Introductory Animal Physiology

McGRAW-HILL SERIES IN ORGANISMIC BIOLOGY

Gardiner *The Biology of Invertebrates*
Gunderson *Mammalogy*
Kramer *Plant and Soil Water Relationships: A Modern Synthesis*
Leopold and Kriedemann *Plant Growth and Development*
Patten and Carlson *Foundations of Embryology*
Phillips *Introduction to the Biochemistry and Physiology of Plant Growth Hormones*
Price *Molecular Approaches to Plant Physiology*
Ralph *Introductory Animal Physiology*
Weichert and Presch *Elements of Chordate Anatomy*

INTRODUCTORY ANIMAL PHYSIOLOGY

CHARLES L. RALPH

Professor of Zoology and Entomology
Colorado State University

McGRAW-HILL BOOK COMPANY

New York St. Louis San Francisco Auckland
Bogotá Düsseldorf Johannesburg London
Madrid Mexico Montreal New Delhi Panama
Paris São Paulo Singapore Sydney Tokyo Toronto

INTRODUCTORY ANIMAL PHYSIOLOGY

1234567890 DODO 78321098

This book was set in Memphis Light by Progressive Typographers.
The editors were James E. Vastyan and James W. Bradley;
the designer was Nicholas Krenitsky;
the production supervisor was Charles Hess;
the drawings were done by Long Island Technical Illustrators.
R. R. Donnelley & Sons Company was printer and binder.

Library of Congress Cataloging in Publication Data

Ralph, Charles L
Introductory animal physiology.

(McGraw-Hill series in organismic biology)
Includes index.
1. Physiology. I. Title.
QP31.2.R33 591.1 77-10416
ISBN 0-07-051156-X

This book is respectfully dedicated to

HOMER R. BOLEN
FRANK A. BROWN, JR.
RICHARD M. FRAPS

in appreciation of their singular
and important influence upon me,
personally and professionally

Contents

In biology one can no longer teach all about any one subject in the time usually allowed; however, one *can* teach what a subject is all about. This book is based on a course with which I experimented for several years, and it represents what I think basic animal physiology is about. It is not intended to be encyclopedic. Neither is the content meant simply to mirror the unevenness of existing information about the functions of organisms; it is easy to be overwhelmed by the great advances in such subdisciplines as muscle and neuron physiology and be unaware of the profound ignorance that exists in others. Rather, this text is intended to outline the functional problems of the living state and, by selected examples, illustrate the strategies that have emerged to solve them.

There has been a deliberate selection of examples from various life forms and at different levels of organization to illustrate the principles advanced. Physiology is not revealed in all its aspects by taking the limited view offered by the cell, or human beings, or plants, or animals. To know what physiology is about requires that no arbitrary limits be placed on our inquiry. However, this book is mostly about animals and largely about the more complex ones. The ultimate expressions of the living state are found in the highly evolved animal forms. Thus, it is appropriate in scanning the numerous life strategies that the highest developments be examined, for incorporated within the highly sophisticated mechanisms are the rudiments of the more primitive ones. Life evolved through minor modifications and gradual, additive refinements.

All organisms, whatever their exact natures, encounter the same problems in coping with the physical environment and in regulating their own components. It is of heuristic value, as well as fascinating, to take account of how various living entities have met specific situations. Furthermore, a problem-oriented approach permits discernment of widely applicable principles. It is the essence of science to formulate such generalities. In the words of Bronowski,[1] "The progress of science is the discovery at each step of a new order which gives unity to what had long seemed unlike."

The topics covered are arranged in a sequence which forms a logical progression. If this book can make any claim to uniqueness, it is in its organization of information from a holistic point of view. After much grappling with the problem of the order of presentation of physiological information to my students, I have arrived at the conclusion that the standard pedagogic approach which starts with "small" things—the cell

[1] J. Bronowski, 1972, *Science and Human Values*, Harper & Row, Publishers, Incorporated, New York.

or, commonly, with elemental biochemistry—and proceeds to the "larger" is neither necessary nor desirable. Such a beginning immediately thrusts the student into an extremely abstract realm of conceptualization. At this level lie the basic roots of understanding function, indeed, but I believe most students are not automatically ready to be enthusiastic about such esoteric fundamentals. A much more logical place to begin is with the organisms. After all, organisms are the functional units with which physiologists always start and, ultimately, to which they must relate their findings and conclusions. Any other level of organization—an organ, cell, or organelle—after all has no reality other than that which relates to the total organism. It is organisms that physiologists, and in final analysis, all biologists, focus upon as units for analysis.

One can proceed away from the organism in either of two main directions. Studies of population or community biology, social structure, and ecology are one direction. They deal with collections of organismic units, as they interact with one another or with their environments. The other direction is toward reduction of organisms into subassemblies—organs, cells, organelles, or molecules. This is the domain of cell physiology and biochemistry. Both departures involve abstract conceptualizations that require, at the outset, knowledge of the nature and workings of organisms, as individuals and as species.

In truth, the organism per se does not exist apart from its surroundings. It is utterly dependent upon the physical environment for exchange of the materials and energy that allow its continued existence. For its continuance as a species, other members of its kind usually are required. The existence of some organisms is tightly interwoven with that of one or more different species, as in the case of parasitic plants and animals. Nevertheless, the organism is as close as we can come to an identifiable, living unit. (This is as true of single-celled organisms as it is of multicellular organisms.) Thus, a most logical place to start the study of living mechanisms—and that is what physiology is about—is with the functions of intact organisms. Once the outlines of organismic functions are grasped, it is appropriate to attempt a progressive disassembly of the very complex organisms into smaller and smaller subassemblies, seeking to analyze mechanisms down to the molecular level. That is the design this book presents.

Following a brief introductory chapter in Part I on the nature of physiology and of life, Part II (Chapters 2 through 8) deals with the ordinary kinds of survival problems that are faced by all animals and several of the more common adaptational strategies that have evolved in response to them.

Part III (Chapters 9 through 12) presents a survey of the mechanisms used by multicellular animals to coalesce their several organs into integrated, coordinated machines that respond appropriately to external and internal changes.

Part IV (Chapters 13 through 16) outlines the main features of electrogenesis, active transport, contractility, and biosynthesis. These are the energy-requiring activities that are common to all cells.

Part V (Chapters 17 through 19) deals with the subcellular and molecular processes that result in acquisition by cells of the chemical energy required to support the processes described in the preceding chapters.

In Part VI, Chapter 20 draws attention to the inseparability of structure and function in living things and alludes to orderliness as a basis of life's uniqueness. A final chapter briefly reviews some speculations about the origin of life and how the beginnings of life, as they were shaped by the physical world, are still reflected in living forms today.

If you, the reader, are a teacher considering the adoption of this book for your class, I encourage you to examine its contents objectively. If you do not accept the order of presentation or do not want to use all of this book, it should be quite feasible to make appropriate selections, arrangements, or adjustments, especially if the six parts, which are fairly independent of one another, are the units that are manipulated.

If you are a student who is about to plunge into this book, I greet you and encourage you to approach the study of physiology with confidence and great expectations. It is an inherently interesting subject that you are going to delve into, and I hope you find this book interesting as well as reliably informative. It is meant to provide a basic understanding of many aspects of physiology which should serve you well if you take more specialized or advanced courses in physiology.

I have generally presented only one or a few interpretations of the current information bearing upon each of the several topics outlined. To present most or all viewpoints is impossible in a book of this kind, but in several places I have pointed out that there is more than one sound way to interpret certain experimental results. Even those things which might be called "principles" should not be regarded as necessarily "true," but as provisional truths.

The success of science depends upon the mutability of its truths. While having faith in their immediate beliefs, scientists (and particularly biologists) have become accustomed to continual revisions. New discoveries are not regarded as catastrophes but as triumphs. Because our understanding of the living state at present is so limited, we can be sure that in physiology there will be many "triumphs" in the future.

I am grateful to Diane Reilly for assistance in the writing of this book. For numerous improvements and corrections, I am pleased to acknowledge the superior assistance of Scott Turner. For her excellent and patient help with the numerous details of assembling the manuscript and coordinating permissions, I specially thank Reta Herbertson.

Charles L. Ralph

Introductory Animal Physiology

I

PROLOGUE

This book begins by considering the nature of physiology and the nature of life. Physiology probes, through a variety of direct and some very indirect means, at the physical and chemical basis of life. It records and measures with numerous devices how hot or cold, how concentrated or dilute some specific things are, or how strong or weak a certain force is. Physiology is very much a measuring science, speaking in moles, microns, degrees, calories, or grams. It also is an analytical science. Starting with a cow, rat, or amoeba, it may focus eventually on a mitochondrion or an enzyme. At all levels of structural organization, physiologists attempt to understand mechanisms and how they function.

For there to be structure there must be *information*—coded, genetic information, in the case of organisms. The story that this book tells is acted out upon the ordered structure of the many kinds of organisms. Much of what follows describes some of that structure. But, gradually and ultimately, interest will turn more to *energy*, for in the dynamics of getting and spending of energy is the real essence of life—and of physiology.

INTRODUCTION 1

In this age of specialization the many kinds of scientists have relatively little communication with one another. Such a state of affairs is inevitable when information is being produced rapidly by specialized groups that do not regularly interact. Many specialists are quite concerned about their own narrowness and very much want to bring their information into a common pool of knowledge. Physiologists have been particularly cognizant of the potential rewards of transfusion of information and methodology from other fields to theirs. In recent years there has been an obvious increase in collaboration of scientists from different disciplines in attacking research problems. For example, groups composed of various combinations of psychologists, pharmacologists, neurophysiologists, and endocrinologists have produced some remarkable advances in understanding learning processes. Engineers and physiologists of several different specialties have cooperated both to analyze living systems by engineering methods and to construct systems whose function is based on living systems. One of the results of such interdisciplinary efforts is the increasing presence of such terms as "control," "feedback," "information," "code," and "memory trace." They have similar meanings for physiologists, psychologists, and engineers. But the words themselves are not important; it is the discovery of the analogous nature of the principles which underlie various phenomena, living and nonliving, that is significant.

1-1 NATURE OF PHYSIOLOGY

From the time of Aristotle there has persisted a school of thought known as **vitalism**, which holds that although many biological phenomena can be explained in physicochemical terms, underlying certain complex processes, such as thought and embryological development, are unknown and perhaps unknowable principles. The real essence of life, according to this view, stems from "vital" forces. Adherents to this philosophy are almost extinct and have been supplanted by **mechanists**, who operate on the hypothesis that life processes are nothing more than the workings of physicochemical systems.

If the organism is a consequence of physicochemical systems performing work, then it is susceptible to analysis in the terms of physical science—like a machine. Perhaps an organism should be regarded as a collection of machines—cellular and subcellular. The several "organelles"—mitochondria, chloroplasts, ribosomes, and others—act as machines; for a specific input there will be a specific output, in both energy and matter. There are even certain of these biological machines that regulate other machines; i.e., the systems are fully automated and employ feedback principles.

The science of controls and information, **cybernetics,** has considerable value to biologists in analyzing the living state in much the same way that an engineer analyzes a complex machine. The rationale for both the biologist and engineer in employing cybernetics is to reduce a complex situation to sets of understandable data about constants and variables, inputs and outputs, and to make testable predictions about what will happen in given circumstances. Although the exact nature of the living machinery is far too complex to be fully understood at present, this is no great deterrent for cybernetic analysis. It is the input and output that are of most concern; exactly what happens in between need not be understood fully, if at all. Thus, a cell or an organism may be regarded as a **black box,** or as sets of black boxes or even as black boxes which contain black boxes. Thus, one can analyze *what* an organism, a cell, a ribosome, or an operon *does* without knowing how it does it (or if it even exists as an entity).

It would be pleasing not to have to regard living systems as black—or very gray—boxes, but note as we go along how little we truly know about the exact workings of biological processes. The major efforts of physiologists are directed at revealing the true nature of those many black boxes. (In Sec. 4-8 the concept of black boxes will be further detailed.)

Another viewpoint useful to the physiologist in interpreting life is to regard organisms of today as the result of an emerging progression which gives the appearance of having been molded by "strategy." This approach, in which one does not lose sight of the fundamental, ubiquitous features of life, may allow one to make some sense out of all the many different mechanisms characteristic of the gamut of life forms. One can take the position that, as the result of a set of fundamental characteristics, life has been endowed with certain strategies. First and foremost there is an energy-binding strategy. It is obvious that all organisms engage in this strategy. They differ, however, in their "tactics"—in the way they initially capture energy, especially, and somewhat in the way they handle energy internally. There is another strategy for continuance, through replication and reproduction tactics. These can vary enormously from one kind of organism to another, but the overriding strategy—heredity—is clearly evident. Thus, one can view the uniqueness and complexity of any organism within the framework of a moderate number of strategies.

If one chooses to regard a life unit as a machine it generally follows that one is a "reductionist," who believes that to understand life it is necessary to dismantle it into smaller and smaller components until they are simple enough to be analyzed by the methods of physics and chemistry, and perhaps to be represented by a mathematical expression. On the other hand, the view of a life unit as a "problem-solving" entity which has emerged from its evolution with strategies for survival and continuance is more "antireductionist" than the machine concept. Both views have considerable merit, are not contradictory, and will be used without apology on different occasions in what follows.

Basically there are two reasons why people are motivated to take up science: to control nature and to understand nature. The latter case may be really only an extension of the former, as understanding generally leads to control. Yet, there are always among us those who apparently take up science for the pure joy of discovery. However, there is no question that the main impetus for advancement of physiological information has been to control nature. Specifically, it has been the desire to be free of diseases and afflictions which has driven men to learn most of what is known about life functions. This is still an overriding motivation, but we now also earnestly seek and place high value on knowledge that is of no immediate application to man's ills.

Modern physiology is organized into several indistinct subdisciplines. One kind of categorization is based upon types of organisms. Thus *animal physiology*, *mammalian physiology*, and *human physiology* are formulated to include the study of function in circumscribed groups of organisms. Another way of dividing the field is to focus on level of structural complexity. For example, *cell physiology* is the study of those processes which occur within individual cells and its adherents deliberately do not limit their interests to mammals or vertebrates or even animals, for a sizable body of their information comes from plant cells and microorganisms. *Comparative animal physiology* is the study of functions at the organ-system level in various types of animals; it seeks to find functional homologies and analogies which transcend specific groups of animals. *Plant physiology* deals with reproduction, nutrition, growth, transport, and other processes as they uniquely occur in plants. *General physiology* is quite comprehensive, dealing with cellular and intercellular functions as they relate to an understanding of the multicellular organism. Textbooks which have drawn together much of the information of these and other specialized kinds of physiology are listed at the end of this chapter.

1-2 OBSERVATION AND EXPERIMENT

Because physiology is based so largely on experiment, some things should be said about the techniques and apparatus employed in obtaining physiological information.

Much, of course, has been learned simply by observation of organisms under conditions that do not require extensive manipulation by the investigator. A splendid example is the 1882 investigation by T. W. Engelmann of the dependence of oxygen production by green plants upon wavelength of light. Using an apparatus designed by Carl Zeiss to project a microscopically small spectrum on algal cells, he demonstrated that motile bacteria in the immediate vicinity of the cells were most active in two bands, one near the red end and, to a lesser degree, in a second band near the violet end. Motility of the bacteria directly reflected oxygen availability. Thus, the relative effectiveness of different wavelengths of light in relation to the production of oxygen by photosynthetic cells was clearly demonstrated (Fig. 1-1).

Such studies, although ingenious, are often limited, especially quantitatively, and therefore the physiologist generally imposes upon the system under study certain experimental procedures in the hope of obtaining the desired information in a quantitative form. Quite often this means attempting to isolate some system or part of an organism for study in vitro (in glass; hence in a tube, chamber, etc.); this is commonly difficult. The kinds of isolation practiced by the physical scientists have few counterparts in biology; isolation generally means death in biological systems. Many times a physiologist will concentrate on one organ or structure in situ (in its natural place) or in vivo (in a living organism) and essentially ignore the remainder of the organism; however, he is fully aware that he is ignoring many things and therefore strives to take this into account when interpreting his findings. It is simply impossible at present to understand the workings and interrelationships of all the myriad parts that compose even the simplest physiological system, but steadily the appropriate techniques for dealing with biological complexity are being developed.

Physiologists have borrowed, devised, and improvised many kinds of apparatus to aid them in resolving the functions of organisms. Among biological sciences, physiology has been especially dependent upon technological advances. Methods for separating and characterizing biological components, such as chromatography, centrifugation, and x-ray diffraction, are borrowed from the domains of chemistry and physics.

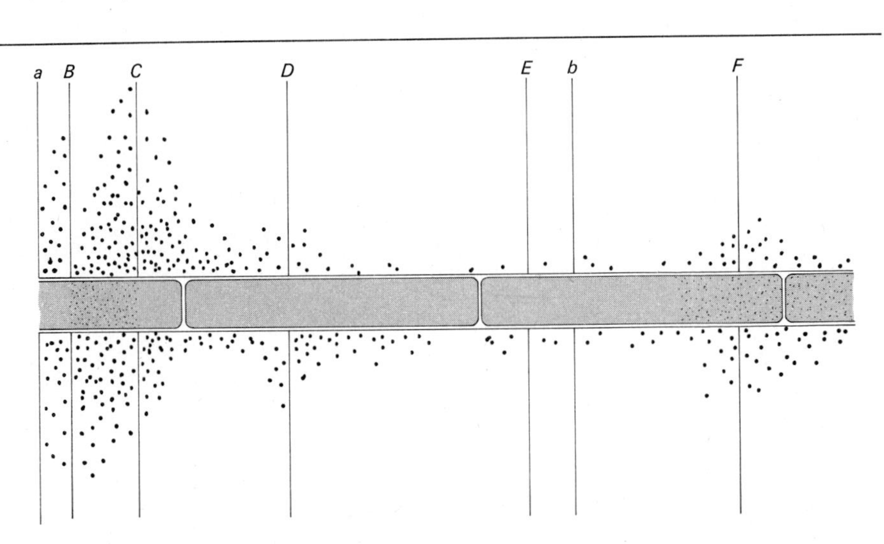

FIGURE 1-1
A demonstration by Engelmann in 1882 of the relative effectiveness of different portions of the spectrum in photosynthesis. Under a microscope a small spectrum of sunlight was projected upon a filament of cells of the alga, *Cladophora*, and the location of accumulations of motile bacteria was noted. The position of Fraunhofer lines is indicated; *B* and *C* are in the red region, and *F* is at the violet end. [*After M. L. Gabriel and S. Fogel (eds.),* 1955. *Great Experiments in Biology*, *Prentice-Hall, Inc., Englewood Cliffs, N.J.*, p. 167.]

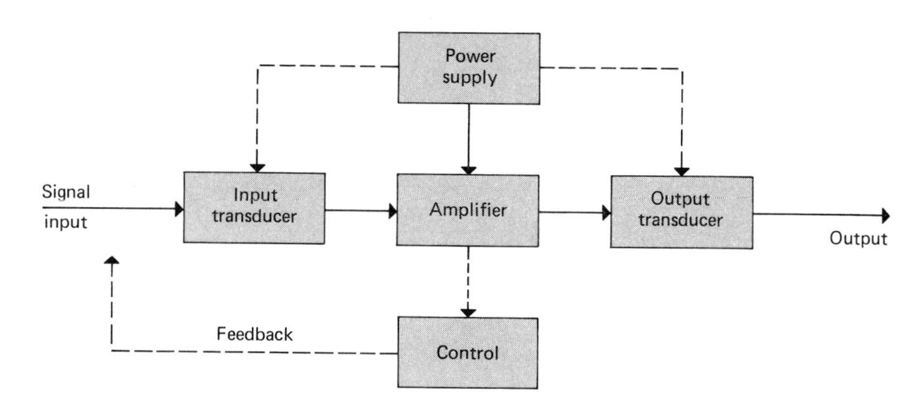

FIGURE 1-2
A typical electronic system which would be adaptable to measurement of a variety of biological phenomena. Some biological reactions produce voltages directly and are relatively easy to measure. Other processes involve a change in resistance, capacitance, or some other property that can be measured. Movement can be detected by a device that responds to mechanical deformation. By changing the components appropriately, principally the specific input transducer, many kinds of biological responses can be converted into an electrical output.

Among the more important tools devised by physiologists are transducer-amplifier devices which permit the detection or measurement of subtle events that occur within living things. For example, if a physiologist wishes to measure dissolved oxygen, let us say in blood, he has several methods from which to choose. One of the most straightforward employs an oxygen electrode, a polarographic device which measures the rate of irreversible molecular transfer of dissolved oxygen to a platinum electrode maintained at a potential negative to the solution. The rate of transfer, and therefore the electrode current, is proportional to the concentration of dissolved oxygen, and the minute current generated may be amplified to drive some kind of read-out meter or mechanical recording device. If appropriate constants and variables of the system are known or can be calculated, a reliable estimate of oxygen content is obtained.

Many pieces of apparatus have been designed to display and record electrical signals from biological material. In fact, progress in neurophysiology clearly has reflected advances in the development of devices for electrical measurements. The minute membrane potentials (50 to 100 mV) of nerve and muscle membranes can be amplified to attain a sufficiently large output to drive a pen-recording device or cathode-ray tube. Amplification is obtained with circuits specifically designed for study of electrogenic cells; uniform responses over a wide range of relatively low frequencies, adjustable bandwidths, and interference filtering are features of such instruments. Cathode-ray oscilloscopes, with built-in amplification, variable time-base rates, and double-beam displays, are in common use even in undergraduate laboratory classes in physiology (Fig. 1-2).

Physiologists admit in good humor that they sometimes have difficulty remembering where the biological preparation is when their various pieces of equipment are ganged about it. An isolated strip of muscle or nerve ganglion can look ridiculously insignificant among the amplifiers, recorders, pumps, and wiring and plumbing that characterize many laboratories. However, from such arrays of devices come the composites of information needed to interpret complex biological phenomena. And though gadgetry is more sophisticated today, ingenious minds of other eras also managed to put together some remarkably clever complexes of devices (Fig. 1-3).

Man has always turned to other living things for ideas about how to improve his lot, and sometimes simply for inspiration. Leonardo da Vinci's design in the year 1505 for a flying machine, inspired by the flight of bats, is a grand illustration. In 1960 a name was given to the art of applying knowledge of living systems to solving man's technical problems: **bionics.** In this field would be those who study the coding and transmission of information by nervous systems in the hope of simulating parts of the processes in communication devices, such as telephones. Visual mechanisms are being analyzed for possibilities of applying their principles to machines that can read. Echolocation of bats might provide ideas for improvements in certain aspects of radar. Beginnings have been made in applying biological designs to a few devices, mostly in information processing, but any notable realization of such enterprises depends upon first deciphering the complexities of living systems.

FIGURE 1-3
Design of apparatus used in the 1880s by du Bois-Reymond to study the discharge of an electric fish. One frog nerve-muscle preparation (G_1) was used to ring a bell when the fish discharged, while a second (G_2) served to disconnect the galvanometer (B) after the first pulse of the train of discharges had been measured. (*From Endeavour, 15,* 218, 1956.)

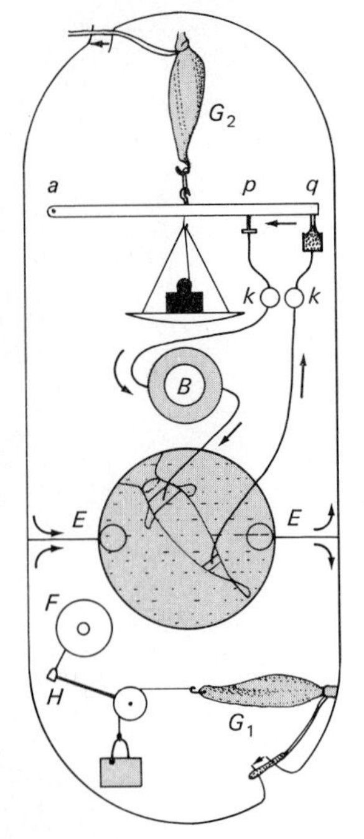

1-3 NATURE OF LIFE

Everyone who studies the fundamentals of biology learns the characteristics that distinguish living from nonliving things. To *define* "life," however, is extremely difficult if not impossible. Often it is easier to study a thing or process than to define it. There are many eminent scientists who have concluded that a precise definition of life is impossible. The highly respected physiologist and biochemist, Szent-Györgyi, has written: "Life as such does not exist: nobody has ever seen it. . . . Whether something is alive or not depends on our idea of what we call 'alive' and what criterion we choose. The noun 'life' has no sense, there being no such thing."[1]

Nevertheless, several remarkable minds have pondered the essential nature of life and reached certain conclusions that at least provide some insight into the nature of the problem. For one thing their thoughts on this subject focus our attention on the fact that the animate and inanimate may imperceptibly merge with one another. That is, living and nonliving things appear to form a continuum at certain levels. Furthermore, it can be argued that the matter of a living thing is expressing inherent properties that are not revealed until incorporated within a complex which we would recognize as "alive."

But what minimum construction can be called "alive," and when does an entering atom or molecule become a part of that which is "alive" or "animate"? And when, in leaving to return to the physical environ-

[1] A. Szent-Györgyi, 1948, *Nature of Life,* Academic Press, Inc., New York, p. 9 and pp. 90–91.

ment, does it become "inanimate" matter again? Horowitz has suggested that the simplest form that living matter can take is a single macromolecule. It is alive, he proposes, if it shows the minimum criteria of mutability, self-duplication, and heterocatalysis. The gene would meet these criteria. But since no living thing can function in a vacuum, Horowitz proposes that "life arose as individual molecules in a polymolecular environment."[1]

Lwoff has placed especially strong emphasis on this latter point. He states: "The living organism is an integrated system of macromolecular structures and functions able to reproduce its kind. . . . Separated from its context—that is, extracted from the cell—any structure, either a nucleic acid or a protein, is just an organic molecule. Such a thing as living substance or living matter does not exist. Life can only be the appanage of the organism as a whole. Only organisms are alive."[2]

Von Bertalanffy comments: "A 'living substance' has often been spoken of. This conception is due to a fundamental fallacy. There is no 'living substance' in the sense that lead, water, or cellulose are substances, where any arbitrarily taken part shares the same properties as the rest. Rather is life bound to individualized and organized systems, the destruction of which puts an end to it."[3]

Many definitions of life select criteria based upon certain kinds of matter-energy relationships. Thus Madison considers: "Life, regardless of what else it may be, is the manifestations of an organism. . . . Organisms can be defined as a group of chemical systems in which free energy is released as part of the reactions of one or more of the systems, and in which some of this free energy is used in the reactions of one or more of the remaining systems."[4] Note in this that we are introduced to one of the ubiquitous features of all biological constructions: coupled reactions involving free energy.

Ehrensvärd says that the true nature of life "is what it is; and it appears to be, in simple outline, a cyclic procession of matter driven by sunlight."[5] Note the emphasis on energy, specifically light. This is a more cosmic view of life than those quoted previously, and the cyclic nature of life is emphasized.

The dynamic nature has been stressed by von Bertalanffy in such statements as: "Living forms are not in being, they are happening; they are the expression of a perpetual stream of matter and energy which passes the organism and at the same time constitutes it."[6]

These various statements on the nature of life place emphasis on somewhat different aspects of matter in the life state, but they all are at-

[1] N. H. Horowitz, 1959, On defining 'life,' in *The Origin of Life on the Earth*, edited by A. I. Oparin et al., Pergamon Press, New York, pp. 106 and 107.

[2] A. Lwoff, 1962, *Biological Order*, The M.I.T. Press, Cambridge, Mass., p. 100.

[3] L. von Bertalanffy, 1952, *Problems of Life*, C. A. Watts & Co., Ltd., London, p. 13.

[4] K. M. Madison, 1953, The organism and its origin, *Evolution*, **7**, 213, 214.

[5] G. Ehrensvärd, 1960, *Life: Origin and Development*, The University of Chicago Press, Chicago, p. 9.

[6] L. von Bertalanffy, op cit., p. 124.

tempts to put into words concepts that severely tax our language. For the present we must be content to study life without knowing what it really is.

A REFLEX[1]
Hear my rigmarole.
Science stuck a pole
Down a likely hole
And he got it bit.
Science gave a stab
And he got a grab.
That was what he got.
"Ah," he said, "Qui vive,
Who goes there, and what
ARE we to believe?
That there is an It?"

READINGS

Selected textbooks in physiology

Davson, H., 1964, *A Textbook of General Physiology*, 3d ed., Little, Brown and Company, Boston. First published in 1951, this textbook remains a classic attempt to explain living phenomena in terms of physics and chemistry.

Florey, E., 1966, *An Introduction to General and Comparative Animal Physiology*, W. B. Saunders Company, Philadelphia. A very good, in-depth examination of selected topics in animal physiology.

Giese, A. C., 1973, *Cell Physiology*, 4th ed., W. B. Saunders Company, Philadelphia. An account of the functions of cells.

Hoar, W. S., 1966, *General and Comparative Physiology*, Prentice-Hall, Inc., Englewood Cliffs, N.J. This book presents animal functions on a framework of phylogeny.

Levitt, J., 1974, *Introduction to Plant Physiology*, The C. V. Mosby Company, St. Louis. One of the relatively few textbooks on this subject; the unique functions of plants, such as photosynthesis, water relations, mineral requirements, and growth processes, are discussed.

Prosser, C. L., 1973, *Comparative Animal Physiology*, 3d ed., W. B. Saunders Company, Philadelphia. A valuable, encyclopedic presentation of physiological information about animals in general. The references serve as guides to most of the literature in this field.

Ruch, T. C., and H. D. Patton (eds.), 1973, *Physiology and Biophysics*, 20th ed., 3 vols., W. B. Saunders Company, Philadelphia. This is an evolutionary product, the original ancestor of which was Howell's *An American Textbook of Physiology*, published in 1896. Although it is primarily a mammalian physiology, its many modern contributors provide a splendid survey of most of general physiology.

Schmidt-Nielsen, K., 1975, *Animal Physiology*, Cambridge University Press, New York. A highly readable introductory textbook with much comparative information.

[1] From *The Poetry of Robert Frost*, edited by Edward Connery Lathem, 1969, Holt, Rinehart and Winston, Inc., New York, p. 468.

On the nature of life

Bertalanffy, L. von, 1952, *Problems of Life: An Evaluation of Modern Biological Thought*, C. A. Watts & Co., Ltd., London. A provocative argument for the oneness of structure and function in living things.

Ehrensvärd, G., 1962, *Life: Origin and Development*, The University of Chicago Press, Chicago. The concept of life as a cycle comprising interaction between organisms and the components of the inorganic environment is well presented.

Grobstein, C., 1964, *The Strategy of Life*, W. H. Freeman and Company, San Francisco. An elementary but cosmic view of life.

Lwoff, A., 1962, *Biological Order*, The M.I.T. Press, Cambridge, Mass. The hereditary order and the functional order of life are the themes of this small volume.

Schrödinger, E., 1945, *What Is Life?* The Macmillan Company, New York. A physicist's attempt to answer this great question.

Waddington, C. H., 1961, *The Nature of Life*, George Allen & Unwin, Ltd., London. Life is here interpreted by a developmentalist.

II

ADAPTATIONS: Animal Specializations

Adaptations are those particulars which result in appropriate and efficient morphological and functional correlations between components of an organism, between individuals of the same species or of different species, and between an organism and its external environment. Collectively, adaptations confer advantages which favor the survival of the individual and the perpetuation of the species.

Internal adaptations involve the development of complex functional unities. Evolutionary changes in interacting organs must be mutual and tightly coordinated. For example, modifications of a skeletal joint must involve mutual accommodations of the articulating bones as well as of the attached muscles. A change in the number or kinds of photoreceptor cells in the retina of the eye (e.g., the acquisition of cone cells of color vision) would necessitate complementary changes in the neuronal connections in the retina and in the brain.

Many adaptations are primarily or, perhaps, exclusively for the internal meshing of physiomorphological mechanisms, but there is no clear demarcation between interior adaptations and adaptations to the external environment. Reproductive, digestive, respiratory, osmoregulatory, excretory, and other systems decidedly reflect the unique features of the particular environment in which an animal exists.

Some truly remarkable interorganismic adaptations have been evolved. Examples would include host-parasite adaptations, adaptations of copulatory devices of the male and female of a species, and the release and detection of pheromones by certain species of animals.

Adaptations permit invasion of an environment which differs from that previously inhabited. Any adaptive changes must be gradual and partial. The evolving animal has to continue to meet the fundamental and general exigencies of the living state while, through successive generations, modifying localized and specialized structures and mechanisms. The species must conduct its "business as usual during alterations."

BEHAVIOR AND SURVIVAL

2

Organisms have to cope with hostile or, at least, uncooperative surroundings. Their survival depends on the maintenance of suitable relations with the physical environment, and they must be prepared to make appropriate responses to other organisms to avoid injury or annihilation, as individuals and as a species. In addition, they must respond to internal signals that indicate physiological states and invoke appropriate measures to ensure proper functioning. These responses may be morphological, physiological, or behavioral—or combinations of the three.

Although it is relatively easy to record morphological and physiological responses to variables and relate them to survival of an individual or a population, it has proved more difficult, and less appealing to biological investigators, to obtain information on the role that behavior may play in survival. It has been the custom until quite recently to regard the study of animal behavior as something largely outside the proper domain of biology. This situation is rapidly changing; a goodly number of pioneering workers has now begun to view organisms in a more total way than has ever been done before. For example, remarkable progress has been realized in analyzing the neuroanatomy, neurochemistry, endocrinology, and behavioral modes supporting regulation of water content, mating success, control of body temperature, and maintenance of nutritional state. Clearly revealed is the fact that behavioral, morphological, and physiological responses are complementary and essential components of processes that protect the animal from numerous hazards, make its life less stressful, and increase its likelihood of survival. Such findings are emerging from the broad-based studies of both the laboratory-bound scientists as well as those engaged in the biological study of behavior under more natural conditions, the ethologists.

In this chapter we will begin to consider how behavior serves adaptive ends, primarily for the individual animal. In the next chapter the interactions of animals and the adaptations of animal groups will be considered. Subsequent chapters will, from time to time, call attention to the contributions that behavioral responses make to certain physiological regulations. However, the distinction between what is properly considered behavior and what is considered physiology is so unclear that it is generally absurd to attempt to make such distinctions, as will become apparent in what follows.

2-1 INNATE BEHAVIOR

Primitively, behavior is largely a response or a sequence of responses triggered by a stimulus. At this level behavior is **stereotyped** and the organism is, to a large extent, **stimulus-bound.** Such behavior is refractory to any significant modification. Because it is essentially an expression of inherited properties of the organism's nervous system, it is said to be **innate.**

Included in the stereotyped, innate modes of behavior are **taxes, kineses, reflexes,** and **fixed action patterns.** These primitive forms of behavior, described below, are the exclusive or predominant modes in those organisms that also are considered relatively more primitive, in terms of phylogeny. Just as one can place organisms in an approximate phylogenetic scheme based upon increasing morphological complexity,

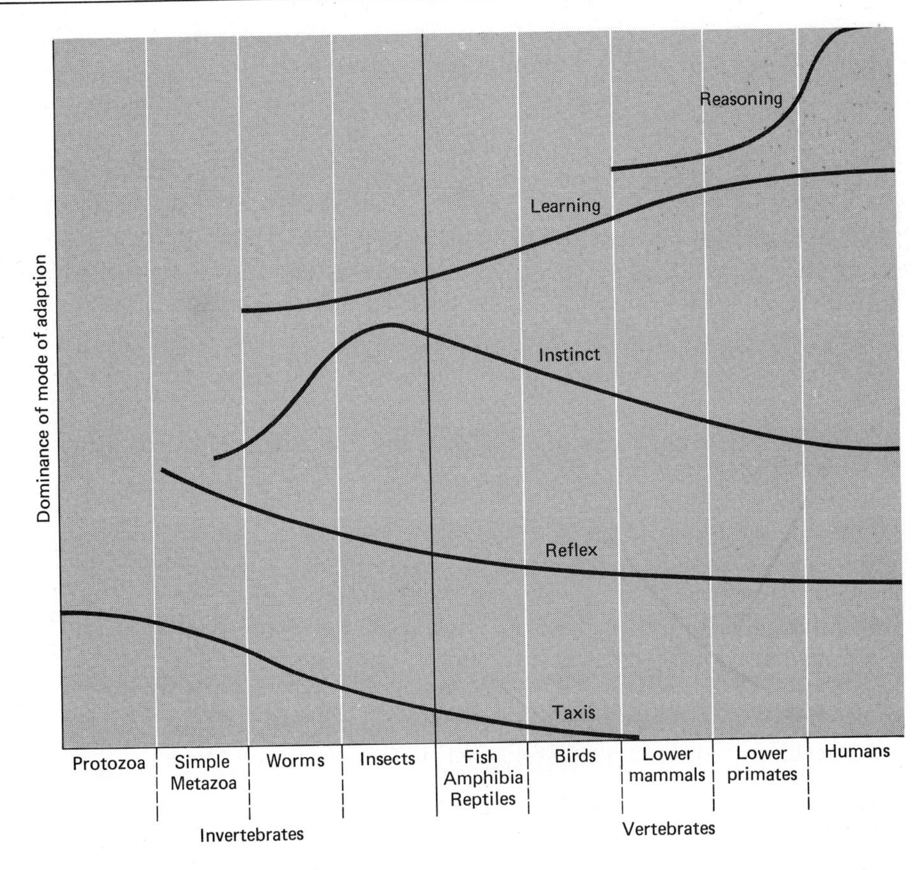

FIGURE 2-1
Schematic portrayal of the changes that take place in the major modes of adaptive behavior in phylogeny. Reading from left to right, it shows the relative development of different modes of adaptation; reading up and down, it shows the relative pattern of modes of adaptation at different levels of the phylogenetic scale. (*After V. G. Dethier and E. Sellar,* 1970, *Animal Behavior,* Prentice-Hall, Inc., *Englewood Cliffs, N.J., p.* 91.

so can animals be arranged with regard to the complexity of their modes of adaptive behavior. Among the so-called higher organisms, stereotyped behavior gives way to **learning ability** and **reasoning,** in parallel progression with the degree of morphological complexity, especially that of the nervous system. However, within each of the major phylogenetic groups there is a considerable range of modifications of the basic sets of behaviors (Fig. 2-1).

Animals below the level of worms do not seem to be capable of learning. They have relatively simple and poorly developed fixed patterns, and are dominated by taxes and reflexes. Insects are relatively poor learners, are dominated by largely unmodifiable patterns, and exhibit several kinds of taxes. The behavioral repertoire of birds consists mainly of instincts and learning, with essentially no reasoning capacity. Mammals display a remarkable range of modes of adaptive behaviors. For example, in the well-studied rat, reasoning is virtually nonexistent but learning is well developed. In human beings the dominant modes of adaptation are reasoning and learning, with very little instinct or reflex that is not considerably modified by experience.

Kineses, Taxes, Orientations

Light plays an important role in the directional orientation of many organisms, and how they respond to it depends upon the nature of their photoreceptors and the organization of their nervous systems. A classification scheme, developed by Fraenkel and Gunn,[1] with emphasis on photoreceptors, distinguishes several simple responses that appropriately serve their possessors.

Kineses are nondirectional velocity changes of motile organisms in response to changes in intensity of environmental stimulating factors. There are two subdivisions of this class of responses:

Orthokinesis **refers to a change in rate of linear movement.**
Klinokinesis **refers to the frequency of change of direction or turning.**

Both types of kineses are dependent upon a receptive system capable of distinguishing intensity differences. It is these responses which, for example, result in organisms *aggregating* at some optimal position in thermal, light, or oxygen gradients. Movement away from the optimum position will result in increased linear velocity and increased rate of turning. As the optimum position is approached both kinds of movements will diminish. The result of the changes of velocity and rate of turning will be that most animals will tend to occupy the optimum position most of the time.

Taxes are the directed orientations of animals which move in their environment. Taxes have been subdivided into three types:

[1] G. Fraenkel and D. L. Gunn, 1940, *The Orientation of Animals*, Oxford University Press, Fair Lawn, N.J., 1940.

Tropotaxis is a smooth, straight orientation toward or away from a stimulus source by *simultaneous bilateral equating* of stimulation. This kind of orientation, exhibited, for example, by planarians and sow bugs (isopods), depends upon a bilateral arrangement of receptors possessing some good degree of directional localization. If one eye is blinded, the animal circles, away from the intact eye if it is negatively phototaxic and toward the intact eye if it is positively phototaxic.

Telotaxis is a smooth, straight orientation toward or away from a stimulus source, without a bilateral comparison of stimulus intensity. Phototelotaxis requires an eye capable of good directional localization, and a mechanism inhibiting response to other stimulating sources simultaneously present. An image-forming eye permits telotaxis.

Klinotaxis is the directed movement toward or away from a stimulus which depends upon frequent bending movements of the head (carrying the receptor) to permit successive *comparisons* of stimulus strength on the two sides. This kind of orientation, as a photoklinotaxis, is found in such organisms as fly maggots and *Euglena*, which have good intensity discrimination but poor directional receptors (Fig. 2-2).

The adaptive value of taxes is illustrated by the grayling butterfly (*Eumenis semele*), which flies toward the sun to escape its predators. *Euglena*, capable of photosynthesis, will tend to remain in illumination by its klinotaxis. Cryptic animals will be inhibited by kineses or taxes from venturing away from concealing shadows.

Two more kinds of orientation, with special regard to light, should be mentioned. The brine shrimp, *Artemia*, may be readily induced to swim upside down by illumination from below. This orientation is called the *dorsal light response*. Many organisms tend to show this reaction if their gravity-sensing units are destroyed, so there is no conflicting information about which direction is "up" and "down." They will orient their dorsal surface toward the light, regardless of where it is placed (Fig. 2-3).

Many organisms orient with respect to a light source but assume a path at some particular *angle* with respect to the source. This **light-**

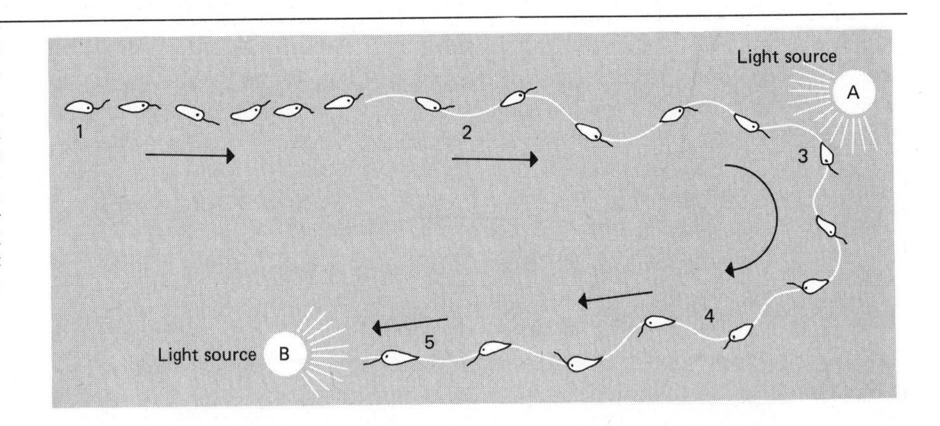

FIGURE 2-2
The phototactic orientation of the flagellate *Euglena*. *Euglena* swims in its characteristic spiral motion toward the source of light (*A*). Upon reaching point 2, the light is changed to *B*. The *Euglena* eventually changes direction and swims toward the new light source.

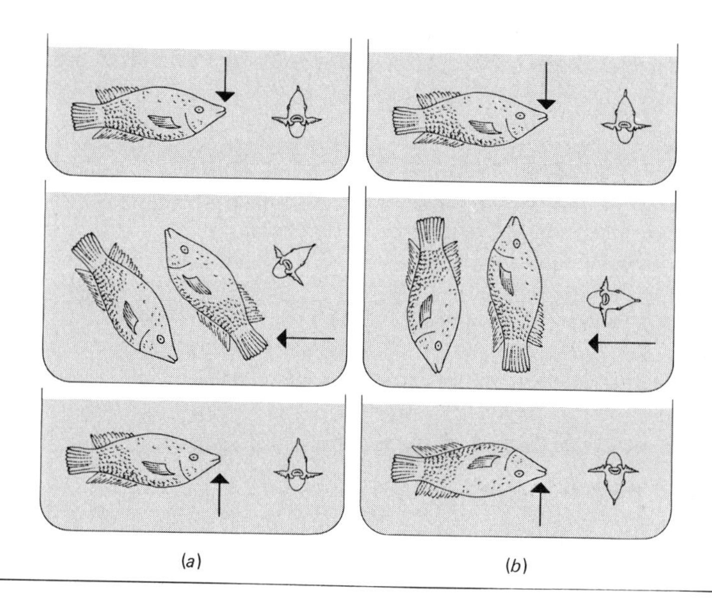

FIGURE 2-3
Reaction of *Crenilabrus rostratus* to light coming from different directions (arrows). In (*a*) the fish is normal; in (*b*) the fish has had its response to gravity eliminated by the removal of the labyrinth of the ear. (*After N. Tinbergen,* 1969, *The Study of Instinct, Oxford University Press, Fair Lawn, N.J., fig.* 82, *p.* 94.)

compass orientation requires the possession of an image-forming eye, upon which the light source may be fixated by any region of the eye. The light-compass reaction is believed to form the basis of celestial navigation for such animals as insects, crustaceans, mollusks, fish, and homing birds. The angular orientation must change at rates which compensate for the earth's rotation relative to the sun, moon, or stars—changes regulated by the so-called *biological clocks* (see later) (Fig. 2-4).

Orientation may involve more than simple kineses and taxes, even though the organism is continuously guided by external stimuli and the

FIGURE 2-4
The effects on initial headings when the biological clocks of mallard ducks are shifted. In all diagrams the open rectangles represent the headings of control birds that had not been subjected to any clock-shifting treatment; these ducks tended to fly northwest. Filled rectangles show headings of birds kept for several days on a shifted light cycle that reset their biological clocks six hours ahead or behind. The sectors at the center of each graph show the angular range within which half the headings were contained. The two graphs on the left show results obtained in daytime when birds had a clear view of the sun; the resetting of biological clocks produced the expected changes in headings. Evidently these birds were using sun-compass orientation.

The two diagrams on the right show the result of the same experiment performed on clear but moonless nights. With the full star pattern available, the directional headings were not significantly affected by resetting the biological clocks. (*After D. R. Griffin,* 1964, *Bird Migration, Dover Publications, Inc., New York, p.* 164.)

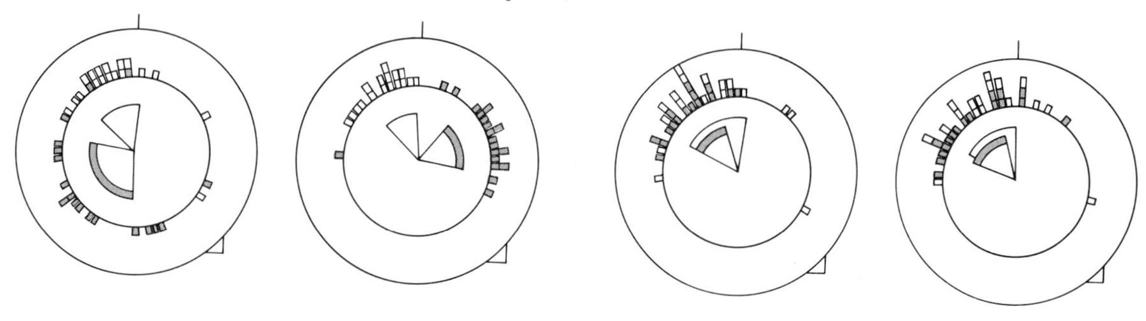

consequences of its own responses to them. The optomotor turning response of the beetle (*Chlorophanus*) to movement is an interesting example. When the beetle is fixed in position inside a drum on which rotating lines of light are projected and its feet grasp a very light-weight Y-maze globe, the beetle's direction and speed of locomotory movements change as a function of the direction and speed of the rotation and the size and brightness of the lines against the background of the drum. For example, as the direction of the rotation is changed from left to right, the beetle accordingly moves along the left or right choices of the Y-maze. Such changes are precisely predictable, as if the beetle were a complex, rigidly programmed machine (Fig. 2-5).

While it is true that, under appropriate conditions, lower organisms will display to external stimuli certain *relatively* fixed and stereotyped reactions of the types described above, even the simplest of organisms may show rather complex behavior. Any organism is responsive to more than one aspect of its environment or internal state at any given instant and may be making several different adaptations at once. Thus, in observing complex patterns of behavior, one can identify component acts that are relatively fixed orientation responses embedded in a more complex whole.

Reflexes

Reflexes are another kind of relatively stereotyped and fixed response to stimuli that are classified as innate because they are the outcome of inherited neural mechanisms. They are difficult to distinguish from taxes, but whereas taxes involve an orientation of the whole body (and include a number of specific reflex responses), the reflexes typically involve responses of only parts of the body, such as flexion of a limb in response to painful stimuli or constriction of the eye pupil to an increase in light intensity. However, reflexes also may involve all or most of the body, as in the startle reflex or righting reflex. The classic righting reflex can be demonstrated in the cat by dropping it blindfolded with its legs pointed upward. The cat turns with almost incredible speed and lights deftly on all four legs. The reflex always begins with rotation of the head to initiate the turn, followed by rotation of the upper body to align it with the head, and this is followed by rotation of the lower body, completing the turn.

FIGURE 2-5
The orientation of a beetle (*Chloropharnus*) on a Y-maze globe it holds with its feet. See text for explanation. (*After R. A. Hinde, Animal Behavior: A Synthesis of Ethology and Comparative Psychology*, McGraw-Hill Book Company, New York, 1966, p. 93.)

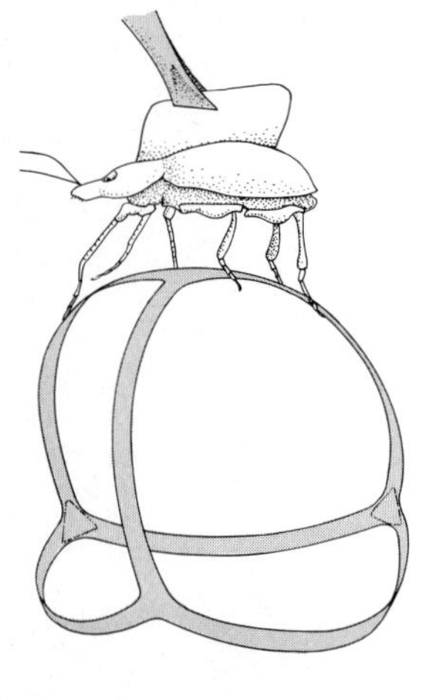

It is obvious that reflexes are adaptive and, in general, relatively invariable. However, there is some variation from species to species, especially among higher vertebrates where reflexes are dependent on "levels" of the nervous system above the spinal cord. Simple reflexes, such as flexion and extension, may be organized within a few segments of the vertebrate spinal cord, but locomotion, which involves a coordinated alternation of flexion and extension, is organized over many segments of the spinal cord and normally requires the influence of the midbrain.

Though the reflex response is one of the major modes of adaptation in the animal kingdom, many kinds of behavior involve something more than complex chaining of simple reflexes with their invariable stimulus-response relationships. In the course of evolution, reflexes became less prominent features of behavior. They were overshadowed by other modes of adaptation and modified by increasingly cephalized nervous systems.

Fixed Action Patterns and Complex Action Sequences

The manner in which a woodpecker drills a hole in a tree, the way a cat arches its back, when threatening, and the defensive posture of the rattlesnake are examples of relatively stereotyped components of behavior which are typical for the members of a species and may be referred to as **fixed action patterns** or *instinctive movements*. The term **instinct** has been applied to such fixed patterns, but as a technical term is nowadays not often used, because it lacks precision and means too many different things to different people.

Many fixed patterns appear so soon after birth or hatching that it must be concluded the underlying physical coordination mechanisms are there at birth. However, strictly speaking, behavior is not "inherited" in the sense of being solely determined by genes. All behavior, whether innate or not, develops—just as do other features—as a result of interactions between an organism's genotype and its environment. The repertoire of complex, species-predictable behavior appears progressively with maturation. Often there is a period during which behaviors are perfected until they reach their final form. In other cases there is little requirement for perfecting and the behavior appears, in complete form, at some stage of the animal's maturation. Behavior patterns that develop through maturation are commonly included within the category of *innate* behavior to distinguish them from behavior which develops as a consequence of learning.

Development of adult courtship and of threat displays seems mainly a matter of maturation. Indeed, male domestic chicks can be induced to crow by injecting them with the testicular hormone, *testosterone*. Except that the voice of the chick is, of course, much higher in pitch, the crowing performance closely resembles that of the adult cock. Precocial sexual behavior can also be induced in male mammals by injections of testosterone. Such findings indicate that the neural substrate for an adult behavior pattern already exists at birth or is acquired early in life, but lies dormant until activated or sensitized—e.g., by hormones.

The adult animal's behavior eventually consists of an interwoven complex of learned and innate components. Maturational processes can be considered as providing only an incomplete "program" for the complex action sequences that must be supplemented by information gained through practice, trial-and-error learning, conditioning, or imitation.

Some good examples of the "shaping" of innate behavior are found in studies of development of song in many songbirds. For example, if some species do not hear the normal adult song during a brief "sensitive period" early in life, they later produce rather abnormal songs—similar to the song of its species but lacking many of the finer details of the normal song. Performance is greatly impaired if they are deafened before they have heard others of their species sing. Thus, exposure to the correct song at the proper time results in the formation of an acquired program or template, which some months later is used in shaping their own singing. Deafening is of little consequence if it is performed after the full song has been mastered.

The outstanding "virtue" of fixed behavior patterns is that they require no experience with the particular situation in which the movements are to be used. The components of an animal's behavior "fit" its natural environment and "preadapt" the animal for its unique niche. Thus, behavior adapted to a particular environmental situation may develop without direct participation in the situation.

Selective responsiveness to a few simple *sign stimuli* appears to be typical of innate behavior. That is, of the many stimuli that an animal perceives simultaneously, most are ignored and it responds only to one or a few. Some illustrations will suffice to indicate the often remarkable match between innate responsiveness and the eliciting stimuli provided by a natural situation.

Experiments with foraging honeybees have shown that the colored dots, lines, or other patterns of flowers are used as "guides" for indicating the location of nectar or the path the insect must follow to gain access to it. (Actually, they respond to the ultraviolet radiation pattern of the flower pigments and do not see some colors as we do.)

Fish-eating birds must swallow their prey head first; the backward-pointing spines prevent them from being swallowed tail first. Newly hatched arctic terns and guillemots, for example, orient fish correctly the first time they are presented with one. Experiments with dummy fish indicate that the sign stimulus in some birds is the position of the eye; in others it is the eye plus the tail fin ray pattern, and in still others the relative width of the head versus the tail serves to signal which is the head end. Other stimuli are not relevant.

Newly hatched gulls peck at the beak of their parent when they are hungry. The beak pecking prompts the parent to provide food—by regurgitation. Naive young will peck at any moving, small conspicuous object within reach of their beaks, but certain features add "releasing value" to such objects. For example, narrow, vertical red and blue objects whose lower end is at eye level and moves from side to side, and especially a conspicuous spot or a break in the outline, enhance the attractiveness of such objects. These features, which elicit "begging" in a chick that has never been with parents, closely agree with aspects of the natural situation (Fig. 2-6).

The young of many gallinaceous (henlike) birds and of waterfowl

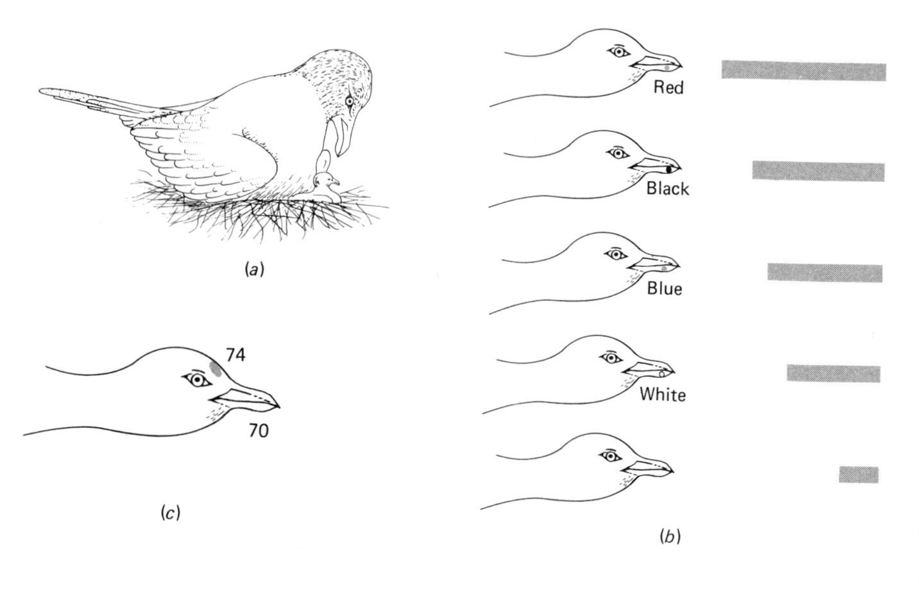

FIGURE 2-6
(a) Adult black-headed gull looking down at a newly hatched chick, the typical situation in which the young starts to beg while seeing the bill of its parent. (b) Models of herring-gull heads used in experiments on the begging response of the young; the strength of the response elicited is indicated by the length of the bar on the right. A red spot is most effective. (c) The begging response of the young gull is directed about as often to the mouth (70 times) as it is to the red spot (74 times), which was shifted in this experiment from the beak to the head of the model. (*B and c redrawn from N. Tinbergen, 1969, The Study of Instinct, Oxford University Press, Fair Lawn, N.J. fig. 24, p. 31.*)

escape or crouch when they see a bird of prey flying overhead. Cardboard models resembling birds of prey elicit escape or crouching responses if they simulate "short-necked" birds, which are like the shape of birds of prey. Models resembling "long-necked" birds flying overhead, which look more like ducks or geese, do not elicit such responses (Fig. 2-7).

Electrophysiological studies of vision have uncovered a number of detector mechanisms which promise to be of importance in explaining innate behavior. The best known of these to date is the so-called "bug detector" in the eye of the frog described by Lettvin and colleagues.[1] The frog retina has some ganglion cells which respond only when a small contrasting object moves erratically in front of the frog's eyes. These units, apparently, inform the frog's brain of the presence of small insects in the visual field. Other units are stimulated by sudden darkening of the visual field and could serve to "warn" the frog of an approaching large predator. They may be a part of the mechanism that makes a frog jump when a predator approaches.

FIGURE 2-7
Model used in testing for escape response in young waterfowl. When sailed overhead with the long "neck" forward it had no effect; when moved with the short "neck" forward it elicited flight and escape.

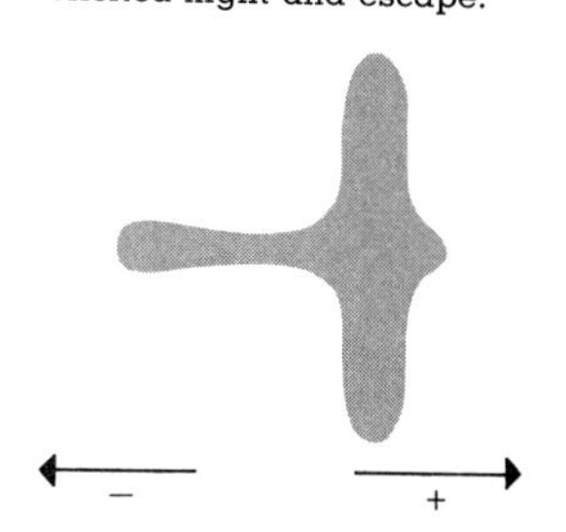

Rhythms and Biological Clocks

Innate activity rhythms are a ubiquitous and important feature in the life of organisms. Most activities occur cyclically, with periodicities ranging from a fraction of a second to years. Many cycles reflect environmental rhythms such as changes of the season, the phase of the moon and the tides, and the day-night cycles.

[1] J. Y. Lettvin, H. R. Maturana, W. H. Pitts, and W. S. McCulloch, 1961, Two remarks on the visual system of the frog, in *Sensory Communication*, edited by W. A. Rosenblith, John Wiley & Sons, Inc., New York.

Under natural conditions the rhythms of plants and animals are synchronized with environmental cues. The best-known examples of this are physiological rhythms that can be "entrained" by light-dark cycles. When the 24-hour light-dark cycle is experimentally reversed in phase, activity rhythms of animals, for example, also will reverse within a few days and be in phase with the new cycle. However, despite this apparent "driving" of the organism's activity by the *exogenous* illumination cycle, the organism has *endogenous* rhythmicity and may not depend on cues from the environment for expression of rhythms.

Female rats are regularly receptive to mating (in heat) every few days. The length of this cycle (the estrous cycle) varies between strains and colonies, but is typically 4 days. During the phase of receptivity female rats are particularly active. When allowed access to a running wheel the female tends to run more at night than during the day, and when in heat she is especially active. Thus, the basic daily, exogenously phased rhythm is modified by an endogenous 4-day rhythm (Fig. 2-8).

Animals placed in isolation so that all fluctuations of light, sound, temperature, and so on are excluded still show physiological cycles with a periodicity of about 24 hours. Such persistence of rhythmicity has been demonstrated in several mammals, birds, reptiles, amphibians, fish, invertebrates, and plants. The cycles tend to be somewhat shorter or longer than 24 hours and thus gradually drift away from earth time. Such maintained rhythms are called **circadian rhythms** (from *circa diem*, about a day). They result from some kind of endogenous, self-sustaining, free-running *biological-clock* mechanism. There is probably no specific location for such clocks, which in fact may be a property of individual, isolated cells.

That biological clocks are innate and not acquired through learning has been demonstrated in experiments in which birds and lizards were hatched in darkness but nevertheless displayed marked circadian rhythms.

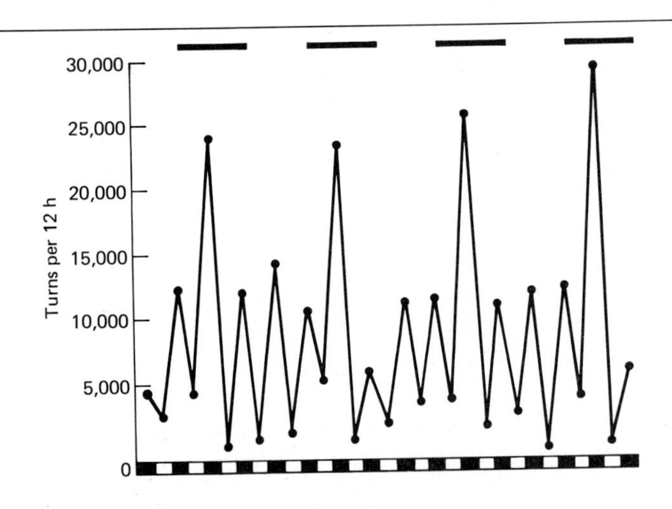

FIGURE 2-8
Activity cycle of a female rat. The ordinate gives the number of turns of a running wheel in 12 hours. Abscissa: the black-and-white pattern represents successive 12-hour light and dark periods. Black bars on top: duration of estrus. (*After L. C. Brownman, in J. Aschoff,* 1958, *Z. Tierpsychol. 15, p.* 19.)

Attempts to imprint periodicities much greater or lesser than 24 hours have been unsuccessful. Animals will, to some extent, conform to imposed photoperiods of unnatural proportions, but when returned to darkness they will revert to a rhythm with a periodicity of about 24 hours.

Some investigators assert that under "constant" conditions the circadian cycles are the result only of an endogenous mechanism operating. Others maintain that endogenous clocks are driven by subtle environmental cues still available to the organism in "constant conditions," but coupled in such a way that phase locking need not occur. Changes in magnetic-field strength or angle may be one of the clues not excluded by standard isolating conditions.

Biological clocks subserve many important functions that have obvious value for survival. They pace the organism's daily, monthly, or yearly activities, so that times of foraging and feeding, sleeping, avoidance of predator, migration, mating, and other functions are optimized. Biological clocks may predict tide changes for littoral animals or time the seasonal cycles of plumage change and breeding of birds. In short, they temporally attune the organism to various rhythmic fluctuations in its environment.

2-2 LEARNING AND BEHAVIOR

Plants and some animals survive and reproduce without any apparent learning, but for vertebrates in general, and especially for the human animal, learning is profoundly significant for biological success. While all types of behavior may be reflexive or spontaneous, it is learning that coalesces them into unified sequences that satisfy biological needs and make these behaviors become *anticipatory*. Learning modifies behavior in ways which are usually adaptive, increasing the likelihood of survival—of the individual and of the species.

Through selection, evolution may produce organisms capable of learning, just as the nerves, muscles, sense organs, and glands which are involved in behavior have evolved. So also may different kinds of learning evolve, trending toward greater complexity. And different kinds of learning may develop within a single organism. Higher mammals appear to have at least three kinds of learning: short-term memory, habituation, and reinforcement learning.

Short-Term Memory

After a few repetitions, motor behavior typically becomes more skilled and effective. This is termed the "warm-up effect" and is the basis of the practice swing of the baseball player and golfer. Sensory and perceptual responses are made more probable by recent stimulation; after seeing or hearing something, one can remember it quite completely for a short period, but this relatively complete memory fades quickly. Short-term memory appears to be effective only for a matter of seconds or, at most, a few minutes. It is postulated that both perceptual behavior and motor behavior activate reverberating circuits (long circular chains of neurons that activate one another) or produce transient changes in neurons, which facilitate behavior. Thus, in **short-term memory** a response or behavior is made *more* likely or vigorous by its recent occurrence.

Habituation

Upon first presentation of a stimulus the evoked response is likely to be relatively vigorous and immediate. If the stimulus is presented again and again the responsiveness may diminish. This effect is termed **habituation.** In contrast to short-term memory, a behavior is made *less* likely or vigorous by its recent occurrence. For example, a noise may elicit an array of behaviors the first time it is heard—the animal becomes alert and looks about, heart rate decreases momentarily, and skin resistance decreases. If the same noise is repeated a few times, certain components of this complex response will become less and less prominent, whereas others will not. The heart rate and skin resistance changes may be more persistent than the tendency to look about, for example.

If there is sufficient lapse of time without presentation of the stimulus, the effects of habituation disappear and the original responsiveness is regained. Recovery from habituation takes minutes, hours, or even days. Thus, the time course for habituation tends to be longer than that for short-term memory. The phenomenon of satiation, for example, may be considered to be a case of habituation in behaviors which are controlled by internal stimuli. A biologically important aspect of habituation is that it increases the relative effectiveness of novel stimuli by decreasing the effectiveness of recurring stimuli.

Reinforcement Learning

Whereas in short-term memory and habituation a behavior is influenced by prior occurrence of the *same* behavior, in **reinforcement learning** a behavior is influenced by prior occurrence of a sequence involving that behavior and a *different* behavior. This different behavior and its stimulus are usually termed *reinforcers*.

The proper temporal sequence is crucial in reinforcement learning: stimuli or responses must be immediately followed by a reinforcer (e.g., reward or punishment). Behaviors which are followed by appropriate reinforcers are protected from habituation. In fact, certain aspects or portions of the behavior produced by the reinforcement stimulus tend to move forward in time—to become anticipatory. The well-known salivation response which begins when conditioned animals see or smell food is an example of an anticipatory behavior. Thus, an animal through conditioning comes to *associate* the sight or smell of food with the consumption of food that will follow. The sight of an object that previously inflicted pain will cause an animal to exhibit avoidance behavior.

The phenomenon of reinforcement learning is found in most, if not all, vertebrates. However, what is positively and negatively reinforcing differs widely among animal species. Food preferences, habitat choice, daily activity cycles, and social status are all influenced by reinforcement.

2-3 LEARNING AND PHYLOGENY

Learning ability differs among animals and tends, more or less, to follow a phylogenetic scale (Fig. 2-1). All efforts, so far, to demonstrate learning in Protozoa have failed to yield conclusive positive results. Simple Metazoa, the Micrometazoa, are capable only of habituation that lasts a matter of minutes. Studies of starfish have successfully demonstrated reinforcement learning. Pacific starfish (*Piaster giganteus*) that habitually rested on the walls of a tank kept in darkness learned in four trials to associate light stimulus with food. When a light was turned on, the starfish descended to the bottom of the tank where food would be presented. Earthworms show clear evidence of habituation and reinforcement learning. Worms can be trained to go to one arm of a T maze leading to a dark, moist chamber and to avoid the other arm, which leads to an electric shock and an irritating salt solution. An average of about 200 trials is required to reach 90 percent correct responses.

Planarian worms (Platyhelminthes) have been used in numerous classical conditioning studies. In one of these a light was turned on, followed 2 seconds later by an electric shock that caused the worms to contract longitudinally. After 150 trials, the worms contracted to the light alone over 90 percent of the time.

Learning in two mollusks is well documented. The sea hare (*Aplysia*), a gastropod, shows rhythms, habituation, and reinforcement learning. Octopuses, cephalopod mollusks, which have the largest brains of all invertebrates, can learn many kinds of visual and tactile discriminations with an accuracy comparable to that exhibited by mammals. By means of rewards (bits of fish or crabs) and punishment (electric shocks) they can be trained to execute a great variety of visual discriminations, and can even master "detour" problems (Fig. 2-9).

Arthropods are capable of complex learning. Visual discriminations, based on color and pattern, have been demonstrated. Ants learn to move through mazes interposed between the nest and food.

In summary, among invertebrates, positive evidence for learning is not found below the level of worms, which possess a bilaterally symmetrical, synaptic nervous system. More consistent and complex learning is characteristic of the cephalopods and arthropods, which have large concentrated neuronal masses (ganglia and "brains"), especially at the cephalic end of their central nervous systems. Compared to vertebrates, however, even the higher invertebrates have relatively inferior capacities for learning. Their behaviors are largely dominated by stereotyped patterns (Fig. 2-1).

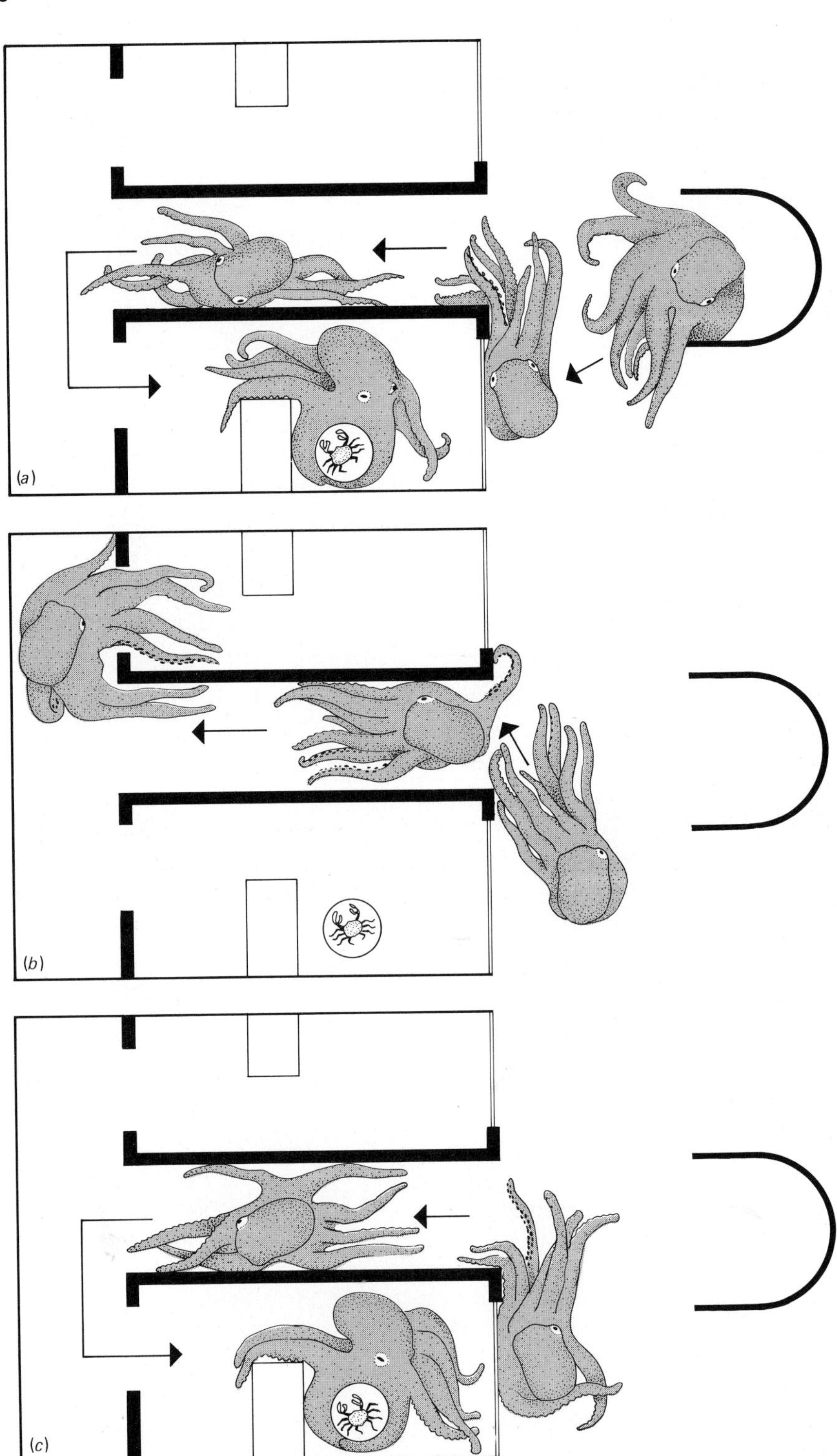

FIGURE 2-9
Apparatus for testing ability of octopuses to solve detour problems. (*a*) Octopus leaves home at right, struggles for 17 seconds against window of feeding area, then passes down opaque corridor and around into feeding compartment. (*b*) Error by octopus blinded in right eye. It found the "wrong" wall and was led astray. (*c*) A correct run by the animal blinded in the right eye. (*After M. J. Wells,* 1964, *J. Exp. Biol., 41,* p. 621–642.)

Fish can master mazes and detour problems much more rapidly and consistently than invertebrates, with the possible exception of ants. Amphibians and reptiles have similar abilities. However, their learning consists of *trial-and-error* learning of relatively simple stimulus-response relationships.

Birds have more facility for learning than lower vertebrates, not only in rate of acquisition, but also in the complexity of problem solving. Mammals are best equipped to learn, being relatively free from simple stimulus-response, trial-and-error learning. Excellent memory from past behavior can be utilized in novel situations. Higher mammals bring to the solution of complex problems rudimentary capacities for reasoning and symbolic behavior (Fig. 2-1).

It should be noted that the learning capabilities have not been tested adequately for any group of organisms. This is true even for mammals, the group whose learning performances have been most thoroughly studied—albeit mostly with genetically uniform strains of a very limited number of species. Kavanau[1] has pointed out that the use of highly inbred domestic animals, where the objective is to minimize individual differences, can reveal only a small portion of the total response spectrum for the species and probably has minimal significance for adaptation and evolution in the wild. The white-footed mice (*Peromyscus*) that he studied were able to master a maze system that was 96 m long, with 1205 90-degree turns and 445 blind alleys in as few as 2 or 3 days, forward and backward, and without extrinsic reward or prior deprivation. Even though such a maze is far more complex than the ones employed in studies with domestic rodents, it is probable that the limits of the learning capacity of white-footed mice were not closely approached by such a challenge.

Genetic information provides the foundation for the behavior of all animals, but in the course of evolution *experiential* information has increased in importance. The nervous system is the machine which gathers and "processes" information and enables an organism to respond to environmental conditions by the appropriate adaptive means. Retention of information increases survival, since it enables the organism to learn from experience and to anticipate the proper adaptive response. Members of the human species, the one best endowed for processing experiential information, add symbolic capacity in the form of languages that allow them to assimilate the past experiences of others of their kind, in addition to their own unique accumulation of experiences.

2-4 POSTULATED MECHANISMS OF MEMORY AND LEARNING

Thermodynamic theory states that there is an exact but inverse equivalence between entropy and information content of a system. Information is thus orderliness, in the thermodynamic sense. However, there are

[1] J. L. Kavanau, 1967, *Science*, **155**, 1623.

many problems to be solved before thermodynamics can be applied to quantitative evaluation of the information content of biological systems. Nevertheless, there are good reasons to believe that the brain of a highly trained organism must possess more order than that of a naive, untrained one. Just where these greater and lesser degrees of order exist is not certain, but that they do exist seems unarguable.

Orderliness accrued in the nervous systems of animals through *learning* expresses itself as memory. Varying degrees of *innate* neural orderliness we might recognize as fixed patterns, as defined above. But how the central nervous system processes innate and experiential information, stores and retrieves long-term memory, and learns as the result of experience are baffling problems. These functions in mammals reside principally in the neocortex, a region containing 70 percent of all brain neurons in human beings and one of unparalleled neuroanatomical complexity.

In 1904 Richard Semon proposed that each stimulus an organism experiences leaves a discrete material trace of some kind within its nervous system. This material trace he called an **engram** and considered that it might be chemical rather than physical.

Several years later the American psychologist, Karl S. Lashley, made a determined "search for the engram" in specific places in the nervous system. He taught rats and monkeys to solve various problems and then made deep incisions in their brains or removed large portions of their brains. Instead of finding a localized engram, he found that his animals showed only "relative" forgetting. That is, if he removed 50 percent of their cortex they forgot about half the problem, and if he removed 25 percent of the cortex they remembered about 75 percent of their training. Lashley concluded from these and other results that storage of experiential information in the brain involves a widespread delocalization of the surviving trace or material which acts as the code.

There is general agreement that some kind of molecular mechanism forms the basis of information processing in nervous systems, but most of the evidence produced to support such a mechanism is very controversial. A brief survey of some current ideas about how learning affects the nervous system and how memory might be coded will serve to indicate the nature of the problem, but may not satisfactorily explain these processes. The various postulates about what transpires in neurons during information processing are not entirely conflicting, but indeed, may be largely complementary.

Electrophysiological Model

During the 1920s and 1930s the brain was considered as a relay organ, like a telephone exchange center, between receptors and effectors so that conditioned responses could easily take place. However, attempts to demonstrate specific pathways in the brain which would be responsible for stimulus-response connections and for the association of new stimuli with already-established memory traces failed. Largely from the ablation studies of Lashley, there emerged the realization that the sought-for "location" of memory was not assignable to some discrete aggregate of neurons. It was concluded that each sensory system of the brain is provided with a large functional reserve, widely distributed through the cortex, and that the *pattern* of neuronal excitation more than the involvement of particular neurons is essential to assure the state of conscious awareness and to account for learning and memory processes.

A Canadian psychologist, D. O. Hebb, formulated in 1947 an electrophysiological model of brain functioning based upon a particular type of patterned modality of neuronal stimulation. He noted that a short-acting electrical stimulation (1 to 2 seconds) on an area of the cerebral cortex causes a prolonged rhythmical production of neuronal discharges (lasting minutes or even sometimes hours). Hebb envisioned "cell assemblies" connected into complex neuronal networks, called "phase assemblies," forming closed loops which, when activated, would permit neural impulses to pass again and again through the same synapses until the circuit fatigued and the circular flow ceased.

In his *dual-trace* model of memory, Hebb suggested that the reverberatory cycling within the neuronal circuits would account for the initial phases of registration of experiental information, or *memory trace*. The establishment of a more durable memory, during the so-called phase of *consolidation of memory*, would result from a structural change in the neuronal network promoted by its having been activated. A *structural* change in the neuronal network was considered to be indispensable for the consolidation of the memory trace.

Clinical studies of head injuries and electroshock treatment indicate that memory does involve at least two storage processes. One process assumes the character of a transient experience, wherein immediate past events may be perfectly clear and complete, only to fade in a matter of seconds if not rehearsed. The other process has the character of a fixed experience, is relatively permanent, and usually can be recalled on demand. The effect of electroconvulsive shock (ECS) on retrograde amnesia has been used to support this memory dichotomy.

If a mouse or rat placed on a raised platform is given a foot shock when it steps off, it will remain on the platform when returned there for the next trial. If ECS follows the foot shock within a brief time interval after the first trial, the animal has retrograde amnesia and steps off the platform on the second trial. Such a result indicates some kind of temporal requirement for "consolidation" of memory. The short-term trace must be converted into a storage form.

A variety of brain manipulations can cause retrograde amnesia, including hypothermia, anesthesia, irritative lesions produced by application of aluminum hydroxide cream to the cortex, a topical application of KC1 on the cortex to produce spreading depression, and direct electrical stimulation of the cortex. All these have the common element of markedly altering electrical activity of the brain.

A large body of evidence can be mustered to support the consolidation theory of memory function. Short-term memory is viewed as essentially the expression of an electrophysiological event involving reverberating circuits of neurons. Memory consolidation needs time to take place and can be disrupted in various ways.

The true nature of the structural change called for in the consolidation phase remains unknown. It has been shown, for example, that the end bulbs of neuron fibers swell as a consequence of activity. Modifications of the content of neurotransmitter affect memory. Facilitation of synaptic transmission is postulated to result from stimulation. Thus, changes in "connectivity" of neurons generally are considered to be the fundamental basis of electrophysiological models of memory and learning.

Biochemical Model

With the discoveries of the macromolecular basis of transfer of *genetic* information, there arose parallel speculations about the possible role of the nucleic acids in processing of *experiential* information. The possibility that synthesis of a new protein within brain cells could occur during a learning experience and represent the material in which the memory trace would be permanently stored, was grasped.

In nerve-cell bodies and extending out into dendrites there is a substance formed into granules and, as revealed by electron microscopy, arranged in a specific way on endoplasmic reticulum. These granules, first seen by light microscopy and named *Nissl substance*, correspond to the microsomal fraction obtainable from other tissues. They are readily stained with basic aniline dyes. Studies in the early 1900s demonstrated that changes in density of staining of Nissl substance occur after neural activity. We now know that these granules are largely ribonucleic acid (RNA) and that they are very active centers of protein synthesis.

It was easy, therefore, for investigators to conclude that in neurons RNA and protein production are modified by "usage." Also, metabolism was shown to increase with increasing demands. Indeed, the neuron can be visualized as a very active, glandlike cell, fulfilling its electrogenic functions while producing proteins and lipoproteins under the modulating control of RNA.

Several pieces of evidence have been assembled to suggest strongly that some relationship exists between RNA and certain types of learning. In the late 1940s Ward Halstead, a psychologist at the University of Chicago, theorized that the "engram" might exist as a molecular change in the nucleic acids of a single cell. An experimental basis for this idea was first provided about 10 years later by the Swedish biologist, Holger Hydén, who carried out a series of experiments that pointed to RNA as the "memory molecule."

Since deoxyribonucleic acid (DNA) stores an organism's "ancestral memories" in coded form, Hydén postulated that RNA might encode an organism's "personal memories." He tested his theory on rats that were divided into two groups. One group was trained to balance on a taut wire in order to reach food; the second group was given passive exercise but did not have to balance on a wire to get food. He then removed large cells individually from areas of the brainstem that were functionally concerned with complicated motor-vestibular functions. He extracted the contents of each cell and, using microanalytical techniques, measured the RNA quantitatively and qualitatively. Hydén found both groups had increased RNA contents, but in the wire-walking rats there had been changes in the base composition of nuclear RNA: viz., an increased adenine/uracil ratio.

Hydén interpreted the change in base composition of RNA as indicating synthesis of a fraction or fractions of RNA in the nucleus with highly specific base composition due to derepressed regions of DNA. Learning is postulated to involve an increase in previously existing messenger RNA (mRNA) or the synthesis of completely new mRNA. His scheme can be summarized as follows: Instructing pattern of impulse → derepression of DNA → new mRNA → new enzyme → new transmitter → increased synaptic efficiency.

It has been found that the gross RNA content in the human brain increases from birth to about age 40 years and then remains relatively constant until about age 60 when it begins to decline. Injections of purified yeast RNA have been reported to have favorable effects on organic memory deficits in aged people! Laboratory rats given purified yeast RNA are reported to be able to acquire behavioral responses faster than control rats. However, these findings have been disputed and remain equivocal.

The molecular changes of the brain occasioned by learning are not restricted to neurons, according to Hydén. Neuroglia, nonneuronal supportive cells of the brain (cf. Chap. 11), transfer their RNA content into neurons during the learning process. Hydén postulates a "symbiosis" between the neuroglia and neurons, with regard to both polynucleotide production and metabolic energy.

Several antibiotics, which inhibit protein synthesis, have been the main tools in the study of memory suppression. *Puromycin*, an antibiotic that prevents protein synthesis by releasing nascent peptides from ribosomes through hydrolysis of the amino acyl link between the most recently added amino acid and transfer RNA (tRNA), is supposed to block memory when bilaterally injected into certain parts of the brain. Likewise, *8-azaguanine*, which will insert into forming RNA in place of guanine and thereby produce meaningless codes, is reported to depress learning when administered to the brain.

B. W. Agranoff, beginning in the 1960s, conditioned goldfish to avoid a shock by swimming back and forth from one compartment to another. When puromycin was injected into their cranial fluid a few seconds after the learning experience, the goldfish forgot the lesson. If the injection is done 1 hour later, however, it remembers the experience. This and similar results have been used to argue that memory consolidation from the trace to the permanent memory is prevented by inhibition of protein synthesis. However, there are conflicting data on the effects of protein inhibitors, especially among experiments with rodents.

Some results indicate that it is not the registration of memory that is affected, but rather *recall*. When memory of maze learning is blocked in mice by puromycin, the block can be removed by intracerebral injection of sodium hydroxide a few days after the training experience and by saline injections at least up to 2 months later. One explanation of these puzzling results is that the relevant effect of puromycin may be to modify the neuronal membrane, especially the synaptic membrane, by causing the formation of abnormal peptides. There is some direct evidence that puromycin interferes with the electrical activity of brain cells. Thus, at least for puromycin, suppression of memory may result from disruption of neuroelectric processes, not from inhibiting nucleic acid or protein synthesis in neurons.

Whatever the molecular mechanism of memory may be, it must involve both neurophysiological and neuroanatomical changes. Presumably, some neurophysiological event in the nerve cells produces the chemical change, which in turn must have some way of altering the function of the nervous system to produce memory. Probably it is at the synapse that the crucial modifications are made.

READINGS

Dethier, V. G., and E. Stellar, 1970, *Animal Behavior*, 3d ed., Prentice-Hall, Inc., Engelwood Cliffs, N.J. This introductory volume approaches animal behavior from an analysis of the complexity of different kinds of nervous systems. Much basic, biologically important information can be found here.

DeWied, D., 1969, Effects of peptide hormones on behavior, in *Frontiers in Neuroendocrinology*, edited by L. Martini and W. F. Ganong, Oxford University Press, London. An accounting of experiments that support the view that peptides of pituitary-gland origin may influence learning and memory.

Honig, W. K., and P. H. R. James (eds.), 1971, *Animal Memory*, Academic Press, Inc., New York. Various models of memory and learning are discussed and their molecular basis is described.

Several Authors, 1971, *Topics in Animal Behavior*, Harper & Row, Publishers, Incorporated, New York. This excellent collection of articles by established authorities deals with innate behavior, learning, imprinting, motivation, and communication.

Ungar, G. (ed.), 1970, *Molecular Mechanisms in Memory and Learning*, Plenum Press, New York. A gathering of reviews of their works by several of the investigators of the molecular mechanisms of learning and memory.

SOCIAL AND REPRODUCTIVE PHYSIOLOGY

3

Social physiology is meant here to encompass those activities which expressly depend upon the interaction of two or more organisms. It includes intra- and interspecific agonistic and affiliative activities, mechanisms that ensure continuance of the species through reproduction and parental care, responses which regulate population size and dispersal, and communication. Much of this subject matter would be classified as social behavior by psychologists and as community dynamics by ecologists. The physiologist is primarily interested in these events in terms of sensory processes, effects on integrating systems, and output responses—i.e., in the mechanisms that underlie the responses of organisms to specific encounters with members of their own or different species.

3-1 AGONISTIC BEHAVIOR

Individuals of a single species tend to aggregate in places that are optimal for their existence. As a consequence there is generally competition for space, food, shelter, and mates. The several kinds of activities displayed during competition, such as fighting or fleeing, hostility or submission, hiding, and "freezing," are called **agonistic behavior.** In intraspecific competition, although there is sometimes serious combat, as when an individual invades the territory of another, more often *threat displays* are invoked. These are ritualized social signals by means of which the stronger (or more strongly motivated) individual tries to dominate its competitor and send it fleeing or force it to accept a submissive role. Intraspecific conflicts seldom result in serious injury to the combatants; their fights are generally *ceremonial.*

Through such combats, animals living in a group establish *social hierarchies* in which the weaker or less aggressive individuals accept dominance by stronger and more aggressive members. Social hierarchies have been identified in a variety of mammalian species, including primates, and in flocking birds. *Pecking order*, so named by Schjelderup-Ebbe in 1913, of domestic chickens results in reducing conflicts arising from competition for food, shelter, or mate. Even among nonsocial animals, such as crayfish and fish, *rank order* is established among individuals sharing a neighborhood. The social position, or the rank of the individual, has profound consequences in such matters as breeding priority and access to food. Also, as we shall describe later, there are significant morphological and physiological correlates of social position.

Fear, rage, and anger are well-known examples of agonistic activities commonly displayed by mammals. These names characterize the outward manifestations of emotional states, and we recognize them by certain stereotyped, *expressive movements* that are typical for the species, such as crouching, freezing, hissing, or baring of teeth. Such emotional arousals involve activation of the autonomic nervous system, which causes changes in rate of breathing, heart rate, blood pressure, sugar metabolism, and brain neural activity. These responses prepare the animal for "fight" or "flight," to use an old cliché.

Certain species are characterized by dominant, *aggressive* traits: the hunting, stalking carnivorous feline and canine species, for example. Others display more *defensive* traits. Herbivorous animals, such as ungulates, to name one group, generally flee when threatened and become aggressive only in a few circumstances. The aggressive species have morphological features, such as sharp teeth and claws and muscular, agile bodies that suit their mode of existence. The defensive species, on the other hand, tend to have legs designed for rapid and sustained running. Here again is the complementary set of adaptive features: behavioral, morphological, and physiological properties matched to the life-style of the organism—or, rather, the form of agonistic behavior, morphology, and physiology is determined by the animal's basic mode of existence.

Endogenous Influences

The appearance of agonistic behavior and sexual maturation are closely correlated in many species. In general, among vertebrates, males are more aggressive than females, and the aggressiveness of the male tends to increase as sexual maturation progresses, becoming particularly pronounced during the breeding season. Male mice confined together in a cage will immediately engage in combat, and the fighting will continue until one establishes dominance over the other. Males castrated prepuberally do not show such aggression. However, when castrated males are injected with *androgens* (male sex hormones, such as *testosterone*), they will engage in combat like normal males. Male mice castrated in adulthood, on the other hand, persist in fighting after castration. Furthermore, castrated males conditioned by testosterone administration continue to be fighters after withdrawal of the hormone.

Circulating levels of androgens thus appear to be causal agents of combativeness. Indeed, administration of testosterone will increase aggressiveness and improve social rank order in female swordtail fish, chickens, ring doves, starlings, female rats, and other animals. On the other hand, female sex hormones, such as *estrogen* and *progesterone*, tend to lower aggression in some species.

Here, then, we have examples of hormones conditioning behavior. To some extent, the opposite may occur: behavior can condition hormone-mediated actions. For example, a male cichlid fish that cannot defend and establish a nesting territory fails to develop its hormonally dependent, nuptial color markings. If, however, the dominant, territory-holding male is removed, the one next in rank order in the social hierarchy will develop nuptial colors. Another example is among chickens, where males that are low in rank in the social hierarchy of a flock show total suppression of sexual behavior, even when the dominant males are removed. When the socially "emasculated" male is put into another flock it still fails to mate. Male rhesus monkeys show a direct relationship between blood level of testosterone and dominance position. When a leader is deposed his testosterone level sinks.

Endogenous influences associated with other types of behavior also may modify aggression. For example, fighting becomes more frequent among animals when they are hungry or thirsty.

Spacing of Animals

A simple type of aggression is commonly associated with the maintenance of "personal space," espcially in the "distance" animals, as contrasted with "contact" animals that much more readily permit proximity and even body contact. Distance animals generally maintain a certain personal space around themselves by limiting the approach of another animal, by withdrawing from it or by driving it away. Personal space has been measured with considerable precision in certain birds. In chaffinches, for example, a male in nonreproductive condition will either attack or withdraw from another male if it approaches closer than about 20 cm. A female can approach a male to within half this distance before it is attacked. Painting the gray-brown breast feathers of the female orange-brown, so as to resemble those of the male, results in the female being attacked at the same distance as males would be.

In animal societies composed of individuals which recognize one another, a stranger elicits aggression more readily than a familiar individual. Recognition may be primarily visual, auditory, olfactory, electrical (among electric fishes), or it can depend upon some combination of these mechanisms.

A fixed space defended by a single animal, a pair, a family, or a larger group of animals is known as a *territory*. The habitats of many animals are mosaics of such territories. Mammals commonly identify their territories by ritualized urination or defecation (e.g., rhinoceros) or by rubbing scent glands on objects in their territories (e.g., rabbits). They mark where they are likely to attack another member of the same species.

Agonistic behavior, then, serves to segregate individuals within a group or a territory in such a way that each can acquire, in accordance with its strength and dominance characteristics, a separate niche for itself.

3-2
AFFILIATIVE BEHAVIOR

Fish tend to aggregate in schools, birds in flocks, mammals in herds, troops, or packs, and some insects into societies. Aggregations may be the result of localized sources of food or water, or optimum sheltering from adverse environmental conditions.

Sexual attractions, seasonally conditioned in most cases, the rituals of courtship, and the cooperative rearing of young may produce affiliative bonds between pairs of male and female animals. The affiliation of sex partners may be ephemeral or may serve as the basis for group formation.

Certain benefits accrue to the individual from belonging to a group. For example, location of food by an individual is signaled to the other members of the group in many species. The hunting and killing of prey, as by a wolf pack, is an example of group aggression. An individual alarmed by the detection of some presumed danger immediately notifies the rest of the group, which can flee or prepare for defense. The *alarm reaction* can involve vocal, visual, or chemical cues (these are discussed in Sec. 3-6).

There are many examples among animals of the initiation or induction of the performance of an act as a result of its exhibition by others. For example, when one bird of a flock suddenly flies off, all the others immediately follow it. Such influences on the group are referred to as *social facilitation*.

The physiological state of individuals is influenced by the organization and size of the group to which they belong. For instance, in mice the mean thickness of the adrenal cortex increases linearly as the local population increases, with a subsequent decrease as population density becomes supersaturated. However, the magnitude of the increase in size of the adrenal gland of individual mice is related to social rank. Those with lowest rank show the greatest increase and, in contrast, the adrenals of dominant mice are comparable in size to the adrenals of mice from small groups. Such changes, presumably, are reflective of an organism's response to *stress*. The adrenal cortex produces steroid hormones which make metabolic adjustments necessitated by exertion or exhaustion. A concomitant effect, either directly or indirectly, is the retardation in growth and fertility of individuals and a consequent fall in population. The latter aspect will be discussed further in Sec. 3-5.

3-3 REPRODUCTION

Survival of most animal species demands sexual propagation and, sometimes, parental care of offspring. Many varied and elaborate mechanisms have evolved to ensure that matings will occur and that young are produced in optimal conditions.

Among most vertebrates and many invertebrates adult sexual competence is a cyclic phenomenon and is usually confined to a certain season of the year. Even among continuous breeders, fertility is only periodic, especially in the female. Consequently, reproduction and the attendant sexual and parental behavior must be precisely timed and coordinated events.

Exogenous Cues for Reproductive Processes

The cyclic maturation of reproductive organs and stimulation of sexual behavior is profoundly influenced by various external conditions, particularly by length of daily illumination, but also by temperature changes, availability of territory, presence of other members of the species, and other factors. The roles of such exogenous stimuli are difficult to separate from the contributions of endogenous stimuli, as is true of any attempt to make such a distinction for any kind of motivated behavior. A number of animals have been observed to have a tendency to exhibit cycles of breeding and nonbreeding while kept under conditions in which such common variables as light and temperature are maintained constant throughout the year. There is no evidence of a precise annual rhythm under such conditions, but there is clearly the predisposition to cycle independently of the more obvious environmental cues. Here, then, we may have an example of *inherited,* innate patterns that normally become interwoven with environmentally induced responsiveness to produce the adaptive annual breeding cycle.

While the breeding season of most species conforms to an annual cycle, there is great variation in duration, synchrony, and responsiveness to environmental cues. Furthermore, in the animal kingdom considered as a whole, reproduction may occur at any time of the year (Fig. 3-1).

Animals inhabiting arid regions commonly are aseasonal breeders, often reproducing in response to rainfall. When rains are seasonal, a more regular cycle is found. Certain birds and amphibians are particularly attuned to such opportunistic modes of reproduction. Their gonads are maintained in a state of readiness for prolonged periods, and within a short time after the onset of favorable conditions they can initiate intense reproductive activity. The red-billed dioch, *Quelea quelea,* of Africa, for example, begins breeding after rain, in apparent response to the availability of green grass with which it weaves its nest. Similarly, the zebra finch, *Peophila castanotis*, of northwestern Australia, after many months with no rain and no breeding activity at all, has been observed to begin courtship behavior in the rain of a thunderstorm.

The reproductive cycles of a few animals are known to be partially regulated by lunar periodicity. The Atlantic palolo worm, *Leodice fucata*, the Pacific palolo worm, *Eunice viridis*, and the grunion fish, *Leuresthes tenuis*, breed annually at certain phases of the moon, with a remarkably accurate timing that is quite predictable.

The Pacific palolo worms only once or twice each year emerge from crevices and holes in coral reefs and swim to the surface. There their bodies separate into two parts, one of which apparently returns to the bottom to conceal itself and regenerate the missing half, while the other segment remains swimming amidst masses of its kind on the surface. This segment is reproductive. Polynesians gather masses of the reproductive segments, eating them raw and collecting baskets full of them for traditional feasts. The average date of swarming is November 27, but the exact date is somehow timed to the lunar cycle, occurring 7, 8, or 9 days after the full moon that falls between November 7 and December 22. The Atlantic palolo is most likely to swarm about 8 days after the full moon in July.

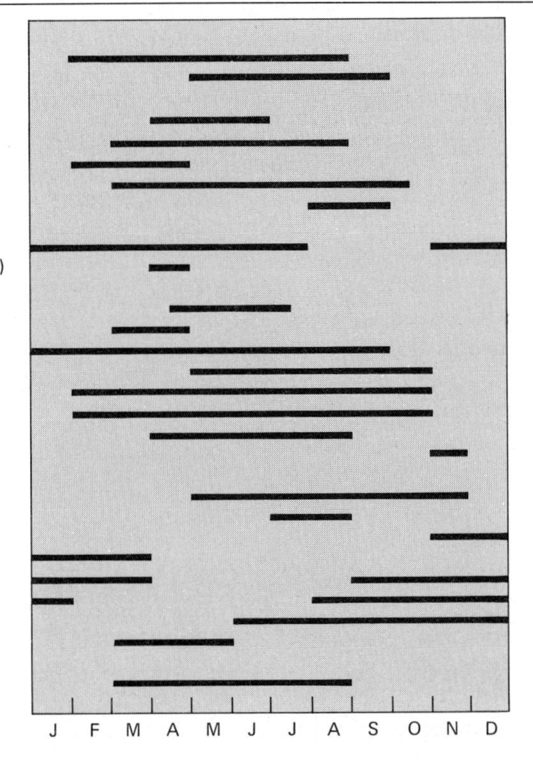

FIGURE 3-1
The season of mating activity (black bars) of some representative mammals. Southern-hemisphere seasons have been converted to Northern-hemisphere equivalents. The records are for several locations. (*After Everett, 1966, from P. Marler and W. J. Hamilton, III, Mechanisms of Animal Behavior, 1967, John Wiley & Sons, Inc., New York, p. 75.*)

Grunions deposit their eggs and sperm in the sand of southern California beaches during a season existing from March through August. However, the heaviest runs are in late spring and summer, coinciding with the highest waters of each tide and with the highest tides of each lunar month. During spawning runs the fish come ashore just after the high tide has begun to ebb. The females wriggle their tails 2 or 3 in (5 to $7^1/_2$ cm) into the wet sand and deposit eggs. Milt from the males runs through the sand or down the sides of the wet females. On the next wave the fish will swim back into the ocean. The eggs develop during 2 weeks in the sand and then hatch when wetted by the next high tide. An individual female makes from four to eight spawning runs during the breeding season; thus, there is a recurring lunar subcycle within the annual cycle.

Day length is the most important external timing stimulus regulating breeding in nonequatorial species. Decreasing light period in the fall is a cue to autumn-breeding animals such as brook trout and ruminant mammals, which begin reproductive behavior in response to shortening days. The combination of lengthening daylight and shortening nights in spring triggers gonadal recrudescence in spring-breeding fish, birds, and mammals. The specific time of breeding for a particular species has probably been selected so that young are produced when the environmental conditions are optimal for their survival. Experimental manipulation of photoperiods has been demonstrated to readily induce reproductive behavior out of season in several kinds of animals.

Hormonal Basis of Reproductive Behavior

A series of well-synchronized events is associated with the maturation of gonads and their products, the gametes, in all species. These include changes in the tracts which will transport the gametes to their place of union, and may include the production of protective coatings (shells) to envelop the ovum in certain species or the production of a brooding chamber within the female parent where development of the young occurs. Certain external morphological changes also accompany gamete maturation in several species. These include the appearance of bright, nuptial plumage in certain species of birds, color patches in fish and reptiles, and courting and ritualized display behavior in most vertebrates.

Maturation of gonads is associated with increased quantities of certain hormones. Some of these trigger sex behavior and thereby ensure that the male and female gametes will be brought together for fertilization. The hormones produced by the gonads during gamete maturation may originally have been involved locally in stimulating metabolism, so that the rather prodigious energy and material demands of gamete formation, especially those of ova, could be met. Only later in the evolutionary history of organisms, according to one hypothesis, did these signal molecules serve to control events in distal parts of the body, including release in the central nervous system of sex behavior. Thus, the production of gametes and the behavior appropriate to favor their efficient usage become linked together through hormones.

Detailed information about the specific patterns of behavior which bring male and female in close enough proximity for successful union of gametes exists chiefly for mammals. The following discussion of this aspect of reproduction, therefore, will be limited to mammals, and much of what follows applies specifically to laboratory rats, the most thoroughly analyzed mammal from the standpoint of reproduction.

Female sex behavior tends to be a passive response: she stops running or trying to avoid the male and stands firmly so the male can mount. Often the perianal region is elevated so that the male has easier access to the vaginal canal; the tail is usually elevated and deflected to the side. This posture, which in many species is readily and unmistakably recognized, is termed **"lordosis."**

(Except in primates, the female mammal usually shows receptivity to the male only during a *specific phase* of the reproductive cycle when she bears mature ova that are ready to be fertilized.)

The sexual behavior pattern of the male is more complex. He must first recognize the receptive female; odor appears to be important. Then he must orient so that mounting occurs with appropriate relationship to the female. Intromission must occur, followed by ejaculation of the male's gametes into the vaginal canal.

If the gonads are removed before an animal attains sexual maturity, the individual never shows sex behavior, implying that substances produced by the gonads are required for sex behavior. If a mature female mammal which is showing receptivity has her gonads removed (a surgical procedure known as *ovariectomy*), she becomes nonreceptive within a few hours and never again shows receptivity. However, if such a castrate female is given a series of injections of hormones produced by ovaries, she will again show receptivity. The reestablishment of normal sexual behavior usually requires a sequential treatment with two ovarian steroids: **estrogen,** for one to several days, followed by **progesterone.** Within 2 to 4 hours following administration of progesterone, the female becomes receptive.

Castration of males (by a surgical procedure known as *orchidectomy*) seriously affects sexual behavior, although sexually experienced males do not show so rapid a time sequence between removal of the testes and the disappearance of the sexual response pattern as in the female. Instead there is a gradual diminution over weeks, months, or years, with disappearance of ejaculation, intromission, and pelvic thrusts, usually in that order. A series of injections of male hormone, **testosterone,** will result in reestablishment of the complete pattern.

The conditions necessary for the production of critical quantities of gonadal hormones that must be present to elicit sexual behavior are fairly well known. Central among the factors that trigger the behavior is the light cycle. Information about the relative amount of light and darkness in a day is somehow transcribed into messages that feed into a central integrative region of the brain and eventually result in the release by the anterior pituitary gland of two gonadotropic hormones. One of these is **follicle-stimulating hormone (FSH),** which stimulates the development in the ovary of egg-containing follicles. Full follicular maturation probably depends upon a second pituitary substance, **luteinizing hormone (LH),** which definitely is required for release of the egg (ovulation).

The pituitary gonadotropins are controlled, not by nervous connections from the brain, but rather by chemicals, called *releasing* and *inhibiting factors* or *hormones*, which are produced by clusters of neurons in the brain—in the hypothalamus specifically—and pass via a blood portal system to the pituitary, where they trigger the release of FSH and LH (Fig. 3-2).

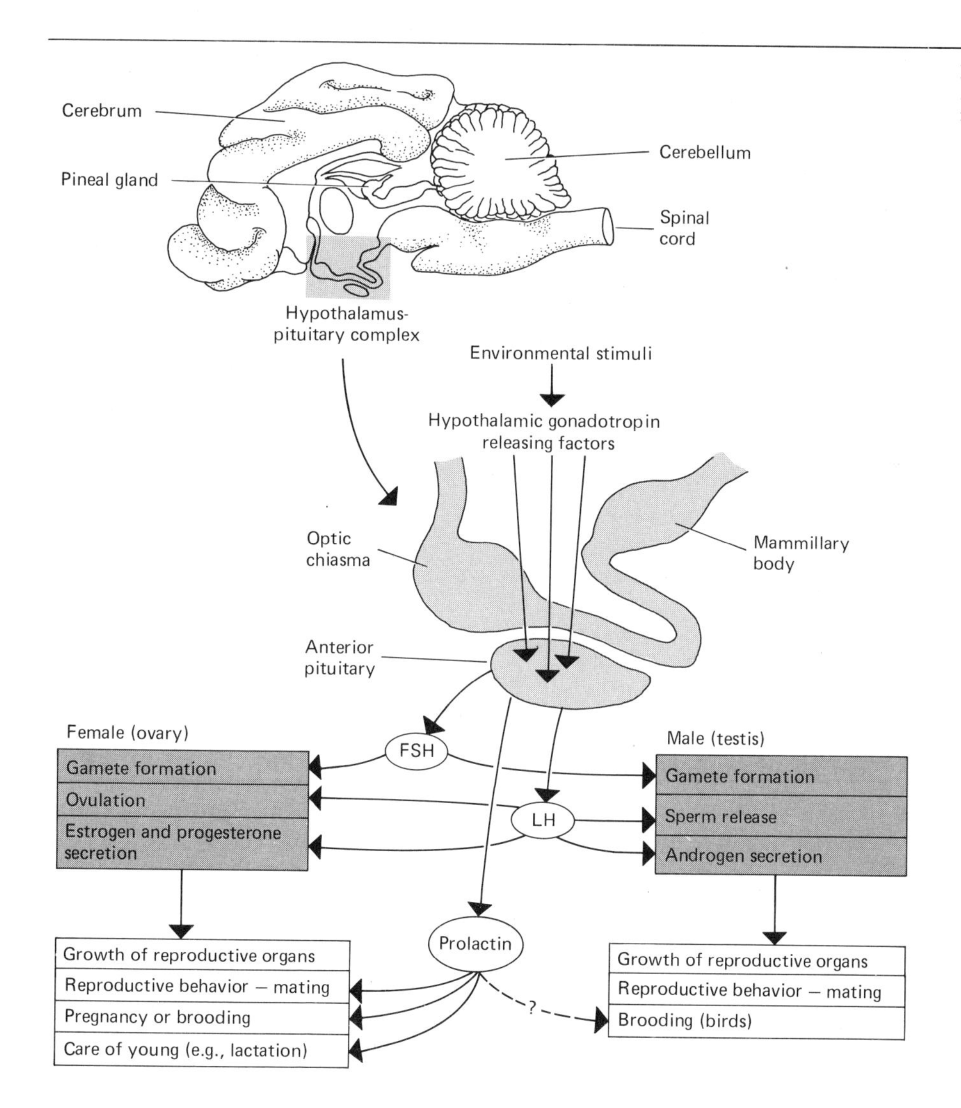

FIGURE 3-2
Neuroendocrine control of reproductive cycles.

It recently has been demonstrated that injection of luteinizing hormone-releasing factor (LRF) into estrogen-primed, hypophysectomized, ovariectomized female rats facilitates the appearance of the lordosis response. Thus, LRF would seem to supplement the copulatory behavior of the female rat that results from estrogen and progesterone action.

The gonadotropins travel via the general systemic circulation and affect the gonads, causing maturation of the gametes. The maturing gonads, in turn, produce hormones. These are steroid molecules: mainly **estrogens** and **progesterone** by the ovary and **testosterone** by the testis. These molecules, in addition to the gonadotropins, are necessary for complete maturation of the gametes. They also have other major functions. The sex steroids are necessary, for example, for the maturation of the reproductive tract and for sexual behavior, as noted above.

The system for regulating gonadal function just described is tightly integrated and rigidly controlled. Detectors in the brain and pituitary continuously monitor the level of sex steroid and regulate, through a chain of processes, the rate of release of gonadotropins. Thus, when the level of sex steroids reaches a predetermined level ("set point," in control-theory terminology, which is genetically determined) the brain sensors reduce the chemical signals to the pituitary; gonadotropin secretion, in turn, is reduced, and steroid production slows down.

The steroid receptors have been localized by implanting minute quantities of sex hormones in various sites of the brain and in the pituitary to see where they are most effective. In the female rat two hypothalamic regions above the pituitary have been found to be steroid-sensitive—a

FIGURE 3-3

Illustration of feedback principle for maintaining a function within "set" limits. (*a*) Thermostat regulating a furnace. (*b*) Analogous system for regulating gonadal function. (*c*) A more detailed model showing where hormone "sensors" are located for regulation of sex behavior and the pituitary-gonad axis (based on studies in the female rat). (*After R. D. Lisk, 1971, in Topics in Animal Behavior, Harper & Row, Publishers, Incorporated, New York, p. 35.*)

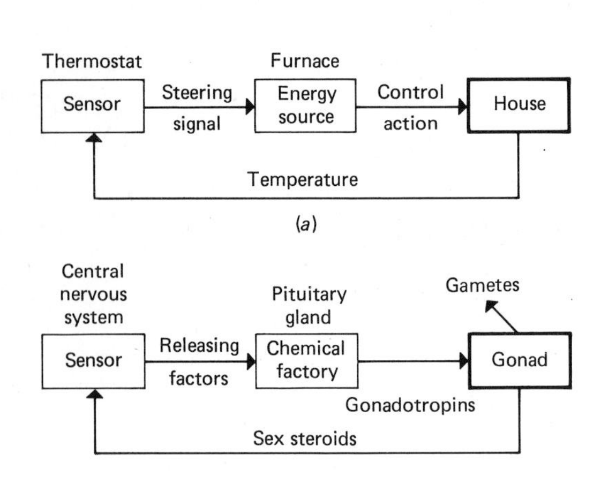

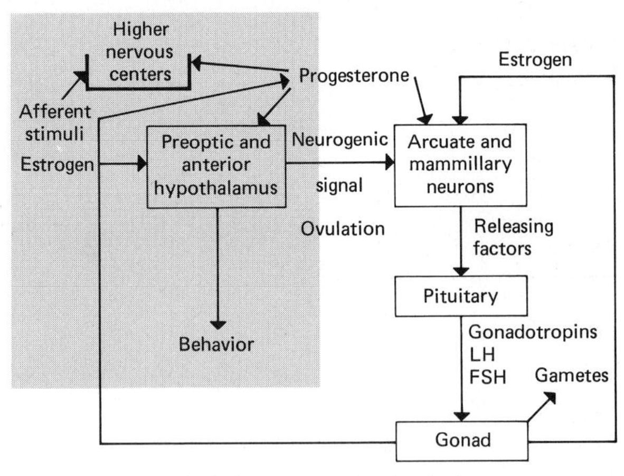

rostral group of neurons in the preoptic and anterior hypothalamus and a neuron cluster more caudal, among the arcuate and mammillary neurons (Fig. 3-3).

Sex steroids applied locally in the hypothalamus of the female can induce receptivity to the male's sexual advances. This has been demonstrated in rats, cats, rabbits, and hamsters. The precise anatomical location within the hypothalamus for the behavioral effect varies from species to species.

Estrous and Menstrual Cycles in Mammals

Within the annual reproductive season, female mammals go through one or more cycles of fertility. These cycles are known as **estrous cycles,** because of the occurrence in each cycle of a period of estrus, or heat, during which the female is sexually aroused and receptive to the male. Males do not show such periodic cycles and are capable of copulation and fertilization throughout the breeding period.

The duration of the female's period of receptivity to males varies from species to species, but generally lasts only a few hours or days. Ovulation and fertilization generally must occur during or shortly after heat. In the absence of fertilization, estrous cycling in polyestrous female mammals continues (Table 3-1).

If fertilization and implantation of the embryo occur, the course of the cycle is broken. There then ensues a series of events which result in changes in the uterine wall and the provision for nutrition, parturition (birth), and suckling of young, together with the associated behaviors directing nest building, defense, and care of the young.

If pregnancy does not occur, there will be a brief pause and then a new estrous cycle will begin. The total length of the cycle, from one period of estrus to the next, varies from a few days to several weeks or months in larger animals.

Primates have a unique kind of reproductive cycle known as the **menstrual cycle.** The length of the cycle is highly variable. In the human female it has commonly been thought that 28 days is typical. Recent sta-

Table 3-1
The Length of the Estrous Cycle and Duration of Heat in Some Animals

Animal	Length of cycle	Duration of heat
Rat	4–5 days	12–15 hours
Guinea pig	16 days	6–12 hours
Dog	3–4 months	7–10 days
Pig	21 days	2–3 days
Sheep	16 days	30–36 hours
Cow	21 days	13–17 hours
Woman	28 days*	No period of heat

* This may or may not be a mean length; cycles vary greatly in length in normal women.

tistical studies demonstrate that only a small percentage of women actually have cycles of this length. What is a "normal" menstrual cycle varies greatly among women, from a very few days to several weeks.

Both estrous and menstrual cycles are regulated by the interactions of the same pituitary and ovarian hormones. They differ in that (1) primates have a menstrual phase in their cycle and (2) sexual receptivity is not confined to a period of heat around the time of ovulation. During the menstrual phase the superficial layers of the uterine endometrium (the glandular interior) are sloughed, with accompanying bleeding; this type of bleeding does not occur in nonprimates.

Four phases of the menstrual cycle can be distinguished: menstrual, proliferative (follicular), ovulatory, and progestational (luteal). The *menstrual phase* is the easiest to recognize, because of menstruation. It corresponds with the formation by the ovary of new *Graafian follicles*, spheres of epithelial cells each enclosing an ovum (Fig. 3-4).

The *follicular phase* is conditioned by estrogen. The endometrium thickens as the estrogen titer rises and becomes glandular and well vascularized.

No conspicuous changes in the endometrium have been seen during the episodic *ovulatory phase*. A distinct rise in body temperature of the human female correlates with ovulation, and it remains high until the onset of the next menstrual period.

During the *luteal phase* the uterus is strongly influenced by both estrogens and progestogens. The endometrium differentiates into a tissue that will allow an embryo to implant. This phase is also characterized by transformation of the follicle, which has just shed its ovum, into an endocrine structure, the *corpus luteum*. In histological sections corpora lutea appear as solid, whitish bodies—hence the name. The corpus luteum, in addition to other sources, produces progestogens (Fig. 3-4).

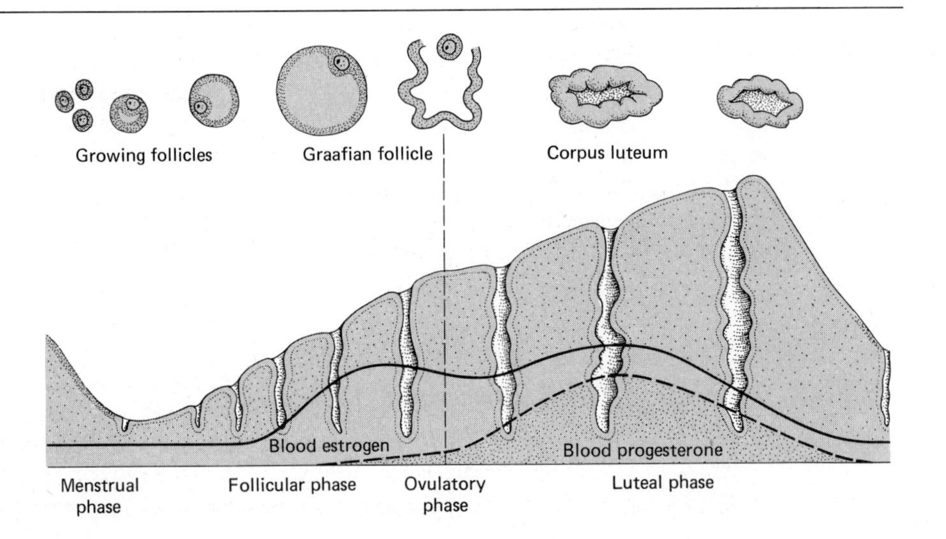

FIGURE 3-4
Diagram showing changes in the endometrium, the ovaries, and the circulating ovarian hormones during a menstrual cycle.

While it is active, the hormones of the corpus luteum inhibit pituitary output of FSH, but with the beginning of menstruation the inhibiting influence is removed. FSH is secreted in increasing amounts and stimulates the growth of young follicles. These, in turn, release increasing quantities of estrogens (Fig. 3-4). Supposedly, the high estrogen content of the blood suppresses FSH production and increases the output of LH, which triggers ovulation when it rises to a proper titer. Following ovulation, the ruptured follicle is transformed under the influence of LH into the corpus luteum, which secretes progesterone and a small amount of estrogen.

If a fertilized egg is not implanted, the corpus luteum begins to degenerate 8 to 10 days following ovulation.

As the corpus diminishes its function, degenerative changes begin in the endometrium. Leukocytes invade the tissue, necrotic changes occur in the stroma, and the uterine glands involute. With the onset of menstruation, the outer portion of the endometrium is lost and there is bleeding into the uterine cavity. The onset of menstrual bleeding correlates with the withdrawal of progesterone and, to a lesser extent, of estrogen from the circulation.

If the ovum is fertilized and implanted, the pituitary continues its output of LH and the corpus luteum increases in size and output of its hormones. The corpus luteum diminishes its hormone output slowly after the fourth month of pregnancy, and the *placenta*, rather than the ovary, then becomes the principal source of progesterone and estrogen for the remainder of the gestation period.

3-4 HORMONES AND PARENTAL BEHAVIOR

Nest building, incubation of eggs, parturition, and rearing of young are dependent upon hormonal activation, but there is relatively little detailed information about the mechanisms by which they are effected. Generally *nest building* occurs among birds at the height of follicular development, and, in species in which the male also participates in nest building, this activity coincides with the height of testicular activity. In night herons, the male and female of which collaborate in nest building, testosterone was found to initiate nest building in both sexes, whereas estrogen had no effect on this behavior in either sex. In ring doves, pretreatment with estrogen induces immediate nest building without the normal antecedent courtship behavior. Injection of progesterone into intact or castrate female mice produces increased nest-building activity, but has no effect on males.

Pituitary leuteotropic hormone (LTH) (also called *prolactin*) evokes *incubation* in broody hens (*Gallus domesticus*). Furthermore, LTH in domestic hens and other birds leads to nurturing the young, including brooding, guiding, and protecting them. Injections of LTH in the domestic cock, which normally does not participate in care of the young, will induce some, though not all, patterns of broody behavior.

Shortly before the start of incubation, many female birds develop one or more *incubation patches* on the breast and belly. The feathers are shed, the epidermis thickens, and there is a great increase in vascularity and swelling (edema) of the underlying tissues. These changes are caused by increased estrogen and LTH production, and perhaps require the participation of progesterone also (Fig. 3-5).

In phalaropes, in which some roles of males and females are reversed, it has been found that testosterone, rather than estrogen, is the gonadal hormone which acts synergistically with LTH to develop the incubation patch. Female phalaropes, which normally do not develop an incubation patch, do so if treated with testosterone and LTH. In a few passerine species both males and females have incubation patches, but the hormonal mechanisms supporting this condition are not known.

The young of some species of birds are well feathered and capable of locomotion immediately after hatching (*precocial young*), whereas others are featherless and incapable of locomotion for several days after birth (*altricial young*). Domestic chicks and ducklings are examples of the first category. The female parent aids the young in finding food, responds to their distress calls, retrieves them when scattered, and spreads her wings around them to provide warmth and protection (brooding).

Parental behavior among parents of altricial young is more complicated. One or both parents may assume the task of feeding the young, depending upon the species. Much of the daily activity of such birds may be consumed in food gathering. The feeding of young is guided in several species by special innate signaling devices. The begging of food by herring-gull chicks, which peck at the color patch on the parent's mandible, is an example (discussed in Chap. 2).

FIGURE 3-5
The incubation patch in a bird. (*a*) Ventral views, showing the position of the patch. (*b*) Stages of the incubation-patch cycle in the female white-crowned sparrow. (*After P. Marler and W. J. Hamilton, III, 1967, Mechanisms of Animal Behavior, John Wiley & Sons, Inc., New York, p. 90.*)

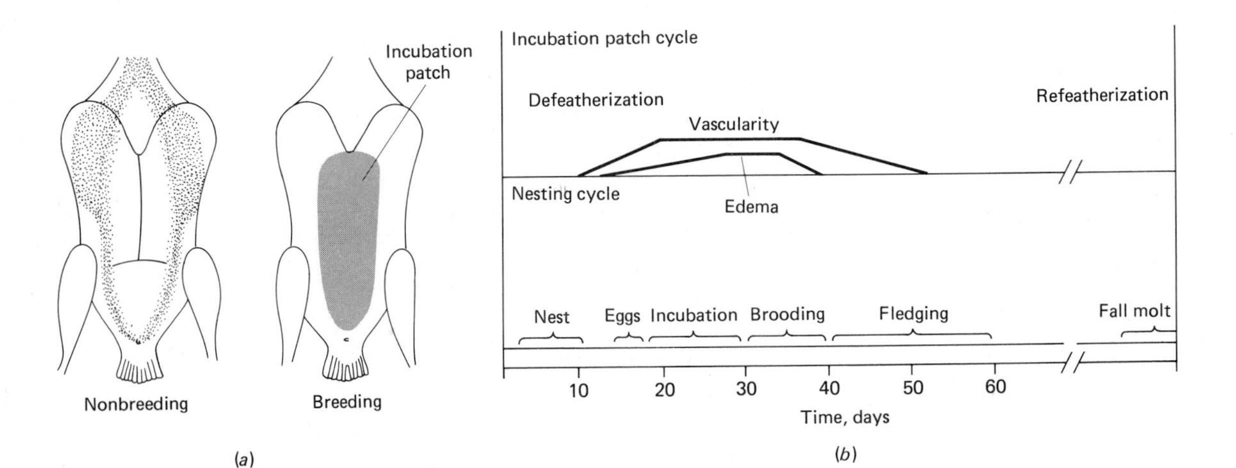

Both sexes of pigeons and doves participate in incubating the eggs and in feeding the young. Cells from the epithelium of their *crop sac*, an expanded portion of the esophagus, desquamate and form a mass called "crop milk"; this is regurgitated and given to the young. Under the influence of LTH, the crop sac begins during the second half of incubation to develop its nutritional function, which continues during the subsequent period of brooding and feeding the young.

The mechanisms involved in the birth process (**parturition**) of the young of mammals are quite complex and poorly understood. The uterus is relatively quiescent during gestation, but as the time of parturition approaches it becomes more sensitive to certain hormones and is more responsive to mechanical stimuli. Also, the amount of actomyosin, the contractile protein of the muscle, increases.

Several hormones have known roles in determining the onset of parturition. Progesterone exerts a pregnancy-stabilizing effect. Until its influence diminishes, birth does not occur. Estrogens increase in amount and effective form near the end of gestation and can promote rhythmic contractility of the uterus. **Oxytocin** from the neurohypophysis, also has a potent stimulatory effect on the uterine muscles, and there is rather good evidence that it may be involved in the expulsion of the young.

Another hormone, secreted by the placenta as well as by the ovary, **relaxin,** softens the cervical canal and relaxes the pubic ligaments in preparation for outward passage of the young through the vagina.

Lactation in mammals begins before parturition, but afterward there is a great increase in milk secretion. The exact nature of the stimulus is not agreed upon. There are indications that estrogens and progestogens act synergistically to inhibit lactation. At parturition these steroids diminish, allowing pituitary and adrenocortical hormones to act upon the mammary glands. Pituitary LTH is extremely effective in inducing lactation, but it undoubtedly works in concert with several other hormones to bring about mammary development and to sustain milk production (Fig. 3-6).

The stimulus of sucking by the young is essential for the maintenance of lactation. Removal of young rats from their mother is followed in a few days by cessation of milk secretion, whereas repeated replacement of maturing rats by younger rats makes possible the extension of the normal 20-day period to 70 days. The daily milking of dairy cows ensures continuous milk production for many months. Among human females, lactation can sometimes be extended for several years by continued nursing. The sensory input accompanying sucking of the nipple is a potent trigger for hormone release. In fact, sucking or simulated sucking can serve to initiate milk secretion in breasts of some women who have never given birth to a child. Milk production in such women can be made still more likely by treatment with lactogenic hormones. Such a possibility makes it feasible for some women to nurse adopted infants.

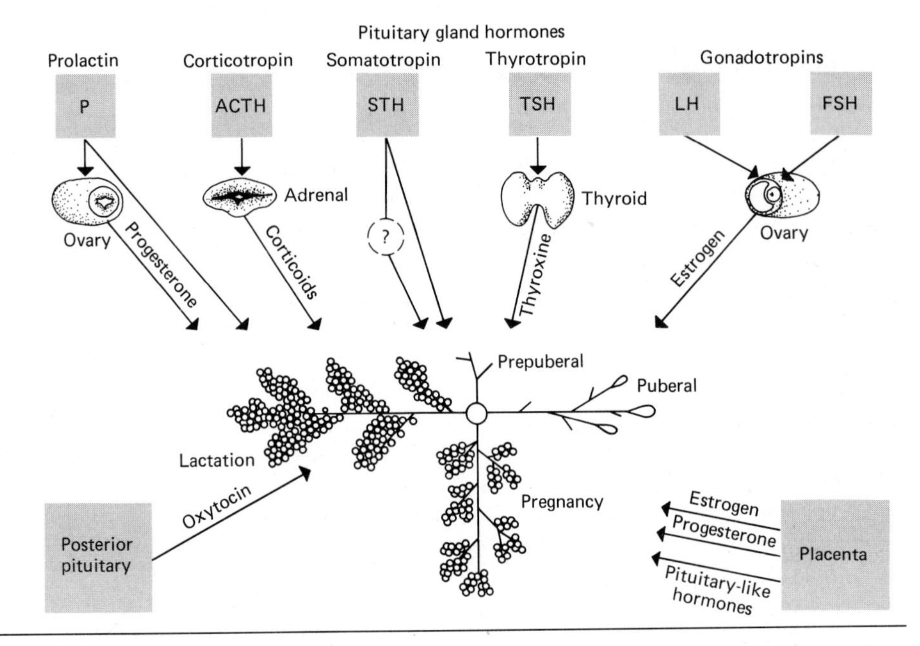

FIGURE 3-6
A simplified diagram showing the action of hormones on mammary-gland growth and lactation. *Upper*, rudimentary gland (prepuberal); *right*, prepuberal to puberal gland; *lower*, prolactational gland of pregnancy; *left*, lactating gland. (*After W. R. Lyons et al.*, 1958, *Recent Prog. Horm. Res. 14*, p. 246.)

3-5
ANIMAL POPULATIONS
Breeding Migrations

The breeding migrations of animals are intriguing phenomena which offer many fascinating and baffling questions for the physiologist to ponder. The eel, *Anguilla*, for example, journeys 3000 miles westward from Europe or eastward from America to its spawning place in the Sargasso Sea (off Bermuda). The young (elvers) return to the freshwater stream of their parents where they grow to sexual maturity. They then return to the sea and breed. A fish having this pattern of migration is called **catadromous.**

Salmon, on the other hand, spawn in freshwater, and the young (smolt) migrate downriver to the ocean. After living in the sea for as long as 5 years they swim back unerringly to the stream of their birth to spawn and then die. This migratory pattern is called **anadromous.** It seems likely that some kind of long-range navigational mechanism brings them to the continental coast and it is the unique odor of their home stream, to which they were conditioned before they went to sea, that they use to identify it. Recent studies suggest that a chemical given off by young salmon upstream is what attracts the older salmon.

To engage in such long-range travels, animals must have specialized mechanisms of several kinds. For example, during the transition from sea to freshwater, or vice versa, there is a radical change in the salinity of the environment. As will be discussed in Chap. 7, Sec. 7-4, the osmoregulatory problems in these two conditions are radically different and demand great changes in response mechanisms, particularly in the way the kidney and ion-exchange cells of the gills operate. Hormones of the gonads, thyroid gland, and adrenal cortex appear to be clearly implicated in these remarkable transitions, but their exact role is but dimly perceived at this time. Such migrations, commonly engaged in by masses of a species, are seasonally timed and, undoubtedly, involve sensory systems that respond to annual meteorological cycles and trigger the hormonal changes that lead to the behavioral, morphological, and physiological changes that make such dramatic journeys possible.

Breeding and seasonal migrations have been extensively investigated in birds. Beginning with the studies of W. Rowan on juncos in Canada during the 1920s, it has been repeatedly demonstrated that changes in gonadal function and fat deposition in many species of birds are regulated by the cycle of the sun and associated factors. In other species other kinds of external stimuli, such as temperature, rainfall, and food supply, are influential in the control of sexual cycles, as noted earlier.

Migration for birds is an energy-demanding, arduous journey. Those crossing large bodies of water face special perils of nonstop, long-distance flying and possible adverse weather. For those species that make such an expensive investment there must be commensurate rewards, for they have been successful in continuing as a species. Some of the rewards are immediate and obvious; some are long-range and less apparent.

The generally most important advantage is the securing of a better environment for living. Birds cannot reproduce successfully, for example, in conditions of intense cold or inadequate food. However, they usually depart from their "wintering" grounds when the environment is more benign than the place to which they fly, and they desert their breeding grounds before adverse conditions set in. Thus, it is not the immediate conditions that cause them to migrate.

By alternately exploiting two different habitats for food, more birds can be sustained. For example, birds that move into the arctic regions to produce their numerous young use the rich food resources available only during the summer. Although the summer is of short duration, the long daylight hours of the Arctic provide a long foraging period each day during the intensive incubating and feeding-of-young phase.

There are advantages also in terms of predator pressure. Commonly the explosive frenzy of breeding, incubating, and catering to the young is well synchronized in massive colonies of birds. The pressure of predation is dispersed, therefore, among a great number of eggs and young, and this results in a tolerable level of mortality. Furthermore, the brief, once-a-year appearance of predator-vulnerable eggs and young denies to the predators the sustained food supply they would need to build up populations large enough to eradicate the birds on which they feed.

In a similar way, it is probable that parasitism and infectious diseases are reduced by the fact that the long, cold arctic winter and small resident animal populations would discourage their survival in infected soil from breeding season to breeding season.

Population Fluctuations

Spectacular increases and decreases in certain populations have been observed and commented upon for centuries. Population fluctuations have been particularly well studied in mammals, and something is known of their physiological basis.

For many years it was assumed that the termination of the explosive growth of populations was precipitated by epizootics, famine, or climatic factors, resulting in often spectacular crashes. However, by the early 1940s it was becoming apparent that the regulation of populations was more intrinsic than extrinsic. As early as 1946, dramatic declines in rat populations were found to be coincident with social disturbances rather than with environmental changes.

The suspicion that social factors were involved led to a search for mechanisms that could regulate the growth of populations in a density-dependent manner. Although social and behavioral phenomena are density-dependent, and become strikingly apparent at high population levels, such factors are also present in low populations, but inconspicuous. Purely ecological factors, such as food and climatic conditions, also affect populations, and, indeed, may prevent populations from reaching a level where social forces become important. However, it is commonly believed by those who work on animal populations that, within broad limits set by the environment, density-dependent mechanisms have evolved within animals which regulate population growth and curtail it short of the point of suicidal destruction of the environment. There is not complete agreement on the mechanisms by which such a result is achieved.

It has been asserted by J. J. Christian and colleagues, on the basis of studies of rodents, lagomorphs, deer, and a few other mammals, that there is a behavioral-endocrine feedback system that is important in the regulation of populations. The feedback is thought to act as a safety device, which limits population growth when other factors fail to do so.

Specifically, changes in certain endocrine organs can be correlated with population size. For example, in experiments with mice in the laboratory, progressive hypertrophy (excessive development) of the adrenal cortex and thymic involution are associated with increasing population size. Somatic growth is suppressed, sexual maturation is delayed, and reproduction ceases at higher population densities. It has been suggested that the progressive inhibition of reproductive function is due to the sterilizing effect of adrenal androgens (sex steroids) in response to pituitary-hormone (adrenocorticotropin) stimulation of the adrenal gland.

It has been observed that the adrenals of wild mice put together in

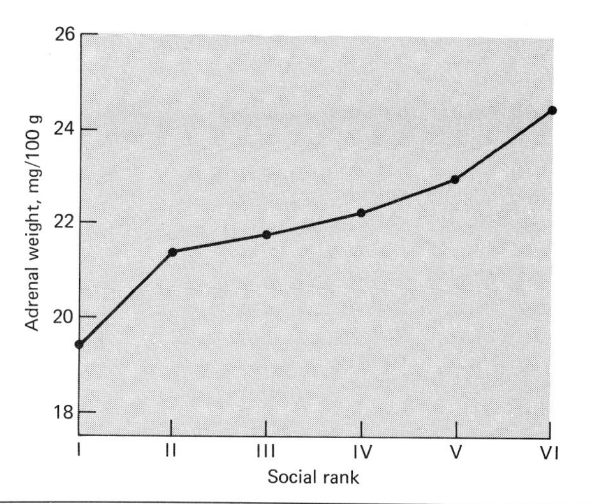

FIGURE 3-7
Relation of adrenal weight to rank. The dominant mice had adrenals about the same size as isolated individuals; the subordinate mice had larger glands. (*After D. E. Davis,* 1966, *Integral Animal Behavior, The Macmillan Company, New York, p.* 83.)

very dense groups were larger than the adrenals of mice kept in isolation. It was also shown that the size of the adrenals was inversely related to social rank: the highest-ranking mice had the smallest adrenal glands, which were about the same size as the adrenals of isolated individuals (Fig. 3-7).

What happens when a population starts to increase rapidly is, theoretically, the following. The frequency of interaction increases, aggression (both real and ritualized) becomes more frequent, and a linked series of physiological events is triggered. Sensory information resulting from social encounters, competition for food or mates, as well as other factors, are integrated in the hypothalamus and are translated into pituitary hormone output. As the population pressures increase there will be a progressive increase in age at sexual maturation, embryos are less frequently sustained, the interval between pregnancies may increase, and mortality among the young is high. The net result is a decrease in population. However, this attractive theory has not yet been adequately supported by findings from wild populations. It appears that in natural conditions the crowding required to produce these effects simply is rarely, if ever, achieved.

3-6 ANIMAL COMMUNICATION

Visual, auditory, and olfactory perception are attributes of almost all complex animals, but there is great variation in the extent to which animals of different groups rely on the different senses for information about their surroundings, and in the extent to which their functions are influenced by specific sensory stimuli. Contrast, for example, the sensory systems of birds and mammals. In most species of birds the olfactory sense is poorly developed, but photoreceptors are supremely important to them. Vision plays a central role in their social and reproductive activities, as well as in food recognition. In mammals, usually, the situation is reversed. Many mammals have mediocre vision, especially color vision, and rely strongly on smell to assess their environment. Smell is a central feature in mammalian social life, and odorous substances produced by mammals are implicated in defense, marking of territory, identification, and in evocation of sexual and aggressive behavior.

Visual Communication

Vision is the most important sensory modality in the intraspecific communication of diurnal insects, fishes, lizards, birds, and primates. Visual signaling can be very complex and subtle (as among rhesus monkeys and baboons which live in cohesive groups and where even facial expressions are involved) or simple and stereotyped (such as the visual signals of many birds, which communicate over greater distances).

Vision is essential for the establishment of social groups in many animals. For example, it is important in the forming and maintaining of schools of fish. Also, numerous kinds of visual signals are used to allow the approach of two individuals, and especially to coordinate reproductive activities. The recognition and feeding of young by parents, especially among birds, frequently rely upon visual cues.

On the other hand, many animals use visual signals to encourage dispersal. Visual displays communicating aggression are legion: the display of weapons (teeth, talons, limbs), color patches or crests, the assumption of a fighting posture.

Specialized visual alarm signals include the action of flight among many species, tail flicking in certain birds, fin flicking in some fishes, and exposure of white rump hair in certain mammals.

Visual signals carry a great amount of information because they may be expressed through combinations of color, posture or form, movement, and timing. They also have the advantage that they can be started or stopped immediately and there is nothing ambiguous about the exact position of the sender.

Visual signals are not of use to most animals at night or in dark places. The fireflies, palolo worms, and other bioluminescent creatures are exceptions. Visual signals convey decreasing amounts of information as the distance increases. It is not possible to increase such a signal by putting more power into it (except for light producers), as is possible with, for example, sound.

Acoustical Communication

There are circumstances in which hearing offers unique advantages, especially in social behavior and in some relationships of the animal to its environment. The orientation of many fishes, especially in dark or turbid waters, to mechanical disturbances depends upon detection by their *lateral line system,* a complex set of canals and sensory *hair cells* along the sides of the body. Insects can sense air movements with their *hair sensilla* and *antennal receptors*. The approach of a predator is often signaled by its sounds. Bats, flying in darkness, use echolocation for navigation and for prey location.

Arthropods utilize friction devices involving wings, legs, or specialized vibrating membranes to make sounds. Vertebrates commonly produce sound signals by forcing air across vibrating membranes of respiratory systems. In addition, rabbits thump the ground, grouse make drumming sounds by beating with their wings, gorillas beat their chests, and woodpeckers drum on hollow trees.

Any sound production is associated with a particular physiological state. For example, many sounds are confined to animals in reproductive condition. Whales produce sonorous and varied sounds during their migrations. They may vocalize at other times also, but they have been studied only during the times when they are together during migrations. It is speculated that the calls serve to keep the herd together.

The sounds of humpback whales (*Megaptera novaeangliae*) have been extensively recorded by Roger S. Payne. This species produces a series of beautiful and varied sounds for a period of 7 to 30 minutes and then repeats the same song with considerable precision. Each individual appears to have its own unique song, and many sing for several hours.

Chemical Communication

Social insects provide some of the most striking examples of intraspecific use of chemical communication. Very specific behavioral patterns are triggered by minute amounts of specific chemicals. The immediate and stereotyped response elicited in such instances has been likened to the action of a hormone released from one site in the body and acting on another. Thus, those chemicals which are secreted by an animal to the outside and cause a specific reaction in a receiving individual of the same species are called **pheromones** (Gr. *pherein*, to carry; *horman*, to excite). Pheromones are not restricted to insects, but are also used by other animals, including mammals, as will be discussed further on.

It seems safe to state that pheromones have a central role in the organization of insect societies. Their remarkable qualities of social life are mass phenomena that emerge from the meshing of simple, individual patterns by means of communication. Edward O. Wilson (see Readings at end of chapter) has recognized about nine categories of response: alarm, simple attraction, recruitment, grooming, exchange of oral and anal liquid, exchange of solid food particles, facilitation, recognition, and caste determination either by stimulation or inhibition. Each of these has been demonstrated to require, to some extent in some social insects, a chemical communication.

Alarm, which is the normal response to intruders in the nest, typically consists of swift, oriented movements. In some species the response is to retreat or even abandon the nest. In others it is aggressive and a fierce defense is staged. A number of insect alarm pheromones have been identified. They are usually fairly small, structurally simple molecules with five to ten carbon members.

Pheromones can act by interfering with endocrine systems. This is the case in *caste control* among social insects. In as short a time as 30 minutes after the mother queen of a honeybee colony is removed, the organized activities of the colony change into disorganized restlessness. One or more worker brood cells are altered to become emergency queen cells, within which potential new queens will be produced. A few days later, some of the workers begin to experience increased ovarian development. These effects are due to the removal of a pheromone present in the queen called *queen substance*. In fact, it appears that she produces at least two inhibitory pheromones plus at least two additional attractive scents in her domination of the colony.

The inhibitory pheromones may act to suppress secretion of "gonadotropic" hormones, and they inhibit ovarian development. One of these is 9-ketodecanoic acid, produced by the queen's mandibular glands, which inhibits both ovarian development and queen rearing. A second inhibitory scent, produced in a part of the body outside the mandibular glands, inhibits ovarian development but not queen rearing. One of the pheromones causing clustering and stabilization of worker swarms is 9-hydroxydecanoic acid.

The mammalian pheromones have been categorized by H. M. Bruce into three groups: *releaser* or *signaling* pheromones, *imprinting* pheromones, and *primer* pheromones. The *releaser* pheromones exert a direct action on the central nervous system, which in turn may produce an immediate behavioral response, such as sexual attraction, or recognition of the receptive female by the male or of the young by their mother. (Sex attractants, trail and alarm substances of insects are examples of pheromones having releaser effects.) Aggressive behavior of unfamiliar male mice is related to a pheromone of urinary origin and to another produced by the footpads.

Secretions from the androgen-dependent chin glands of the Australian rabbit are used in marking localized features of its territory and repelling or inhibiting signals from other males. Dominant males have larger chin glands than subordinate males. An anal gland also secretes a pheromone, which, along with urine, serves to establish the rabbit's overall territory. The largest anal glands are found in dominant individuals.

The *imprinting* pheromones, perhaps an aspect of primer pheromone activity, are considered to apply to the olfactory environment at and shortly after birth in the young mammal. For example, in mice and rats, social behavior in the adult may be modified by olfactory experience during the period of suckling. Female mice reared in the absence of the father show a loss of discrimination in sexual selection when adult, and the same deficiency develops if the olfactory atmosphere of the nest is artificially altered by spraying it every day with perfume.

The *primer* pheromones bring about a delayed response to prolonged stimulation mediated through the nervous and endocrine system, including the pituitary gland. Those of mice have been most thoroughly studied. In female mice caged together the estrous cycles tend to be disturbed. Estrous periods are interrupted by pseudopregnancies or by extended periods of diestrus (Lee-Boot effect). This effect disappears when the females are subjected to olfactory lobotomy. It also has been shown that the introduction of a male mouse into a group of female mice accelerates the attainment of estrus and shortens the cycle (Whitten effect). The responsible pheromone is present in the urine of males; it disappears if the male is castrated, but reappears if the castrate male is given testosterone therapy. Thus, the pheromone is thought to be either an androgen metabolite or the product of an androgen-dependent tissue.

If a newly mated female mouse is removed from the stud male and

FIGURE 3-8
A chain of neuroendocrine reactions triggered by pheromones. Pregnancy may be blocked in a newly impregnated mouse by exposure to a strange male of a different strain (Bruce effect). 1, The volatile pheromone is perceived by the olfactory epithelium; 2, impulses are relayed through the olfactory lobes and cerebrum to the hypothalamus; 3, gonadotropin-releasing factors in the median eminence are conveyed over the hypophysial portal veins to the anterior lobe of the pituitary; 4, the releasing factors regulate the output of pituitary gonadotropin; 5, gonadotropins condition the production of steroid hormones by the ovaries; 6, the ovarian hormones are deficient or of the wrong kind and a pregnancy-type uterus cannot be developed and maintained; 7, young embryos fail to implant, thus terminating pregnancy. (The size and shape of the pituitary gland are exaggerated in this figure for the sake of clarity.) (*After C. D. Turner and J. T. Bagnara, 1971, General Endocrinology, W. B. Saunders Company, Philadelphia, p. 13.*)

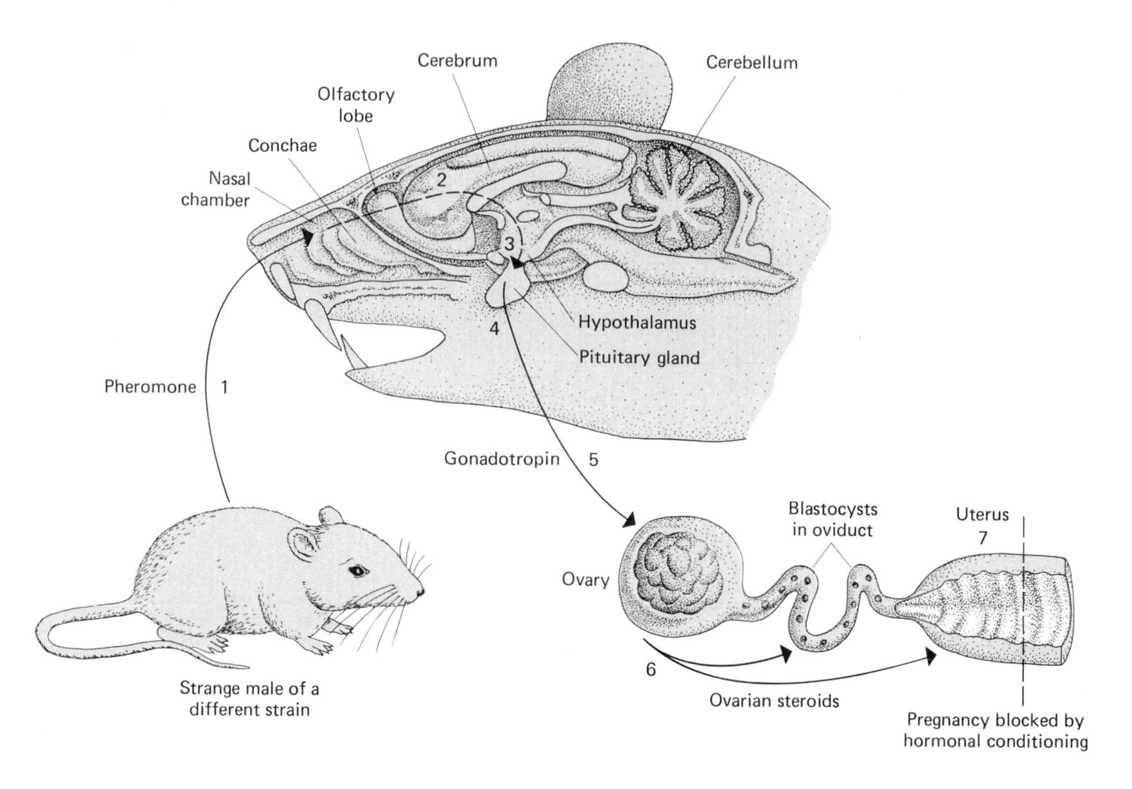

exposed for 2 or 3 days to other males, pregnancy is blocked to a large extent (Bruce effect). Contact between the sexes is not necessary. Even exposing the female to litter freshly soiled by a strange male will prevent successful pregnancy.

The causative pheromone is present in urine, and castration of the adult male abolishes the pheromone. Its action is to prevent successful implantation because the ovaries fail to produce the hormones required to build up the uterine endometrium. Prevention of the effects of the pheromone are provided by injecting prolactin or progesterone during the period of exposure to the strange male. Thus, stimulation of the olfactory tract appears somehow to influence the neuroendocrine apparatus that controls anterior pituitary gonadotropins and, eventually, the ovaries (Fig. 3-8).

Thus, in a manner somewhat analogous to the way in which one endocrine gland influences another gland or a nonglandular tissue of the same organism, so certain exocrine glands of animals influence target structures in other organisms. Thus, "exocrinology," seems to be practiced on a large scale by arthropods and mammals in their complex interactions with members of their own species.

In this chapter we have moved along the interfaces between behavior, ecology, and basic organismic physiology. Within these zones reside many yet-to-be-understood mechanisms that enhance the likelihood of survival and continuance of the various species of animals. Some of these are subtle but others are obvious. Though physiologists are concerned primarily with the organism as a functional unit, they must also be cognizant of the fact that an organism is enmeshed in complexly interdependent relationships with other organisms. Its very survival depends upon being properly attuned to the opportunities and limitations that such organization imposes.

DEPARTMENTAL[1]

An ant on the tablecloth
Ran into a dormant moth
Of many times his size.
He showed not the least surprise.
His business wasn't with such.
He gave it scarcely a touch,
And was off on his duty run.
Yet if he encountered one
Of the hive's enquiry squad
Whose work is to find out God
And the nature of time and space,
He would put him onto the case.
Ants are a curious race;

[1] From *The Poetry of Robert Frost,* edited by Edward Connery Lathem, 1969, Holt, Rinehart and Winston, New York, pp. 287–289.

One crossing with hurried tread
The body of one of their dead
Isn't given a moment's arrest—
Seems not even impressed.
But he no doubt reports to any
With whom he crosses antennae,
And they no doubt report
To the higher up at court.
Then word goes forth in Formic:
"Death's come to Jerry McCormic,
Our selfless forager Jerry.
Will the special Janizary
Whose office it is to bury
The dead of the commissary
Go bring him home to his people.
Lay him in state on a sepal.
Wrap him for shroud in a petal.
Embalm him with ichor of nettle.
This is the word of your Queen."
And presently on the scene
Appears a solemn mortician;
And taking formal position
With feelers calmly atwiddle,
Seizes the dead by the middle,
And heaving him high in air,
Carries him out of there.
No one stands round to stare.
It is nobody else's affair.

It couldn't be called ungentle.
But how thoroughly departmental.

READINGS

Altman, J., 1966, *Organic Foundations of Animal Behavior*, Holt, Rinehart and Winston, Inc., New York. This book, a biological approach to psychology, is an original synthesis of biochemical, neurological, and endocrinological information. The reference list is extensive.

Marler, O., and W. J. Hamilton, III, 1967, *Mechanisms of Animal Behavior*, John Wiley & Sons, Inc., New York. A wealth of basic information about behavior and underlying processes is contained in this book. It is, to quote the authors, "about the processes that determine when behavior will occur and what form it will take." It is an interesting physiology of behavior with numerous references for each chapter.

Wilson, E. O., 1975, *Sociobiology, The New Synthesis*, Belknap Press of Harvard University Press, Cambridge, Mass. This book is a comprehensive review and synthesis of literature on social organization, ecology, and evolutionary genetics of all animal groups whose members exhibit some forms of social behavior in addition to mating. It is a monumental work that is destined to become a classic.

ECONOMIC PHYSIOLOGY 4

In the preceding chapter the associations and interactions of animals that primarily ensure the continuance of the *species* were considered. With this chapter we begin examining several of the special mechanisms that serve primarily to preserve the *individual*, as it contends with the physical environment and other organisms as well. In other words, the management of its household—its economic physiology—is our subject. Many of these mechanisms are devices for supporting on-going, maintenance functions, such as those relating to nutritional state or thermoregulation. Others cope with the annual changes of environment encountered by organisms in temperate latitudes. Still others deal with special, stressful, or emergency situations. The total aggregate of these mechanisms serves to ensure the well-being of the individual organism.

An organism that has satisfactorily secured its own survival, of course, has the potential to contribute to the survival of the species, because it may have the opportunity to participate in reproduction or support of a social unit. Anything which increases the probability of survival of the individual organism for some optimum period of time—especially until it engages in reproduction—should have positive selection value. In fact, it probably is legitimate to speculate that the survival of the species is what determines the selection of mechanisms that preserve the individual. Life for the individual organism, then, could be considered a secondary consequence of propagation of the species.

Biologists sometimes refer to the constellation of unique mechanisms possessed by the organisms of a species which enable them to survive under the special environmental conditions in which they are found, as **adaptations.** For a proper consideration of the full meaning of adaptation and its basis, it would be necessary to go into some details about inheritance and evolution, and that is beyond the scope of this treatise. Here we shall consider only some selected examples to illustrate the kinds of adaptations that organisms employ in coping with their surroundings.

4-1 DEFENSIVE AND OFFENSIVE SECRETIONS

In contrast to *pheromones* (Sec. 3-6), which are used in intraspecific communication, a vast array of substances serves for the transmission of information between members of *different* species. Among these are several kinds of agents that serve as chemical defenses against predators. Again unlike pheromones, which are often produced in minute quantities and usually can be detected only by sensitive bioassays, defensive secretions may be detected with relative ease since they often are strongly odorous and discharged in substantial amounts.

Certain mammals discharge fetid secretions when threatened. Examples are found in several families. Some of the Viverridae (civets and mongooses) have perineal glands that discharge a stinking secretion when danger threatens. It is in the skunks (Mustelidae), however, that this mode of defense reaches its greatest development. They have large anal glands that can be protruded to shoot their secretion with astonishing precision to a distance of up to about 4 m. The vile odor is due mainly to a mixture of crotyl mercaptan, isopentyl mercaptan, and methycrotyl disulfide. Contrary to previous reports, *n*-butyl mercaptan is not the active ingredient. Before discharging their defensive secretion, skunks assume "threatening attitudes" that vary widely according to species. Some stamp on the ground with their hind feet; others, such as the spotted skunk, walk for a few seconds on their front feet with the hindquarters elevated. When sprayed with the skunk's secretion, would-be predators exhibit nausea, confusion, and temporary blindness if the eyes are hit.

Skunks, like civets and mongooses that use stink glands, have conspicuous patterns of coat coloration. The contrasting black-and-white coloration of skunks makes them conspicuous and unique among mammals, which are usually clothed in subdued coloration to provide camouflage. Skunks boldly advertise their almost invulnerable defensive mechanism.

Musk glands discharged in fright are found in a wide variety of mammals. It is not readily apparent that this behavior has any defensive utility, and it would seem likely that several intermediate stages exist between the fear-induced secretion of certain cutaneous glands and their intentional utilization for defense against an aggressor.

Among arthropods, chemical defenses, comparable to that of the skunk, are common. In fact, in terrestrial animals, chemical defenses against predators are probably more diverse and widespread among arthropods than in any other group. The glands responsible for the defensive secretions of arthropods are integumental organs. They are essentially infoldings of the body wall, but are quite varied in morphological detail. Typically, a gland consists of a saclike reservoir in which the secretion is stored, and of a glandular tissue that may be a part of the wall of the reservoir itself or may lie apart from the reservoir and be joined to it by one or more special ducts. The number and distribution of glands is so variable as to suggest that they have arisen many times independently in the course of evolution.

The modes of discharge of the glands have been categorized as three by Thomas Eisner. In one type, present commonly in millipedes, the secretion simply oozes from the glands onto the animal's own surface. In a second, found in some caterpillars and beetles, the secretion is "aired" by evagination of the gland as a whole. The third and most spectacular method includes those glands that discharge their contents as a spray, sometimes to a distance of a few meters. Arthropods that spray include cockroaches, earwigs, stick insects, stink bugs (Hemiptera), notodontid caterpillars, grasshoppers, certain beetles, scorpions, and millipedes. Commonly the ejected spray is precisely aimed at the attacking predator to ensure maximum effectiveness of the discharge.

The mechanisms for forceful expulsion of the defensive secretion vary. In some cases muscular compression of the reservoir is involved, or it may be squeezed by vascular fluid pressure. Some glands are connected to respiratory tracheae and rely on air pressure for discharge. A most remarkable mechanism is present in the bombardier beetles (*Brachinus* spp.), large carabid beetles having a pair of glands, opening at the tip of the abdomen, that explosively discharge a strong repellent spray. They can repeatedly and rapidly fire off these glands and, through rotation of the abdominal tip, they can aim the spray in virtually any direction (Fig. 4-1).

Defensive secretions, in general, are very effective weapons for deterring attack from numerous types of predators. Arthropods eject their

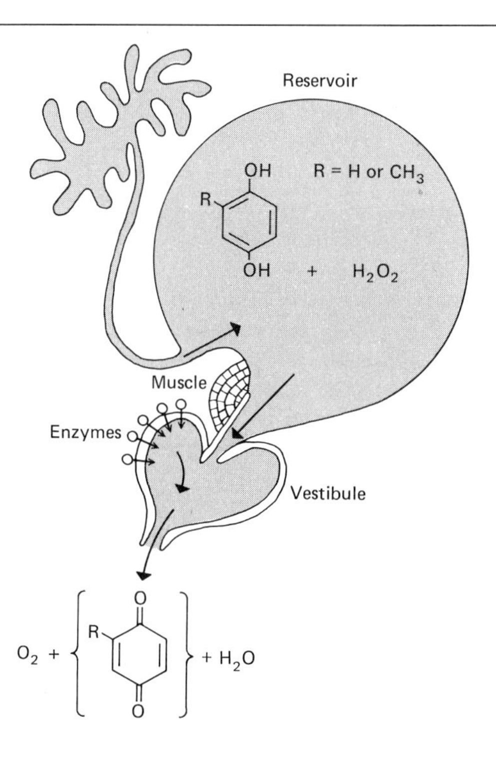

FIGURE 4-1
Diagram of a gland that produces a defensive secretion in the bombardier beetle. The reservoir contains an aqueous solution of hydroquinones and hydrogen peroxide, obtained from a duct that drains an outlying cluster of secretory tissue. The outer compartment (vestibule) contains a mixture of catalase and peroxidase enzymes that are secreted by cells on the wall of the vestibule itself. To effect a discharge, the beetle squeezes some reservoir fluid into the vestibule through the valve, controlled by muscles, between the two compartments. The catalases promote the decomposition of hydrogen peroxide, while the peroxidases force the oxidation of the hydroquinones to their respective quinones. Under pressure of the free oxygen, the mixture of repellent quinones is explosively discharged.

secretions during the initial phases of the attack, causing the predator to desist from the attack and engage in vigorous cleansing activities.

It is intriguing to note that the active components in the defensive secretions of arthropods also occur, almost without exception, as secondary substances of plants. Compounds that are commonly found in plants and are used by arthropods as defensive sprays include 2-hexenal and α-pinene. The fact that the active principles in the defensive secretions are effective repellents for predators raises the possibility that these same substances, as secondary constituents of plants, serve to protect plants against herbivores.

Ants (family Formicidae, order Hymenoptera) produce a number of exocrine secretions which are of prime importance to their social organization and individual specialization. By their behavior they must be judged as among the most highly evolved Insecta. Ducted glands, producing secretions that may be categorized as venoms, attractants (mostly pheromones), and repellents, enable the ants to kill or paralyze the prey that forms its food or the food of its young; to convey messages to others of its species concerning food sources, mating, and the presence of enemies; and to discourage or prevent those enemies from interfering with its social pattern (Fig. 4-2).

Brown, Eisner, and Whittaker[1] have proposed for the field of "chemical ecology" the grouping of interspecific agents into two categories:

[1] W. L. Brown, Jr., T. Eisner, and R. H. Whittaker, 1970, Allomones and kairomones: transspecific chemical messengers. *BioScience*, **20**: 21–22.

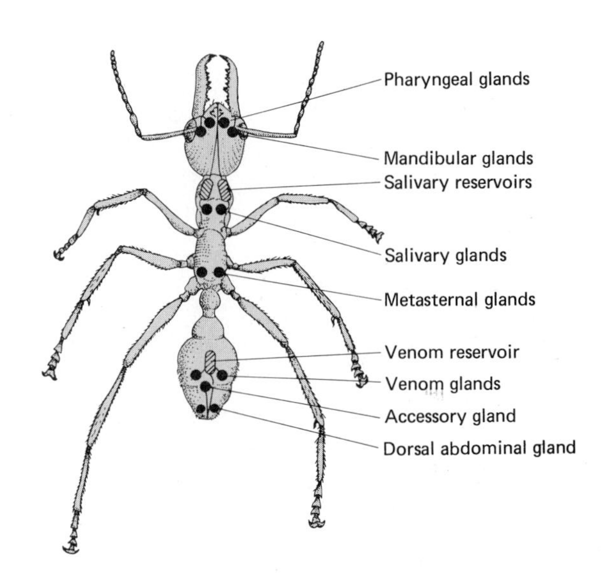

FIGURE 4-2
Exocrine gland system of the bull ant, *Myrmecia gulosa* (Fabr.), subfamily Myrmeciinae. The pharyngeal glands open into the pharynx and have a digestive function. The mandibular glands are a pair of sacs which open into the base of the mandibles and probably secrete an alarm substance. The function of the metasternal glands is unknown; they were thought earlier to play a role in determining nest odor. The accessory gland (Dufour's gland) is a part of the venom apparatus but is used as an odor-trail substance in at least one group. The dorsal abdominal glands are of unknown function but perhaps are scent glands. (*After G. W. Cavill and P. Robertson,* 1965, *Science, 149,* p. 1341.)

Allomones. Chemical substances, produced or acquired by an organism, which, when they contact an individual of another species in the natural context, evoke in the receiver a behavioral or physiological reaction adaptively favorable to the emitter.

Kairomones. Chemical substances, as above, but the adaptive benefit of which falls on the recipient rather than the emitter. They are, in fact, commonly nonadaptive or maladaptive to the transmitter. They include attractants, phagostimulants, and other substances that mediate the positive responses of predators to their prey, herbivores to their food plants, and parasites to their hosts. Kairomones, obviously, are a diffuse lot of substances. They are very largely undefined chemically, and undoubtedly are variable mixtures of substances in many cases. Some workers in the field of animal communication do not believe such a loose assemblage deserves to be dignified by a special name, but there is decidedly some value in having a term that conveys the concept.

These terms are intentional parallels to the terms "pheromones" and "hormones." Indeed, the four major categories of chemical messengers—pheromones, allomones, kairomones, and hormones—are not mutually exclusive. For example, the sex pheromone of the bark beetle, *Ips*, also serves as a kairomone that attracts certain other beetles that prey on *Ips*. The secretion discharged by ants when the nest is disturbed acts as a social alarm pheromone, but it may act as a repellent allomone against aggressors. Even a hormone may serve also as an allomone. In the wood-eating roach, *Cryptocercus punctulatus*, the hormone **ecdysone** regulates the growth and molting cycle, and it also serves as an allomone by inducing the sexual cycles of the symbiotic wood-digesting Protozoa that inhabit the gut of the roach. A hormone also may act as a kairomone. The rabbit flea depends for maturation of its eggs on ingestion of blood laden with the proper amount of corticosteroid hormones from the pregnant host.

Chemical agents thus carry messages that relate one part of an organism to another, individuals of a species to one another, and members of different species to a community. Of tremendous importance to the survival of any organism is its ability to intercept a wide variety of physical and chemical stimuli from the environment and to respond in some appropriate manner to whatever is relevant.

4-2 ADAPTIVE COLORATION

Concealment and Camouflage

The appearance of an animal is connected with its survival. It can seek protection both through concealment (cryptic coloration) and through displaying bright pigments (warning coloring). There are several ways by which animals camouflage themselves. These include:

Markings that blend with the landscape
Concealed contours
Camouflage through masking
Camouflage through resemblance
Camouflage through mimicry

In the first category are animals with markings which "run on" into the background. By use of lines, spots, and stripes the animal's body appears to dissolve into its surroundings, a concealing effect called *somatolysis*. Snakes use this kind of concealment very effectively. Many ground-dwelling birds do likewise, particularly the young. The effectiveness of the camouflage usually depends upon the ability of the animal to remain still, an inherited behavior that commonly goes with this type of morphological pattern. In many instances concealment is employed in aggression: the rigidly still and camouflaged predator waits until its unsuspecting prey approaches close enough and then captures it with a rapid strike.

The often remarkable pairing of adaptive coloration and adaptive behavior is exemplified by the bittern (*Botaurus lentiginosus*), a bird that inhabits marshes. When the bittern is alarmed it tilts up its head, pointing its beak to the sky. A bittern's long neck, with russet and black plumage, blends with the surrounding reeds. Should a gust of wind cause the reeds to sway, the bird will sway rhythmically with them.

Concealment of contours is effected, for example, by reducing or eliminating a shadow. The animal can crouch close to the ground to reduce its shadow. Its body may be structured so that it is flattened or expanded by lateral flaps or projections to minimize shadowing. Another method of doing away with the shadow effect is *countershading*. This is seen in numerous animals where the upper part of the body (toward the sunlight) is a darker color and the underside is lighter, so that the body's own shade balances the light intensity and the body outlines become dissolved.

Good examples of masking are found among certain crabs. Some pick algae and sponges, cut them to shape, add a glue from their mouth, and then stick them onto their carapace. The sponge crab holds a cut piece of sponge in place with a specially modified last pair of legs.

The white froth of many sucking insects (Homoptera) is another kind of mask. The predatory larva of the lacewing fly (*Chrysopa*) covers itself with the remains of dead ants.

Camouflage through resemblance is quite diverse and widespread. Only a few examples will be mentioned here. Winged insects often mimic leaves, not only in color and venation, but even to including apparent imperfections such as fungal growth, discolorations, and eaten edges. The seas abound with organisms that resemble objects in their surroundings, including the long, slender pipefish among grasslike plants and the elaborately formed and colored frogfish that looks like seaweed.

Camouflage through mimicry is fairly common among insects. The subject was first defined by Henry W. Bates, an English explorer who spent 11 years in the jungles of the Amazon. He observed that certain slow-flying butterflies, which should have been ideal prey, were usually ignored by birds. Bates suspected that birds were sickened by some of their secretions, and once a bird had captured one and become sick, it no longer preyed upon that particular kind of butterfly. Other butterflies that resembled the unpalatable insect also, it was observed, were left alone. *Batesian mimicry* is now the term which refers exclusively to the imitation of offensive properties. The predator does not instinctively avoid the prey, but learns for itself how distasteful or unpleasant the model is and then avoids others of the same species, as well as mimics of that species.

As early as 1867, Alfred Wallace expressed the opinion that once a brightly colored insect had repelled a bird through repulsive taste, the bird learns to associate the insect's distinctive coloration with the previous unpleasant experience and thus tends to avoid it.

Chromatophores and Color Change

Many species can change color through movement of pigments within certain integumentary cells or organs. This ability has been observed for numerous cyclostomes, fishes, amphibians, reptiles, crustaceans, cephalopods, and leeches, as well as a few insects, echinoderms, and polychaetes.

Special pigment-containing cells, called **chromatophores,** bring about the color changes. Although they are not confined to the surface, but are found in deeper tissues as well, it is mostly the integumentary ones that effect the color changes to be considered here. Chromatophores can concentrate pigment into a small mass (punctate condition) and have it contribute little to the gross coloration of the individual, or disperse it over a large surface (reticulate condition) and thereby impart its tint to the animal. This mechanism of concentration and dispersion of pigment is referred to as **physiological color change** (Figs. 4-3 and 4-4).

The response of the chromatophores is predominantly, although not exclusively, to color or shade of background, suggesting that color change contributes significantly to the concealment of the animal for protection or aggression, and hence to its chance for survival. However, there are few demonstrations that this is so. Among these is a study in which it was shown that if fishes were given time to change their coloration, a smaller percentage were seized by predatory birds than if not allowed to adapt. Similar results were obtained in a study of insect predation by birds. Furthermore, fishes which are black-adapted tend to select a black background over a white background when given a choice.

Chromatophores appear also to serve as a shield against bright illumination which could be deleterious. In some animals the pigments are dispersed in bright illumination, regardless of background. Such a response might be considered primitive, with the obliterative role appearing later in the evolution of chromatophore systems.

FIGURE 4-3
The cephalopod chromatophore. A central uninucleate cell contains pigment within an intracellular sac. Radiating out from the cell in the plane of the skin are 6 to 20 or more uninucleate smooth-muscle fibers which usually contract simultaneously, stretching out the central cell into a disc, as in *b*. When the radiating fibers relax, the elastic sac of the central cell restores it to a small, spherical shape, as in *a*. The muscle fibers are caused to contract by neural excitation. The general coloration of the animal can be altered rapidly by the many chromatophores of the skin dispersing or concentrating their different pigments.

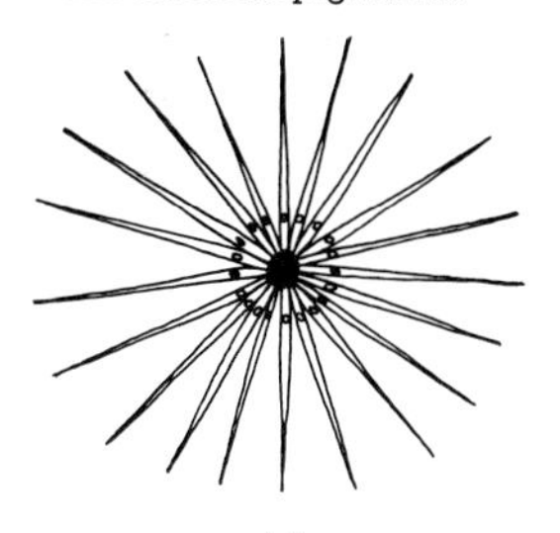

(*a*)

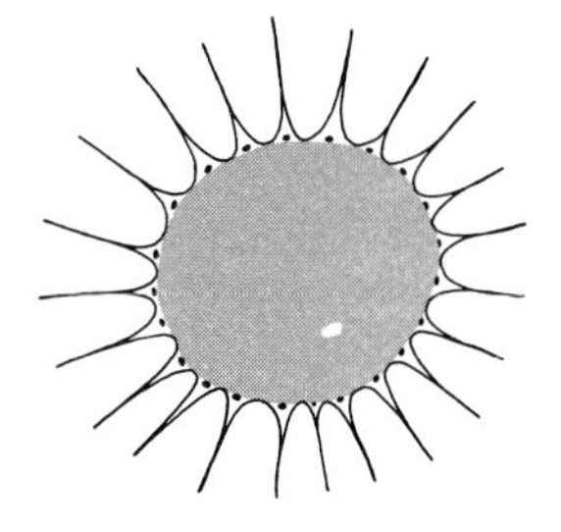

(*b*)

FIGURE 4-4
Photograph of two *Rana pipiens*, one of which (*left*) has been on a white background for several minutes and the other (*right*) on a black background for several minutes.

Another function attributed to chromatophores is thermoregulation. Among crabs, frogs, and, particularly, reptiles, a thermoregulatory role for chromatophores has been supported. This appears to be clearly documented, for example, in the desert lizard, *Phrynosoma*. It tends to be light at night and during midday, but dark during the early morning and late afternoon. The animal apparently is adaptively controlling heat absorption and radiation at the various times of day by chromatophore pigment manipulation (Fig. 4-5).

Finally, chromatophores play a role in the color displays that some animals use in mating. For example, in the lizard, *Anolis*, the males show a striking change from green to brown during pairing. Certain fishes and cephalopods also have color displays associated with breeding.

Morphological color change involves actual change in *quantity* of pigment within the animal or its integument. It may involve both an increase in the amount of pigment within chromatophores and the number of functional chromatophores per unit area of integument (Fig. 4-6).

There appears to be a close functional relationship between physiological and morphological color changes. Maintained concentration of pigment within chromatophores seems usually to be correlated with a reduction in quantity of that pigment and, conversely, pigment dispersion seems to stimulate pigment production. In other words, the mechanisms controlling pigment movement within cells also control pigment synthesis. Presumably, this would involve hormones, particularly melanocyte-stimulating hormone (see Chap. 9, Sec. 9-6).

The Uses of Color by Birds

Birds as a group are remarkably varied in their external pigmentation and patterns of coloration. Their feathers, scales, and skin commonly contain both dark colors (black through reddish browns, which are melanins that are synthesized by the birds themselves) and bright pigments

(a)

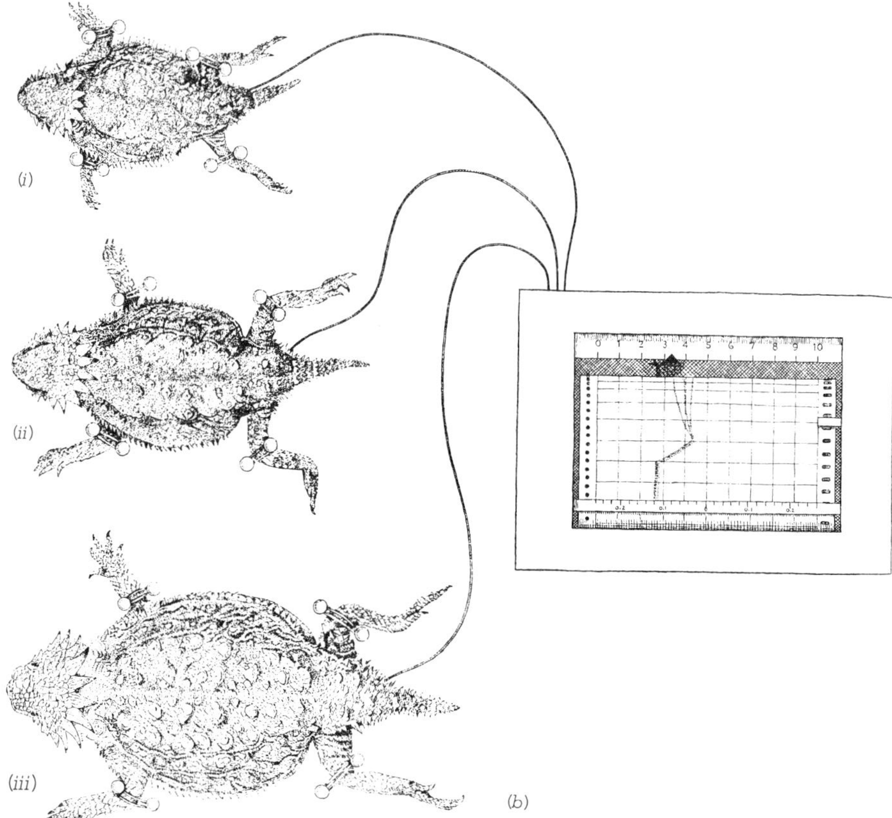

(i)

(ii)

(iii)

(b)

FIGURE 4-5
(*a*) Texas horned lizard (*Phrynosoma cornutum*). When its body temperature is low, the animal is quite dark (*left*), but the same animal becomes paler (*right*) after it has been exposed to higher temperatures. (*Photograph, courtesy of C. M. Bogert.*) (*b*) Regal horned lizards (*Phrynosoma solare*), weighing respectively 12.4 g (*i*), 29.4 g (*ii*), and 85.5 g (*iii*), were exposed to the midday sun while their body temperatures were continuously measured by means of thermocouple wires from the cloaca of each to a recording potentiometer. The colors of the lizards were nearly the same at start but then changed, the smallest lizard becoming palest. (*From C. M. Bogert, 1959, Sci. Am., April, p. 108.*)

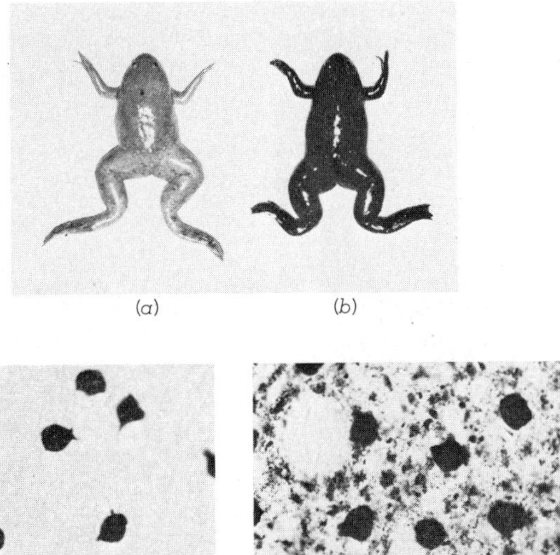

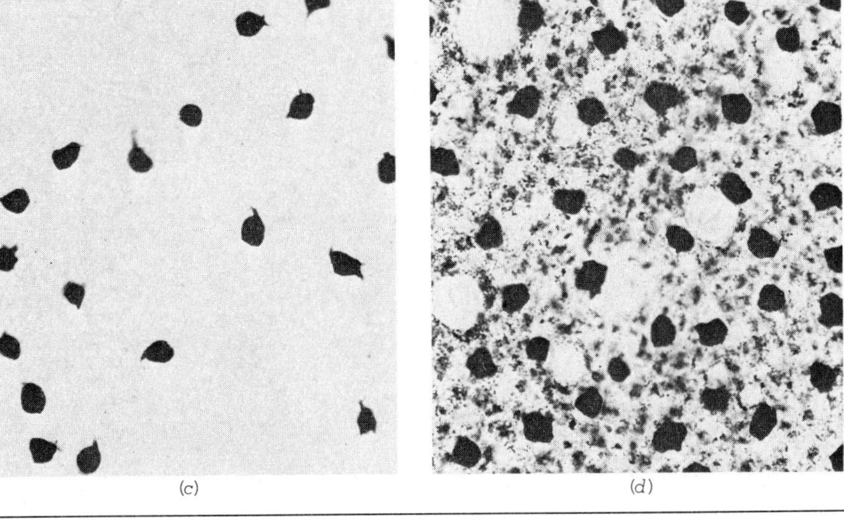

FIGURE 4-6
Two adult *Xenopus* that have been kept on a white (*a*) and a black (*b*) background for 6 weeks. Skin preparations beneath indicate that the white-background–adapted frog has few epidermal melanophores (c), whereas the black-background–adapted frog has many melanophores (*d*). The melanophores in both preparations are punctate because the skins were maintained in Ringer's solution prior to fixation. (*Photograph, Courtesy of M. E. Hadley.*)

(red, orange, and yellow compounds obtained from food). These colors are mainly important for their effects on the behavior of other animals, including other birds. They also have intrinsic values. For example, pigmented feathers are sturdier and do not wear out as fast as unpigmented ones. Pigments also absorb radiant energy and thus are important in thermoregulation. Pigmented feathers serve as a barrier to ultraviolet radiation. Some birds that have translucent feathers, on the other hand, have black, melanin-containing skin, which presumably stops ultraviolet radiation and serves to absorb heat.

Many species of birds use colors for concealment. Some, such as the snowy owl (*Nyctea scandiaca*) and the Greenland gyrfalcon (*Falco rusticolus*) are white and resemble their arctic surroundings. The willow ptarmigan (*Lagopus lagopus*) changes its plumage from a winter white to a summer brown to match the seasonal change in the environment. Many ground-dwelling birds are rendered inconspicuous by their brown colors, whereas birds of the forest canopy, such as vireos and warblers, often resemble green, sun-flecked foliage. The eggs of birds which nest on the ground are generally colored with markings that make them difficult to see. Likewise, the precocial chicks of such birds commonly blend so perfectly with the background that they are quite inconspicuous. The chicks assume cryptic postures when alarmed, squatting low, often with the head and neck flat on the ground to eliminate shadows.

Conspicuous coloration is employed by birds for quick recognition, often at a great distance, of friend or foe, and can serve to hold flocks together. Color may be used for threat or warning, thus sometimes allowing birds to avoid the energy costs and hazards of actual combat. A bright patch of color has the disadvantage of attracting the attention of potential predators, but one must assume that it plays a part of sufficient importance to outweigh this undesirable aspect.

Coloration is used by many birds in courtship ceremonies to attract mates, to stimulate them into sex readiness, and to synchronize the male and female sexual reproductive states. Color patterns of birds in the breeding condition, particularly in the males, are often quite spectacular. Gaudy feathers, brightly colored combs, wattles, neck pouches, or legs and feet may be displayed in courting rituals by the male to stimulate the female. Mating success hinges upon successful combinations of exact color patterns and their proper display in an appropriate environment at the right time. The development of the pigmentation and release of mating behaviors are conditioned and orchestrated by complex neuroendocrine mechanisms, centering principally about pituitary gonadotropic hormones and gonadal steroids.

4-3 SLEEP

Although much is known about sleep and different kinds have been described and analyzed in great detail, sleep still defies definition. No satisfactory explanation for the necessity of sleep has been offered, and no certain function has been assigned to it. Sleep is hazardous in that it renders an animal defenseless for a third or more of its life. And yet, if sleep were not essential it presumably would have been eliminated by natural selection. Sleep deprivation results in reduced efficiency, and prolonged deprivation eventually kills. Most organisms can survive food and water deprivation for much longer periods than sleep deprivation.

Although the level of consciousness is reduced during sleep, it is an organized physiological state. Somatic activity is decreased, and there is a reduction of muscle tone. The autonomic reflexes continue unaltered, and digestion, respiration, and circulation continue normally throughout sleep, although the heart rate decreases and blood pressure falls, reflecting reduced activity of the voluntary somatic muscles.

One of the most instructive ways to study sleep is to use an electroencephalogram (EEG), a technique introduced by Berger in 1930. The EEG is a graphic record of the small voltage changes observed at the scalp surface. In the *awake* state in the adult human with the eyes closed, the EEG is dominated by a so-called *alpha rhythm*, consisting of a series of waves between 8 and 13 per second, having an amplitude between 30 and 50 μV. As the individual becomes drowsy, the alpha rhythm diminishes. Upon entry of the subject into the sleep state the EEG displays waxing-waning bursts (*spindles*) of regular waves of 12 to 14 cycles/second with a low-voltage background. As sleep becomes deeper, large-amplitude waves of 1 or 2 cycles/second or slower appear and the spindles progressively disappear. The stages of sleep in humans have been classified by a few different investigators. One such scheme is shown in Fig. 4-7.

The EEG recordings from other mammals show comparable patterns, and although there is great variation among species, a sequence of patterns can be established for each of them. It is evident that the divisions between stages are arbitrary, as the stages form a temporal continuum, but the divisions are useful in characterizing depth of sleep

FIGURE 4-7
Examples of the recorded tracings of EEG stages of sleep for a human, classified according to the scheme of Dement and Kleitman. *Awake* state is characterized by a mixture of frequencies above 8 cycles/second; often there are dominant, continuous sinusoidal fluctuations of 8 to 13 cycles/second, called the alpha rhythm. *Stage 1:* a low-voltage record with a mixture of both slow and fast frequencies; the alpha rhythm is much reduced; sleep spindles, generally present in other sleep stages, are absent. *Stage 2:* characterized by the presence of sleep spindles (waxing-waning bursts of regular waves of 12 to 14 cycles/second) with a low-voltage background. *Stage 3:* an intermediate stage characterized by the appearance of moderate amounts of large-amplitude waves of 1 to 2 cycles/second or slower; sleep spindles generally present. *Stage 4:* the record is dominated by large-amplitude slow waves. (*Redrawn from W. C. Dement, K. C. Fisher, 1968, in Mammalian Hibernation III, American Elsevier Publishing Company, Inc., New York, p. 179.*)

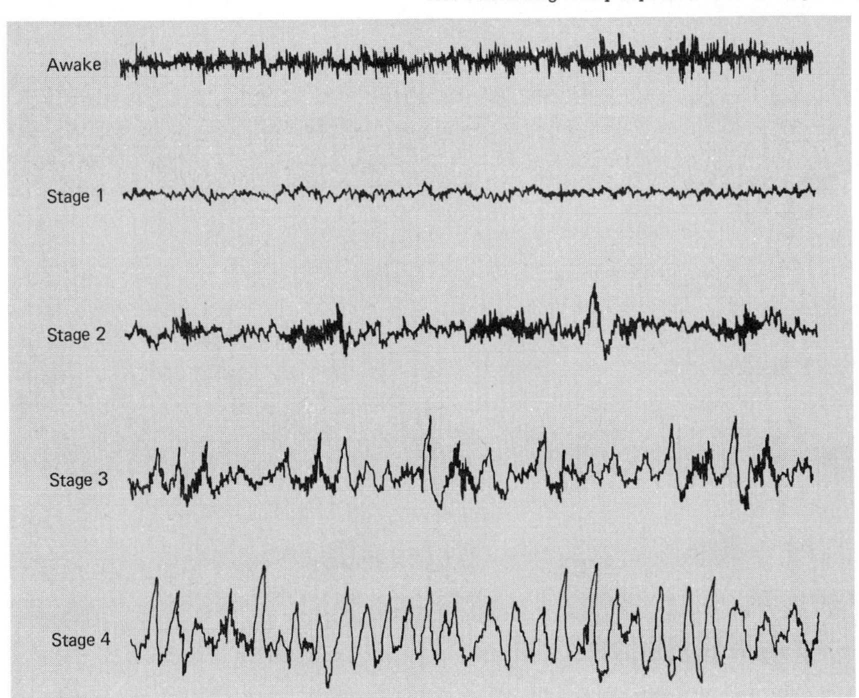

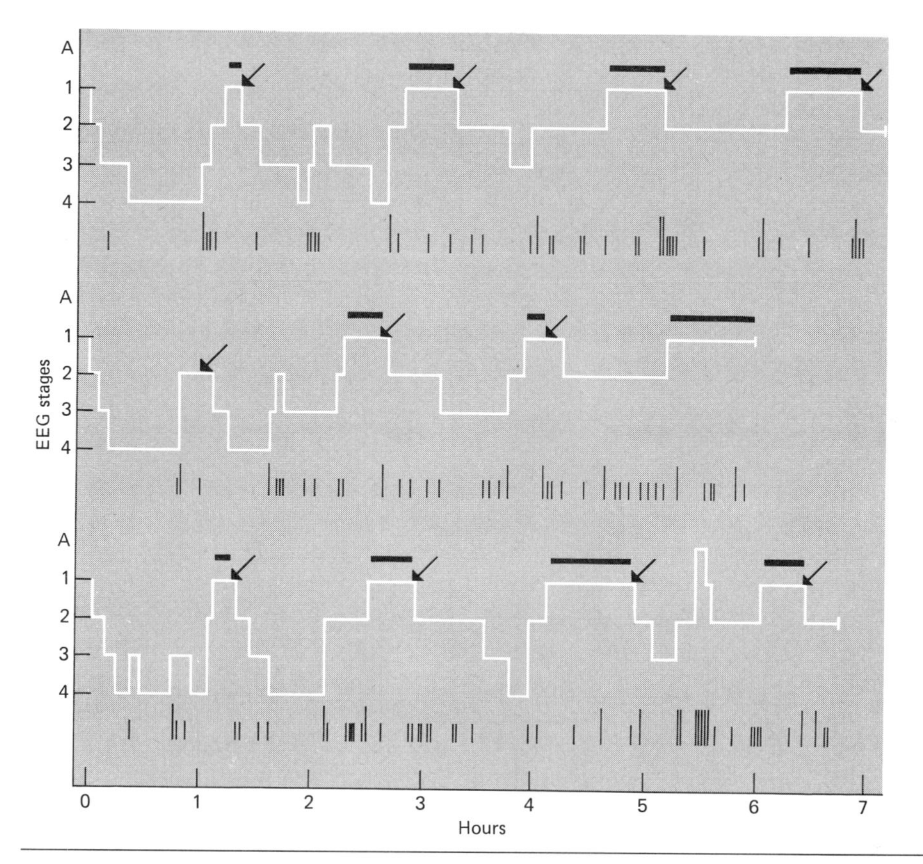

FIGURE 4-8
Continuous plots of the EEG stages of a human during three representative nights. The thick bars immediately above the EEG lines indicate periods during which rapid eye movements were seen. The vertical lines below stand for body movements. The longer lines indicate large movements, e.g., changes in position of the whole body; the shorter lines represent smaller movements. The arrows indicate both the end of one EEG cycle and the beginning of the next. (*Redrawn from W. C. Dement and N. Kleitman, 1957, Electroencephalogr. Clin. Neurophysiol., 9, p. 680.*)

states. When entering into and emerging from deep sleep, the animal passes through these stages in sequence.

While studying spontaneous movements during sleep, Kleitman and associates discovered (in the mid-fifties) that there are several times during a night when *rapid eye movements* (REMs) occur. It was found that this appears in the stage 1 EEG sleep pattern when the subject is ascending from the stage 2 pattern. Dreaming occurs during such arousals. From EEG records it appears that sleeping subjects oscillate back and forth through the sleep states. They progressively enter deeper states and then return to a more aroused state, with REM sleep occupying a minor portion of the total sleep period (Fig. 4-8).

The function of these stage changes is quite unclear, but it appears that to be most beneficial to humans a proper proportion of REM sleep and non-REM sleep is required. When a human subject is deprived of REM sleep and then allowed to sleep without interference, the proportion of REM sleep will be increased, when total sleep is held constant, until the deficit is repaired. However, animals deprived of REM sleep for long periods of time do not seem obviously impaired.

4-4 THE THERMAL FITNESS OF THE ENVIRONMENT

The chemical systems of organisms, based on carbon compounds, are confined in their functioning to a temperature range of about 50°C. They become too inactive to support life near the freezing point of water. Proteins usually begin to denature at 45 to 50°C, although a few algae exist at temperatures above 70°C. Thus, in this narrow thermal spectrum in which almost all living things exist, animals either find environments which are neither too hot nor too cold or else they maintain internal temperatures that do not transgress the thermal limits.

Proper thermal environments for life are commonplace on the earth. This is especially true for bodies of water which, because of the thermal characteristics of water, are eminently suitable for sustaining life. The "fitness" of the physical environment to the conditions of temperature appropriate to life as we know it is remarkable.[1] The temperature-stabilizing properties of water are paramount in making the earth a benign place for life.

The radiant energy which impinges on the surface of the earth—about 10^{15} cal/km^2 at the equator—would produce lethally high temperatures during the day and temperatures below the freezing point of water at night if the surface were entirely made up of rock and soil. Because there are large masses of water on the earth and water vapor in the atmosphere, some of the incoming radiation is absorbed by the water with its high *specific heat* (the number of calories required to raise the temperature of 1 g of a substance 1°C; for water it is 1.0, a value exceeded only by that of liquid ammonia). Thus a body of water can lose or absorb a considerable amount of heat with relatively little change in temperature.

The lakes and oceans, which occupy a majority of the earth's surface, provide a very large *heat capacity* (the number of calories required to raise the temperature of the entire body 1°C), and thus serve as stabilizers or buffers that resist temperature change. They absorb radiant energy during the day and reradiate it at night. Furthermore, the heat reradiated from the land masses is absorbed by water vapor of the atmosphere, preventing the earth from cooling at night as rapidly as it otherwise would. This is the so-called *greenhouse effect*.

Additional contributions of water to temperature stabilization result from its high *heat of vaporization* (the number of calories required to change 1 g of liquid to vapor; 539 cal at 100°C for water, one of the highest known). Large amounts of the sun's heat are absorbed during evaporation from the vast surface of the oceans. Upon condensation of water vapor, heat is liberated, thereby counteracting the coolness of regions in which precipitation occurs. This may happen far from the orig-

[1] See L. J. Henderson, 1913, *The Fitness of the Environment*, The Macmillan Company, New York; reissued in 1958 by Beacon Press, Boston.

inal source of the absorbed heat and thus serve to distribute heat more uniformly across the earth's surface.

Water's high *heat of fusion* (the number of calories required to convert 1 g of solid at the freezing point to 1 g of liquid at the same temperature; for water it is 80 cal, a value greater than that for most other substances) requires a considerable loss of heat before ice is formed and a comparable gain of heat before it melts. Thus, massive gains and losses of heat can occur in large bodies of water at 0°C without change of temperature.

An additional thermostabilizing contribution stems from the fact that water is at *maximum specific gravity at 4°C*. Thus, water below 4°C rises and freezing begins at the surface. One interesting consequence is that ice floats, and icebergs formed at polar regions melt as they move toward the equator. Were it not for this unusual feature of water's specific gravity, the lakes and oceans would presumably be permanently frozen solid.

Finally, water is a very much better *heat conductor* than any other common liquid. Consequently, absorbed heat is readily distributed in water bodies.

All these thermal characteristics of water that so importantly provide automatic temperature stabilization in the external environment likewise serve in similar ways to stabilize the temperature of organisms. Like the surface of the earth, the composition of the bodies of organisms includes more water than anything else.

4-5 STRATEGIES FOR BODY-TEMPERATURE STABILIZATION

Animals can be categorized, approximately, by the way in which they control their body temperature. The term "**poikilotherm**" refers to those with a relatively variable body temperature, and "**homeotherm**" (homoiotherm) applies to those with a relatively constant body temperature. However, these names convey no information about mechanisms, and they suggest too simplistic a separation of what is a truly complex situation. Homeothermy is a feature of birds and mammals, but many of them, under certain conditions, behave more like poikilotherms. And through behavioral and physiological means, many poikilotherms can keep their body temperature relatively constant under a variety of conditions and for prolonged periods of time.

Body temperature is the result of two processes: heat gain and heat loss. All animals produce heat, but commonly the rate of heat production is so low, thermal conductance is high, and the surrounding environment has such a large capacity as a heat sink that there is no perceptible effect on their temperature. Consequently, in most animals body temperature can be independent of heat produced internally and be determined exclusively by heat acquired from the environment. Such animals are described as "**ectotherms,**" which includes the great majority of animals. However, a few atypical forms produce sufficient heat by oxidative metabolism and have low enough thermal conductance that the heat which contributes most to their body temperature is internally produced. Such animals are called **endotherms.** Only certain birds and mammals are continually endothermic.

Some large reptiles and a few fast-swimming fishes are at least partially endothermic. Certain insects, such as honeybees and lepidopterans, are endothermic during periods of activity and can maintain their body temperature well above the ambient temperature. Even some plants maintain temperatures 10°C or more above exogeneous temperatures by heat produced during flowering.

The term **"heterotherm"** was originally applied to birds and mammals that are not continuously homeothermic. However, it can be applied to members of any vertebrate or invertebrate species which during activity are capable of endothermic temperature regulation, but which do not sustain a constant level of body temperature.

It will become apparent that the combinations afforded by thermally variable animals and thermally complex environments allow for several different thermoregulatory capacities. By appropriate combinations of the terms "endothermic," "ectothermic," "poikilothermic," "homeothermic," and "heterothermic" one can describe fairly accurately the patterns of body temperature of most animals.

Large bodies of water, particularly the open oceans, are the most stable and uniform environments in the biosphere. Aquatic animals, except mammals and birds, tend to have body temperatures that do not vary significantly from that of the water in which they are suspended. Most fish, for example, producing heat by the muscle activity associated with swimming, will have a core temperature the same as the temperature of the surrounding water. Any heat produced is conducted into the bloodstream. The blood inevitably moves to the animal's surface, including the extensive gill structures. The gill apparatus is adapted for efficient exchange of dissolved gases between the fish and the environment by having large surface areas intensively vascularized in such a way as to maximize exchange. The same construction that favors gas exchange also facilitates heat exchange. It is virtually inevitable that gill-breathing aquatics are poikilotherms (but some interesting exceptions will be mentioned later).

Aquatic poikilotherms are commonly very mobile and can avoid extremes of temperature. They can recognize and select water masses with temperatures appropriate for their physiological functions (Fig. 4-9).

A whale or a porpoise swimming in arctic waters of about 0°C can maintain a core temperature like that of other mammals—about 38°C. This is possible for several reasons. In the first place these pelagic mammals are air breathers and do not have extensive heat-losing vasculature exposed to the water the way gill breathers do. Most of the body surface is insulated by a thick layer of blubber. Only the flippers and tail flukes are thin-skinned and poorly insulated, and while they are well vas-

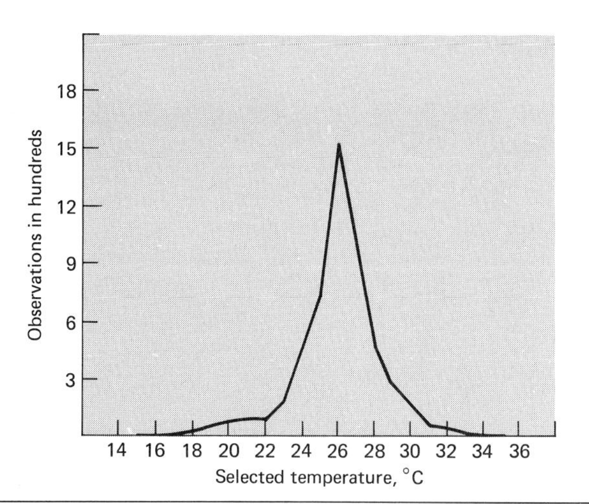

FIGURE 4-9
Temperature selection by 39 individuals of shore fish, the California opaleye (*Girella nigricans*). (*After K. S. Norris, 1963, The functions of temperature in the ecology of the percoid fish* Girella nigricans *(Ayres). Ecol. Monogr. 33, p. 48.*)

cularized and well suited for dissipating heat, the blood vessels in these structures are arranged so as to function as excellent heat exchangers. The principle of **countercurrent exchange** is employed. That is, in the limbs heat in outgoing arteries is taken up by the incoming veins which lie next to the arteries. In fact, a special network of blood vessels, called *rete mirabile* (wonderful net), consisting of counterflowing arteries and veins acts as a heat trap (Fig. 4-10).

The retia are commonly situated at the places where the trunk of the animal is joined to the extremities—limb, fin, or tail—and efficiently return heat to the trunk. Consequently, the extremities generally operate at temperatures cooler than that of the central body. Retia have been described in whales, seals, long-legged wading birds, manatees, and a few land-dwelling forms, including the sloth.

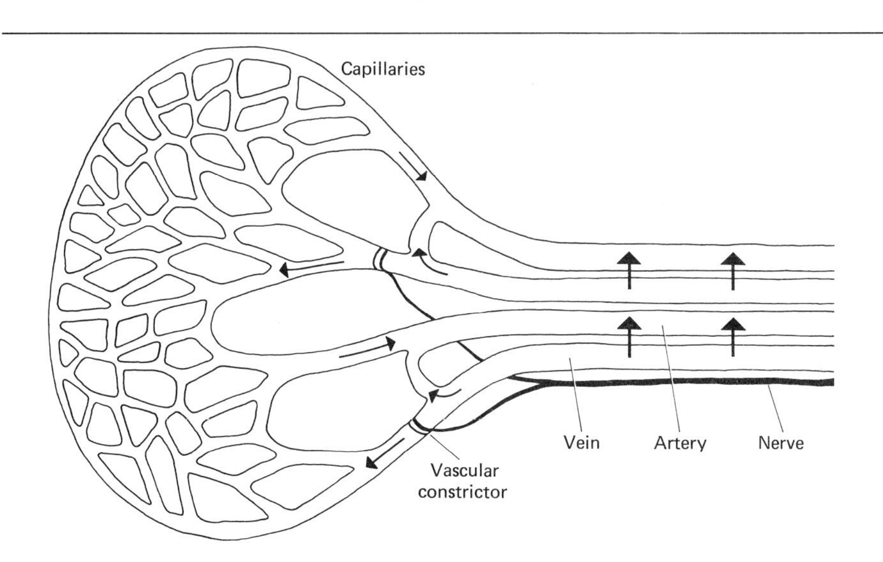

FIGURE 4-10
Adaptation to cold environmental temperature can involve an exchange of heat between arterial and venous blood. The cold venous blood returning from an extremity acquires heat (vertical arrows) from an arterial network. The outgoing arterial blood is thus cooled. Hence the countercurrent exchange helps to keep heat in the central body and away from the extremities when the animal is exposed to low temperatures. The effect is enhanced by the fact that blood vessels near the surface constrict in cold.

As noted, gill-breathing prevents most aquatic animals from maintaining a significant difference between their core temperature and that of the surrounding water. Nevertheless, by use of countercurrent heat exchangers, certain members of two quite unrelated families of fast-swimming fish have evolved effective mechanisms for endothermy. The tuna, a bony fish, and the mako and porbeagle sharks, cartilaginous fish, can maintain central temperatures 10 to 15°C above ambient water temperature. Because these fish must swim continuously in order to respire, they have a continuous supply of muscle-generated heat. Heat is conserved by interposition of countercurrent heat exchangers in the circulatory system between the axial muscles and the gills. The advantage of endothermy to fishes is probably enhanced power generation. For every 10°C increase in the temperature of vertebrate muscle there is a tripling of the power output (Fig. 4-11).

All terrestrial animals are ectotherms, except for mammals, birds, and a few insects. Behavioral regulation is the common mode of thermoregulation in terrestrial ectotherms. Given the opportunity, they will select habitats with preferred temperatures. Some ectotherms are *heliotherms;* i.e., they acquire heat from solar radiation and raise their temperature. Bullfrogs, for example, control their body temperature by changes of location and posture. Solar and thermal radiation are used as heat sources. Pond water serves as a heat source on overcast days and as a heat sink on clear sunny days. Thus, by proper interaction with its environment, a bullfrog in the summertime can maintain its body temperature between 26 and 33°C during daytime activity.

Reptiles employ patterns of behavioral and postural thermoregulation that allow remarkably precise control of body temperature. Lizards such as horned lizards (*Phrynosoma* spp.) can control their rate of heat

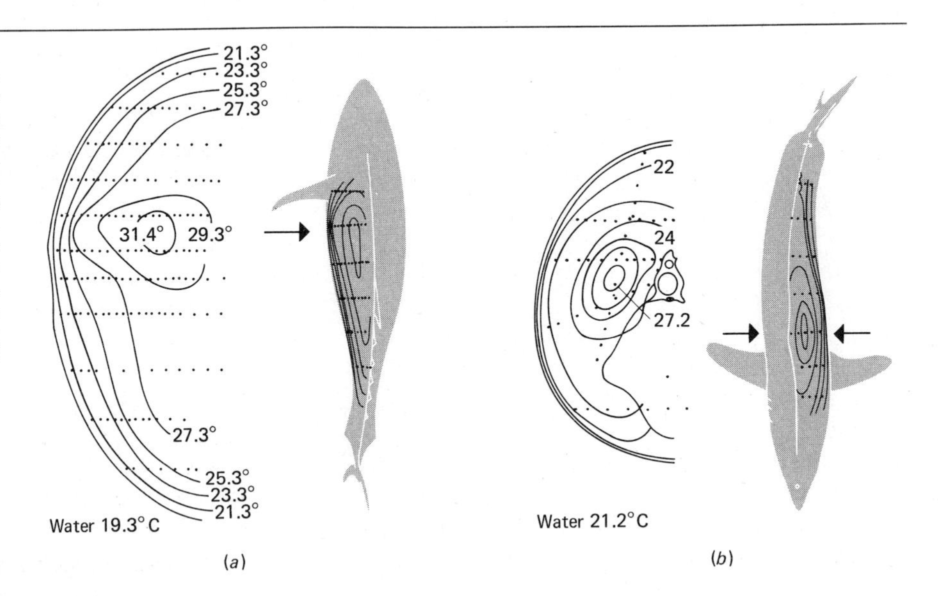

FIGURE 4-11
(*a*) Temperature distribution in a bluefin tuna, *Thunnus thynnus.* Shortly after death, temperatures were measured with long thermistor needles at points indicated by dots. Isotherms are drawn on 2°C contours. (*b*) Temperature distribution in a mako shark, *Isurus oxyrinchus;* 1°C isotherms. (*After F. G. Carey et al.,* 1969, *Comp. Biochem. Physiol. 28, pp.* 200 *and* 206.)

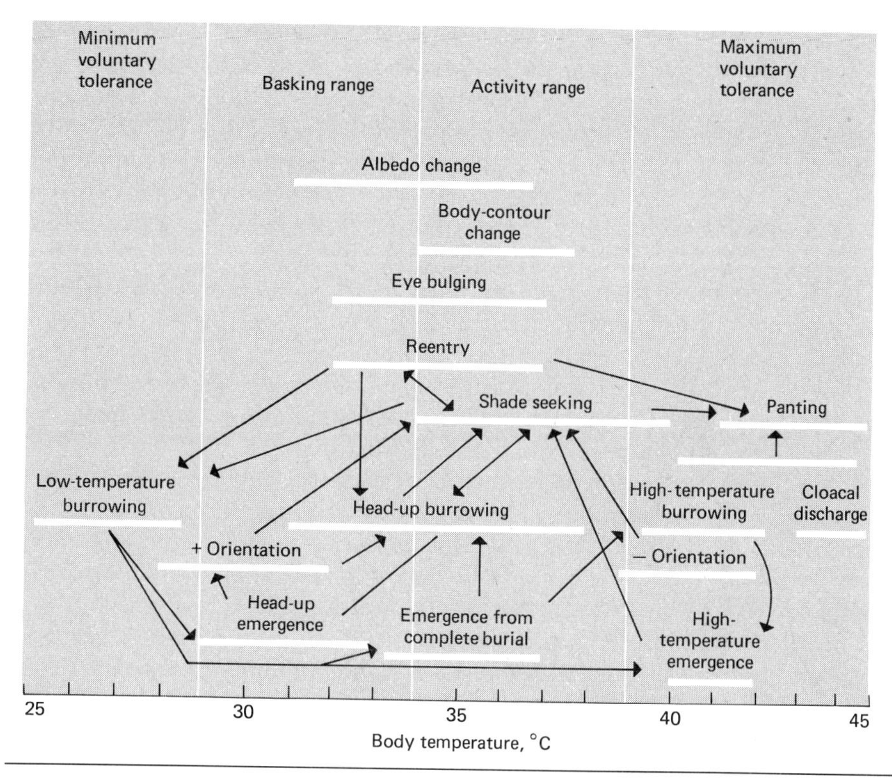

FIGURE 4-12
Relations between different patterns of behavioral thermoregulation and body temperature in the horned lizard (*Phrynosoma coronatum*). (*After J. E. Heath, 1965, Univ. Calif. Berkeley Publ. Zool., 64, 97.*)

gain from solar radiation by variation in orientation with reference to the sun and by changes in body shape. Additionally, they can seek shade or burrow to escape the sun, and they can change the coloration of their body to absorb more or less radiant energy (Fig. 4-5).

Complex combinations of all these devices are triggered at certain *set points* in the neural system of the lizard, which monitors its temperature. (The meaning of the term "set point" is further discussed in Sec. 4-8.) Some indication of the elaborate pattern of thermoregulatory behavior employed by *Phrynosoma* is shown in Fig. 4-12.

Very useful thermal adjustments are made, usually seasonally, by ectotherms which allow their metabolic rates to have some independence from temperature. In a matter of days, and sometimes hours, there will be a thermal acclimation so that the expected proportional changes in metabolic rate with changes of temperature are somewhat compensated for. Enzyme activity often is calibrated to an animal's temperature optimum. For example, a metabolic "fit" with the thermal environment has been found in lizards, where the temperature dependence of ATPase activity of muscle myosin correlates with the lizard's preferred body temperature.

It should now be apparent that ectotherms are not simply passive objects reflecting the temperature of their ambient environment. They can maintain a closely regulated level of body temperature during periods of activity. They do this by behavioral means and appear to have hypothalamic temperature sensors which are involved in behavioral regulation. Many also use cardiovascular control, evaporative cooling, and endogenous heat production to control temperature. Reptiles, which have highly sophisticated mechanisms for thermoregulation, are quite similar to endotherms, except that they lack an insulated body covering and have a much lower metabolic rate.

Several kinds of insects can attain thoracic temperatures 10 to 20°C above ambient temperature by retaining some of the heat produced by the contraction of the wing muscles. The primary functional benefit of elevated temperature is an increase in the power output and the frequency of wingbeat. Preflight warm-up (movement of wings while staying in place) has been demonstrated in butterflies and beetles. In nocturnal beetles this is a particular advantage, since they have little opportunity to elevate body temperature by behavioral thermoregulation. In addition, heat produced by flight muscles also can be important in processes not involved with flight, such as brooding in bees and stridulation of katydids.

Birds and mammals have largely escaped the thermal constraints imposed by the environment which in the ectotherm limit the timing and spatial patterns of their activities. The thermoregulatory system of birds and mammals has as its central element a high rate of endogenous heat production coupled with control of heat production and heat exchange. These are the unique features of the endotherm.

One can readily appreciate the fundamental nature of endothermy by examining the relationship between metabolic rate and environmental temperature. Rate of oxygen consumption is a conventional index of metabolic rate. The relationship is not so simple and direct as in an ectotherm. Within a certain range of temperature, endotherms show a rate of oxygen consumption that is minimal and virtually independent of temperature. This is referred to as the **thermal neutral zone** (Fig. 4-13).

In the thermal neutral zone an animal is expending a minimum of energy on thermoregulation. On either side of this zone an animal must do metabolic work above the minimum to maintain a constant body temperature. Thus, the thermal neutral zone is the range over which a homeotherm can vary its thermal conductance in an energetically inexpensive manner. This is done by varying changes in the supply of blood to superficial areas, by increasing or decreasing the insulation afforded by fur or feathers through fluffing or compressing, or by simple postural adjustments. These changes allow regulation of body temperature through control of heat flow with negligible energy cost. At all temperatures a homeotherm produces a large amount of heat that is continuously being lost to the environment.

Below its critical temperature (Fig. 4-13) the animal can no longer decrease its thermal conductance enough to compensate adequately for

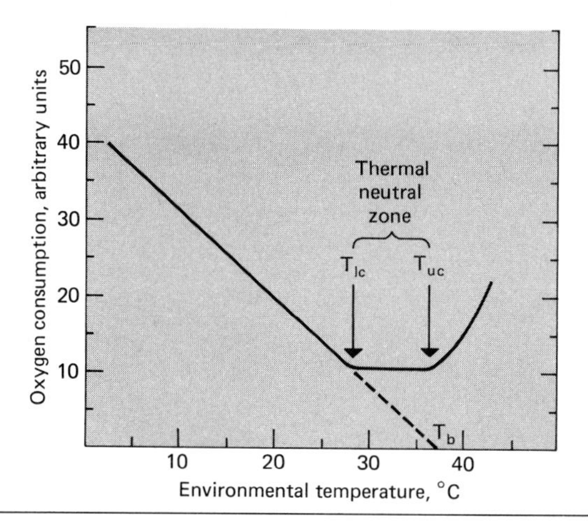

FIGURE 4-13
Relation of oxygen consumption to environmental temperature in a hypothetical mammal. T_{lc} = lower critical temperature; T_{uc} = upper critical temperature; T_b = core body temperature.

the rate of heat loss as the difference between body and ambient temperature increases. Thermal conductance is minimal. Therefore, it must increase the rate of heat production to prevent a decrease in temperature.

At the upper critical temperature (Fig. 4-13) thermal conductance is maximal. This point is by necessity slightly below the operational body temperature; otherwise heat would not be lost to the environment. When ambient temperature exceeds body temperature an animal can no longer passively lose heat to the environment through conduction, convection, or radiation. It must do so by means of water evaporation. Work is required to mobilize the water used in evaporative cooling, as readily seen in the increased oxygen consumption associated with increased environmental temperature (Fig. 4-13).

The adaptations of endotherms to temperature extremes are varied and interesting. The most direct and metabolically inexpensive method for dealing with low ambient temperatures is to increase the effectiveness of insulative mechanisms and thereby diminish thermal conductance. This can involve seasonal morphological changes that are under endocrine control, short-term changes in cardiovascular patterns, and behavioral changes, including postural changes. Some animals simply avoid low environmental temperatures. They may migrate to a warmer climate or may build nests or shelters. Others avoid the problem, to some extent, by allowing their body temperature to drop 20 to 35°C below their active level and enter an estivation or hibernation state, as discussed below. Some birds and mammals practice *regional heterothermy;* i.e., extremities are not maintained at the same warm temperature which characterizes the body core. By means not at all understood, tissues of such animals that are allowed to cool to just a few degrees above 0°C still function in a normal manner (Fig. 4-14).

An active, and energetically expensive, response to a cold environment is increased heat production to offset heat loss, after further reduc-

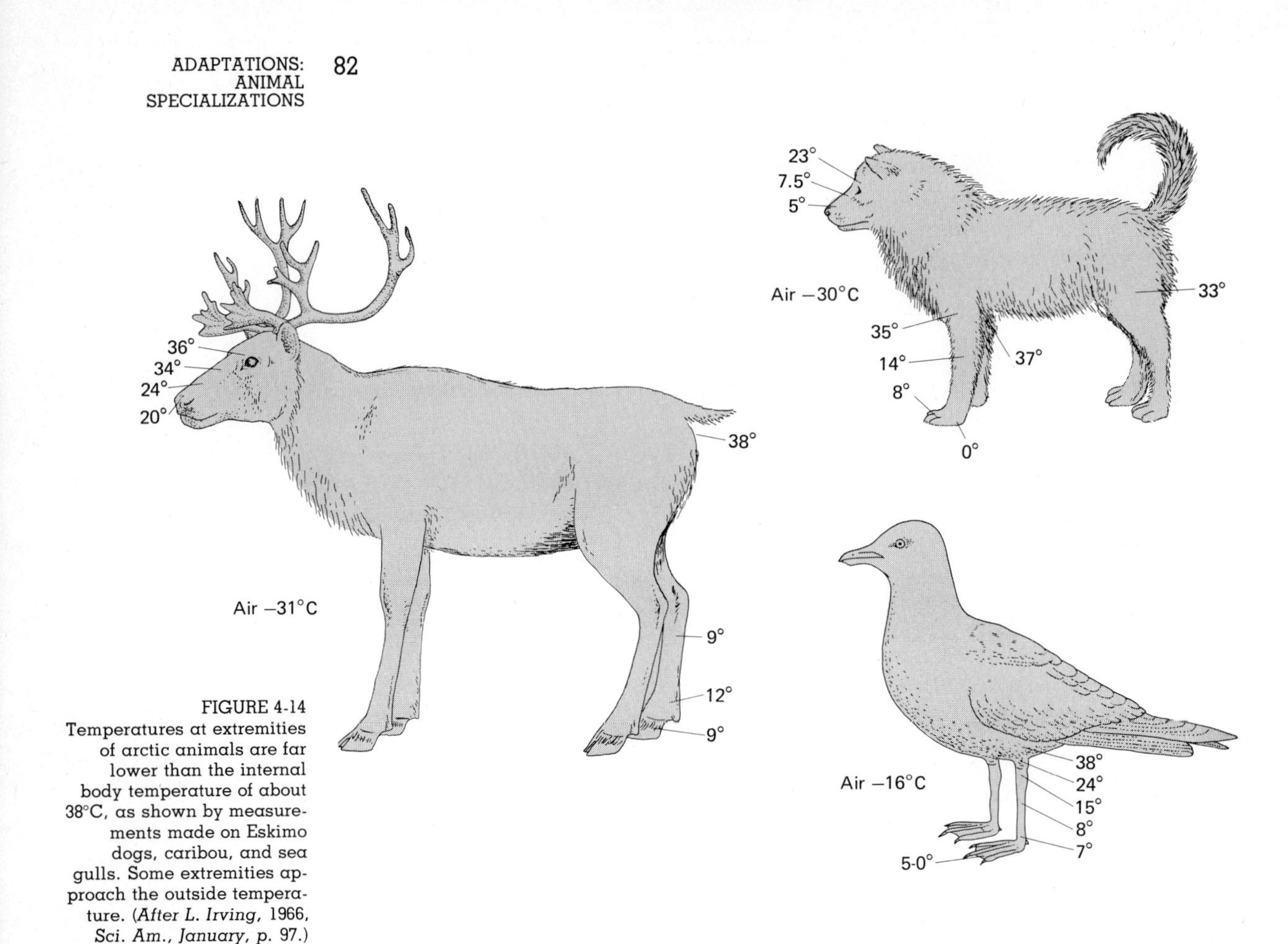

FIGURE 4-14
Temperatures at extremities of arctic animals are far lower than the internal body temperature of about 38°C, as shown by measurements made on Eskimo dogs, caribou, and sea gulls. Some extremities approach the outside temperature. (*After L. Irving, 1966, Sci. Am., January, p. 97.*)

tions in conductance are no longer possible. The control mechanisms of intermediary metabolism ultimately are involved, and seasonal and short-term adjustments are made. However, the most conspicuous mechanism by which endotherms increase heat production is by increased muscular activity, either by locomotion or by *shivering*.

In several kinds of mammals heat production can be augmented without increased muscular activity, i.e., by *nonshivering thermogenesis*. Apparently, with cold adaptation there are shifts in the pathways of intermediary metabolism which result in greater than usual heat production. It is probable that there is increased utilization of lipids. Thyroid hormones are involved in the long-term development of nonshivering thermogenesis.

An important site of nonshivering thermogenesis is *brown fat*, a highly vascular adipose tissue that is characteristic of the young of most species of mammals. As noted below, brown fat also is particularly well developed in mammals that hibernate, and it usually increases in mass during prolonged exposure to low ambient temperature. Because brown fat depots are localized around the neck, thorax, and major blood vessels, the heat generated in them is transported to the heart and brain, structures for which temperature is particularly critical.

4-6 ADAPTIVE HYPOTHERMIA

Deriving heat for regulating body temperature from oxidative metabolism is energetically expensive. The continuous maintenance of a high and uniform body temperature (homeothermy) requires sustained high levels of energy utilization and frequent food consumption. These conditions are difficult to meet, especially for small animals. Very small birds and mammals often must consume food equal to their body weight every day to maintain a constant body temperature. A major reason for this demand for energy is that *relative* heat loss of a body increases with decreasing size. Smaller animals have a relatively greater surface area than larger ones (surface area of a mammal is given by a constant × weight$^{0.67}$), since heat production is proportional to body mass but loss is proportional to body surface; they thus have relatively greater potential for heat loss through greater thermal conductance. There are several factors that modify thermal conductance—including rate and pattern of blood circulation, posture, piloerection, and rate of evaporative water loss, as noted above—but, in general, smaller homeotherms have greater heat production and a higher thermal conductance than larger ones.

Considering the energy cost, it is not surprising that a few small birds and many small and medium-sized mammals will abandon homeothermy for intervals of sometimes hours or several days, allowing their body temperature to fall to levels equal to or only slightly above those of their surroundings. The lowered body temperature is not a symptom of failure of the animal's thermoregulatory machinery, however, but rather the result of a switch to a clearly adaptive, precisely monitored state of *hypothermia* from which the organism can arouse by endogenous heat production. Rates of heat production and levels of body temperature are controlled on a much more flexible basis than in those birds and mammals whose thermal set points are at a relatively fixed position.

Hibernation, or *winter dormancy*, of mammals in cool temperate and boreal regions is the most familiar of the adaptive hypothermia patterns. However, this is but one form of energy-saving hypothermia. Other animals have daily periods of **torpidity.** Still others have a *partial torpidity* characterized by a few degrees of hypothermia, usually referred to as *seasonal lethargy* or *partial hibernation*.

All the kinds of adaptive hypothermia share the following attributes: (1) body temperature drops to within 1°C or less of ambient temperature, (2) oxygen consumption is greatly reduced, to as little as 5 percent of the basal metabolic rate, (3) breathing slows, often to one exchange or less per minute, (4) heart rate slows considerably, (5) a condition of torpor or dormancy develops that is much more profound than deep sleep, and (6) a return to high body-temperature levels characteristic of normally active endotherms can be effected by invoking arousal mechanisms that lead to endogenous heat production and the conservation of body heat. The specific form of each of these features, as well as the conditions that trigger the hypothermic responses, varies considerably among animals.

Daily torpor is exhibited by some small birds and mammals that have relatively high metabolic rates and store only small amounts of energy as fats. They are always only a few hours from death by starvation, particularly at low ambient temperature. One solution is daily torpor; i.e., while active they are warm-bodied and thereby gain the advantages of independence from the environment, but when inactive they lower their body temperature and reduce their energy expenditure. Such a combination allows them to exploit habitats that otherwise would not be available to them. The routines of small bats and hummingbirds are examples of such a life-style. Both have restrictive feeding adaptations that limit their periods of foraging (daytime in hummingbirds, nighttime in bats). Both have developed the same strategy: daily periods of torpor.

The energy saving made possible by torpor is quite great. For example, if the California pocket mouse (*Perognathus californicus*) were to maintain torpidity for 10 hours at 15°C the energy cost of the entire torpor cycle has been calculated to be only 10 percent of that required for maintenance of its normal body temperature for the same time. In this and other desert-dwelling mice either food or water restriction can trigger torpidity.

Daily torpor is a fairly widespread phenomenon, occurring in hummingbirds, bats, and a variety of rodents. It may also intergrade with *seasonal dormancy*, or have a strong seasonal component. Some bats show daily torpor during the winter but not during the summer. Other bats have daily cycles of torpor during the summer and prolonged ones during the winter. Some rodents intergrade between daily and seasonal torpor. The birch mouse (*Sicista betulina*) shows daily cycles of torpidity during summer and early fall but hibernates during the winter.

Hibernation is characteristic of certain small- or medium-sized mammals inhabiting environments where winters are cold. It is a prolonged period of hypothermia and dormancy that requires a period of preparation and adjustment. The pattern of development and regulation of hibernation is complex and variable among species. It appears to have originated independently among different groups of mammals. Enormous amounts of body fat are deposited prior to entry into torpor, there is an increased lethargy and a series of increasingly lower and more prolonged periods of hypothermia, followed by return to normal body temperature.

In contrast to the passively determined body temperature of the torpid state, in hibernation temperature is loosely regulated, or, at least, monitored. If the body temperature starts to fall to lethal levels, there is arousal and the animal warms to its normal body temperature. Regular arousal is, in fact, characteristic of hibernators. Although they may remain relatively inactive and not leave their den for months at a time, they apparently arouse every week or two, maintain a warm body temperature for a few hours or days, eliminate accumulated metabolic wastes, and then return to dormancy (Fig. 4-15).

When arousal from hibernation is effected, mechanisms for maximal heat production are activated. Both *shivering* and *nonshivering thermogenesis* are employed, and heat production in the depots of highly vascularized deposits of brown fat contribute importantly to the elevation of temperature. As noted earlier, the heat generated is usually sequestered in the anterior parts of the body, where it warms the heart and the central nervous system. This is the result of the particular localization of the brown-fat depots and appropriate cardiovascular shunting.

Estivation is a form of summer dormancy used by animals to escape prolonged periods of heat, drought, or some other lengthy and recurrent pattern of seasonal stress. The advantages of seasonal dormancy that accrue to winter hibernators apply equally well to estivators. Seasonal dormancy is a broadly distributed response pattern that is not restricted to cold-latitude hibernators. This is clearly indicated by the situation demonstrated by ground squirrels of the genus *Citellus*, which has representatives in arctic, temperate, and tropical regions. Those living where winters are cold hibernate; where there are seasonal droughts they estivate; they both estivate and hibernate, with the two merging, in northern arid regions where precipitation is restricted to winter and spring, but not

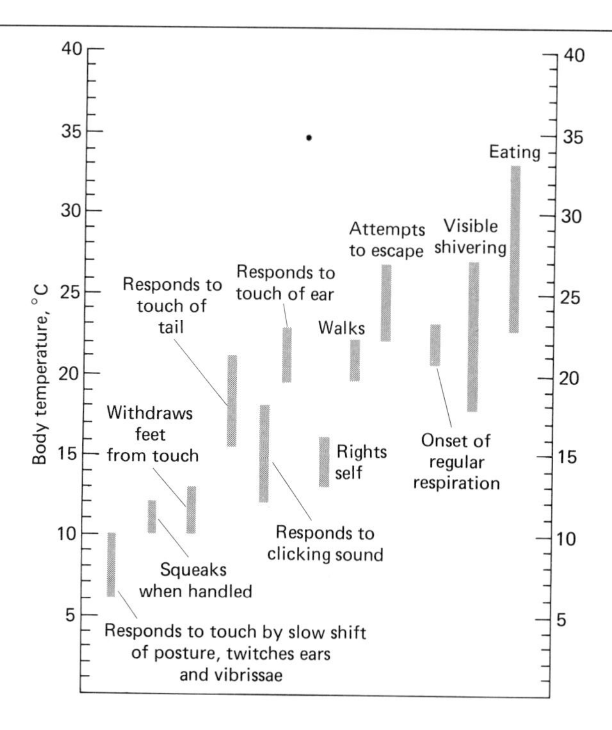

FIGURE 4-15
Range of minimal body temperatures for various patterns of behavior by the little pocket mouse during arousal from hibernation. (*After G. A. Bartholomew and T. J. Cade, 1957, J. Mammal., 38, p. 67.*)

in areas of regular summer rainfall; others inhabiting warm temperate or tropical areas neither hibernate nor estivate.

Estivation occurs at relatively moderate environmental temperatures, and unless its body temperature is measured, an animal may practice hypothermia and go unnoticed. Estivating mammals commonly behave in a normal-appearing manner even though their body temperature is near 25°C.

Estivation is fairly common in rodents, and has been recorded in insectivores and marsupials. Swifts and goatsuckers, two groups of aerial-feeding insectivorous birds, exhibit hypothermia and torpidity that are similar to mammalian estivation during inclement weather.

4-7 HOMEOSTASIS

Living systems at all levels not only can react to changes in their surroundings, they also can control and regulate their reactions. This fact became clearly apparent to the great French physiologist of the late nineteenth century, Claude Bernard. He recognized three forms of life: *la vie latente*, *la vie oscillante*, and *la vie constante*. In his classification latent life was that exemplified by spores and encysted microorganisms, in which the influence of the external environment is so dominant that the manifestations of life are arrested completely. Oscillatory life was that of poikilotherms, which, as we have seen, do not always conform to external temperature so passively as Bernard believed. Constant life was that of homeotherms, in which life, so Bernard assumed, is carried on at a constant temperature. *Vitalism* (see Sec. 1-1), a strong philosophy of the time, held that homeotherms are controlled in some way by an internal vital principle which endowed them with their constancy and which would be beyond the reach of scientific analysis. Bernard rejected that view and believed that the constancy is a result of regulatory processes which operate according to analyzable physiological principles and which continuously maintain this constancy by responding to the factors tending to upset it.

Bernard proposed that the maintenance of a stable steady state of the internal milieu by organisms was a requirement for their survival. His views have profoundly influenced subsequent developments in animal physiology and have been extended to apply not only to the various mechanisms regulating the internal composition of the body, but also to behavior, to population dynamics, and to other areas as well. However, it is the more classical concept of constancy of the internal environment of organisms that we are interested in now.

The regulating mechanisms of higher vertebrates have developed from simpler mechanisms through the operation of natural selection. The higher an animal is on the evolutionary scale, the greater are its regulatory capacities. Birds and mammals regulate the largest number of internal factors with the greatest precision. The operation of these mechanisms was embodied in the concept of **homeostasis,** formulated by W. B. Cannon in 1929.

The constant conditions which are maintained in the body might be termed *equilibria*. That word, however, has come to have fairly exact meaning as applied to relatively simple physico-chemical states, in closed systems, where known forces are balanced. The coordinated physiological processes which maintain most of the steady states in the organism are so complex and so peculiar to living beings—involving, as they may, the brain and nerves, the heart, lungs, kidneys and spleen, all working cooperatively—that I have suggested a special designation for these states, *homeostasis*. The word does not imply something set and immobile, a stagnation. It means a condition—a condition which may vary, but which is relatively constant.[1]

Homeostasis includes the regulation of such variables as temperature, salt and water content, pH, and nutrient concentrations.

4-8 BIOLOGICAL CONTROL SYSTEMS

The maintenance of a constant level of some physical or chemical quantity against disturbances from the outside, whether in a machine or in a living organism, suggests the presence of a "servomechanism." Such things as blood pressure, rate of blood flow, and body temperature are automatically controlled by such mechanisms. Just as in man-made devices, the chain of control begins with a *sensing unit*, which measures the variable and in turn relays information about it to a *controller*. In the controller a comparison is made between the current value of the variable and a *set point* to which the variable is to be held. If there is not a proper match, then an *effector* mechanism is activated to bring the variable into accord with the set point (Fig. 4-16).

[1] W. B. Cannon, 1939, *The Wisdom of the Body*, W. W. Norton & Company, Inc., New York.

FIGURE 4-16
(*a*) Representation of a simple system containing an input, an output, and a set of rules governing the input-output relationship. (*b*) Conventional symbols used to represent various parts of systems. Square or rectangular boxes indicate the system under analysis. Lines represent information-flow pathways; i.e., electrical wires, nerve cells, or other mechanisms which convey information, either as energy or matter, from one part of the system to another. Arrows show the direction of information flow. Circles represent sites in the system where several lines of information merge. Open circles are mechanisms where manipulation or control of activity occurs. Quartered circles are sites at which algebraic addition of signals takes place, often with reference to some set-point level of information. Such comparators may or may not be included as part of the controlling system. These components are used, for example, in the system of Figs. 4-17 and 4-18.

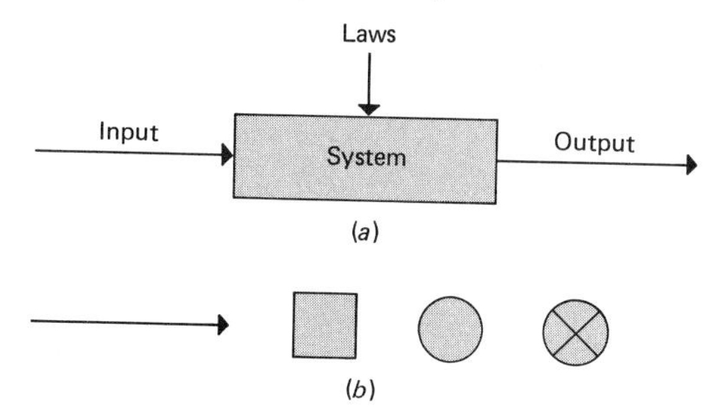

Homeostatic systems contain one or more negative-feedback loops. **Feedback,** a basic idea borrowed from cybernetics (see Chap. 1), is defined as the influence of the output of a system on its input. *Positive feedback* occurs when the output, or some function of the output, is *added* to the input—resulting in greater output. Certain autocatalytic processes and some amplifying neurosensory pathways employ positive feedback. Such systems can become explosive (in a literal sense) and require eventual suppression, but they can provide stability under certain conditions. *Negative feedback* occurs where the output, or some function of the output, is algebraically subtracted from the input of the system, resulting in an increasing or decreasing output about some set level so that the entire system is controlled (Fig. 4-17).

A system that exhibits regulation must contain a *negative*-feedback loop. Any stimulus (disturbance) will affect the output and thus the feedback signal, depending on whether the output is larger or smaller than the predetermined value (*set point*). Thus, the level of the feedback signal alters the value of the output signal and control is achieved (Fig. 4-18).

Because most of the information received by receptors, including feedback sensors, is relayed to the central nervous system for processing, it is commonly assumed that this is the site of set-point comparisons for many control systems. In a few instances, such as for thermoregulation, this is certainly the case. The location of the controller is in the hypothalamus. It monitors as well as regulates the temperature of the blood flowing through the brain. It is part of a *feedback* system which causes heat dissipation and thus keeps the body from overheating under normal conditions. Whereas this system is rather well analyzed, the system which steps up metabolic heat production and keeps body temperature from falling below optimum is not at all well elucidated.

How to analyze and describe complex systems, where many different components and functional levels are involved and where many variables affect the system, has been a chronic problem in biology. One approach that has been rather successful in providing logical presenta-

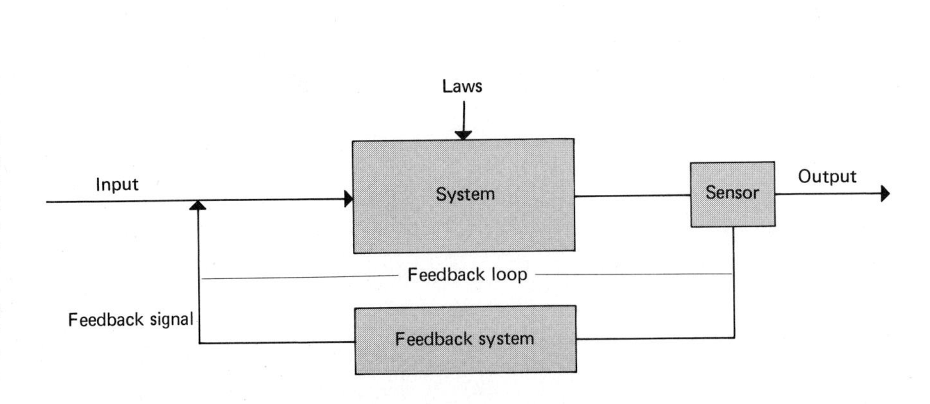

FIGURE 4-17
An input-output system as previously diagramed (Fig. 4-16), but with a closed loop (negative-feedback loop) added. Part of the output is sensed by an element of the feedback system, and a signal is sent back to the input. In negative-feedback regulating systems, the feedback is algebraically subtracted from the input, causing an appropriate increase or decrease in the system's output—thereby achieving a relatively constant output.

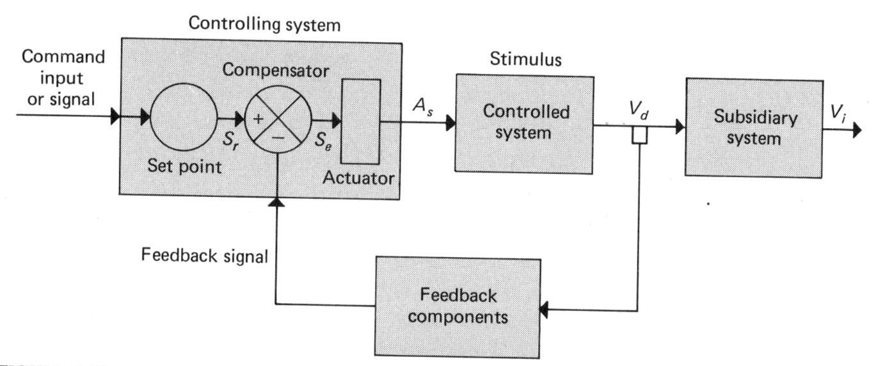

FIGURE 4-18
Schematic of a negative-feedback system and its components. Each of the required components is shown as a separate entity, but in the organism many of the mechanisms required for feedback control in homeostasis may be combined in one cell or tissue. In most cases the actual location and mode of operation of the feedback-system components in the organism are not known. Note that there are inputs and outputs for each subsystem and that information flow is an important factor in the system. The controlling system includes an actuator, responsible for increasing or decreasing the value of the controlled-output variable. The actuator signal or forcing signal (A_s) is determined by the strength of the error signal (S_e), derived by comparing the feedback signal with a reference signal (S_r) in the component known as the comparator or compensator. The set-point device selects the required value of the controlled-output variable; it can be activated by a command signal so that a reference signal results. A controlled system is the part of the feedback system which is to have one or more output variables controlled. The directly controlled variable (V_d) often is the input to a subsidiary system, whose output is then indirectly controlled (V_i). The input labeled "stimulus" at the level of the controlled system may alter the output, which must then be regulated to overcome the effect of the disturbance. The feedback loop contains components to sense the magnitude of the controlled output and to relay information via a feedback signal to a controlling system.

tions and a few widely applicable generalizations about biological mechanisms is *systems analysis*. Although systems analysis is a mathematically based methodology, a verbal description will be used in the brief account here.[1]

Systems analysis is not always readily applied to organisms. It was in engineering, where the interest is in how to *build* complex systems, that systems analysis originated. Quite a different situation is presented to the biologist who is faced with trying to *analyze* an established, complex system that is the result of millions of years of evolution. Instead of putting together components to make complex functional systems as the engineer does, the biologist must astutely dismantle the functioning organism to gain some information about its component parts.

Systems analysis permits one to regard organisms as entities built of smaller systems such as organs, cells, or subcellular units. Whereas in the most general sense a *system* is any part of the universe a scientist wishes to study, in systems analysis a system is defined as a collection of objects or components functioning together or coordinated according to a set of relationships or laws. Systems possess at least the following parts: (1) one or more inputs, (2) relationships between the input and output, and (3) a surrounding, or in biological terms, an environment. Systems can be represented by diagrams which may be simple or complex, de-

[1] See J. H. Milsum, 1966, *Biological Control System Analysis*, McGraw-Hill Book Company, New York, for a quantitative presentation.

pending upon the amount of detail which is to be shown, such as those shown in Figs. 4-16, 4-17, and 4-18.

Organisms and parts of organisms may be treated as systems. The term "living system" is used to refer to any or all of the various structural-functional components of an organism. The biological investigator must by necessity limit his interest to some arbitrarily delimited system, such as an organ, a cell, or an enzyme system.

Systems may be classified into two general types: **closed systems** and **open systems.** Closed systems do not exchange matter with their surroundings, and they are capable of attaining true thermodynamic equilibrium. An example of a closed system would be a chemical reaction occurring in a flask during which no matter is lost or gained. Energy may be lost or gained by the system, however. Living things and their component parts are open systems; they do exchange matter with their surroundings. They are not usually in true thermodynamic equilibrium, but rather in a dynamic *steady state* in which the rate of formation of a given component is exactly counterbalanced by an equal rate of removal or breakdown. Thus, the open system may give the appearance of being stationary in concentration but is actually engaged in exchange with the environment. Nutrients, oxygen, carbon dioxide, and water are exchanged on a very large scale by living, open systems.

Some of the energetic aspects of living systems will be considered near the end of this book. Our more immediate concern will be with how living systems respond to *information* inputs about changes in the external and internal environments and make output responses to these changes. Concepts such as communication, control, and regulation will be continuing themes through the next several chapters.

One of the most discernible attributes of the living state is irritability, the ability to react to changes in the environment. The change which provokes the response is called the *stimulus.* The stimulus can be any of various kinds in terms of energy form, intensity, duration, or pattern and can be simple or complex. By analyzing the nature of the stimulus (input) and the causally related response (output), information can be obtained about the system. The system can be an animal population, a single organism, an organ, a cell, or a molecular complex. Such analysis makes use of the **black-box concept** of Norbert Wiener. A black box is a system whose components and mechanisms are unknown but for which a stimulus-response relation can be established. This is particularly useful for the physiologist, who works with extremely complex systems which are presently very largely enigmatic. By appropriately stimulating his black box—a neuron, an endocrine gland, liver cells, or other system—under controlled conditions, he hopes to gain insight into the nature of its contents and also obtain clues which will suggest other more productive stimulus probes to apply.

Sometimes an investigator employs the **white-box concept.** A white box is a collection of known components, which might be a computer simulation of the black box, or it might be a hardware *model* of a well-analyzed black box that is designed to have the same stimulus-response relationships as the black box. Whatever its nature, the white box is used to learn more about the system of interest. Models are, of course, simplified representations of real, complex systems.

Models may be mathematical rather than of the hardware kind. Mathematical models, though they cannot truly represent the multivariate and nonlinear character of complex living systems in toto, are very useful. They can be rigorous and concise, and they use a universal language. Modeling, using both mathematical constructs and physical representations, has been especially prominent in the study of conductance of excitable cell membranes.

In the next few chapters, which deal with the subjects of nutrition, respiration, circulation, osmoregulation, and excretion, several kinds of regulatory systems will be illustrated. Each of these involves specific and numerous intricacies, but they become less formidable if they are dealt with in the generalizations of control systems, such as have been outlined here. Through manipulating the system in a controlled manner, information can be gained about the mechanisms that reside therein.

READINGS

Among the many articles to be found on defensive and offensive secretions, those by G. W. K. Cavill and P. L. Robertson, 1965, Ant venoms, attractants and repellents, *Science,* **149:**1337–1345, and by T. Eisner and J. Meinwald, 1966, Defensive secretions of arthropods, *Science,* **153:**1341–1350, will be found especially interesting.

Bagnara, J. T., and M. E. Hadley, 1973, *Chromatophores and Color Change,* Prentice-Hall, Inc., Englewood Cliffs, N. J., deals with various aspects of animal pigmentation; contains excellent illustrations and numerous references.

Gordon, M. S., 1972, *Animal Physiology,* The Macmillan Company, New York. Chapter 8 of this textbook is an excellent account of body-temperature regulation and energy metabolism.

Kalmus, H. (ed.), 1966. *Regulation and Control in Living Systems,* John Wiley & Sons, Inc., London. An anthology of articles that illustrates the application of the principles of control theory, as well as its limitations, to various biological systems.

Portman, A., 1959, *Animal Camouflage,* The University of Michigan Press, Ann Arbor, is a nontechnical but authoritative introduction to the techniques used by animals to conceal themselves.

Whittow, G. C., 1970–1973, *Comparative Physiology of Thermoregulation,* 3 vols., Academic Press, Inc., New York. In this treatise thermoregulation is dealt with by taxonomic groups in a very thorough manner.

NUTRITION 5

"It is usual to suppose that living organisms are material systems competing for energy, whereas it seems more likely that they are energy systems competing for materials."[1]

The biochemical basis of chemical needs is the subject of this chapter. Specific chemical needs have been more or less defined for about 200 animals and 300 plants; i.e., about 0.4 percent of the estimated kinds of organisms. Known best are the requirements for organisms that are pathogens, cause damage to materials of human importance, are sources of human food or drugs, are important in manufacturing (e.g., the fermentation and silk industries), or are popular as experimental subjects for biological research.

Nutrition is the general term used for the sum of the various processes by which an animal or plant acquires and incorporates required substances from the external environment. The disposal of terminal waste products is called **elimination.**

One of the great accomplishments of physiological chemistry in the nineteenth and first half of the twentieth centuries was the identification of proteins, carbohydrates, and lipids as both the major structural and nutritional constituents of organisms, as well as the discovery of the detailed structure and metabolic fates of these compounds.

5-1 CONSTITUENTS OF ORGANISMS

The predominant chemical elements in the crust of the earth, in grams per ton, are:

Oxygen*	466,000	Potassium*	25,900	Sulfur*	520
Silicon	277,200	Magnesium*	20,900	Strontium	450
Aluminum	81,300	Titanium	4,400	Barium	400
Iron*	50,000	Hydrogen*	1,400	Carbon*	320
Calcium*	36,300	Phosphorus*	1,180	Chromium	200
Sodium*	28,300	Manganese*	1,000	Chlorine*	200

These elements, with the exceptions of strontium and barium, are members of the first four periods of the periodic table; i.e., they are among the lighter elements and, therefore, tend to be more common at the surface of the stratified lithosphere. What is the biological significance of this information?

The elements marked * above are known to be major functional

[1] A. E. Needham, 1959, *Q. Rev. Biol.*, **34**, 202.

components of organisms, and if only nitrogen is added to the list, they would account for at least 99.9 percent of the mass of organisms. The remaining 0.1 percent is composed of various "trace" elements found in varying amounts in different organisms. These include silicon and aluminum from the above list, as well as cobalt, copper, zinc, iodine, fluorine, bromine, vanadium, and perhaps a few others.

Only three of the predominant crustal elements listed above have no known biological role. Titanium in nature is found only in higher states of oxidation and does not appear to form ions. Barium is toxic to higher organisms. Strontium is not known to perform any specific function in organisms but seems to accompany or displace calcium.

Those elements which have functional roles in the biosphere tend to be the most soluble in water and as a consequence are found widely distributed in the earth's crust. Also, they are either weakly electronegative, meaning that they tend to lose electrons rather easily and form cations (Ca^{2+}, Mg^{2+}, K^{+}, Na^{+}), or they are strongly electronegative and consequently tend to acquire electrons and become anions (S^{2-} and SO_4^{2-}, P^{3-} and PO_4^{3-}, Cl^{-}, and CO_3^{2-}).

Nitrogen is conspicuously rare in the lithosphere, yet it comprises 78

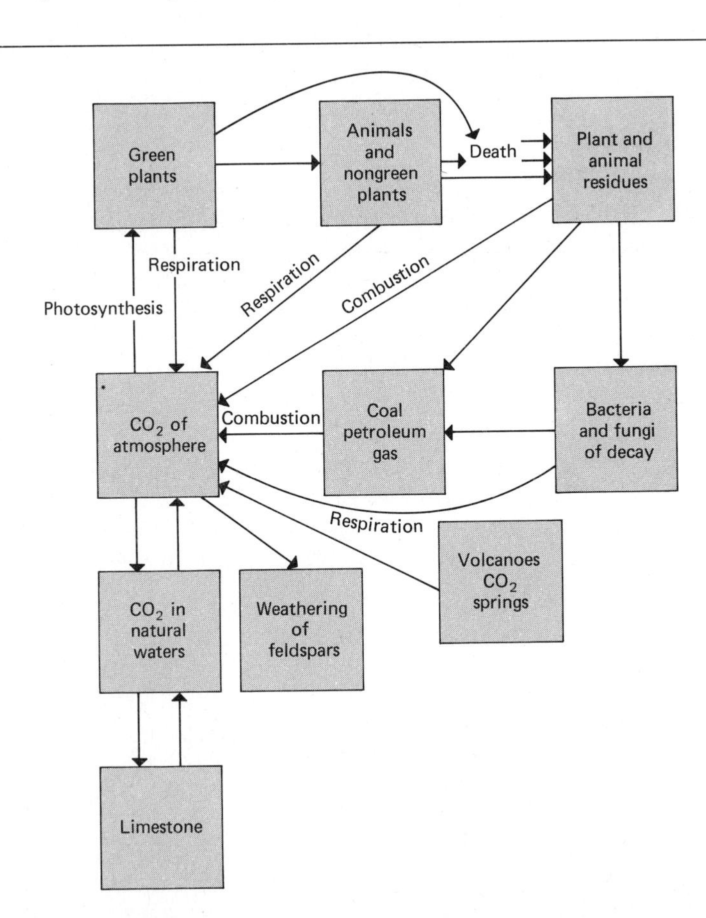

FIGURE 5-1 The carbon cycle. Carbon dioxide in the atmosphere (and dissolved in water) forms the principal inorganic reservoir of carbon. The photosynthetic reactions of green plants incorporate atmospheric carbon dioxide into organic compounds. Animals, bacteria, and fungi are largely responsible for the respiration that releases carbon dioxide back to the air reservoir.

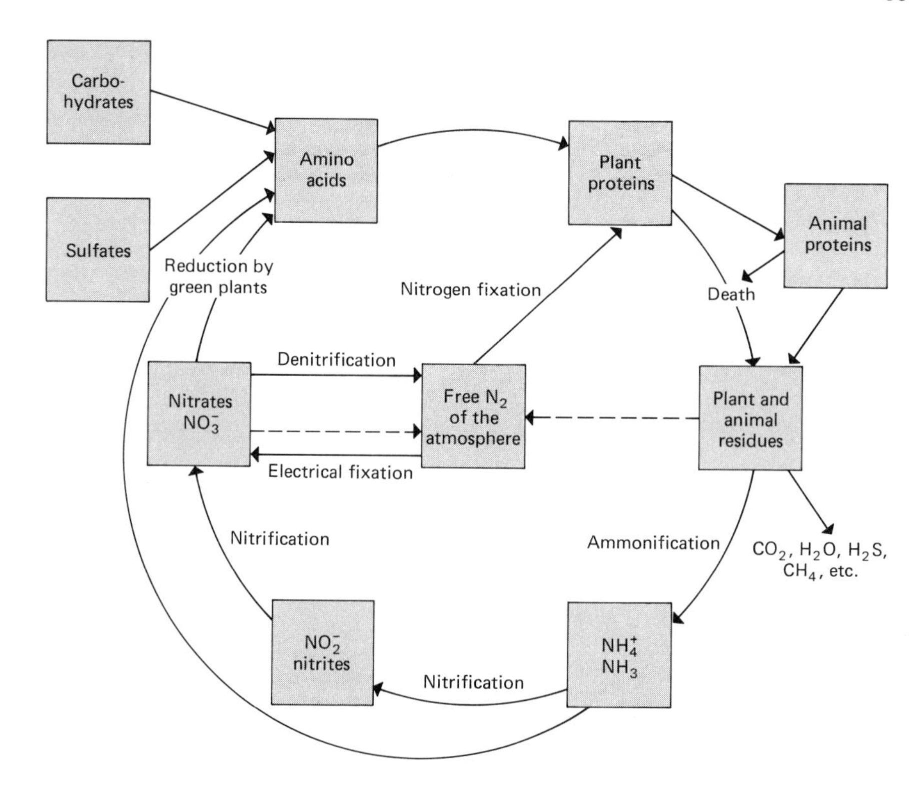

FIGURE 5-2
The nitrogen cycle. About 78 percent of the atmosphere is nitrogen (N_2), the largest gaseous reserve of any element. Nitrogen-fixing bacteria and blue-green algae convert N_2 into forms, especially nitrates (NO_3^-), that are usable by plants. Some, but not all, of the nitrogen-fixing bacteria exist symbiotically in nodules on legume plants. The host plant or adjacent plants can absorb the nitrates through their roots and use them in protein synthesis. Decomposition of plant and animal material releases ammonia (NH_3), which can be converted by nitrite bacteria into NO_2^-, and still another group of bacteria (nitrate bacteria) converts the nitrites to NO_3^-. This nitrate is also available for plant use, but it also can be acted upon by denitrifying bacteria, which return N_2 to the atmosphere. Only a small portion of the nitrogen fixed from air is nonbiological (electrification by lightning).

percent by volume of the atmosphere. It is an essential element for life, but relatively few organisms can capture it as a gas. The nitrogen content of soils is usually less than 1 percent and of this only about 1 percent may be available as inorganic salts. It is, therefore, striking that the nitrogen content of organisms is considerable, being about 3 percent (dry weight) for green plants and 12 percent for mammals. Nitrogen in the biosphere, as will be noted below, is extremely mobile, cycling in a variety of complex processes.

As a summarizing generalization, then, we can state that organisms are composed of the lightest, most water-soluble, most ionic, most mobile, and most abundant elements of the earth. Life is not merely *on* the earth; it is *of* the earth!

The total amount of carbon in the biosphere is estimated to be about 2.8×10^{17}g. The total mass of the biosphere is about 10 times this, or 3×10^{18}g. Photosynthesis produces about 2.5×10^{17}g of glucose annually, consuming about 3.6×10^{17}g CO_2 in the process. Thus, the amount of carbon mobilized by the biomass is prodigious (Fig. 5-1).

Cycles of a similar magnitude occur for several of the other elements which compose living material (e.g., sulfur, phosphorus, oxygen, nitrogen), and they are driven primarily through biological actions. Nonbiological physical and chemical forces contribute only in a relatively minor way to these events (Fig. 5-2).

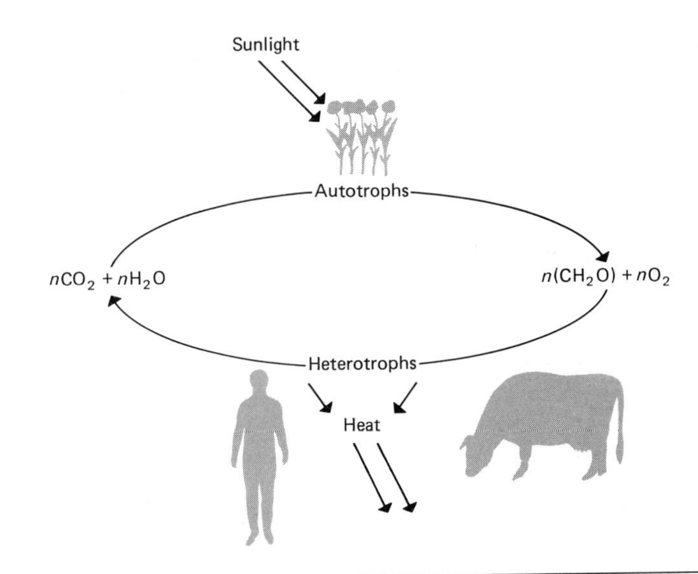

FIGURE 5-3
Matter and energy cycles in the biosphere, as exemplified by the overall net formulas of photosynthesis and respiration. Note that the matter cycle is "closed" and that the energy cycle is "open."

Ehrensvärd's assertion that life is ". . . a cyclic procession of matter driven by sunlight" is a succinct and appropriate statement about the nature of life. Green plants capture the energy of light quanta and use it to synthesize several high-energy compounds from carbon dioxide, hydrogen, and a few additional elements. The former are used structurally and nutritionally by the plants that make them. These compounds also serve as food for organisms that feed on the plants, and indirectly as food for organisms that feed on the organisms that fed on the green plants—and so on. Thus, matter circulates through organismal components of the biomass in a *closed* cycle that is powered by energy derived from solar photons which are captured in photosynthesis and converted to chemical-bond energy. Eventually the energy is degraded to heat and lost to the universe. The energy cycle, then, in contrast to that of matter, is an *open* cycle (Fig. 5-3).

5-2 ACQUISITION OF NUTRIENTS

Respiration

The utilization of organic substances as sources of energy commonly requires the availability of *oxygen*, obtained from air or from water. An array of lung and gill devices is used by animals to gain access to oxygen and to release carbon dioxide. The nature of these structures, as well as mechanisms for transport of the gases throughout the body, will be dealt with in the following chapter.

Most air-breathing animals normally do not need to seek out oxygen, as it is ubiquitous in the surrounding atmosphere. Certain places, however, can be oxygen-deficient, including underground chambers or high altitudes. These are avoided to prevent oxygen deprivation.

Seals, dolphins, whales, diving birds, turtles, and other air-breathing animals become oxygen-hungry during prolonged submergence and will seek access to air to relieve their deficiency. In certain pathological states, such as pneumonia, and when drowning, animals engage in violent activity in an effort to obtain oxygen.

Water breathers commonly encounter wide variations in oxygen availability. The more mobile and active ones, such as fish, seek out places that have the higher oxygen content. Some fish will gulp air when the oxygen content of the water falls too low to sustain them.

Feeding

Food is obtained by a diversity of mechanisms. Within any one taxonomic group there tends to be a common pattern for feeding. That is, certain orders of birds all eat seeds and have appropriate beaks for seed cracking. Ungulate mammals are grazers. Certain groups of bugs all suck plant juices, and so on. However, within some other groups there are adaptations for dealing with quite a variety of foods.

One can classify feeding mechanisms in a number of ways. The classification of C. M. Yonge, originally intended for invertebrate mechanisms and later expanded to include vertebrates, is satisfactory for illustrating the range of mechanisms that deal with acquisition of organic matter. It is based on character of the food, especially the size of its elements. There are two main groups: the **microphages,** animals that feed on small particles; and the **macrophages,** animals that feed on large-size food. A third, and much smaller, group includes animals that feed on fluids or on dissolved food. Subdivisions of the three principal groups can be made according to the nature of the mechanisms that handle the food (Table 5-1).

Microphages feed on particles that are too small to be sensed or seized individually and are, therefore, taken up in bulk. Macrophages sense and seize the material on which they feed. Many of them reduce it to smaller masses by *trituration* before swallowing. Fluid feeders must take in the water medium along with the contained food, as the latter is in either a dissolved or a finely dispersed state.

Feeding Behavior

Feeding in higher vertebrates, especially mammals, in which the process has been well studied, may be divided into phases: *appetitive*, *consummatory*, and *satiation*. In the appetitive phase the animal exhibits behavior appropriate for the procurement of food. When food is acquired the animal engages in the consummatory phase, or ingestion. If the obtained food is not sufficient (or is rejected), the cycle of food procurement and ingestion is repeated, until satiation is achieved. Satiation terminates the cycle temporarily.

The appetitive phase usually begins with random activity, combined with various manifestations of "hunger." The exact nature of the internal signals that create "hunger" and trigger feeding are not well understood. In the early 1900s several *peripheral theories* of hunger were postulated. The subjective experience of hunger in the human is associated with *stomach contractions*. It was demonstrated that lowering of blood-glucose level increased stomach contractions and that elevating glucose decreased contractions. Additionally, *stomach distension* was shown to play a role in registering food intake. Inert substances placed in the stomach produce a reduction in food ingestion like that occurring after similar amounts of food have been placed there. Animals with

Table 5-1
Classification of Feeding Mechanisms Used by Animals

Mechanisms for dealing with food	Examples
Microphages	
Pseudopods	Radiolarians, foraminiferans
Cilia	Ciliates, sponges, many worms, brachiopods, bivalves, anuran tadpoles (e.g., *Xenopus*, *Bufo*, and others)
Mucus, especially in form of filtering sheets	Several gastropods, tunicates, *Amphioxus*, Ammocoetes
Setae and similar filtering structures	*Daphnia*, copepods, and other crustaceans, basking shark, some teleosts, flamingo, whalebone whales
Macrophages	
Mechanisms for swallowing the surrounding medium (mud, sand, earth, etc.)	Many burrowing and digging forms in sediments in the sea and fresh waters, earthworms
Mechanisms for seizing prey	Most coelenterates, many polychaetes, most nonmammalian vertebrates, some mammals
Mechanisms for seizing and masticating prey, and for biting, rasping, grazing, etc., often combined with mastication of food	Many gastropods, cephalopods, crustaceans, and insects; cyclostomes, some birds and most mammals
Feeders on fluids and dissolved food	
Mechanisms for sucking fluids	Trematodes, nematodes, leeches, parasitic copepods, several insect groups, young of mammals
Mechanisms for absorbing dissolved food through external surfaces: parenteral food uptake	Many parasites, aquatic animals

Source: C. M. Yonge, 1928, *Biol. Rev.*, **3**, 21–76; H. J. Jordan and Hirsch, G. C., 1927, in A. Bethe et al. (eds.), *Handbuch der normalen und pathologischen Physiologie*, vol. 3, Springer-Verlag OHG, Berlin, part II, pp. 24–101.

surgically produced esophageal fistulas, which prevent ingested food from reaching the stomach, tend to eat longer and more often than normal animals. The role of peripheral information, however, cannot be crucial in regulating food intake. The removal of the entire stomach in humans does not interfere with hunger and feeding. Also, when the gastrointestinal tract of rats and dogs is denervated, they still show normal feeding phases.

When the importance of the brain's hypothalamus for many "visceral" regulations came to be appreciated, the peripheral theories of food intake were largely abandoned in favor of *central theories*. There are, apparently, monitoring cells in the hypothalamus which register blood-glucose concentration. Decreased utilizability or availability of glucose produces increased food intake. Mice injected with glucose to which gold is attached (goldthioglucose) develop obesity. They also have lesions in the ventromedial region of their hypothalamus, suggesting that those cells are the ones with glucose receptors, and that they are destroyed by the abnormal metallic glucose which they bind. However, goldthioglucose produces lesions in several brain regions. Furthermore, the correlation between glucose availability and eating is less than perfect.

The peripheral and central theories have been amalgamated by some investigators into a *multifactor theory* of the regulation of food intake. Accordingly, both peripheral and central mechanisms gauge the food requirements in a complementary manner. Indeed, some of the peripheral signals from the gastrointestinal tract may be registered in the hypothalamus.

The observation that animals will selectively feed on substances in which their diet has previously been deficient, has spawned much experimentation and speculation. This phenomenon, called *specific hunger*, permits animals to satisfy nutritional needs beyond the simple replenishment of caloric losses. It has been known that domestic animals given a free choice of nutrients fare as well as ones whose feed has been selected for balanced nutrition. Food selection conditioned by specific hunger can be remarkably selective. Rats allowed to become deficient in a B vitamin consume large quantities of this substance when it is presented to them in crystalline form. However, this kind of food "intelligence" does not extend to all substances, and different animals show wide variability in selective ability.

5-3 TROPHIC REQUIREMENTS

The biochemical basis of chemical needs can be considered functionally from the standpoint of several different basic requirements for life processes; viz., maintenance of the internal milieu, sources of free energy, structural organization, catalysis, and certain other specialized activities. Though these needs are rather obvious and basic, it is to be noted that other kinds of needs lie outside the traditional scope of nutrition. For example, there are needs of tissues for chemical substances supplied by other tissues of the same organism (hormones, organizers of development, auxins in plants); the need of one member of a species for substances supplied by members of the same species (pheromones; cf. Chap. 3); and the need of a member of one species for substances emitted by a member of another species (allomones and kairomones, cf. Chap. 4).

Requirements for Maintaining the Internal Milieu

For the early and simple forms of life that presumably originated in the sea, exchange of materials with the environment could have occurred directly between their cells and the watery solution surrounding them. As organisms became larger and more complex, it probably became necessary to provide channels for moving the watery medium through their bodies to facilitate exchange between the more internal cells and the now more remote environment. Eventually, the openings of the channels presumably were closed and some kind of internalized circulatory system was effected.

A closed circulatory system affords a high degree of independence from the vagaries of the environment, but it also imposes problems, for materials moved into and out of the organism must pass through barrier membranes. If the body fluids of an organism are to differ in solute composition from that of the external environment, as is commonly the case, then there are osmotic problems. Substances will tend to diffuse along their concentration gradients, or along electrochemical gradients for charged particles. Differential rates of movement of solutes, principally electrolytes, through membranes can generate osmotic (and electrical) imbalances. But there still must be exchange between organism and environment across the barrier membranes—exchange with the environment is a hallmark of a living system—and so methods had to be devised to control the transmembrane movement of solutes and water.

Selective absorption and a number of other physiological devices evolved to meet the needs of the various kinds of organisms—marine, freshwater, brackish-water, terrestrial—and provide relative degrees of independence from the environment. A major advance, for example, was the adoption by some groups of plants and animals of osmotically active organic solutes, allowing an almost complete independence from the obligatory high nutritional requirement for maintenance of the original marine electrolytes.

The nutritional requirements for the maintenance of the internal environment of animals may be summarized as specific requirements for:

1 Maintaining a near-neutral pH. Land vertebrates average a pH of about 7.4, and all others between about 7.0 and 7.8. These values are maintained by buffering substances which, with the major exception in some instances of the proteins, are practically identical with the materials that maintain osmotic balances.

2 Maintaining osmotic and electrolytic balance. This is generally achieved through regulation of the cations sodium, potassium, calcium, and magnesium. Chloride, bicarbonate, sulfate, phosphate, organic acids, and proteins act as anions. The bicarbonate component of the plasma of all animals is supplied metabolically, so it is not a dietary requirement.

Special mention will be made later (Chap. 7) of the organic nutritional requirements for osmoregulation in elasmobranchs, in which trimethylamine and urea, supplied by intermediary metabolism, play an important role. In insects amino acids and other organic solutes achieve such high levels in the hemolymph that they also contribute importantly to osmotic balance.

Requirements for Nutritional Sources of Free Energy

The most *immediate* source of free energy for all known life forms resides in the covalent bonds of monosaccharides. There is, thus, an obvious unity of life in selection of monosaccharides as the common and primary energy source. Almost every cell of most plants and animals examined has a "main line" of metabolism that makes use of the monosaccharides, or other reduced forms of carbon which are directly convertible to them. Metabolism will be detailed in Part V of this book.

There is remarkably little variation among organisms in the way they secure their energy status. One of the conventional and logical ways to classify organisms—on the basis of the nutritional source of carbon and of free energy—makes this clear. Among the photosynthetic algae, bacteria, and higher plants, the captured light photons provide energy for the reduction of carbon dioxide to saccharides, which are required for immediate energy. As a group these organisms are called **photosynthetic autotrophs** (Gr., *autos*, self; *trophos*, feeder). Certain bacteria, called **chemosynthetic autotrophs,** can use the free energy from oxidation of inorganic substances for the reduction of carbon dioxide to the requisite monosaccharides. The third and final group in this classification consists of the **heterotrophs,** those that feed on "others" (Gr., *heteros*). These are the etiolated plants, some algae, many bacteria, and fungi, most protozoans, and all known higher animals. They require carbon compounds of varying kinds, but at least as complex as acetate. Heterotrophs, obviously, are nutritionally dependent upon autotrophs (Fig. 5-3).

Other bases for nutritive classification include specific requirements for preformed compounds, especially the nitrogenous ones. Table 5-2 is one such scheme, along with a classification based upon energy and carbon sources.

As indicated in Table 5-2, metatrophic heterotrophs must acquire organic carbon and organic nitrogen compounds from their environment. These organisms will ordinarily grow only when provided with a complex mixture of compounds, which generally include (1) carbon compounds with at least three carbon atoms in the chain, (2) a mixture of

10 or more amino acids, and (3) an assortment of several vitamins. Heterotrophs can utilize most, if not all, aliphatic compounds, including their derivatives, as energy sources, when structural details do not make the compounds prohibitively toxic. The same is true for aromatic compounds. However, all heterotrophic organisms do best with those compounds that are the most closely related to amino acids, lipids, and carbohydrates.

Organisms do not produce or consume energy; they can only convert energy among its various forms. Ideally, biologically useful energy should be available from a variety of sources, since, given proper conditions, any form of energy may be converted into any other form. However, there are limitations to the types of conversions that are practical for organisms. They cannot change thermal or heat energy into any other energy forms. Efficient utilization of thermal energy requires large temperature differences, while organismic processes are essentially isothermal (constant-temperature). Mechanical energy, electrical energy, and chemical-bond energy are converted readily to heat, but the reverse

Table 5-2
A Nutritive Classification of Organisms

Requirement for nitrogenous compounds	Source of energy		
	Inorganic reactions provide energy for reduction of CO_2: *Chemotrophic*	Sunlight provides energy for reduction of CO_2 by means of chlorophyll: *Phototrophic*	Energy can be derived from carbon compounds at least as complex as acetate: *Heterotrophic*
Molecular nitrogen can be utilized: *Prototrophic*			Nitrogen-fixing bacteria
Nitrogen must be combined as nitrate, nitrite, or ammonia, but organic nitrogen compounds can be formed from these combinations: *Autotrophic*	Iron and sulfur bacteria	All green plants, many green flagellates, certain bacteria	Many colorless flagellates, many bacteria
Nitrogen must be combined as amino acids, but one or a very few amino acids will suffice: *Mesotrophic*		Many green flagellates	Many colorless flagellates, many bacteria
A complex mixture of amino acids must be available: *Metatrophic*		Certain green flagellates	All animals except certain flagellates, many bacteria

Source: Based on W. L. Doyle, 1943, *Biol. Rev.*, **18**, 119.

does not occur. Thus, the heat which results from the dissipation by organisms of other energy forms is degraded energy, and it is lost to the universe (Fig. 5-3).

Unlike photosynthetic autotrophs, the heterotrophic animal derives no energetic support directly from the sun, because it lacks the appropriate energy-coupling mechanisms. The only form of energy which animals can utilize is that inherent in specific molecular configurations of certain ingested substances: the chemical free energy of food. This limitation is imposed by the possession of the structural and physiological devices which allow for the transduction of the free energy of chemical bonds into the other energy forms essential for life. Relatively large, highly ordered, and thus energetically rich, molecules are ingested by animals. Heat and simple molecules of low energy content are returned to the environment as wastes. Between ingestion and excretion are a variety of metabolic processes involved in energetic conversions.

It is well established thermodynamically that the energy available from any chemical reaction is solely determined by the difference in energy content of the end products and initial substrates, and is independent of the particular chemical pathway by which the reaction proceeds or where it occurs. To cite a useful example, the oxidation of glucose is described by the stoichiometric relation:

$$\underset{100\ g}{C_6H_{12}O_6} + \underset{74.7\ l}{6O_2} \longrightarrow \underset{74.7\ l}{6CO_2} + \underset{0.06\ l}{6H_2O} + 381\ \text{kcal}$$

This equation applies equally to the direct and simple process of combustion as well as to the multistep, complex biological oxidation of glucose. The total energy release is the same.

Chemical energy is ordinarily measured as heat and expressed as *calories*. A **calorie** is the amount of heat required to raise the temperature of 1 g of water from 14.5 to 15.5°C. A thousand calories are called a **kilocalorie.** (Since electrical measurements can be standardized more accurately than heat measurements, the calorie is now officially defined as 5.1833 international joules.)

The technique employed in determining the heat content of substances is known as **calorimetry,** and the device used is called a **calorimeter.** The heat of combustion of fuels and foods is usually measured by a *bomb calorimeter*. The sample is placed on a nickel or platinum dish and suspended inside the sealed inner metal chamber of the calorimeter. An ignition wire is placed in contact with the sample. The chamber is sealed and then filled with oxygen at 25 or 30 atm. The chamber (= bomb) is immersed in a weighed amount of water, and the whole system is encased in an insulated jacket. An electric current is sent through the ignition wire, which burns and in turn ignites the sample. By taking a series of temperature measurements of the water which surrounds the combustion chamber before and after ignition, one can calculate the heat of combustion. When related to the amount of material combusted, the caloric value in kilocalories (kcal) per gram can be calculated.

In applying bomb calorimetric predictions to animal metabolism, certain corrections must be made. For one thing, the bomb calorimeter burns all foods to their final oxidation products, but this is not necessarily true for animals, particularly in the case of proteins. The nitrogenous end products of protein metabolism—urea, uric acid, and ammonia—are not the nitrogen gas or oxides of nitrogen characteristic of bomb combustion. That is, heterotrophs do not oxidize proteins to their lowest energy state. Thus, the energy released by the calorimeter is greater than that released when an equal amount of protein is biologically oxidized.

However, bomb calorimetric values can be corrected by simply determining the caloric content of appropriate amounts of metabolic end products of protein metabolism. By combusting urea, uric acid, or ammonia in the bomb calorimeter, one can obtain correction values for subtracting from the heat of the total oxidation of protein to find the in vivo caloric value.

Another possible source of difficulty in applying calorimetric measurements to animal nutrition arises from the fact that not all substances ingested are necessarily absorbed from the digestive tract. If ingested substances are not absorbed, of course, they cannot contribute to metabolic energy. For example, cellulose oxidized in a calorimeter will release heat in an amount similar to other carbohydrates, but many animals do not digest cellulose, and for them it has no energy value.

Carbohydrates, proteins, and lipids constitute the only significant energy sources in the foods of animals. The remainder of ingested substances, although they may be important in particular aspects of metabolism, contribute little to energy intake. Inorganic materials undergo no chemical transformations that contribute to the available energy. Other organic matter, such as vitamins, is present in amounts simply too small to make a significant energetic contribution.

Carbohydrates Although not all carbohydrates have identical energy values, they are sufficiently similar for humans that a single figure is commonly used for all foods in this category. Taking the common carbohydrates in the proportion in which they normally occur in our diet gives a mean value of about 4.1 kcal/g of carbohydrate oxidized. Because the carbohydrates are almost completely absorbed, the "physiologic" caloric

Table 5-3
Energy Content and Physiologic Value of Food for Human Beings

Food	Energy, kcal/g		
	Bomb calorimeter	Human oxidation	Physiologic value
Carbohydrate	4.1	4.1	4.0
Protein	5.4	4.2	4.0
Lipid	9.3	9.3	9.0

value—the number of kilocalories per gram *ingested*—is only slightly lower: about 4 kcal/g (Table 5-3).

Proteins An average value from bomb calorimetry of proteins is 5.4 kcal/g, but since energy remains in excretory products when proteins are oxidized in vivo, as noted above, the potential available energy is only 4.2 kcal/g absorbed. However, lack of complete absorption further reduces the available energy to 4 kcal/g ingested (Table 5-3).

Lipids This class of compounds is the most "energy dense" of common foodstuffs. Animal fats average 9.4 kcal/g, and vegetable fats and oils average 9.3 kcal/g for a net mean for all fats in a human diet of about 9.3 kcal/g. Failure to absorb all consumed fat reduces the physiologic value to 9 kcal/g ingested (Table 5-3).

Requirements for Structural Organization

It is a patent fact that life is contained in specific shapes or forms. This is true at all levels of organization. From the distinctive features by which we recognize and name organisms, to the smallest details that have been visualized by electron microscopy, there are distinctive morphological characteristics, unique to each entity. Although there are varying degrees of difference, there is much constancy of form among organisms. Mitochondria, for example, although quite variable in size and shape, are generally recognizable in electron micrographs made from cells of organisms as divergent as carrots and cows.

Most cells have a similar, distinctive set of organelles and are enclosed in plasma membranes. Basically, all the recognizable components of cells seem to be built on a modular plan: many small units assembled to form a larger unit, which in turn forms a still larger unit, etc. Thus, mitochondria seem to be an ordered montage of a few kinds of respiratory assemblies and enzyme aggregates. The membranes forming the endoplasmic reticulum of eukaryotic cells incorporate ribosomes. The plasma membrane contains numerous characteristic enzymes and "receptor" proteins. The implication of this is that the "structure" that one sees in cells is not inert fabric, but is, in fact, the very functional machinery of the cell.

An important point is that molecules are only temporarily stabilized in "structure"—they are in a dynamic, steady flux. Matter constitutes the organism while passing through it in a continuous stream. Furthermore, many of those component molecules that are classifiied as structural can also be used nutritionally for energy. Thus, while sugars are the main nutrient molecules, as polymers they form the major structural fibers of plant cell walls (mainly cellulose) and arthropod exoskeletons (including chitin) (Fig. 5-4). Likewise, amino acids can be used to provide energy to support metabolism, as well as be polymerized into structural proteins, such as keratins.

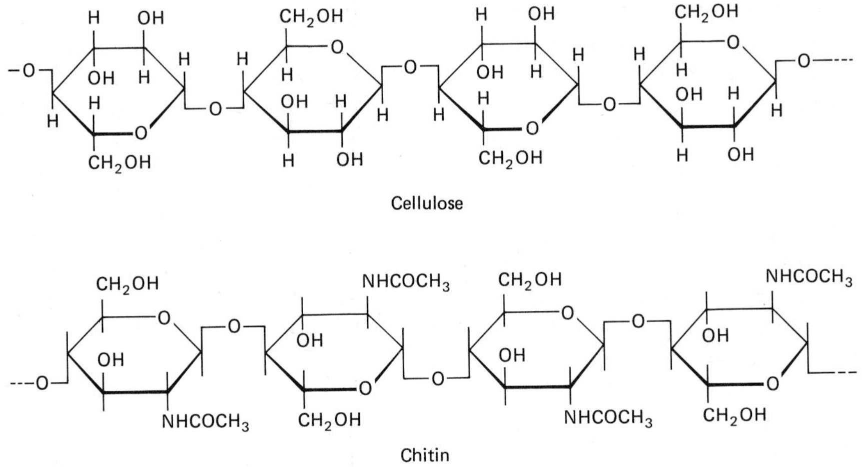

Cellulose

Chitin

FIGURE 5-4 Structures of the polysaccharides cellulose and chitin.

Structuring permits order and prevents chaos among the various metabolic processes that occur simultaneously in cells, through "compartmentalizing" them. It also ensures that certain reactions or processes will occur, or that they will occur efficiently, by placing the interacting participants in close proximity. Protection also results from enclosing sensitive or fragile mechanisms inside barrier structures. Finally, structured nucleoproteins provide for recording exact ancestral and experiential information.

The chemical substances that satisfy the structural requirements are generally either very high-molecular-weight organic compounds (polymers) or relatively insoluble inorganic crystallites. The former may include proteins, polysaccharides, nucleic acids, and lipids.

The structural proteins are composed, for the most part, of all or most of some 20 different α-amino acids. Several other very rare amino acids may occur in a few proteins. The total amino acid (or protein) content of the diet varies with a number of factors, but for most vertebrates it is about 20 percent.

Green plants, other than flagellates, do not require amino acids for growth, for they can synthesize their own. A few green flagellates, however, will grow only if supplied with one or more specific amino acids. The same is true for many species or strains of fungi, bacteria, and protozoans (Table 5-2). Genetic analyses of fungi (*Neurospora*) have shown that specific amino acid requirements arise when a gene mutation occurs which interferes with a specific step in the synthesis of that amino acid. In insects and vertebrates many amino acids are required. This is the result of the gradual accumulation of gene mutations which, through evolution, have involved progressive losses of synthetic ability (see Table 5-4).

The chemical needs for polysaccharide synthesis have scarcely been investigated. This is partly because a portion of the same carbohydrate "pool" that is available for free energy and as carbon skeletons for amino acid biosynthesis is used in the formation of structural polysaccharides in all organisms. Thus, it is difficult to segregate this portion of the total requirement. Indeed, there may be no specific monosaccharide nutritional requirement at all. However, there is some evidence that polysaccharide structure may be remarkably complex and contain unusual or esoteric sugars. Furthermore, the common polysaccharide-synthesizing enzymes often appear to *require* a small amount of polysaccharide as a template to initiate synthesis. The requirements for nitrogen, phosphorus, and sulfur are increased by the formation of some polysaccharides, especially the mucoproteins (or aminopolysaccharides), and the glycolipids (or lipopolysaccharides).

Chemical requirements for structural *lipid*, *lipoprotein*, and *liposaccharide* synthesis are unknown. Lecithins and cephalins, major lipid constituents of tissues, require ethanolamine, serine, and choline moieties.

The precisely ordered polymeric structures known as *polynucleotides* have several biological functions, in addition to those associated with structure, including the genetic mechanisms involving DNA and RNA and certain catalyst-carrier functions. Also, the uses of the base moieties of nucleotides, the purines and pyrimidines, are many and diverse. In addition to the commoner purines (adenine and guanine) and pyrimidines (cytosine, thymine, and uracil), a large number of other

Table 5-4
Amino Acid Requirements and Utilization by Major Groups of Organisms

Amino acid	Higher plants	Fungi	Yeasts	Bacteria	Algae	Green flagellates	Other protozoa	Insects	Vertebrates
Leucine	N u	r u	r u	r u	N	N u	r u	R*	R
Lysine	N u	r u	u	r u	N	N u	r u	R	R
Methionine	N u	r u	r u	r u	N	r	r u	R	R
Phenylalanine	N u	r u	r u	r u	N	N u	r u	R*	R
Threonine	N u	r u	u	r u	N	. . .	r u	R*	R
Histidine	N u	r u	r u	r u	N u	r u	r u	R	R
Isoleucine	N u	r u	r u	r u	N u	. . .	r u	R	R
Tryptophan	N u	r u	u	r u	N u	N	r u	R*	R
Valine	N u	r u	u	r u	N u	N u	r u	R*	R
Arginine	N u	r u	r u	r u	N u	N u	r u	R	r
Cystine-cysteine	N u	r u	u	r u	N u	. . .	u	R*	N
Glutamic acid	N u	r u	u	r u	N u	N u	u	r	r
Glycine	N u	r u	u	r u	N u	N u	r u	r	r
Proline	N u	r u	u	r u	N	N u	r u	r	r
Serine	N u	r u	u	r u	N	N u	r u	r	N
Alanine	N u	u	u	r u	N u	N u	u	r*	N
Aspartic acid	N u	u	u	r u	N u	r u	u	r*	N
Tyrosine	N u	r u	u	r u	. . .	N	r u	N	N

* Some substitutions are possible. Thus phe will substitute for thre, try for val, val for leu, tyr for phe, met for cys, asp or glu for ala, glu for asp, in some species.

Note: N—not required by any; u—used by some; r—required by some; R—required by all tested.

Source: From B. T. Scheer, 1963, *Animal Physiology*, John Wiley & Sons, Inc., New York, p. 179.

methyl and amino derivatives occurs in nucleotides. Most interestingly, the chemical needs for purines and pyrimidines appear to vary inversely with the degree of phylogenetic development: requirements are common in plants, occasional in protozoans, rare in insects and nonexistent in vertebrates. Vertebrates can synthesize all their bases from simple organic compounds. (See Fig. 16-7 for nucleotide structure.)

Specific but poorly defined mineral requirements exist in most phyla for construction of a skeletal framework. Even in plants where cellulose fibers mainly provide rigidity, silicates also have a structural role. Several plants are considerably more fragile when grown without silicon, and although it is not essential for growth, it serves an important structural function in them. Among the protozoans are several that have highly organized skeletal structures composed of calcareous or siliceous materials. Different species of sponges have skeletal structures made primarily of either calcium carbonate or silica. Calcium carbonate constitutes 87 to 97 percent of the shells of mollusks. Calcium is essential for the formation of the tests of echinoderms and the exoskeletons of crustaceans. The vertebrate endoskeleton incorporates about 99 percent of the calcium, 80 percent of the phosphorus, and 70 percent of the magnesium of the body. They are bound in the hydroxyapatite crystallites of bone and teeth.

Requirements for Catalysis

In the nineteenth century it was ascertained that foodstuffs were composed essentially of three main classes of organic substances: proteins, fats, and carbohydrates, together with various mineral salts and water. In chemical analyses, these components quantitatively accounted for approximately 100 percent of the materials that organisms required. Therefore, at the beginning of the twentieth century it was the accepted practice to define the nutritional value of any foodstuff in terms of these substances only. Subsequently, it became increasingly clear that adequate nutrition involves very small amounts of certain substances, in addition to the major foodstuffs.

Vitamins Lunin, in 1881 in Switzerland, demonstrated that mice could not survive on a diet of carbohydrate, fat, protein, milk salts, and water, but the addition of fresh milk kept them alive. This, together with some dietary findings by Eijkman (1890) in Java, and notably Hopkins (1912) in England, forced the conclusion that there were other things that mattered nutritionally in foods in addition to the well-recognized main components. The newly detected ingredients, present in foods in extremely minute amounts, but nevertheless essential for the maintenance of health, came to be known as **vitamins.** This term, originally spelled "vitamine," was coined by Funk (1912) at the Lister Institute in London and was intended by him to imply first that they were "vital" for the survival of the living organism and secondly that some of them at least were basic in chemical character, or "amines."

No vitamins had yet been isolated or chemically characterized when they were named. However, one by one the need for vitamins was recognized, each responsible for preventing some different disorder in humans or other animals. Subsequently many were separated in a pure state, identified, and synthesized, and their mode of action was determined. Methods for estimating the amount present in various foodstuffs were devised and the requirements determined for humans and other animals. Because they are distributed in foodstuffs in such relatively minute amounts, the chemistry of vitamins is a relatively recent chapter in the nutritional literature. Today more than 20 vitamins are known, and at least a dozen of them are known to be needed by humans.

The first convincing indication of the multiplicity of vitamins came in 1915 when McCollum and Davis in America showed that, in rats, at least two accessory factors were needed for growth. One, present in fatty foods, was named "fat-soluble A"; the other, in nonfatty foods, was called "water-soluble B."

At the suggestion of Drummond (1920), "water-soluble B" was subsequently named *vitamin* **B**, the terminal "e" now dropped to avoid the possibly unwarranted implication that it was an amine. It was agreed that this was the substance with properties of the antineuritic (antiberiberi) factor. Similarly "fat-soluble A" became *vitamin* **A**. The most noticeable effects of its deficiency were failure of growth and increased susceptibility to infection, especially of the respiratory system.

It was deduced that there was a third and entirely different chemical that was an antiscurvy factor. It was accordingly given the next letter in the alphabet and became *vitamin* **C**. Shortly thereafter, it was proved that experimental rickets in dogs was due to the absence of a fat-soluble vitamin (Mellanby, 1918) and, extending the alphabet, it was named *vitamin* **D**. Evans and Bishop (1922), in California, described a factor needed to ensure normal reproduction in rats. This antisterility substance was called *vitamin* **E** (now also known as **α-tocopherol**).

The multiplicity of vitamins required by humans became astonishingly apparent when it was realized that vitamin B was a complex, not a single substance. In 1926 Goldberger and colleagues proved that pellagra was associated with the lack of a vitamin. The effective substance had a distribution somewhat similar to that of the antiberiberi factor but was more heat-stable. The antiberiberi vitamin was renamed *vitamin* $\mathbf{B_1}$, and the "heat-stable component" was called *vitamin* $\mathbf{B_2}$. Subsequently **riboflavin** was isolated and for a time was called "vitamin B_2." However, it soon became apparent that B_2 was also a *complex* of substances. A second component, now named **pyridoxine,** was identified in 1934. And it turned out that the pellagra-preventing factor was still a third substance, identified as **nicotinamide** (or nicotinic acid or niacin).

Through the 1930s and 1940s more than a dozen additional vitamins were found and characterized. At least three of these—vitamin **K, folic acid,** and vitamin B_{12}—definitely are important in human nutrition. Others are required by microorganisms, rats, mice, or chicks, and may be important, in some cases, for humans. These include vitamins **F** and **P, pantothenic acid, biotin, choline,** and **inositol.** The better-known vitamins have been isolated, characterized chemically, and synthesized in the laboratory.

Vitamin B_{12}, also known as **cyanocobalamin,** is one of the most recently synthesized vitamins. It is the largest and most complex vitamin yet discovered, and it also has the distinction of being the only known vitamin to contain a metal ion, in this case cobalt. Lack of B_{12} causes pernicious anemia, a condition in which red blood cells are not properly formed. Pernicious anemia is generally caused by the absence of a *gastric intrinsic factor,* a glycoprotein needed to absorb the vitamin from the intestine.

With the discovery that various vitamins of the B group function as moieties of coenzymes, the true nature of the "need" for many of the vitamins was envisioned. (See Sec. 18-3 for explanation of the term "coenzyme".) *Nicotinamide* is a part of the coenzymes nicotinamide-adenine dinucleotide (NAD) and nicotinamide-adenine dinucleotide phosphate (NADP) (see Fig. 18-7), which function in electron transfer between reduced and oxidized metabolites, or between a reduced metabolite and flavin adenine dinucleotide (FAD). FAD and another coenzyme, flavin mononucleotide (FMN), both of which contain *riboflavin* (see Fig. 18-10), also function in electron transfers in biological oxidation-reduction. *Pantothenic acid* forms a part of coenzyme A (see Fig. 18-9), which is involved with transport of the acetyl group as an intermediate in many important steps in carbohydrate and lipid metabolism. **Lipoic acid** (see Fig. 18-11) functions as the cocarboxylase conjugate in the transfer of acyl groups between thiamine pyrophosphate and coenzyme A.

Ascorbic acid (vitamin C) is apparently involved in many metabolic oxidations. It is not a known coenzyme for any specific enzyme systems, however, and its action is usually nonspecific. **Thiamine** (vitamin B_1), in the form of the pyrophosphate called cocarboxylase, functions in the α-decarboxylation of certain organic acids. **Biotin** participates in a number of processes in which CO_2 is fixed into larger organic molecules or released from them. Vitamin **K** is an essential cofactor for synthesis by the liver of prothrombin and other plasma factors that function in blood clotting. **Folic acid** participates in several reactions that lead to the formation of deoxyribonucleic acids. Thus, although the complete function of any one of the vitamins is not known, it can be stated, as a generality, that vitamins are essential in maintaining metabolic and biosynthetic processes at normal levels in animals and certain microorganisms.

Trace elements Although a wide range of functions has been established for essential elements, nearly all of them are believed to have one or more catalytic functions in the cell. Catalysts from living organisms are called **enzymes,** and all enzymes are protein. (This subject is more fully explored in Chap. 18.) Many kinds of enzyme molecules contain metals. The known catalyst-associated elements are Ca, Cl, Co, Cu, Fe, K, Mg, Mn, Mo, Na, P, S, and Zn.

Two subclasses of metal-containing enzymes have been distinguished: metal-activated enzymes and metalloenzymes. Using *M* as a symbol for the metal concerned, their characteristics are differentiated as follows:

Metal-activated enzymes	Metalloenzymes
M reversibly bonded to enzyme	*M* firmly bonded to enzyme
Dissociation constant measurable	Dissociation constant very small
M/protein ratio variable	*M*/protein ratio a small integer
M/enzyme activity ratio variable	*M*/enzyme activity ratio constant
M not necessarily unique	*M* unique
Enzyme activity may exist without *M*	No enzyme activity without *M*

For many *metal-activated enzymes,* in experimental situations, the activating metal is not unique and can be replaced by another divalent metal, but the efficiencies of the metals as activators differ widely. The catalog of metal-activated enzymes continues to grow rapidly, and only a few will be mentioned here in passing. *Magnesium,* the most common enzyme activator, is especially important in activating phosphate transferases and a number of decarboxylases. *Manganese* resembles magnesium in activating a number of phosphate transferases and decarboxylases, notably those participating in the tricarboxylic acid (Krebs) cycle. *Calcium* activates a few enzymes, including animal α-amylase. *Sodium* and *potassium* activate an ATPase that functions in active transport across cell membranes (discussed further in Sec. 14-6). Potassium also activates pyruvate kinase and some proteinases. *Iron* activates a number of oxidases.

In *metalloenzymes* the active metal is firmly bound in a constant stoichiometric ratio to the protein. Frequently the metal is contained in a small molecule called the "prosthetic group." (See Sec. 18-3 for the meaning of this term.) Metalloenzymes containing *copper, iron, molybdenum,* and *zinc* are rather numerous. It is interesting that the free metal alone is often a catalyst for the reaction concerned, although much less efficient than the enzyme itself.

One manner of classifying metalloenzymes is according to the prosthetic group involved. Thus, the following categories:

Metalloprotein enzymes have no prosthetic group; the metal is bound directly to the protein. Many copper, zinc (see Fig. 18-3), and some iron enzymes are of this type.

Metalloporphyrin enzymes have the metal chelated by a porphyrin prosthetic group. Included are many oxidases containing iron bound in a heme group.

Metalloflavin enzymes contain a flavin prosthetic group, but the mode of attachment of the metal is not known. They include many oxidases or dehydrogenases containing iron, copper, magnesium, manganese, molybdenum, or zinc together with FAD or FMN.

The trace elements have several other functions which are not catalytic but which relate to the several mechanisms supporting energy capture, transfer, and utilization. A brief and incomplete listing of these follows:

Magnesium forms stable complexes with adenosine triphosphate (ATP). It is thought that most of the ATP in cells is present as the Mg^{2+} complex (see Fig. 17-3).

Magnesium is incorporated in the porphyrin structure of chlorophyll.

Iron porphyrins form the functional core of cytochromes (see Fig. 19-8), hemoglobin (see Fig. 6-24), and myoglobin.

Calcium (and *magnesium*) are involved in muscle contraction, as will be discussed in Chap. 15.

Cobalt, as noted above, is incorporated into the porphyrin of cyanocobalamin, better known as vitamin B_{12}.

Copper is present in the blue blood of mollusks and certain arthropods, as will be noted in Chap. 6.

The known chemical needs of organisms, only briefly sketched here, indicate the patterns that exist throughout the biosphere, but we are far from defining the total requirements of organisms, even of the more common ones.

5-4 DIGESTION

Digestive systems of heterotrophs make organic matter available to the body cells. Digestion also eliminates possible antigenic properties of food molecules. These results are accomplished by (1) mechanical disruption and breakdown of foodstuffs entering the digestive tract, (2) extra- and intracellular enzymatic cleavage of large, complex molecules into smaller diffusible compounds, and (3) transfer of these diffusible compounds through the membranes of the digestive tract into the body fluids. These are the three stages of *digestion*.

The digestive system of the higher Metazoa are morphologically similar, consisting of a *gastrointestinal tract*, which is basically a tube that extends between the mouth and the anus, and associated *digestive glands* (Fig. 5-5).

The digestive tract is usually composed of three main segments: the *esophagus*, *stomach*, and *intestine*. The stomach is often compart-

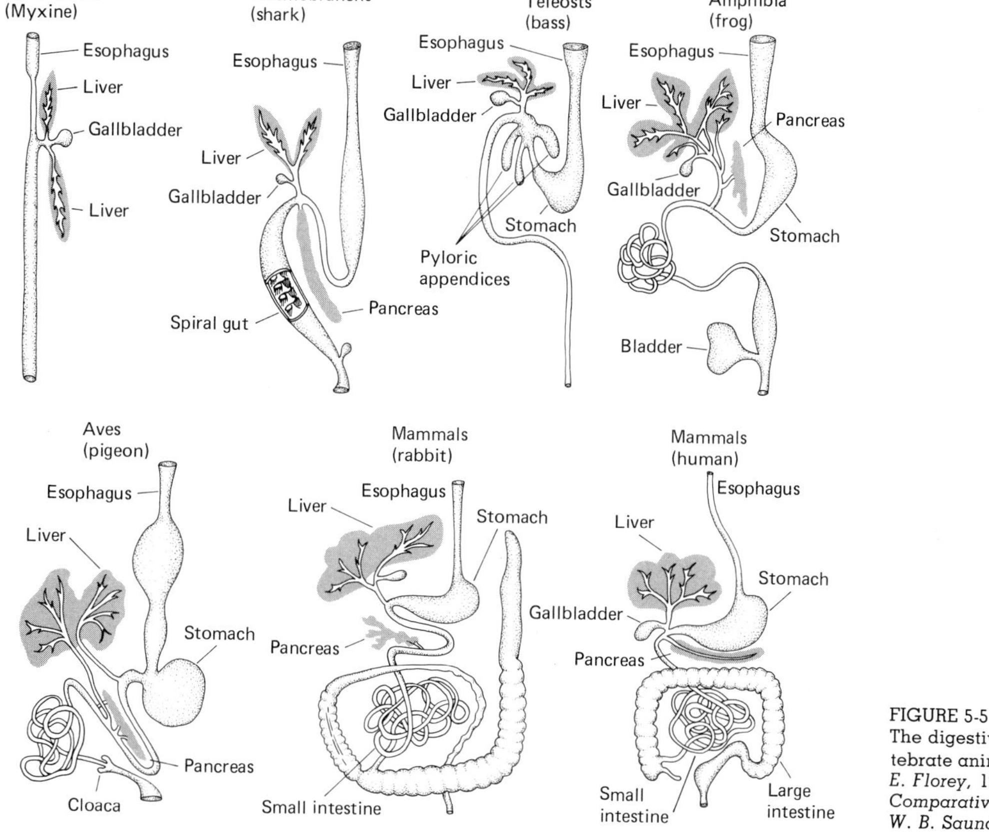

FIGURE 5-5
The digestive tracts of vertebrate animals. (*After E. Florey, 1966, General and Comparative Physiology, W. B. Saunders Company, Philadelphia, p. 242.*)

mentalized, and the intestine may consist of several functionally different parts. The digestive glands are either diffusive glandular cells in the wall of the gastrointestinal tract or separate aggregates of gland cells that deliver their secretions to the tract through connecting ducts. Ducted glands opening into the esophagus (or pharynx) near the mouth are called *salivary glands*. Their secretions usually contain some combination of lubricating mucus, digestive enzymes, and toxins, but sometimes only one of these. The digestive juice of the salivary glands is known as *saliva*.

The other digestive glands or gland cells produce digestive enzymes, mucus, emulsifying agents, or acid. Each of these is generally produced by different gland cells, which, by their spatial arrangement, provide a specialized sequence of action on foodstuffs passing through the digestive tract.

Digestion occurs in some organisms as an *intracellular* process following *phagocytosis* of particulate matter by cells of the digestive tract. Intracellular digestion is considered to be a primitive condition, taken over from protozoans. It is the only method of digestion in Porifera (sponges), which lack a digestive tract. Intracellular digestion occurs in coelenterates, many platyhelminths, and *Limulus*, organisms considered to be phylogenetically primitive.

Extracellular digestion, in the lumen of the gastrointestinal tract, is characteristic of more advanced animals. In crustaceans, insects, cephalopods, tunicates, and all vertebrates, digestion is practically completed before absorption takes place by the cells of the digestive tract.

It is interesting that feeding on particulate food is often associated with intracellular digestion. Microphagous groups, such as brachiopods, rotifers, bivalves, and cephalochordates, practice intracellular digestion. Macrophagous feeding is not, however, uniformly associated with pronounced extracellular digestion. Often there is a predigestion and breakdown in the intestinal lumen of the food mass to particles of suitable size for intracellular uptake. This is the case in coelenterates, platyhelminths, and *Limulus*, which, as noted above, engage in intracellular digestion.

The great advantage of extracellular digestion over intracellular digestion is that it allows a *regional* differentiation for secretion of enzymes, digestion, and absorption of digested products. Indeed, the uptake of food, its temporary storage in the intestine, mechanical disintegration, sequential digestion, absorption, and elimination as feces are unique and efficient functions of the several specialized segments of the digestive tract of the more advanced animals.

Digestive Enzymes

All known digestive enzymes belong to the general group called **hydrolases.** The reactions they catalyze can be represented by the following scheme, in which R represents part of a carbohydrate, protein, or lipid:

$$R - R + H_2O \rightarrow R - OH + H - R$$

Enzymatic hydrolysis of organic molecules does not liberate significant amounts of energy. This is in striking contrast to the highly energetic reactions catalyzed by intracellular enzymes in the metabolic pathways.

Like other enzymes, the digestive enzymes are protein. They are classified by the chemical reactions they catalyze, not by their structure. Such categorization is necessitated by the fact that it is generally impractical, at present, to determine the exact structure of every enzyme and give it a specific name. Enzymes are large, complex proteins, and those that catalyze any one reaction tend to differ from species to species in pH or temperature optima, in reaction rate, and in other ways. Thus, enzyme names are *class* names, not names of specific compounds.

Digestive enzymes of animals are classified in the following outline, which, though not an exhaustive list, names the more common enzymes.

PROTEASES

Endopeptidases (hydrolyze proteins to polypeptides)

- **Pepsinases** (pH optimum 1.5 to 2.5)
- **Tryptases** (trypsin, chymotrypsin) (pH optimum in the alkaline range)

Exopeptidases (hydrolyze polypeptides to peptides and amino acids)
- **Polypeptidases**
- **Dipeptidases**

CARBOHYDRASES

Polyases or **polysaccharidases** (hydrolyze high-molecular-weight carbohydrates, such as starch, glycogen, and cellulose, to oligosaccharides, disaccharides, and monosaccharides)

Amylases (hydrolyze starch and glycogen into α-maltose)
- **Cellulases** (hydrolyze cellulose to simple sugars)
- **Chitinases** (hydrolyze chitin to simple sugar; these enzymes usually derived from flora of the digestive tract and not produced by animals themselves)

Oligases or **oligosaccharidases** (hydrolyze trisaccharides and disaccharides to simple sugars)
- **α-Glucosidases** (substrate: maltose, saccharose, α-glucoside)
- **β-Glucosidases** (substrate: cellobiose, gentiobiose, β-glucoside)
- **α-Galactosidases** (substrate: melibiose, raffinose, α-galactoside)
- **β-Galactosidases** (substrate: lactose, β-galactoside)
- **β-Fructosidase** (substrate: saccharose, raffinose, gentianose)

ESTERASES (hydrolyze esters)

Lipases (hydrolyze the long-chain esters formed by glycerol with long-chain fatty acids)

Esterases proper (hydrolyze short-chain, common esters)

These three classes of digestive enzymes occur in all animals, as far as is known. They correspond to the three major classes of organic foodstuffs: proteins, carbohydrates, and fats. The relative abundance in any one species of the corresponding three kinds of enzymes is related to the nature of the food utilized by that species. That is, where protein predominates in the diet, the organism has a relatively great abundance of proteolytic enzyme in its digestive tract. If it exists mainly on carbohydrate, then it will probably have very active carbohydrases and relatively few proteases or esterases.

Proteases split protein and peptide molecules by hydrolysis, according to the general equation:

$$NH_2-\underset{\substack{|\\H}}{\overset{\substack{R_1\\|}}{C}}-\overset{\substack{O\\\|}}{C}-\overset{\substack{H\\|}}{N}-\underset{\substack{|\\H}}{\overset{\substack{R_2\\|}}{C}}-\overset{\substack{O\\\|}}{C}-OH + HOH \longrightarrow$$

$$NH_2-\underset{\substack{|\\H}}{\overset{\substack{R_1\\|}}{C}}-\overset{\substack{O\\\|}}{C}-OH + NH_2-\underset{\substack{|\\H}}{\overset{\substack{R_2\\|}}{C}}-\overset{\substack{O\\\|}}{C}-OH$$

The acid-amide bond which is split is called a *peptide linkage* or *peptide bond*.

Proteins are macromolecules (i.e., of 5000 mol. wt. or more, the number 5000 being quite arbitrary). They are composed of hundreds or thousands of amino acids that are joined together by peptide bonds. Assemblies of amino acids of less than 5000 mol. wt. are called *polypeptides* (Gr. *poly*, many). When the number of amino acids is very small and known, these compounds are given numerical names; thus, dipeptides—two amino acids; tripeptides—three amino acids; etc.

Apparently, digestion of protein is carried out in the same way throughout the animal kingdom. There are three main stages of degradation:

1 **Proteins are broken into smaller units by peptide-peptidohydrolases, formerly known as proteinases, but now called *endopeptidases* (see classification above). These enzymes hydrolyze peptide bonds at various places along peptide chains but do not attack bonds of terminal carboxyl or amino groups. Trypsins are a widely distributed type of endopeptidase.**
2 **Whereas few free amino acids or small peptides are produced by endopeptidases, a second class of proteases, the *exopeptidases*, hydrolyze terminal peptide bonds and commonly produce small peptides, including dipeptides and free amino acids. There are two types, one attacking the peptide bonds at the carboxyl-bearing end of peptide chains, and the other hydrolyzing peptide bonds at the amino terminal ends of the peptide.**
3 **The last stage of protein digestion is handled by *polypeptidases* and *dipeptidases*, which finally yield free amino acids.**

It should be pointed out that, in addition to the relative substrate "preference" of proteases indicated by the above scheme, there are varying degrees of still greater substrate *specificity* among some of these proteases. For example, mammalian trypsin hydrolyzes peptide bonds in which the carboxyl group belongs to a basic amino acid, especially lysine or arginine. Chymotrypsin cleaves peptide bonds in which the carboxyl group belongs either to an aromatic *l*-amino acid or to one of a few other amino acids. Pepsin hydrolyzes peptide bonds joining aromatic or dicarboxylic *l*-amino acid residues (Fig. 5-6).

Carbohydrases hydrolyze glycosidic bonds that join together the monosaccharides forming polysaccharides. *Starches* are the predominant carbohydrate in plant food, and *glycogen* is the major carbohydrate obtainable from animals. Both are *polysaccharides*. Starch is composed of two kinds of molecules: *amylose*, an unbranched chain molecule composed of hundreds of glucose units, and *amylopectin*, a branched chain also composed of hundreds or thousands of glucose molecules. The structure of glycogen resembles that of amylopectin. Partial structures of these molecules are shown in Fig. 16-6.

The polysaccharidases that hydrolyze starch and glycogen are known as *amylases*. They result in the splitting off of disaccharides (*mal-*

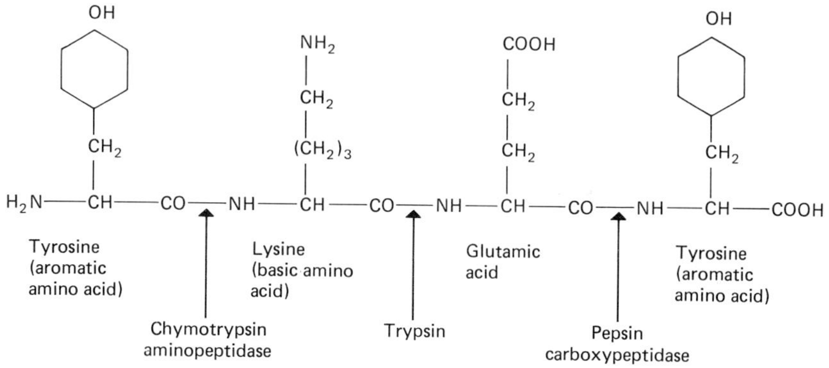

$H_2N—CH(CH_2-C_6H_4OH)—CO—NH—CH((CH_2)_3-CH_2-NH_2)—CO—NH—CH(CH_2-CH_2-COOH)—CO—NH—CH(CH_2-C_6H_4OH)—COOH$

Tyrosine (aromatic amino acid)

Lysine (basic amino acid)

Glutamic acid

Tyrosine (aromatic amino acid)

Chymotrypsin aminopeptidase

Trypsin

Pepsin carboxypeptidase

FIGURE 5-6
Examples of the specificities of some digestive enzymes.

tose) and, to a limited degree, glucose. There is an amylase of mammalian saliva, sometimes called *ptyalin*.

Cellulose, the plant structural polysaccharide, is produced in tremendous quantities. Like starch and glycogen, it is composed of branched chains of glucose units, but whereas these two compounds are formed from α-glucose units, cellulose is made up of β-glucose isomers (Fig. 5-4).

Despite its abundance, many organisms cannot effectively exploit cellulose as a nutritional substance. Those that can, rely upon symbiotic bacteria or protozoa to provide the digestive enzymes, *cellulases*. This subject will be explored later with regard to cellulose digestion by ruminant mammals.

Chitin, a structural material of the exoskeleton of arthropods, is a condensation product of acetylglucosamine. *Glucosamine* is one of the amino sugars—hexoses in which amino groups are substituted for hydroxyl groups. Chitin molecules are chains of acetylglucosamine units linked in a way similar to that of the glucose units in cellulose (β-1,4 linkage) (Fig. 5-4). Apparently, a few animals produce *chitinases*, but it is believed that, as in the case of cellulose, intestinal flora and fauna are largely responsible for the digestion of chitin.

The di- and trisaccharides, both naturally occurring and those produced by polysaccharidase digestion, are grouped under the heading of *oligases* or *oligosaccharidases* (Gr. *oligo*, small) and then subdivided into five classes according to substrates, as outlined above. There are probably other oligosaccharidases in addition to those named. These enzymes produce the simple sugars, which may then be absorbed.

Esterases and *lipases* act upon the esters formed by glycerol and fatty acids, which are the constituents of fats or lipids. The distinction between esterases and lipases is rather arbitrary, the term "lipase" generally referring to an enzyme that digests long-chain esters and "esterase" to an enzyme that hydrolyzes shorter-chain esters.

5-5 ABSORPTION

Emulsifiers, although they are not enzymes, play an important role in increasing the efficiency of both the digestion and the absorption of fats. A secretion of the liver of vertebrates, called *bile*, contains, among several other things, *bile salts*. The bile salts are chemically based on cholesterol and have an acid group in the side chain, making them detergents. Bile salts emulsify large-diameter fat droplets into many small-diameter droplets. Pancreatic lipase acts upon the surface of the small droplets, forming free fatty acids and mixtures of mono- and diglycerides, with only small amounts of free glycerol. As digestion proceeds, the bile salts promote aggregation of the liberated free fatty acids, monoglycerides, and cholesterol into water-soluble *micelles* of up to 10 nm in diameter. It is from these micelles that fatty acids and monoglycerides enter the intestinal cells. It is not known whether individual molecules or whole micelles enter the cell. If it is the latter, it is speculated that this may be effected through *pinocytosis by microvilli* of the absorptive cells which form the surface of the intestinal *villi* (Fig. 5-7).

Although fatty acids are the primary form of fat entering the epithelial cells, very little free fatty acid is released into the circulation. During their passage through the epithelial cells, fatty acids and monoglycerides are resynthesized into triglycerides, and it is the triglycerides which are released into the circulation. However, the triglycerides enter the *lacteals* (lymph vessels) (Fig. 5-7) rather than the capillaries, primarily in the form of small lipid droplets.

Carbohydrates were previously believed to be absorbed only in the form of monosaccharides. It is now widely accepted that the intestinal cells absorb significant amounts of disaccharides before they are split by intracellular enzymes. The monosaccharides absorbed from the intestine include hexoses and pentoses. It is believed that most, if not all, monosaccharides enter the epithelium bound to a carrier molecule. (Carrier-mediated transport will be discussed in Chap. 14.)

In general, only amino acids freed from proteins are absorbed. However, dipeptides appear to be absorbed to some extent and then hydrolyzed intracellularly in the intestinal epithelium. Some amino acids, including glutamic acid and aspartic acid, may diffuse passively through the intestinal wall, but others appear to be taken up by active mechanisms. There seems to be one transport mechanism for neutral amino acids, and another mechanism may be responsible for the transport of the basic amino acids.

The coordination of the mechanisms that deal sequentially with the progressive degradation of foodstuffs is a function, in part, of the nervous system and, in part, of certain endocrines. (The gastrointestinal hormones are discussed in Chap. 9.) As the contained material moves through the various specialized regions of the intestine, digestive enzymes act upon

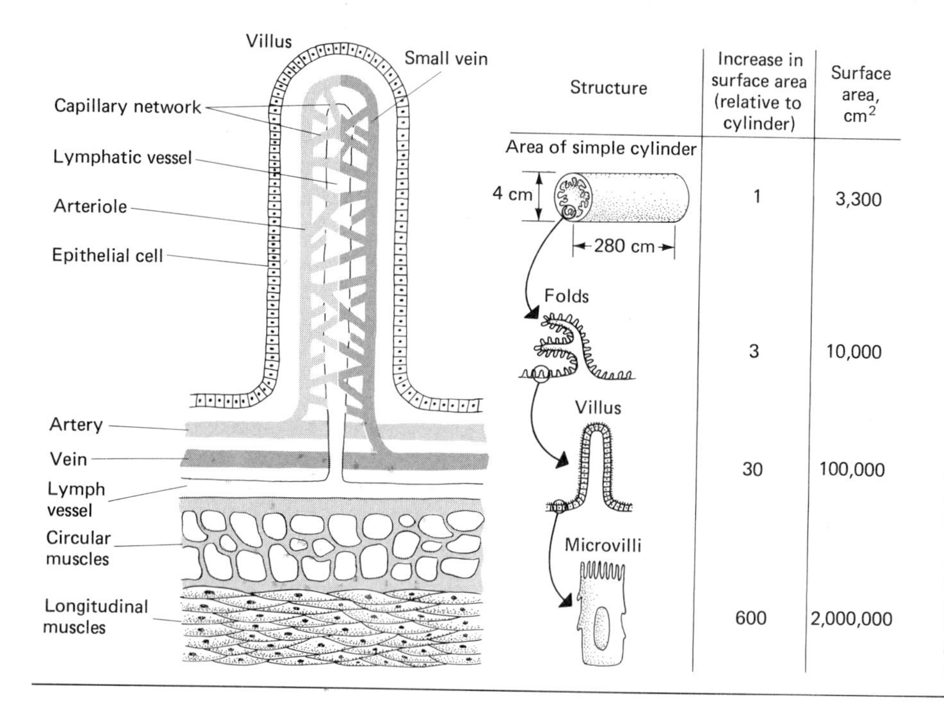

Structure	Increase in surface area (relative to cylinder)	Surface area, cm^2
Area of simple cylinder	1	3,300
Folds	3	10,000
Villus	30	100,000
Microvilli	600	2,000,000

FIGURE 5-7
(Left) A diagram of the microscopic structure of an intestinal villus. *(Right)* A table of the structures by which the surface area of the mammalian intestine is increased. (*After T. H. Wilson, 1962, Intestinal Absorption, W. B. Saunders Co., Philadelphia.*)

progressively simpler products of earlier splittings, until the simple absorbable molecules are all that remain (Fig. 5-8).

The efficiency of the digestive process raises an interesting question: what prevents the digestive enzymes, particularly the proteases, from digesting the cells that produce them and the other cells that form the surface of the digestive tract? If these same cells were part of the food, they would be rapidly destroyed as their proteins were broken down. Of course, cells of the intestine do die and are replaced, just like any other part of the body, but they are not eroded away by the action of the digestive enzymes, although any animal produces enough enzymes for ready digestion of itself.

Just how the organism is protected from its own enzymes is not entirely clear. For one thing, pepsin and trypsin are secreted as inactive forms. The inactive (*proenzyme* or *zymogen*) form of pepsin, *pepsinogen*, is converted to *pepsin* by splitting off a small fragment of the molecule. Hydrochloric acid secreted by parietal cells of the stomach wall initiates the process, and once pepsin is formed, it can act upon other molecules of pepsinogen to form more pepsin (i.e., it is autocatalytic). For *trypsinogen*, the primary event in activation involves the hydrolysis (by *trypsin*) of an amino terminal hexapeptide in the bovine enzyme and an amino terminal octapeptide in the porcine enzyme. These and other proteolytic enzymes are activated in the intestine and do not have hydrolytic activity when they are intracellular. However, as noted above there are some intracellular dipeptidases. These, it would seem necessary to postulate, are somehow "compartmentalized" in the cell so that they do not destroy it.

Considering the high concentration of acid alone, it is most curious that the stomach does not digest itself. Several mechanisms protect the walls of the stomach and intestine. For one, the surface of the mucosal cells is covered by a slightly alkaline mucus which serves to neutralize hydrogen ions in the immediate area of the epithelial cell layer. The stomach surface also is relatively impermeable to hydrogen ions. It is thought that unique structures, called *Brunner's glands*, through their secretions somehow protect the duodenal mucosa by forming a protective coating.

In some cases, the barrier is penetrated, and erosions (ulcers) of the gastric or duodenal mucosa may occur. About 10 percent of the population of the United States are found at autopsy to have ulcers, 10 times more frequently found in the walls of the duodenum than in the stomach itself.

5-6 SYMBIOSIS AND ENDOPARASITISM

The great abundance of green plants offers the possibility of widespread development of herbivorous fauna. However, the enzymes required to

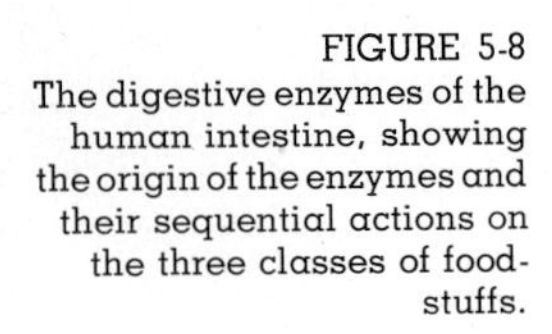

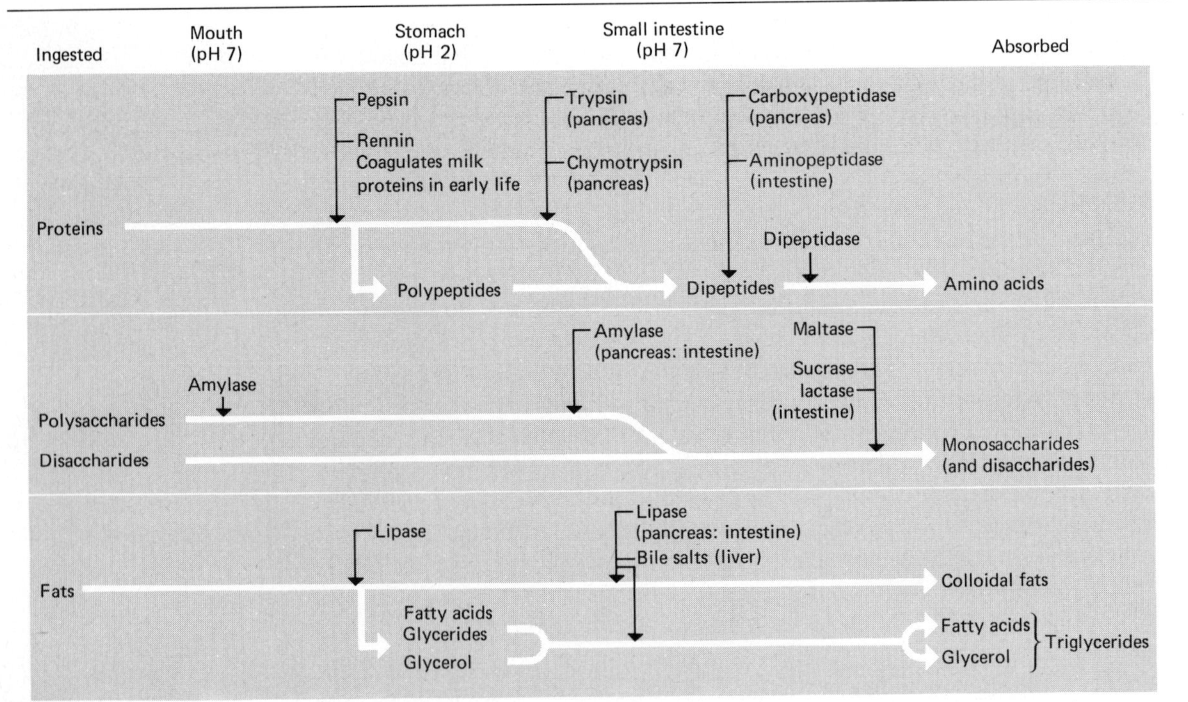

FIGURE 5-8 The digestive enzymes of the human intestine, showing the origin of the enzymes and their sequential actions on the three classes of foodstuffs.

cleave the β-glycosidic bonds of *structural* polysaccharides (cellulose, xylans, pectin, and lignin) have rarely been developed among animals. The herbivorous mode has only developed in certain groups and is based upon one of several specializations: (1) secreted cellulase and related hydrolases; found so far only in a very few animals and never in vertebrates or tunicates; (2) harboring of cellulolytic microorganisms as gastric symbionts, fermenting food prior to final digestion (i.e., the ruminant mode); (3) use of those symbionts for a secondary fermentation in the hindgut after the main digestion has occurred conventionally, as in the horse and elephant; (4) coprophagy (i.e., reingestion of feces) so that secondary bacterial attack can be exploited, as in the rabbit; or (5) use of microorganisms for secretion of ancillary enzymes in the gut for hydrolysis without fermentation, as in many of the phytophagous insects. The structural polysaccharides, generally indigestible by other groups of animals, are degraded to varying degrees in these classes of herbivores.

Symbiont-aided digestion has reached its highest development in the ruminants and other groups of ruminantlike animals. Among those which ruminate are the camel, llama, hippopotamus, sloth, deer, and cattle. The ruminants owe their name to their habit of returning some of the stomach content to the oral cavity for further chewing. The chewing increases the surface area of the food available to the action of digestive enzymes.

The true ruminants have a multicompartmental stomach. Large-scale symbiont digestion of insoluble and resistant polysaccharides is space- and time-consuming. Often several days are required to break down plant material. The greatly expanded and specialized stomach of the ruminant provides a proper set of chambers for these protracted events (Fig. 5-9).

The largest stomach compartment, the *rumen*, and the associated compartment, the *reticulum*, maintain a continuous anaerobic culture of bacteria and protozoa. No digestive enzymes are secreted by the reticulorumen. Structural polysaccharides are partially digested by the microorganisms, and the products and other digestible carbohydrates are fermented, producing volatile fatty acids, H_2CO_3, and CH_4.

Thus, although the initial stages in the breakdown of the food in the rumen are carried out by hydrolases, the sugars produced are not absorbed by the host but instead are largely absorbed and utilized by the microorganisms themselves, to produce fatty acids mainly. Acetic acid appears to be the major product, but large amounts of propionic and butyric acids also are produced. It is these fatty acids that constitute the host's share of the carbohydrate in the food. The blood-sugar level of adult ruminants is only about one-half that of other mammals. Ruminants synthesize glucose from propionic acid.

The proteins of food also are split by proteolytic enzymes secreted by microorganisms. Some of the freed amino acids are absorbed through the walls of the reticulorumen, but most of them appear to go to the competing microorganisms to be used for their own growth and multiplication. The protein is not lost by the ruminant, however, since the microorganisms themselves eventually pass from the reticulorumen to the *abomasum*, of the stomach proper, in which normal digestion occurs. Here the microorganisms are digested by the host. They are an important nutritional source of protein. In fact, most of the true food of the ruminant is its own mircroorganisms.

The rumen microorganisms also can synthesize proteins from ammonia and urea. The latter is produced by protein catabolism and, by diffusion from the blood, can enter the reticulorumen. On a protein-poor diet, the synthetic capacity of the microorganisms can be very important in the protein economy of the ruminant.

In addition to the advantages to the host of symbiont-aided digestion already mentioned, there is the benefit of vitamin synthesis by the microorganisms. Those of the B group, especially, are synthesized. Thus, dietary vitamin needs are restricted almost entirely to vitamins A and D.

Certain groups of animals have, in adapting to a highly specialized feeding habit, lost the ability to synthesize some digestive enzymes. For example, aphids, which feed on the phloem sap of plants, lack proteases and polysaccharidases. Leeches supposedly have no endopeptidases in the gut, but rely on gut flora to provide them.

Endoparasites often represent the extreme of such adaptations. The Cestoda and Acanthocephala have no digestive tract and no digestive enzymes. They absorb metabolites from the host through their body surface. Other endoparasites combine digestion by their own enzymes of host foodstuffs with direct absorption of simple nutrients through their body surface.

Symbiosis with unicellular algae (zooanthellae and zoochlorellae) occurs in many invertebrate groups, and in some cases, is linked to digestive adaptations. Intracellular photosynthesizing algae provide

FIGURE 5-9

The stomach of the ruminant animal is multicompartmented. The true stomach, or abomasum, is preceded by several other compartments that serve as fermentation chambers in which microorganisms digest polysaccharides. (*After K. Schmidt-Nielsen, 1975, Animal Physiology, Cambridge University Press, New York, p. 184.*)

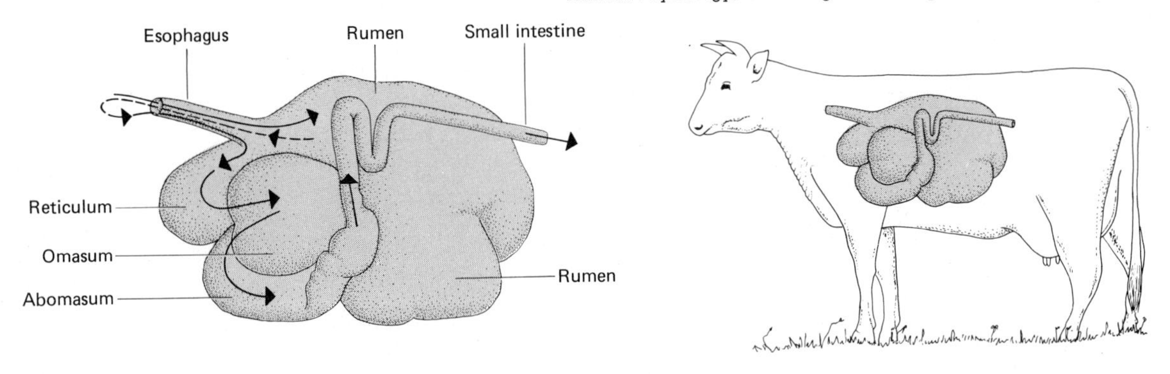

sugar and glycerol to some species of sea anemones, corals, and hydra. In the turbellarian, *Convoluta roscoffensis*, the digestive tract disappears in the adult as zoochlorellae multiply in its cells. In certain mollusks the superficial tissues contain numerous photosynthesizing algae which apparently provide nutrients to them.

The attine ants, commonly known as fungus-growing ants, provide a spectacular example of a symbiotic association of two very different types of organisms. They cultivate fungus gardens, the viability of which depends directly upon the presence of the ants. Their fecal material is incorporated into the substrate of plant material (or insect carcasses in some cases) on which the fungus grows, and it is applied continuously to the growing garden. The fecal material contains *proteolytic enzymes* which compensate for a deficiency of such enzymes in the fungus. In addition, the nitrogenous components in the fecal material facilitate the initial growth of the fungus. The ants contribute their enzymes to degrade protein, and the fungus is a cellulose-degrading organism. Carbon and nitrogen metabolism of the two organisms have been integrated. Thus, there is a joining of *complementary* metabolic capabilities and deficiencies in this natural symbiotic relationship.

5-7 STORAGE AND MOBILIZATION OF NUTRIENTS

Many animals, especially the larger and more complex ones, tend to feed episodically. The gastrointestinal tract is filled at certain times and empty at other times. Feeding-fasting cycles may occur several times daily, weekly, or less often, depending upon the species. Some animals feed only for a few weeks or months of the year. Interrupted food intake necessitates some kind of mechanism by which nutrients can be stored following absorption and then doled out to all the body cells in the interim between feedings. Thus, in higher animals one speaks of the *absorptive state*, during which ingested nutrients are entering the blood from the gastrointestinal tract, and the *postabsorptive* (or *fasting*), *state*, during which the gastrointestinal tract is empty and nutrients must be supplied by the body's endogenous stores.

Let us consider the *absorptive* state as it occurs in the human. The average meal might contain approximately 65 percent carbohydrate, 10 percent fat, and 25 percent protein. Their digested products enter the blood and lymph as monosaccharides, triglycerides, and amino acids, respectively. All the blood leaving the intestine goes directly to the *liver* (via the *portal-system* vessels), where the composition of the blood is altered before the blood flows to the remainder of the body (Fig. 5-10).

Glucose, as well as some galactose and fructose, which the liver converts to glucose, is incorporated into the polysaccharide glycogen, or transformed into fat by liver cells. Note in Fig. 5-10 that glucose can provide both the glycerol and fatty acid moieties of triglycerides. (The interconvertibility of the three major classes of foodstuffs—carbohydrates,

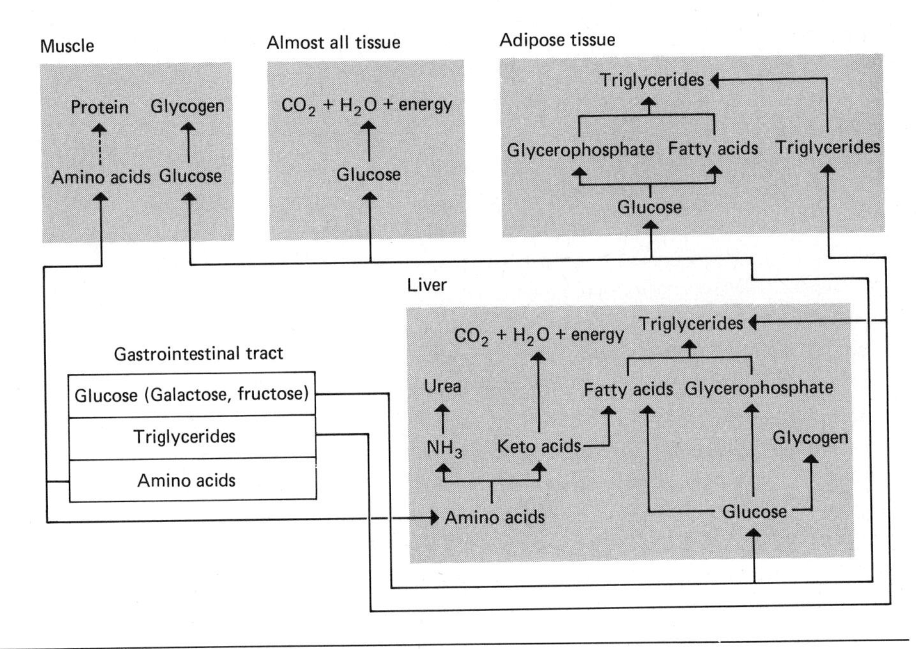

FIGURE 5-10 Major metabolic pathways of the absorptive phase of mammals. (*After A. J. Vander et al.*, 1970, *Human Physiology, McGraw-Hill Book Company, New York*, p. 307.)

fats, and proteins—should be noted in this and the following figure.) Some of the fat synthesized by the liver may be stored there, but most of it is transported by the blood to adipose-tissue cells. Much of the absorbed glucose which is not picked up by liver cells can enter adipose-tissue cells and there be transformed into fat. Another fraction of the glucose is stored as glycogen in skeletal muscle and certain other tissues. A very large fraction enters the various cells of the body, where it is oxidized. Glucose is the body's major energy source during the absorptive state.

Almost all ingested fat is absorbed into the lymph, and not into the blood, as fat droplets, containing primarily triglycerides, which enter adipose-tissue cells, where they are stored. Some of the fat is also oxidized during the absorptive state and provides energy. (This is not shown in Fig. 5-10.) Just how much is utilized energetically depends upon the content of the meal and the subject's nutritional status.

Amino acids may enter liver cells and be converted into carbohydrate (keto acids) following deamination (NH_3 removal). The ammonia is converted to urea in the liver, from which it diffuses into the blood and is excreted through the kidneys. Some of the keto acids provide most of the energy for liver cells in the absorptive state. Any remaining keto acids can be converted to fatty acids and thus contribute to synthesis of fat.

Amino acids not trapped by the liver may enter various cells, in addition to those of muscle, as indicated in Fig. 5-10, and be synthesized into protein. However, excess amino acids are not stored as protein. In the figure the dashed line is intended to call attention to this fact. Excess amino acids are converted into carbohydrate or fat. Excess "calories" are mainly stored as fat. Glycogen is a quantitatively less important storage form for carbohydrate. Because fat contains more than twice the calories of an equal mass of protein or glycogen, and because there is very little water in adipose tissue, fat is an excellent energy-storage form for mobile animals.

During the *postabsorptive* state, no glucose is being absorbed from the intestine, yet the plasma glucose concentration must be maintained because the nervous system is an obligatory user of glucose. (This is not absolutely true, since, after 4 to 5 days of fasting, brain cells begin using ketone bodies, as well as glucose, as an energy source.) Neurons normally use no other nutrient for energy, and if glucose falls too low in the blood, especially if the decrease is rapid, severe neural disturbances occur, including coma and eventual death if glucose deprivation lasts several minutes. Thus, a central necessity of the postabsorptive state is the maintenance of the blood-glucose concentration. The mechanisms for effecting this are shown in Fig. 5-11.

During fasting, glycogen stores of the liver release glucose, but they are adequate only for a short time (usually for about 4 hours). Glycogen in muscles, and to a lesser extent in other tissues, provides an amount of glucose about the same as that from the liver. However, muscle cells lack the necessary enzyme to form free glucose from glycogen. Instead, through the glycolytic pathway, glycogen yields pyruvate and lactate, which are liberated into the blood and then converted in the liver into free glucose. Catabolism of triglycerides yields glycerol and fatty acids, as shown in Fig. 5-11. Glycerol, liberated into the blood, can be con-

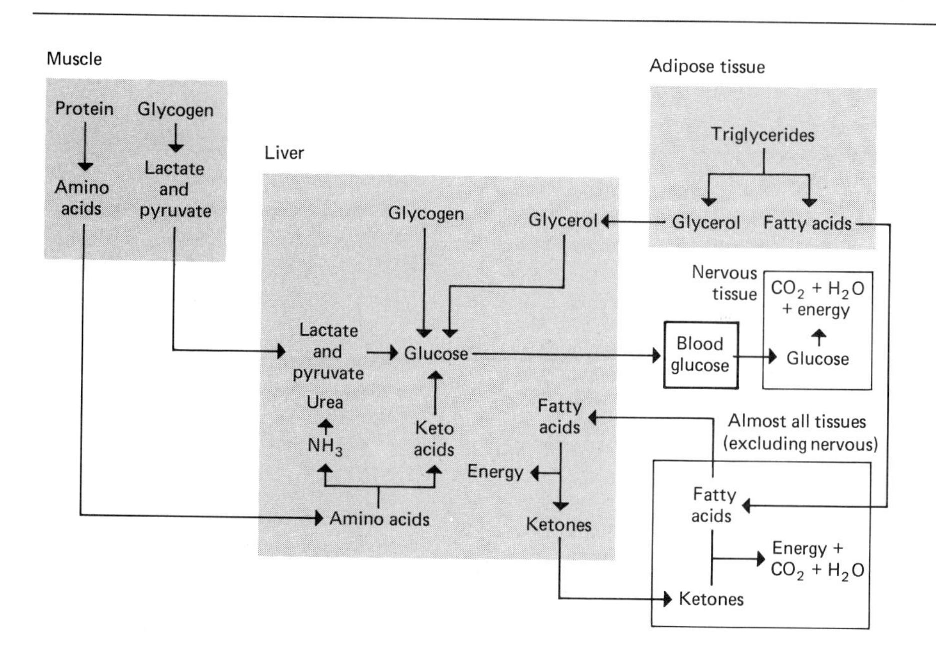

FIGURE 5-11
Major metabolic pathways of the postabsorptive (fasting) phase in a mammal. The central focus is regulation of the blood-glucose concentration. (*After A. J. Vander et al.,* 1970, *Human Physiology, McGraw-Hill Book Company, New York, p.* 408.)

verted to glucose by the liver, but fatty acids are used directly as energy sources by most nonneural tissues (and even by neural tissues during prolonged fasting).

Protein serves as the major source of blood glucose during *prolonged* periods of fasting. Considerable quantities of protein can be withdrawn from muscle and a few other tissues without serious loss of function. The amino acids freed are converted in the liver to glucose. However, beyond a limit of protein loss, severe, debilitating failures of physiological mechanisms occur and will, if not reversed, lead to death. The adaptation of brain cells switching to ketone-body utilization during prolonged fasting, as mentioned above, has survival value in that protein stores can be protected longer while adipose stores are being utilized.

Not only is energy storage important in animals that feed discontinuously; vitamins and inorganic ions also must be held in reserve. The *fat-soluble* vitamins, A, D, K, and E, are stored in considerable quantities, especially in the liver. The liver of the polar bear stores so much vitamin A that it is toxic to humans. The remainder of the vitamins are *water-soluble* and are not readily stored. Deprived of a dietary source, animals quickly become deficient in water-soluble vitamins.

Inorganic ions are stored in the form of insoluble crystalline deposits within endoskeletons (bone, spicules, cartilage), exoskeletons, and calcareous shells. In vertebrates, absorption of monovalent ions occurs throughout the digestive tract. Divalent and trivalent ions, in contrast, do not penetrate readily. Specific active-transport mechanisms for the accumulation of calcium ion and ferrous and ferric ions from the intestine are indicated.

5-8
ENERGY METABOLISM OF WHOLE ANIMALS

The overall energy processes of organisms are commonly called **metabolism.** However, this term is used in several different ways. It can be used to specify the physiological processing associated with an element or a class of molecules, as in "calcium metabolism" or "carbohydrate metabolism." It is sometimes expanded, as in "intermediary metabolism," to mean all the biochemical transformations that take place in the organism. Thus, although it is a very useful term, it is somewhat imprecise.

Metabolic rate refers to energy metabolism per unit of time. The metabolic rate of an animal eventually must relate to the energy cost of all its activities, which can be enormously variable and diverse. To measure metabolic rate in a meaningful way, the conditions which affect it must be taken into account. Otherwise, meaningful and comparative data are unlikely to be obtained. Environmental temperature, digestion, time of day and year, muscular activity, age, and many other factors influence rates of metabolism.

Metabolic rate of whole animals can be measured in several ways. One may determine by calorimetry the difference between the energy content of all food ingested and that of all excreta. However, this method is cumbersome and will not detect changes in storage of materials or growth. Another method is to measure total heat production—the so-called *direct calorimetry* technique. An animal is placed in an insulated chamber, and the rise in temperature of the medium surrounding the chamber is measured as heat is lost from the animal's body. This method can be quite reliable, but it becomes a complex procedure if one attempts to increase its accuracy. Furthermore, direct calorimetry places stringent restrictions on the environment and behavioral condition of the animal being measured.

Measurement of metabolic rate based on determination of oxygen consumption, carbon dioxide production, or combinations of the two is the most frequently employed method. Provided one knows which substances are being oxidized, and there is no anaerobic metabolism or external work performed, this method, called *indirect calorimetry*, is fairly accurate, versatile, and technically convenient.

The practicality of the method stems from the fact that the amount of heat produced for each liter of oxygen consumed in metabolism is relatively constant, regardless of whether carbohydrate, fat, or protein is being oxidized (Table 5-5). The difference between the highest value (carbohydrate, 5.0 kcal/l O_2) and the lowest value (protein, 4.5 kcal/l O_2) is only 10 percent. An average value of 4.8 is conventionally used in metabolic rate calculations.

Whereas 1 liter of oxygen provides similar amounts of energy for all three major foodstuffs, a comparison of the values in Table 5-3 with those in Table 5-5 will lead to the obvious conclusion that different volumes of oxygen are required in the oxidation of equal quantities of these compounds. Indeed, more oxygen per gram is used in protein oxidation than in the oxidation of carbohydrate, and still more is used in the oxidation of fat. The energy derived from the oxidation of 1 g fat is more than twice that derived from the oxidation of 1 g carbohydrate (Table 5-3). This has important biological implications for the use of fat as a compact form of energy storage.

The ratio, known as the **respiratory quotient** (RQ), between the carbon dioxide released in metabolism and the oxygen used provides information about the nature of the fuel being consumed in respiration. Usually the RQ is between 0.7 and 1.0. If it is near 0.7 it probably means that fat is being used primarily. If it is near 1.0 it suggests that mostly car-

Table 5-5
Heat Produced in the Metabolism of Common Foodstuffs and the Respiratory Quotient

Food	Kilocalories per liter O_2	RQ
Carbohydrate	5.0	1.00
Protein	4.5	0.81
Fat	4.7	0.71

bohydrate is being oxidized. Intermediate values would indicate that some mixture of two or all three kinds of foodstuffs is being used. If the amount of protein utilization is determined from excreted nitrogen, that fraction of oxygen consumed and carbon dioxide produced by protein metabolism can be subtracted from the totals. The remainder is due to carbohydrate and fat, and the relative amounts of each can then be calculated. However, when measurements of standard and basal metabolism are made, the animal is fasting and relatively little protein is being oxidized. Hence, the RQ is usually not corrected for its protein component.

The differences in RQ ratios for the three kinds of foodstuffs reflect their differences in chemical composition. Carbohydrate oxidation results in an equal consumption of O_2 and CO_2 evolution. (See formula for glucose oxidation in Sec. 5-3, above.) Protein and especially fat, containing relatively less oxygen than carbohydrate, produce lesser amounts of CO_2 in proportion to the O_2 used in their oxidation.

For comparative purposes, one has to measure metabolic rate under standard, repeatable conditions. As a starting point it is necessary to determine the rate under the simplest and least psychologically and physiologically stressful conditions. The animal should be fasting, at rest, and under no thermal stress. Having established, as best as possible, these conditions, one measures the animal's energy metabolism over a sufficient duration to determine a stable minimum level. In the case of mammals, including humans, this stable minimum rate of energy metabolism is conventionally called the **basal metabolic rate** (BMR). The term "basal" is appropriate for homeotherms but is ambiguous for the majority of animal species. Consequently, BMR is restricted primarily to avian and mammalian (especially the human) species. Because the metabolic rate of poikilotherms is temperature-dependent, there is no metabolic state that could be called "basal" for the animals. Instead, the minimum metabolism of fasting individuals at a given environmental temperature is referred to as the **standard metabolic rate** (SMR) for that temperature.

Though it is perfectly logical to predict that, in general, larger animals will have greater metabolic demands than smaller animals of a given kind, it is less obvious that the larger animal should have a lower metabolic rate *per unit mass* than the smaller animal. For example, the oxygen consumption per gram of harvest mouse is almost 40 times that per gram of elephant, about 10 times that per gram of human, and about 4 times that per gram of cat. Similar trends can be seen with size variations among any group of animals. A sufficient set of such comparisons would reveal that the logarithm of the SMR (or BMR) is a linear function of the logarithm of body weight, or to say this another way, the SMR is proportional to some exponential function of body weight. This relationship is described by:

$$M = aW^b$$

where M = the SMR (or BMR) of the whole animal

a = the metabolic rate per unit weight
W = the weight
b = the exponent

Dividing by W gives the metabolic rate relative to weight:

$$\frac{M}{W} = aW^{(b-1)}$$

Whereas M, a, and W can be objectively determined, the value assigned to the exponent b has been very controversial. Its average value for all organisms lies between 0.7 and 0.8, and 0.75 is regarded by some authorities as the best fit of the data. What this says is that, in general, for each doubling in body weight, standard metabolism increases on the average by about 75 percent (and not by 100 percent).

Why this is so is not at all clear. Because the early studies of animal energetics almost exclusively employed mammals, an interesting explanation, proposed more than 100 years ago, was based on the correlation between basal metabolism and body surface. A small animal has a larger body surface relative to the body weight than a large animal. An equation of the same form that predicts BMR also predicts surface area:

$$\textit{Surface area} = aW^{0.67}$$

The rate of heat loss is proportional to surface area, as noted in Sec. 4-6.

Small mammals, because of their larger relative surface area, must therefore produce a greater amount of heat per unit of body weight in order to maintain a constant, warm body temperature. However, this pleasing "surface rule" fails to explain why frogs, fish, and other poikilotherms, as well as plants, which do not produce heat to thermoregulate, show the same relationship. Thus, while relative surface area could be one of the important determinants of metabolic rate, especially in homeotherms, it appears not to explain why bigger beasts have lower unit energy costs than smaller ones.

Metabolic rate changes must match the variation in performance of organisms. Rarely will any animal be in "basal" state. Any physical activity—walking, running, swimming, flying—increases metabolic rate. However, there is not a direct coupling between oxygen consumption and release of energy for muscular work. Intense muscular activity can consume energy at a rate several times the maximum that can be provided by oxidation. However, very high rates of energy expenditure can be sustained only for short durations. Much of the energy utilized is derived from anaerobic processes, resulting in a local accumulation of metabolites, such as lactic and pyruvic acids, which will later be aerobically oxidized (to be described in Part V). An animal in such a state has an *oxygen debt*, which must be repaid by sustaining a level of oxygen consumption above the resting level during a recovery period following the period of intense activity.

It is useful to know the maximum amount of energy that an animal can release over and above the amount required just to maintain its physiological machinery. The difference between minimum and maximum metabolic rates under a given set of conditions is called **metabolic scope.** It conveniently serves to indicate the capacity for activity. Another similar concept is the **index of metabolic expansibility,** which is used with mammals to define the ratio of metabolic rate at peak effort during sustained muscular work to basal metabolism.

Human beings are ideal experimental animals for determining metabolic scope and the energy costs of various physical activities, because they usually are very cooperative and highly motivated. A person in good physical condition can increase O_2 consumption 15 to 20 times the basal rate, and, during a few seconds of maximal activity, energy expenditure can be as much as 100 times resting level.

The ability of other mammals and birds to increase aerobic metabolism is also impressive. Increases of 3 to 10 times are common. These values are all the more impressive when it is recalled that homeotherms have high resting rates of oxygen consumption. Furthermore, the capacity to incur large oxygen debts allows truly remarkable increases in energy expenditure. The latter capability is an especially important key to success among diving mammals and birds.

In this chapter we have seen how the total complex of feeding, digestive, absorptive, and storage mechanisms is related to the type of food used by the animal and to its metabolic needs. Nutrition occupies a central position in the life of organisms. Animals truly can be viewed as energy systems that must continuously or regularly search for the materials that will support their metabolic machinery.

READINGS

Altman, J., 1966, *Organic Foundations of Animal Behavior*, Holt, Rinehart and Winston, Inc., New York. An account of theories of the control of hunger and feeding can be found here.

Beerstecher, E., Jr., 1964, The biochemical basis of chemical needs, in *Comparative Biochemistry*, vol. 6, edited by M. Florkin and H. S. Mason, Academic Press, Inc., New York. Most of Sec. 5-3 is based on this article.

Bowen, H. J. M., 1966, *Trace Elements in Biochemistry*, Academic Press, Inc., New York. The uptake and function of essential elements is dealt with in considerable detail.

Kleiber, M., 1961, *The Fire of Life*, John Wiley & Sons, Inc., New York. Life as a combustion process and food as the fuel are the predominant themes of this classic book.

Needham, J. (ed.), 1970, *The Chemistry of Life*, Cambridge University Press, New York. This book, subtitled "Eight Lectures on the History of Biochemistry," contains interesting accounts of the discovery of vitamins and enzymes.

Prosser, C. L., 1973, *Comparative Animal Physiology*, W. B. Saunders Company, Philadelphia. Here can be found, in the chapters on nutrition and digestion, a wealth of information about physiological and biochemical adaptations for feeding, digestion, and absorption.

CIRCULATION AND RESPIRATION 6

As noted in the preceding chapter, multicellular organisms contain a fluid milieu which is in an exchange relationship with the external environment and also with the cells it bathes. Nutrients, respiratory gases, and waste metabolites may be exchanged with this extracellular fluid. There are no special arrangements for circulating it in many kinds of small or sluggish organisms. Coelenterates and flatworms have achieved considerable size and complexity, but because they have a highly branched and ramifying gut or gastrovascular cavity which combines some of the functions of a circulatory system with the digestive structures, they require no special internal transport system. Echinoderms have such a low rate of metabolism that an efficient circulation is apparently not required. However, the majority of multicellular animals, and plants too, because of mass, complexity, or activity and the associated metabolic demands, require a continuous and directed circulation of body fluids.

Circulation primarily enhances diffusion and thereby increases the efficiency of exchange. Depending upon the particular species, animal circulatory systems can support many functions, including nutrition, excretion, respiration, osmotic and acid-base regulation, heat transfer, and communication. In animals without skeletons or with little rigid structure, and in many plants, the body fluid can serve hydraulic functions, such as maintaining posture, giving form, and providing means of movement and locomotion. (Fig. 6-1).

6-1 AMOUNT AND COMPOSITION OF BODY FLUIDS

As noted in a previous chapter (Sec. 4-7), organisms regulate, to varying degrees, their internal environment through several *homeostatic* mechanisms. The more complex the organism, the more factors regulated and the more closely they are controlled. At the very core of these regulations is the maintenance of a stability of the body fluids. The concentrations of oxygen, carbon dioxide, nutrients, wastes, and inorganic ions all must remain within certain ranges in the fluid bathing the body's cells.

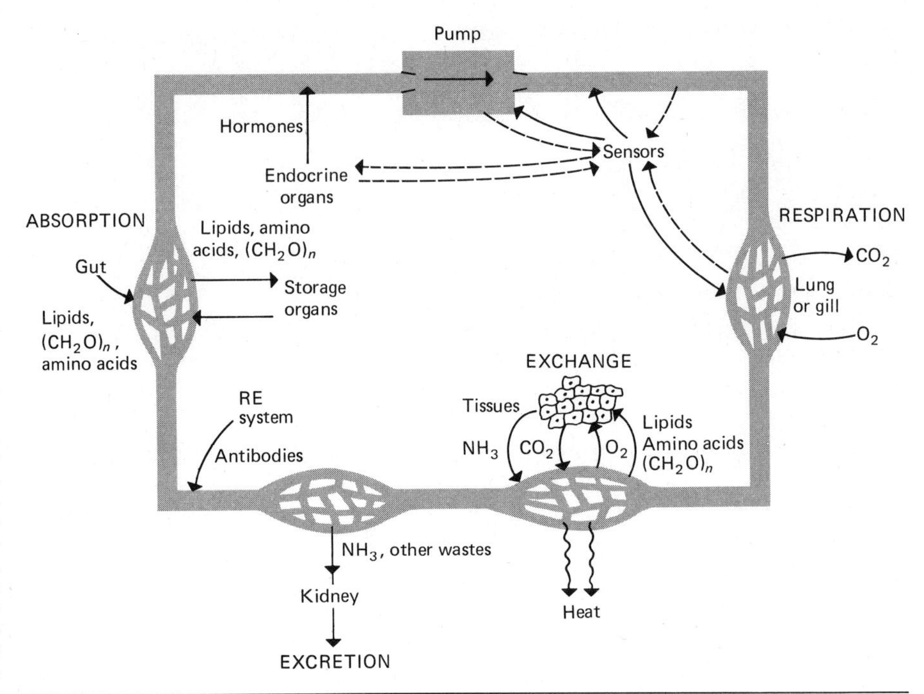

FIGURE 6-1
Diagram representing the general functions of circulatory systems. Some animals do not make significant use of the circulatory system for respiratory exchange, as is the case in arthropods with tracheal respiratory systems. In all animals the circulatory system serves as a medium for metabolite transport and probably for transmission of hormones. Certain cells or elements in the circulation can serve importantly in phagocytosis, clotting, and wound healing. In larger animals the circulatory fluid commonly plays a vital role in heat distribution or dissipation. In many of the soft-bodied animals it has a hydraulic function, serving as a hydrostatic skeleton.

Water is by far the most abundant component of animals. In humans it constitutes about 60 percent of the total body weight, but there is considerable variation among individuals (40 to 80 percent), primarily due to the amount of body fat. Because adipose tissue contains relatively little water, the percentage of water is relatively less in fatter people.

Body water can be said to exist as **intracellular fluid** (inside cells) and **extracellular fluid** (outside cells). The extracellular fluid is the "internal environment" referred to in homeostatic regulation. It constitutes approximately 20 percent of body weight. Usually about twice as much water is intracellular fluid; i.e., about 40 percent of the body weight.

The extracellular fluid of vertebrates can be subdivided into three, and perhaps four, so-called *compartments:* the fluid portion of the blood, or **plasma;** the **coelomic fluid** in such spaces as the peritoneal, pericardial, and pleural cavities; and the **intercellular** or **interstitial fluid,** which is in diffusion equilibrium with blood, coelomic, and intracellular fluid. The intercellular fluid diffuses and is filtered into blind-ended ducts, where it is called **lymph** (except in fishes, other than the teleosts, which lack proper lymphatics). At several places the system of lymph vessels opens into the blood-vascular system. The **cerebrospinal fluid,** which occupies the ventricular spaces of the brain and spinal cord, could be considered still another fluid compartment.

In the open circulatory system of invertebrates (see later) the distinction between blood and lymph or tissue fluids cannot be drawn. The same fluid moves through vascular channels and tissue spaces and is, therefore, commonly called **hemolymph** to emphasize this commonality.

Volume

The amount of body fluid, which varies greatly from animal to animal, can be estimated in various ways, using the concentration of a substance which either is naturally present or has been introduced. For example, if a colored pigment such as hemoglobin is present, a small sample of blood, diluted by a known amount, is matched in color with the total hemoglobin (which can be extracted by bleeding and then washing the minced tissues), also diluted by a measured amount. Thus, if a sample of 0.1 ml is diluted by 10 ml and then is matched in color, and therefore in hemoglobin concentration, with the total extractable blood when it is diluted to a volume of, say, 60 ml, then the blood volume should be 0.6 ml.

In another method a nontoxic dye, such as Evans's blue (T-1824), is injected and allowed to become completely mixed in the circulation. It is believed that the dye forms a complex with albumin and leaves the blood slowly. Usually, at 30 min after administering T-1824 the rate of its fall in plasma is steady because mixing is complete. By extrapolating a line back to zero from the slope formed by several samples, one can estimate what the concentration of the dye would have been if complete mixing had occurred at zero time, and thereby an estimate of the volume dilution is obtained. Thus,

$$\text{volume} = \frac{\text{amount injected}}{\text{concentration}}, \text{ since concentration} = \frac{\text{amount injected}}{\text{volume}}$$

Among annelids, the blood constitutes 30 to 50 percent of their weight. In the arthropods, the values range from 8 percent of body weight in spider crabs (*Maia*) to 37 percent in *Carcinus*. Among insects, values of 20 to 40 percent are common. In mammals, birds, and reptiles blood volumes range between 6 and 10 percent of their body weight. Fish tend to have lower volumes—1.5 to 3 percent of body weight.

The relatively small amount of blood of vertebrates, as compared with invertebrates, is in a closed system of vessels, as contrasted to the open circulatory system of invertebrates. Blood also is commonly more rapidly and thoroughly circulated in vertebrates than in invertebrates. Furthermore, the small volume of blood of the vertebrate is backed up by a large volume of tissue fluid which is in continuous interchange with the blood.

Composition

Body fluids vary greatly in their make-up from group to group of animals. All are viscous, watery solutions of salts, nutritive materials, and wastes. Some contain proteins and pigment-protein complexes involved in gas transport. Unique kinds of cells that function in protective or gas-transport mechanisms are present in the body fluids of some organisms. The constituents of human blood are presented in Fig. 6-2.

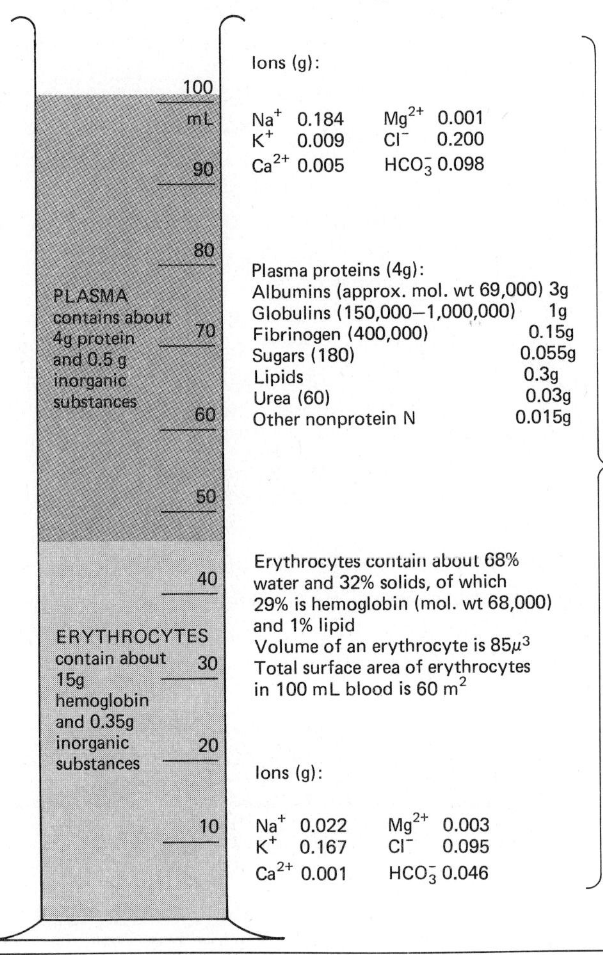

FIGURE 6-2
Pictorial representation of the amounts of the main constituents of human blood. The quantities given are necessarily approximate. (After G. Chapman, 1967, *The Body Fluids and Their Functions*, Edward Arnold, Edinburgh, p. 8.)

The ionic composition of the body fluids will be considered in the following chapter in the context of osmoregulation. Pigmented proteins will be discussed in Sec. 6-4 of this chapter. The role of blood cells in immunity and protection will be dealt with in Chap. 8.

6-2 MOVEMENT OF BODY FLUIDS

The body fluids are transport media, and to serve the requirements of nutrition, gas exchange, excretion, and chemical communication an adaptive and well-integrated circulation is advantageous. Mechanisms for propulsion of blood, especially, are several and quite diverse. *Hemodynamics* is the name given to the study of mechanisms for generating blood pressure and maintaining blood flow.

William Harvey (1578–1657) provided the first recorded demonstrations of how a mammalian circulatory system works. Prior to his studies the conceptions about blood and circulation were nebulous and generally incorrect. By simple experiments and logical deductions he proved the existence of a completely closed circulatory system. Although he could not see the connecting links, the *capillaries*, he argued from the direction of flow in the larger vessels and from calculation of the volume of blood ejected from the heart that the vascular system was continuous.

In the larger, more complex animals, two patterns of development of mesodermal cavities are found, which lead to two quite different patterns of body-fluid circulation. Within the mesoderm, in addition to a series of blood spaces and channels, there is a space provided within which large visceral organs can achieve independence of movement and through which excretory products and reproductive cells may find ready exit. These spaces become important compartments for body fluids. In one developmental pattern the major perivisceral space—referred to as the **primary body cavity**—is, in fact, a persistent blastocoel which is not obliterated by the expanding mesoderm. This space in arthropods, for example, becomes an enlarged blood sinus, the **hemocoel.** The coelomic spaces proper (cavities in the mesoderm) are restricted to the gonads and excretory organs (Fig. 6-3).

In the second pattern, an extensive **secondary body cavity** (true **coelom**) develops within the mesoderm and expands to obliterate the primary cavity (Fig. 6-3).

Thus, two distinct arrangements of circulatory channels character-

FIGURE 6-3
Cross-sectional diagrams to show the relations of the body cavities in annelids and arthropods. (*a*) Early embryonic condition. (*b* and *c*) The blastocoel persists only in the dorsal and ventral blood vessels, and the coelom becomes the secondary body cavity. (*d* and *e*) The coelom remains small, and the blastocoel persists as the hemocoel, or primary body cavity. (*After W. S. Hoar, 1966, General and Comparative Physiology, Prentice-Hall, Inc., Englewood Cliffs, N.J., p. 138.*)

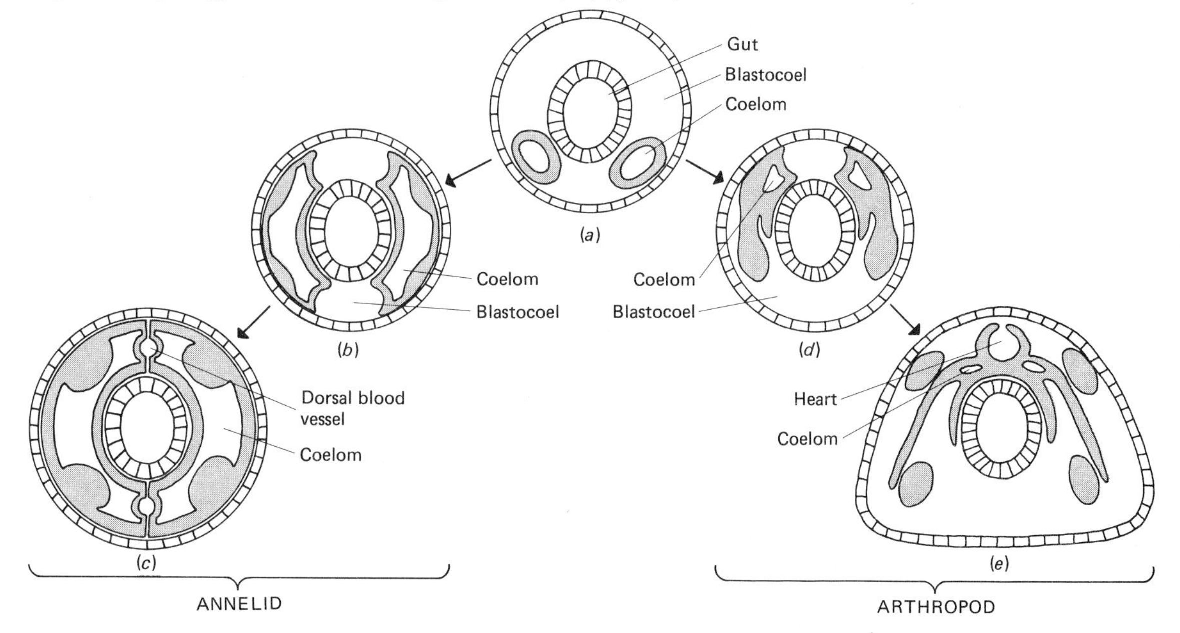

ize the more highly organized groups of animals. Arthropods, most mollusks, and several lesser groups have an **open system** in which there is no completely closed network of vessels. There are vessels, channels, and hearts which give the blood a directed flow, but eventually arterial blood enters sinuses (large spaces) or lacunae (small spaces) where the body fluid bathes the major organs and tissues. The organs lie directly in the blood-filled hemocoel or in a primary body cavity.

In the second type of system, found in vertebrates, annelids, and a few other groups of invertebrates, the blood channels form a **closed system.** A continuous network of minute capillaries unites the arterial and venous vessels. Tissues and visceral organs do not lie in the blood but are bathed in fluid compartments that exchange materials with the vessel-enclosed blood.

Two basic devices for moving fluid are employed: *cilia* (or *flagella*) and *muscles*. Flagella only are used in sponges. Muscles alone are employed by nematodes and arthropods, but both cilia and muscles are used by all other groups. In the echinoderms, for instance, the fluid in the various coelomic spaces is propelled by currents created by a ciliated epithelium. Additionally, fluid contained in the water-vascular system, a set of structures serving locomotion, is circulated both by ciliary action and by muscle contractions. Volume changes in this system serve indirectly to disturb and circulate the contents of the perivascular fluid.

Locomotion and the associated body movements provide sufficient stirring of the body fluids in coelenterates, flatworms, and some of the small members of more advanced phyla. It is also the main propulsive force or an important adjunct in the hemodynamics of larger animals. For example, the massaging action of muscles in the appendages on thin-walled veins and lymphatics in higher vertebrates aids in the return of venous blood. Locomotion is not an adequate driving force for the blood of bulky animals with different kinds of highly specialized tissues. Generally paralleling the taxonomic position of animals is the complexity of the mechanisms related to hemodynamics. Muscle-reinforced tubes, pumps, valves, and control by the nervous system become collectively more sophisticated in the higher groups.

The movement of viscous fluids through extremely small and lengthy blood vessels requires considerable force. Pressure must be maintained to overcome resistance to flow. As will be noted later in this section, hydrostatic pressure also is responsible for perfusing cells by driving fluid out of the capillaries. The same thing applies to filtration in the kidney glomeruli, as is discussed in the following chapter. Continuous flow at high pressures is the hallmark of the highly evolved circulatory system.

Pumps

Pulsating vessels *Peristalsis* is a widespread, primitive mechanism for the movement of the contents of tubular organs. In annelids, many blood

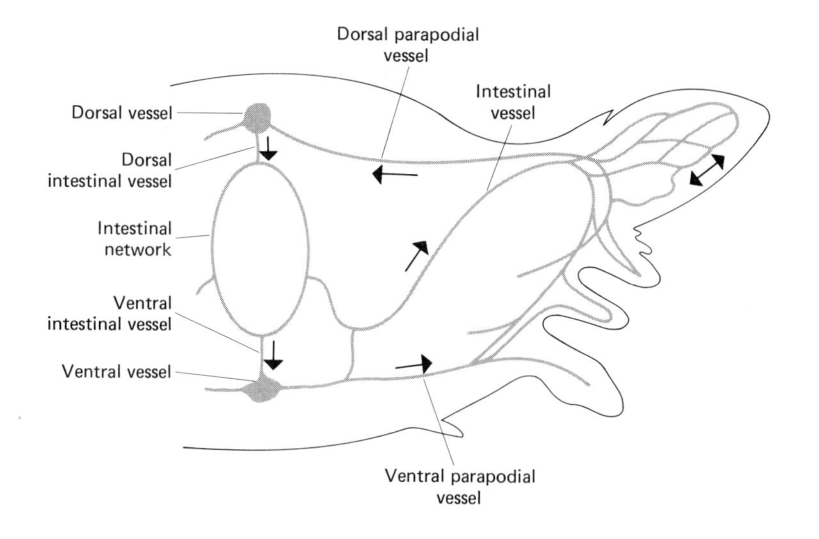

FIGURE 6-4
Cross section through *Nereis* at the level of a parapodium to show the arrangement of the major blood vessels. (*After W. S. Hoar, 1966; General and Comparative Physiology, Prentice-Hall, Inc., Englewood Cliffs, N.J., p. 142.*)

vessels show rhythmic peristalsis. In *Nereis*, a polychaete annelid, contraction waves in the dorsal blood vessel pass from the posterior to the anterior end; waves in the connecting segmental vessels drive blood ventrally into an intestinal network. By way of several possible routes, through the parapodia and intestinal network, blood finds its way into the noncontractile ventral vessel, which carries it passively caudalward. (Fig. 6-4).

In the earthworm, *Lumbricus*, contractions in the dorsal vessel move blood forward. It then passes through several pairs of lateral "hearts" to the contractile ventral vessel.

Pulsating vessels are found in other annelids, some holothurians, and in protochordates. In some higher vertebrates, pulsating veins, as in bat wings, aid in propelling blood, but only as accessories to the main pump, the heart.

Tubular hearts The systemic heart of most arthropods is a specialized area of a dorsal vessel which contracts rhythmically. It is a single chamber with heavily reinforced walls containing striated muscle. The major artery is directed anteriorly and carries hemolymph forward; it returns through laterally paired, valved *ostia* and sometimes through posterior veins. The heart may be tubular and extend for a considerable length of the body, as in Insecta, or it may be a pulsating muscular sac as in Crustacea (Figs. 6-5 and 6-6).

The heart of tunicates, a delicate, nonvalved tube lying in a pericardium, pumps for a time in one direction and then reverses, so that blood does not flow in the same circuit continuously. Thus, the main blood vessels serve alternately as arteries and veins. The stimulus for the reversal of the direction of the heart pulsations is assumed to be an increase in pressure in that part of the vascular system serving as arteries.

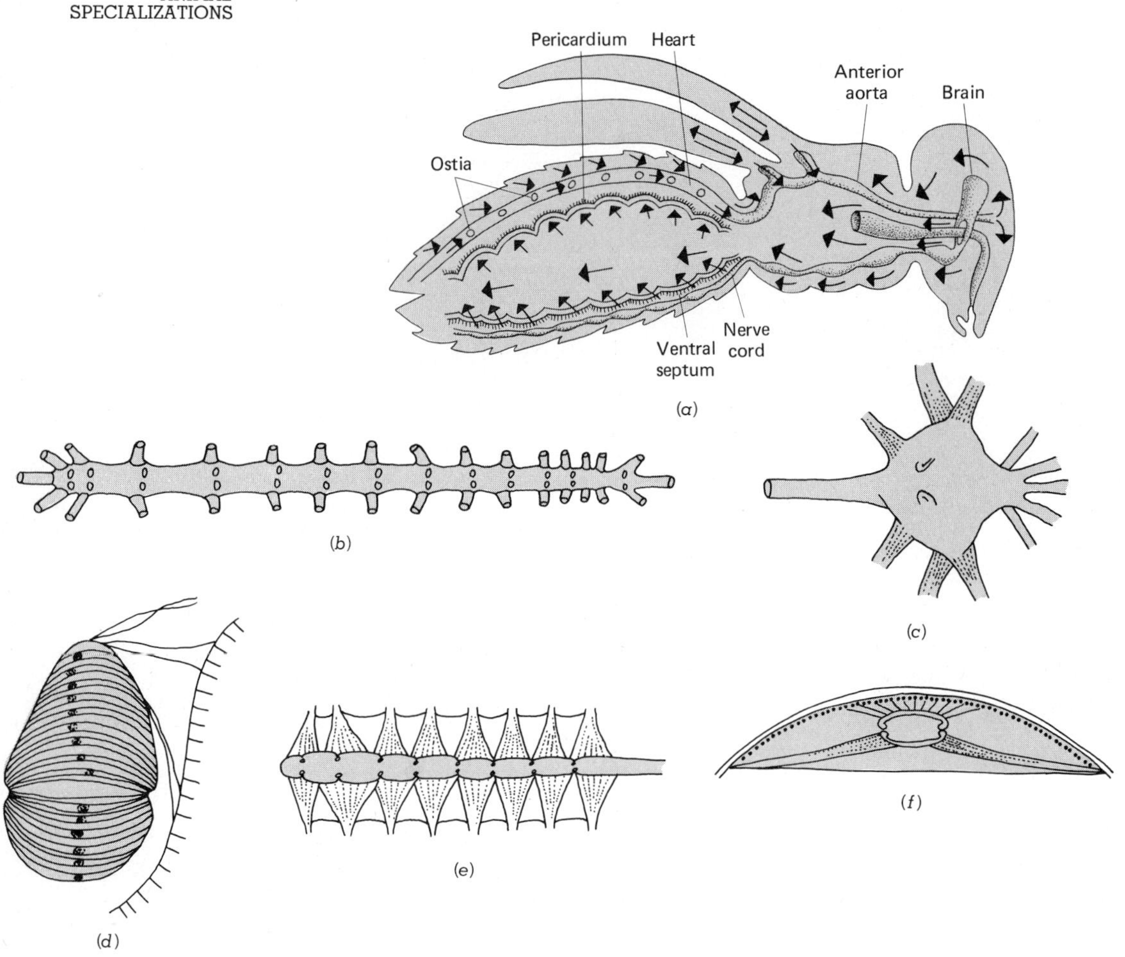

FIGURE 6-5
(*a*) Generalized insect circulatory system. Types of crustacean hearts: (*b*) *Squilla,* (*c*) *Astacus,* (*d*) *Daphnia.* Insect hearts: (*e*) a beetle, (*f*) an insect heart in transverse section to show the alary muscles and the dorsal suspensory fibers.

Accessory or auxiliary hearts Auxiliary hearts are found in a number of invertebrates and vertebrates. Pulsatile ampullar organs are common in insects, particularly at the base of antennae and at the attachment of wings and the legs. Cephalopod mollusks have branchial hearts for propelling the blood through the paired gills. Fishes, amphibians, and reptiles have lymph hearts, which are contractile enlargements of lymph vessels that aid in the movement of lymph into veins. They are composed of striated anastomosing fibers and have valves which prevent backflow.

Chambered hearts These are characteristic of most chordates and mollusks. Chambered hearts may be one-sided with two chambers, or two-sided with three or four chambers. Most fishes have a single atrium and ventricle. Blood must pass through gill vessels before entering the aorta and the systemic circulation. Thus, the heart must provide pressure for traversing two capillary beds. However, the primitive cyclostomes (hagfishes) have a portal, a cardinal, and a caudal heart in addition to the systemic pump. They have a partially open, low-pressure system.

The vascular pumps and general circulatory arrangements of the major vertebrate groups are illustrated in Fig. 6-7.

The two-sided heart appears in the air-breathing fishes (dipnoans). The return of blood from the air bladder is separate from the return of the blood from the rest of the fish. Amphibians, with lung and skin breathing, have a divided atrium (but a single ventricle); the pulmonary vein is separate from the vena cava. Some reptiles (crocodiles) have a two-sided ventricle, but lizards, snakes, and turtles have an incomplete ventricular septum. Birds and mammals maintain a continuous circulation of their body fluids at high pressure with two strong, muscular pumps—one for the systemic circulation and the other for the pulmonary circuit.

The chambered hearts of mollusks consist of one or two atria (four in tetrabranch cephalopods) and one ventricle. Cephalopod mollusks utilize two auxiliary pumps (branchial pumps) to push blood through the gills (Fig. 6-8).

Valves

Efficient circulatory systems require valves at appropriate places to direct the flow and control the volume of the fluid. They are particularly vital to the sustained high pressures and continuous flows of the blood of vertebrate species. However, valves are common in invertebrates as well, particularly in hearts. A flaplike valve in the dorsal vessel of annelids ensures the forward movement of the blood. Several types of valves have been described in crustaceans, associated with the heart and the arterial and venous sinuses.

Paired flaplike valves are found in the veins and lymphatics of the vertebrates, particularly in the limbs. When muscles squeeze the veins,

FIGURE 6-6
Diagrams of hemolymph flow in crustaceans. Solid arrows, flow in vessels; broken arrows, flow in unbound sinuses. (*After W. S. Hoar, 1966, General and Comparative Physiology, Prentice-Hall Inc., Englewood Cliffs, N.J., p.* 145.)

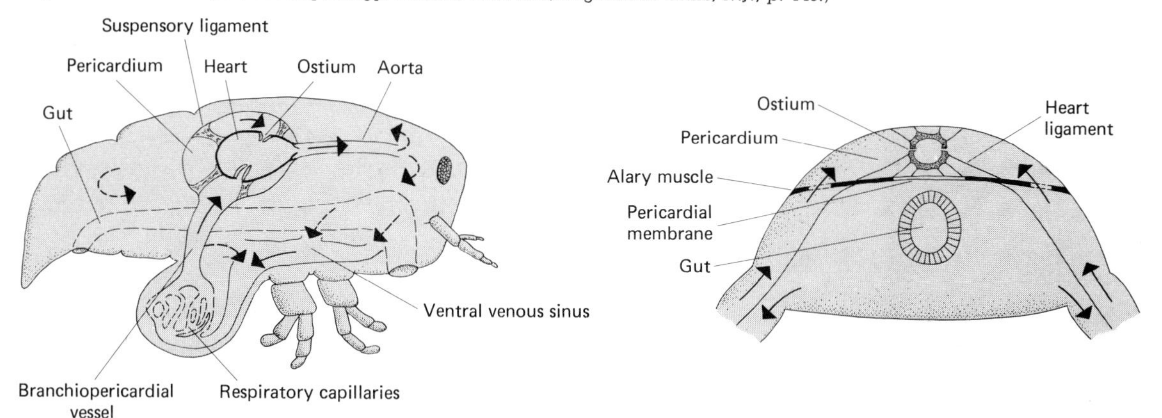

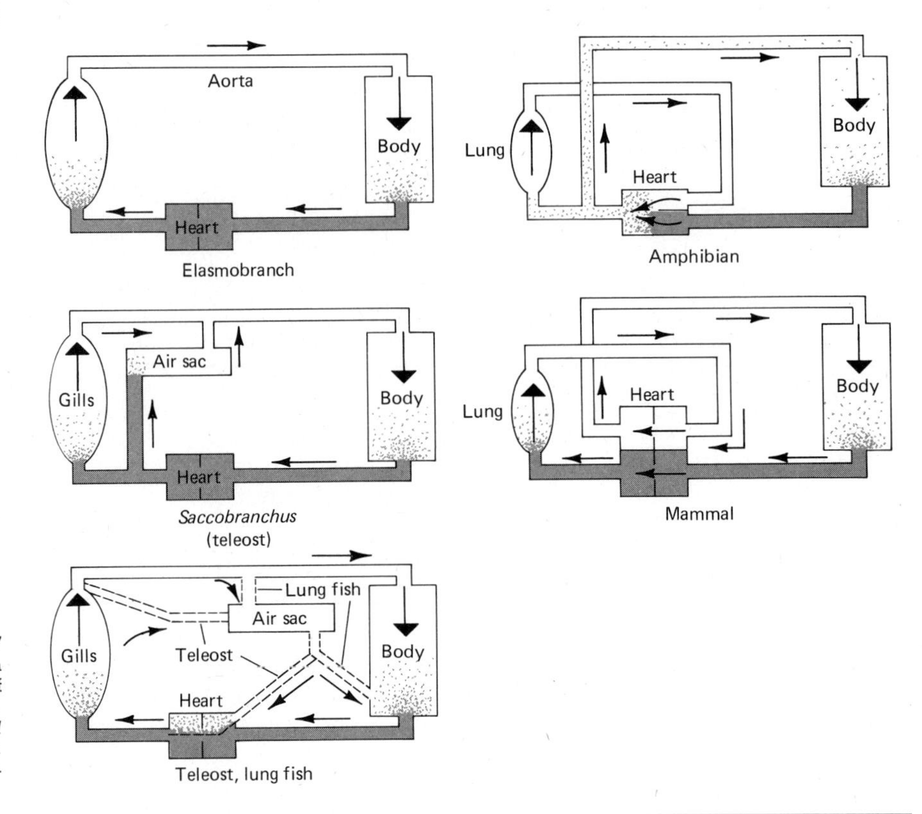

FIGURE 6-7
Diagram of circulation in several different groups of vertebrates. (*After W. S. Hoar, 1966, General and Comparative Physiology, Prentice-Hall, Inc., Englewood Cliffs, N.J., p. 123.*)

FIGURE 6-8
(*a*) Diagram of the central vascular system in the octopus. (*b*) A pressure record from the aorta cephalica with time marks at 2 seconds. (*After W. S. Hoar, 1966, General and Comparative Physiology, Prentice-Hall, Inc., Englewood Cliffs, N.J., p. 147.*)

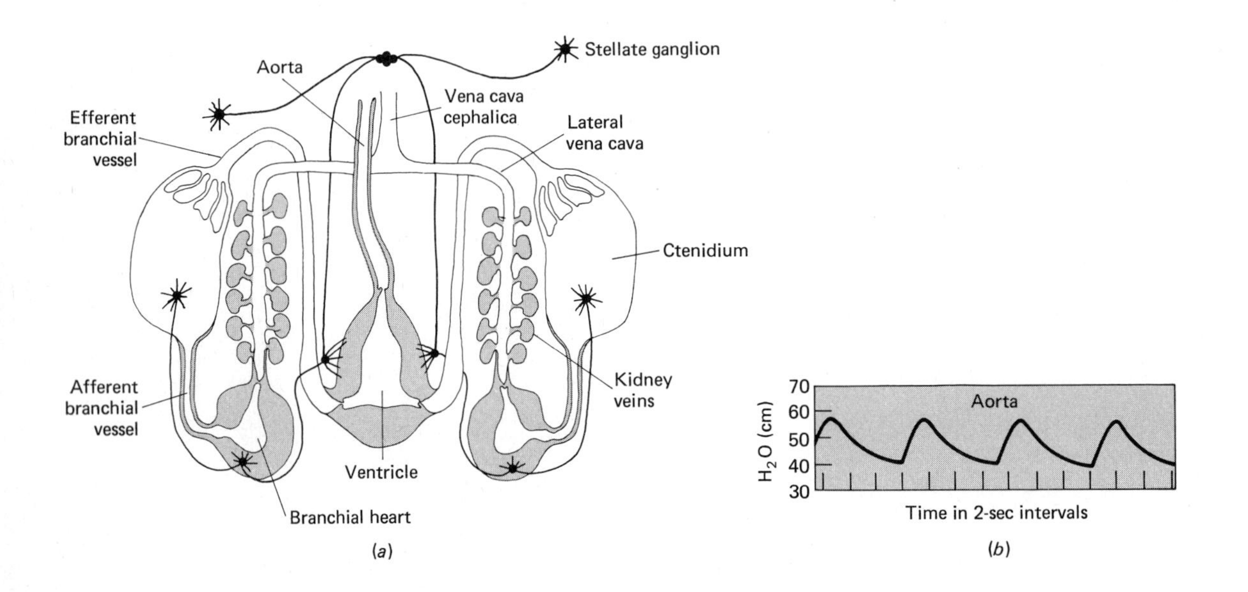

blood is forced toward the heart because the valves prevent backward flow (Fig. 6-9).

Chambered hearts of vertebrates and invertebrates possess valves as an essential feature of the primer pump arrangement (Fig. 6-10).

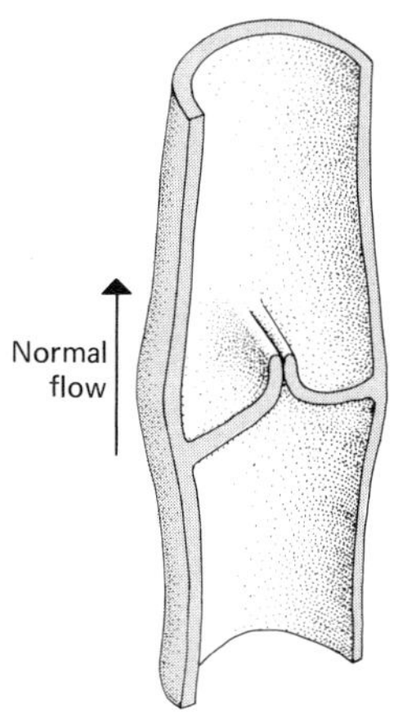

FIGURE 6-9
A venous valve. Retrograde flow tends to press the valve leaflets together and prevent blood from reversing direction of normal flow.

Nervous Control of Cardiovascular Systems

With few exceptions, the pumps and muscular vasculature of circulatory systems are under nervous control. Tunicates lack nervous control, and the cyclostome systemic heart appears to be without innervation, although the accessory hearts are, at least in part, under nervous control. Neural elements control the diameter of certain blood vessels by altering the tonus of their musculature as well as the frequency and amplitude of the contraction of the main heart and accessory pumps.

Pulsating blood vessels, accessory hearts, and main hearts may have either direct or indirect nervous control. In the case of *direct* control, impulses are initiated in neurons, and rhythmic contractions of the musculature result from the periodic output of the neurons. The driving neurons are usually collected together in a compact *cardiac* or *heart ganglion*. The ganglion in turn is modulated by nerves emerging from the central nervous system. The latter control consists of a regulation of the frequency and duration (or intensity) of the activity periods of the cardiac ganglion.

Pulsating organs in which muscle contraction is initiated by nerve impulses are known as *neurogenic* organs. The hearts of decapod crustaceans, *Limulus* (Xiphosura), and scorpions are of this type (Fig. 6-11).

Other pulsating organs are controlled *indirectly* by nerves emerging from the central nervous system, but, as a rule, do not initiate contractions. Instead, excitation that leads to muscle contraction arises from within the organs themselves. They contain modified muscle cells, called **pacemakers,** that spontaneously produce rhythmic "excitations,"

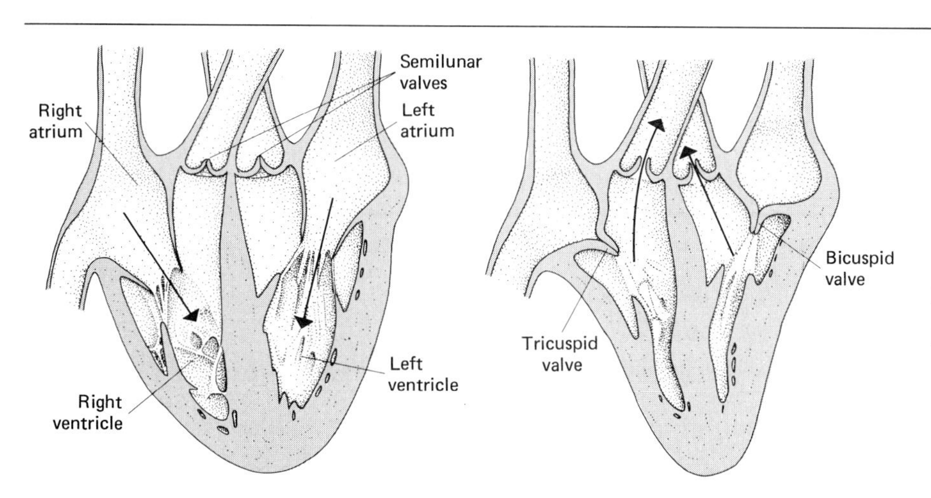

FIGURE 6-10
Diastole and systole in a mammalian heart. (*a*) Diastole, in which the ventricles relax and blood flows into them from the atria. Inlet (tricuspid and bicuspid) valves of the ventricles are open; outlet (semilunar) valves are closed. (*b*) Systole, in which the ventricles contract, closing the inlet valves and forcing blood out through the outlet valves.

or impulses, that sweep across the organ from cell to cell and thereby trigger contractions. Pulsating organs with "muscular" pacemakers are known as *myogenic* organs. The vertebrate and tunicate hearts are typical examples. The hearts of insects also may be primarily myogenic.

It is common to find in both neurogenic and myogenic hearts a double innervation—one set of nerves causing an acceleration and augmentation, the other a slowing and diminishing of the contractions (Fig. 6-11). However, in several cases only an inhibitory innervation has been found (elasmobranchs, cladoceran crustaceans) or only an acclerator nerve is known (certain lamellibranch mollusks).

Heartbeat coordination and vascular reflexes are best known for mammalian hearts, and a very brief summary will illustrate the underlying principles involved. Several areas of the adult mammalian heart have the capacity to develop membrane depolarizations spontaneously.[1] The one with the fastest inherent rhythm of discharge becomes the *pacemaker* for the entire heart. A small mass of specialized myocardial cells embedded in the right atrial wall, called the **sinoatrial node** (SA) is the normal heart pacemaker. Excitation spreads through the atria from myocardial cell to cell, there is no specialized conducting system within the atria. In less than 0.1 second in mammalian hearts all parts of the atria are involved in the spreading excitation, and both atria contract essentially at the same time.

At the base of the right atrium very near the interventricular septum, there is a second small mass of specialized cells, the **atrioventricular** (AV) **node.** Excitation of the atria is transmitted from the AV node along specialized myocardial fibers which extend down the interventric-

[1] The plasma membrane is electrically polarized, negative on the interior. Excitation involves a reduction or reversal of normal polarization, or depolarization. The depolarization in a cell or group of cells can trigger depolarizations in neighboring cells, thus leading to a spreading excitation.

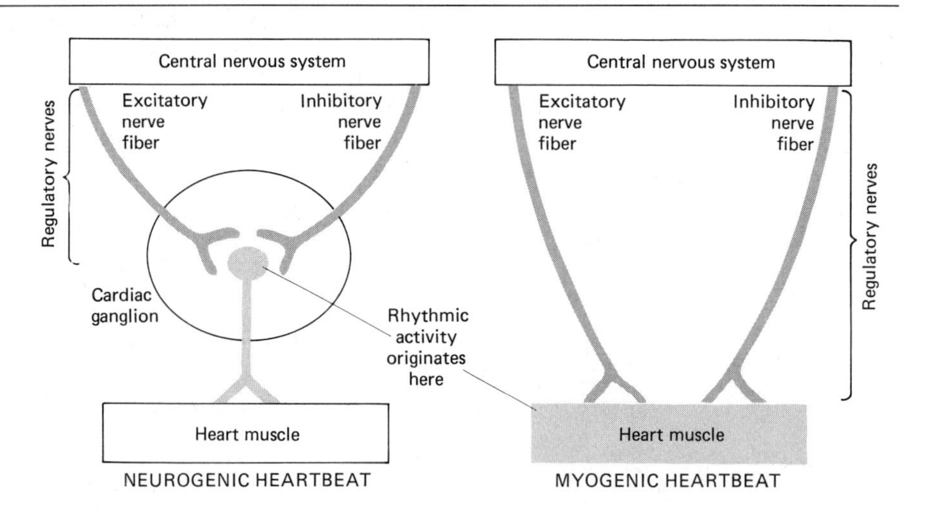

FIGURE 6-11
Conceptual diagram to explain the difference between a neurogenic and a myogenic heart and its regulatory nerve supply. (*After E. Florey, 1966, An Introduction to General and Comparative Animal Physiology, W. B. Saunders Company, Philadelphia, p. 219.*)

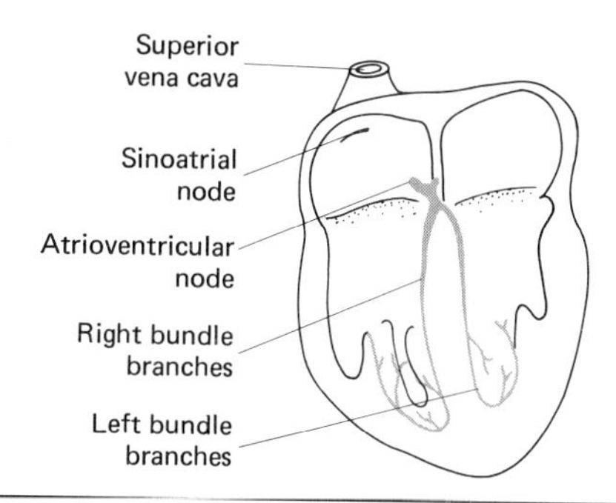

FIGURE 6-12
Conducting system of the human heart. Note the absence of specialized conducting tissue between the sinoatrial and atrioventricular nodes.

ular septum as bundles which eventually make contact with myocardial cells. The excitation spreads from cell to cell throughout the ventricles. The rapid conduction along these fibers and the highly diffuse distribution of the excitation from cell to cell cause depolarization of all right and left ventricular cells more or less simultaneously and ensure a single coordinated contraction (Fig. 6-12).

Sensitive electronic recording devices can detect the electrical events of the cardiac muscle at the surface of the body. The action potentials in the heart may be viewed as batteries which cause current flow throughout the body fluids. A minute fraction of these currents can be detected in metal plates placed on the surface of the body at different places. A record of these minute electrical changes is called an *electrocardiogram* (ECG, or EKG more commonly, from the German terminology). The EKG has great clinical usefulness in health care. The intervals and patterns of waves differ in various cardiac pathologies (Fig. 6-13).

The orderly process of depolarization that triggers contraction of the atria and then the ventricles generates fluid pressures, which, by the

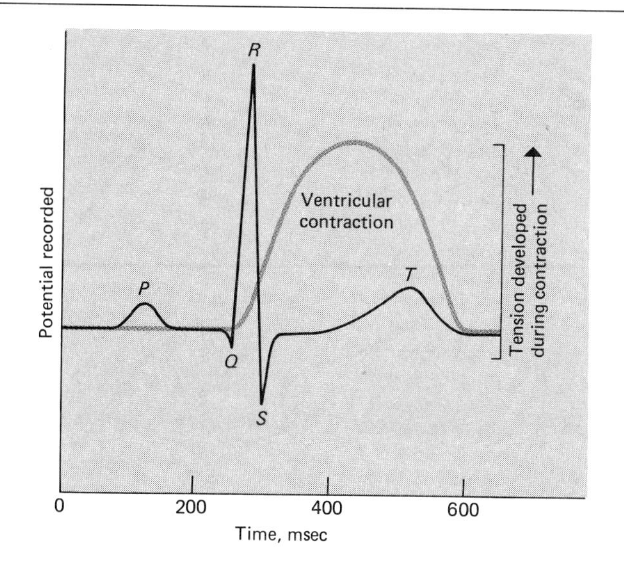

FIGURE 6-13
A typical electrocardiogram recorded as the potential difference between the right and left wrists. The first wave, P, represents atrial depolarization. The second complex, QRS, corresponds to ventricular depolarization, and the T wave is ventricular repolarization.

arrangement of heart valves, direct blood flow through both sides of the heart. The events of the *cardiac cycle* are shown in Fig. 6-14.

The rhythmic discharge of the SA node of mammalian hearts occurs spontaneously in the complete absence of any nervous or humoral influence. However, it can be modified both by its autonomic innervation and by hormones. A large number of parasympathetic and sympathetic fibers terminate in the SA node, as well as on other areas of the con-

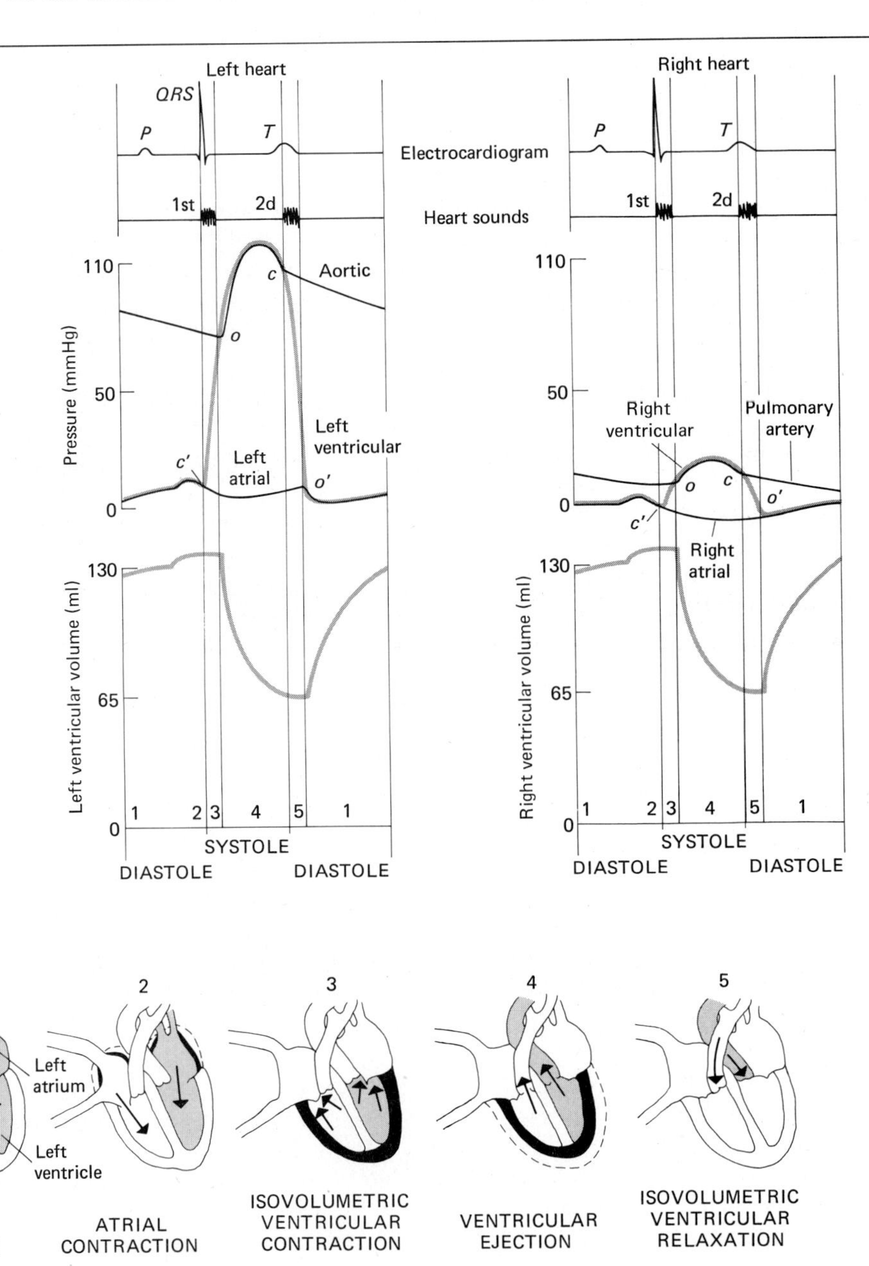

FIGURE 6-14
(*Left*) Summary of events in the left side of the heart and the aorta during the cardiac cycle. At c′ the AV valve closes; at o′ it opens. At o the aortic valve opens; at c it closes. The contracting portions of the heart are shown in black. (*Right*) Summary of events in the pulmonary artery and right side of the heart during the cardiac cycle. At c′ the AV valve closes; at o′ it opens. At o the pulmonary valve opens; at c it closes. (*From A. J. Vander et al.*, 1970; *Human Physiology, McGraw-Hill Book Company, New York, p.* 256.)

ducting system. As noted above (Fig. 6-11), there is a dual nervous control, excitatory and inhibitory. Stimulation of the *parasympathetics* (or local application of the parasympathetic transmitter, *acetylcholine*, to the SA node) causes slowing of the heart. Cutting the parasympathetics causes the heart rate to increase. Stimulation of the *sympathetic* nerves (or local application of the sympathetic transmitter, *norepinephrine*) increases heart rate. Cutting the sympathetics slows the heart. Thus, the antagonistic sympathetic-parasympathetic innervation maintains a control over the autonomous discharge of the myogenic pacemaker, the sympathetic component causing a more rapid development of the membrane potential and subsequent depolarization, whereas the parasympathetics cause a retarding of the rate of potential development.

Maintenance of Blood Pressure and Volume

Arterial pressure in homeotherms is maintained relatively constant despite changes in activity, local metabolic demands, and blood-volume changes. This constancy is largely achieved by altering of cardiac output and peripheral resistance. The controls are neural, humoral, and physical in nature. A partial list of some regulatory mechanisms that have been studied in common mammalian laboratory animals follows:

1 Sensory elements known as *pressoreceptors* (or baroreceptors), located in the walls of the aortic arches and carotid sinuses, detect increases in arterial pressure (*hypertension*). They transmit excitation to cardiovascular centers of the brainstem, which in turn increases parasympathetic outflow to the heart (through the *vagus* nerve), and thereby slow the heart. The parasympathetics also inhibit sympathetic activity to peripheral arterioles and veins. Thus, through reduction of cardiac output and of peripheral resistance, arteriole pressure approaches the norm. The converse effects are elicited by *hypotensive* (low-pressure) states.

2 In hypotensive states the sympathetic system is activated. Cardiac output, peripheral resistance, and venous tone are increased. Another consequence is the release of catecholamines from the adrenal medulla. Adrenal catecholamines (epinephrine, or adrenalin, principally) further facilitate the effects of indirect sympathetic stimulation.

3 Hypotensive states, especially when characterized by low-oxygen and high-carbon-dioxide levels in the blood, evoke further pressoreceptor-type sympathetic adjustments and accelerate breathing. The detectors involved are chemoreceptors such as in the *carotid* and *aortic bodies*. Venous return to the heart is augmented under these conditions by the mechanical effects of deep and rapid breathing.

4 *Hypovolemia* (low volume) resulting from hemorrhage is partially counteracted by bulk filtration into the capillaries from interstitial fluid occasioned by the lowered hydrostatic pressure. (See below for an explanation of fluid exchanges effected by opposing osmotic and hydrostatic pressures.) This physical adjustment is a significant mechanism for maintenance of blood volume and pressure in emergency situations.

5 Hypovolemia also triggers mechanisms that conserve fluid volume. Renal vascular resistance rises and blood flow diminishes, so that kidney glomerular filtration is reduced. Urine volume and sodium-ion excretion fall, indicating both water and electrolyte conservation. Additionally, hormones augment these actions. Antidiuretic hormone, released through the mediation of volume receptors, favors water recovery in the kidney. The renin-angiotensin system (described in Sec. 9-1) is activated, with the consequent release of aldosterone from the adrenal cortex. Sodium reabsorption in the kidney tubules is promoted by this hormone, further augmenting electrolyte and water conservation. Additionally, angiotensin acts to augment the sympathetically mediated rise in peripheral vascular resistance.

Fluid Exchange

In the human, approximately 5 percent of the total circulating blood flows through the *capillaries* at any one time. This small amount is the only blood in the cardiovascular system which engages in the primary function of the system: the exchange of nutrients and metabolic end products. All other vessels of the vascular network subserve the objective of maintaining adequate blood flow in the capillaries. No cell is more than 0.1 mm from a capillary. Thus, the diffusion distance is short and exchange between the blood and cells is highly efficient.

Although each individual capillary is only about 1 mm long, it has been estimated that an adult human has about 60,000 miles of capillaries. However, circulation is not going on in all of them all the time. Indeed, the majority may be closed at any given moment.

The capillary wall lacks the connective tissue and smooth muscle that enclose other parts of the vascular system. It is thin (a single layer of cells) and behaves as if it were perforated by small pores through which water and solute particles smaller than protein readily move. In fact, it acts as a filter through which protein-free plasma moves by *bulk flow*, known as **ultrafiltration**, under the influence of a hydrostatic pressure gradient. The magnitude of the bulk flow is directly proportional to the hydrostatic pressure difference between the inside of the capillary (blood pressure) and the outside (interstitial-fluid pressure). Normally, the former is much larger in the arterial end of capillaries and the filtrate is driven out. However, plasma proteins are largely retained in the capillary and exert an osmotic effect. [The term *colloid osmotic pressure* is used to designate the effect caused by these large proteins, in contrast to the *crystalloids* (salts and glucose), which move freely through the capillary and do not contribute significantly to the osmotic pressure.] This causes the osmotic content of the blood to be greater than that of the tissue fluids, and produces a countering osmotic pressure. Fluid filtered out in the arterial end where hydrostatic pressure exceeds colloid osmotic pressure is replaced in the venous end, because there the hydrostatic pressure is less than the osmotic pressure. Thus, two major opposing forces act to regulate movement of fluid across the capillary wall: (1) the hydrostatic pressure gradient and (2) the osmotic-pressure gradient. In

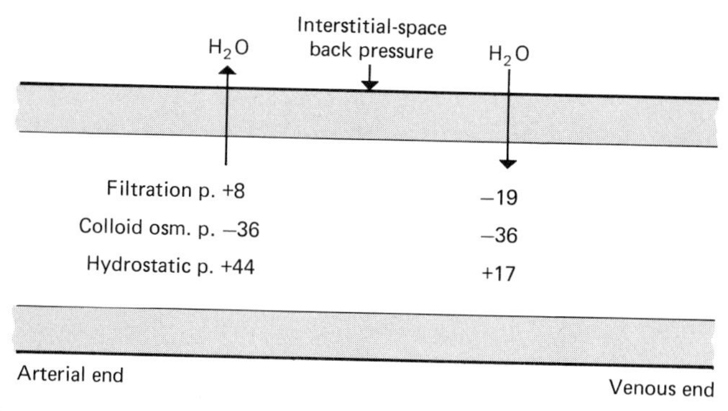

FIGURE 6-15
Diagram of fluid exchange across the wall of a capillary. Blood pressure (hydrostatic pressure) forces fluid out through the semipermeable capillary wall (ultrafiltration) at the arterial end. The plasma proteins remain in the capillary, providing a constant colloidal osmotic pressure. As hydrostatic pressure diminishes along the length of the capillary, the osmotic pressure due to proteins causes a flow of fluid back into the capillary in the venous end. The net pressure difference, or filtration pressure, thus changes from positive to negative.

addition, there is a certain mechanical resistance (tissue pressure) to the flow of fluids from the capillaries into the tissue spaces. Figure 6-15 depicts these forces acting across the capillary wall.

6-3 EXTERNAL RESPIRATION

Diffusion

Gases exchange across respiratory membranes by **diffusion.** The process of diffusion is a basic phenomenon of profound importance in many biologic mechanisms. It results in the spontaneous movement of matter from regions of higher to regions of lower concentration, or, to be more precise, from regions of higher to regions of lower free energy. Animals that use their circulatory system for transport of oxygen and carbon dioxide take advantage of diffusion for exchange of O_2 and CO_2 between their blood and the atmosphere or the water surrounding them and between the blood and the tissues of the body.

Gases will physically dissolve in water. Except for CO_2, which reacts chemically with water, gases take up places between the molecules of water or aggregates of these. When air and water are at equilibrium, the *partial pressures* and the *free energies* of the gases are identical in the two phases. The partial pressure of a gas dissolved in water is often termed the gas *tension*.

Each gas dissolves in water independently of the others and in proportion to its partial pressure. *Partial pressure* is equal to the total pressure times the volume proportion of the particular gas. At 1 atm (atmosphere), the content and partial pressures of the constituent gases of air are: O_2, 20.93 percent, 0.2093 atm; CO_2, 0.03 percent, 0.0003 atm; inert gases 79.04 percent, 0.7904 atm. At equilibrium the concentration (C) of each gas dissolved in water is equal to the product of its partial pressure (p) and its *solubility coefficient* (α), $C = p\alpha$. The solubility coefficient of a gas expresses the ability of water to accommodate the gas molecules.

Exchange of a gas will occur by diffusion between two gases, two fluids, or a fluid and a gas, provided the two are not separated by an impermeable barrier. The gas will move from the phase where its partial pressure is higher to where it is lower. For nonphotosynthesizing organisms, which continuously use O_2 and produce CO_2, there will always be a partial pressure difference of these two gases between the organism and its environment. Biological materials are quite permeable to gases. Hence, diffusion of gases between an organism and its environment occurs continuously. In higher animals, sites of exchange are largely restricted to the respiratory organs: lungs, gills, or regions of the skin.

Contrary to some earlier beliefs, it is today widely accepted that diffusion is the only biological mechanism of gas exchange. Largely because of the brilliant studies of August Krogh (1910), it has been concluded that gas exchange by diffusion can account for the entire gas exchange in the lungs. No active, energy-requiring exchange process is involved.

Air and Water as Respiratory Media

Water has a much greater density and viscosity than air. Thus, it requires more energy to ventilate an organ by water than by air. Furthermore, water equilibrated with air contains only 0.03 percent as much O_2 as air, and gases diffuse about 10^6 times slower in water than in air.

Only dissolved gases and a few other volatile substances are exchanged between the blood of air-breathing animals and the gaseous environment. Water breathers, however, must deal with the possibility that *all* diffusible substances could be exchanged between their blood and the bathing water, although to varying degrees. CO_2 and O_2 penetrate biological membranes at least 1000 times faster than ions, but exchange of ions in aquatic organisms is still significant, as attested by the fact that fish gills have salt-secreting cells in the epithelium. Similar mechanisms for controlling the composition of the internal environment are found among other water breathers.

Water breathers expose their blood to an external fluid of the same *heat capacity*. In contrast, the blood of air breathers is exposed to a medium which has a 3000 times lower heat capacity (on a volume basis). Furthermore, whereas 100 volumes of air contain 20 volumes of O_2, 100 volumes of water equilibrated with air contain only one-thirtieth this amount at 10°C. Thus, in order to bring an equal amount of O_2 to the respiratory surface, water breathers must expose their blood to a heat sink which has 90,000 times larger capacity than air. This difference readily explains why true homeothermy is not found among water breathers. As noted earlier (Sec. 4-5, Fig. 4-11), only a few of the fishes maintain central body temperatures above ambient water temperatures. Air breathing provides at least the opportunity for homeothermy; water breathing largely ensures the continued thermal enslavement of the organism. Thus, gas exchange with air is less expensive, energetically speaking, and more compatible with the maintenance of constant composition and temperature of the organism than is aquatic respiration. Consequently, air breathers can more readily attain homeostatic regulation of several parameters and gain more freedom from vagaries of the environment. The independence accruing from air breathing, in fact, could be considered the major key to the development of mental capacity!

Air breathing does bring with it one great disadvantage: water loss. Membranes adapted to the exchange of respiratory gases inevitably permit evaporative loss of water. An average man, working lightly at 25°C and 50 percent relative humidity, can lose about 300 g of water a day through expiration. Losses through the skin also can be quite significant for many organisms. Water loss can be minimized by structural and behavioral adaptations. The fact that the lungs are cavities with restricted openings is, of course, a major adaptation. Evaporative loss is reduced in small mammals and birds by a kind of *countercurrent* heat-exchange mechanism in the respiratory passages. Inspired air is warmed and humidified on its passage through the upper part of the respiratory tract, and the walls of the passages are correspondingly cooled. Expired air, which is initially saturated with water vapor at body temperature, is cooled during its return passage, and some of its water is condensed. The nocturnal and fossorial habits of small desert mammals, reptiles, and amphibians illustrate behavioral adaptations for avoiding low humidity and high temperatures.

Aquatic Gas Exchange

In simple forms, such as protozoa and platyhelminths, gas exchange occurs without respiratory organs, circulatory systems, or respiratory pigments. Exchange by diffusion across the surface without any special provision is sufficient in animals below a certain size.

Sponges have a simple ventilatory system which enhances respiratory exchange. Water is moved by flagella through numerous pores and irrigates a large internal surface. Coelenterates, ctenophores, and roundworms have some minor specializations for respiration. Echinoderms have respiratory organs and a primitive type of circulatory system, but no respiratory pigment in their internal fluid. The circulatory fluid is moved by cilia.

An increased demand for oxygen with increase in size and activity requires the development of specialized respiratory organs. These are customarily called **gills** when they take the form of evagination from the surface, as in most aquatic forms. Many mollusks, crustaceans, and annelids have ventilated respiratory gills. Furthermore, many have respiratory pigments. Although they have open circulatory systems, they often have capillaries in their gill structures.

Most aquatic mollusks have gills specialized for gas exchange only. However, in bivalve mollusks (clams, oysters) the gills also are devices for filtering and transporting food particles, and in cephalopods (squid, cuttlefish, octopus) ventilation serves to propel water for locomotion. The molluskan gills are highly folded and thus have a large respiratory surface. They are supplied with blood by an artery and drained by a vein. The external medium is pumped across the gills, but the local currents of the exchange surface are controlled by cilia. There is a *countercurrent* circulation so that blood flows inside the flat lamellae in an opposite direction to the flow of water across the lamellae. Such an arrangment permits the arterial partial pressure of O_2 (pO_2) to approach that of inflowing water (Fig. 6-16).

A respiratory pigment is present in many mollusks, but others have no pigment at all. Furthermore, the heterogeneity of the phylum is reflected in the fact that while most have *hemocyanin* pigment, some have *hemoglobin*, and some have both. The presence or absence of respiratory pigments is a most important respiratory variable relating to adaptations for various habitats. (The structure and function of the several kinds of respiratory pigments will be discussed in the following section.)

Among crustaceans there are highly specialized respiratory organs. Their gills are not ciliated but are covered with a thin layer of chitin, which gives them considerable mechanical strength. The laterally paired gills are encased by a hard, protective carapace, and water is pumped across them by special paddlelike appendages. Most crustaceans have a respiratory pigment, either *hemoglobin* or, more commonly, *hemocyanin*.

Aquatic gas exchange is most highly developed in fish, which are characterized by the most specialized gill and ventilatory structures, a closed circulatory system, and a respiratory pigment. Both teleost and elasmobranch gills are countercurrent exchangers. Typically in teleosts the gill arches have two rows of filaments, on the surface of which are numerous thin-walled, parallel, and evenly spaced lamellae. Water passing caudalward across the gill apparatus is exposed to a very large surface, beneath which blood in a thin film flows rostralward (Fig. 6-17).

The lamellae of some fish gills have cells of a distinctive secretory type, which are thought to be specialized for salt transport.

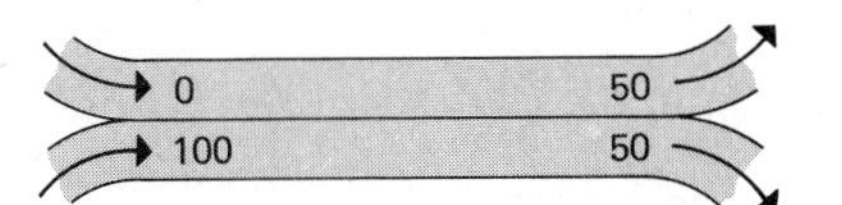

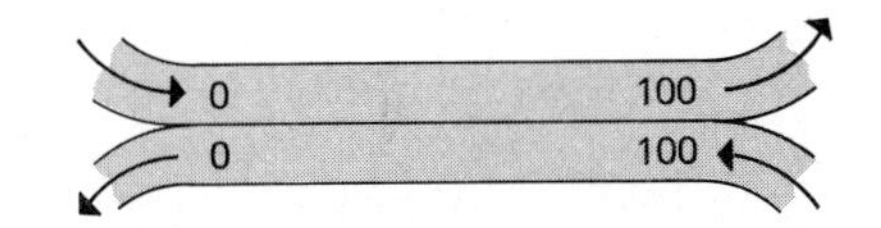

FIGURE 6-16
Schematic representation of the difference between an exchange system with concurrent flow (*left*) and countercurrent flow (*right*). The flow is the same in all tubes, and the numbers represent pO_2. The outflow pO_2 is under conditions of maximum exchange.

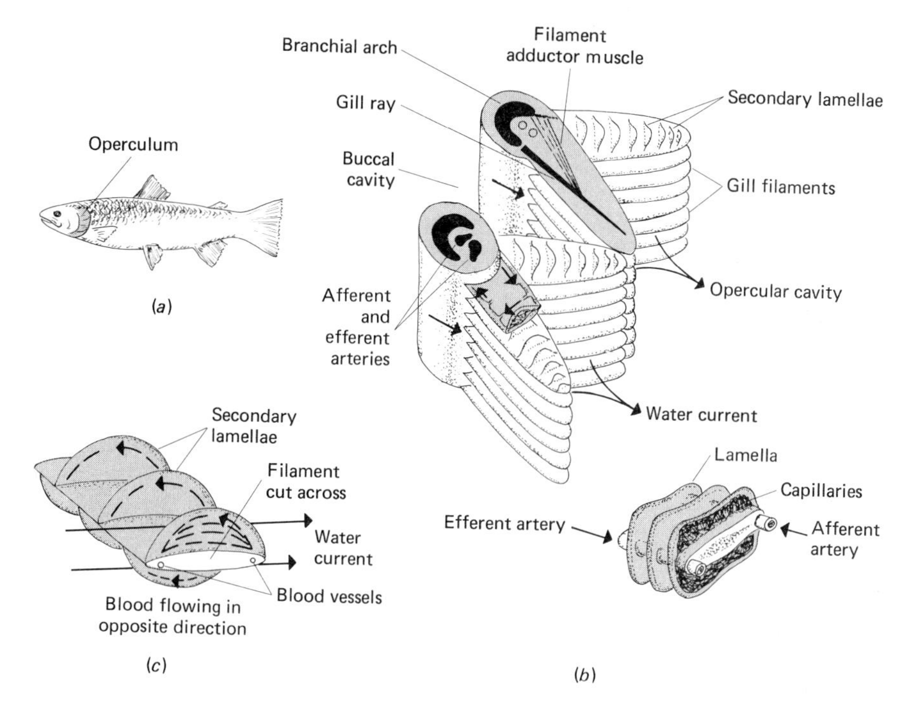

FIGURE 6-17
The teleost gill as a countercurrent exchange system. (a) Diagram to show the position of the four gill arches beneath the operculum of the left side of a teleost fish. (b) Parts of two of the gill arches with the filaments of adjacent rows in contact at their tips. The blood vessels which carry the blood before and after its passage over the gills are shown. (c) Part of a single filament with three lamellae on each side. The flow of blood is in the opposite direction to that of the water.

Transitional Breathing

Several animals have mechanisms that allow them to breathe in either water or air. The majority of these are water breathers with accessory organs for air breathing. Only a few air breathers have accessory organs for water breathing. Secondary adaptation to an aquatic mode of life has mainly involved either *bubble* or *plastron* breathing in insects, or a *diving* syndrome as in reptiles, birds, and mammals (discussed later).

Rarely does the same organ serve both air and water breathing. Gills usually depend on being supported in water and collapse when in air to such an extent that their respiratory surface area is greatly reduced. Lungs are more suited for use in air. However, some animals with rigid lamellae do use gills for air breathing.

There are a few fish, such as the common eel, *Anguilla vulgaris*, which can respire in air without any apparent special modifications. It uses its typical teleost gills for aerial respiration. While traveling across land through wet grass, the eel inspires and keeps the gas in the gill cavity until the pO_2 has reached about 100 mmHg (pO_2 of air at 1 atm is 159.1 mmHg). Then the opercula open, gas is expelled, and a new volume of air is taken in. At cool temperatures such aerial gill respiration, augmented by gas exchange across the skin, adequately supports the metabolic needs of the eel. At higher temperatures, however, the eel develops metabolic acidosis and an O_2 debt.

The majority of air-breathing fishes show structural adaptations for gas exchange with air. At least four different types of mechanisms have evolved:

1 *Mouth breathing;* increased vascularization of mouth and pharynx
2 *Intestinal breathing;* increased vascularization of the gastrointestinal tract
3 *Bladder breathing;* swim bladder provided with postbranchial blood
4 *Lung breathing;* lung provided with prebranchial blood

The South American fish, *Symbranchus marmoratus*, has, in addition to gills, a richly vascularized mouth roof and gill cover. During air exposure it fills its mouth cavity with air and expels it when O_2 content falls to a certain level. Both gills and mouth cavity take part in this gas exchange.

The electric eel, *Electrophorus electricus*, is an obligate air breather. It has very reduced gills; there are no proper lamellae, and the exchange area is small. The mouth cavity has an elaborate system of extremely well-vascularized papillae (Fig. 6-18).

Some fishes belonging to the Gobiidae that inhabit tidal swamps have become so adapted to aerial respiration that they cannot respire in water. In *Periophthalmus schlosseri*, for example, the gill lamellae have coalesced and the gill chamber must be periodically ventilated with air.

Intestinal breathing occurs in some fish where part of the stomach or part of the small intestine has been modified for gas exchange. When the pO_2 of the ambient water falls low enough or when the fish migrates over land, it swallows air rhythmically and expires through either the mouth or the anus.

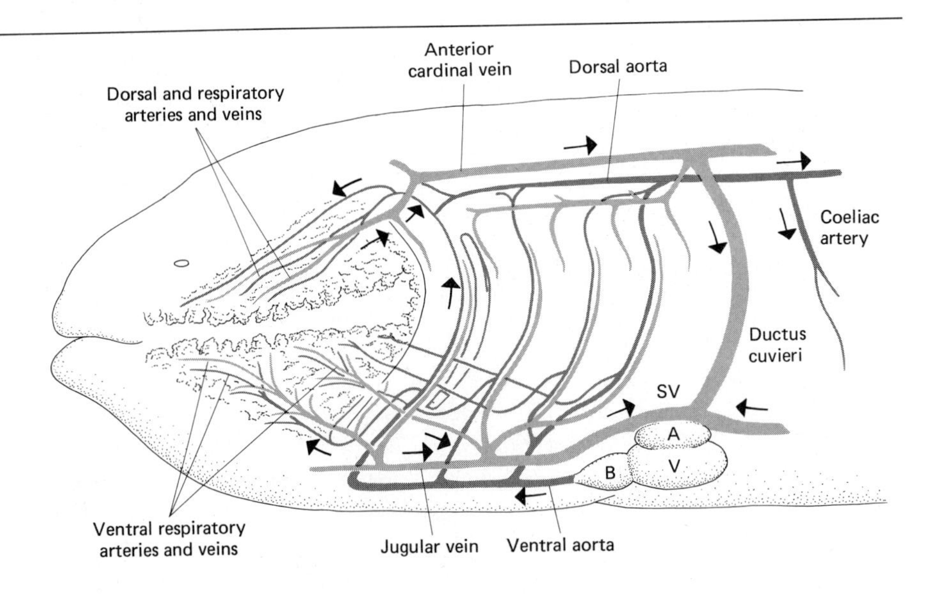

FIGURE 6-18 Schematic drawing of gills and mouth cavity of the electric eel. *SV*, sinus venosus; *A*, atrium; *V*, ventricle; *B*, bulbus.

Although the *swim bladder* serves mainly as a hydrostatic organ for regulating buoyancy, in some fish it also serves as an accessory organ for gas exchange during situations of low water pO_2. Bladder breathing in the common eel increases when it is exposed to air and may account for as much as half the total O_2 uptake. However, the bladder of the eel is not open to the air for ventilation; its O_2 is derived from the arterial blood. Thus, the bladder probably serves as an O_2 store to be drawn upon during hypoxic conditions.

There are other teleosts that can ventilate their swim bladders and utilize them as a "lung" (e.g., *Saccobranchus*). However, these fishes are distinguished from the *lungfishes* by the fact that the bladder is supplied by a branch from the dorsal artery that drains into the systemic venous circulation, whereas in lungfishes the lung is supplied by a branch from a gill artery emptying directly into the heart (Fig. 6-7). Thus, the bladder breathers do not separate venous blood from oxygenated blood anatomically, as lungfishes do.

The bowfin, *Amia calva*, has well-developed gills and a vascular bladder which can be ventilated. It is an active swimmer, but if its habitat dries up it can estivate in pockets of moist soil. Thus, under appropriate conditions it can rely almost entirely on gills or bladder, although it normally uses both simultaneously. When one of the respiratory organs dominates, the blood is to a considerable extent shunted to bypass the exchange vessels of the other. This is appropriate because if, for example, the fish were in O_2-poor water it could lose O_2 picked up in the bladder via the gills.

A few bony fishes are characterized by the presence of a true *lung*, an organ serving for gas exchange primarily; i.e., a unique organ, not a modification of a preexisting one. The Dipnoi, an order of the subclass Choanichthyes, are commonly referred to as *lungfishes*. They are represented today by three genera, one each in Australia, Africa, and South America. They have survived only in regions where we find conditions of seasonal drought similar to those believed to have been present in the Devonian period when the Dipnoi proliferated.

Typical lungs also are present in *Polypterus*, a chondrostean fish (subclass Actinopterygii) found in Central Africa, where it lives in much the same environment as the lungfish of that continent. Lungs appear to have been present in all primitive bony fishes, although today such structures usually have been lost or converted into swim bladders.

A lungfish can, in contrast to most air-breathing fishes, use the lungs and the gills simultaneously in their respective respiratory media. A most important corollary development to the lung is a separate return of pulmonary blood, not to the general venous system, but directly to the heart. A division of the heart has started, resembling that of primitive tetrapods, so that deoxygenated systemic blood and the oxygenated pulmonary blood are partially separated (Fig. 6-7).

Aerial Gas Exchange

Respiratory exchange with air can be categorized approximately into three typical mechanisms (but intermediate ones are frequently found). (1) exchange across the general body surface with circulation and respiratory pigment; (2) exchange through tubes which lead directly to the tissue; (3) exchange with lungs and a circulatory system with respiratory pigments.

The first type of respiratory mechanism is exemplified by many annelids, which have a closed circulatory system and hemoglobin. There are no special respiratory structures; gas exchange occurs across the body surface. The O_2 capacity of earthworm blood is high, but the amount exchanged by the pigment is relatively low, suggesting that the high O_2-capacity blood may serve as an O_2 reservoir, much like myoglobin (see later) of higher animals.

The fact that some salamanders exist without lungs or other respiratory organs illustrates that cutaneous gas exchange can indeed be efficient. Among amphibians, in general, body surface exchange makes important contributions to respiration, but more commonly is a supplement to gill or lung mechanisms.

Gas exchange through tubes, which give the tissues direct access to air, is found among terrestrial arthropods. In these animals the respiratory function of the circulatory system is not required. The network of channels that provide the anatomical basis for this style of respiration is known as the **tracheal system.** Although its structure varies greatly from species to species, the tracheal system basically consists of rigid, branching, air-filled tubes ending blindly in the tissues. At the body surface the tubes open through pairs of *spiracles* (Fig. 6-19).

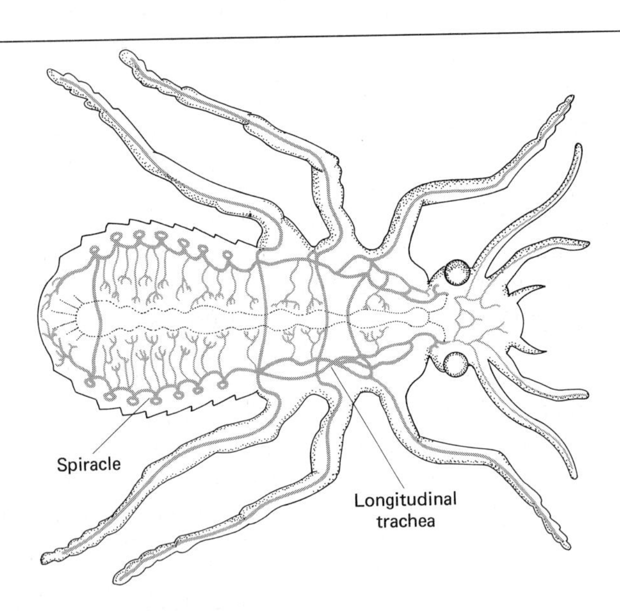

FIGURE 6-19
Tracheal system in a generalized drawing of an insect. This respiratory system depends upon simple diffusion of gases through complex tube systems throughout the body. The tubes open to the surface through spiracles for exchange of gases with the external atmosphere.

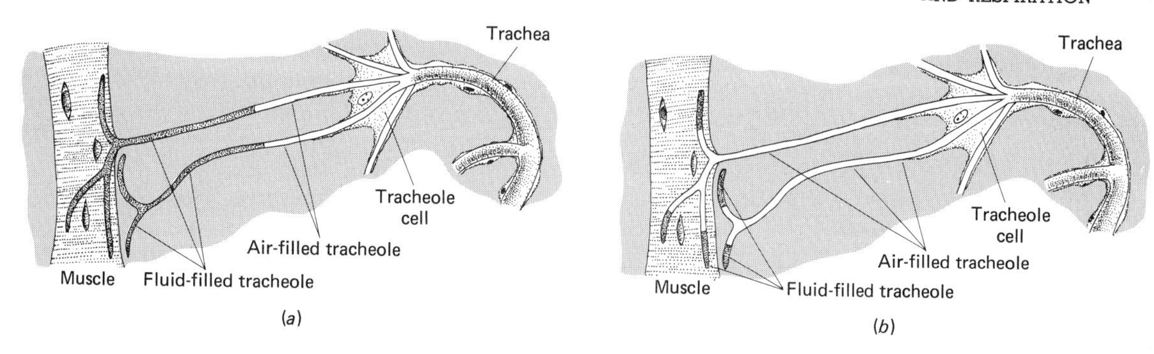

FIGURE 6-20
Diagram of a portion of a tracheal system showing the movement of liquid in the terminal tracheoles. Tracheae are tubes which lead from the spiracles of the integument. They contain chitinized spiral thickenings (taenidia) which permit stretching of the tubes but prevent their collapse. The walls of the tracheae become progressively thinner as the tubes become smaller. Finally the tracheae pass into the air capillaries or tracheoles less than 1 μm in diameter. It is the latter which are the physiologically important structures of gas exchange. They commonly are partially filled with fluid when the animal is resting (*a*), but when the animal is active the fluid is withdrawn from the tracheoles (*b*), bringing the air supply closer to the cells. The mechanism by which this to-and-fro movement is regulated remains unknown. (*After V. B. Wigglesworth,* 1930, *Proc. R. Soc. London B, 106* p. 231.)

In primitive insects the spiracles are simple holes whose aperture cannot be regulated. In most insects, however, the spiracles are complex valves that can be opened or closed according to respiratory demand or can be used in sequences that facilitate ventilatory movements of air in the tube network. Active ventilation is effected through movements of the body wall which rhythmically compress the air spaces and force the air out of them. Dorsoventral flattening of the abdomen (grasshoppers and beetles) or telescoping movements of the abdominal segments (bees and flies) are the more common means. In the grasshopper the flow of air is unidirectional, so that dead air space is reduced; the thoracic spiracles are used primarily for inspiration, while expiration takes place through the abdominal spiracles. However, only the larger trunks are actively ventilated in this manner; smaller branches always depend on diffusion and the movement of fluids in the *tracheoles* (Fig. 6-20). The efficiency of ventilation is greatly improved in some species by the *air sacs*. These are balloonlike dilatations of the trachea which range in size from small vesicles in the legs to very large sacs in the body cavity.

The greatest defect of the tracheal system is the ease with which water can be lost through the open spiracles. A *diffusion control* of respiration by the opening and closing of the spiracles is an effective and important compromise between water loss and the requirements for gas exchange. In the flea, for example, only two pairs of spiracles operate when the animal is at rest. They open and close rhythmically at rates which depend on the temperature. However, when the flea is most active, eight pairs of spiracles remain open until the activity has ceased and metabolism has returned to resting conditions. At intermediate levels of activity, the two pairs of spiracles may remain continuously open and the others may rhythmically open and close or remain open for long durations.

Thus, there is regulation of the air supply to meet prevailing demands while providing for the conservation of water.

Lungs are any kind of invagination from the surface, or cavity which serves a respiratory function. Animals that engage in respiration by lungs depend upon a circulatory system to transport O_2 and CO_2 between the respiratory organ and the tissues. This is highly efficient when a gas-carrying pigment is present.

There are basically two kinds of lungs: *diffusion* lungs and *ventilation* lungs. The former, found in smaller mollusks and arachnids, rely upon diffusion alone for movement of gases. Ventilation lungs, found in larger mollusks and in all vertebrates, facilitate gas diffusion by forcefully moving air across the lung surface. However, the distinction between diffusion and ventilation lungs is not sharp. The important feature is that small animals with low metabolism can get by with lungs that need not be ventilated, whereas larger or more active animals must increase the efficiency of their lungs by ventilation.

Among spiders are found paired diffusion lungs known as **book lungs.** A series of blood-filled, parallel, thin plates separated by narrow air spaces is enclosed in an invagination of the abdominal wall. The air space is connected to the exterior by a spiracle which, in some species at least, opens and closes at a rate related to activity.

The most primitive diffusion lungs are found in snails and slugs. In these animals the mantle cavity is developed into a lung. The surface area is increased by being numerously ridged and well vascularized.

In aquatic mollusks the lung serves as a hydrostatic organ. Pulmonate snails can regulate their buoyancy, rising to the surface to fill their lungs with air when the O_2 content falls to a low level and then diving for a prolonged period. Some mollusks ventilate their lungs in response to O_2 tension or changing air volume. Thus, the distinction between diffusion lungs and ventilation lungs becomes quite blurred in many mollusks. Pulmonary oxygen uptake is supplementary to cutaneous exchange, especially in aquatic snails; and in these forms the lung is probably quite insignificant for carbon dioxide elimination (CO_2 diffuses readily across the epithelium into a medium in which it is highly soluble).

Only the energetic vertebrates have developed air-breathing lungs with mechanisms for regular ventilation. However, even they are not permitted to exploit fully the consistently high pO_2 of the medium. The tidal ventilation mechanism (in and out through the same tube) results in a marked dilution of inspired air with residual air, and alveolar pO_2 does not normally exceed 105 mmHg (the pO_2 of air at 20°C, 760 mmHg, and 50 percent relative humidity is 157.4 mmHg). However, the appropriate adjustment of the loading tension of the hemoglobin (see next section) and the relative low cost of ventilation in air (see above) combine with the superior blood/gas equilibration condition in the lung to compensate for the loss of about one-third of the potential pO_2 gradient.

The ventilation mechanism of the amphibian lung has only recently been understood, largely as a result of an investigation on the bullfrog, *Rana catesbeiana*, by deJongh and Gans (1969). There are two fundamental breathing cycles. One involves only the buccopharyngeal region. With the glottis closed and the nostrils open, air is pumped in and out of the buccal chamber at a frequency of 50 to 100 cycles/minute. These movements are called *oscillatory cycles*. Periodically a more complex *ventilation cycle* occurs. First, air passively leaves the lungs through the open glottis and nostrils. Then the nostrils close, the buccal floor is strongly contracted, and a sharp rise in buccal pressure causes a reinflation of the lungs. The cycle ends when the glottis is closed, the nostrils are reopened, and buccal pressure falls. The buccal floor drops when its muscles are relaxed, and then begins the next sequence of oscillatory cycles (Fig. 6-21).

There is a third kind of cycle in frog respiration, called *inflation cycles* by deJongh and Gans. These consist of a series of ventilation cycles, interrupted by a pause in breathing at the end of the fourth phase of a ventilatory cycle. The intensity of the ventilatory cycles increases before this pause and decreases immediately thereafter.

The lungs of amphibians are quite elastic, expanding upon being filled with air by the buccal force pump and deflating passively by relaxation of their stretched tissue. A similar mechanism of lung ventilation is found in the lungfishes, except that the mouth is opened to fill the buccal cavity with air prior to deflating and reinflating the lung. Neither lungfishes nor amphibia involve the ribs in ventilation. The reptiles, however, have a radically different method of inflating the lungs: a costal suction pump.

A triphasic pattern of respiratory movements has been described in the lizard *Lacerta*. With the glottis open, contraction of abdominal muscles and of the smooth muscles of the lungs produces an expiration. Inspiration is effected through expansion of the rib cage by contraction of intercostal muscles. The glottis closes, and then follows a movement of the abdominal viscera which, because the glottis remains closed, causes the pressure of the pulmonary air to rise slightly and to be sustained above atmospheric pressure for a brief moment. Then the glottis opens and expiration ensues.

In mammals there are still further improvements in pulmonary performance because of a much more elaborate lung structure and refinements of the ventilation mechanism. The costal pump is supplied with two sets of antagonistic *intercostal muscles*. The external intercostals produce an active enlargement of the rib cage for inspiration. Contraction of the internal intercostals causes forced exhalation, but during resting states expiration is largely due to the elasticity of the lung itself; only when high ventilation volumes are required do the internal intercostals become important. Additionally, a muscular *diaphragm*, which is anchored around the circumference of the lower thorax and separates the

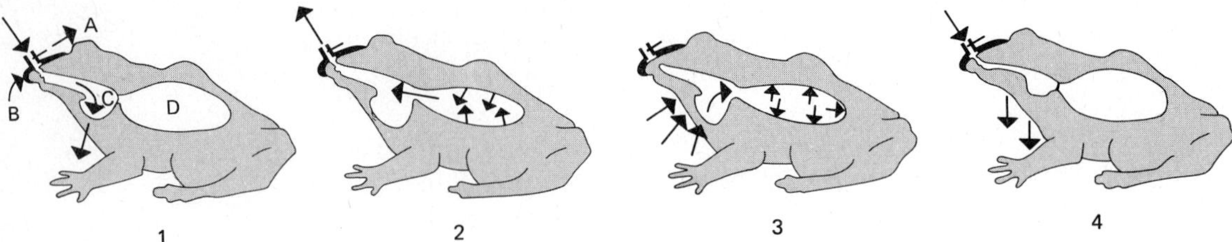

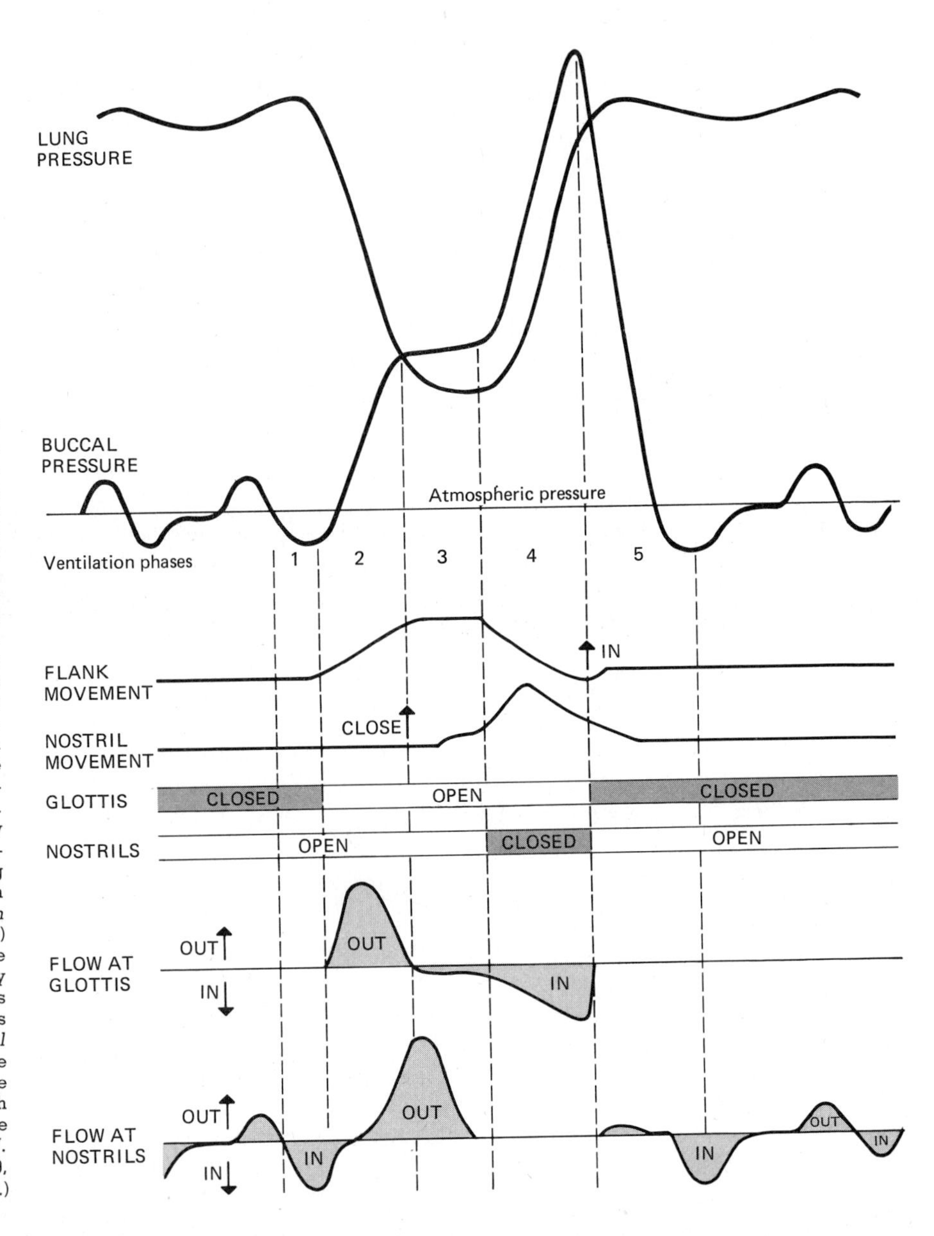

FIGURE 6-21
(*a*) Successive stages of airflow during the ventilatory cycle in *Rana catesbeiana*. The gas-sampling probe leading to the respiratory mass spectrometer (*A*) is supported by a plastic collar in the rubber mask (*B*). The arrows indicate the approximate sequence of gas flow into and out of the buccal cavity (*C*) and lung (*D*). Stage 2 shows how the pulmonary efflux apparently bypasses the inhaled gases which rest in the posterior portion of the buccal cavity. The glottis is closed in stages 1 and 4; it is open in stages 2 and 3. (*After C. Gans et al.*, 1969, *Science*, pp. 163, 1224.) (*b*) Diagram summarizing some of the mechanical events observed in respiration of *R. catesbeiana*. Oscillatory cycles, consisting of rhythmical raising and lowering of the floor of the mouth with open nostrils (*bottom line, extreme left and right*) introduce fresh air into the buccal cavity. Ventilatory cycles, one of which is shown with its five phases (*separated by vertical dashed lines*), involve opening and closing of the glottis and nostrils with renewal of a portion of the pulmonary gas. (*After H. J. deJongh and C. Gans*, 1969, *J. Morphol.*, pp. 127, 279.)

thoracic and abdominal cavities, is primarily responsible for ventilatory actions in many mammals, including humans. When the muscle of the diaphragm contracts, the mobile central portion of the sheet moves downward, much as a piston moves in a cylinder (Fig. 6-22).

The activities of the costal and diaphragmatic muscles cause suction pumping of the elastic lungs. There is a simple rhythmic cycle of pressure changes in the pleural cavities associated with inspiration and expiration. There is no closure of the glottis.

The lungs of mammals are highly complex, consisting of air-containing tubes, blood vessels, and elastic connective tissue. The *conducting portion* of the respiratory system is a series of highly branched hollow tubes which become smaller in diameter and more numerous at each branching. The smallest of these tubes end in tiny blind sacs, the *alveoli*, which are the sites of gas exchange with the blood (Fig. 6-23).

Control of Breathing in Humans

The diaphragm and intercostal muscles depend upon cyclic stimulation by nerves to maintain the rhythmic ventilatory cycle. The nerves control-

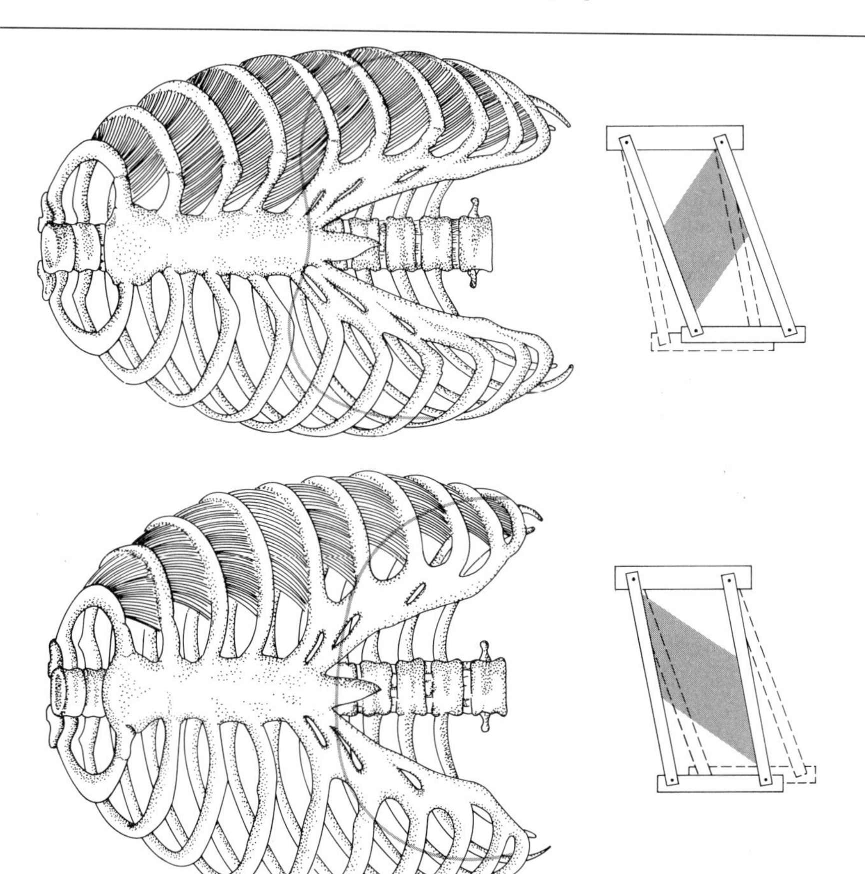

FIGURE 6-22
Mechanical events in ventilation of the mammalian lung. During inhalation the thorax expands (*top left*), principally because of contraction and flattening of the diaphragm (*line*) but also because of expansion by contraction of the external-intercostal muscles of the rib cage (*detail at bottom left*). Exhalation occurs when passive relaxation reduces the size of the thorax (*top right*), thereby raising the gas in the lungs to greater than atmospheric pressure. The internal intercostal muscles (*detail at bottom right*) aid in forced exhalation under conditions of stress.

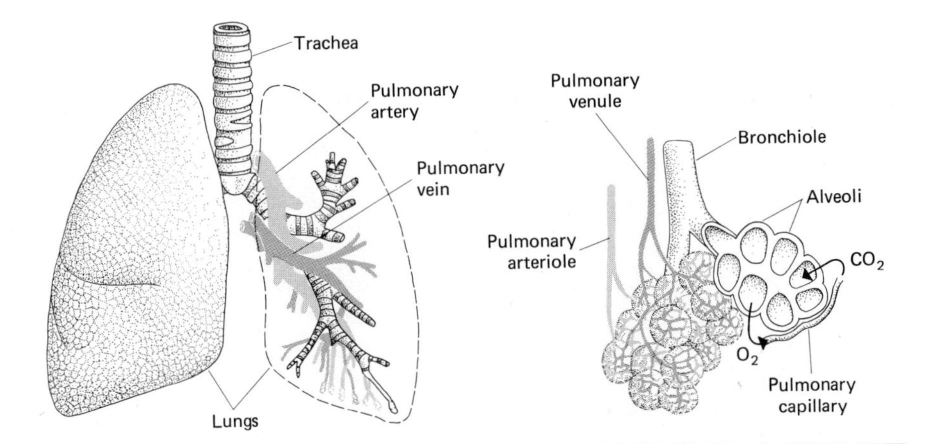

FIGURE 6-23
Structure of the mammalian lung, showing the relationship between respiratory airways and blood vessels (*After A. J. Vander et al.,* 1970, *Human Physiology, McGraw-Hill Book Company, New York,* p. 303.)

ling breathing movements originate in the spinal cord at the levels of the neck and thorax. Destruction of the phrenic nerve to the diaphragm and the intercostal nerves results in complete paralysis of the respiratory muscles and death, unless some form of artificial respiration is instituted.

Neural control of the respiratory muscles resides primarily in neurons with cell bodies in the lower portion of the brainstem, the medulla, which also houses the cardiovascular control centers. However, in addition to the respiratory system of the medulla, other controlling influences exist in the upper part of the spinal cord. The common textbook diagram showing inspiratory, pneumotaxic, and expiratory centers in the lower brainstem belies the actual complexity of the system that generates rhythmical breathing. It is, in truth, poorly understood.

We do know a fair amount about factors in mammals that influence breathing rate and depth. Experiments with humans demonstrate that there is a response to reduced plasma pO_2. However, it is essential that plasma pCO_2 be held constant in such experiments, because with reduced oxygen the ventilation rate will be increased and this will lower plasma pCO_2, which will inhibit ventilation and thereby counteract the stimulatory effect of reduced pO_2. The receptors stimulated by lowering oxygen are located at the bifurcation of the common carotid arteries and in the arch of the aorta, quite close to, but distinct from, the pressoreceptors mentioned in Sec. 6-2. These receptors, the *carotid* and *aortic bodies*, send afferent nerves into the medullary centers. Lowered dissolved oxygen in the plasma (not *total* blood oxygen, most of which is combined with hemoglobin) increases the rate of receptor firing. If the afferent nerves from the carotid and aortic bodies are severed, the usual increase in ventilation resulting from exposure to low pO_2 does not occur, demonstrating that the stimulatory effects of lowered oxygen are completely mediated by this pathway.

It is interesting that because the oxygen "detectors" respond only to oxygen in solution, carbon monoxide poisoning and anemia do not cause stimulation of ventilation. Carbon monoxide reacts with the same sites on the hemoglobin as oxygen and thus reduces the ability of oxygen to combine with hemoglobin. Since it does not affect the amount of oxygen which can physically dissolve in blood, the pO_2 is unaltered, the carotid and aortic bodies are not stimulated, and the victim can lose consciousness and die of oxygen lack without ever increasing ventilation. Similarly, the decreased hemoglobin content in the anemic condition does not affect blood pO_2. These two conditions are quite in contrast to those which do result in lowered plasma pO_2, such as lung disease or high altitudes, and therefore do result in respiratory stimulation.

It is the pCO_2 of plasma that plays the predominant role in the control of ventilation. Normally atmospheric air contains very little carbon dioxide (about 0.03 percent). An increase of just 5 mmHg in alveolar pCO_2 will cause a human to increase ventilatory rate by 100 percent. Conversely, if a subject is asked to breathe as rapidly and deeply as possible (hyperventilation) for a few minutes and then told to breathe naturally, there is often a period of 1 to 2 minutes when breathing stops completely (apnea). The reason is that during hyperventilation carbon dioxide is blown off faster than it is produced, plasma pCO_2 falls and ventilation ceases until metabolically produced CO_2 returns to normal. The normal value of arterial blood pCO_2 is 40 mmHg.

The effects of carbon dioxide on ventilation may be due not to carbon dioxide itself but to the associated changes in hydrogen-ion concentration. Indeed, it can be demonstrated that changes in hydrogen-ion concentration can alter ventilation when pCO_2 remains unchanged, or even changes in the opposite direction. CO_2 in solution in water forms carbonic acid, which in turn almost completely ionizes, resulting ultimately in bicarbonate and hydrogen ions, according to the following equation:

$$CO_2 + H_2O \rightleftharpoons H_2CO_3 \rightleftharpoons HCO_3^- + H^+$$

The critical hydrogen-ion concentration appears to be not that of the arterial blood but instead that of the cerebrospinal fluid—at least this is the currently popular hypothesis.

Although it may at first seem strange that the control of breathing is primarily responsive to brain hydrogen-ion concentration, it actually is quite appropriate. Since there is a fixed relationship between hydrogen ion and pCO_2, and there is a close relationship between oxygen consumption and CO_2 production, monitoring of hydrogen ion and regulating ventilation accordingly would serve to ensure adequate oxygen and a constant pCO_2. Furthermore, it is important that brain pH be carefully controlled as brain neurons are extremely sensitive to hydrogen-ion concentration and even small changes could induce serious malfunction.

The carotid and aortic bodies which are responsive to low oxygen also are sensitive to changes in plasma ion concentration, but their contribution to the overall ventilatory response to changes in carbon dioxide and hydrogen ion is quite small compared to that of the brain.

6-4 GAS TRANSPORT

Possibly more important than the development of efficient circulatory systems and respiratory organs for ensuring an adequate quantity and partial pressure of oxygen at the tissue level is the development of respiratory pigments. The presently recognized pigments of proved respiratory function may be grouped into these four classes: hemoglobins, hemocyanins, chlorocruorins, and hemerythrins. A fifth category, the *erythrocruorins*, was formerly recognized but because these porphyrin-based pigments of invertebrates are now known not to differ essentially from the hemoglobins of mammals this term is no longer used.

Respiratory Pigments

Hemoglobins are characteristic of vertebrates and, with the notable exception of some antarctic fishes and occasional specimens of adult amphibia (e.g., *Xenopus*), are present in the blood of all normal adults. Hemoglobins consist of an iron-porphyrin (heme) bound to a protein, globin. The protein moiety varies considerably in size, amino acid composition, solubility, and other physical properties from species to species, but all appear to have the same heme. They are invariably found in the blood corpuscles (erythrocytes) of vertebrates. In addition, hemoglobins, designated *myoglobins*, are found in muscle cells of mammals and birds and occasionally in teleosts and elasmobranchs.

Whole human blood can absorb about 75 times more O_2 than the same volume of solution without hemoglobin. Hemoglobin has the unique chemical property of combining reversibly with both O_2 and CO_2. In the lungs, where the partial pressure of O_2 may be as high as 100 mmHg, hemoglobin becomes 98 percent saturated. In the tissues, where O_2 partial pressure may be 24 to 40 mmHg, as much as 60 percent of the O_2 in the blood will be released. This ability of combining reversibly with oxygen is a first requisite of an oxygen-transport substance.

Respiratory pigments occur very randomly among invertebrates. In some taxonomic groups they are very common, whereas in others only a few genera or the members of only one species have them. Annelida are fairly consistent in having dispersed (noncellular) hemoglobin in the plasma of many families with closed circulatory systems, and in some cases erythrocytes are found in the coelomic fluid as well. Holothurians, echiuroids, and phoronids, which lack a closed circulatory system, have hemoglobin in corpuscles. Large quantities of dispersed hemoglobin also occur in the Pogonophora. In the remaining phyla the occurrence of hemoglobin is erratic. It is present in some crustacea, a few pelecypods, one gastropod, a few nemertines and nematodes, and in the larvae of a number of chironomid midges.

Why some animals have hemoglobins that are carried in solution and others have hemoglobins contained in blood corpuscles is an interesting problem. Animals with high hemoglobin concentrations always have blood corpuscles. Hemoglobins carried in solution have large molecular weights, whereas the hemoglobins located in corpuscles have low molecular weights. There are advantages to having low-molecular-weight hemoglobins within corpuscles: (1) Enclosing the hemoglobin in corpuscles appears to be a requisite for a reduction in molecular weight; high concentrations of hemoglobin will increase O_2 transport only if the molecular weight of the hemoglobin is low, because of the oxygen-binding properties of the molecule. (2) Freely dissolved molecules must have a large molecular weight if they are not to be lost via excretory structures; animals with low-molecular-weight hemoglobins do not have filtering excretory systems. (3) If large amounts of small-molecular-weight hemoglobin were to be in solution, the plasma osmotic pressure would be intolerably increased (threefold in mammals). Thus, the evolution of corpuscles appears to be a requisite for having an efficient, large-capacity, O_2-transporting system.

Chlorocruorins, which are closely related to hemoglobins, are found only in solution in the plasma and appear to be confined to four polychaete annelid families. Different species of the serpulid *Spirorbis* have chlorocruorin or hemoglobin or neither, and the genus *Serpula* has both in its blood.

Hemocyanins have not been found in corpuscles or in tissue cells. These blue, copper-containing pigments are next in importance to hemoglobins, as judged by their occurrence. They are found among mollusks (amphineurans, cephalopods, and some gastropods), arthropods (many malacostracans and crustaceans), *Limulus*, and a few arachnids. Many mollusks, e.g., the whelk *Busycon* and the amphineuran *Cryptochiton*, have hemocyanin in the blood and myoglobin in some muscles.

Hemerythrins are iron-containing pigments with a very restricted distribution. They are known in small groups in four unrelated phyla: *Magelona* (a polychaete worm), probably in most sipunculids, in some priapulids, and in two brachiopods. Hemerythrins, which are violet, are found in corpuscles; despite the name, the iron is not contained in a heme porphyrin.

All four classes of pigments are conjugated proteins (long polypeptide chains bearing a non–amino acid prosthetic group). The prosthetic group contains a metal, except in the case of hemocyanin.

The structural configuration of hemoglobin is rather well established, largely because of the x-ray diffraction studies of Kendrew and Perutz. In addition, the sequence of amino acids in the primary chains of a number of different forms of hemoglobin is known. Heme, the prosthetic group of hemoglobins, is a metalloporphyrin with an atom of iron at its center (Figs. 6-24 and 6-25).

One molecule of oxygen is reversibly bound to the single ferrous iron of hemoglobin by a coordinate bond on one side and to the polypeptide chain by the imidazole group of a histidine residue on the other side. A second coordinate bond links the iron atom to the polypeptide chain,

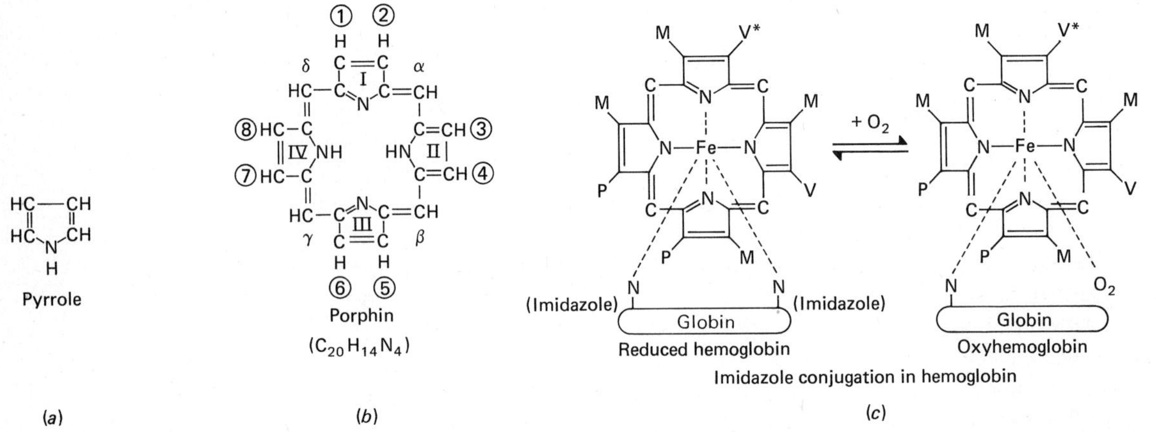

Pyrrole

(a)

Porphin

($C_{20}H_{14}N_4$)

(b)

Reduced hemoglobin

Oxyhemoglobin

Imidazole conjugation in hemoglobin

(c)

FIGURE 6-24

The porphyrin structure of hemoglobin (a) Pyrrole. (b) Skeletal structure of the porphin molecule (tetrapyrrole ring) with the conventional numbering of positions and rings. (c) Imidazole conjugation in hemoglobin. *M*, the methyl group; *V*, the vinyl group; *P*, the propionic group. In chlorocruorin, position 2 (marked* in hemoglobin) is filled by the formyl group.

also by an imidazole group of a histidine residue. When oxygen is gained and lost, the metal remains in the ferrous state. The electrons are merely redistributed, so in this system oxygenation and deoxygenation are not oxidation and reduction processes (Fig. 6-24).

Chlorocruorin differs from heme in that on one pyrrole ring a vinyl

FIGURE 6-25

Representations of the tertiary and quaternary structures of hemoglobin. Vertebrate hemoglobins are tetramers of polypeptide chains, represented here by layers of electron-density "contours" as determined by x-ray diffraction analysis (α chains, white; β chains, black). The aggregation is held together by hydrophobic interactions between unlike chains ($\alpha_1\,\beta_2$, $\alpha_1\,\beta_1$, $\alpha_2\,\beta_2$, $\alpha_2\,\beta_1$). Heme groups, seen here as gray discs in α_1 and α_2 only, lie in pockets facing outwards on the hydrophilic surface of the aggregate. Oxygenation of one heme group results in a subtle but complex conformational change in the corresponding chain, which in turn influences the conformation of the corresponding residues in the other chains. These stereochemical changes could account for the Bohr effect and for heme-heme interactions. (*After J. D. Jones, 1972, Comparative Physiology of Respiration, Edward Arnold, Edinburgh, and Crane Russak, New York, p. 93.*)

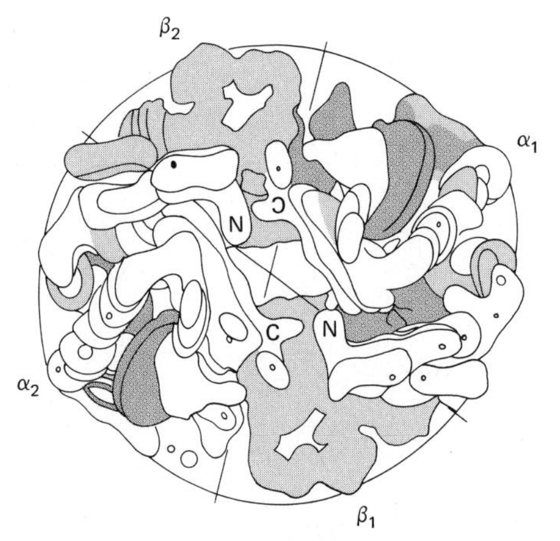

Hemoglobin

chain is replaced by a formyl (Fig. 6-24). The amino acid chains of chlorocruorin resemble in proportion those of invertebrate hemoglobins.

In hemocyanin the copper is bound to the protein and does not occur in a prosthetic group. The nature of the attachment residue is not known with certainty, but —SH groups are probably involved. It appears that one molecule of oxygen is bound by two copper atoms. Both are in the cuprous state before the attachment of oxygen, but in oxygenation one may lose an electron to become cupric.

Oxygenation is a function of the pO_2 with which the pigment is in equilibrium. However, the proportion of the total hemoglobin (or other pigment) molecules in the oxygenated (oxyhemoglobin) state (percentage saturation) is related to the pO_2 in a relatively complex manner, compared with the relation between pO_2 and dissolved oxygen concentration, which is linear. In the first place, the chances of an oxygen molecule's striking an unoccupied binding site on the pigment are increasingly less as the pO_2 increases. Thus, because an increasingly large number of oxygen molecules is required to effect each increment in the saturation of the pigment, an asymptotic curve must result. Secondly, in virtually all vertebrate bloods the functional hemoglobin molecule is an aggregate of four units (Fig. 6-25), each of which consists of a prosthetic group and polypeptide chain. The binding of oxygen at any one site may influence the ease of subsequent binding at other sites. Indeed, later bindings are facilitated by earlier ones. This influence, known as *heme-heme interactions*, causes the *binding curve* (or what is more commonly called the *dissociation curve*) to be steeper than it would be otherwise. The resulting degree of sigmoidness is a measure of the intensity of these reactions (Fig. 6-26).

The basic pattern of oxygen affinity of any respiratory pigment, however, is a species-specific property, determined by the influence of groups in the polypeptide chain upon the equilibrium constants of the reactions at individual binding sites. Hemoglobins differ greatly in their

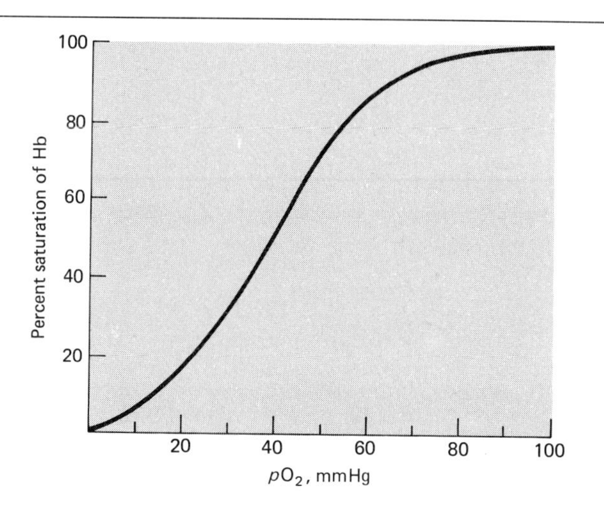

FIGURE 6-26
Oxygen-hemoglobin dissociation curve (whole human blood: pH, 7.4; 38 °C). Samples of normal blood are equilibrated with gas mixtures having various partial pressures of oxygen. The amount of oxygen per unit volume of blood is then determined and percent of saturation is calculated.

oxygen affinities, and bloods differ in the amounts of oxygen they carry when saturated (Fig. 6-27).

Though the basic oxygen affinity of any pigment, with or without the modifying effect of prosthetic group interactions, is genetically determined, there are additionally several normal environmental variables which often modify the relationship between percentage saturation and pO_2. Increase of *temperature* is almost universally found to displace the dissociation curve to the right. This is the simple consequence of the exothermic nature of oxygen-binding processes. Of much greater interest and functional significance is the influence of pCO_2 on oxygen affinity. Called the *Bohr effect* (after one of its discoverers), this is known to be a consequence of change in *pH*, as noted earlier. An increase of carbon dioxide or other acidic metabolites, such as lactic acid, in most species moves the oxygen-dissociation curve to the right. Free H^+ from the dissociation of carbonic acid or other acidic metabolite is bound at some site on the polypeptide chain, thereby reducing negative charge on the protein molecule and causing a change in the oxygen affinity of the binding site (Fig. 6-28).

The reciprocal of the Bohr effect is the facilitation of CO_2 loss from the blood on oxygenation and facilitation of CO_2 uptake in tissues on deoxygenation. This is known as the *Haldane effect*. It results from the fact that oxyhemoglobin is a stronger acid than deoxygenated hemoglobin. Thus, as O_2 is given off in the tissues, the unloaded hemoglobin becomes a better buffer, and in the lungs, as O_2 is taken up, there is an increase of negative charge and equivalent displacement of HCO_3^-, thus facilitating CO_2 loss. In the capillaries, particularly those in active tissues such as contracting muscles, the blood tends to give up its oxygen. Because the pO_2 in the tissues is lower than the pO_2 in plasma and the red

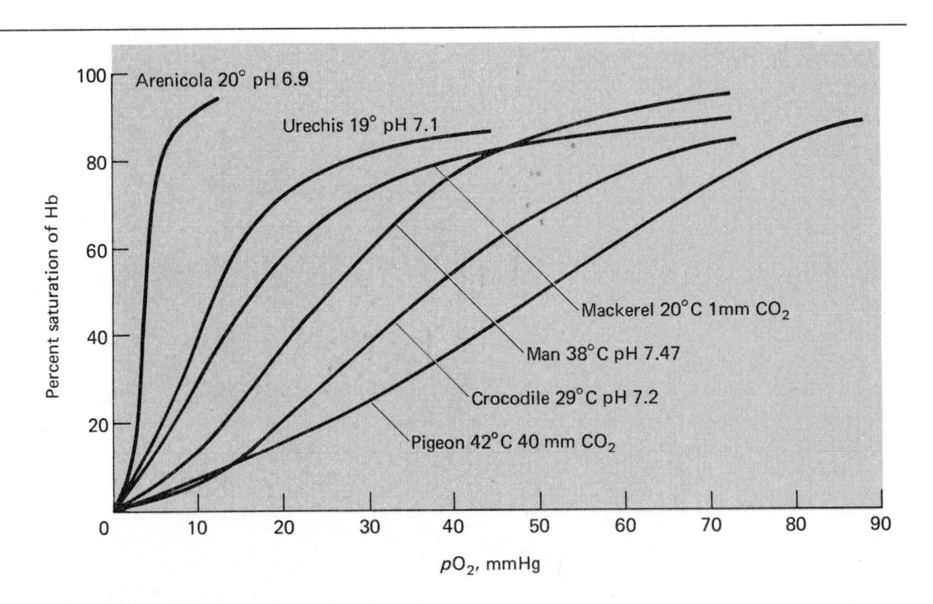

FIGURE 6-27 Oxygen equilibrium curves in percent saturation of hemoglobin as a function of oxygen pressure in millimeters of mercury in a variety of animals. (*After C. L. Prosser, 1973, Comparative Animal Physiology, W. B. Saunders Company, Philadelphia, p. 339.*)

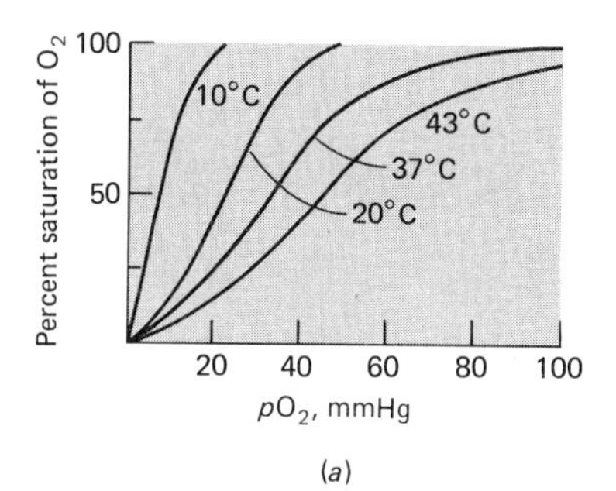

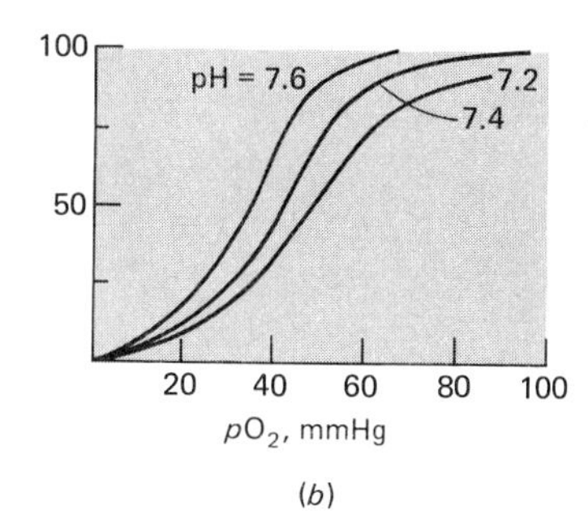

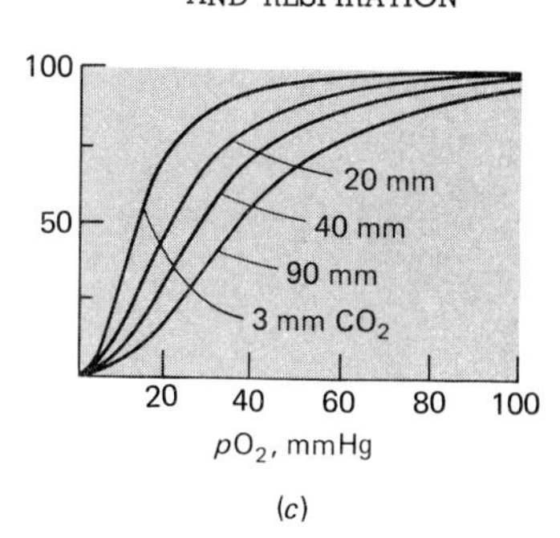

FIGURE 6-28
Oxygen-hemoglobin curves, showing effect of (a) temperature, (b) pH and (c) pCO_2.

cells, oxygen moves according to this pressure gradient, and dissociation of oxygen and hemoglobin occurs.

The circulation rate in humans can locally increase by about 10 times during severe muscular effort. Additionally, a redistribution of blood takes place, and a larger portion of the cardiac output flows through muscles. Not only is there a faster flow of blood, but more capillary networks open up as well. Thus, the blood flow in active muscles may be as much as 30 times that when they are at rest. Furthermore, the coefficient of oxygen utilization (the fraction of the blood's total oxygen content that is given up in passage through a capillary bed) rises during exercise, so even though the blood may spend a shorter time in a given capillary channel, it actually unloads more oxygen.

In contrast to oxygen, which is almost entirely transported in association with hemoglobin, only between 2 and 10 percent of the carbon dioxide transported by the blood of humans is bound in a similar loose chemical combination with hemoglobin. A still smaller amount combines with plasma protein to form carbamino compounds. Approximately 5 percent is carried in simple solution in plasma. The remaining 85 to 90 percent is carried as bicarbonate ion.

Because tissue pCO_2 is higher than arterial pCO_2, carbon dioxide diffuses from the tissues into the arterial blood. Some carbon dioxide dissolves in the plasma, a very small amount of which reacts slowly to form carbonic acid and then dissociates:

$$H_2O + CO_2 \rightleftharpoons H_2CO_3 \rightleftharpoons H^+ + HCO_3^-$$

Most of the carbon dioxide carried by the blood passes into erythrocytes, some of it remaining in solution and some of it combining with hemoglobin to form carbamino compounds, as follows:

$$RNH_2 + CO_2 \longrightarrow RNHCOO^- + H^+$$

This is a rapid, uncatalyzed reaction. The remainder of CO_2 in the cell is rapidly combined with water, through the influence of the enzyme *carbonic anhydrase* to form H_2CO_3 and then $HCO_3^- + H^+$. Only about 0.1 percent of absorbed carbon dioxide is in the form of carbonic acid (Fig. 6-29).

The blood phosphates play a minor part in buffering the blood against changes of pH. However, the blood proteins are much more important. These molecules are amphoteric, and since blood pH is normally on the alkaline side of the isoelectric point, the molecules act as proteinic acids. Also, as just noted, CO_2 can make direct combinations with free amino groups on the blood proteins, forming carbamino compounds.

The HCO_3^- formed inside the red cell diffuses into the plasma. The membrane of the red cell is not readily permeated by cations, and anions from the plasma diffuse in and preserve electrical neutrality. The major ion that replaces bicarbonate is Cl^-. The exchange is called the *Hamburger interchange*, after the Dutch physiologist who discovered it. To maintain osmotic balance, water also diffuses into the red cells, with a consequent slight swelling in venous blood; this is reversed in the lungs.

In humans the blood is exposed in the lungs to oxygen at partial pressures of approximately 100 mmHg (Fig. 6-29). When the blood leaves the lungs it carries 19 vol/100 ml of O_2 at 80 mmHg, and the hemoglobin is 98 percent saturated. In the capillaries, the blood passes through tissues where the O_2 pressure is 30 mmHg or less (Fig. 6-29). Here 25 to 30 percent of the O_2 is unloaded.

FIGURE 6-29

Exchange of respiratory gases in tissues. Hb = hemoglobin. CO_2 diffuses from the cells, through the interstitial fluid and capillary wall into the plasma. Some of the CO_2 reacts slowly with H_2O in the plasma to form H_2CO_3, which in turn ionizes and liberates H^+. Most of the CO_2 diffuses into the red corpuscles, where it combines with water as it does in the plasma, but the reaction (*) is rapid because it is catalyzed by carbonic anhydrase. The H^+ that is ultimately formed is taken up by HbO_2^- to form HHb and O_2. Some of the CO_2 combines with various forms of hemoglobin, the most important reaction being with HbO_2, for it releases O_2. CO_2 combines with amino groups of hemoglobin to form carbamino hemoglobin (HbO_2^-). O_2 diffuses out of the red cell into the tissues. The excess HCO_3^- that diffuses out of the cell into the plasma is electrically counterbalanced by Cl^- moving into the cell.

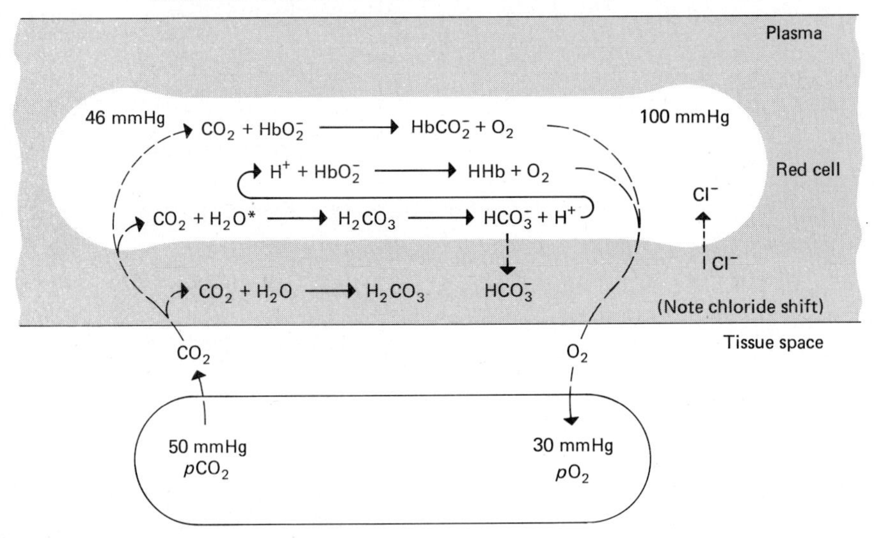

Myoglobin (muscle hemoglobin) has a greater affinity for O_2 than erythrocyte Hb. Consequently, it can transfer oxygen from blood to cell enzymes. This points to another important function of hemoglobin: facilitation of O_2 diffusion. In experimental situations it can be demonstrated that the rate of diffusion of O_2 is several hundred times greater in the presence of the pigment.

Most respiratory transport pigments become saturated at lower partial pressures of oxygen than that present in air at sea level. The most adaptive of the differences among the hemoglobins of different animals are the partial pressures at which they load and unload oxygen (Fig. 6-27). Unloading pressures set the upper limit of tissue oxygen pressure and the lower limit of environmental oxygen for function of the pigment. These pressures determine the range of usefulness of the pigments. From *oxygen equilibrium* curves, the affinity can best be expressed as the P_{50}, the pressure for *half-saturation*. As is true for the Bohr effect, elevating temperature, lowering pH, or increasing pCO_2 shifts the P_{50} upward; i.e., high P_{50} means low affinity and vice versa.

Mammals that make quick movements (mouse, cat) tend to have higher P_{50} values than animals that are slow and steady (dog). The P_{50} decreases with increasing body size, and larger animals tend to have a smaller Bohr effect than do smaller ones. Burrowing animals (prairie dog) have lower P_{50} values than do surface or arboreal mammals. Both deep-diving and high-altitude mammals have high affinities.

In general, hemoglobins of birds require higher O_2 pressures than do hemoglobins of mammals to reach saturation. Amphibians that spend much of their time on land have oxygen equilibrium curves to the right of those of aquatic amphibian species. In the adult bullfrog there is a pronounced effect of pH on oxygen affinity, but there is none in the tadpole. Fishes that live in relatively stagnant water have a low P_{50}, and the Bohr effect does not shift the dissociation curve out of the useful range. Fishes (such as trout) with a high P_{50}, on the other hand, are so responsive to CO_2 that the curve may move so far to the right that the fish suffocates in ample O_2; i.e., CO_2 which favors O_2 unloading in the tissues can prevent loading in the gills.

Many air-breathing animals survive prolonged periods of submergence. Some turtles are able to survive 12 hours of submersion. The capacities of some mammals to stay submerged are compared in Table 6-1.

Diving Vertebrates

Diving vertebrates have developed physiological adjustments that are not unique adaptations but rather a perfection of a general defense mechanism found in all vertebrates, with one exception. That exception is the defense against the effects of decompression.

The main O_2 stores of the body are the blood, the muscle pigments, and the lungs. Diving animals have, relative to their body weight, about twice the blood volume of nondivers. The O_2 capacity of their blood is generally also higher. Not only do divers tend to have high blood O_2 affinity; most of them also have a more pronounced Bohr effect than nondivers. Commonly they have high myoglobin concentrations. However, the lung volume of divers is not significantly different from that of nondivers, and, in fact, may be rather small in deep-diving species. This may be advantageous, as large lungs would constitute a cumbersome float during the dive. Divers do have a large tidal volume (volume of one breath) relative to lung volume. This permits a quick renewal of gas between dives.

Preparatory to a dive seals expire rather than inspire. This is thought to relate to defense against decompression sickness (formation of gas bubbles in blood and tissues during decompression). During a dive the lungs of whales are compressed; at 100-m depth the lung volume is at most one-tenth of what it was at the surface. The lung endothelium shrinks and thickens, and supersaturation of the blood is slowed. The total effect is to create an anatomical dead space where diffusion of gas into the blood is reduced, provided the descent is rapid.

The cardiovascular reactions during diving are characterized by reduced heart rate (*bradycardia*) and redistribution of blood. A reduction to one-tenth of normal heart rate during a dive is common. Because resistance to blood flow is increased in proportion to the decreased cardiac output, blood pressure is not seriously reduced.

Table 6-1
Duration and Depth of Diving of Mammals

Species	Duration, min	Depth, m
Platypus	10	
Mink, *Mustela vison*	3	
Harbor seal, *Phoca vitulina*	20	
Walrus, *Odobenus rosmarus*	10	80
Steller's sea lion, *Eumetopias jubata*		146
Gray seal, *Halichoerus grypus*	20	100
Weddell seal, *Leptonychotes weddelli*	43	600*
Bottle-nosed whale, *Hyperoodon ampullatum*	120	Deep
Sperm whale, *Physeter catodon*	75	900
Blue whale, *Sibbaldus musculus*	49	100
Harbor porpoise, *Phocaena phocaena*	12	20
Bottle-nosed porpoise, *Tursiops truncatus*	5	
Beaver, *Castor canadensis*	15	
Muskrat, *Ondatra zibethicus*	12	
Most humans	1	
Experienced skin divers	2.5	161†

* From G. L. Kooyman, 1966, *Science*, **151**, 1553–1554.

† Greek sponge diver. Stotti Georghios (1913).

Source: L. Irving, 1964, In W. O. Fenn and H. Rahn (eds.), *Handbook of Physiology*, sec. 3, vol. 1, American Physiology Society, Washington, D.C.; and M. S. Gordon et al., 1972, *Animal Physiology*, The Macmillan Company, New York.

Bradycardia is a vagal reflex; sectioning of the vagi or administration of atropine (which blocks the action of acetylcholine, the transmitter of the vagi) eliminates this response.

During a dive the large skeletal muscles, such as the pectorals and the gastrocnemius, receive reduced blood supply. There is commonly renal, splanchnic, and cutaneous vasoconstriction. Peripheral vasoconstriction is accomplished by constriction of the larger supply arteries through sympathetic innervation. On the other hand, the heart and brain have supernormal flows. These organs obtain energy by aerobic metabolism. The others must rely largely upon anaerobic processes. Consequently, lactic acid concentration increases during the dive and pH values of arterial blood below 7 are not unusual. Thus, even the heart and brain may be exposed to severe acidosis during diving.

The noncirculated organs accumulate metabolites during submersion, and upon surfacing there is a large increase in the lactic acid content of the blood. The total energy production during the dive is much reduced, in keeping with the strict economy of using limited O_2 stores. Some divers (seals and ducks), if not all, are less sensitive to arterial pCO_2 than are nondivers.

Some interesting developments relating to respiration have resulted from attempts to find substitutes for blood plasma that can also provide adequate oxygen transport. Certain perfluorochemicals—organic compounds in which all hydrogens have been replaced with fluorine—can transport oxygen and, in conjunction with a simulated blood plasma, perform many functions of whole blood. Partial or complete replacement of blood with these compounds has been achieved temporarily in rodents. However, the propensity of perfluorochemicals to accumulate in body tissues, with unpredictable effects, will probably continue to make their use in humans unsuitable.

Oxygen is highly soluble in liquid perfluorochemicals, a fact dramatically demonstrated by Leland Clark at the University of Cincinnati in 1966 when he submerged mice in inert liquid perfluorochemicals for extended periods. The animals were able to obtain sufficient oxygen by breathing the liquid, and, upon removal, showed no apparent ill effects from the experience. Clark also has shown that breathing such liquids can protect from the effect of rapid decompression and suggests that they would be useful for such applications as escape from submarines and deep-sea diving.

READINGS

Burton, A. C., 1972, *Physiology and Biophysics of Circulation*, 2d ed., Year Book Medical Publishers, Inc., Chicago. This is a fairly detailed account of mammalian, principally human, hemodynamics and cardiovascular physiology.

Chapman, G., 1967, *The Body Fluids and Their Function*, St. Martin's Press, Inc., New York. This booklet provides an outline of the main features of circulating body fluids.

Jones, J. D., 1972, *Comparative Physiology of Respiration*, Edward Arnold (Publishers) Ltd., Edinburgh, England. This excellent monograph deals with respiratory organs, respiratory pigments, and gas transport.

Prosser, C. L., 1973, *Comparative Animal Physiology*, 3d ed., W. B. Saunders Company, Philadelphia. Chapters 5 and 8 are very useful sources of information on such subjects as environmental adaptations for respiration, diving animals, and the function of respiratory pigments.

Steen, J. B., 1971, *Comparative Physiology of Respiratory Mechanisms*. Academic Press, Inc., New York. Essential features of the respiratory mechanisms of animals are summarized in this small monograph.

Vander, A. J., et al., 1970, *Human Physiology*, McGraw-Hill Book Company, New York. Chapter 10 on circulation and Chap. 11 on respiration provide good treatments of these functions in mammals.

OSMO-REGULATION, IONIC REGULATION, AND EXCRETION

Water makes up 50 to 95 percent of the total mass of animals. It is the medium in which the metabolic reactions of cells occur, it surrounds and supports cells and organs, and fertilization can take place only in water. For aquatic organisms, it is the major component of their ambient environment. Water profoundly shapes the basic nature of life. In Chap. 4, the thermal properties of water were related to the living state; here we take into account water's solvent and ionizing aspects, for they explain much about the physical basis of life.

Dissolved or suspended in the watery fluids surrounding and invading the parts of organisms are the various ions, molecules, and particles which pass into and out of the living machinery. Some of the mechanisms that maintain the body and cellular fluids in steady-state compositions are the central subjects of this chapter. Additive processes related to acquisition of nutrients and respiratory exchanges have been considered in the two previous chapters. Now the devices that maintain ionic composition and extract metabolic wastes from body fluids will be featured.

7-1 PROPERTIES OF SOLUTIONS

As a solvent, water is without parallel. This fact and its importance to life have been eloquently argued by L. J. Henderson in his *Fitness of the*

Environment.[1] Most inorganic substances—acids, bases, salts—are exceedingly soluble in water. Water also is a remarkably good solvent for most organic compounds, as well as for oxygen and carbon dioxide. Only nonpolar compounds, such as fats, do not dissolve in water.

Water is the best solvent largely because it has a *dielectric constant* higher than most substances. Such a value is a measure of the amount by which a substance decreases the electrostatic forces between two charged bodies. Electrolytes readily ionize in water because water molecules have an unequal charge distribution and are, therefore, electric dipoles, or polar molecules. The four valence electrons shared with the two hydrogens are nearer to the oxygen than the hydrogens, leaving the region around the latter with a net positive charge. The two other pairs in the eight-electron shell of oxygen are directed more outward from the O—H bond. Thus, the water molecule is electrically asymmetrical, and when a salt is dissolved "shells" of water form around the ions and tend to keep them apart (Fig. 7-1).

Furthermore, *hydrogen bonds* form between the water molecules and tend to stabilize shells around and between solute molecules. It is believed that formation and breakage of hydrogen bonds are responsible for most of the unique thermal and solvent properties of water. The negative region around the oxygen is attracted to a net positive hydrogen. A group of water molecules, therefore, can form a tetrahedral[2] configuration about the oxygen atom (Fig. 7-1).

It is conjectured that the molecules of liquid water are held together in groups of a few to several molecules by hydrogen bonds. These bonds are dynamic, breaking and reforming continuously. The existence of hydrogen-bonded clusters gives liquid water a partially crystalline character. Lowering temperature increases the number and size of such clusters until, at the freezing point, the most stable state is that of a continuous crystal, or ice. Raising the temperature breaks hydrogen bonds, but even at 100°C the high heat of vaporization of water suggests some quasicrystalline structure. Hydrogen bonds are believed to be quite important in determining biological structure, particularly that related to proteins. Though the energy of formation or breakage of hydrogen bonds is relatively low (4.5 kcal/mol of bonds), as compared to covalent bonds (110 kcal/mol of bonds), it is sufficient to determine configuration and influence solubilities.

Hydrogen bonding of water accounts for its *high surface tension*—the highest for any natural liquid. This property, combined with the excellent *wetting property* of water, ensures that it will move readily through capillary spaces and pores, such as in soils. Therefore, water is rapidly and evenly distributed.

Even *ionization* of water is unique: it is the only species of molecule

[1] Henderson, L. J., 1913, *The Fitness of the Environment*, The Macmillan Company, New York; reissued in 1958 by Beacon Press, Boston.

[2] The angle of the bonds is 105° instead of 109° 29″ as in a perfect tetrahedron.

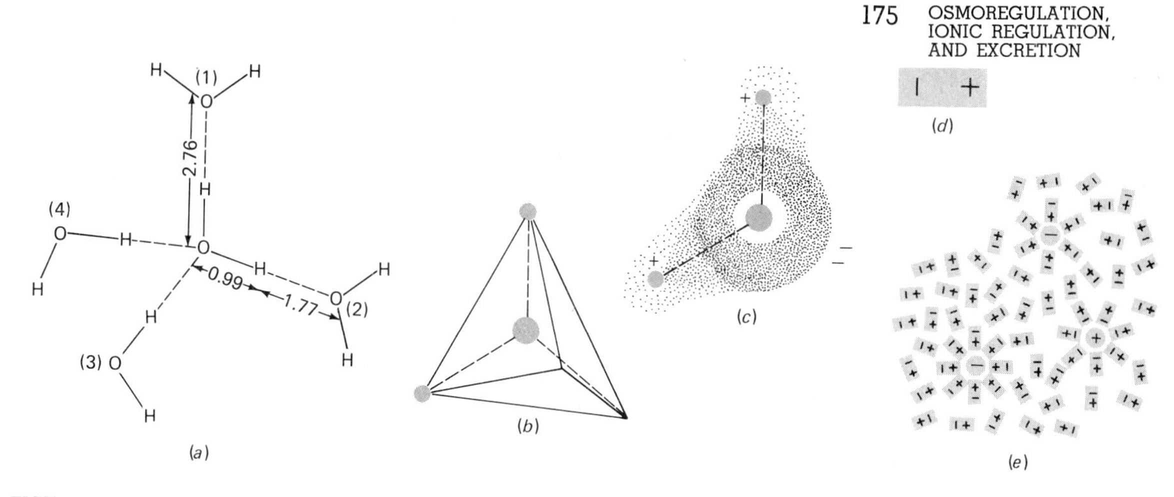

FIGURE 7-1
(a) Tetrahedral coordination of water molecules in ice: molecules (1) and (2), as well as the central H_2O molecule, lie entirely in the plane of the paper; molecule (3) lies above this plane, and molecule (4) below it. Solid lines, covalent bonds; dashed lines hydrogen bonds. (b) A tetrahedron described about the oxygen atom of water, with the H nuclei at two of the corners; the other two corners define the direction of regions of high-electron-density distribution, due to the unshared electrons (centers of negative charge). (c) The polar nature of water symbolized by the density of the "electron cloud." (d) An abstraction of water as an electric dipole. (e) Symbolic representation of the hydration of ions through formation of "shells" of polar water molecules.

that produces equal amounts of hydrogen[1] and hydroxyl ions. It is, therefore, a neutral substance. Short-lived hydronium ions are formed when hydrogen protons "hop" into an association with the negative oxygen atom. Hydrogens appear to move back and forth in their relationships with oxygen, in a kind of resonant state (Fig. 7-2).

The state of water in cells is not well understood. Nuclear magnetic resonance measurements indicate that more of the water in tissues is in a "crystalline" structure than in a liquid form. Free or solvent water, as contrasted with bound water or water of hydration, varies according to protein content. The amount of solvent water also varies according to whether inorganic ions or nonelectrolytes serve as solutes, in part because organic solutes may form multiple hydrogen bonds.

Thus, the water content of a solution, either a biological fluid or a physical one, provides no direct indication of the effective water activity, in a thermodynamic sense. However, for simple considerations, biological fluids are commonly regarded as dilute solutions. The total effective concentration of all solutes present, or the *osmotic concentration*, is often expressed in *osmoles*—the total number of moles of solute per liter of solvent. The osmolal concentration of a solution can be determined by its *colligative* properties. In fact, any one of the colligative properties of a solution may be calculated from any of the others. The higher the concentration of solute, the greater is the osmotic pressure; the higher the boiling

[1] The term "hydrogen ion" is a convention for referring to what is actually the hydronium ion, H_3O^+, in the case of water. The hydrogen, H^+, is actually hydrated by being combined with a molecule of water.

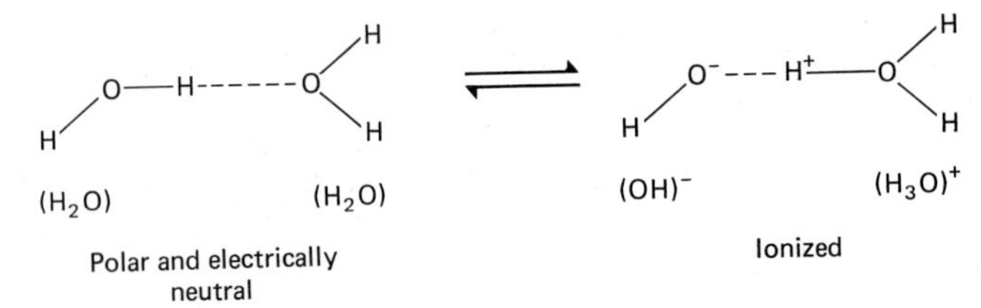

$H-O-H \cdots O(H)(H) \rightleftharpoons H-O^{-} \cdots H^{+}-O(H)(H)$

(H_2O) (H_2O) $(OH)^-$ $(H_3O)^+$

Polar and electrically neutral

Ionized

FIGURE 7-2

Mechanism for the ionization of water. The proton at intervals of a very small fraction of a second may "hop" from one oxygen to its neighbor. In a kilogram of water at 25°, which contains 55.5 mol of H_2O, there are 10^{-7} mol of OH^- and 10^{-7} mol of H_3O^+. Since by definition $pH = -\log(H^+)$, the pH of pure water is 7. Dashed lines: hydrogen bonds; solid lines: covalent bonds.

point, the lower is the vapor pressure, and the larger is the depression of the freezing point of a solution, and so on.

Osmotic pressure, that pressure which would be necessary to prevent net flux of water across a semipermeable membrane (one which permits only solvent to pass), is not a very useful biological concept, because strictly semipermeable membranes rarely if ever exist in organisms. Certain solutes pass all membranes that have been examined, and no situation is known where pure water occurs within organisms.

Surprisingly, biologists were responsible for much of the early theory of solutions. From studies in the late 1800s of plant cell *plasmolysis* (shrinking of the cytoplasm of a cell due to loss of water; discussed in Chap. 14) came van't Hoff's theory that dilute solutions behave like gases. It had been demonstrated for gases that:

$$P = \frac{n}{V} RT$$

where P = pressure
n = number of moles of gas
V = volume
R = the universal gas constant
T = absolute temperature

By analogy, the osmotic pressure (π) of solutions should equal the osmolal concentration (C) times the gas constant ($R = 0.082$ at 1 atm pressure) and the absolute temperature (T). Thus:

$$\pi = CRT$$

However, the botanist DeVries soon realized that solutions of salts produced more osmotic pressure than did equimolar solutions of sugars. From this finding, Arrhenius formulated the theory of electrolytes, thus modifying the van't Hoff equation to:

$$\pi = iCRT$$

where i is an isotonic[1] coefficient. (Despite the fact that weak electrolytes in dilute solution are completely dissociated, i is less than 2 for univalent

[1] *Isotonic* (Gr. *isos*, equal; *tonikos*, tension) means having the same or equal osmotic pressure.

salts and less than 3 for salts that form three ions because of ionic interactions; *i* values must be determined empirically.)

Because determination of freezing point is relatively easy to do, osmotic concentrations are commonly calculated from the freezing point of a solution. A 1-osmolal aqueous solution freezes at $-1.86°C$; hence, the depression of the freezing point of water by a solute is :

$$\Delta_{fp} = -1.86\, iC$$

and since a 1-osmole solution has an osmotic pressure (π) of 22.4 atm,[1]

$$\pi = \Delta_{fp} \frac{22.4}{1.86} = 12.06\, \Delta_{fp}$$

(The symbol Δ means "change of," or in this instance lowering of the freezing point.)

Solutes move under the influence of several forces. Two of the most important, for biological considerations, are *chemical potential* (concentration) and, if the solute bears a charge, *electrical potential*. The movement of solutes in biological systems is strongly modified by membranes, with their barrier characteristics which restrict and modify the diffusion of solutes. These complications are given more consideration in Chap. 14.

7-2 PHYSIOLOGICAL SALINES

Physiological salines are water solutions whose ionic concentration and osmotic pressure allow cells to temporarily survive in them. Their composition is instructive, especially in that they empirically define minimal requirements for the media that surround cells in vivo. However, they are at best only a temporary substitute for the natural body fluids. Normally animal tissues are bathed by blood or an ultrafiltrate of blood and, thus, the most appropriate composition for a physiological saline would seem to be one based on analyses of blood composition. Indeed, some of the most satisfactory and recently developed salines are designed accordingly, but allowances usually have to be made for the absence of charge-modifying polyvalent proteins present in blood.

In his classical studies, Sydney Ringer, Professor of Medicine at University College, London,[2] demonstrated that a bathing solution in which the concentration or ratios of the constituent ions were grossly incorrect had immediate deleterious effects on a tissue. When the balance of ions was appropriate, an organ might survive for hours or days in the solution. By empirical experimentation, Ringer established the foundation for in vitro culture of cells and organs. He defined the essential requirements for physiological salines and provided some insight into ion antagonisms.

The biologically important properties of ions are those which depend upon the atomic number, valency, and tendency to form complexes

[1] At 0°C the volume of 1 mol of an ideal gas is 0.082×273 (R × °K) = 22.4 liters.

[2] S. Ringer, 1882, *J. Physiol.*, **3**, 380; 1883, **4**, 29, 222; 1895, **18**, 425.

with water and organic molecules. Lockwood,[1] in his comprehensive review, points out that these properties determine such diverse ion functions as (1) the stabilization of proteins in solution and regulation of their colloidal properties, (2) the activation of enzyme systems, (3) the development of electrical excitability, (4) the dynamic maintenance of isotonicity between cells and body fluids, and (5) the control of the permeability of membranes. Maintenance of isotonicity between cells and body fluids will be discussed in this chapter; the other functions of ions will be dealt with later.

A brief consideration of certain specific functions of some of the more abundant ions that are found in biological fluids and in cells will be useful in understanding the composition of salines. *Potassium* is a major cation of cells. Some intracellular K is presumably ionized, since it can be shown to migrate under the influence of an electric field. As a free ion it would contribute to the osmotic pressure of the cell. Much of it, however, is bound in certain cells to organic components. Many enzyme systems are activated by K, a function presumably requiring the binding of some of the ion with enzymes or substrates. Potassium plays a unique role in being the major contributor of the resting potential in most electrogenic cells (e.g., neurons and muscle cells).

Sodium is the principal cation of extracellular fluids. The depolarization of the cell membrane of electrogenic cells (e.g., as in the passage of an action potential along a neuron membrane) is intimately associated with changes in Na conductance across the membrane. Replacement of Na in the extracellular fluid, while maintaining osmotic pressure constant by substituting other particles, results in a loss of membrane excitability.

Calcium is somehow very important in regulating cell-membrane permeability. Calcium-deficient media cause muscles and nerves to develop unstable membrane potentials and eventually to lose excitability. Ca is also associated with the activation of the contractile filaments of muscle cells. In the absence of Ca, cell masses tend to separate. Ca antagonizes magnesium in several situations. For example, Mg decreases the release of acetylcholine at nerve endings, whereas Ca increases the release.

Magnesium behaves in a manner similar to that of Ca in many reactions. Like Ca, Mg alters cell permeability, and it is also effective in maintaining intracellular cement. Mg decreases activity in contractile tissues, and its absence from the bathing medium results in hypersensitivity. It activates a number of enzyme systems.

The anions *phosphate* and *bicarbonate* provide buffering capacity in the blood. In contrast to the alkali metal cations, few specific roles are known for these ions. HPO_4 is intimately related to the nucleoside phosphate energetics of cells. If HCO_3 is absent from the saline bathing tissue slices, they soon lose their enzymatic activities. CO_2 and HCO_3 are essential to the functioning of most ventilatory mechanisms of respiratory systems.

[1] A. P. M. Lockwood, 1961, *Comp. Biochem. Physiol.*, **2**, 241.

Chloride is the principal anion of extracellular fluids. In concert with K, Cl plays a role in the development of resting potential in electrogenic cells. Cl activates at least one enzyme, amylase.

Hydrogen-ion concentration must be closely regulated in physiological saline, as the tissues of many animals are markedly affected by deviations of only 0.1 to 0.2 pH units from normal body fluid values. The pH of the bathing medium can affect rate of ion transport, colloidal osmotic pressure, enzymatic activity, imbibition of water by gels, and the chemical properties of proteins, since they are amphoteric molecules. The pH of the bloods of most animals falls within the range of 6.8 to 7.6.

Ringer's 1883 physiological saline (Table 7-1) was developed for the isolated frog heart. It is a balanced concentration of the chlorides of Na, K, and Ca, with a small amount of $NaHCO_3$. Locke (1901) modified Ringer's solution for use with mammalian tissues by raising the osmotic concentration and adding glucose for energy support (Table 7-1). Tyrode (1910) made further modifications by adding magnesium and phosphate for use with the mammalian gut (Table 7-1). Tyrode's solution works well with bird tissues also.

Salines have been designed specifically for only a very few heterothermic vertebrates and invertebrates, although the ionic composition of the blood of several species is known and artificial solutions could be concocted accordingly.

Insects present a particularly interesting case in that the composition of their body fluid (called **hemolymph,** since it combines the functions of the vertebrate blood and lymph) has the greatest variation found in the animal kingdom. Their tissues also are commonly very tolerant of quite wide ranges in concentration of ions. A large proportion of the total osmotic pressure of holometabolous insects is contributed by sugars and amino acids, the latter constituent being much higher than is found in other classes of animals. Salines based only on ion composition of insect hemolymph tend to have too low an osmotic pressure.

In the use of salines for maintaining living tissues in vitro, one must not only attend to their initial chemical composition, but also have regard for their pH stability, adequacy of metabolic support, aeration and, particularly if the tissue is from a homeotherm, the appropriateness of the temperature.

Table 7-1
Composition of Some Physiological Salines, g/l

Authority	NaCl	KCl	$CaCl_2$	$MgCl_2$	$MgSO_4$	$NaHCO_3$	NaH_2PO_4	KH_2PO_4	Glucose
Ringer (1883) (frog)	6.5	0.14	0.12			0.2			
Locke (1901) (mammal)	9.0	0.42	0.24			0.2			1.0
Tyrode (1910) (bird and mammal)	8.0	0.20	0.20	0.05		1.0	0.04		1.0
Krebs and Henseleit (1932) (mammal)	6.92	0.35	0.28		0.15	2.1		0.16	2.0

7-3 COMPOSITION OF BODY FLUID

Seawater contains about 3.5 percent salt. The major ions are sodium and chloride. Magnesium, sulfate, calcium, potassium, and bicarbonate constitute most of the remaining salt content. Although it varies somewhat from place to place and time to time, depending upon rate of evaporation, inflow from freshwater streams, rainfall, and other factors, seawater remains rather constant in its composition (Table 7-2).

Freshwater has a much more dilute concentration than seawater and also has a more variable solute concentration. Depending upon what materials the water passes over or through, it may contain virtually no solutes ("soft" water) or it may have dissolved in it relatively large amounts of solutes ("hard" water, particularly if the solutes include calcium or magnesium salts). The composition of various types of freshwater is given in Table 7-3.

Brackish water occurs in coastal regions where seawater is mixed with freshwater. Estuarine zones at the mouths of rivers that empty into the sea have greatly variable salinities, often from nearly fresh to almost undiluted seawater as the tides ebb and flow. Although it is not easy to define exactly the geographical boundaries between seawater and brackish water or between freshwater and brackish water, because of a continuous gradient with barely perceptible changes, a commonly accepted way of defining brackish water is to say that salinities between 3.0 and 0.05 percent are brackish. Brackish water is a physiologically challenging environment, forming a barrier to the distribution of both marine animals on one side and of freshwater animals on the other, while at the same time forming a gradual transition between marine and freshwater habitats. However, it covers less than 1 percent of the earth's surface.

The simplest forms of life are those small organisms that exist in seawater. They exchange materials directly with the surrounding

Table 7-2
Composition of Seawater*

	In 1 liter seawater		In 1 kg H_2O	
Ion	(mmol)	(g)	(mmol)	(g)
Sodium	470.20	10.813	475.40	10.933
Magnesium	53.57	1.303	54.17	1.317
Calcium	10.23	0.410	10.34	0.414
Potassium	9.96	0.389	10.07	0.394
Chloride	548.30	19.440	554.40	19.658
Sulfate	28.25	2.713	28.56	2.744
Bicarbonate	2.34	0.143	2.37	0.145

* In addition to the ions listed, seawater contains small amounts of virtually all elements found on earth.
Source: From W. T. W. Potts and G. Parry, 1964, *Osmotic and Ionic Regulation in Animals*, Pergamon Press, New York, p. 90.

medium, taking nutrients and oxygen and returning waste products. More complex animals do not engage in such direct exchange; instead their cells are bathed in some kind of body fluid that is the medium for exchange. Various kinds of devices provide for exchange of materials between the extracellular body fluids and the physical environment. However, because the volume of the extracellular fluid is usually smaller than that of the cells it encloses, many complex mechanisms have developed to regulate the composition of the fluids. Through a variety of devices, employing feedback principles, homeostasis of varying degrees, depending upon the species, is maintained for the body fluids. Respiration provides oxygen and removes carbon dioxide; feeding and digestion provide nutrients; and osmoregulation controls the volume and composition of body fluids. Lungs, gills, guts, and kidneys are among the devices which add and extract substances from these private ponds (Fig. 6-1).

The extracellular fluids of most *marine* animals are in osmotic equilibrium and almost in ionic equilibrium with the surrounding seawater. Some of the concentrations are almost identical to seawater, but others differ considerably. Magnesium and sulfate, particularly, are much lower in certain marine species than they are in seawater. Such great differences indicate a nonequilibrium situation and suggest the existence of regulatory mechanisms for maintaining a steady-state concentration (Table 7-4).

The most striking thing about the ionic composition of those animals listed in Table 7-4, however, is not the differences from seawater but the great similarities to it. They obviously have few, if any, problems of an osmoregulatory nature.

The development of an internal medium, with its associated control systems, has facilitated the differentiation of freshwater and terrestrial

Table 7-3

Typical Composition of Soft Water, Hard Water, and Inland Saline Water, mmol/kg H_2O (Listed in Same Order as in Table 7-2)

	Soft lake water*	River water†	Hard river water‡	Saline water§	Dead Sea¶
Sodium	0.17	0.39	6.13	640	840
Magnesium	0.15	0.21	0.66	6	2302
Calcium	0.22	0.52	5.01	32	583
Potassium		0.04	0.11	16	152
Chloride	0.03	0.23	13.44	630	6662
Sulfate	0.09	0.21	1.40	54	8.4
Bicarbonate	0.43	1.11	1.39	3	Trace

* Lake Nipissing, Ontario.

† Mean composition of North American rivers.

‡ Tuscarawas River, Ohio.

§ Bad Water, Death Valley, California.

¶ Dead Sea, Israel. This water also contains 118 mmol/kg H_2O of bromide.

Source: From K. Schmidt-Nielsen, 1975, *Animal Physiology*, Cambridge University Press, p. 373.

animals which maintain body fluids that are quite different from the composition of their surrounding medium. The body fluids of *freshwater* animals show many similarities to seawater but have a lower osmotic concentration. Sodium and chloride are generally the most abundant ions, while the concentration of potassium is very much lower. Calcium is present in lesser amounts than in seawater, but usually it is proportionately less reduced than the total ions. Very little magnesium is found in the blood of freshwater animals. Sulfate concentration is extremely low (Table 7-5).

The solute concentrations of the body fluids of tetrapods are approximately equivalent to one-fourth or one-third that of seawater. In most *terrestrial* animals, the greater part of the osmotic pressure of the blood is accounted for by sodium and chloride ions. The concentration of potas-

Table 7-4

Concentration of Common Ions, mmol/kg water, in the Blood of Some Marine Animals

Animal	Na	Mg	Ca	K	Cl	SO_4	Protein, g/l
(Seawater)	478.3	54.5	10.5	10.1	558.4	28.8	
Jellyfish (*Aurelia*)	474	53.0	10.0	10.7	580	15.8	0.7
Polychaete (*Aphrodite*)	476	54.6	10.5	10.5	557	26.5	0.2
Sea urchin (*Echinus*)	474	53.5	10.6	10.1	557	28.7	0.3
Mussel (*Mytilus*)	474	52.6	11.9	12.0	553	28.9	1.6
Squid (*Loligo*)	456	55.4	10.6	22.2	578	8.1	150
Isopod (*Ligia*)	566	20.2	34.9	13.3	629	4.0	
Crab (*Maia*)	488	44.1	13.6	12.4	554	14.5	
Shore crab (*Carcinus*)	531	19.5	13.3	12.3	557	16.5	60
Norwegian lobster (*Nephreps*)	541	9.3	11.9	7.8	552	19.8	33
Hagfish (*Myxine*)	437	18.0	5.9	9.1	542	6.3	67

Source: From W. T. W. Potts and G. Parry, 1964, *Osmotic and Ionic Regulation in Animals*, Pergamon Press, New York, p. 95.

Table 7-5

The Osmotic Constituents of the Blood of Some Freshwater Animals

	Concentrations in m*M*/kg water or m*M*/l blood						
Animal	Na	K	Ca	Mg	Cl	HCO_3	Other ions
Frog (*Rana esculenta**)	109	2.6	2.1	1.3	78	26.6	Lactate, 3.5
Trout (*Salmo trutta*†)	161	5.3	6.3	0.93	119	n.d.	Phosphate, 1.0
Crab (*Potamon niloticus*†)	259	8.4	12.7	n.d.	242	n.d.	
Crayfish (*Astacus fluviatilis*†)	212	4.1	15.8	1.5	199	15	
Alderfly (*Sialis lutaria*†)	109	5	7.5	19	31	15	Amino acids, 152
Freshwater clam (*Anodonta sygnaea**)	15.6	0.5	6	0.2	11.7	12	Amino acids, 0.2

* m*M*/kg water.

† m*M*/l blood.

Note: n.d. = no data.

Source: From W. T. W. Potts and G. Parry, 1964, *Osmotic and Ionic Regulation in Animals*, Pergamon Press, New York, p. 169.

sium is usually small, so that the ratio of Na/K is 20 or more. As in freshwater forms, magnesium is in very low concentration, and sulfate is so low that it is difficult to detect (Table 7-6).

As noted in the preceding section, the composition of insect hemolymph is peculiar in having unusually large amounts of sugars and amino acids. It is also unusual, particularly in herbivorous insects, in having a very high concentration of potassium, while at the same time having a relatively low sodium concentration, so that the Na/K ratio may be less than one (see *Bombyx* in Table 7-6). Insect hemolymph also is notable for a high concentration of magnesium and, generally, for a low concentration of chloride (Table 7-6).

7-4 SOLUTE AND WATER REGULATION

The literature dealing with osmotic and ionic regulation in animals is very large. Fortunately, excellent reviews and syntheses are available. In 1939 August Krogh, of the University of Copenhagen, published his classic *Osmotic Regulation in Aquatic Animals*. Twenty-five years later, *Osmotic and Ionic Regulation in Animals* by Potts and Parry provided an extensive summary of more recent discoveries based on the use of modern methods such as microchemical analyses and radioisotope techniques. (See references at the end of the chapter.)

Osmotic situations can be categorized roughly into relatively few types. First, there are the marine invertebrates with essentially an independence from osmotic problems because, as illustrated in Table 7-4, their fluids closely approximate seawater. However, they may regulate the concentration of specific ions so that they are noticeably different from the concentrations in seawater. *Ligia* for example, has three times more calcium and about half as much magnesium in its blood compared to seawater, and *Loligo* has twice as much potassium (Table 7-4).

Table 7-6
Composition of the Blood of Some Land Animals

Animal	Blood composition, m*M*/kg or m*M*/l							
	Na	K	Ca	Mg	Cl	HCO_3	Amino acids	PO_4
Isopod (*Ligia oceanica*)	586	14	36	21	596			5
Spider (*Tegenaria atrica*)	207	9.6			193			
Thysanuran (*Petrobius* spp.)	208	5.8			194			
Dragonfly (*Aeschna* spp., larva)	135	5.4	7.5	6.0	120			
Cockroach (*Periplaneta* spp.)	156	7.7	4.2	5.4	144			
Silkworm (*Bombyx mori*, larva)	3.4	41.8	12.3	40.4	14			
Snail (*Helix pomatia*)	113	4	11	13.4		24.2		
Frog (*Rana temporaria*)	104	2.5	2	1.2	74.3	30	0.7	3
Rat (*Rattus rattus*)	140	6.4	3.4	1.6	119	24.3	3	2.3

Source: From W. T. W. Potts and G. Parry, 1964, *Osmotic and Ionic Regulation in Animals*, Pergamon Press, New York, p. 227.

Freshwater invertebrates and freshwater fishes and amphibians, on the other hand, are presented with osmotic problems in that they generally have a much higher total solute concentration, as well as a much higher concentration of specific ions, than those found in freshwater (compare Table 7-3 with Table 7-5). The range of osmotic concentrations is considerable, but the blood of freshwater animals is almost always less concentrated than the blood of marine ones.

Marine teleosts have a severe osmotic problem because their blood has much lower solute concentrations than that of the bathing seawater. Most interestingly, the osmotic pressures of the blood of freshwater and marine teleosts as well as terrestrial vertebrates are all about the same. Examples of these groups are represented in Fig. 7-3, along with the osmotic pressures of some other groups for comparison.

Most of the major animal phyla have species in both sea- and freshwater, although the number that are in the sea is by far the greatest. Principal exceptions are the echinoderms and cephalopod mollusks (octopus and squid), which are not found in freshwater.

Let us now define certain osmotic problems in more detail and examine strategies for their solution. A bit of useful vocabulary should be acquired first, however.

Since most marine invertebrates have body fluids with the same osmotic pressure as seawater they are said to be **isoosmotic** or **isosmotic** (Gr. *isos*, equal) with the external medium (e.g., marine mollusks and crustaceans, Fig. 7-3). Marine teleosts are **hyposmotic** (Gr. *hypo*, under, beneath) to seawater, but freshwater teleosts are **hyperosmotic** (Gr. *hyper*, over, above) to freshwater (Fig. 7-3).

Euryhaline animals (Gr. *eurys*, wide, broad, *halos*, salt) can tolerate wide variations in the salinity of the water in which they live. For example, a marine animal that can survive in brackish water is euryhaline. It even may tolerate a limited exposure to freshwater. A freshwater animal that can endure a considerable increase in salinity also is referred to as euryhaline. Other animals have limited tolerance to variations in the concentration of their aquatic medium and are called **stenohaline** (Gr. *stenos*, narrow, close). A stenohaline animal, whether marine or freshwater, can tolerate only small changes in the salt concentration of its ambient water. There is no clear-cut separation of euryhaline and stenohaline animals; these terms represent idealized extremes, whereas there actually are varying grades of tolerance. The osmotic limitations to distribution of a group of animals can be obtained by observing their response to **osmotic stress.** There are two extreme patterns of response. Animals may be osmotically labile (dependent) and their body fluids will change with the medium; these are **osmoconformers.** Others may be osmotically stable (independent) and, as the medium changes, their internal concentration remains fairly constant; these are **osmoregulators.** The terms **poikilosmotic** and **homoiosmotic** are sometimes applied respectively to conforming and regulating animals. Of course, there are gradations between the extremes of lability and constancy.

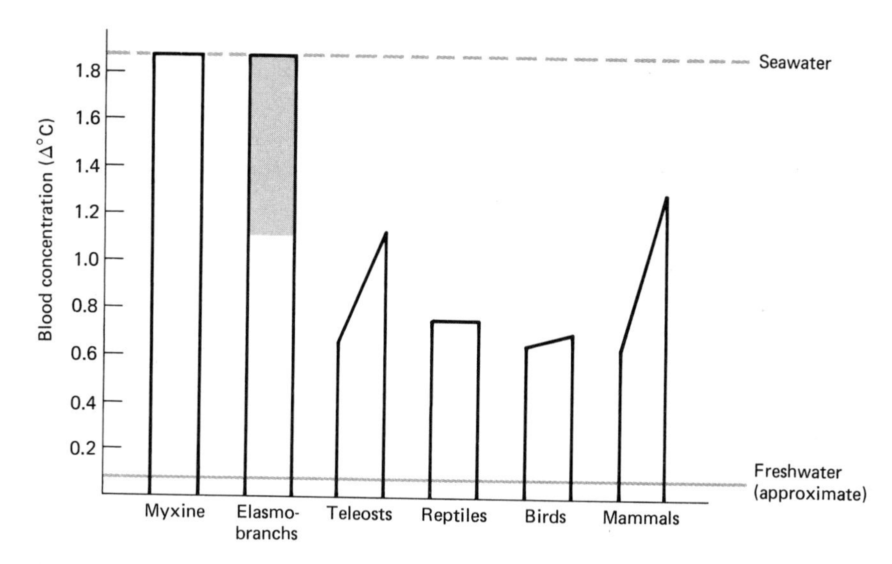

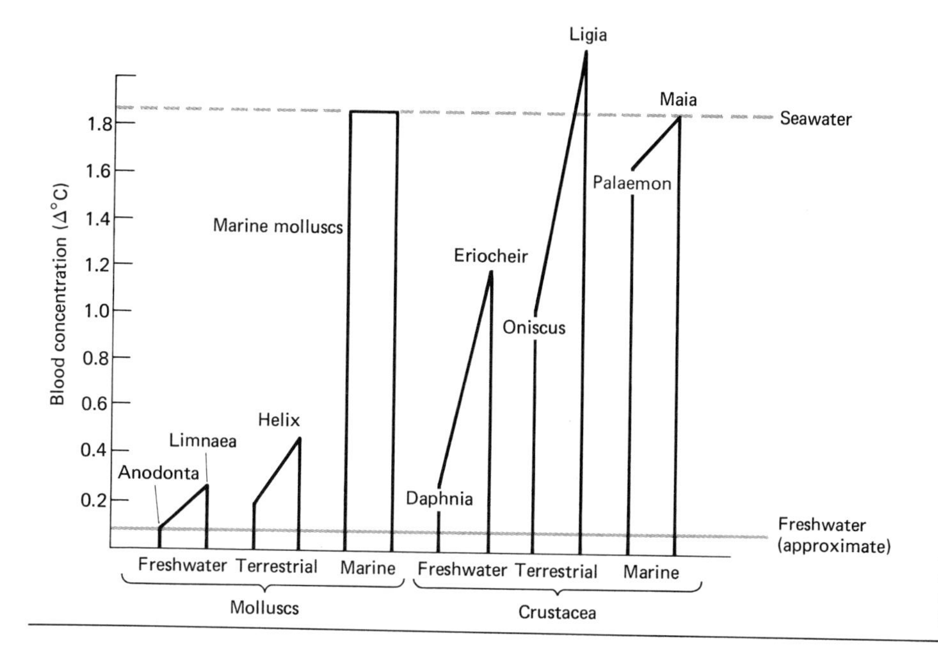

FIGURE 7-3
(*Above*) The approximate osmotic concentrations of the blood of various marine vertebrates. Only the elasmobranchs and *Myxine* have blood nearly isosmotic with seawater. Note the considerable contribution of urea and trimethylamine oxide (indicated by shading) to the total osmotic concentrations of elasmobranch blood. (*Below*) The osmotic concentrations of the blood of mollusks and crustaceans from various environments.

Osmotic changes may result in gain or loss of water and consequent volume changes. If appropriate solute changes also occur, the body volume remains constant while the animal's concentration changes with the medium.

Marine Animals

Most marine invertebrates obviously are *osmoconformers* and do not have to regulate water movement. However, as stated before, some may have a *solute* concentration that is different from seawater and thus engage in *ionic regulation*. Others, such as echinoderms and coelenterates, show little or no evidence of ionic regulation.

Marine animals that penetrate into brackish water can be of two types: osmoconformers or osmoregulators. Again, there are varying degrees of osmoconformity and osmoregulation. An osmoconformer, such as the oyster living in an estuary, may tolerate considerable dilution. However, it can avoid some dilution by keeping its shell closed during the periodic ebbing of the tide.

Nereis diversicolor is a euryhaline, nereid polychaete that can penetrate brackish water. In dilute seawater *Nereis* remains hyperosmotic to the medium, indicating osmoregulation, but is not able to maintain the body fluid concentration it had in seawater. Oxygen consumption increases initially on transfer to dilute seawater, partly because of muscular resistance to swelling. It takes up ions, including chloride, by an active process, while passive permeability to water and ions decreases when the external medium is diluted (Fig. 7-4).

The European shore crab, *Carcinus*, although a good osmotic regulator, generally will not survive exposure to brackish water more dilute than about one-third of normal seawater. *Eriocheir*, the Chinese mitten crab, is much more tolerant, being able to penetrate into freshwater. However, it must return to the sea to reproduce and thus is a marine animal that has excellent regulating capacity in freshwater (Figs. 7-3 and 7-4).

The genus *Gammarus* includes marine, brackish, and freshwater species. Some are restricted in salinity range; others are extremely euryhaline. Marine species show some hyperosmotic regulation in dilute

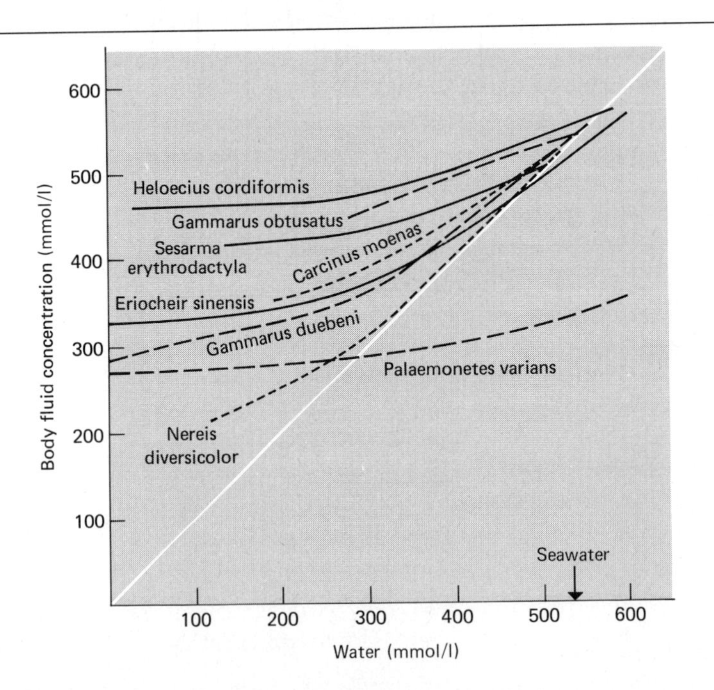

FIGURE 7-4
Relation between the concentration of body fluids and medium in various brackish-water animals. Full-strength seawater is indicated by an arrow. Diagonal line indicates equal concentrations in body fluid and medium (*After L. C. Beadle*, 1943, *Biol. Rev.*, 18, 174.)

medium, but regulation fails at relatively high saline concentrations. *Gammarus duebeni*, a brackish-water species, has an efficient excretory organ that produces a hyposmotic urine in dilute medium and, by some unknown mechanism, can actively concentrate sodium from the external medium (Fig. 7-4).

Freshwater Animals

Freshwater animals behave osmotically much like successful osmoregulators in brackish water, but there are great differences in the concentrations at which they maintain their body fluids. In Table 7-5, note that the crab, *Potamon*, maintains a much higher osmotic concentration than the clam, *Anodonta*. However, both are hyperosmotic, and no freshwater animal is known that has body fluids as dilute as the water in which it lives.

An animal that is hyperosmotic to its medium has two osmotic problems: (1) it tends to lose solutes to the dilute medium, and (2) water tends to flow into it. The common strategies for dealing with these problems are: (1) make the exterior surfaces highly impermeable, to reduce salt loss and water gain, (2) expel the flood of water by secretion of a dilute urine, and (3) compensate for salt loss above that gained by dietary intake through active uptake of ions from the medium.

It is not possible for an animal to make itself completely impermeable. For instance, respiratory surfaces which must be extensive enough and thin enough to permit adequate diffusion of gases also will leak salt and water. However, it is possible to restrict such permeable surfaces to a minimum. Water that enters the animal due to osmotic driving force can be pumped out as urine, but no animal is known that can produce a urine that is pure water; thus, salts are lost through urine excretion. Uptake of ions to compensate for losses occurs against a concentration gradient by an *active* transport process, commonly in some specialized gill cells (Fig. 7-5*a*).

The organ responsible for ion transport is not always known. In some freshwater animals it is assumed that the general body surface is capable of active ion uptake. In crustaceans there is conclusive evidence that the gill is the organ of active transport. In certain larval insects the "anal gills," which probably have no respiratory function, are involved in ion uptake. However, there often is an association of respiratory surfaces and ion-transport sites, as in crustaceans and also in fish.

It is assumed that at some early stage vertebrates must have entered freshwater by penetrating into estuaries and subsequently moving into rivers and lakes. Adaptation to freshwater existence would have involved a reduction in the osmotic pressure of their body fluids. However, despite the lowered solute concentration, a marine origin is still reflected in the general ratios of ions in the body fluids of all the higher vertebrates, even in those that have become fully terrestrial.

Freshwater teleosts are, in a sense, in continuous danger of drowning because of osmotic influx of water. *Marine teleosts*, on the other hand, are faced continuously with the possibility of desiccation due

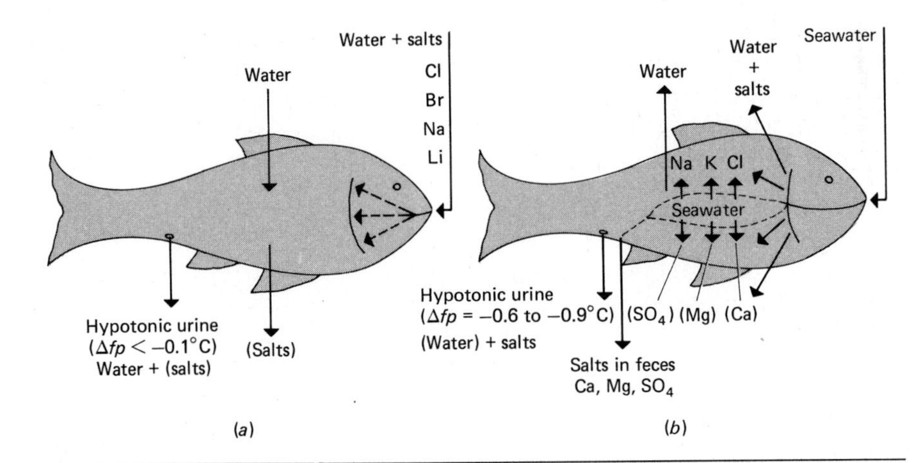

FIGURE 7-5
(*a*) Osmotic regulation in freshwater teleost fishes. (*b*) Osmotic regulation in marine teleost fishes.

to osmotic loss of water. Some early freshwater bony fishes presumably returned to the sea to give rise to modern-day marine teleosts, encountering thereby quite different osmotic problems. The internal osmotic concentration of marine bony fishes is similar to that of their freshwater ancestors and, therefore, is below that of the external medium. Thus, there is an osmotic efflux of water. Marine fishes compensate by drinking seawater. However, in taking in seawater they acquire a large quantity of monovalent ions which pass into the blood from the intestine. The excess is lost through the gills, chloride being actively extruded by so-called *chloride cells*. Sodium may passively accompany chloride efflux, or sodium may be actively transported as well; at present the exact situation remains to be clarified. Most of the divalent ions remain in the intestinal lumen and are lost in the feces. Smaller amounts are excreted in the urine (Fig. 7-5*b*).

Migratory teleosts, such as salmon, or euryhaline teleosts, such as the killifish (*Fundulus*), which pass from sea- to freshwater or vice versa, adapt their osmoregulatory mechanisms appropriately to the particular medium in which they are existing. *Catadromous* fish breed in the sea and mature in freshwater. For example, eels (*Anguilla*) breed in the Sargasso Sea of the mid-Atlantic, and the young elvings migrate either to the American or European continent. In seawater, their gills actively excrete Cl^-; in freshwater the gills absorb Cl^- and some ions are obtained in food (Table 7-7).

Anadromous fish, such as salmon, spawn in freshwater, and the young then migrate to the sea as smolt to mature. The blood of salmon freezes at $-0.76°C$ when at sea and at $-0.67°C$ in the freshwater spawning grounds. When in freshwater they also produce a more dilute urine (Table 7-7).

Elasmobranchs

Elasmobranchs (sharks and rays) are, as a group, more committed to the marine environment than are bony fishes. They have had an evolutionary history that has long been independent of teleosts and have met their osmotic problems in the sea in a unique way. While the ionic composition of their body fluids is somewhat greater than that of teleosts (Table 7-7), they have considerably elevated their internal osmotic pressure by accumulating urea and trimethylamine oxide (Fig. 7-3).

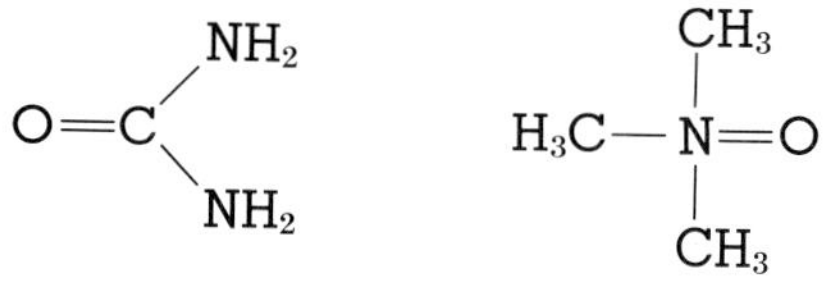

$$O{=}C\begin{matrix} / \; NH_2 \\ \backslash \; NH_2 \end{matrix} \qquad \begin{matrix} & CH_3 \\ & | \\ H_3C{-} & N{=}O \\ & | \\ & CH_3 \end{matrix}$$

Urea — Trimethylamine oxide

Urea, as will be discussed in Sec. 7-7, is an end product of protein metabolism in certain vertebrates and it is excreted by the kidney of mammals. In contrast, the shark kidney retains most of this compound and returns it to the blood. Trimethylamine oxide is a compound found in many marine organisms, but its origin and metabolism are poorly understood. Whether elasmobranchs produce it or obtain it from their diet remains uncertain. The net result of accumulation of these organic solutes

Table 7-7
Concentrations of Major Solutes, mmol/l, in Blood Plasma of Aquatic Vertebrates

Vertebrate	Habitat*	Na	K	Urea†	Osmotic concentration, mOsm/l
(Seawater)		~450	10	0	~1000
Cyclostomes					
Hagfish (*Myxine*) (*a*)	M	549	11		1152
Lamprey (*Petromyzon*) (*b*)	M				317
Lamprey (*Lampetra*) (*a*)	F	120	3	<1	270
Elasmobranchs					
Ray (*Raja*) (*a*)	M	289	4	444‡	1050
Dogfish (*Squalus*) (*a*)	M	287	4	354‡	1000
Freshwater ray (*Potamotrygon*) (*c*)	F	150	6	<1	308
Coelacanth					
Latimeria (*a,d*)	M	181		355‡	1181
Teleosts					
Goldfish (*Carassius*) (*a*)	F	115	4		259
Toadfish (*Opsanus*) (*a*)	M	160	5		392
Eel (*Anguilla*) (*a*)	F	155	3		323
	M	177	3		371
Salmon (*Salmo*) (*a*)	F	181	2		340
	M	212	3		400
Amphibia					
Frog (*Rana*) (*e*)	F	92	3	~1	
Crab-eating frog (*R. cancrivora*) (*f*)	M	252	14	350	830§

* *M* = marine, *F* = freshwater.
† When no value is listed for urea, the concentration is of the order of 1 mmol/l and osmotically insignificant.
‡ Also contains trimethylamine oxide.
§ Values for frogs kept in a medium of about 800 mOsm/l, or four-fifths the concentration of normal seawater.
Source: From K. Schmidt-Nielsen, 1975, *Animal Physiology*, Cambridge University Press, New York, p. 355; (*a*) Bentley, 1971; (*b*) Robertson, 1954; (*c*) Thorson, Cowan, and Watson, 1967; (*d*) Lutz and Robertson, 1971; (*e*) Mayer, 1969; (*f*) Gordon, Schmidt-Neilsen, and Kelly, 1961.

is that elasmobranchs are isosmotic with seawater. They still require specific regulation of ions, however. The sodium concentration, for example, is maintained at about half that of seawater (Table 7-7). Excess sodium is handled in part by the kidney, but a special excretory structure in the posterior part of the intestine, the **rectal gland,** is probably more important. The gills also may actively extrude sodium.

The great majority of elasmobranchs are sea creatures, but there are a few that enter rivers and lakes, and some may take up permanent residence in freshwater. The ones that can make the transition have blood solute concentrations that are lower than those in the strictly marine forms. Urea concentration is strikingly lower in those that have colonized freshwater (Table 7-7, freshwater ray).

The crossopterygian fish, *Latimeria*, which was thought to be extinct for 75 million years until one was caught off the coast of southwest Africa in 1938, solves its osmotic problems in the same way as elasmobranchs. It has a high urea and trimethylamine oxide concentration (Table 7-7).

The *cyclostomes*, eels and hagfishes, are considered to be the most primitive of living vertebrates. Lampreys occur both in the sea and in freshwater. The hagfishes are strictly marine and stenohaline. They are the only vertebrates which have body fluids with a salt concentration similar to that of seawater (Table 7-7). However, ionic regulation does not occur in hagfishes. The lampreys, whether marine or freshwater, have osmotic concentrations similar to those of teleosts (Table 7-7).

Amphibians

Most amphibians are aquatic or semiaquatic and, with regard to osmotic regulation, are quite similar to teleost fishes. Virtually all amphibians are freshwater inhabitants. In the adult the skin serves as the main organ of osmoregulation. Water flowing in through the skin is eventually excreted as a highly dilute urine. Loss of salt, both in the urine and across the skin, is balanced by active uptake of salt by cells in the skin from the highly dilute medium (Fig. 7-6).

In Southeast Asia there is a crab-eating frog (*Rana cancrivora*) which lives in coastal mangrove swamps and seeks its food in full-strength seawater. It employs the elasmobranch strategy of adding urea to its body fluids (Table 7-7). It does not seem to reabsorb urea in the

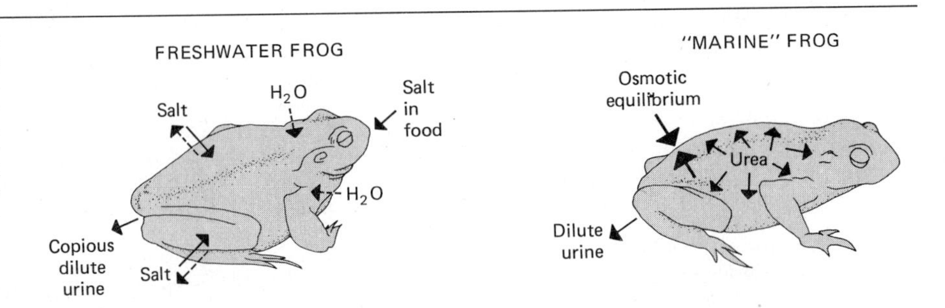

FIGURE 7-6
Schematic representation of mechanisms of osmoregulation in freshwater and crab-eating "marine" frogs. Dashed lines indicate passive osmotic movements; solid lines show active energy-requiring processes.

kidney as sharks do, but effects urea retention primarily by a reduction in urine volume when in seawater (Fig. 7-6).

All amphibians must contend with the problem of an integument that is ineffective as a barrier to water loss. Temperate-zone frogs evaporate water from their body surfaces at the same rate as from free water surfaces of the same area at the same temperature. For this reason, most amphibians are restricted to freshwater bodies or to microhabitats where freshwater is available. Behavioral and physiological mechanisms for conserving and obtaining water are highly evolved in amphibians, especially in those living in arid or desert habitats.

Amphibians do not take up water as vapor, even from a saturated atmosphere. Terrestrial and semiterrestrial species absorb water from a moist surface. The ventral skin which can be adpressed against the substrate is particularly effective in water uptake. By osmotic uptake, water is moved across the skin. Desert species, such as spadefood toads, burrow into the soil, where they can encounter soil water. The water tension is low enough for the water to be osmotically extracted and absorbed across their skin.

Some of the more terrestrial amphibians can tolerate a fair amount of evaporative water loss. Many of the desert species use their bladders and perhaps their lymph spaces as accessory reservoirs for water storage during dry periods. However, amphibians do have some control of water loss, primarily through the kidneys. The few that have been studied show a marked decrease in rate of urine production when they are out of water. Furthermore, the solute concentration increases considerably when they are in freshwater. Whereas ammonia will be the principal nitrogenous waste when water is abundant, the less toxic compound, urea, is the major form when amphibians are deprived of water. Some African and South American frogs which live in arid regions have been reported to excrete 60 to 80 percent of their waste nitrogen as the relatively insoluble uric acid. This is the characteristic nitrogen excretory form of the most terrestrial vertebrates, particularly reptiles and birds.

Terrestrial Animals

The great physiological advantage of terrestrial life is easy access to oxygen; the principal physiological threat to life on land is desiccation. Only two phyla have been truly successful in colonizing the land: arthropods and vertebrates. Evaporative water loss is minimized by development of a relatively impermeable surface and by placement of respiratory surfaces in cavities (gills or lungs). Excretory water loss is reduced by the development of efficient kidneys.

Terrestrialism required many kinds of adaptations. Because toxic substances, such as ammonia, cannot be disposed of across gill surfaces into an abundant supply of water in land animals, profound changes in protein metabolism have taken place, and kidneys have assumed a role in nitrogenous waste secretion. The closed egg and viviparity are part of the reproductive strategies of land animals. Temperature variations tend to be greater on land than in water, and high temperatures with low humidity cause great evaporative water loss. So thermal adaptations also evolved.

Hormones

It is quite remarkable that mammals can maintain a relatively constant composition of body fluids in an environment where access to food and water is often highly variable. The homeostatic regulation of ions and water is largely the result of mechanisms employing two different hormones. (The source, control, and chemical nature of these hormones will be further elaborated in Chap. 9.) One of the hormones is an octapeptide known as **antidiuretic hormone** (ADH), which is released from the posterior lobe of the pituitary gland. As the name implies, it restricts the secretion of urine in the kidney. The absence of this hormone results in a pathologic condition known as *diabetes insipidus*, characterized by the excretion of a large quantity of urine. As much as 30 liters of water a day must be drunk by an adult human with this condition. The hormone produces its effect by acting upon the kidney tubules (described in Sec. 7-6) to cause an increase in water uptake. The regulated release of this hormone makes it possible for mammals to adjust their urine output to the availability of water.

In some amphibians, ADH affects water permeability of the skin, bladder, and renal tubule. It may exert little or no influence in the strictly aquatic amphibians (e.g., *Xenopus*), but has the greatest effect in primarily terrestrial groups such as bufonid toads. However, even in the latter group, water balance is nearly normal in the absence of the posterior lobe of the pituitary. Thus, ADH probably became incorporated into the homeostatic water-balance mechanisms of animals early in the adaptation to terrestrial life, but the essential action of the hormone developed only later in birds and mammals.

The second important factor regulating body fluids in mammals originates in the cortex of the adrenal gland. The adrenocortical hormones are steroid molecules. Among them are a set, called **mineralocorticoids,** that exert their effect primarily on ion metabolism. Principally concerned with this action is the one called **aldosterone.** In the absence of mineralocorticoids, mammals generally do not survive. Sodium is lost excessively through the urine, and as the sodium level in the blood falls, there is a fall in chloride and bicarbonate and some rise in the blood-potassium level. Associated with these ionic changes are an increased loss of water in the urine and a consequent hemoconcentration that may be the primary cause of death. The syndrome resulting from mineralocorticoid deprivation is called *Addison's disease*.

Whereas ADH acts by facilitating the passive movement of water down osmotic gradients, the mineralocorticoids, by contrast, act by facilitating the active transport of sodium in the kidney tubules.

7-5 SALT GLANDS AND OTHER EXCRETORY ORGANS

A kidney is not uniformly the most important organ of excretion among animals. In fact, mammals are the only class of vertebrates in which kidneys are always the major organ of excretion. Excreting organs commonly are, in addition to being devices for ridding animals of metabolic wastes, mechanisms used in ionic and osmotic homeostasis. To maintain their constant composition, organisms must rid themselves of materials equal to the amount taken in; thus the necessity of having excretory mechanisms that can be closely controlled to eliminate variable amounts of a great variety of substances. In addition to maintenance of a proper concentration of ions, excretory systems also play a major role in maintenance of proper water volume and, thus, osmotic concentration, as well as in removal of metabolic end products, and foreign substances (e.g., drugs) or their metabolites.

One of the major metabolic products is carbon dioxide. As described in Chap. 6, carbon dioxide is primarily excreted via respiratory devices. Most other metabolic end products are removed by specialized excretory structures. Thus, the main role of excretory organs is subtractive—helping to maintain a steady-state composition by balancing inputs of materials acquired through the gut (Fig. 6-1).

Contractile vacuoles are excretory organelles present in many protozoa and in the different cells of freshwater sponges. The contractile vacuole fills up gradually with fluid and then collapses, discharging its content to the outside. The primary role of the contractile vacuole may be in osmoregulation. Facts which support this contention are (1) the vacuole occurs, generally, in freshwater forms, (2) it is absent in many marine and parasitic species, (3) injection of distilled water into amoebas increases the rate of pulsation and water output of the contractile vacuole, (4) the osmotic concentration of the vacuole fluid is hypotonic to the remainder of the cell, (5) the rate of pulsation bears a reciprocal relation to the salinity of the outer medium in some marine and parasitic forms, and (6) contractile vacuoles appear de novo in some marine amoebas following a decrease in salinity of the medium. Therefore, it appears that the contractile vacuole serves to bail out water which enters by osmotic flux. However, in a marine habitat, where there is no significant osmotic difference between the organism and the external medium, the slowly pulsating contractile vacuole is assumed to be not primarily an organelle of osmoregulation but an excretory device.

Most curious is the fact that no specific excretory organ has been identified in coelenterates (Cnidaria) or echinoderms. This is especially interesting in that there are some freshwater coelenterates which are quite hypertonic to the medium and they undoubtedly gain water by osmotic uptake. How this water is drained off is not known. Echinoderms, which are only marine, have no osmoregulatory problem, being isosmotic with seawater.

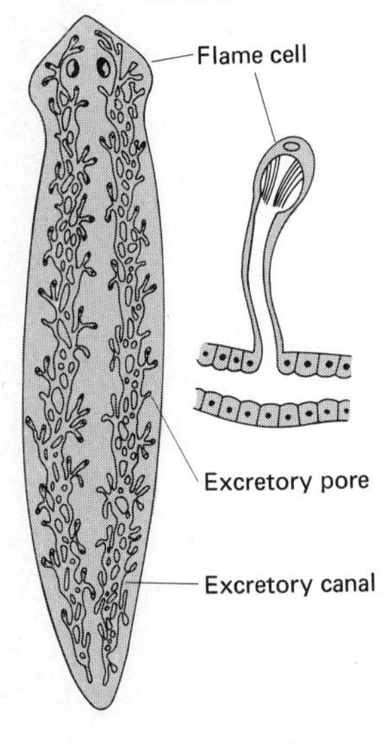

FIGURE 7-7
The protonephridial excretory system of planaria (Platyhelminthes). The highly branched system forms excretory fluid in terminal *flame cells*, so called because they contain an undulating tuft of cilia which flickers like the flame of a candle. (If there is a single cilium, the terminal cell is called a *solenocyte*.) The nephridial canals discharge the fluid through excretory pores.

With the development of bilaterally symmetrical animals, true organs of excretion appear. The most common type, widely distributed among invertebrates, are the **nephridial organs.** They are simple or branching tubes which open to the outside through pores—the nephridial pores. In flatworms (Platyhelminthes) the tubules have enlarged blind ends, containing one or several long cilia, and are called *protonephridia*. This type of nephridium is characteristic of animals lacking a coelom (Fig. 7-7).

Metanephridia are found in annelids. This type of nephridium occurs as a pair in each body segment, and they open into the coelom as ciliated funnels called nephrostomes. Fluid from the coelom drains into the nephridium through the nephrostome, and as it courses through the extensively looped duct, its composition is modified. Upon entering the nephridium the fluid is isotonic, but salts are removed through the terminal parts of the nephridium, and what is discharged is a dilute urine. The metanephridium, therefore, is a filtration-reabsorption kidney (Fig. 7-8).

Mollusks have rather elaborate nephridial organs which are often called "kidneys." An initial fluid is formed by *ultrafiltration*. Ultrafiltration, you will recall from the preceding chapter, is the process in which hydrostatic pressure forces fluid through a semipermeable membrane which allows water and small molecules to pass but does not allow cells, proteins, and similar large molecules to penetrate. Thus, the ultrafiltrate contains, with the major exception of proteins, the same solutes as are present in the blood serum. These include not only substances to be excreted, but useful substances such as sugars and amino acids as well. The loss of these valuable materials is greatly diminished by having them selectively reabsorbed before the urine is discharged. The molluskan excretory organ also can actively *secrete* substances to be added to the urinary fluid.

The chief excretory organ of crustaceans is the so-called **antennal gland** or **green gland.** A pair of these structures is located in the head. They consist of an internal end sac, an excretory tubule, and an exit duct that is sometimes enlarged into a bladder. Urine is formed by ultrafiltration and reabsorption, with tubular secretions added (Fig. 7-9).

The excretory structures of insects are known as **Malpighian tubules.** They may number from two to several hundred, depending upon the species. Malphigian tubules have blind ends which lie in the body cavity (hemocoel) and open at the other end into the intestine at the junction between the midgut and hindgut (Fig. 7-10).

Ultrafiltration does not appear to occur in urine formation in insects. Instead, Malphighian tubules function in the following manner. Potassium is actively secreted into the lumen of the tubule, and water follows passively under osmotic drive. A potassium-rich fluid enters the hindgut, where solutes and much of the water are reabsorbed, and uric acid (the main nitrogenous waste of insects, which enters the tubule as water-soluble potassium urate) is precipitated.

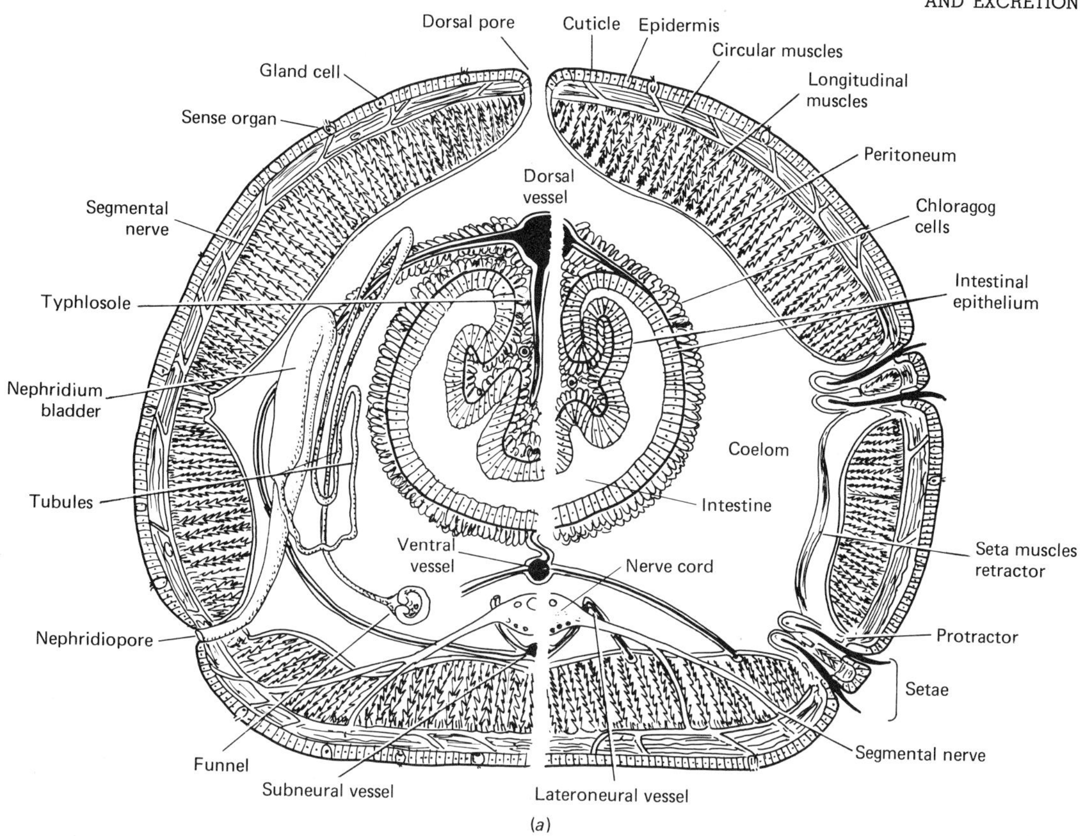

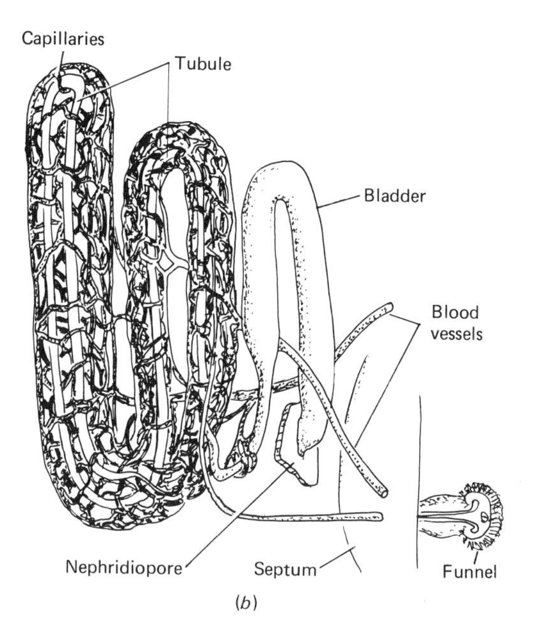

FIGURE 7-8
(*a*) Diagramatic cross section of an earthworm. The left half shows an entire *nephridium*. The right half, at a level in which the nephridium is absent, shows *chloragog cells*, a tissue surrounding the gut which is thought to accumulate and store excretory materials. (*b*) An entire earthworm nephridium. (*After T. I. Storer et al.*, 1972, *General Zoology*, *McGraw-Hill Book Company*, *New York*. *pp.* 514 *and* 516.)

As mentioned in the preceding section, elasmobranchs have a special gland, the **rectal gland,** which excretes excess sodium. This small gland, which opens by a duct into the rectum, secretes a fluid with a high sodium and chloride concentration. Its concentration can be somewhat higher than that of seawater.

So-called **salt glands** are found in certain marine turtles, lizards, snakes, and birds. These glands, located in the head, excrete excess salt which their kidneys cannot handle. Salt glands are specialized, auxiliary excretory organs, producing a highly concentrated fluid containing primarily sodium and chloride.

The Galápagos iguana, *Amblyrhynchus cristatus,* which lives near the surf and feeds on seaweed, has salt glands which empty into the nasal cavity, from which a sudden exhalation will expel a spray of droplets through the nostrils. Marine turtles, both plant-eating and carnivorous, have a large salt-excreting gland in the ocular orbits. The duct from

FIGURE 7-9
(*a*) The structure and relations of the renal organ, the antennal gland or green gland, in the crayfish, *Astacus pallipes.* (*b*) The concentration of chloride and the formation of hypoosomotic urine in the antennal gland of the crayfish, *Astacus fluviatilis.* The gland is shown below as if it were stretched out so that the effect of different portions on solute concentration can be illustrated by the graph above. (*After W. T. W. Potts and G. Parry,* 1964, *Osmotic and Ionic Regulation in Animals, Pergamon Press, New York, pp.* 69 *and* 72.)

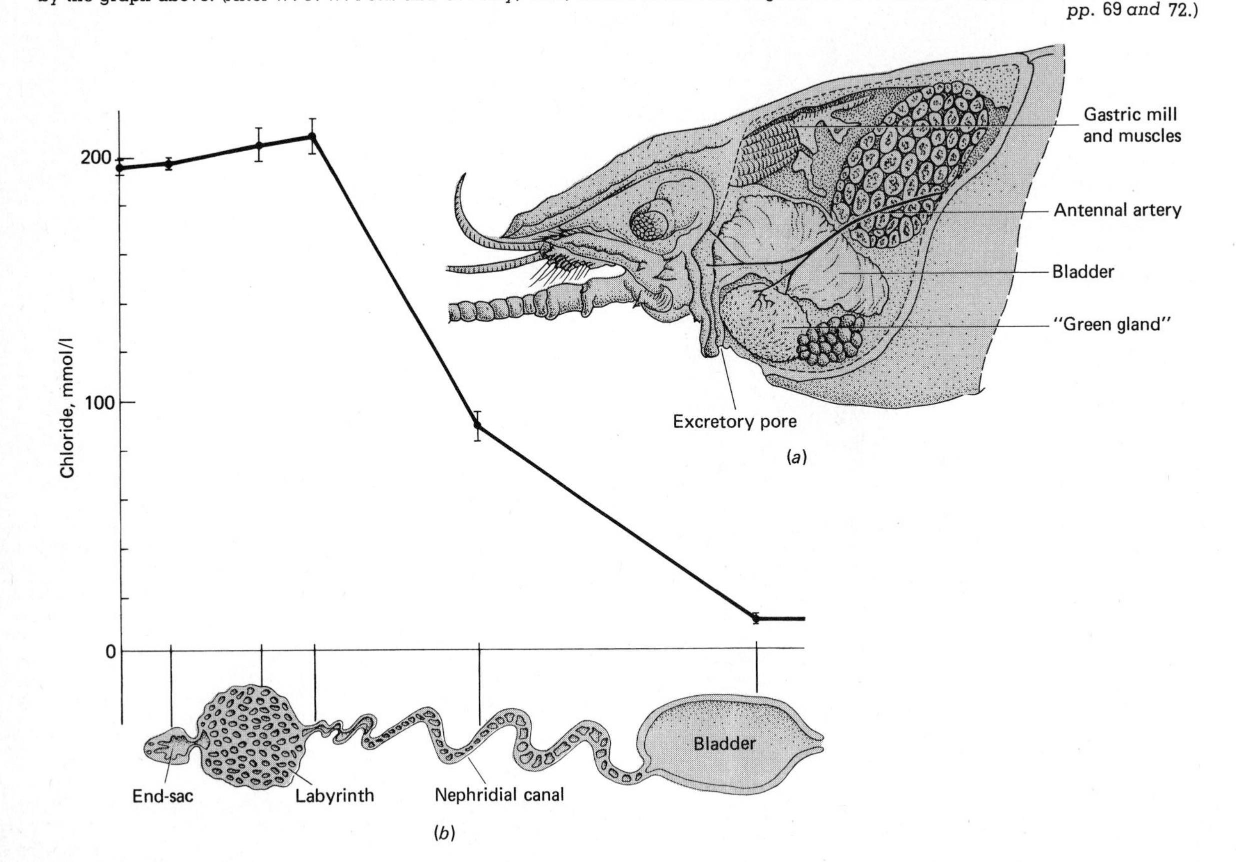

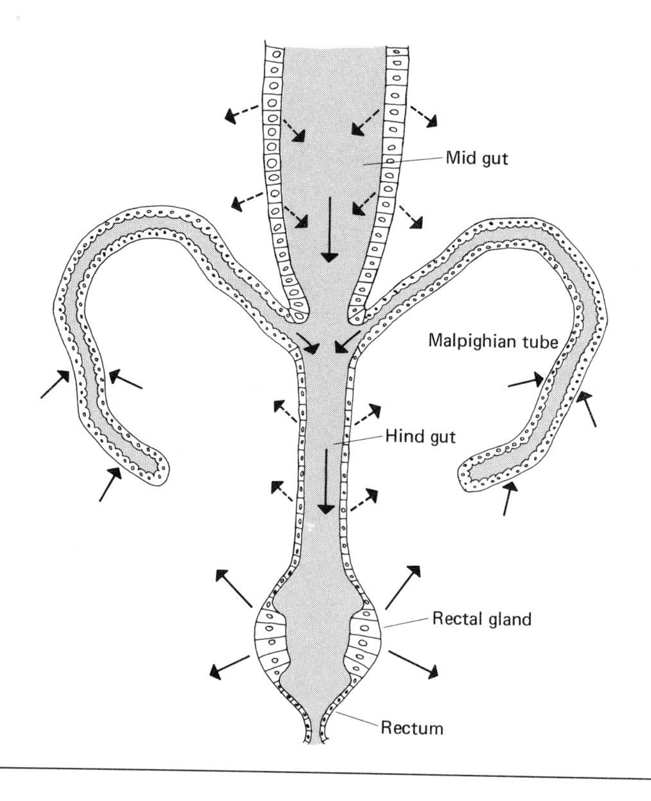

FIGURE 7-10
The insect excretory system. The arrows indicate the direction of water movement. (*Based on V. B. Wigglesworth*, 1931, *J. Exp. Biol.*, 8, *pp.* 411–451.)

the gland opens into the posterior corner of the orbit, and a turtle ridding itself of a salt load "cries" salty tears. Sea snakes have salt glands that empty into the oral cavity.

Bird salt glands have been studied in great detail, especially by Schmidt-Nielsen. Marine birds, in particular, have very large salt glands. Some marine birds are coastal, but others are truly pelagic, such as the albatross. The latter have an especially well-developed capacity to secrete salt. All marine birds have a pair of nasal salt glands, most frequently located on top of the skull above the orbit of the eyes (Fig. 7-11).

Unlike kidneys, which are continuously active, the salt glands are usually inactive and start secreting only in response to an osmotic stress, as when seawater or salty food is ingested. The secreted fluid contains mostly sodium and chloride in rather constant concentrations. Salt glands have an extraordinary capacity to secrete salt. The arrangement of capillaries and secretory ducts in the salt gland is such as to suggest that it employs the *countercurrent principle* to concentrate salt (Fig. 7-12).

7-6 KIDNEYS

With the exception of a few teleost fishes that have *secretion-type kidneys*, all other vertebrates—most fish, the amphibians, reptiles, birds, and mammals—have kidneys of the *ultrafiltration-reabsorption* type.

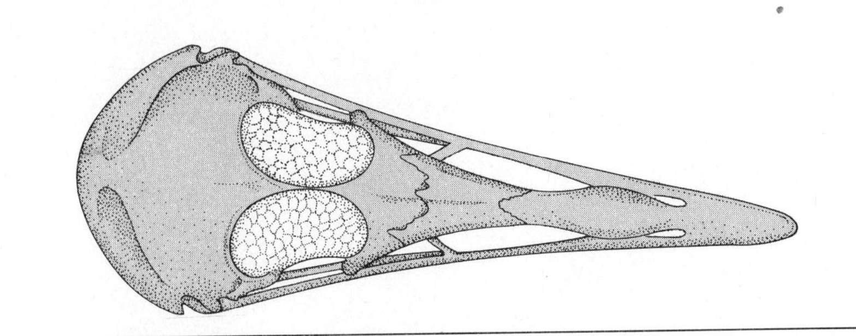

FIGURE 7-11
The salt-secreting glands of the sea gull are located on top of the head, above the eye, in shallow depressions in the bone. (*After K. Schmidt-Nielsen*, 1975, *Animal Physiology, Cambridge University Press, New York*, p. 431.)

The initial filtrate contains all the compounds present in blood, except large molecules, which are mainly protein. Reabsorption mechanisms recover glucose, amino acids, vitamins, salts, and water.

The advantages of nonselectively filtering the blood and then selectively returning most of the solutes and water can be questioned logically. It would seem to be more efficient to use selective secretion only and to avoid ultrafiltration. However, there is one unique consequence of ultrafiltration: any substance that has been filtered will remain in the urine, unless it is reabsorbed. Thus, the filtration-reabsorption kidney probably permits an organism to penetrate new environments and utilize novel food, for example, without development of special mechanisms to secrete "new" substances. The secretion-type, vertebrate kidney is limited to a small number of marine fish which live in a dependably constant environment, suggesting that, indeed, it is ill-suited to the vagaries that characterize the habitat of the great majority of vertebrates.

All vertebrates can produce urine that is *isotonic* or *hypotonic* to the blood. Freshwater fishes have kidneys that produce *hypotonic* urine;

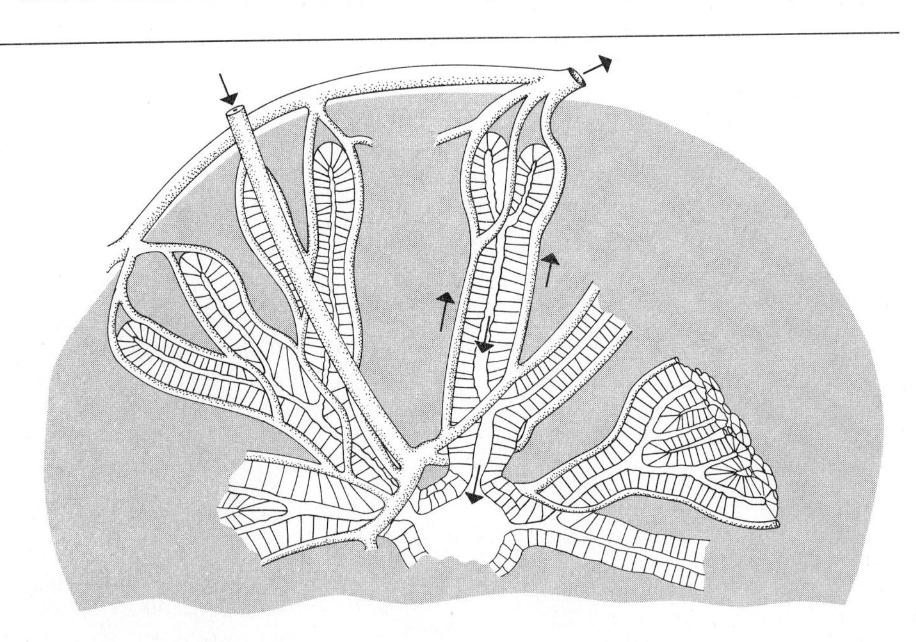

FIGURE 7-12
The arrangement of the blood vessels and capillaries in the lobes of the salt gland indicates that the capillary blood flow is countercurrent to the flow of secreted fluid in the secretory tubule. (*After K. Schmidt-Nielsen*, 1960, *Circulation*, pp. 21, 964, *by permission of the American Heart Association.*)

they bail out excess water while withholding solutes. Marine teleosts produce a more concentrated urine, but it is still *hypotonic* and they must add special glands for excess salt secretion, as noted earlier. Only birds and mammals can form a urine more concentrated than the body fluids. Marine and terrestrial mammals have kidneys with excellent concentrating ability, and they handle excess salt strictly by renal excretion, producing a *hypertonic* urine.

To describe properly the concentrating capacity of the filtration-reabsorption kidney, as found in birds and mammals, the concept of *osmolality* is quite useful. As noted in Sec. 7-1, 1 osmole of particles exerts an osmotic pressure of 22.4 atm. For biological purposes the osmole is too large a unit; instead, the milliosmole (mOsm, one-thousandth of an osmole) is usually used. The osmolality of human blood plasma is about 300 mOsm/l. The human kidney produces urine that is 4 to 5 times as concentrated, or about 1400 mOsm/l. Certain desert rodents, which do not need to drink water but satisfy their requirements from metabolic water, can do far better. The African sand rat (*Psammomys*), for example, can produce urine that is 6000 mOsm/l from a plasma filtrate of 300 mOsm/l, a twentyfold concentration. Its urine has an osmolality equivalent to almost 140 atm!

To understand the workings of the vertebrate kidney requires a detailed acquaintance with its geometric arrangement. The kidney is an organ in which structure and function are very closely correlated. By an ingenious anatomical arrangement of blood capillaries and kidney tubules, plus active transport of *sodium* at specific loci, a highly concentrated urine can be produced.

The **nephron,** the basic unit of the kidney, consists of a **glomerulus** and a long **tubule** along which a number of sharply demarcated segments can be distinguished (Fig. 7-13).

The concentrating capacity of the mammalian kidney is closely related to the *length* of **Henle's loop,** which is the hairpin loop formed by the thin segment of the tubule. Most mammalian kidneys have two types of nephrons: ones with long loops and ones with short loops. Those animals that produce the most highly concentrated urine have only long-looped nephrons. The beaver and pig have only short loops; their urine is only about twice as concentrated as the blood plasma.

The tubule loop is not present in fish, amphibians, and reptiles; they produce urine no more concentrated than blood plasma. Bird kidneys have only a few mammal-like tubules with a medullary loop (analogous to the loop of Henle). The vast majority of the nephrons are reptilelike, with short loops that are confined to the outer region of the kidney.

In each human kidney there are approximately 1 million nephrons. The glomeruli lie in a clearly demarcated outermost zone called the **cortex.** Here also are the greater part of the proximal and distal tubules, as well as the beginning of the collecting ducts. Inside the cortex lies the **medulla,** containing the loops of Henle and the larger collecting ducts.

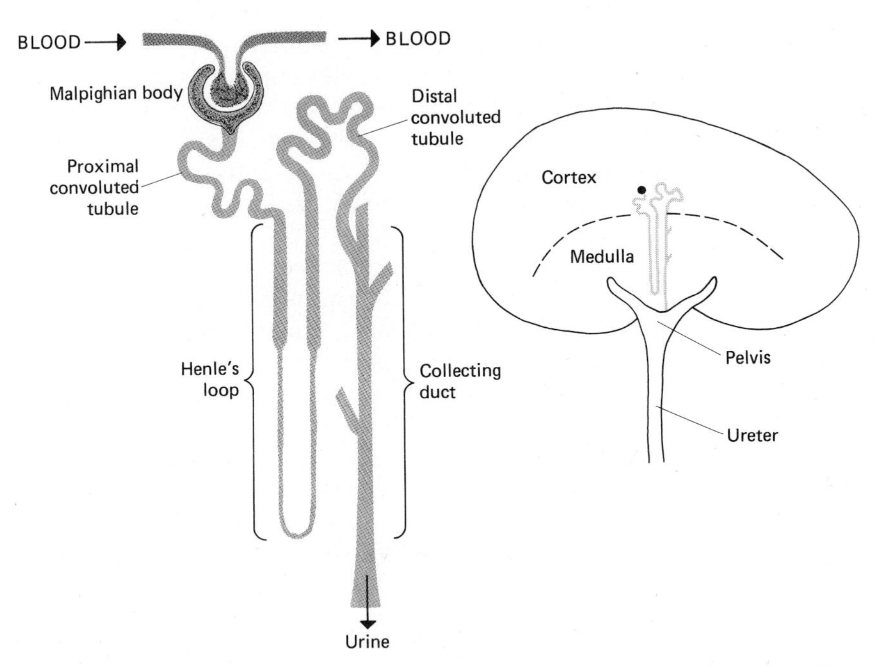

FIGURE 7-13
Schematic diagram of a mammalian kidney on the right. The kidney contains a large number of single nephrons (up to several million). Only one nephron is indicated in this diagram, and is shown enlarged to the left. The outer layer of the kidney, the *cortex,* contains the *Malpighian bodies* and the proximal and the distal convoluted tubules. The capillary network within the Malpighian body is known as the *glomerulus.* The inner portion, the *medulla,* contains Henle's loops and *collecting ducts.* The urine is initially formed by ultrafiltration in the Malpighian bodies. The filtered fluid is modified and greatly reduced in volume as it passes down the renal tubule and into the collecting ducts, which empty the urine into the *pelvis,* from which it is conveyed to the bladder via the *ureter.* (*After K. Schmidt-Nielsen,* 1975, *Animal Physiology, Cambridge University Press, New York, p.* 460.)

Blood vessels, including a finely branched capillary plexus, are found in each zone. The inner medulla projects into the **pelvis** of the **ureter,** through which urine drains to the **bladder** (Fig. 7-13).

Marcello Malpighi, in the middle of the seventeenth century, first described the renal corpuscles and correctly surmised that they were concerned with urine formation. In 1842 Bowman described and demonstrated the anatomical relationship of the **Malpighian body** (Fig. 7-13) to the rest of the tubule. The afferent arteriole divides to form a network of anastomosing capillaries, the **glomerulus,** which reunite to form the outgoing or efferent arteriole. The glomerular capillaries are closely apposed to the inner epithelial lining of **Bowman's capsule,** and the outer layer of the capsule is continuous with the proximal end of the tubule (Fig. 7-14).

The function of the glomerulus is to produce an almost protein-free ultrafiltrate of the blood plasma which enters the cavity of Bowman's capsule and then passes into the proximal segment of the tubule. The driving force for the filtering process is the blood pressure within the capillaries, which is probably about 70 to 75 mmHg. The kidneys of humans together receive about one-quarter of the cardiac output, or about 1300 ml of blood per minute. This volume contains about 700 ml of plasma, the rest being cells, and of this, about 125 ml/minute are filtered through the glomeruli. Thus, total daily filtration volume is in the neighborhood of 180 liters—more than 40 gal! However, only about 1.5 liters of urine are normally produced. Thus, 178.5 liters of the filtrate are reabsorbed by the tubules.

The study of the function of the renal tubules has been approached in several ways. The proof that the renal corpuscles produce an ultrafiltrate came in 1924 when Richards and associates published the results of their micropuncture studies. They withdrew fluid from the capsule of the frog and mudpuppy (*Necturus*) and showed that it contained glucose and chloride in about the same concentration as the plasma, but was normally free of protein. Later, pressures were measured and the dynamics of the corpuscles were characterized in terms of blood pressure, osmotic pressure, and pressures of the capsular fluid.

Renal tubules secrete as well as reabsorb. One of the proofs of this comes from the study of certain fishes (*Lophius, Opsanus*) which have

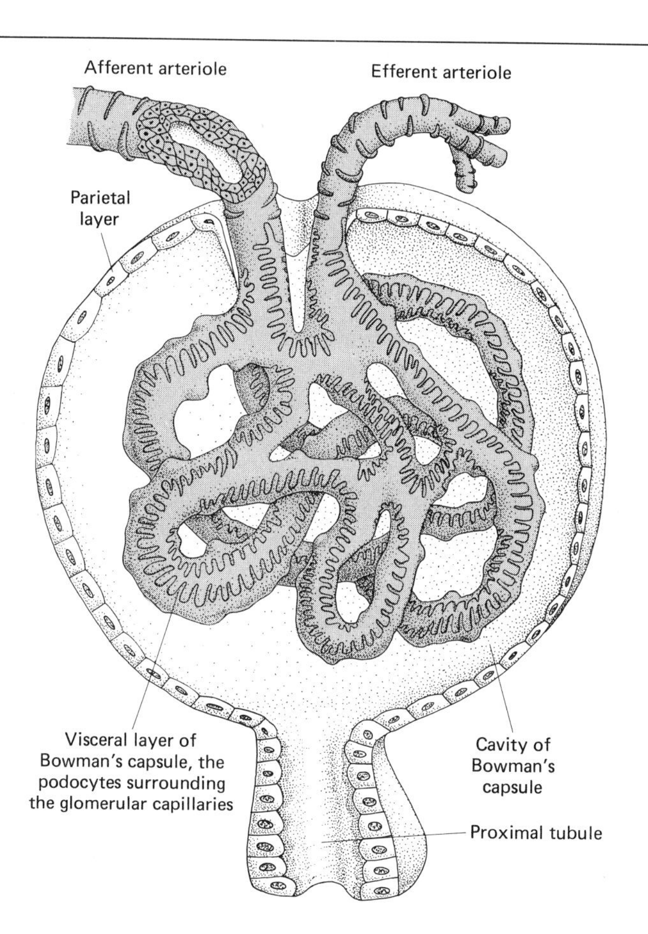

FIGURE 7-14
Diagram of Bowman's capsule and the glomerular tuft of capillaries. (*After D. B. Moffat, 1971, The Control of Water Balance by the Kidney, Oxford Biology Reader, Oxford University Press, p. 5.*)

secretion-type kidneys with aglomerular tubules. Only the tubular epithelium is involved in urine formation. The aglomerular kidney was shown to excrete various nitrogenous wastes, ions, and a variety of foreign substances, including certain dyes. However, glucose, inulin (a large polysaccharide excreted only by filtration), and several other substances never appear in the urine. Active secretion was thus demonstrated in a vertebrate nephron, which is considered to be homologous with the proximal tubule of higher forms.

Radioisotope tracer studies have shown that sodium is actively transported by tubular cells and passed into the capillaries, accompanied by appropriate anions, especially chloride. Analysis of fluids obtained through microcatheter and puncture, as well as direct cryoscopy of kidney slices, has shown that the urine is isotonic in the proximal tubule, becomes progressively more hypertonic as it descends the loop of Henle, and then gradually becomes less hypertonic in the ascending limb. As noted earlier, *aldosterone* influences the sodium-transporting mechanism.

All these and other pieces of information have been meshed into a fairly comprehensive understanding of how urine formation occurs. In the human, the greater part of the 125 ml of filtrate per minute is reabsorbed by the proximal tubule, leaving only 15 to 20 ml/minute to enter

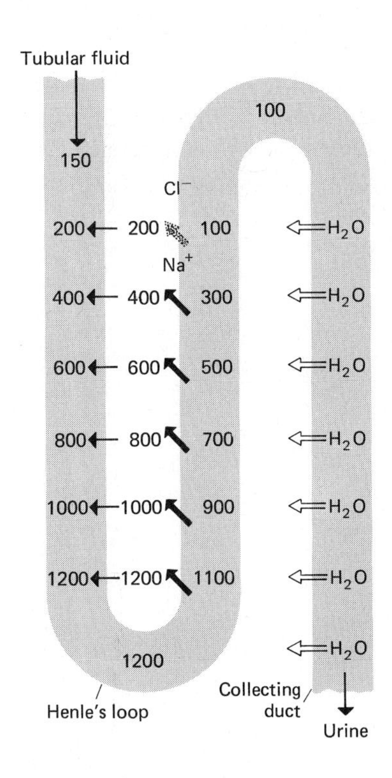

FIGURE 7-15
Diagram of the countercurrent multiplier system in the mammalian kidney. The figures refer to sodium concentrations, arbitrarily chosen to indicate the increase in concentration toward the bottom of the loop. Heavy black arrows refer to active sodium transport, thin arrows to passive diffusion of sodium. Stippled arrow means active chloride transport. Open arrows indicate passive movement of water due to osmotic forces.

the loops of Henle. This remainder is reabsorbed to a varying extent, so that the final urine volume ranges between 0.3 and 20 ml/minute. The fluid entering the loop has a sodium concentration like that of the plasma, about 150 mmol/l. Sodium is actively transported out of the *distal* or *ascending* limb of the loop, resulting in a concentration of sodium in the surrounding tissue. The *descending* limb is permeable to sodium. Therefore, sodium diffuses in and establishes an equilibrium with the interstitial tissue. (The ascending limb tends to be impermeable to water; otherwise water would passively follow sodium.) Thus, the osmolality of the loops of Henle and of the interstitial tissue (including the blood vessels) increases toward the tip of the loop.

A recent major addition to the traditional explanation of kidney function has been forced by the discovery that a portion of the mammalian tubule actively transports *chloride* but probably not sodium. While active sodium transport is the basic process in proximal tubules, distal convoluted tubules, and the collecting ducts, the thick ascending limb of Henle's loop (Fig. 7-13), sometimes called the *diluting segment*, actively absorbs chloride from the tubule lumen but sodium movement is probably passive. Water permeability is low in this segment, thus accounting for the dilution of urine in this segment (Fig. 7-15).

There is extensive reabsorption of other cations in the thick ascending portion of the tubule, including calcium, magnesium, and potassium. The driving force for uptake of cations, in general, could be provided by the electromotive gradient generated by active chloride transport. A comparable portion of the proximal tubule in amphibians likewise actively transports chloride.

The collecting ducts must traverse the zone of increased osmolality in passing their content to the pelvis of the ureter. The walls of the collecting ducts are permeable to water, and here urine can be further concentrated. The tubule and collecting duct, by virtue of their geometry and the active movement of sodium out of the ascending limb, form a **countercurrent multiplier system** (Fig. 7-15).

If ADH is present, water accompanies sodium out of the distal tubule and also is osmotically withdrawn from the collecting duct, thus producing a more concentrated urine. In the absence of ADH, more water remains in the distal tubule and in the collecting duct, since the walls are impermeable to water. The result is a dilute urine.

This account of urine formation is quite abbreviated and simplified. The mechanisms involved are more complicated and controversial than this description would suggest. However, it is almost universally accepted that the principle of countercurrent multiplication underlies tubular function and that active transport of sodium occurs somewhere in the ascending limb. The anatomical peculiarities of the blood supply to the medulla suggest that kidney function may depend partly upon vascular arrangements. Also, the precise mechanism by which ADH increases permeability to water remains to be determined.

7-7
NITROGEN EXCRETION

Most of the food consumed by animals consists of carbohydrates, fats, proteins, and lesser amounts of nucleic acids. The metabolism of carbohydrates and fats results in the liberation of carbon dioxide and water. When proteins and nucleic acid are oxidized they also yield carbon dioxide and water, but in addition, nitrogen-containing compounds are produced. As noted in Sec. 5-7, deamination of proteins results in the removal of the amino group (—NH_2) and the appearance of ammonia (NH_3). Removal of ammonia is one of the major tasks of excretion. Ammonia is highly toxic and is always removed in solution—frequently in some detoxified form. The particular nitrogenous product excreted by an animal is closely related to the environment in which it normally lives, and in particular with the availability of water.

Ammonia

Ammonia is excreted as such only when there is an abundance of water for its rapid removal. It is extremely toxic and normally never accumulates in living cells or their surrounding medium. Marine invertebrates and freshwater vertebrates and invertebrates excrete ammonia.

Animals which excrete ammonia as the main end product of amino nitrogen metabolism are said to be **ammoniotelic** (Gr. *telos*, end). Water, however, is commonly too limited to allow pollution with ammonia. Such is the case with marine fishes and terrestrial forms. In their desiccating environments they have discovered how to make less toxic forms of ammonia, such as urea (**ureotelic** animals) or uric acid (**uricotelic** animals). In truth, most animals produce a mixture of nitrogenous wastes. For example, urea and uric acid can be found in the urine of both freshwater and marine invertebrates. However, there is characteristically only one main form of nitrogen waste.

Urea

The production and osmotic use of urea by elasmobranchs and other vertebrates have already been mentioned. Lungfishes excrete ammonia when they are active, just as other freshwater fish do, but when estivating they synthesize urea. Thus, they become ureotelic when water supply is restricted but do not bother with urea synthesis when an abundant supply of water is available for dilution of their ammonia. Such an option is probably of considerable advantage, because the synthesis of urea involves a considerable cost of chemical-bond energy. Urea synthesis is by no means a simple process (Fig. 7-16).

The tadpoles of frogs and toads excrete mostly ammonia; the semiterrestrial adults commonly excrete urea. Where the adult remains aquatic, as in the South African frog, *Xenopus*, only ammonia is normally excreted. However, if it is kept out of water for several weeks, *Xenopus* will accumulate urea in the blood. The change to urea excretion

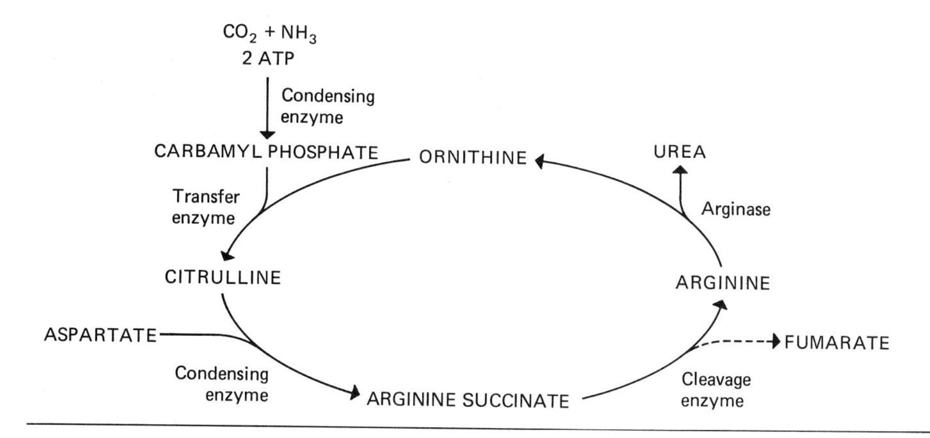

FIGURE 7-16
Synthesis of urea by the ornithine cycle. Ammonia and carbon dioxide condense with the amino acid, ornithine. Through several additional steps arginine is formed, which, with the aid of the enzyme arginase, splits off urea. Ornithine is thereby regenerated, and it can again enter the cycle.

at the onset of metamorphosis in the semiterrestrial forms is associated with marked increases in the activities of the liver enzymes of the ornithine cycle.

Uric Acid

Uric acid, a member of the purines, is the third conspicuous nitrogenous waste. It is highly insoluble and precipitates readily from supersaturated colloidal solution, permitting nitrogen removal in a solid form without loss of water. All the successful groups of arid-living animals (pulmonate snails, insects, birds, saurian reptiles) are uricotelic. As noted in Sec. 7-4 above, a few amphibians of arid regions also can excrete uric acid. They form it by a complex series of reactions (Fig. 16-8).

Joseph Needham has suggested that the difference between the higher vertebrates that form urea (mainly mammals and amphibians) and those that form uric acid (mainly reptiles and birds) is primarily related to their mode of reproduction. The amphibian egg develops in water, and the mammalian embryo exchanges wastes through the placenta with the blood of the mother. Reptiles and birds develop, in contrast, in a closed egg, the so-called *cleidoic* (Gr, *kleistos*, closed, from *kleis*, key) egg, where excretory products (except gases) must accumulate. In the limited water supply of the cleidoic egg, ammonia would be too toxic and urea would accumulate in solution to intolerable levels. Uric acid, by being precipitated, is essentially eliminated from the egg. Nitrogen excretion patterns closely follow phylogenetic lines (Fig. 7-17).

It would be erroneous to believe that ammonia is excreted only by aquatic animals. Ammonia is a normal constituent in the urine of terrestrial animals and serves to regulate the pH of the urine. It is added to the urine to balance excessive acidity. However, this ammonia is produced in the kidney from the amino acid glutamine and has no direct connection with the ammonia produced in the liver by the deamination of amino acids.

Urea in the mammalian kidney is filtered in the glomerulus and is

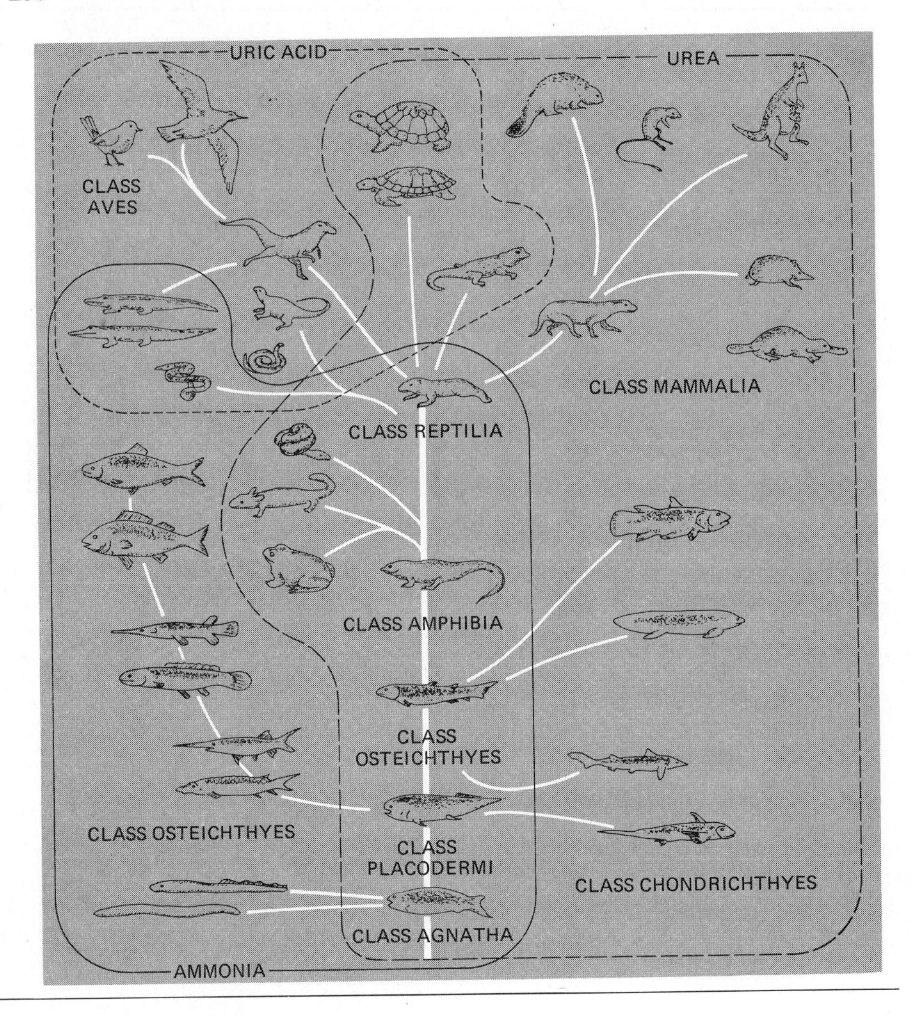

FIGURE 7-17
Nitrogen excretion in relation to the phylogeny of vertebrates. The lines enclose groups of animals which use ammonia, urea, and uric acid, respectively, as the major nitrogenous excretory product. This is a general scheme, and there are some exceptions, as noted in the text. (*After Schmidt-Nielsen, 1972, Nitrogen Metabolism and the Environment, edited by J. W. Campbell and L. Goldstein, Academic Press Inc., New York, p. 80.*)

presumed to be excreted passively. However, there is some evidence that urea may not be treated entirely passively, there being a well-regulated control of urea excretion. In elasmobranchs, urea is filtered in the glomerulus and reconvered from the tubule by active reabsorption. In amphibians, the tubules actively add urea to the urine. Thus, several kinds of animals actively transport urea, but they move it in different directions, depending upon the end result to be achieved.

READINGS

Baldwin, E., 1949, *An Introduction to Comparative Biochemistry*, Cambridge University Press, New York. In this small, very readable volume the basic osmotic and excretory problems are defined and the strategies for their solution are sketched.

Korgh, A., 1939, *Osmotic Regulation in Aquatic Animals*, Cambridge University Press, New York. Also available as a 1965 republication by Dover Publications, New York. This is a classic treatise, still well worth reading or consulting.

Pitts, R. F., 1968, *Physiology of the Kidney and Body Fluids*, 2d ed., Year Book Medical Publishers, Inc., Chicago. The intricacies of kidney function are presented in detail.

Potts, W. T. W., and G. Parry, 1964, *Osmotic and Ionic Regulation in Animals*, Pergamon Press, New York. A comprehensive, standard work in the field, covering various aspects of osmotic and ionic regulation, including that in terrestrial animals.

DEFENSE AND IMMUNE MECHANISMS

For cells to carry on their intricate processes, it is appropriate that they be protected, commonly to a great degree, from any large changes in their immediate surroundings. In the preceding chapters, many of the mechanisms that ensure some degree of homeostatic regulation of nutrition, osmolarity, respiration, and temperature have been indicated. Continuing to outline the standard patterns which animals have evolved to cope with environmental and internal instability, we now direct attention mainly to the interface between organisms and their environments. Those structures and systems which initially absorb the impact of physical variations and which protect against invasion by other organisms will be considered.

A living thing must exchange materials with the environment, yet do so selectively. It must protect itself against the loss of valuable substances, as well as against the entry of harmful agents. Thus, an organism must have well-designed barriers enclosing it but still be able to engage in selective commerce with the outside world. Failure of the peripheral lines of defense will lead to disruption of steady states in the protected community of cells and organs and, if severe enough, to the death of the organism (Fig. 8-1).

8-1 EXTRAORGANISMIC BARRIERS

The formation, secretion, or fashioning of some kind of protective, nonliving case, shell, capsule, or nest is a very common and effective way of providing protection against physical dangers and biological assaults. Plants, for example, employ *shells* to surround seeds, *bark* to protect stems, and *sporangia* for their asexual spores. The larvae of caddis flies (Order Trichoptera) construct distinctive, portable *cases* of leaves, plant stems, sand, or gravel that are bound together by a lining of silk secreted by glands opening near the mouth.

To protect its rather soft body, the hermit crab inserts the posterior

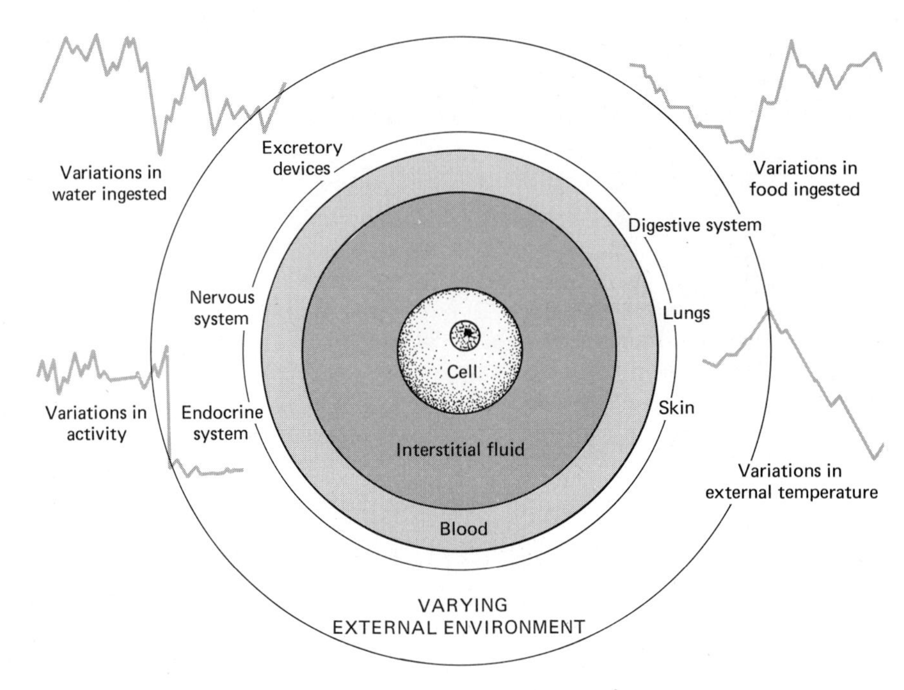

FIGURE 8-1
The cell, represented as being surrounded by layers of fluids and structures which "insulate" it from the hostile and variable physical environment. Each layer, from outside to inside, progressively buffers, or damps, any changes so that the cell is protected from major variances. (*After F. L. Strand,* 1965, *Modern Physiology, The Macmillan Company, New York, p.* 31.)

end of its body into an abandoned *shell* of a snail and carries it about as a refuge into which to retreat when danger threatens. Some hermit crabs place sea anemones on their shell, thereby providing camouflage and an improved defense with the nematocysts of the anemones. Such crabs have been reported to transplant the sessile associates to a new shell when they have outgrown the old one and are forced to change.

The *nests* of social insects provide a relatively stabilized physical environment and help to control temperature, humidity, and air flow at more optimal values, as well as to protect from predators. Nests of primitive social insects are often merely burrows in the soil or in dead wood. In the more advanced forms, elaborate structures may be built of glued earth particles. Others use plant material, such as wood particles and plant fibers (termites, ants, wasps), excretion (termites), or secretions from special glands (wax of bees). The nests of some social insects may serve for the storage of food or the cultivation of fungi (see Sec. 5-6). The social homeostasis attained through nest construction enables some insects to inhabit otherwise unfavorable places. For example, the mound nests of ants and termites permit them to invade grasslands or swampy regions that are periodically subject to flooding. Also, termites, which are susceptible to desiccation, can inhabit deserts because of the constructions of nests and shelter tubes.

The *nests* of birds hold the eggs and protect the bird's young. These structures range from a few pebbles or bits of grass on the ground, as in shore birds, to elaborately woven nests of leaves, as in the weaverbirds or orioles. Chimney swifts construct nests from small twigs, glued together with saliva. In Malaya and the South Sea Islands, some swifts built nests almost entirely of saliva. Those nests built by swifts of the genus *Collocalia* are collected and used in making the luxurious concoction called bird's-nest soup.

The female bird secretes a *calcareous shell* around the egg during its passage down the oviduct. This protective enclosure retards water loss but permits gaseous exchange. At the end of the incubation period, the chick must liberate itself from its limestone shell. In preparation for this ordeal, the maturing chick develops a short, pointed, horny "egg tooth" at the tip of its upper mandible. Also a special set of muscles, located largely on the upper side of the neck and head, and supposedly used to force the head and egg-tooth outward against the shell, are maximally developed just prior to hatching. Both these structures disappear shortly after emergence. The length of time required for the chick to hammer its way out of the shell is from a few hours to a few days, varying greatly from species to species. The egg shell must be thick enough to be sturdy, yet thin enough to permit the chick to break out of it.

Burrowing is a common way to avoid predators and environmental changes. Bivalve mollusks make use of their hydrostatic skeleton in the foot for such a purpose. Water is ejected from the mantle cavity to loosen the sand, and then the foot is protruded under pressure; the end of the foot is dilated to provide a pedal anchor, and the siphons close. The valves (shell) then partly close, and retractor muscles of the foot pull the shell downward. The valves then open more, forming a shell anchor and the foot is retracted (Fig. 8-2).

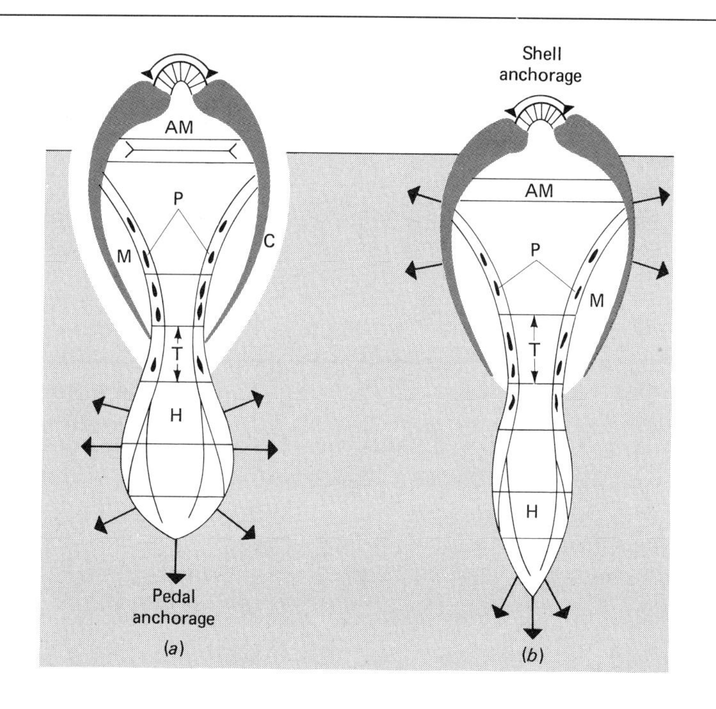

FIGURE 8-2
Diagram of stages of burrowing of a bivalve mollusk; pedal anchorage and shell anchorage are shown by arrows. (*a*) Valves adducted by muscle (*AM*), producing pedal dilatation and a cavity (*C*) in sand around valves. (*b*) Valve reopened by elasticity of ligament, holding shell fast while contraction of protractor (*P*) and transverse muscles (*T*) causes pedal protraction. *H*, pedal hemocoel; *M* mantle cavity. (*After E. R. Trueman*, 1966, *Biol. Bull., 131*, p. 374.)

8-2 SHELLS, CUTICLES, AND SKINS

Mollusks are endowed with remarkable body coverings. Their *shells* provide excellent protection, but the disadvantages are that they impede locomotion and require much energy and material to make. The gastropod mollusks produce a shell that may be coiled (helicoid or turbinate) or may be in the form of a flattened spiral or a short cone. Some cephalopods form chambered, coiled shells, as in *Nautilus*. Bivalve mollusks, such as clams, mussels, and oysters, form bilaterally symmetrical, calcareous *valves* which are united by an elastic hinged ligament.

The *shell*, or *valve*, is secreted by an outer part of the mantle. It is mainly made up of calcium carbonate crystals enclosed by a covering of protein fibers. The crystals may be of calcite or of aragonite. The flexibility or brittleness of the shell corresponds to differing relative amounts of protein and crystalline calcium carbonate. The three layers composing the shell are shown in Fig. 8-3.

The *integument* of *arthropods* not only serves as a protective barrier between the outside environment and the internal organ systems, it also is an **exoskeleton** from which organs are suspended and on which attached muscles do work. In addition, the exoskeleton bears a variety of sensory devices. This type of skeleton is size-limiting, becoming inconve-

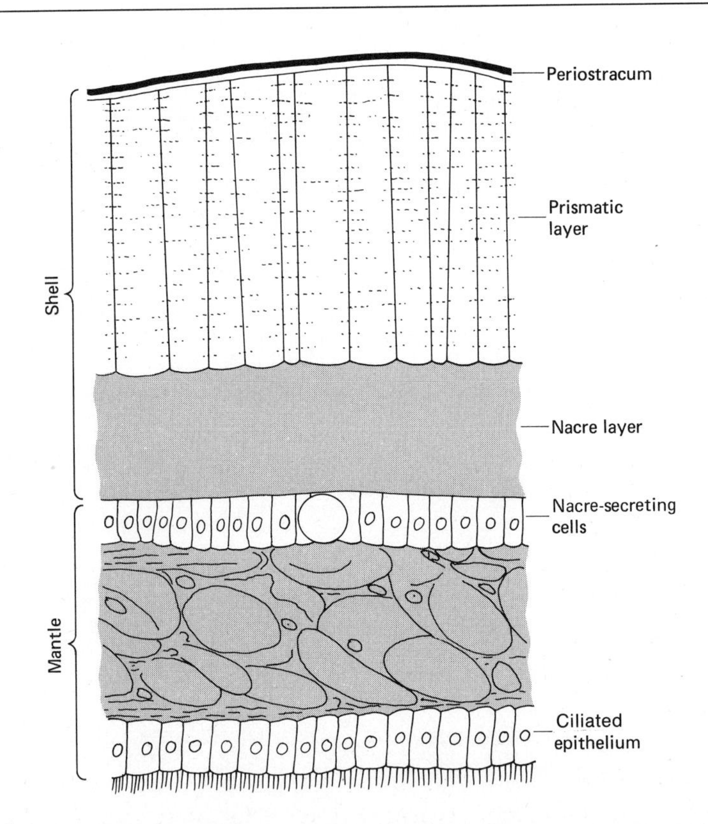

FIGURE 8-3 Enlarged cross section of shell and mantle of a fresh-water clam. The external *periostracum* is a thin, colored, horny covering that protects the inner layers from being dissolved by carbonic acid in the water. The *prismatic layer* is composed of crystalline calcium carbonate. The inner *nacre*, or "mother of pearl," formed of many thin layers of calcium carbonate, has a slight iridescence. The shell grows in area at its margin by secretion from the *mantle*, and in thickness by successive deposits of nacre. (*After T. I. Storer et al., 1972, General Zoology, McGraw-Hill Book Company, New York, p. 495.*)

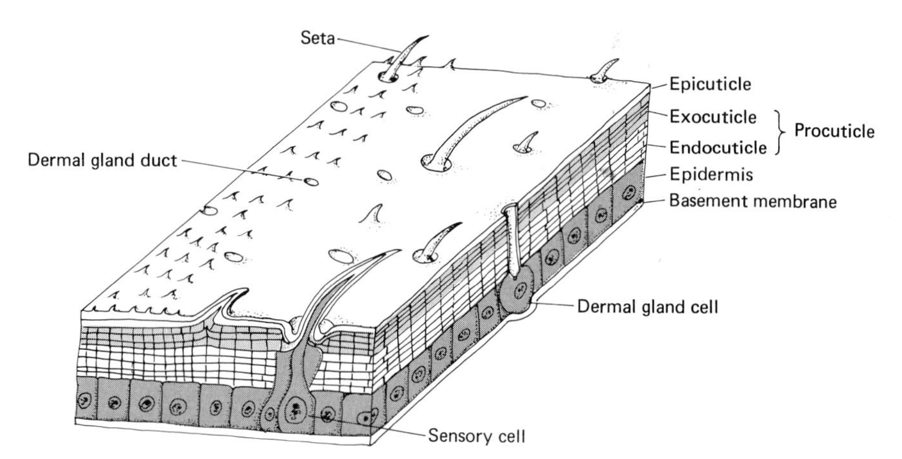

FIGURE 8-4
Digrammatic representation of a typical section of insect integument. (*After R. L. Patton, 1963, Introductory Insect Physiology, W. B. Saunders Company, Philadelphia, p. 190.*)

niently heavy with increasing size. It must be shed, episodically, to permit bursts of growth. The period of shedding and renewal is a hazardous one, for then the animal is vulnerable to physical damage, and there is always the possibility of failure in the precise timing of one or more of the events involved in the complex process. Also, exoskeleton renewal is a costly process in energy and materials. The arthropod integument completely invests the body externally and extends internally to line such structures as the foregut and hindgut, and the air tubes of insects.

The exoskeleton consists of three zones. There is a *basement membrane,* which lines the integument and separates the body cavity from the cellular *epidermis.* The epidermis includes the glands of the integument and is responsible for secreting the nonliving layers. The layer next to the epidermis is called the *procuticle.* It consists of many fused lamina, forming a massive structure that contributes to the mechanical properties of the exoskeleton (Fig. 8-4).

One of the unique constituents of the procuticle is **chitin,** a polysaccharide which resembles cellulose except that an acetylamine is substituted on the number 2 carbon in place of the hydroxyl group (cf. Fig. 5-4). The procuticle also contains proteins. The *epicuticle,* the outermost layer, is responsible for most of the barrier features of the exoskeleton. This layer is nonwettable, is almost impermeable in a physicochemical sense, and provides protection against microorganisms. In insects, the epicuticle has an outermost layer of polymerized lipoprotein, secreted from pore canals of the epidermal cells which extend through the procuticle and the epicuticle. Immediately below this is a *polyphenol layer,* then a *wax layer,* and finally a cement layer composed of *cuticulin.*

The procuticle remains flexible but nonelastic in such places as joints of the arthropod limbs and body. To make it more protective, most of the procuticle is hardened and becomes rigid. In the crustaceans, this is accomplished by the deposition of calcium carbonate in the middle layer of the procuticle. Another method, used to some extent by all arthropods, involves the so-called *tanning* of the protein component in the procuticle layer. This involves cross-linking of the protein chains by orthoquinones and incorporation, in addition to proteins, of polyphenols and polyphenol oxidases.

Although the integument greatly retards evaporative water loss, some insects, ticks, and mites have the unique property of being able to absorb water from an atmosphere which is humid but less than saturated. The mechanism of absorption of water vapor is unclear. Water moves inward three times more readily than it moves outward. The absorption occurs only in living insects.

Vertebrate **skin** is a complex structure, consisting of an avascular *epidermis*, a vascular *dermis*, and subcutaneous *fat*. The epidermis may be regarded as the protective coat of the animal, but, unlike the cuticle of arthropods, it is entirely cellular, being made up typically of an outer,

FIGURE 8-5

(*a*) The components of the skin are illustrated schematically to indicate their anatomical relations. (*b*) Cells in the germinal layer of the epidermis divide, and the daughter cells migrate toward the surface, specializing into flat cells of the stratum corneum. (*c*) The skin layers vary greatly in thickness in various regions of the body. (*After R. F. Rushmer et al.*, 1966, *Science*, *154*, p. 344.)

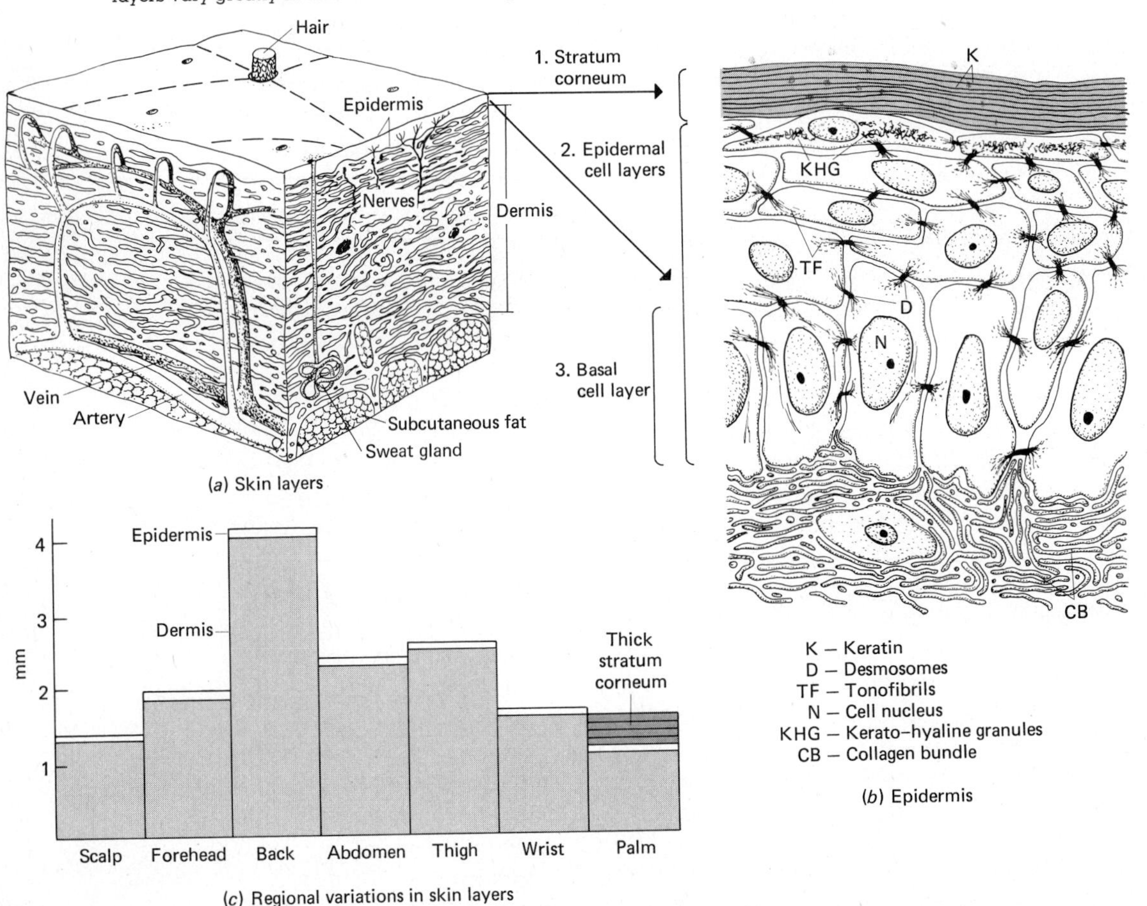

horny, *stratum corneum* and an inner *stratum germinativum*. The corneum is continually shed and replaced as the cells of the germinativum become "keratinized" into cells of the corneum (Fig. 8-5).

Keratin is a scleroprotein synthesized by the epidermis. It consists of protein strands of amino acids held together by hydrogen bonds. It also is found in accessory structures of the skin, including nails, hair, horn, and feathers. Because keratin is elastic the skin is quite distensible.

The skin has four major functions: (1) *protection:* against invasion by foreign organisms, such as bacteria and viruses; against mechanical injury; against dehydration; and against some wavelengths of electromagnetic radiations; (2) *perception:* the detectors of changes in the environment are located in the skin; (3) *thermoregulation:* the skin is one of the components of the thermoregulatory system; across it heat loss can be either increased or decreased (e.g., by changing degree of piloerection or blood flow); and (4) *excretion:* glands in the skin excrete salts, nitrogenous end products, and other wastes, making the skin a site of exchange with the environment. Sweat excretions of mammals evaporating from the skin accelerate heat loss. In some vertebrates the skin may secrete offensive substances that repel predators.

The skin extends around and into the body to line the lungs, digestive tract, and excretory-genital organs, providing a mechanical, chemical, and physiological barrier between the vertebrate and the environment. Human skin is remarkably effective in retarding diffusion of gases, water, and many other types of chemicals. Body constituents are effectively retained within the skin. It is not a perfect barrier, however. Most irritants and cellular poisons are unable to penetrate it, but there are a few notable exceptions, including mustard gas and the irritant of poison ivy. Several kinds of bacteria can cross the skin, and several kinds of fungi can form colonies on the surface of it. Nevertheless, the skin does offer a rather high degree of protection against invasion.

Long-chain fatty acids and soaps give the skin an "acid coat" (pH about 3.5). Other antimicrobial factors, such as *lysozymes*, provide autodisinfection mechanisms by breaking down bacterial cell walls. Tears also contain these enzymes. Furthermore, skin is commonly differentiated into mucous membranes, such as those lining respiratory cavities, that serve to trap particulate matter. Commonly mucous membranes are ciliated, giving them the capability of ridding themselves of trapped debris. Mucous secretions, such as tears, saliva, and bile, often serve in mechanical cleansing. Mucus also serves as waterproofing for fish and amphibians.

The principle of use of polymer fibers as a major strengthening element is employed, as we have seen, in the shells of mollusks, the integument of arthropods, and the skin of vertebrates. In mollusks and vertebrates, protein fibers are used. In arthropods, the fibers are polysaccharide. Plant cell walls also employ a polysaccharide, cellulose in this case, as the main strengthening material. Mainly due to the properties of cellulose, plant cell walls have low elasticity, reduced flexibility, and very high tensile strength (comparable to that of spring steel). Very conveniently, the cell wall is also transparent, permitting chlorophyll-containing cells to engage in photosynthesis.

8-3 VENOMS AND TOXINS

Many animals use poisonous secretions to make themselves offensive to potential predators or to incapacitate or kill prey. Sometimes these two categories of function are served by the same devices and substances.

The **nematocysts,** or stinging capsules, of cnidarians discharge neurotoxins. These poisons appear to be of several kinds, but they are probably all protein.

Many of the larger nemertines (ribbon worms) are armed with one or more stylets on the proboscis, which are used in feeding. It is thought that some have toxic secretions to aid in its use.

The sting of bees, a modified ovipositor present only in workers and in queens, delivers from a large median poison sac a venom that can kill small invaders. The poison of spiders can quickly kill invertebrates, and large spiders can even kill small vertebrates. Several hundred human cases of poisoning by black widow spiders have been recorded, with about 5 percent mortality. Some scorpions use a powerful neurotoxic venom.

The pufferfishes (Tetraodontidae) contain a potent nerve poison (*tetrodotoxin*), found mainly in the liver and roe. Evidently it is identical with the toxin in newts (*Taricha*). Tetrodotoxin has been widely employed by neurophysiologists in recent years because it selectively blocks the action potentials of neurons but does not alter the generator potential.

The skin of amphibians contains poison glands, sometimes in special groups—warts on toads, dorsolateral folds on frogs, and the upper surface of the tail of some salamanders. The toxin of frogs is a thick, whit-

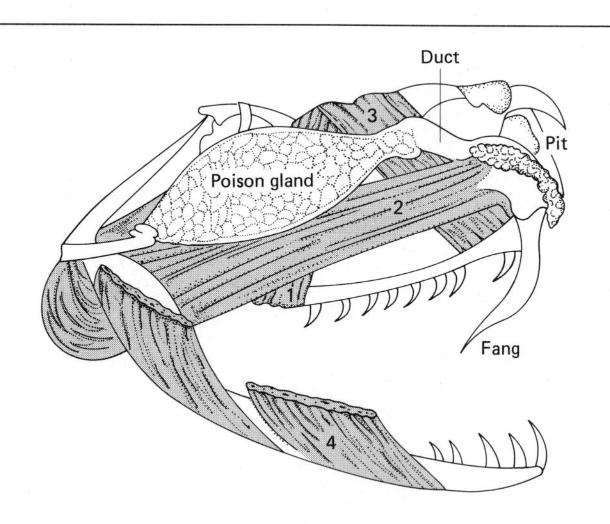

FIGURE 8-6
Rattlesnake: mechanism of the head used in striking. The sphenopterygoid muscle (1) contracts to rock the fang forward; after the fang enters the victim, the external pterygoid (2) and sphenopalatine (3) muscles contract to draw the fang more deeply; then the anterior temporal muscle (4) draws up the lower jaw and compresses the poison gland to force venom through the duct and fang into the wound. (*After T. I. Storer et al., 1972, General Zoology, McGraw-Hill Book Company, New York, p. 727.*)

ish, granular alkaloid that causes a burning sensation and protects the animal to some degree from predators. Some amphibians have buccal glands that produce toxic materials.

The skin secretion from a small, brightly colored frog, *Phyllobates aurotaenia*, has long been used by Indians of the Choco rain forest of Colombia to prepare the deadly darts for their blowguns. The active principle, called *batrachotoxin*, is a pyrrolecarboxylic ester of a novel steroidal base with unique and selective actions on a variety of electrogenic membranes. It blocks neuromuscular transmission irreversibly, induces convulsions, and causes muscle contracture.

Only two species of lizard, *Heloderma horridum*, the beaded lizard, and *H. suspectum*, the Gila monster, are venomous. Their venom is as poisonous as that of a rattlesnake, but they deliver it effectively only to captured prey on which they are chewing. All other poisonous reptiles are snakes. They use venom for the capture of small prey and defensively against large animals. Venom is secreted by a pair of glands, one on either side of the upper jaw, and delivered to the victim by paired fangs thrust out from the front of the upper jaw (Fig. 8-6).

Some of the snake venoms contain a phospholipase, capable of hydrolyzing lecithin to produce a powerful hemolytic agent, lysolecithin. Others contain proteolytic enzymes which act as thromboplastic substances and cause extensive intravascular clotting. The venoms of rattlesnakes and vipers (Viperidae) usually act mainly on the circulatory system, causing breakdown of capillary walls and disruption of blood cells. That of the cobra and its relatives (Elapidae) acts upon the nervous system, particularly on respiratory centers, causing death by asphyxiation.

8-4 BODY-FLUID BUFFERING AND CLOTTING

In Chap. 6, the functions of body fluids and circulatory systems were outlined. Some of the functions are the result of unique features, such as gas transport by special pigments. Most of the functions, however, can be attributed to the fact that the fluid is aqueous; the properties of water are capitalized upon. For example, blood and other kinds of body fluids transmit force. This is useful in such actions as burrowing of mollusks, as discussed earlier, in earthworm locomotion, and in ultrafiltration in the kidney glomeruli. Body fluids also dissipate force and reduce the possibility of injury to an animal. Pressure applied in some localized region, causing compression, will be transmitted hydraulically throughout the body fluids and be relieved by expansion of elastic regions elsewhere in the body.

Because body fluids are mostly water, they afford thermal stabilization. This is simply because, as noted earlier (Sec. 4-4), water has a very high specific heat. The proportionately large volume of water contained in animals serves as a heat sink with a large capacity for metabolically generated heat or for heat from external sources. There is, correspondingly, a slow fall in body temperature during heat loss because of the large heat capacity of water.

Body fluids, especially the rapidly circulated blood of birds and mammals, transport heat from deeper organs to the body surface for dissipation. This is especially important during vigorous use of muscles, as in long-distance running or flying. Surface capillaries can be expanded (as indicated by flushed skin in the exercising human) to dissipate heat more rapidly, while the circulatory system acts as a heat pump, carrying heat from skeletal muscles and trunk organs to the skin.

Generally, body fluids have the capacity to *buffer* internal changes in pH. Most of this capacity is due to chemical buffering, but there is also simply a dilution effect in the large volume of the body fluids. Similarly, there is dilution of nutrients and wastes in the body fluids. Thus, rapid changes in the concentration of substances put into or taken from the body fluids are prevented. Where fluids are pumped and circulated, there is even better buffering against changes in chemical composition because of rapid mixing with the entire volume of body fluid.

An inherent characteristic of many circulatory systems is the ability to retard or prevent the loss of body fluids. The most primitive arrangement depends solely on the contractility of body musculature and blood vessels. For soft-bodied invertebrates these are adequate *hemostatic mechanisms*. For hard-bodied animals, such as arthropods, many echinoderms, and mollusks, with even moderate blood pressure, additional mechanisms are employed. Their vascular fluids contain cellular elements and, in some species, special jelling proteins which form plugs at injury sites. In the more primitive groups, including most echinoderms, some arthropods, and some mollusks, only blood cells are involved in *clot* formation. The cells of crustaceans produce, in addition, fibrous proteins which serve to entangle other cells and thus increase the size and strength of the clot. The fibrous material is not biochemically similar to the fibrin of vertebrates.

Several mechanisms help prevent blood loss from ruptured blood vessels of vertebrates. The loss of blood following puncture of a vessel, if severe enough, will lead to a decrease of blood pressure and thereby reduce the flow of blood to the damaged region. Also, the muscular blood vessel automatically contracts when damaged and reduces blood flow. However, the most effective hemostatic mechanism is plug formation by coagulated proteins and blood cells.

The *coagulation* process has been well studied in mammals, especially in humans. An understanding of the clotting process is obviously of great medical importance. If the blood does not properly form clots, death to the individual may result. The well-known condition of *hemophilia* is characterized by failure of clot formation. It is essential also that the size and duration of the clot be rigidly controlled. Once initiated, the clotting process has the potential for spreading excessively through the vascular channels and blocking circulation. Normally this does not happen. The coagulation mechanism is an example of remarkable autoregulation.

For a long time it was puzzling as to why the process of blood clotting in the mammal was so complicated. An extraordinarily large number of factors—at least 12—play a part in the human process. In 1964, McFarlane proposed a rather simple explanation; the clotting factors in human blood appear to form a chain of *proenzymes*. When one is activated, it in turn activates the next proenzyme in the sequence. These factors are usually specified by Roman numerals—I, II, III, etc. Some also bear special names, such as Christmas factor (IX). (An English boy with the surname of Christmas was injured in a bombing during World War II and proved to be deficient in a clotting factor; the factor found in normal blood was named after him.)

The clotting process is initiated when the blood is exposed to an unusual surface, as that resulting from injury. The final steps in the long sequence of proenzyme to enzyme conversions are as follows:

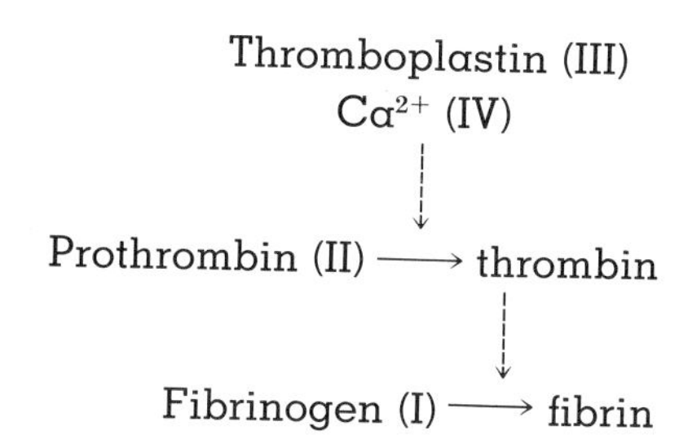

McFarlane refers to this sequence of events as a cascade of proenzyme-enzyme transformations and recognizes its role as a biochemical amplifier in which enzymes are analogous to photomultiplier or transistor stages. In terms of control theory, the system could be represented as:

$$X \longrightarrow \boxed{G} \longrightarrow Y$$

where X is the initial activation by surface contact; the gain or transfer function, G, is provided by the several proenzyme-enzyme stages; and the output, Y, is fibrin, an essential substance for the clot. In such a system, a very small input can result in an enormously large output.

Consider the difference between a "zero-gain" system and the clotting amplifier system. The glycolytic process of cell respiration is a zero-gain system and could be represented as:

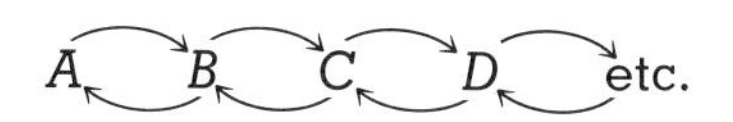

This system can yield an output no greater than the input. For contrast, consider the following:

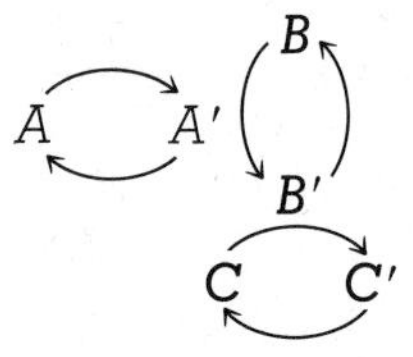

$$A \rightleftarrows A' \quad B \rightleftarrows B' \quad C \rightleftarrows C'$$

If A' and B' represent catalysts which transform other catalysts to active states, such a system might involve any degree of amplification. It is by this latter design that blood clotting works.

It is important that a good clotting system respond quickly when the input signals for clot initiation. It also must act vigorously. The latter requirement seems to be provided for by a positive feedback (or feedforward) effect. Thrombin appears to accelerate the reaction associated with the conversion of perhaps two different proenzymes of the clotting sequence. It is equally important that the output cease quickly when the stimulus for clotting is no longer present. This seems to be achieved by (1) rapid inactivation or decay of the clotting factors (thrombin appears to have a mean lifetime of 24 seconds), and (2) negative feedback provided by fibrin, which seems to accelerate the disappearance of other clotting factors. Thus, the size of the clot and its duration are closely regulated and directly related to the magnitude of the injury.

8-5 IMMUNE SYSTEMS

The first line of defense of an animal is its skin or integument, as noted above. **Phagocytosis** by motile or sessile *reticuloendothelial* (RE) *cells* is often referred to as the second line of defense. These cells can ingest and digest bacteria, viruses, and defunct cells. The phagocytic systems of the human include the motile *leukocytes* of blood and lymph, the microglial cells of the central nervous system, and the reticulum cells or Kupffer cells of the liver, bone marrow, spleen, lymph nodes, and other sinusoidal tissues.

Blood clotting or infection is often followed by **inflammation,** another form of defense, especially for isolating a wound. In an inflamed region blood vessels constrict, and flow becomes sluggish. The vessel walls become sticky for leukocytes which attach and form a coat. Vessels then become dilated, and leukocytes crawl through the wall into the injured or infected region. Plasma fluid may leak through the vessel wall and form a clot at the periphery of the inflamed area. Escaping infectious agents are removed from the lymph by the RE cells of the lymph nodes.

The protective systems mentioned so far are commonly called *nonspecific* defense systems. We now come to *specific* defense systems.

Interferons are a class of proteins produced in animals in response to various viral or nonviral substances, which have the ability to make cells resistant to viruses. Infection by viruses is the natural inducer of interferons. Interferon responses may occur in all vertebrates, since in vitro cultured cells of fish, turtle, chickens, and various mammals produce interferons in response to viral infection. The interferon response differs in several ways from the immune response (discussed later). The immune response is confined to a specialized cell system, whereas most, if not all, cells in the body can make interferon. Furthermore, the interferon response is rapid (measured in hours) compared to the immune system (measured in days). The induced interferon has a very short half-life, however—several hours or a few days—and, again unlike the immune response, a second dose of an inducer does not raise the level of interferon.

There are two kinds of interferons: low-molecular-weight (20,000 to 30,000) that are synthesized by all kinds of cells; high-molecular-weight (50,000 to 100,000) that are synthesized only by the RE system. Interferons of either low or high molecular weight protect cells from a broad spectrum of DNA and RNA viruses by conferring on infected cells the ability to inhibit replication of the virus. As a consequence, spread of the virus is reduced or stopped. Interferons are effective only in the species that produce them and, therefore, are not useful as protective agents between different animal types.

Many groups of animals have the special capacity for removing or inactivating specific bacteria, viruses, and large organic molecules (usually mol. wt. 10,000 or more.) These mechanisms depend upon the elaboration of specific proteins, **immunoglobulins,** which agglutinate, precipitate, neutralize, or dissolve foreign organisms and substances. The agent which evokes such a response in an organism is called an **antigen,** and the protein produced to counteract its presence is called **antibody.** The process generally is referred to as the **specific immune system.**

The *spleen* and *lymph nodes* are the major sources of cells involved in the immune system. The cells are (1) the *lymphocyte* (immunologically competent cell), (2) the *plasma cell,* and (3) the *macrophage.*

One subclass of cells of the immune system, called *T lymphocytes,* depend for normal differentiation and development on the *thymus gland.* The thymus is active only in the very young animal, so this portion of the immune system has its unique properties permanently established early in an organism's life. A polypeptide hormone that can be prepared from thymus glands, **thymosin,** will cause cells to be transformed into T lymphocytes. The graft-versus-host reaction, transplant rejection, immune surveillance of tumor cells, delayed hypersensitivity to foreign antigens, and resistance to viruses all depend on this portion of the immune system. These are the provinces of so-called *cell-mediated* immunity.

T lymphocytes also participate in the regulation of *humoral immu-*

nity, the province of *B cells* (for bone marrow), the second subclass of lymphocytes. When properly stimulated by antigens, B cells differentiate into antibody-secreting cells, the *plasma cells*. The antibodies produced are large proteins (mol. wt. 150,000 to 800,000) of a class called γ-immunoglobins. Antibodies are effective against a number of kinds of antigens, including bacteria and viruses. The T lymphocytes may exert their regulation through *helper T cells*, which are needed for synthesis of many antibodies. There also are *suppressor T cells*, which repress antibody production.

Proteins, either as part of an organism or in solution, are usually what evoke production of antibodies, but large polysaccharides also can be effective. The first exposure to the antigen causes a *primary response*. Although antibodies are produced, the response is usually not overtly detectable. However, a second exposure to the same antigen after the primary response has decayed usually leads to a *secondary response*. In the secondary response, there is rapid and vigorous production of antibodies. In fact, there may be a violent reaction, producing a variety of overt symptoms which, in some instances, may lead to death (Fig. 8-7).

The antibodies seem to act by bridging together foreign particles, clumping them together. The clumps may then be dissolved or be phagocytized by RE cells. These reactions are remarkably specific, especially in mammals. Exposure to an antigen may cause development of immunity only to that one foreign antigen or to a small group of very similar antigens.

The proper result of the sequence of actions in the specific immune response is protection of the integrity of the organism against invasion and accumulation of foreign and, therefore, potentially harmful organisms or substances.

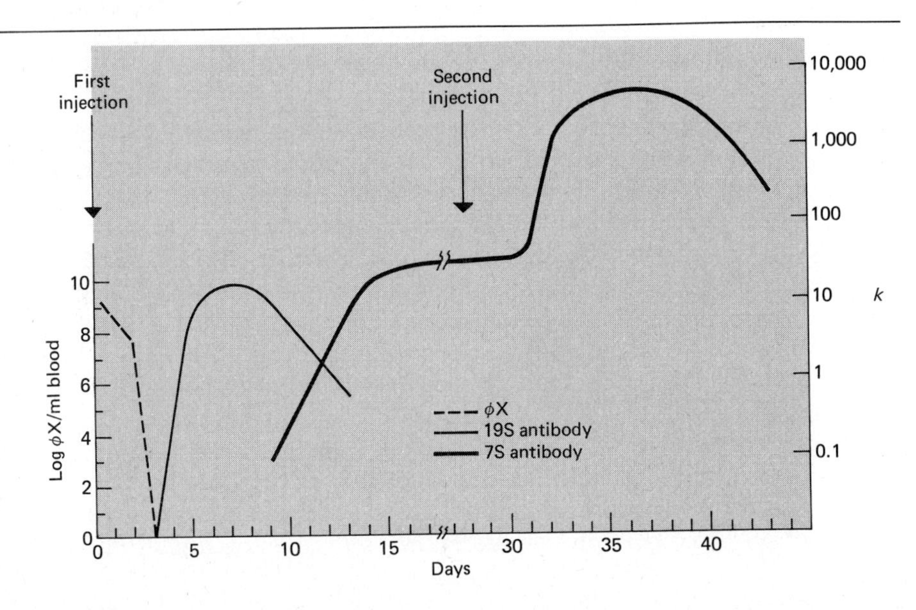

FIGURE 8-7
Sequence of antibody response to bacteriophage (ϕ X 174) in guinea pig given two intravenous injections of phage administered 1 month apart. After a latent period, antibodies belonging to the γM-immunoglobulins, with a sediment coefficient of 19S, are produced for a few days. About 1 week after this injection, antibodies of the γG-immunoglobulin class (7S sedimentation coefficient) start appearing. The second injection causes a great increase in 7S antibody production. Both 19S and 7S antibodies apparently are directed against the same antigen. (*After J. W. Uhr, 1964, Science, 145, p. 458.*)

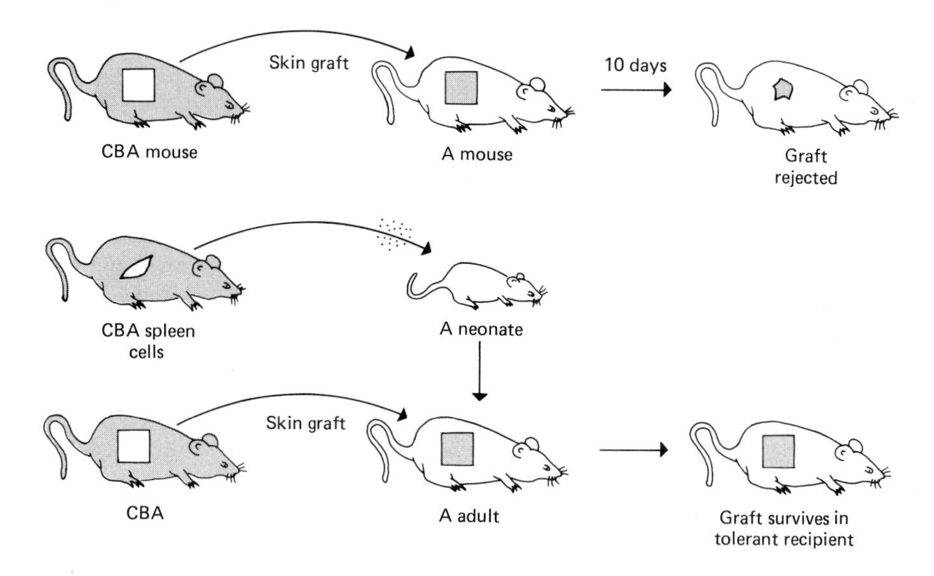

FIGURE 8-8
Production of immunological tolerance to foreign skin grafts by injection of cells into neonatal mice. (*After S. Leskowitz, 1968, Bioscience, 18, p. 1030.*)

The immune system does not normally attack the body's own tissues, but sometimes this tolerance for self is lost and *autoimmunity* results. Several debilitating diseases, including rheumatoid arthritis and systemic lupus erythematosus, are thought to be autoimmune conditions. Some investigators think that suppressor T cells normally prevent the production of autoantibodies.

Immunological tolerance is obviously of great importance in human health. Although it is surgically possible to transplant almost any organ from one organism to another, it is commonly not feasible because of the virtually inevitable rejection of such grafts by the immunologic response of the recipient to the donor's foreign antigens. It appears that the *macrophage* cells play a major role in the rejection response.

Experimentally, animals can be rendered unresponsive to potentially antigenic tissue grafts. Medawar, Billingham, and Brent reported in 1955 success with skin grafts between different strains of mice. Normally such grafts are destroyed by an immunologic reaction in a matter of days. However, injection of living spleen cells from one strain of mouse (CBA) into neonatal mice of another strain (A) produced a situation in which, at maturity, the recipients did not reject grafts of skins from the donors (Fig. 8-8).

The basis of the immune response is not fully understood, at either the cellular or molecular level. How immunity and tolerance are delicately balanced, how tolerance to self is established and maintained, and how an organism can adapt its immune responses to constantly changing kinds and amounts of antigens are among the intriguing problems that are still far from being solved.

READINGS

Barrett, J. T., 1974, *Textbook of Immunology*, The C. V. Mosby Company, St. Louis.

Bücherl, W., E. E. Buckley, and V. Deulofeu (eds.), 1968, *Venomous Animals and Their Venoms*, Academic Press, Inc., New York, 3 vols. The venoms of mammals, snakes, lizards, frogs, fishes, and invertebrates are discussed in detail.

Champion, R. H., et al. (eds.), 1970, *An Introduction to the Biology of the Skin*, Blackwell Scientific Publications, Ltd., Oxford.

Nour-Eldin, F., 1971, *Blood Coagulation Simplified*, Butterworth & Co. (Publishers), Ltd., London.

III

INTEGRATION: Organismic Regulatory Mechanisms

In Part II, attention was focused mainly on adaptive units of function. Lungs, hearts, kidneys, intestines, and other devices were analyzed as components in respiratory, nutritional, circulatory, excretory, thermoregulatory, and protective systems. All these systems, however, are abstractions from the still more complex total organism. It is obvious, for example, that respiration is causally related to nutrition through metabolic chemistry and that excretory output reflects preceding metabolic processes. Mechanisms that allow the animal to behave as a whole organism are the subjects of the next four chapters.

It will become apparent that the multitude of processes which together integrate the activities of the diverse parts of living things actually are varieties of a few common mechanisms. Furthermore, it will be seen that within classes of mechanisms, integrative processes which appear quite unrelated when observed at the more superficial level of the total organism, upon closer examination turn out to be basically very much alike.

CHEMICAL INTEGRATION 9

Underlying virtually all the physiological activities as well as the development of higher organisms, there is a complex regulatory and integrative network which uses specific compounds as chemical signals emitted by certain cells and interpreted by others. . . . The development, the functioning and the survival of complex organisms would hardly be conceivable, were it not for the existence of these regulatory chemical integrations between cells, tissues, and organs.[1]

Some of the signal molecules used by organisms are metabolic intermediates or end products; others are nonmetabolic, specific molecules synthesized for export and tailored to trigger certain activities in distant cells. Several of the integrative, informational molecules are produced by localized clusters of cells (glands), whereas others come from widely scattered sites within an organism. When one attempts to classify these agents into logical categories, many exceptions are found and numerous qualifications must be added.

When in 1902 Bayliss and Starling, two English physiologists, introduced the term "hormone" (with special reference to a substance they called **secretin,** which is produced by cells of the duodenal mucosa), they were primarily interested in delineating a type of integration that could occur without assistance from the nervous system. **Hormone** (Gr. *hormon,* exciting, setting in motion) was meant to refer to a "chemical messenger," but since those uncluttered years at the turn of the century many blood-borne chemical messengers have been discovered that are not hormones. Additionally, carbon dioxide and urea can enter the vascular system from numerous loci and produce physiological responses at various sites. Such substances, which have general origins, in contrast to the hormones of specific glandular sources, are commonly known as **parahormones.**

In addition to hormones and parahormones, other chemical messengers include **phytohormones** (plant hormones) of several different kinds, **neurohormones** and **neurosecretions** that are secretions of neurons (see following chapter), and **pheromones** (discussed in Chap. 3). These are not entirely satisfactory categories, nor are others that are currently used, and undoubtedly we shall see the terminology in this field evolve in the future.

[1] J. Monod, 1966.

9-1 PARAHORMONES

There are several compounds whose action in animals is like that of hormones but which are not hormones by definition, because of their diffuse origin. A few of the better-known ones will be briefly characterized.

Angiotensin

That there is an important set of relationships between mammalian kidney function and blood pressure and volume regulation has been known for over 100 years. Diseases of the kidneys commonly are correlated with chronic high blood pressure, or *hypertension*. Experimentally, it was demonstrated several years ago that extracts of kidney tissue would cause acute hypertension. More recently it has been shown that if flow of blood through the kidneys is reduced by such a method as applying an adjustable clamp to the renal artery, producing a condition called *ischemia*, hypertensive states can be produced. This procedure and several others which result in a deficiency of oxygen in the renal tissues can cause the blood pressure to rise.

If a kidney is transplanted into the neck by joining the renal artery with the carotid and the renal vein with the jugular, and then is made ischemic by clamping the artery, typical hypertension ensues. Thus, a kidney completely devoid of nervous connections must be exerting its effect on the vascular system by means of a chemical agent.

It appears that the key informational molecule is **renin**, an enzyme produced by the juxtaglomerular cells of the kidney and released into the blood under certain conditions, including ischemia. This triggers a series of reactions, beginning with the conversion by renin of **renin-substrate**, a globular protein in blood that originates in the liver, to **angiotensin I**. The amino acid sequence of angiotensin I is:

Asp-Arg-Val-Tyr-Iso-His-Pro-Phe-His-Leu

Another blood protein, known as *converting enzyme*, cleaves histidine and leucine from this decapeptide to produce the octapeptide, **angiotensin II**. It is this molecule which, by constricting arterioles, causes the hypertension of renal origin (Fig. 9-1).

The principal action of angiotension II is to cause a general constriction of smooth muscle, but it apparently acts also on the zona glomerulosa of the adrenal gland to cause **aldosterone** to be released. Aldosterone, as noted in Chap. 7, is the principal agent of the adrenal that enhances sodium and water resorption by the kidney. The blood pressure—volume changes effected by the principal actions of angiotensin influence the juxtaglomerular cells in their production of renin, thus completing a functional "loop."

There are other peptides that largely duplicate the physiological effects of angiotensin. For example, **vasopressins**, the antidiuretic hormones of mammalian neurohypophyses, cause elevation of blood pres-

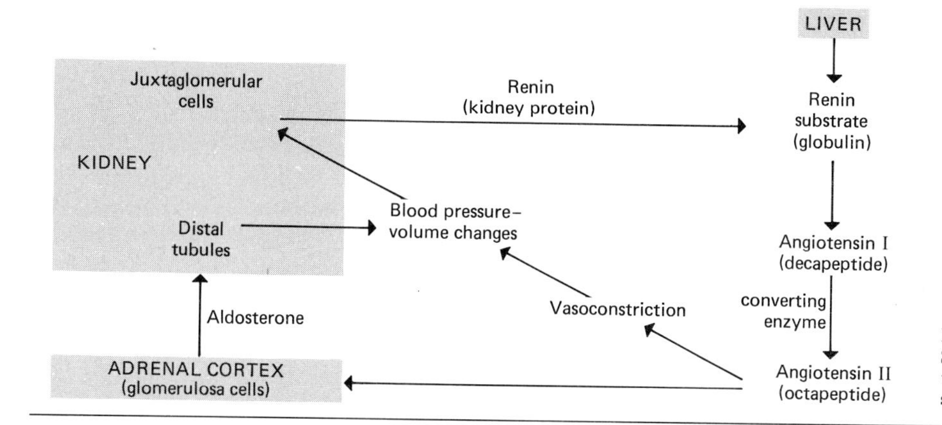

FIGURE 9-1
Schematic representation of the renin-angiotensin system.

sure and influence water resorption in the kidneys, as also noted in Chap. 7. Another neurohypophyseal hormone, **oxytocin,** causes smooth-muscle contraction. (These hormones will be further discussed later on.)

It recently has been demonstrated in mammals that angiotensin acts at specific brain sites to induce drinking. This action, its role in blood-pressure regulation, and, consequently, its effects on sodium and water resorption through action on the adrenal cortex add up to a complex set of physiological and behavioral responses that are directed to solving problems related to homeostatic maintenance of the composition and volume of the blood. Thus, embodied in this one informational molecule is the potential for triggering such diverse actions as stimulating muscle contraction, releasing a hormone, and evoking complex behavior, all directed to the central function of osmoregulation.

Erythropoietin

Hypoxia (abnormally low oxygen pressure) is an effective stimulant for the production of red blood cells and hemoglobin. In mammals, at least, a glycoprotein hormone, **erythropoietin,** has been implicated in this response. The search for the source of this substance has indicated that the kidney is a major site of production, but other organs also release it. Its action is to promote proliferation of red blood cells by the bone marrow.

Bradykinin

Like angiotensin, **bradykinin** is a linear polypeptide, that is released from plasma globulins and acts on smooth muscle. It derives from its precursor (bradykinogen) in mammalian plasma by proteolytic (esterolytic) enzymes such as trypsin and plasmin, or by similar enzymes present in snake venoms. When released it gives rise to a fall in blood pressure due to vasodilatation, but produces a slow contraction of certain smooth muscles and can cause pain sensations.

Physiological roles for bradykinin are not securely established, but because it is a most potent endogenous vasodilator substance and is stored in large quantity in circulating blood, it seems quite probable that it is of importance in some situations. In particular, it is speculated that it mediates the phenomenon of *reactive hyperemia*. That is, if arterial inflow to a tissue is partly occluded, that tissue is deprived of a normal supply of blood, leading to ischemia. The response to ischemia is local arteriolar dilatation which increases blood flow. Bradykinin may act locally to effect the dilatation.

Prostaglandins

Prostaglandins (PGs) are a group of lipid compounds which were first detected in human semen in the early 1930s. Since then a variety of them have been found in various mammalian tissues, including brain and spinal cord, lung, kidney, placenta, and thymus, and they are suspected of being present in a number of other tissues. Thus, even though seminal fluid is one of the richest sources of these compounds, the term "prostaglandins" is something of a misnomer in that it implies they are resident only in the prostate gland.

The prostaglandins are all 20-carbon fatty acids having the same basic skeleton, prostanoic acid, and are derived from essential fatty acids by cyclization and oxidation. They are subdivided into series having common chemical groups, and these are known as the A, B, E, and F prostaglandins. The first two isolated in pure crystalline form, PGE_1 and $PGF_{1\alpha}$, were obtained by Bergström in 1957 from sheep seminal vesicles (Fig. 9-2).

FIGURE 9-2 Structural formula of prostaglandins PGE_1 and $PGF_{1\alpha}$. Their empirical formulas are $C_{20}H_{34}O_5$ and $C_{20}H_{36}O_5$, respectively. More than a dozen of these C_{20}, lipid-soluble, unsaturated hydroxy acids have been isolated and identified chemically.

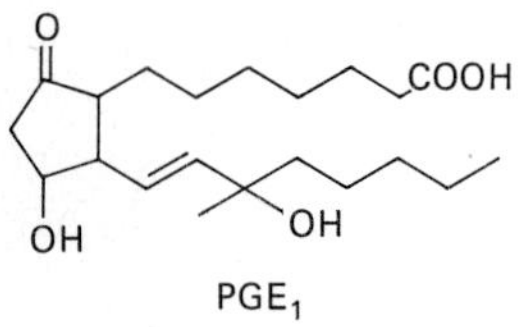

PGE_1

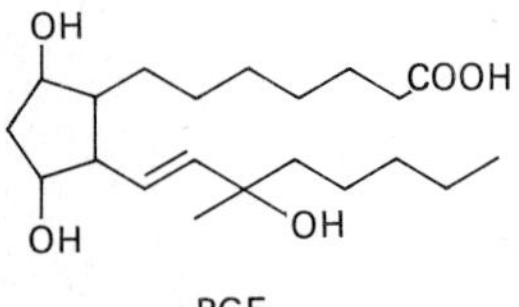

$PGF_{1\alpha}$

Prostaglandins as a group have a wide spectrum of activities, but individual ones differ in their actions, which are often dissimilar and sometimes opposed. Thus, the PGEs are vasodepressor agents, and they also decrease the motility of the human uterus at ovulation, whereas the PGFs have the opposite effects. In contrast, both PGEs and PGFs tend to increase cardiac output, probably as a result of increased return of blood to the heart, and to induce contraction of isolated gastrointestinal muscle and contraction of the iris. The PGAs are like the PGEs in their actions on vascular tissue, but are virtually inactive as gastrointestinal-contracting agents.

PGEs have been used to prevent duodenal ulcers in dogs and rats and may be useful in the same capacity in humans. However, prostaglandins have attracted the most attention as potential agents in the control of fertility and induction of parturition in mammals. Intravenous infusion of certain prostaglandins tends to increase labor. They also have been used as abortifacients for terminating pregnancy and as contraceptives to prevent pregnancy. In contrast, there seems to be a correlation between seminal content of PGE and fertility in males; infertile males tend to have low prostaglandin content.

The bewildering variety of prostaglandin actions perhaps will be explained by their having primary action on cell membranes which leads to changes in intracellular levels of *cyclic 3′,5′-adenosine monophosphate*, the so-called "second messenger" of many hormones. This compound will be discussed later in this chapter.

9-2 HORMONES

The word "hormone" is usually reserved for those chemical agents which are synthesized by circumscribed parts of the body and carried by circulation to other parts of the body where they effect systemic adjustments by acting on specific tissues and organs. More specifically, it applies best to the regulatory products of specialized, ductless glands, the so-called **endocrine glands** (or to endocrine-type cells, where a "gland" cannot be defined).

The glands that collectively are commonly referred to as the **endocrine system** are quite diverse in origin, being derived from ectoderm, mesoderm, or endoderm. They are greatly variable in structure, complexity, and function, and range from autonomous, semi-independent to elaborately regulated systems (Fig. 9-3).

Hormones are physiological regulators, affecting the rate of biological functions in many instances. For example, epinephrine causes vertebrate heart muscle to contract with greater force; antidiuretic hormones retard the rate of water loss in the vertebrate kidney. In other cases, hormones evoke differentiation and growth, as in the development of sex organs. Generally, hormones are effective in extremely small quantities. Plasma concentrations are commonly expressed in microgram (10^{-6}g), nanogram (10^{-9}g), or picogram (10^{-12}) quantities. In the vertebrate hypothalamus, where many hormones act in feedback systems, femtogram (10^{-15}g) quantities are effective. The amount of hormone any particular gland secretes can be conditioned by such factors as age of the organism, season of the year, and physiological state. For instance, as related in Chap. 8, Sec. 8-5, the *thymus gland* and its secretion, *thymosin*, are active only in the very young animal when its immune system is being established. Also, hormones from the pituitary gland, gonads, and other organs vary considerably with reproductive state, particularly in the female. Thus, hormones are not secreted as a steady output; their secretion is modulated according to "need." In fact, with the advent of *radioimmunoassay* techniques, which permit sampling of blood at frequent intervals, because the amount required is so small, it has been shown that many hormones, particularly those of the pituitary gland, are secreted in brief surges. For example, in humans a regular period of growth-hormone secretion occurs during the first several hours after the onset of sleep. A secondary secretory period occurs in the late afternoon, or about 12 hours apart from the primary surge. Furthermore, there are brief, sporadic secretory episodes in between. Some investigators have proposed that it is only the occasional surges of hormone secretion that are physio-

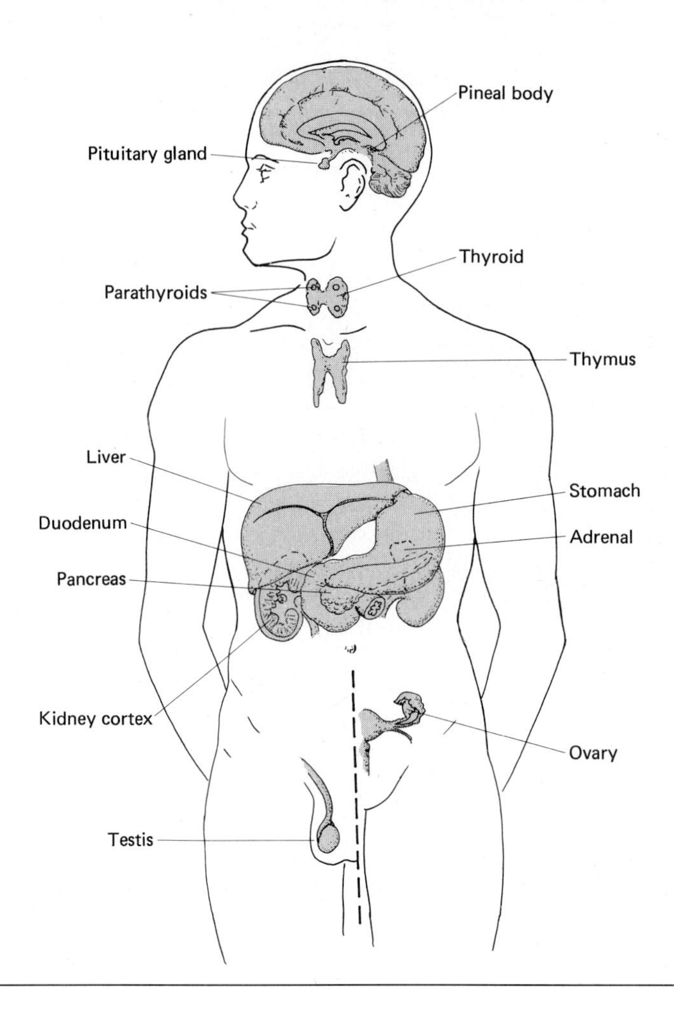

FIGURE 9-3
Approximate locations of the endocrine glands of the human. Although the liver and kidneys add important materials to the blood, they are not usually considered to be endocrine glands.

logically meaningful concentrations, rather than the low basal concentrations.

Conventional hormones are secreted into the blood (or equivalent body fluid in invertebrates) and are transported by the circulatory system to target organs. Some glands release *prohormones* which require extracellular conversion. As noted above, the kidney secretes *renin,* a proteolytic enzyme, which converts renin-substrate to angiotensin I. Insulin and glucagon, for example, are secreted as prohormones which are subsequently cleaved into active principles and residual peptides.

Chemically, hormones include modified amino acids (e.g., catecholamines) small peptides, large peptides, proteins, glycoproteins, steroids, cyclic fatty acids, and possibly mucopolysaccharides. Peptide and protein hormones are synthesized on ribosomes of the rough endoplasmic reticulum and packaged into membrane-enclosed vesicles by the Golgi apparatus for later release. Low-molecular-weight peptides appear to be synthesized by nonribosomal cytoplasmic enzymes. Synthesis of steroids and catecholamines involves both cytoplasmic and mitochondrial enzyme systems. Catecholamines are packaged into granules, but steroid hormones are not.

9-3 GASTROINTESTINAL HORMONES

At the beginning of the twentieth century, largely because of the brilliant studies of the physiology of digestion by Pavlov, it was supposed that the coordination of the processing of food in the alimentary canal was provided by the nervous system. Indeed, not only was the general importance of the antagonistic components of the autonomic nervous system in visceral functions well established at this time, but Pavlov had demonstrated the direct involvement of nerves in the control of the secretion of pancreatic juice. Gastric juice, which is acidic, or else a 0.5 percent solution of HCl, when introduced into the stomach of an experimental dog could routinely be shown to initiate flow of pancreatic juice. The stimulus for the latter was presumed to be transmitted to the pancreas via the vagus nerve.

Not surprisingly, since logic so often fails when too carefully applied in interpreting complex biological systems, Pavlov's explanations of gastrointestinal coordination were shortly to receive some rude blows. Popielski, in the early 1900s, demonstrated that the stimulating effect of acid upon pancreatic secretion could still be demonstrated even after all the neural connections of the alimentary tract had been destroyed. He concluded that the Pavlov effect must be mediated by local reflexes involving peripheral rather than central neurons. This interpretation was discredited by the landmark studies of Bayliss and Starling.

In 1902, using an anesthetized dog, Bayliss and Starling removed the ganglia of the solar plexus, cut both vagi, tied off a segment of jejunum at both ends, and cut the mesenteric nerve supplying it; the resulting segment was connected to the rest of the body only by arteries and veins. With a cannula inserted into the pancreatic duct, and with blood pressure being monitored from the carotid artery, they introduced 20 ml of 0.4 percent HCl into the duodenum. A well-marked pancreatic secretion of one drop every 20 seconds began after 2 minutes and lasted 6 minutes. Acid introduced into the semi-isolated jejunal segment also evoked pancreatic secretion. Bayliss and Starling concluded, eventually, that acid stimulates the intestinal mucosa to release some chemical excitant into the circulation which, in turn, stimulates the pancreas. Indeed, they found that a crude extract of the intestinal mucosa when injected into the jugular vein would cause pancreatic flow.

They gave the name **secretin** to the substance produced by the *intestine* and at this time introduced the name "hormone" as a general term for such chemical messengers. Secretin has a specific site of origin, in that it is produced in a restricted segment of the intestine, but the cells that produce it are apparently scattered among other cells of the mucosa and remain to be identified. Secretin has been purified from hog gut and found to be composed of 27 amino acids.

The main effect of the injection of secretin preparations into the bloodstream is to promote an increase in volume of *pancreatic fluid* with a relatively low enzyme content. This fact caused considerable consternation for several years, and the physiological importance of secretin was questioned. Different methods of preparation of pancreatic extracts finally led to the revelation of a second hormone contained in the duodenum. Whereas secretin influences mainly the volume of fluid secreted, the second hormone, called **pancreozymin,** strongly stimulates the output of digestive enzymes.

Cholecytokinin is the name formerly given to a substance extractable from the *intestinal mucosa* which causes contraction of the muscularis of the gallbladder. Presumably, it is responsible for the emptying of the gallbladder at the time when bile is needed during digestion of a meal in the intestine, and its secretion is stimulated especially by contact of fats and fatty acids with the duodenal mucosa.

Pancreozymin and cholecystokinin once were thought to be two distinct substances but now appear to be a single hormone. In addition to acting on the pancreas to cause release of digestive enzymes, **cholecystokinin-pancreozymin** also stimulates release of the pancreatic hormones, insulin and glucagon.

The secretory response of the stomach to feeding takes place in two distinct places. The first, the so-called *nervous* or *cephalic* phase, is controlled reflexly through the parasympathetic division of the autonomic system. It occurs either in response to the stimulation of taste buds by ingested food or as a conditioned reflex, with higher nervous centers participating, in response to stimulation of other receptors such as the optic or olfactory. Triggering stimuli are carried by the vagus nerve. The resulting gastric flow is very rich in the proteolytic enzyme, *pepsin.*

The nervous phase is followed in mammals by a *chemical phase,* which can be subdivided into pyloric and intestinal stages. The pyloric stage is believed to be regulated through a hormonal mechanism. Either mechanical or chemical stimulation evokes the release of a hormone, known as **gastrin,** from the mucous membrane of the pyloric region of the stomach. Gastrin then is thought to evoke the secretion of acid, which in turn participates in the conversion of the inactive to the active form of pepsin. A hormone of the duodenum, **enterogastrone,** has been postulated to play a role in the termination of gastric digestion (Figs. 9-4 and 9-5).

Duocrinin is a hormone that has been claimed to increase secretion by *Brunner's glands* in the submucosa of the duodenum. Brunner's glands are unique structures that lie between the circular muscles and the muscularis mucosae. It is thought that their secretion somehow protects the duodenal mucosa by adhering to it and forming a protective coating against the acid contents arriving from the stomach as noted in Sec. 5-5. The stimulus for duocrinin release is believed to be the presence of acid or lipid in the duodenum.

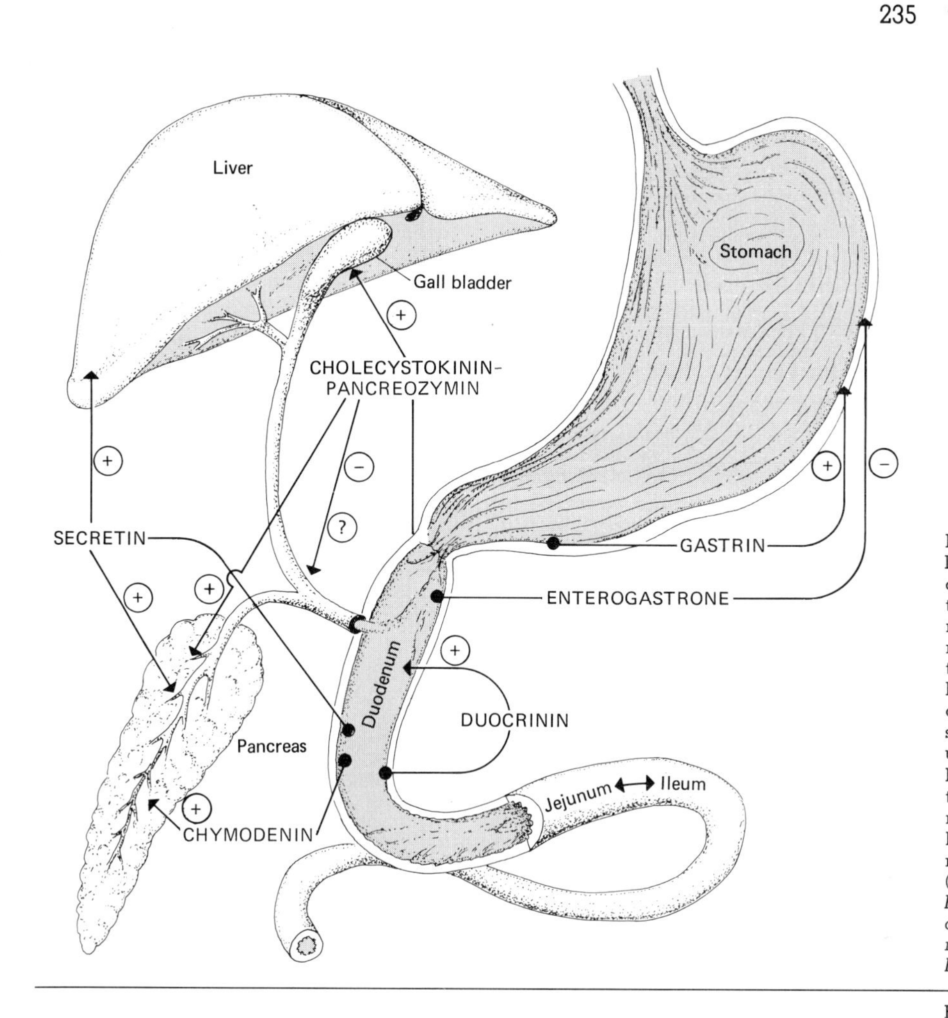

FIGURE 9-4
Diagram showing the place of origin and the site of action of gastrointestinal hormones. Although they originate and act entirely within the digestive tract, these hormones are carried from one place to another by the systemic circulation. A stimulatory action is indicated by a plus sign; an inhibitory action is indicated by a minus sign. Cholecystokinin and pancreozymin are now thought to be identical. (*After A. Gorbman and H. A. Bern, 1962, A Textbook of Comparative Endocrinology, John Wiley & Sons, Inc., New York, p. 194.*)

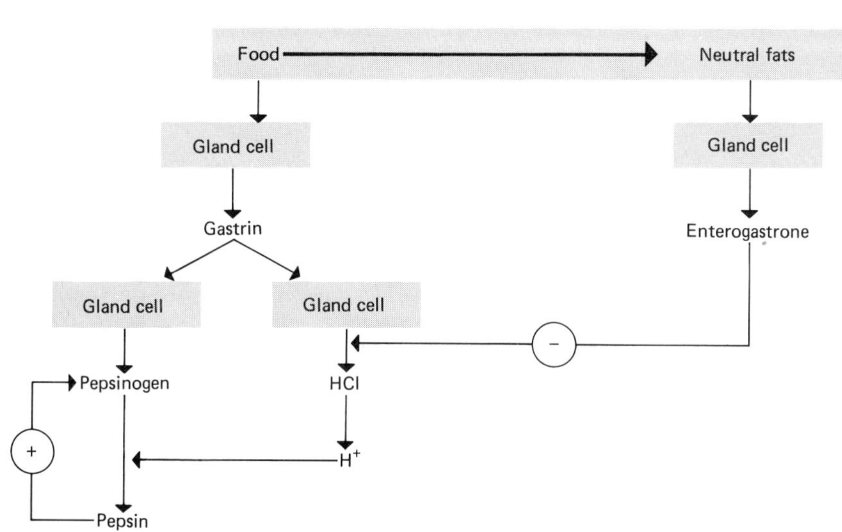

FIGURE 9-5
A simplified scheme indicating the mode of action of gastrin and enterogastrone in the control of the proteolytic enzyme of the stomach pepsin. In the presence of acid the enzymatically inactive pepsinogen is converted to the active pepsin. Gastrin is believed to stimulate the secretion of the glands producing these agents. Enterogastrone, apparently of duodenal origin, inhibits gastric secretion. The inhibition appears to be largely, if not exclusively, confined to the hydrochloric acid releasing mechanism. As indicated, neutral fats are particularly effective in evoking enterogastrone secretion, but other substances will also cause its release.

Chymodenin (named after the enzyme whose secretion it stimulates and its tissue of origin) is recently reported to be a peptide of the *duodenum* which can selectively elicit rapid increases in secretion of *chymotrypsinogen*, the precursor of the digestive enzyme, *chymotrypsin*. Unlike other hormones which trigger en masse secretion of enzymes, chymodenin elicits secretion of the single enzyme.

9-4 REGULATION OF BLOOD GLUCOSE

The discovery that the pancreas secretes hormones (in addition to digestive enzymes) came about, as so many other discoveries did, as the result of attempts to understand an abnormal physiological condition. A normal human filters blood glucose into the kidney tubule. If, however, the level of blood sugar exceeds a certain critical point (about 160 mg/100 ml), the tubules are unable to reabsorb all the glucose, and as a result some of it passes out in the urine. Such a condition is called *diabetes mellitus* (L. *mellitus*, sweetened with honey), because the urine tastes sweet. However, diabetes mellitus is a condition that results from a profound disturbance in metabolic mechanisms, and sugar in the urine is only one symptom of a severe malfunctioning of certain homeostatic mechanisms related to carbohydrate metabolism. The blood-sugar level can become unusually high, and there is a corresponding reduction in glycogen content of the muscles and liver. A general feeling of malaise and muscular weakness accompanies such a condition. **Glycogenesis** (formation of glycogen) is reduced, **glycogenolysis** (breakdown of glycogen) is increased, and there is a decreased ability to oxidize glucose. The explanation of these effects lies in an understanding of how certain hormones regulate the activity of enzymes. But first, let us identify the endocrine products of the pancreas that are so importantly implicated in the control of carbohydrate metabolism.

The mammalian pancreas consists predominantly of *zymogen cells* that secrete digestive enzymes, but in addition there are small islands or groups of cells, called the *islets of Langerhans*, that produce hormones. It was mainly the work of the Canadians, Banting and Best, which established that a hormone of islet tissue, now known as **insulin,** is in large measure responsible for regulation of carbohydrate metabolism. Extracts containing insulin, and later the synthesized molecule, have relieved the diabetic condition and extended the lives of thousands of persons.

The exact source of insulin is well established. Two types of cells, termed α and β cells, are probably present in the pancreatic tissue of all vertebrates from fish to mammals. (There appears to be a third cell type in some organisms, but its significance is uncertain.) Only the β cells synthesize insulin. This is demonstrated by (1) a positive response of β cells to the histochemical test for sulfhydryl groups—which are a feature of the insulin molecule, as we shall see; (2) cytological changes occasioned by injection of a large load of glucose into the bloodstream; β cells become degranulated and vacuolated, suggesting stimulation of secretion rate, while other cell types remain unaffected; and (3) administration of *alloxan*, a substance that selectively destroys the β cells while leaving the other types relatively undisturbed, results in development of a diabetic condition.

By 1926 Abel had isolated insulin in crystalline form. It was shown to be a protein. In a period of about 10 years—1945 to 1955—Frederick Sanger, at Cambridge University, through painstaking analytical work revealed its structure. Though today it is a rather routine, semiautomated task, determination of amino acid sequences of proteins was at the time of Sanger's efforts a tedious, demanding undertaking, requiring innovation and insight. Sanger depended mostly on controlled hydrolysis of the insulin molecule by acid or proteolytic enzymes, followed by separation and identification of the resulting small fragments by paper chromatography and ion-exchange chromatography. The insulin molecule, with a molecular weight of 6000, was found to be composed of two polypeptide chains linked together at two places by the disulfide bridges of cysteine residues (Fig. 9-6).

Insulin molecules vary from species to species in their amino acid sequences, but, nevertheless, as far as is now known, all insulins possess the same characteristic biological activities.

Glucagon is a second hormone of the pancreas that plays a role in carbohydrate metabolism and utilization. It is a polypeptide, but smaller than insulin, containing 29 amino acids and having a molecular weight of 3485 (Fig. 9-6).

Glucagon is apparently secreted by α cells, since it can be extracted from the pancreas when the enzyme-secreting tissue has been destroyed by ligation and the β cells have been obliterated by alloxan. Furthermore, α cells contain granules that react positively to histochemical tests for tryptophan, a constituent of glucagon that is absent from insulin, and negatively for cysteine, a component of insulin that is not found in glucagon.

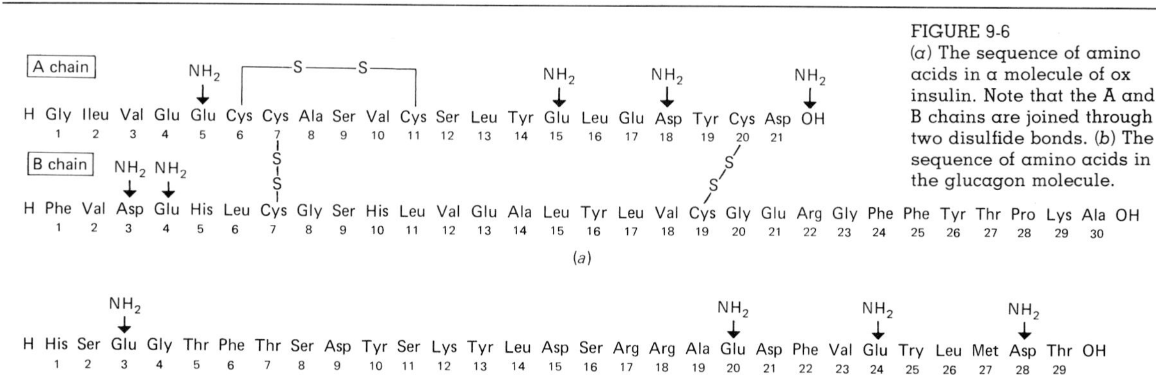

FIGURE 9-6
(*a*) The sequence of amino acids in a molecule of ox insulin. Note that the A and B chains are joined through two disulfide bonds. (*b*) The sequence of amino acids in the glucagon molecule.

The overall action of glucagon is to decrease stores of glycogen, principally in the liver, and to elevate blood-sugar level. In addition, it facilitates the conversion of noncarbohydrate molecules to carbohydrates, which subsequently can be used in conventional carbohydrate metabolism. To understand the actions of both insulin and glucagon better we will need to consider their specific metabolic effects.

It is well established that in metabolism, carbohydrate transformations constitute the central pathway or main line of events and that protein and lipid components are peripherally linked to this main line. Furthermore, the metabolic sequences are essentially all reversible—but for some steps there are two enzymes, one acting in one direction and the other acting when the reaction flows in the opposite direction. (For the student unfamiliar with these concepts, a brief study of Figs. 19-2 and 19-3 will be helpful. Note that the interconversion of glucose and glucose 6-phosphate involves two oppositely biased enzymes, as does the fructose 6-phosphate and fructose 1,6-diphosphate transformation.)

The formation of carbohydrate from noncarbohydrates (i.e., proteins and lipids) is termed **gluconeogenesis.** The principle sites of gluconeogenesis in the mammal appear to be the liver and kidney cortex, with the kidney perhaps being more active on a cellular basis. However, the liver is the principal organ because of its greater mass.

An enzyme-linked, metabolic sequence, such as glycolysis, obviously can be rate-limited by the slowest reaction. By controlling the activities of just a few key enzymes, a complex set of metabolic reactions can be manipulated. That is, to regulate the rate of formation of any end product it is not necessary to regulate each of the several enzymes participating in the sequential reactions that lead to its formation; it is sufficient to manipulate only a few "bottlenecks" to regulate the "traffic," since all molecules in a sequence must pass through the same "bottlenecks" to become that end product.

This is the principle used by pancreatic hormones to favor or suppress certain metabolic sequences. Involved also in these regulations are certain of the steroid hormones emanating from the cortex of the adrenal gland, the so-called **adrenocortical hormones** or **corticoids.** Additionally, **epinephrine** (or **adrenalin**), a catecholamine synthesized and released by the adrenal medulla, plays a role in increasing blood level of glucose by stimulating *glycogenolysis* (breaking of glucosidic bonds of glycogen by a process of phosphorolysis, in this instance). (The corticoids will be discussed subsequently in more detail and in other contexts, and the mechanism of action of epinephrine will be dealt with in the last section of this chapter.) The actions of the several hormones that influence the central pathway of carbohydrate metabolism are indicated in Fig. 9-7.

The hormones regulating metabolism act in some instances, it appears, through activation of preexisting enzymes and in other cases by influencing synthesis of enzymes or both. Additionally, there appear to be effects on cellular transport mechanisms. For instance, when insulin is

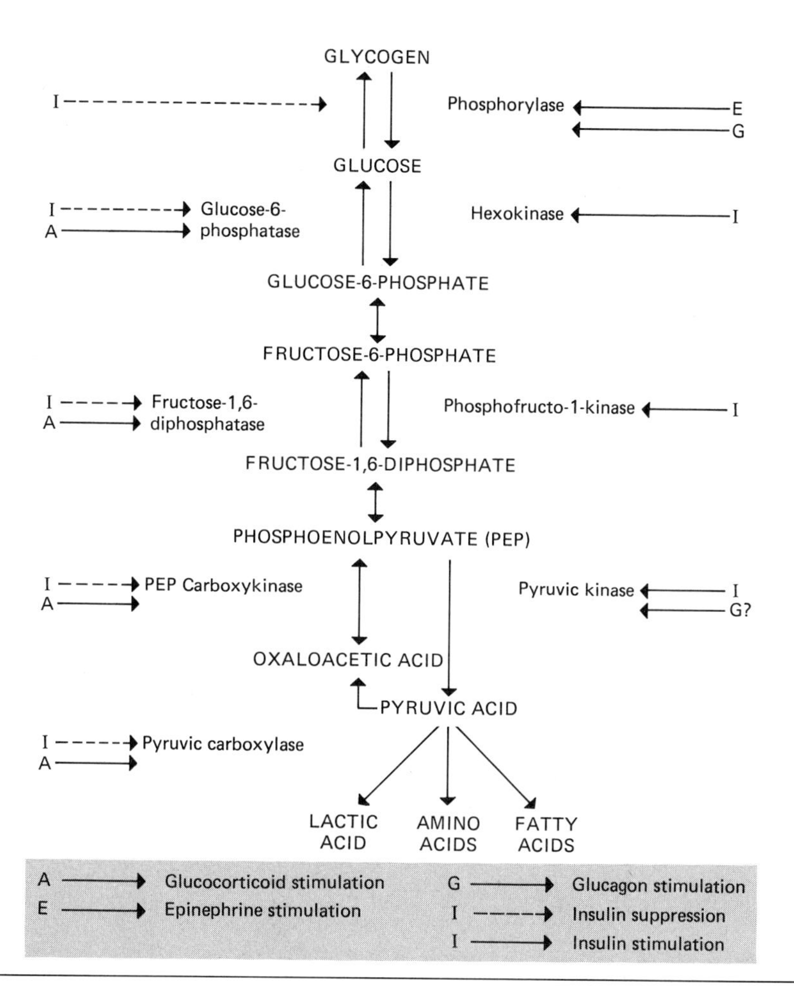

FIGURE 9-7
A highly simplified scheme showing some of the sites of action of certain hormones that regulate gluconeogenic enzymes in the glycolytic pathway.

added to the medium in which isolated pieces of rat diaphragm are being maintained, uptake of glucose, as well as glycogenesis, is stimulated.

Whatever the exact mechanisms of their action, there seems to be no question that the main line of carbohydrate metabolism—the very core of the energy-yielding machinery—is normally tightly regulated by a rather small set of hormones.

The actions of insulin and glucagon are by no means limited to those just described. Insulin, among other effects, increases cellular uptake of glucose by skeletal muscles, adipose tissue, and liver when blood-glucose concentrations are elevated, as after a meal. Thus, the potential for high blood-sugar levels (*hyperglycemia*) is suppressed by insulin and by storage of glucose as glycogen or, through conversion, as fats is promoted.

The pancreatic hormones have been thought to play their role in an essentially autonomous way, responding to changing levels of blood glucose as sensed in the islet tissue. Within the last few years, however, it has been established that the sympathetic nervous system exerts an effect on insulin dynamics. Thus, pancreatic secretions may be evoked both directly, in response to circulatory levels of glucose, and indirectly, through mediation of the nervous system. The invoking of adrenal hormones in sugar regulation, on the other hand, seems exclusively to involve the nervous system. The adrenal *medulla*, the source of epinephrine, is in fact a modified portion of the sympathetic system. The adrenal *cortex*, in contrast, is influenced by a pituitary hormone, which in turn, is regulated by a secretion of the brain. Such relations where the nervous and endocrine systems interact, forming "neuroendocrine" mechanisms, will be examined in the next chapter.

9-5 REGULATION OF CALCIUM AND PHOSPHORUS

The concentration of *calcium* in body fluids is relatively low and is highly regulated. Calcium ions play a vital role in maintaining neuromuscular excitability, membrane permeability, enzyme activity, blood clotting, and the regulation of blood pH. One of the more spectacular effects of subnormal levels of circulating Ca^{2+} (*hypocalcemia*) is *tetany* (tonic spasms of the skeletal muscles). Death from asphyxia occurs when the spasms involve the muscles of the larynx and diaphragm. The isolated heart of a frog will become arrested in diastole in the absence of Ca^{2+} and will show sustained contraction, or rigor, if the concentration is too great. Sudden elevation of calcium level (*hypercalcemia*) will cause the heart to be arrested in systole (calcium rigor).

In contrast to the small but vital amount in the body fluid, calcium is the most abundant cation in terms of the total composition of vertebrates. About 99 percent of this ion in the human is present in the mineral crystals (hydroxyapatite) of bone. An average adult contains about 1300 g of calcium, most of which is in the skeleton. About 12 g is in the soft tissues, and less than 1 g is in the body fluids. There are large and variable movements of the ion between these compartments, but the level of serum calcium is held remarkably constant by the action, principally, of **parathyroid hormones.**

Though the precise roles of calcium in cellular functions remain to be adequately demonstrated, it has been established that the proper excitability of electrogenic membranes, such as those of neurons and muscle cells, depends upon calcium. Calcium also serves as a coupling factor during excitation and contraction in muscles. Additionally, it seems certain that calcium participates in the early stages of the action of several hormones.

The total inorganic and organic *phosphate* of the adult human body amounts to about 500 to 600 g. About 85 percent of this is in the skeleton. Blood serum contains 5 to 7 mg of phosphorus per 100 ml. This small concentration is closely regulated, chiefly by parathyroid hormone, in a very dynamic steady-state relationship with the much larger tissue reservoirs.

The phosphate radical is present in many cellular constituents, including the nucleotides and phosphorylated compounds of the energetic pathways. Inorganic phosphate has a key role in the regulation of glycolysis and metabolism. It is structurally important in the phospholipids of membranes and as an ingredient in bone salts. Several key metabolic enzymes are transformed from the inactive to the active state by the acquisition or removal of a phosphate group. It is not surprising, therefore, that calcium and phosphorus are hormonally regulated in quite precise ways. The **parathyroid glands** are most central in the regulation of these minerals.

As is true for most of the information on hormones, we know about parathyroid glands primarily from studies of eutherian mammals. A bit is known of these glands in birds, and a few studies have been done with amphibians and reptiles. Fish appear not to have parathyroids, but this does not preclude the possibility that they have parathyroid-type hormones. Indeed, a principle that has physiological activities like one of the mammalian parathyroid hormones has been extracted from the **ultimobranchial bodies,** which are derivatives of the fifth pair of pharyngeal pouches. The parathyroids of mammals originate from the dorsal halves of the third and fourth pair of pharyngeal pouches. Life for the great majority of vertebrates is either precarious or impossible without parathyroid hormones.

The term "parathyroid" is unfortunate, as it implies a relationship to the thyroid gland. Parathyroids do have an intimate anatomical association with the thyroid in certain animals (Fig. 9-3), but their physiological role is distinctly different. The parathyroid hormones are the principal agents involved in *regulating the metabolism of ions,* such as calcium, phosphate, pyrophosphate, citrate, and magnesium, and in regulating the metabolism of bone and its organic constituents.

From bovine and porcine parathyroid glands a straight-chain polypeptide consisting of about 75 amino acid residues has been isolated. This substance, called **parathyroid hormone** (PTH), has both calcium-mobilizing and phosphaturic (increasing phosphate in the urine) activities. That is, PTH *increases* serum calcium and *decreases* serum phosphate by a direct action on bone, kidney, and probably the intestinal mucosa.

A second hormone, called **calcitonin** (CT), also has been isolated from parathyroid glands (and demonstrated in the ultimobranchial bodies of sharks, amphibians, lizards, and chickens, as well). It is a polypeptide chain composed of 32 amino acid residues. The best-established action of CT is the lowering of serum-calcium level by suppressing the resorption of bone, but there also are indications that it may be somehow related to the metabolism and transport of phosphate.

The secretion of parathyroid hormones is influenced by levels of plasma calcium and phosphate. There seem to be no neural components or pituitary factors operating in such responses; they are purely local, autonomous responses to the ionic concentration of the blood which is continuously monitored by the cells of the parathyroid. As proof of this contention, it has been demonstrated that when parathyroids are cultured in a low-calcium medium, the output of PTH is increased; in a high-calcium medium, the output of the hormone is diminished. The rate of PTH secretion responds also to fluctuating levels of phosphate in organisms in which the serum-calcium level is held relatively constant. The secretion of CT is responsive, so it appears, only to the level of plasma calcium.

The current concept of the physiological action of PTH is that it acts principally upon bone collagen and apatite crystals to release both calcium and phosphate into the circulation. It also acts upon the kidney tubules to increase the resorption of calcium, but to decrease the resorption of phosphate. There is some contradictory evidence that PTH also stimulates the absorption of calcium and phosphate through the intestinal mucosa.

Somehow, the *D vitamins*, a group of sterols, are involved in the mineral metabolism of bone and gastrointestinal tract. It has long been known that vitamin D–deficient animals have abnormal epiphyseal plates in their bones, leading to a syndrome called *rickets*. Vitamin D promotes absorption of calcium from the small intestine and plays a permissive role in enabling PTH to exert its normal action upon bone resorption.

One current hypothesis to account for the action of PTH and CT is that of Talmadge. He believes that the essential action of PTH is to facilitate a rapid transport of Ca^{2+} from mineralized bone across the membranes of the *osteoblasts* (cells that actively synthesize bone collagen) into the extracellular fluid compartment (Fig. 9-8).

CT, it is established, decreases the rate of bone resorption. It has been suggested that CT has effects upon bone accretion and that it is involved in transport of phosphate. Talmadge postulates that CT may act through an enzyme system which promotes the release of organic phosphate. The latter then combines with calcium, reducing thereby the amount of calcium available through the osteoblast membrane. The result would be a *hypocalcemia* effect on blood, while the concentration of calcium and phosphate ions in the areas where new bone is being formed would be elevated and made available for bone growth.

Other hormones that influence calcium and phosphorus metabolism include adrenal glucocorticoids, which depress calcium uptake by the intestine and inhibit its reabsorption by the kidney. Both actions induce hypocalcemia, which leads secondarily to stimulation of the parathyroid glands. Glucocorticoids also directly inhibit mitosis of osteoblast progenitor cells. Somatotropic hormone, on the other hand, stimulates osteoblast formation, promotes phosphorus retention and calcium excretion in the kidney, and increases calcium absorption from the intestine. Estrogens, in some poorly understood ways, influence calcium and phos-

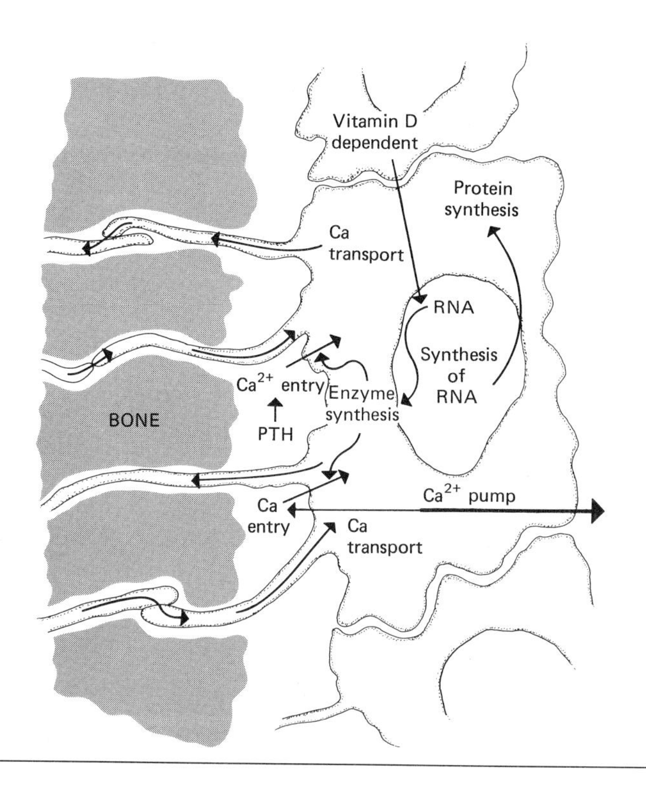

FIGURE 9-8
Diagram illustrating some factors involved in the mechanism of action of PTH on bone cells. The hormone is thought to promote the entry of calcium ions (Ca^{2+}) into cells of the osteoblast membrane, and a calcium pump on the opposite side of the cell passes the ions into the extracellular-fluid compartment. Note the open channels existing between contiguous osteoblast cells; these should permit fluid and ions to leak back into the mineralized compartment. (*Redrawn from R. V. Talmage, 1969, Clin. Orthop., no. 67, p. 212*)

phorus metabolism, as well as bone growth. In birds, estrogens mobilize calcium and phosphorus for egg-shell formation. Androgens stimulate growth of bone, and thyroid hormone promotes bone maturation.

We need now to learn more about the various hormones just alluded to, and so we are led to an examination of that very central gland, the pituitary body.

9-6 THE PITUITARY GLAND AND SOME RELATED HORMONES

The **pituitary gland** (or **hypophysis cerebri**) lies close to the floor of the brain, to which it is attached by a stalk (Fig. 9-3). This relationship is more than fortuitous, as the functional capacity of the pituitary body depends upon its neural and vascular connections with the hypothalamus. The pituitary body is actually an assemblage of glands which not only have a hybrid origin but also exhibit diverse controlling mechanisms and participate in a multitude of unique and varied functions. Removal of the pituitary (*hypophysectomy*) generally causes severe functional deficits, and in many instances will lead to death of the animal unless special procedures are initiated to sustain it.

The pituitary is of ectodermal origin but is derived from two different sources. An outpocketing of the brain, the **infundibulum**, gives rise to the **pars nervosa** or **neurohypophysis**. The neurohypophysis remains permanently connected to the hypothalamus by the stalk formed during its development and is, in fact, an appendage of the brain. From an outgrowth arising in the roof of the future mouth, called **Rathke's pouch,** the remainder of the pituitary, the so-called **adenohypophysis,** is formed. Rathke's pouch fuses to the infundibulum and loses its connection to the buccal epithelium. It differentiates in various ways in different vertebrates, but commonly forms a major body called the **pars distalis,** as well as a **pars intermedia** that lies between the distalis and the neurohypophysis. Additionally, a **pars tuberalis** can be identified in many vertebrates as a thin epithelial plate extending from the pars distalis around the more basal and anterior part of the infundibular stalk. The pars intermedia (when present; some vertebrates, including birds, do not have this lobe) generally is intimately associated with the pars nervosa, and together they are called the **posterior pituitary.** The remainder, including the distalis and tuberalis parts, is called the **anterior pituitary.** Figure 9-9 illustrates the origins of a typical pituitary gland and shows a common pattern of its differentiation into adult form.

The *pars nervosa* consists of branching cells called *pituicytes* and thick networks of fine, unmyelinated nerve fibers that are the terminations of the *hypothalamohypophysial tract*. It appears that it is the secreted products of these neurons, not the pituicytes, that are the hormones of the nervosa. The cell bodies of the neurons lie in clusters (called *nuclei*) within the hypothalamus of the brain, and it is there that the hormones of the nervosa are synthesized. From the cell bodies the hormone molecules travel, in aggregated clumps containing a large carrier protein, along the axons to their terminus in the neural lobe, where they can be freed from the carrier and released into blood capillaries (Fig. 9-10).

Hormones produced by neurons are called **neurohormones,** and the process of synthesis and release of such substances is often called **neurosecretion.** The special structural relationship between neural ter-

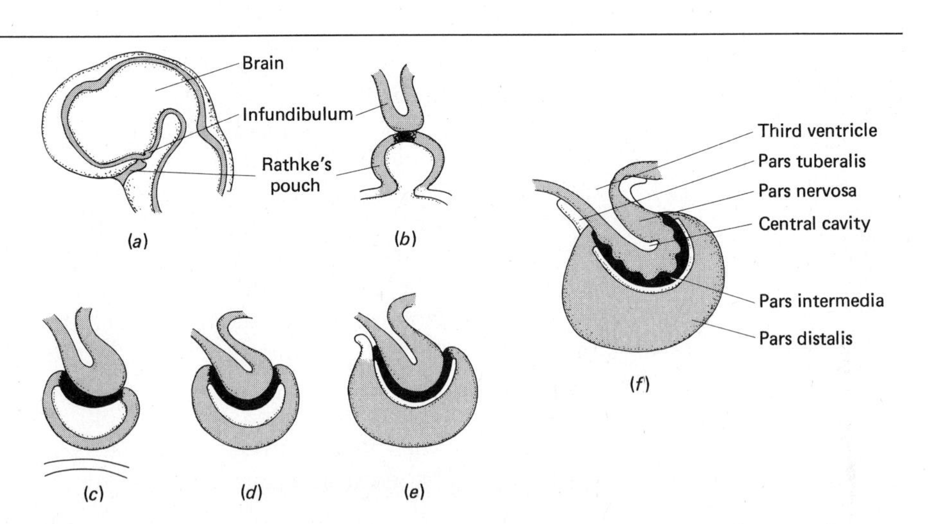

FIGURE 9-9
Diagrams showing progressive stages (*a* through *f*) in the embryonic development of the pituitary gland. Rathke's pouch becomes detached from the oral epithelium at stage c. [(*a*) *After C. A. Vilee et al.*, 1963, *General Zoology, 2d ed., W. B. Saunders Co., Philadelphia; (b)–(f) after C. D. Turner and J. T. Bagnara,* 1971, *General Endocrinology, W. B. Saunders Company, Philadelphia,* p. 78.]

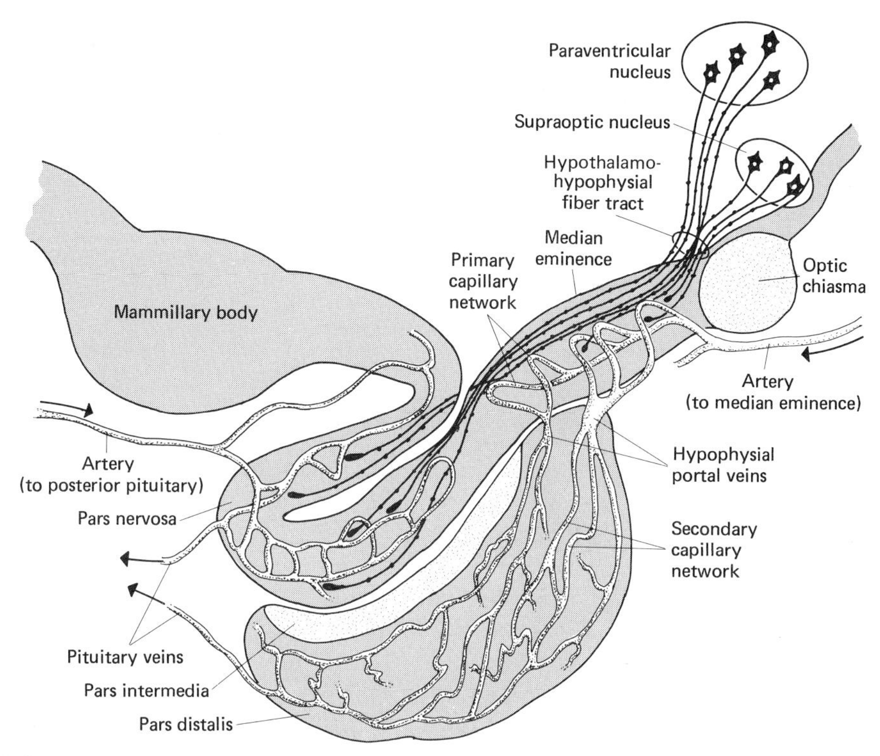

FIGURE 9-10
Diagram of the anatomic connections between the hypothalamus and the pituitary gland. Neurosecretory cells are present in certain hypothalamic nuclei: some of the secretory axons pass down the infundibular stalk and terminate near blood vessels in the pars nervosa; others terminate in close proximity to the capillary loops of the median eminence. The hormones of the neurohypophysis (vasopressin and oxytocin) are the products of hypothalamic neurosecretory cells and are stored in and released from the pars nervosa (a neurohemal organ). The hypophysial portal venules start at the primary plexus of the medium eminence and convey blood downward to the sinusoids of the pars distalis. There are strong indications that the hypothalamic axons of the median eminence liberate multiple releasing and inhibiting factors (probably peptide in nature) into the portal vessels and that these neural factors are concerned with the regulation of anterior pituitary functions. It is apparent that the whole pituitary gland is predominantly subservient to the hypothalamic portion of the brain and has partly evolved from it. (See also Fig. 3-2.)

minations and capillaries, such as found in the pars nervosa, produces what has been called a **neurohemal organ.** Additional information about neuroendocrine structures and their functions is contained in the next chapter.

All the well-characterized hormones that are contained within the neural lobe are octapeptides. Different species may have slightly different hormones, but they all belong to a family of polypeptides in which there is a ring, formed by six amino acid residues closed by an S-S linkage between two cysteine residues, and a side chain containing three amino acid residues (Figs. 9-11 and 9-12).

Little is known about the function of most of the neurohypophysial peptides. **Vasotocin** and **vasopressin** act as water-retention hormones in tetrapods and, accordingly, are said to be **antidiuretic hormones.** (The action of these hormones in the regulation of body fluids is discussed in

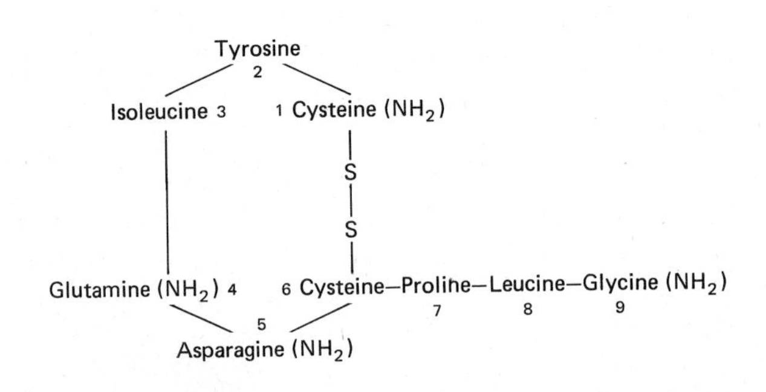

FIGURE 9-11
The arrangement of amino acids in a molecule of oxytocin, a hormone synthesized by neuron cell bodies in the hypothalamus and delivered by the axons to the pars nervosa, from which release into the blood can occur. Oxytocin, like all neural-lobe secretions, is an octapeptide; i.e., it consists of eight amino acids. The two sulfur-linked cysteine molecules are generally counted as one cystine molecule.

Chap. 7.) **Oxytocin** has been postulated to act in mammals to promote uterine contractility, presumably facilitating the ascent of spermatozoa in the female tract after intromission, and also expelling the fetus at parturition. Posterior-lobe peptides also effect oviposition in the hen by inducing contraction of the oviduct and cloaca, and they stimulate contractions of the oviduct of turtles.

Recently evidence has been presented for a new peptide, called **coherin,** in the neurohypophysis that influences gastrointestinal motility. Its structural configuration has not been determined at the time of this writing.

The *pars intermedia* (Figs. 9-9 and 9-10) is a poorly understood portion of the pituitary. Though it is absent in birds and in certain mammals (whale, Indian elephant, armadillo), it has been identified in most of the

FIGURE 9-12
Amino acid sequences in the natural neurohormones of the pars nervosa. All contain a pentapeptide ring to which is attached a side chain composed of three amino acids. The ring is closed by an S-S linkage between two cysteine residues. Substitutions occur at positions 3, 4, and 8 to produce peptides with different biological potencies. The animals in which the molecules have been identified are indicated.

OXYTOCIN (mammals, birds, some elasmobranchs; possibly some other tetrapods)	H-Cys-Tyr-ILE-GLU-Asp-Cys-Pro-LEU-Gly-NH_2 1 2 3 4 5 6 7 8 9
ARGININE VASOPRESSIN (most mammals)	H-Cys-Tyr-PHE-GLU-Asp-Cys-Pro-ARG-Gly-NH_2
LYSINE VASOPRESSIN (pigs)	H-Cys-Tyr-PHE-GLU-Asp-Cys-Pro-LYS-Gly-NH_2
ARGININE VASOTOCIN (all nonmammalian vertebrates)	H-Cys-Tyr-ILE-GLU-Asp-Cys-Pro-ARG-Gly-NH_2
MESOTOCIN (amphibia, reptiles, lungfish; possibly some other fish)	H-Cys-Tyr-ILE-GLU-Asp-Cys-Pro-ILE-Gly-NH_2
ISOTOCIN (teleosts)	H-Cys-Tyr-ILE-SER-Asp-Cys-Pro-ILE-Gly-NH_2
GLUMITOCIN (some elasmobranchs)	H-Cys-Tyr-ILE-SER-Asp-Cys-Pro-GLU-Gly-NH_2
ASPARTOCIN (some elasmobranchs)	H-Cys-Tyr-ILE-ASP-Asp-Cys-Pro-LEU-Gly-NH_2
VALITOCIN (some elasmobranchs)	H-Cys-Tyr-ILE-GLU-Asp-Cys-Pro-VAL-Gly-NH_2

mammals, reptiles, amphibians, and fishes that have been examined. Its chief, or only, type of hormone is **melanocyte-stimulating hormone** (MSH) (Fig. 9-13).

The only function established with some certainty for MSH is its influence on skin-color change in certain fishes, amphibians, and reptiles. It causes the dispersion of pigment within *melanophores* (chromatophores containing melanin pigment) and influences the distribution of pigments of other chromatic cells as well. These are rapid changes (occurring in a few to several minutes) and are employed in response to environmental stimuli to subserve such functions as adapting a color pattern to the background and thus rendering the animal inconspicuous, signaling other individuals a state of sexual readiness, protecting against excessive radiation, and aiding in thermoregulation by altering integumentary heat and light absorption (Figs. 4-4 and 4-5). MSH also can increase the amount of pigment within chromatophores, as well as the number of chromatophores in some vertebrates (Fig. 4-6).

In amphibians, where the control of the pars intermedia has been most studied, it appears that MSH release is under inhibitory control by the hypothalamus. The nature of this inhibitory action on the pars intermedia is not yet resolved. Adrenergic (catecholamine-containing) neurons, as well as neurosecretory-type neurons, invade the amphibian intermediate lobe. There is strong evidence that the hypothalamic inhibition of MSH release is mediated by monoamines which are released from the adrenergic neurons. However, the pars intermedia of some lizard species is not innervated. Also, some studies indicate that the hypothalamus secretes small peptides which act as inhibitors of MSH release, while others suggest that an MSH-releasing factor, as well, is involved in the control of MSH secretion.

FIGURE 9-13
Structural comparison of MSH, ACTH, and LPH from various species. Only partial sequences for ACTH and β-LPH are shown. All contain a common sequence of seven amino acids (enclosed by shaded rectangle) plus proline in the same relative position.

α-MSH	Acetyl-Ser-Tyr-Ser-Met-Glu-His-Phe-Arg-Try-Gly-Lys-Pro-Val-NH_2
MSH (dogfish)	H-Ser-Met-Glu-His-Phe-Arg-Try-Gly-Lys-Pro-Met-(NH_2)
MSH (dogfish)	H-Tyr-Ser-Met-Glu-His-Phe-Arg-Try-Gly-Lys-Pro-Met-(NH_2)
β-MSH (cow)	H-Asp-Ser-Gly-Pro-Tyr-Lys-Met-Glu-His-Phe-Arg-Try-Gly-Ser-Pro-Pro-Lys-Asp-OH
β-MSH (pig)	H-Asp-Glu-Gly-Pro-Tyr-Lys-Met-Glu-His-Phe-Arg-Try-Gly-Ser-Pro-Pro-Lys-Asp-OH
β-MSH (monkey)	H-Asp-Glu-Gly-Pro-Tyr-Arg-Met-Glu-His-Phe-Arg-Try-Gly-Ser-Pro-Pro-Lys-Asp-OH
β-MSH (human)	H-Ala-Glu-Lys-Lys-Asp-Glu-Gly-Pro-Tyr-Arg-Met-Glu-His-Phe-Arg-Try-Gly-Ser-Pro-Pro-Lys-Asp-OH
β-MSH (horse)	H-Asp-Glu-Gly-Pro-Tyr-Lys-Met-Glu-His-Phe-Arg-Try-Gly-Ser-Pro-Arg-Lys-Asp-OH
ACTH (res. 1-17)	H-Ser-Tyr-Ser-Met-Glu-His-Phe-Arg-Try-Gly-Lys-Pro-Val-Gly-Lys-Lys-Arg . . .
β-LPH (ovine) (res. 37-60)	. . . Ala-Glu-Lys-Lys-Asp-Ser-Gly-Pro-Tyr-Lys-Met-Glu-His-Phe-Arg-Try-Gly-Ser-Pro-Pro-Lys-Asp-Lys-Arg . . .

In the pars intermedia there is sometimes found a chemically related hormone, **adrenocorticotropic hormone** (ATCH), which shares with the MSHs a common sequence of seven amino acids (Fig. 9-13), and also has some chromatophoric activity. ACTH is generally and primarily found in the pars distalis, however, and its best-known action, as the name implies, is to stimulate the adrenal cortex (see later).

This is an appropriate time to mention the **pineal body** (Fig. 9-3), a structure derived from the roof of the diencephalon (in contrast to the infundibulum, which emerges from the floor). Its best-known secretion is an amino acid derivative, *N*-acetyl-5-methoxytryptamine (**melatonin**). The best-characterized action of melatonin is concentration of pigment granules in the melanophores of larval amphibians, resulting in what is known as the "body-blanching reaction." That the reaction is mediated by the pineal body in larval amphibia seems fairly certain, but this response may be limited to larvae and not persist in adult animals. There is no other clearly established functional role for melatonin, although it has been implicated in inhibition of gonadal function in certain mammals. However, it now seems likely that an agent of the pineal body—melatonin or something else—acts on certain brain sites and thereby influences the secretion of pituitary hormones, thus accounting for the effect of the pineal body on reproductive organs, and on other organs as well. Further adding to the confusion surrounding the pineal body are the reports of extractable peptide agents that have hormonal activities. Here, then, is a situation in which several candidates have been nominated as pineal hormones, but for which functions must be secured in order to accord them such a status.

The *pars tuberalis* (Fig. 9-9), although it is commonly well vascularized and receives many sympathetic fibers, has no known endocrine function. Several years ago it was proposed as the site in anuran amphibians of production of a substance that could cause pigment concentration in melanophores (the so-called "W" substance of Hogben), but this idea is no longer seriously entertained largely because of recent findings that allow new interpretations of the data that led to the original formulation of this hypothesis. We do not know, surprisingly, what (if any) secretory material is produced by the pars tuberalis.

The *pars distalis* (Figs. 9-9 and 9-10) synthesizes a number of polypeptide or protein hormones—at least eight. Cytologically, the distalis can be demonstrated to contain several kinds of cells, suggesting that the different hormones are produced in specific cell types. The hormones of the distalis have been named for a principal function, but, in truth, they commonly contribute to several actions. **Somatotropic hormone**[1](STH), or **growth hormone**), for example, is named for its capacity to stimulate

[1] The adjectival combining form, *-tropic*, is derived from Gr. *tropos*, meaning "turning," which, in the case of hormones, suggests changing or influencing. Other authors commonly use *-trophic*, from Gr. *trophikos*, meaning "nursing," as the combining term for naming hormones, which would suggest that they "feed" or "support" certain organs. Thus, somato*tropin* is also called somato*tropin;* gonado*tropins* are also called gonado*trophins*, etc. "Tropin" is becoming the preferred term and will be used here.

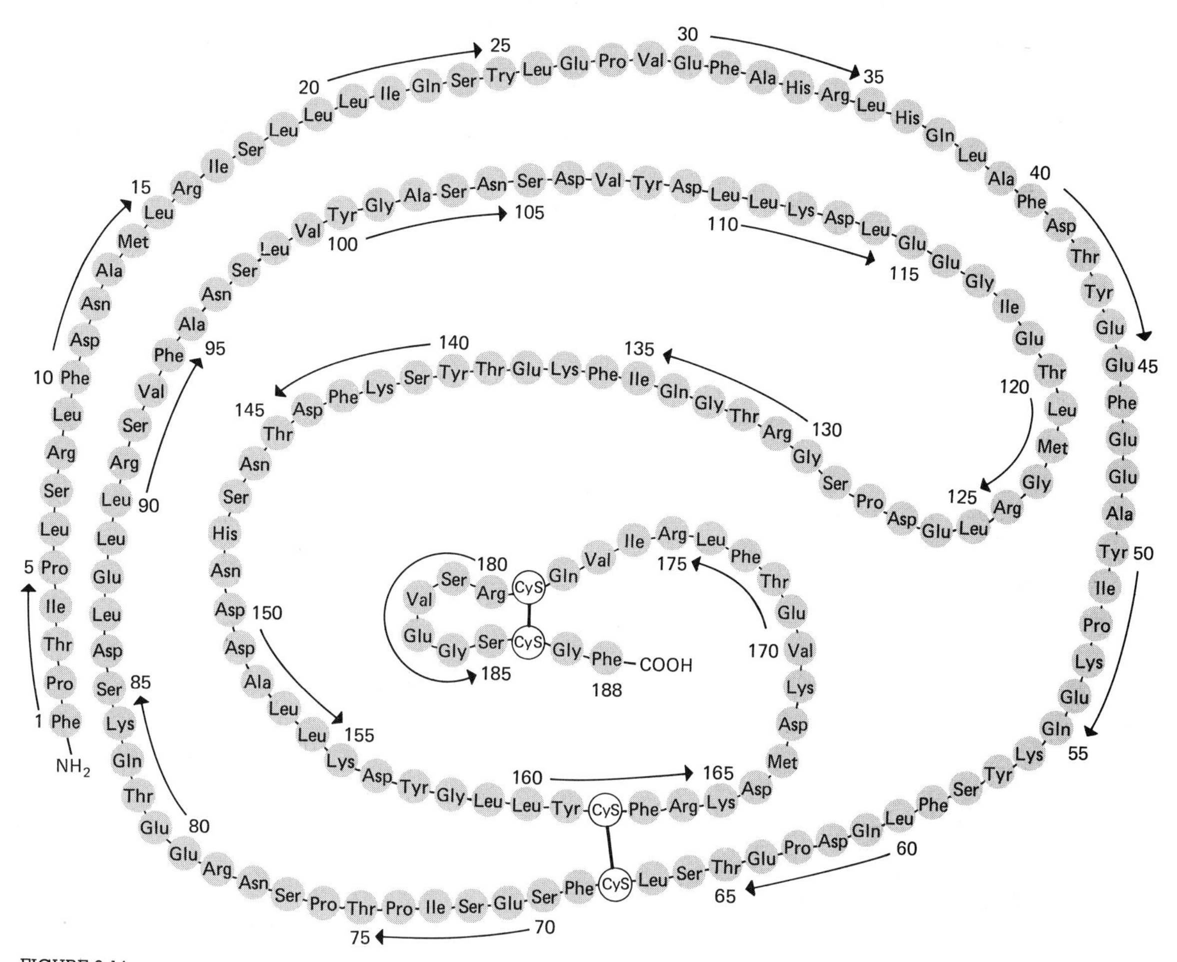

FIGURE 9-14
The complete amino acid sequence of human STH molecule. (*After C. H. Li,* 1969, *Recent studies on the chemistry of human growth hormone, in La Specificité des zoologique des hormones hypophysaires et de leurs activités, edited by M. Fontaine, Centre National de la Recherche Scientifique, Paris.*)

skeletal growth, but it also affects metabolism and the immune response, as well as interacting with other hormones, such as inhibiting the action of insulin. Primate STH also influences the development and functioning of the lactating apparatus of the female mammal.

STH, like most adenohypophysial hormones, is not chemically the same from species to species. Immunological evidence and isoelectric point measurements show distinct species differences for this hormone. However, somatotropins appear to have similar amino acid compositions and have molecular weights of about 22,000. The amino acid sequence for human STH is known, and partial sequences for somatotropins of other species have been determined. There is some interspecies reactivity, but commonly the STH of one species has little or no effect on another (Fig. 9-14).

STH affects the growth of many tissues, including the skeleton. Hypophysectomy of the young animal results in attenuation of linear growth, largely because of impaired development of the skeleton.

Evidence recently gathered indicates that the effects of growth hormone on rapidly growing cartilage are exerted indirectly through a substance found in the blood of mammals called **sulfation factor** (because it stimulates sulfate incorporation into newly forming protein) or **somatomedin** (because it mediates the effect of somatotropin). The factor may be made in the liver, and its production is probably elicited by STH stimulation. Actions of STH other than that on cartilage growth may not be mediated by somatomedin. Furthermore, no one has yet been able to produce growth by administering somatomedin.

Leuteotropic hormone (LTH) (also called prolactin or lactogenic hormone) has a multiplicity of diverse actions among the various vertebrates. It stimulates the crop sacs of pigeons and doves to produce the "crop milk" on which their young feed, acts synergistically with estrogen to induce brood patches in birds (Fig. 3-5), elicits lactation in mammals (Fig. 3-6), favors return from land dwelling to water and subsequent spawning in salamanders, and may have a role in osmoregulation in fishes. It is the most versatile of the adenophypophysial hormones!

In 1962 the presence of a protein hormone in the human placenta which has lactogenic properties and is chemically similar to pituitary STH and LTH was demonstrated. It is called **human placental lactogen.**

It is interesting to note here that new "uses" seem to have been "found" in more recently evolved species for hormone molecules that existed in their more ancient progenitors. LTH would appear to be an interesting example of this. LTH activity has been demonstrated in the pituitaries of all major vertebrate phyla, yet lactation is an attribute only of the relatively recent Mammalia. The molecule (in basic composition, but not necessarily with exactly the same amino acid sequence) existed before the lactational function was "found" for it.

LTH's functions are best known in the mammal. In addition to stimulating mammary-gland growth and development (which it does synergistically with ovarian steroids) and inducing milk production, LTH also promotes the secretion of **progesterone** by the corpora lutea of the ovary (see below)—the so-called *luteotropic effect*. As already noted, there is considerable overlap in actions between STH and LTH in promoting lactation; they are both complementary and redundant. In addition to these two hormones, there are several other hormones, including ovarian, placental, thyroid, and adrenocortical, that are orchestrated to bring about development of the mammary glands and to promote and maintain milk production (Fig. 3-6).

Another interesting example of overlapping function was discovered by Li in the process of purifying ACTH from sheep pituitaries. A component was obtained that had distinctive properties quite different from those of ACTH. When assayed by the rabbit fat-pad method for lipolytic activity, it proved as effective as ACTH, and because of this action it was designated **lipotropin** (β-lipotropic hormone, more precisely, or β-LPH). When its structure was revealed it was found to resemble both ACTH and MSH. It appears that β-LPH is a functional adenohypophysial hormone with biological properties between those of ACTH and β-MSH (Fig. 9-13).

Recently, oligopeptides that have sequences of amino acid residues identical to portions of β-LPH have been isolated from brain tissue and the pituitary gland. They act like morphine sulfate in stimulating prolactin and growth-hormone release. These morphinomimetic (or opioid) peptides have been called **endorphins** (from "endogenous" plus "morphine"). Three endorphins have been isolated from hypothalamic tissue. These small peptides, named **α-endorphin, β-endorphin,** and **γ-endorphin,** are respectively identical to the sequence β-LPH-(61–76), β-LPH-(61–91), and β-LPH-(61–87). β-LPH may be the *prohormone* for endorphins, in a manner analogous to angiotensin formation. Indeed, the brain contains endopeptidases that have the capacity to produce such fragments. The significance of the opiatelike peptides in the physiological control of LTH and STH remains to be defined. It is speculated that endorphins may play a role in natural pain killing and may be involved in narcotic addiction.

ACTH has been purified from pituitaries of cows, sheep, pigs, and humans. In all four species, the hormone is a straight-chain polypeptide of 39 amino acid residues. There are minor variations in a few sequences (positions 25 through 33), but they all appear to be equivalent in their actions on adrenal cortices.

The adrenal cortex is the primary site of action for ACTH. It enlarges the adrenal cortices of normal animals and restores the atrophic cortices of hypophysectomized ones. ACTH stimulates the output of cortical secretions, which, chemically, are steroid hormones. The steroids, in turn, produce specific metabolic changes, especially affecting carbohydrate, protein, and mineral metabolism. Although nearly 50 steroids have been obtained from the cortical component of the mammalian adrenal gland, only a few of these are considered to be released and biologically functional.

The ones that have been detected in circulation are all derivatives of pregnane and contain a total of 21 carbon atoms. Although their actions overlap, the adrenal steroids may be grouped as either *glucocorticoids* (if they primarily influence carbohydrate metabolism) or *mineralocorticoids* (if they influence mineral metabolism). The 11-oxygenated corticosteroids, **cortisol, corticosterone, cortisone,** and **dehydrocorticosterone,** are the most important natural steroids in affecting carbohydrate and protein metabolism, but they have relatively little effect on water and electrolyte metabolism. Corticoids that lack oxygen at position 11 (**deoxy-**

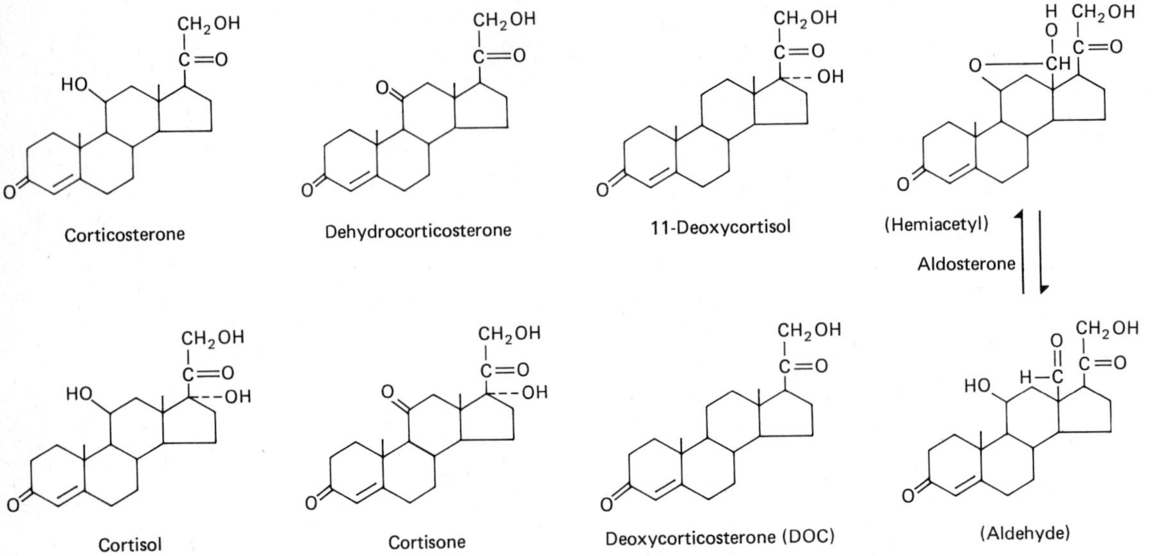

FIGURE 9-15
Some biologically active C_{21} adrenosteroids. Aldosterone, when in solution, exists in aldehyde and hemiacetyl forms.

corticosterone, deoxycortisol) and **aldosterone** have major effects on electrolyte and water metabolism (see Chap. 7) but do not have much effect on energy metabolism (Fig. 9-15).

The role of the adrenal steroids in stress responses has been discussed in Chap. 3, and in regulation of body fluids and electrolytes in Chap. 7.

Three hormones of the pars distalis are *glycoproteins* (i.e., they contain carbohydrate moieties): **follicle-stimulating hormone** (FSH), **luteinizing hormone** (LH), and **thyrotropin** (or **thyroid-stimulating hormone,** TSH). In addition, these three hormones are also alike in that they have similar amino acid compositions, their molecular weights are about 30,000, and they have a complex subunit structure.

It should now be apparent that the hormones of the pituitary gland, although several and diverse in function, belong to only a few chemical families. Indeed, pituitary hormones provide a particularly useful model for studying protein evolution. Since the primary structure of a protein or (in most cases) a polypeptide is directly related to the base sequence of the gene which directs its biosynthesis, the amino acid sequence gives information about that gene. Each hormone is seen to be one of a family, the members of which are related in structure although not necessarily in function. Thus, all the pituitary hormones that have been mentioned above may be grouped into four families:

1 The neurohypophysial octapeptides (oxytocin, vasopressin, and analogues)

2 Corticotropin (ACTH), melanotropin (MSH), and lipotropin (LPH)

3 Growth hormone (STH), LTH, and placental lactogen

4 Thyrotropin (TSH) and gonadotropins (LH and FSH)

9-7 REPRODUCTION

The principal action of FSH in female mammals is to stimulate the development of ovarian follicles, and in males to stimulate development of the seminiferous tubules. However, LH is required for full gonadal development and for production of mature eggs and sperm, as well as for the development of certain sex accessory structures. Thus, FSH and LH act synergistically to stimulate gametogenesis and support reproductivity (Fig. 3-2).

There is much variation among vertebrate species in the manner in which the gonads function, so that it is difficult to make generalizations. For present purposes, only a few endocrine-influenced features will be mentioned to illustrate some principles of chemical integration by hormones.

The *testes* produce masculinizing steroid compounds known as **androgens.** Androgens also are elaborated by ovaries and, as noted below, by adrenal cortices, from biosynthetic pathways like those in testes. **Testosterone** and **androstenedione** are the main circulating androgens of testicular origin (Fig. 9-16).

Since androgens are essential for spermatogenesis, the testis must be regarded as a target acted upon by its own hormones. The accessory system of male ducts and glands is, morphologically and physiologically, dependent upon the production of androgens. Secondary sex characteristics of males, such as the combs of chickens, the blackening of the beak and production of bright-pigmented plumage of certain birds in the breeding season, the presence of horns in sheep and deer, vocalization of toads, and nuptial pigmentation in several species of fishes, are conditioned by male sex hormones. In addition, as discussed below, the adenohypophysis and brain are targets for gonadal steroids; they act in "feedback" control of gonadotropin output.

The *ovaries* of vertebrates elaborate the steroids **estrogens, progestogens, androgens,** and a nonsteroid hormone called **relaxin.** The predominant natural estrogens of the human are **estradiol**-17β, **estrone,** and **estriol** (Fig. 9-17).

Estrogens exert direct, local effects upon the ovary that lead to hypertrophy of that organ and also stimulate gamete maturation. Growth of the vagina, thickening of the uterus, and preparation of mammary glands for milk secretion result from estrogen stimulation. In addition, in mammals exhibiting an estrous cycle, sexual receptivity typically coincides with estrus, a period during which the ovaries are secreting large

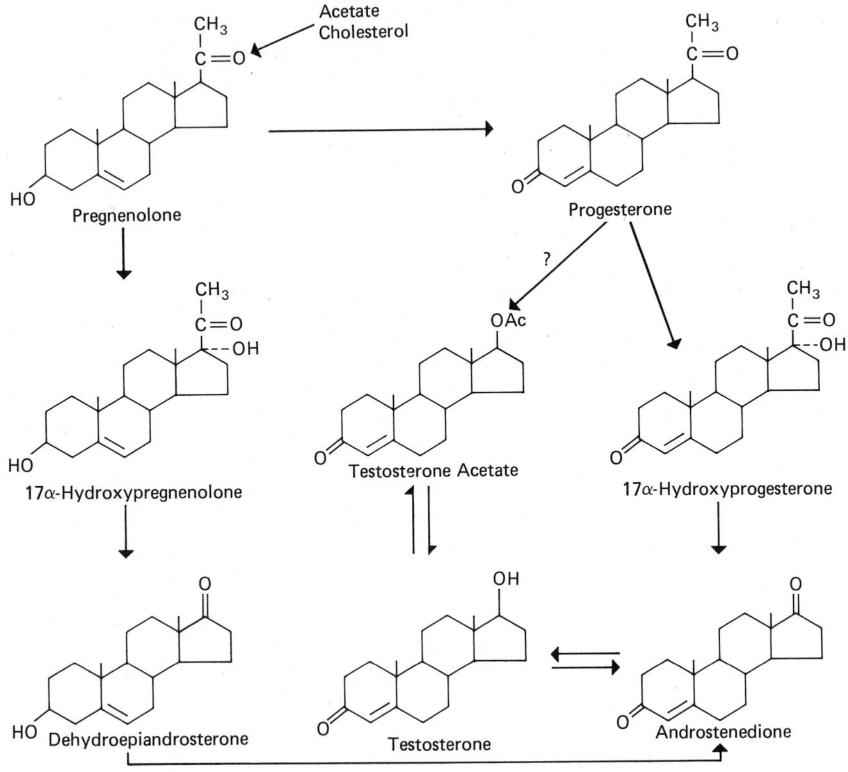

Acetate
Cholesterol
Pregnenolone
Progesterone
17α-Hydroxypregnenolone
Testosterone Acetate
17α-Hydroxyprogesterone
Dehydroepiandrosterone
Testosterone
Androstenedione

FIGURE 9-16 Probable pathways of androgen biosynthesis in the gonads.

amounts of estrogen. Full mating behavior generally depends upon both estrogen and progesterone (see Chap. 3, Sec. 3-3).

Progestogens are substances producing changes in the uterus that prepare it for implantation of blastocysts, maintain pregnancy, and regulate accessory organs during the reproductive cycle. They often act synergistically with estrogens (Fig. 3-4), but also, under other conditions, act antagonistically. **Progesterone,** the most prominent of these C_{21} steroids, is present in the ovary, testis, adrenal cortex, and placenta (Fig. 9-18).

Relaxin is a peptide present in various mammalian species during pregnancy. The placenta and uterus have been suggested as possible sources of relaxin, but the best-known source is the corpus luteum (Fig. 3-4) of the pregnant sow. In 1976 it was determined that relaxin is produced by the human corpus luteum. Operating in conjunction with other hormones, especially estrogen, it causes a marked pelvic relaxation during late pregnancy. The pelvic bones become less rigid and the birth canal is enlarged, thus facilitating parturition.

As a generality, steroid hormones are potent stimulators of metabolic machinery in virtually every tissue. It is probably no coincidence that certain ones are involved in reproductive processes which, metabolically, are formidable, especially in the female. Some interesting findings

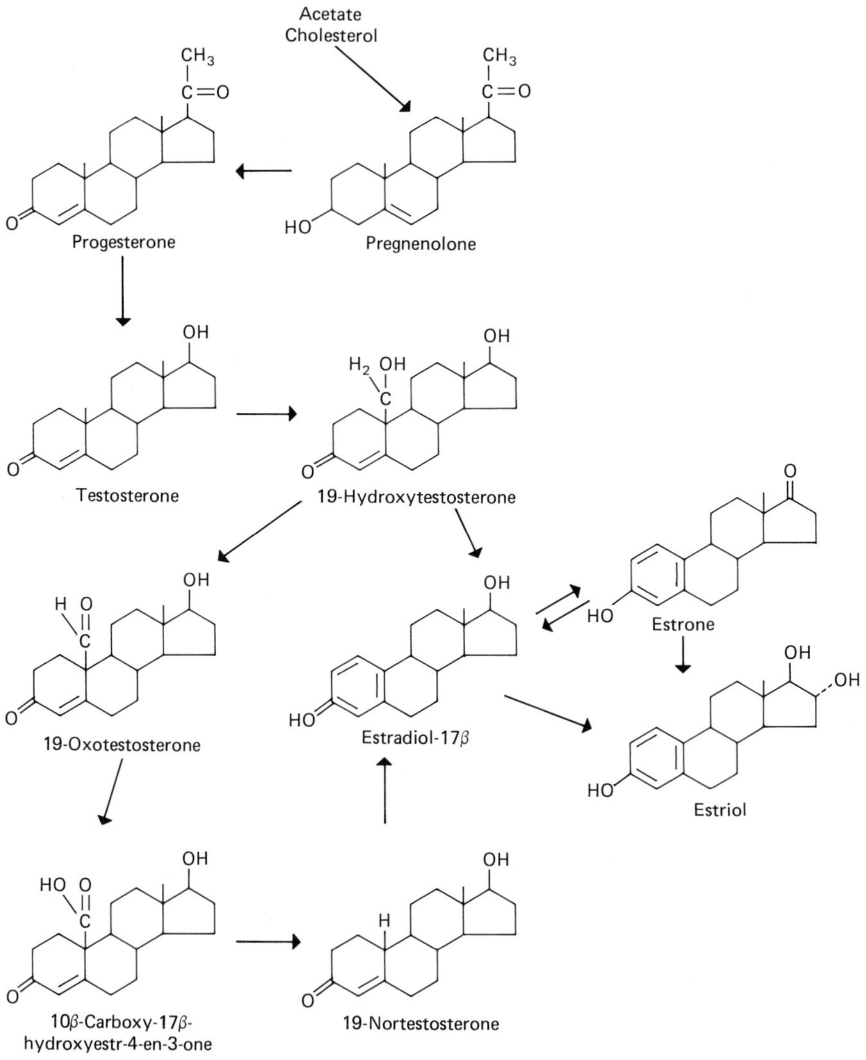

Acetate
Cholesterol

CH_3 C=O

Progesterone

Pregnenolone

HO

OH

Testosterone

H_2 OH C

19-Hydroxytestosterone

H O C

19-Oxotestosterone

Estradiol-17β

Estrone

Estriol

HO O C

10β-Carboxy-17β-hydroxyestr-4-en-3-one

19-Nortestosterone

FIGURE 9-17
Probable pathways in the biosynthesis and metabolism of estrogens.

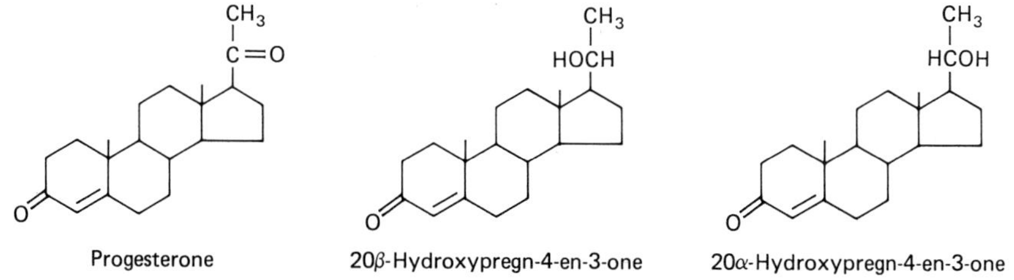

Progesterone

20β-Hydroxypregn-4-en-3-one

20α-Hydroxypregn-4-en-3-one

FIGURE 9-18
Naturally occurring progestogens.

from the study of fish underscore the fact that the pituitary hormones we refer to as gonadotropins—because of their marked action on gonads of homeotherms—were perhaps originally metabolic regulators. Some factors in the teleost pituitary that stimulate its thyroid tissue appear to be identical to gonadotropic hormones of mammals. Accordingly LH and FSH of mammals have the capacity to stimulate the thyroid of teleosts, but not that of mammals. Thus, the glycoprotein hormones of the pituitary, all similar in size and chemical composition, may share the common property of being, originally and primarily, metabolic regulators.

In addition to the glycoprotein hormones of the pituitary gland, there are, in the placenta of many (and perhaps most) eutherian mammals, at least one gonadotropic hormone, called **chorionic gonadotropin** [human chorionic gonadotropin (HCG)] and one **thyrotropin.** The properties of HCG are similar to those of pituitary LH. HCG is responsible for maintaining the function of the corpus luteum during pregnancy. The thyrotropin of the placenta has not been fully characterized, and although it appears to resemble pituitary TSH quite closely, its function is not known.

9-8 REGULATION OF METABOLISM, GROWTH, AND DEVELOPMENT

Pituitary TSH is the main factor controlling thyroid tissue in homeotherms. After ablation of the pituitary, the thyroid atrophies and its secretory capacity is reduced to a minimum. The synthesis of hormone by thyroids depends upon the accumulation of diffusible iodide, a process stimulated by TSH. It is generally believed that the gland engages in a transcellular active transport of iodide from the blood into thyroid-gland follicles. The iodide is incorporated into **thyroglobulin,** the principal storage form of the thyroid hormones. This large protein is hydrolyzed to yield several iodinated amino acids. Only two of these, 3,5,3′**-triiodothyronine** (or T_3) and **thyroxine** (or 3,5,3′,5′**-tetraiodthyronine,** or T_4), are known to have any biological activity. Monoiodotyrosine and diiodotryosine do not leave the follicle; they are rapidly deiodinated within the thyroid cells, where the iodine is reutilized for a recycling synthesis of thyroglobulin (Fig. 9-19).

Many kinds of environmental stimuli influence the release of thyroid hormones. Exposure of warm-blooded animals to cold environments increases thyroid activity. On the other hand, physical stressors, such as trauma and hemorrhage, evoke prompt and prolonged inhibition of thyroid secretion.

In homeotherms, the most characteristic effect of thyroid hormones is to increase energy production and oxygen consumption of most tissues, an effect called *calorigenesis*. Thyroid hormones, however, are not alone in promoting increased metabolic rates, as adrenocortical hormones, ACTH, and STH are calorigenic also. Thyroid hormones have a number

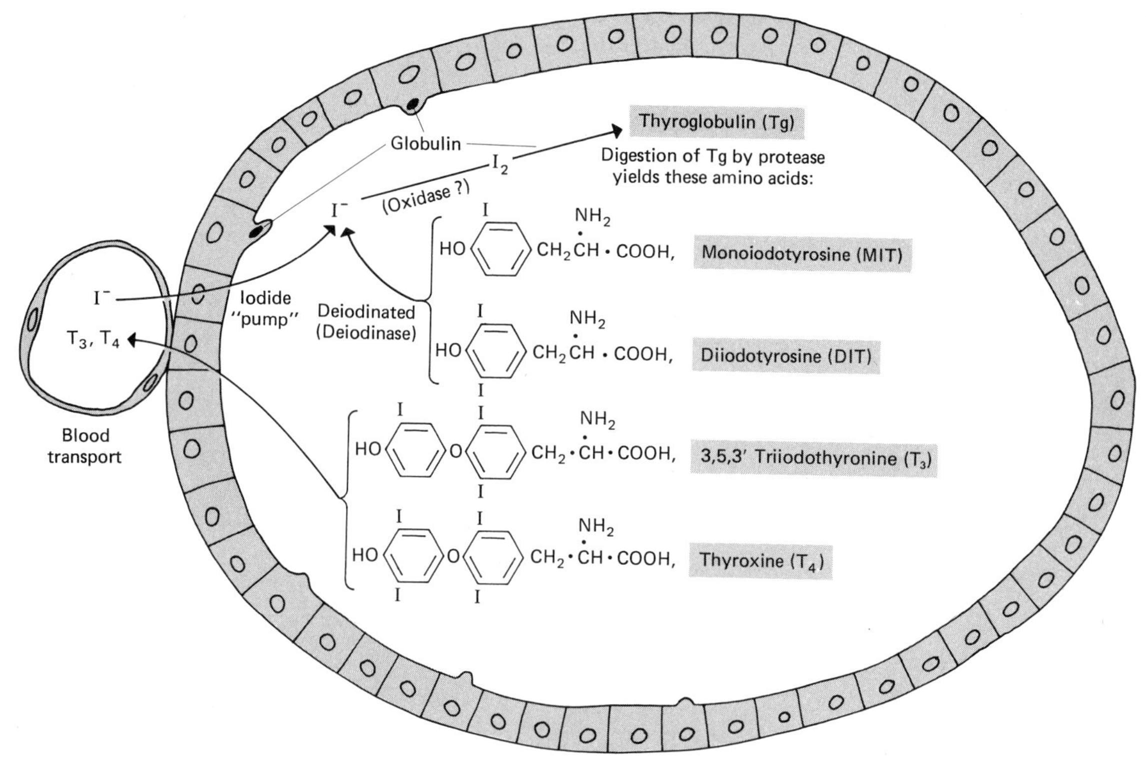

FIGURE 9-19
Diagrammatic summary of the biochemical events in the synthesis of thyroid hormone. (*Modified from A. Gorbman, and H. A. Bern, 1962, A Textbook of Comparative Endocrinology, John Wiley & Sons, Inc., New York, p. 122.*)

of diverse effects on carbohydrate, fat, and protein metabolism, as well as Na^+, Cl^-, and water retention.

The second major category of actions of thyroid hormones is promotion of growth and development. The molting of hair in mammals, of feathers in birds, and of skin in amphibians is strongly conditioned by thyroxines. The best-known function of the thyroid hormones among poikilotherms is the control of metamorphosis. This has been especially well studied in amphibia. It is thought that thyroxines rise slowly during metamorphosis, progressively reaching threshold levels for a series of target tissues whose differentiation to adult condition permits the orderly transformation of the larval form to the adult amphibian.

In several species, especially in certain mammals, it has been shown that if portions of the thyroid gland are removed surgically, causing blood levels of thyroxines to fall, the pituitary will augment its release of TSH, the thyroid will hypertrophy, and thyroxine levels will come up to normal levels. After electrolytic lesions are used to destroy specific bilateral regions of the hypothalamus, the compensatory hypertrophy of the thyroid remnant does not occur. Thus, hypothalamic lesions prevent the elevated output of TSH, and the thyroid is not stimulated. The implication of these observations is that the hypothalamus somehow controls the output of TSH by the pituitary, which in turn controls the thyroid. Indeed, from studies employing electrical stimulation and hormone implantation, as well as lesioning, it is generally agreed that a neural mechanism concerned with TSH control is located in the median eminence region of the hypothalamus (Fig. 9-10). The hypothalamic mechanism appears to influence TSH secretion through release of one or more chemical agents, which reach the appropriate pituitary cells by way of the hypophysial portal vessels. One such hypothalamic factor, called **thyrotropic hormone-releasing factor** (TRF) has been isolated from hog brain and identified. Porcine TRF is known to be L-(pyro)GluHisPro (NH_2).

The level of circulating thyroxine is believed to have a regulatory effect on the output of TSH; thus, there is "feedback." However, it does not appear to be a simple closed-circuit system. Blood levels of thyroid hormone seem to affect the release of TSH by way of the hypothalamic mechanism (long-loop feedback) as well as by direct action on the pituitary gland (short-loop feedback).

9-9 ENDOCRINE INTEGRATION

The neuroendocrine control mechanism for the thyroid just described is but one of the several operating in the control of pituitary hormones and their associated endocrine glands. It has been well documented, beginning in the 1940s, that the pituitary is subject to nervous control. All the pars distalis hormones appear to be controlled by so-called **hypophysiotropic factors** (or **hypophysiotropic hormones**) arising from hypothalamic neurons. These **neurosecretions,** as distinct from those destined for the pars nervosa, are delivered to the primary plexus of the hypophysial-portal system, which then transports them into the pars distalis (Figs. 9-10 and 9-20).

The first of these factors to be isolated and identified was TRF, but the first to be detected was a factor associated with the release of ACTH and called **corticotropin-releasing factor** (CRF). As of this writing, it appears that there may be two CRFs, with molecular weights respectively of 1300 and 2500, but the amino acid sequences of these presumed peptides have not been determined. Other hypophysiotropic factors have been found, and they are now the subjects of vigorous investigation. Though most appear to be *releasing* agents, some of them in mammals, including one that controls LTH, act to *inhibit* secretion. Also, a factor called **somatostatin,** a tetradecapeptide that is synthesized in the hypothalamus, and possibly the pancreas, inhibits growth-hormone secretion. It also inhibits glucagon and insulin secretion from isolated rat pancreatic islands. Whether or not there are specific hypophysiotropic factors controlling LH and FSH individually is presently contested. Many experimental results suggest that there may be only a single factor, a gonadotropin-releasing hormone, controlling both these pituitary hormones.

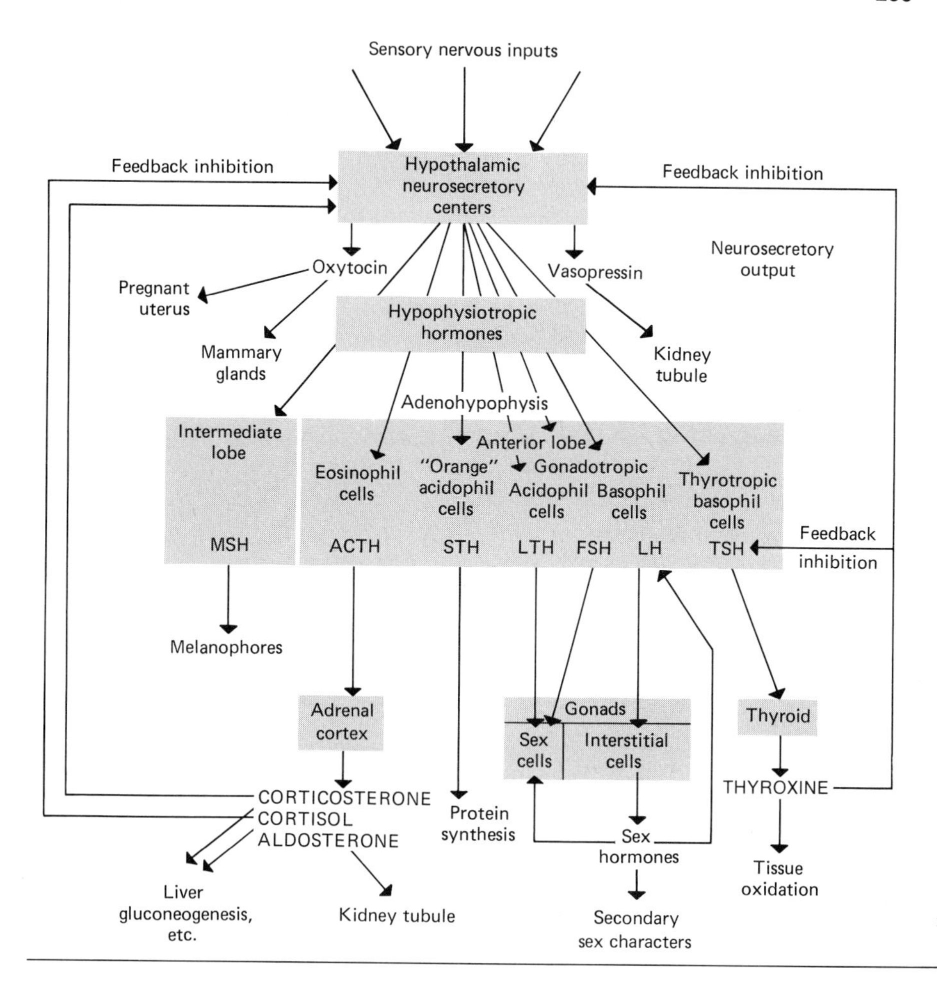

FIGURE 9-20
A generalized scheme for indicating the interrelationships of the hypothalamohypophysial neurosecretory-endocrine system in vertebrates.

The exact identity of these substances is not yet certain in all cases, but all of them appear to be very small peptides. Furthermore, all the distalis hormones appear to be regulated by similar control loops employing feedback, principally to the hypothalamus, but perhaps also to the pituitary. Fairly discrete aggregations of hypothalamic neurons are concerned with each of the pituitary hormones. Their secretions are directed toward the median eminence, from which they are conveyed in common to the pituitary by the portal vessels. Thus, the hypothalamic neurosecretory centers, the adenohypophysis, and several endocrine glands are linked together in a highly regulated, multifunctional, integrated system (Fig. 9-20).

9-10 CYCLIC NUCLEOTIDES

In the 1950s, evidence was obtained by Sutherland and his coworkers that glucagon and epinephrine exert their glycogenolytic action on liver and muscle tissue by increasing the activity of the enzyme glycogen

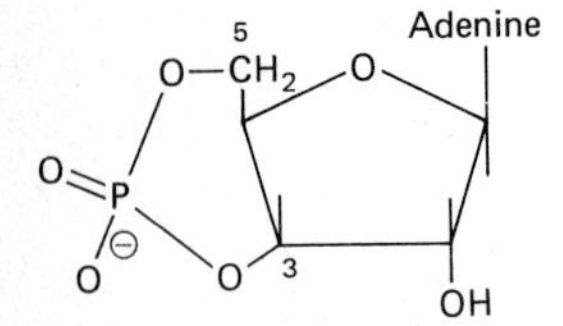

FIGURE 9-21
The structure of cyclic 3′5′-adenosine monophosphate (cyclic AMP).

phosphorylase. Later they showed that the hormones exert their action through altering the level of **cyclic adenosine monophosphate** (cAMP), the product of adenosine triphosphate (ATP) cyclization (Fig. 9-21).

The only enzymatic reaction resulting in the formation of cAMP which has been demonstrated is that catalyzed by *adenyl cyclase*. Cyclic AMP in turn causes the conversion of inactive phosphorylase to the active form. This conversion has been shown to involve dimerization and phosphorylation of enzyme residues with ATP. The reverse reaction is accomplished by phosphate removal. A phosphorylase kinase and a phosphorylase phosphatase are thus involved (Fig. 9-22).

A great number of polypeptide hormones, in addition to glucagon, also appear to exert effects through cAMP. These include ACTH, TSH, LH, FSH, MSH, and vasopressin. However, some nonpeptide hormones (and a few peptide hormones) do not appear to act through cAMP. In addition to epinephrine, other biogenic amines, including norepinephrine, histamine, and 5-hydroxytryptamine, enhance cAMP formation. Cyclic AMP has been implicated in the release of neurotransmitters at synapses and neuromuscular junctions.

FIGURE 9-22
The mechanism of action of epinephrine and glucagon. Glycogen breakdown to glucose phosphate is controlled by the enzyme phosphorylase, which exists in two forms—an active form called phosphorylase *a*, and an inactive form, phosphorylase *b*. The effect of epinephrine or glucagon is to increase the proportion of active phosphorylase, which may increase several hundred times during hormone treatment. Conversion of phosphorylase *b* to *a* is catalyzed by the enzyme dephosphophosphorylase kinase, which in turn is activated by cyclic adenosine monophosphate. This compound is generated from adenosine triphosphate in the presence of an enzyme known as *adenyl cyclase*. Epinephrine and glucagon somehow activate the cyclase and thus set in motion a series of reactions that lead to an increase in phosphorylase activity and glycogen breakdown. Muscle tissue lacks the enzyme glucose 6-phosphatase, which is required to liberate free glucose from glucose phosphate. In muscle, therefore, the effect of epinephrine is to accelerate degradation of glycogen to carbon dioxide and water, or to lactic acid under anaerobic conditions of exercise. Liver contains glucose 6-phosphatase, so the net effect of epinephrine or glucagon on liver is to stimulate secretion of glucose. (*Modified from B. E. Frye*, 1967, *Hormonal Control in Vertebrates, The Macmillan Company, New York*, p. 57.)

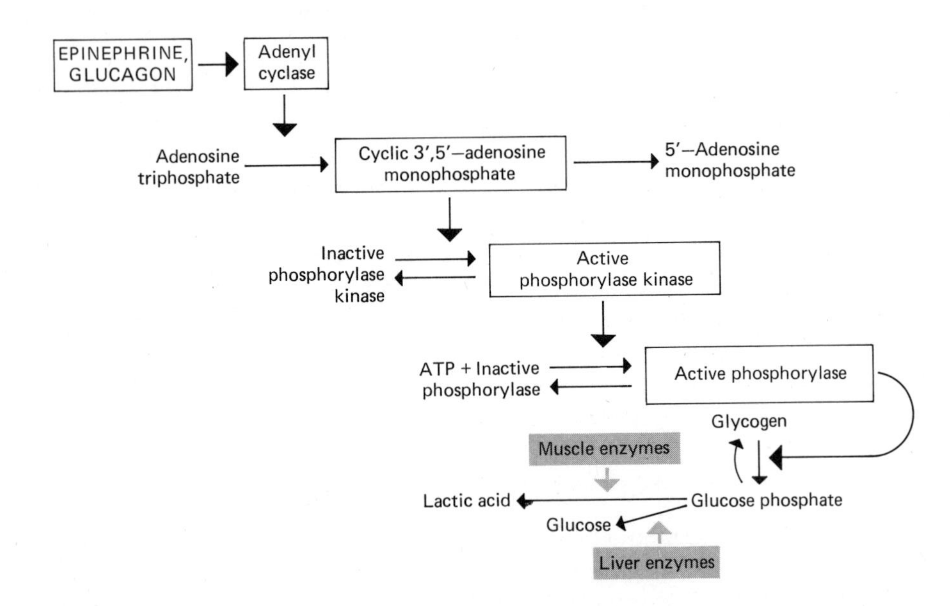

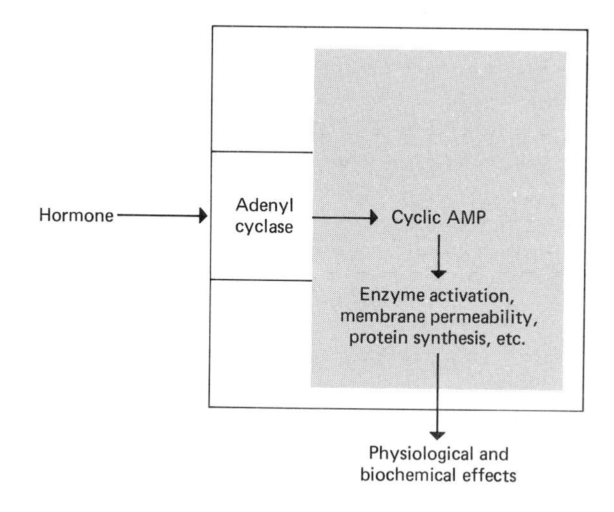

FIGURE 9-23
Diagramatic representation of the concept of hormone action on cells through cyclic AMP.

The enzyme responsible for cAMP synthesis, adenyl cyclase, is bound to the plasma membrane of nearly all animal cells examined, whereas the enzyme responsible for its degradation, phosphodiesterase, usually is located in the interior of the cell. Primarily from this evidence, Sutherland developed the "second-messenger" concept. That is, the "first messenger," an external stimulus specific for the particular cell type, such as a hormone, interacts with the membrane-bound adenyl cyclase, leading to increased cAMP synthesis within the cell. Then cAMP, acting as a second messenger, activates one or more processes or enzymes. Remarkably diverse physiological responses result from the increase in cAMP within different cells. Apparently, what kind of response is evoked depends upon the nature of the cell stimulated (Fig. 9-23).

Cyclic guanosine monophosphate (cGMP), formed from GTP, is known to exert some actions opposite to those of cAMP. For example, cGMP concentrations rise in response to acetylcholine stimulation in heart, brain, and other tissues, whereas cAMP concentrations are increased following epinephrine stimulation. High doses of ACTH elevate cAMP while depressing cGMP, whereas hypophysectomy induces changes in the opposite direction. However, there are presently enough inconsistencies in the responses of the two monophosphates to manipulations to prevent the drawing of the general conclusion that they are somehow acting in a "push-pull" type of control mechanism.

In the operation of the second-messenger monophosphates, there is perhaps revealed to us a magnificently simple control system that may be a universal mechanism for integrating certain intracellular responses of animal cells to specific external stimuli, a system that has permitted the adaptation of cellular activities to novel stimuli as encountered through evolution.

READINGS

Barrington, E. J. W., 1963, *An Introduction to General and Comparative Endocrinology*, Clarendon Press, Oxford. This useful textbook contains interesting historical accounts of the discovery of some of the hormones.

Frye, B. E., 1967, *Hormonal Control in Vertebrates*, The Macmillan Company, New York. This small volume is a concise and well-organized survey of the principal aspects of endocrinology.

Gorbman, A., and H. A. Bern, 1962, *A Textbook of Comparative Endocrinology*, John Wiley & Sons, Inc., New York. A unique collection of information on mammalian and nonmammalian forms that is to be found in no other textbook.

Turner, C. D., and J. T. Bagnara, 1976, *General Endocrinology*, W. B. Saunders Company, Philadelphia. A splendid textbook providing a comprehensive introduction to endocrinology and containing excellent sets of references at the end of each chapter.

NEURO-ENDOCRINE INTEGRATION

10

In terms of rapidity and complexity of response to a disturbance in the internal state or the external environment, the multicellular animals are without peer. Their exquisite and varied repertoire is achieved by having control systems that detect and transmit information efficiently between components that may be spaced widely apart within the organism.

Somewhat arbitrarily, the bulk of the animal's integrating control mechanisms has been categorized as two major systems: the endocrine system and the nervous system. One of the more recent and intellectually pleasing developments in physiology is the recognition that such a separation cannot be adequately defended, especially if one views an adequate representation of multicellular organisms. It is now very well established that the nervous and endocrine systems complement one another and that they also are interrelated in varying degrees. They are, in fact, one supersystem. The nervous system is involved in the control of endocrine secretion, hormones influence the functioning of the nervous system, and certain components of the nervous system produce hormones.

10-1 NEUROENDOCRINE REFLEXES

The transition from the study of endocrine to neural modes of integration is easily effected, as there is an intermixing of neural and endocrine components within organismic control systems such that a continuous spectrum exists from purely endocrine to purely neural, with various admixtures between. This can be conveniently communicated by reference to Fig. 10-1.

The basic plan of all neuroendocrine systems incorporates sense organs for error detection at the input, integration of the generated afferent signals by the central nervous systems (CNS) (except for a few situations exemplified by the A diagram), and output of an efferent signal, either neural or humoral, to the effectors. A few specific examples will illustrate the kinds of relationships that the various components of organismic control systems can assume.

The control mode diagramed in Fig. 10-1*a* is exemplified by the endocrine reflex of the pancreas in the regulation of blood glucose. As documented in the previous chapter, the pancreatic islet cells are both error detectors and the source of efferent outputs, insulin and glucagon. These two signal molecules communicate with their widely scattered target cells (collectively, the effector) via the vascular system. Likewise, parathyroid hormone and calcitonin are involved in a similar mode of operation.

FIGURE 10-1

The position of hormones in the integration of stimulus and response. Several possible kinds of relationships are illustrated. (A) The endocrine gland is also the sensory unit and responds directly to a stimulus (change in internal chemistry) by secreting a hormone, which in turn elicits a response by the effector organ. (B) Sensory information is transmitted to the central nervous system, where it is integrated with other sensory input into the central nervous system, which then directs an appropriate efferent signal. (B_1) A typical neural reflex arc in which the efferent signal is a motor-nerve impulse. (B_2) The efferent signal is a hormone produced by neurosecretory cells in the central nervous system. The position of the neurosecretory cells is analogous to that of the motor nerve in the overall reflex. (C) The efferent signal is propagated in two stages: a motor impulse emanates from the central nervous system and stimulates an endocrine gland to secrete a hormone that elicits the final response. (D_1) Analogous to C, but both stages of the efferent pathway are hormonal. Neurosecretory cells secrete a hormone that activates an endocrine gland. (D_2) A further level of complexity in which, beginning with the neurosecretory hormone, three different hormones are successively involved in the efferent pathway. (*After B. E. Frye, 1967, Hormonal Control in Vertebrates, The Macmillan Company, New York, p. 35.*)

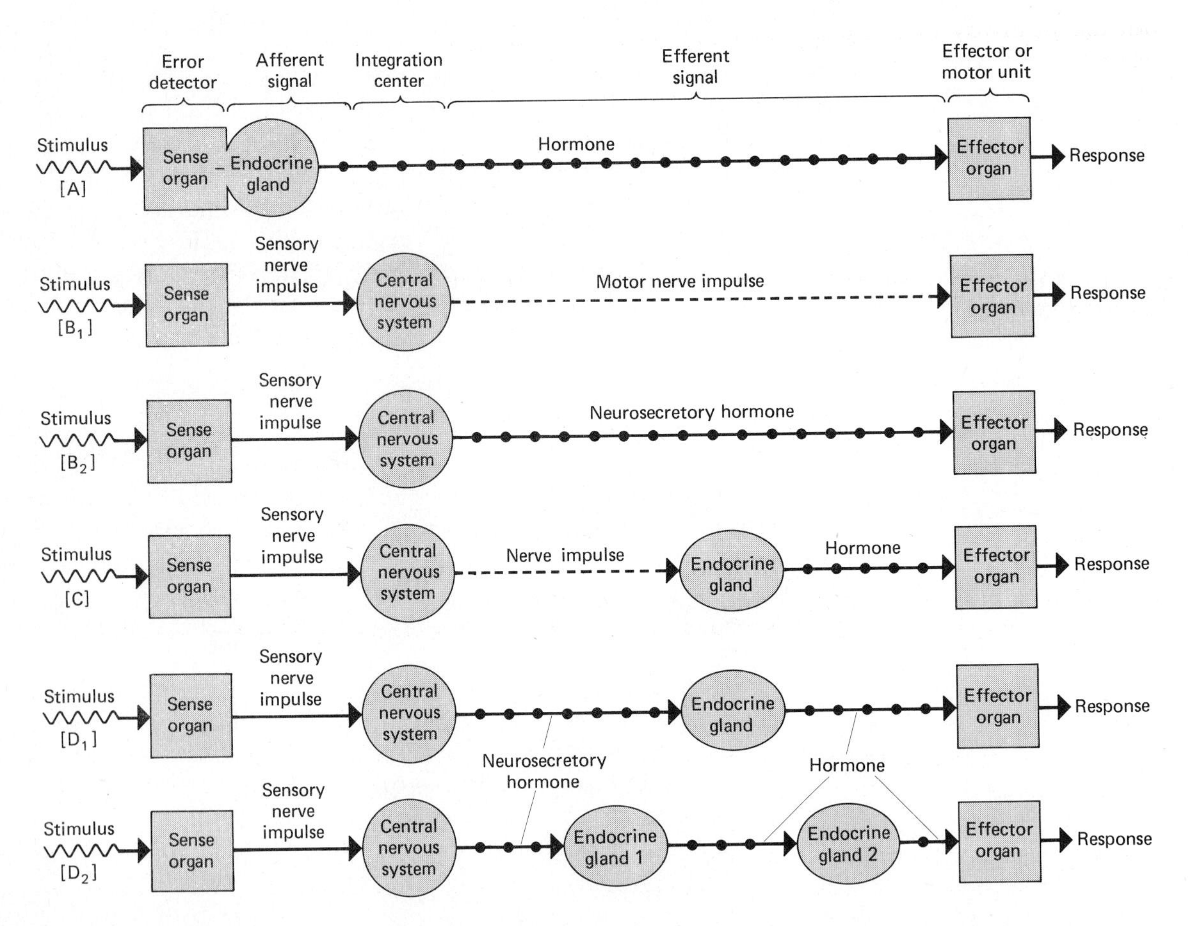

Diagram B_1 in the figure is a representation of the classic neuromotor reflex arc. In such a mechanism a sense organ (e.g., a stretch receptor) initiates a neural input which is transmitted by sensory neurons to the CNS, wherein an output is initiated, which then is transmitted via motor neurons to the effector organ (e.g., skeletal muscle). The details of this and other types of neural integrative mechanisms will be dealt with in the next chapter.

In the mechanism exemplified in diagram B_2 the efferent signal is a neurosecretory hormone, but otherwise it is like the neuromotor reflex described above. Neurosecretions are products of neurons that act as hormones and commonly are released into the circulatory fluids, by which they are conveyed to their targets. The nature of these hormones and their origins will be discussed in the next section. For now, some examples of how this particular kind of reflex arc functions will be sufficient.

In the ejection of milk from the lactating mammary gland, sensory input from the nipple (the sucking stimulus) is referred to the brain (hypothalamus, eventually), and the release of oxytocin (a neurosecretory hormone) from the neurohypophysis into the blood results in contraction of the myoepithelial cells of the mammary alveoli (the effector organ). Note that the pathway of communication between the brain and the mammary gland is vascular rather than neural.

A second example of this type of mechanism is found in the control of kidney diuresis. An increase in osmotic concentration of the blood (hypertonicity), a decrease in blood volume (hypovolemia), as by loss of blood through hemorrhage, as well as several other conditions, will stimulate receptors in the brain, which in turn activate the release of ADH from the neurosecretory endings in the posterior pituitary. Again, communication with the effector organ, the kidney in this instance, is by way of the blood.

The C-type mechanism of Fig. 10-1 is illustrated by the "alarm reaction," in which sensory input to the CNS results in stimulation of the sympathetic division of the autonomic nervous system, which activates the adrenal medulla by its preganglionic sympathetic innervation, causing the release into the bloodstream of a mixture of epinephrine and norepinephrine from the chromaffin cells. (Some details of autonomic-system organization are included in the chapter to follow.)

The type of mechanism illustrated by D_1 in the figure typifies the manner in which certain pituitary hormones are controlled. For example, the release of somatotropic hormone (STH) from the pars distalis of the pituitary is presumably effected by a hypophysiotropic hormone of hypothalamic neurons (the neurosecretory hormone of the figure). STH then travels via the blood to target tissues (e.g., epiphysial discs of long bones).

Another possible example is the background adaptation response of anuran amphibians (although, as noted in Chap. 9, it is not certain that a neurosecretory hormone controls MSH release).

Mechanisms controlling development in certain insects seem to have invoked this same mode of operation. Cell bodies of neurosecretory

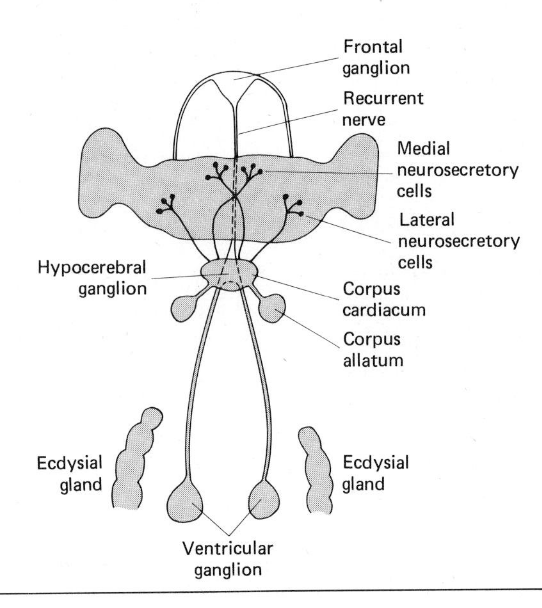

FIGURE 10-2
Diagram of the neuroendocrine system of the head in insects.

cells in the insect brain produce an **ecdysiotropic hormone (brain hormone)** which is transported through paired neural tracts to the **corpora cardiaca** (which are presumed to be neurohemal organs). From these bodies the hormone is transmitted, presumably through the blood (hemolymph), to the **ecdysial gland (prothoracic gland)** from which **ecdysone**, a molt-inducing hormone, is released (Fig. 10-2).

Integrative mechanisms that involve several of the pars distalis hormones are variants of the mechanism illustrated by the D_2 scheme of Fig. 10-1. Certain neurons of the diencephalon produce and release neurosecretory hormones (hypophysiotropic hormones) which travel via the portal system of vessels from the median eminence to the pars distalis ("endocrine gland 1" in the diagram), wherein they influence release (and synthesis, in some cases) of certain pituitary hormones. These pituitary hormones, the several "tropic" hormones, in turn control other endocrine glands ("endocrine gland 2" in the diagram). The secretions of the latter glands (e.g., thyroid, gonad, adrenal cortex) have a variety of specific targets, as discussed in the previous chapter.

Many, perhaps most, of the neuroendocrine reflexes are closed-loop systems employing feedback. For example, secretions of the adrenal cortex, gonad, and thyroid gland serve as inputs to the error detectors of the CNS, or to the pituitary directly, in some cases. They exert an inhibitory effect that results in a reduction of the output of specific tropic hormones from the pituitary (Fig. 9-20).

10-2 NEUROSECRETIONS

As indicated in the previous section, there is a class of chemical mediators of neural origin which are like endocrine factors in that they reach distant "targets" by way of the circulatory system. Thus, they have been called **neurohormones.** The neurons of the special glandular kind that produce these "hormones" are referred to as **neurosecretory cells,** and, hence, their secretions are also sometimes called **neurosecretions.**

Whereas neurohormones operate over some distance, in the conventional mode of integrative activities within the nervous system chemical mediators operate between closely apposed cell membranes, traversing a distance of only a few nanometers, or within the membrane of the cell of origin. **Neurohumors,** also called **neurotransmitters,** are the messenger substances involved in transfer of signals between cells of the nervous system. The site of transfer is a highly specialized junction between the membranes of two neuronal cells, called a **synapse,** wherein neurohumors arising in the presynaptic neuron participate in eliciting a very localized, characteristic response of extremely short duration in the postsynaptic neuron. Such messenger substances generally lack essential features of endocrine factors—in particular, access to and use of vascular pathways. They also are quickly and locally inactivated. These and other characteristics of neurohumors will be the subject of the next section.

The neurons which produce neuro*hormones* are not ubiquitous

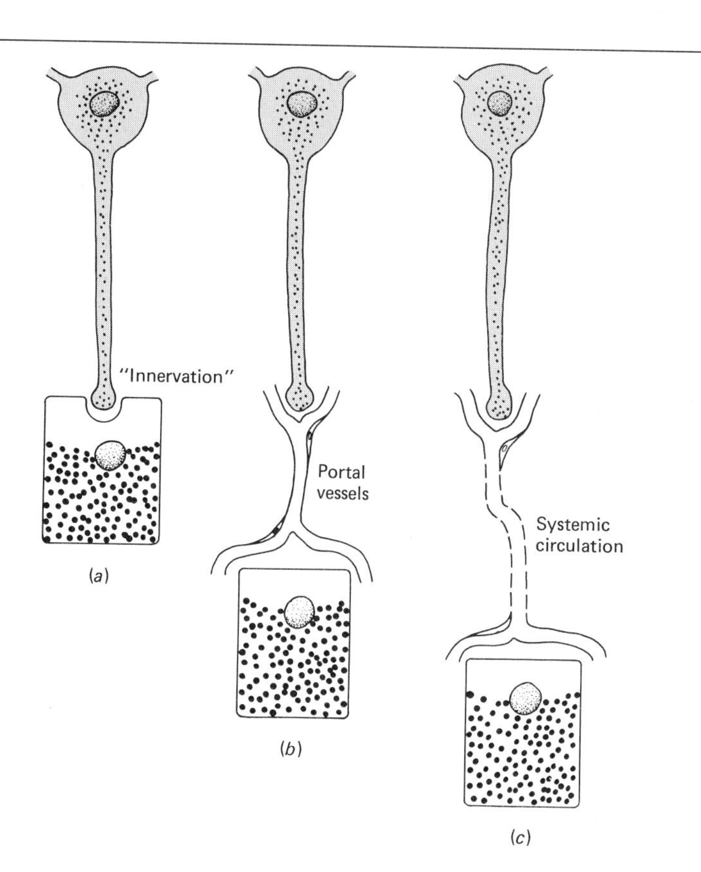

FIGURE 10-3
Possible relations between neurosecretory neurons and target cells. (*a*) Neurosecretomotor innervation represents possible control (often inhibitory) over endocrine cells by neurosecretory neurons, such as may occur in the fish pars intermedia or the insect corpus allatum. (*b*) The portal situation illustrates the nature of the hypothalamic control over hormone secretion by the pars distalis of the hypophysis in tetrapod vertebrates; neurohormones serve as releasers or inhibitors of adenohypophysial secretion. (*c*) Neurohormones released into the systemic circulation act directly upon their targets, as do hormones from non-neurosecretory endocrine tissues; sometimes, as with the insect ecdysial (prothoracic) gland, the target organ may be another endocrine gland. (*After H. A. Bern and F. G. W. Knowles, 1966, in Neuroendocrinology, vol. 1, edited by L. Martini and W. F. Ganong, Academic Press, Inc., New York, p. 143.*)

within nervous tissue, as are those that produce neuro*humors*, but rather tend to form highly specialized assemblies. Furthermore, although some appear to establish synaptic contact with contiguous effector cells, they more commonly terminate in close proximity to vascular channels (Fig. 10-3).

Neurons of this special glandular kind appear to engage in prodigious secretory activity. This is reflected in a decidedly glandular appearance, closely resembling that of nonneuronal cells which produce chemically similar secretory products. Typically, these neuronally derived compounds are proteinaceous, another characteristic distinguishing them from the known neurotransmitters, which are small nonprotein molecules. The secretory products can be made conspicuous by selective staining procedures for light microscopy. In electron micrographs these substances appear as membrane-bounded, more or less electron-dense granules in several characteristic size ranges (Fig. 10-4).

Neurosecretory granules reach the fiber terminal by proximodistal axonal transport and accumulate there in a bulbous enlargement, an important site of release of neurosecretory hormones. Groups of axon terminals, in which neurosecretory substances are stored before being released, often make contact with blood vessels or hemocoels to form **neurohemal organs** (Fig. 10-5).

The best-known neurohemal organ is the posterior pituitary (pars nervosa). As noted in Chap. 9, the posterior pituitary is the site of release of neurohormones (oxytocin and vasopressin) that are synthesized in the perikarya of neurons in the rostral hypothalamus. These octapeptides are bound to a carrier protein **(neurophysin)** during transport, and it is

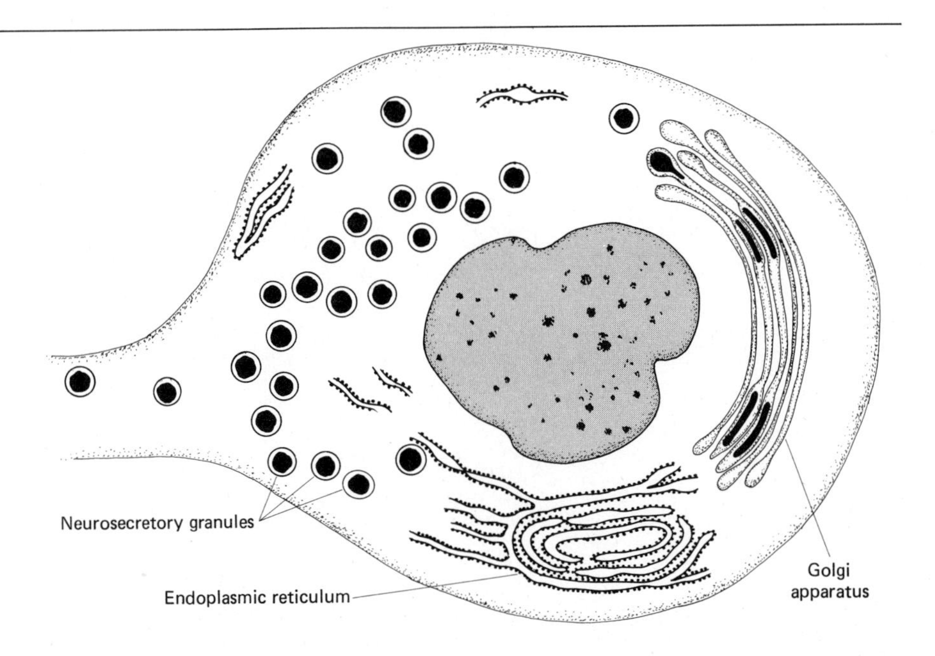

FIGURE 10-4
Diagram of presumed perikaryal synthesis of neurosecretory material, with endoplasmic reticulum contributing protein "raw material" to the Golgi apparatus for packaging into elementary neurosecretory granules. (*After E. Scharrer and S. Brown, 1963, Gen. Comp. Endocrinol. 2, p. 2.*)

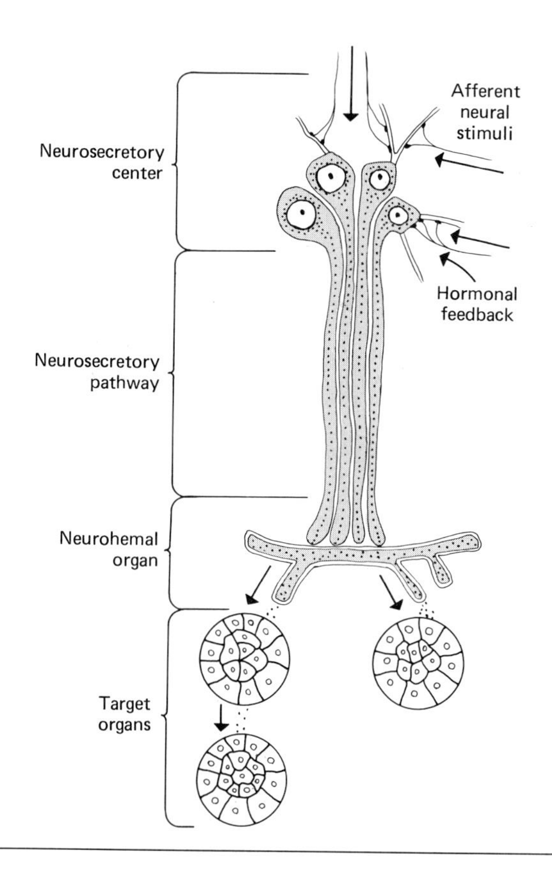

FIGURE 10-5
A common pattern of arrangement of neurosecretory neurons, in which impinging neural stimuli control the output of their neurohormone through a neurohemal organ. (*After E. Scharrer and B. Scharrer, 1963, Neuroendocrinology, Columbia University Press, New York, p. 22.*)

perhaps the latter that gives the stainability which is characteristic of the neurosecretory cell.

The hypophysiotropic factors, or hormones, which regulate hormones of the adenohypophysis are also neurohormones. The axon terminals of these neurosecretory neurons pass their products into the primary plexus of the portal vascular system, which in turn conveys them to the endocrine cells of the pars distalis (Fig. 9-10).

The central nervous systems of all animals, from worms to mammals, have groups of nerve cells that exhibit features of gland cells. Many of these have been demonstrated to be sources of neurohormones. For others, the evidence that they are neurosecretory remains circumstantial, based on cytological criteria. In addition to the known mammalian neurohormones, among insects, crustaceans, mollusks, and annelids several neurosecretions have been characterized in terms of source, site of action, and possible function. Among the functional roles assigned presently to invertebrate neurohormones are antidiuresis, control of blood-sugar level, maturation of oocytes, color change, and regulation of rhythmic muscle contractions.

The occurrence of neurosecretory neurons in so many phyla indicates their functional importance. They serve basically as a link between the nervous and endocrine apparatus. Each system functions in characteristic ways, but the two have a high degree of effective interaction. The special role played by neurosecretions in the neuroendocrine complex is that of generating signals for sustained control mechanisms. The neurosecretory cell, with properties both of a neuron and a gland cell, is made to order for this function. It seems to be uniquely endowed with the capacity for receiving and processing information from neural centers and passing it on to the endocrine system in appropriate language (Fig. 10-6).

Typically, neurohormones reach their "targets" by way of either systemic circulation or locally restricted vascular pathways, such as the hypothalamohypophysial portal system. However, ultrastructural analysis of a variety of neurosecretory systems has revealed that not all the terminals in which secretory granules are stored are associated with the vascular system. For example, there are axons of neurosecretory neurons which invade endocrine structures—the pars intermedia of vertebrates and the corpus allatum of insects. In certain insects, neurosecretory processes associate with muscle cells. Still others appear to make direct contact with other neurons, forming what appear to be "synaptoid areas," with intercellular spaces measuring about 20 to 30 nm. These situations invalidate the standard distinctions between ordinary and neurosecretory neurons (see again Fig. 10-3*a*).

When neurosecretory material is delivered directly to target tissues and not to the circulatory fluid, it could hardly be called a neurohormone, and yet the material is derived from a cell showing all the cytological characteristics of a neurosecretory cell, rather than those of a conventional neuron. Here, then, is an intermediate situation which bridges two modes of neuronal function. Thus, while it is true that all neurohormones are derived from neurosecretory neurons, not all the products of such neurons are hormones in the strict sense of the word.

There is a second kind of situation which further blurs the distinction between neurosecretory and neurohumoral neurons. There are a few neurosecretory neurons which appear to discharge material into the circulatory system or elsewhere that is not a peptide but rather seem to produce large quantities of catecholamines. Therefore, chemically they are like conventional adrenergic neurons, producing the same kind of neurotransmitter, but functionally and cytologically they resemble conventional neurosecretory neurons, differing only in the chemical nature of their product.

Tentatively, those classical neurosecretory neurons producing peptide secretions, regardless of the site of release, are called *peptidergic*. Those neurosecretory neurons producing catecholamines are called *aminergic*. These terms are consonant with the standard terminology which chemically classifies conventional neurons, as detailed in the next section.

FIGURE 10-6
Diagram to illustrate the possible neuronal "circuitry" in which a neurosecretory neuron can be involved, based on electrophysiological and cytological information. *It is to be emphasized that several of the relations indicated here are conjectural only.* Illustrated are axodendritic junctions (1); axosomatic junctions (2); axoaxonic junctions, in which the presynaptic fiber is "ordinary" (3) or neurosecretory (4); possible control of hormone release from the axon terminal by electrotonic or synaptic transmission (not established) from "ordinary" (5) or neurosecretory (6) fibers. In addition, hormonal factors can act upon the neurosecretory cell directly (7), as in feedback control of secretion. (*After H. A. Bern and F. G. W. Knowles, 1966, in Neuroendocrinology, vol. 1, edited by L. Martini and W. F. Ganong, Academic Press, Inc., New York,* p. 143.)

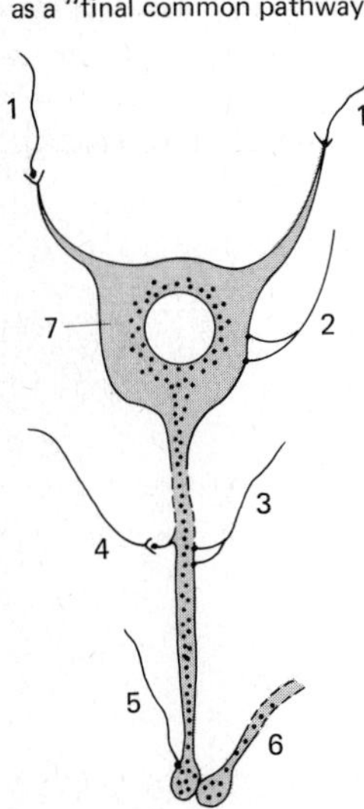

10-3 NEUROHUMORS

It now is apparent that all neurons (with very few known exceptions) are secretory cells. Those neurons that make neurohumors—the very great majority—are now our subjects for consideration. Their products are secreted in the junctions between neurons and thus chemically link chains of neurons together for transmission of information into, within, and out from the nervous system.

An understanding of the neurohumors—their chemical nature and modes of action—is essential to any understanding of how nervous systems function. It will become apparent that there is yet much to be learned about neurohumors, especially with regard to neurons of the central parts of nervous systems. Most of what is known about chemical transmission between neurons has come from the study of peripheral neural junctions. A considerable body of information has been gathered from studying the junctions motor neurons make with skeletal and cardiac muscle.

Acetylcholine was the first substance demonstrated to have a role in junctional transmission. Otto Loewi in 1921 showed that a frog heart would stop beating when its vagus nerve was stimulated, because the nerve secreted an inhibitory substance. Over the next few years Loewi and his colleagues characterized the inhibitor pharmacologically and, to some extent, chemically, and decided that it was acetylcholine. Confirmation of this conclusion came in the early 1930s through the work of Dale and Feldberg, who established that acetylocholine was liberated at a number of mammalian peripheral synapses. We now know acetylcholine is functional in the autonomic nervous system as well as at the terminals of motor neurons on skeletal and cardiac muscles. To this day, acetylcholine remains the best understood and most thoroughly characterized of the synaptic transmitters. Neurons secreting acetylcholine are called *cholinergic*.

The biosynthesis of acetylcholine involves acetyl-coenzyme A and choline, a quaternary base, in the following reaction:

$$\underset{\text{Acetyl-coenzyme A}}{CH_3\overset{\overset{O}{\|}}{C}-S\ CoA} + \underset{\text{Choline}}{(CH_3)_3\overset{+}{N}CH_2CH_2OH} \longrightarrow \underset{\text{Acetylcholine}}{CH_3\overset{\overset{O}{\|}}{C}-O-CH_2CH_2\overset{+}{N}(CH_3)_3} + \underset{\text{Coenzyme A}}{CoASH}$$

The enzyme catalyzing the reaction, *choline acetylase*, is widely distributed in the nervous system, being most active in cholinergic neurons and much less active in noncholinergic neurons.

A second enzyme, *acetylcholinesterase*, is responsible for inactivating acetylcholine by hydrolyzing it to acetate plus choline.

Acetylcholine in the mammalian heart produces inhibition by promoting an increase in potassium permeability. In skeletal neuromuscular junctions it increases small cation permeability, in general, with the net effect being depolarization above excitation threshold, so that action potentials are initiated. Thus, the transmitter can have opposite physiological effects. It is the nature of the receptor site which determines which of the two effects—excitation or inhibition—will ensue.

The chemistry of receptor sites for neurohumors is largely a great mystery. It appears that the sites are accessible to neurohumors on the outside of the postsynaptic cell. It is generally assumed that they are protein. De Robertis has purified a proteolipid from electric tissue of fishes which has a high affinity for binding acetylcholine. Receptor sites can be selectively interfered with by use of agents such as *curare* (e.g., in skeletal neuromuscular junctions) or *atropine* (e.g., in the heart) so that combination with transmitter is diminished or blocked. However, such blocks do not interfere with the action of acetylcholinesterase, as the enzyme will still rapidly hydrolyze available acetylcholine.

Drugs such as *physostigmine* (*eserine*) or *neostigmine* (Prostigmine) can be shown to reduce or prevent destruction of acetylcholine selectively but not to directly affect receptor sites. Thus, it appears that acetylcholinesterase and acetylcholine receptors occupy two distinct sites within the postsynaptic membrane. However, as would be expected, inhibiting either the esterase or the receptor site usually blocks transmission at cholinergic junctions. In the first instance the transmitter is removed from the receptor site only by diffusion, and prolonged permeability changes are a consequence. In the latter situation the receptor is not available and permeability changes cannot occur.

Norepinephrine, a catecholamine, is the second best characterized transmitter substance. Because norepinephrine is synthesized and stored in sympathetic neurons of vertebrates, is released by them in significant amounts during sympathetic-nerve stimulation, and has effects that simulate those of nerve stimulation, it has been generally accepted that norepinephrine is a neurochemical transmitter of sympathetic-nerve endings. Also, drugs that antagonize the action of exogenous norepinephrine also block the effect of sympathetic-nerve stimulation of the organ in question. Neurons secreting norepinephrine are called *adrenergic*.

Sympathetic nerves have all the enzymatic machinery necessary for the synthesis of norepinephrine from tyrosine. In fact, the enzyme that catalyzes the first steps of the reaction, *tyrosine hydroxylase*, may be confined to neurons and specialized tissue, such as adrenal medulla, that contain norepinephrine. This speculation is supported by the fact that the enzyme disappears from tissues deprived of adrenergic innervation. The other two enzymes of the pathway, however, are found in a number of tissues. The synthesis of norepinephrine and related amines is depicted in Fig. 10-7.

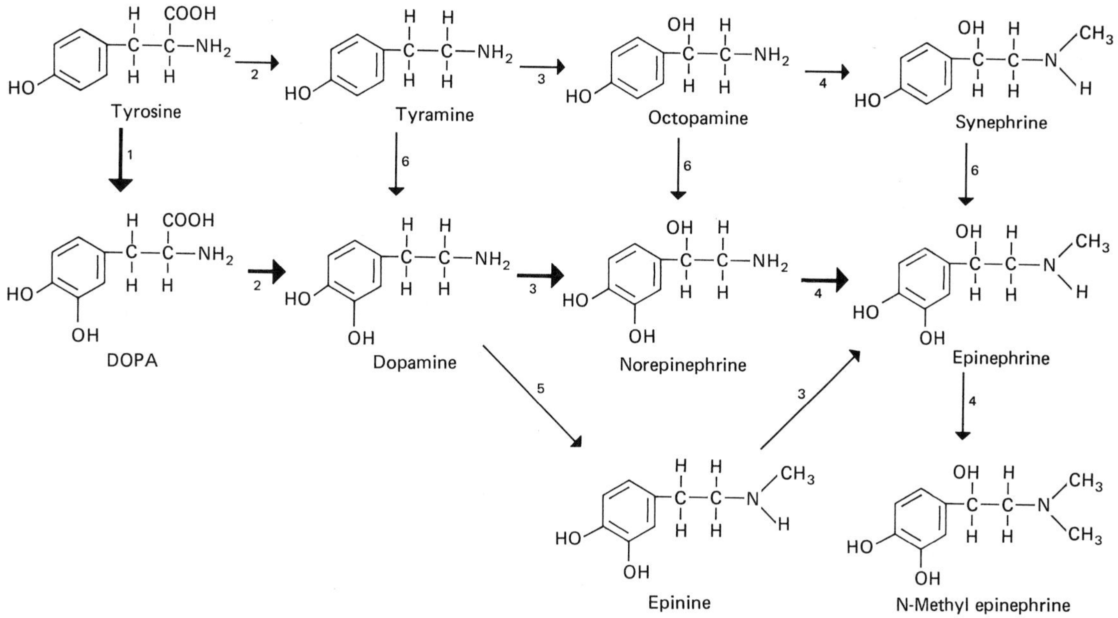

FIGURE 10-7
Primary and alternative pathways in the formation of catecholamines: (1) tyrosine hydroxylase, (2) aromatic amino acid decarboxylase, (3) dopamine-β-oxidase, (4) phenylethanolamine-*N*-methyl transferase, (5) nonspecific *N*-methyl transferase in lung, and (6) catechol-forming enzyme.

The evidence that norepinephrine acts as a neurotransmitter in the CNS is not so compelling as that for the sympathetic system. In the CNS, neurons are not so amenable to cellular investigations because they exist in a totally heterogeneous system of chemical outputs in close proximity to one another, making it difficult to discern events taking place in any one particular neuron ending. However, brain slices, and isolated spinal cord when electrically stimulated, release some norepinephrine into the bathing medium.

The degradation of norepinephrine is considerably slower than that of acetylcholine and occurs in quite another way. The major mammalian enzymes of importance in inactivating the amine are *monoamine oxidase* (MAO) and *catechol-O-methyltransferase* (COMT). MAO converts catecholamines to their corresponding aldehydes, which in turn are usually oxidized by aldehyde dehydrogenase to acids. COMT catalyzes the transfer of a methyl group from S-adenosyl-methionine to the catechol. However, neither MAO nor COMT seems to be the primary mechanism for terminating the action of norepinephrine liberated at sympathetic terminals. It is believed that inactivation is mainly effected by uptake (or re-uptake) by the neurons that release norepinephrine. (This will be discussed in Sec. 10-5.)

It is very interesting to recall that norepinephrine is a hormone of the adrenal medulla, which is a component of the sympathetic nervous system. In fact, norepinephrine is also called *noradrenalin*. Thus, norepinephrine is a *neurohormone* as well as a *neurohumor*.

It can be stated with considerable certainty that acetylcholine and norepinephrine are neurohumors in the peripheral nervous system, and that these agents appear also to serve similar roles in the CNS. With much less certainty it can be said that there also are other substances that seem to be functioning in the CNS as neurohumors, but we presently know little about them, primarily because they do not seem to be operative in peripheral nerves and, hence, cannot be subjected to the kind of analyses that have been used with acetylcholine and norepinephrine in the periphery.

Dopamine, a catecholamine (see Fig. 10-7), is found in mammalian CNS, and its distribution differs markedly from that of norepinephrine, suggesting that it is not exclusively serving as a precursor in norepinephrine synthesis. **Epinephrine (adrenalin)** concentration in the mammalian CNS is relatively low (5 to 17 percent of the norepinephrine content), but certain areas such as olfactory structures contain substantial amounts of the enzyme that forms epinephrine (see Fig. 10-7, enzyme 4), suggesting that it may play a role as a neurohumoral agent. There is good evidence that **5-hydroxytryptamine (serotonin),** an indoleamine which occurs in small quantities in nervous tissues, is an important synaptic transmitter in the brain. However, the central pharmacology of 5-hydroxytryptamine is poorly elucidated and the locations of the neurons that may employ it are not yet defined. **γ-Aminobutyric acid** (GABA) has been detected in brain and spinal cord of vertebrates and in nerves of invertebrates; much evidence has accumulated supporting the hypothesis that it is an inhibitory transmitter at the crustacean myoneural junction. There is also some evidence that the amino acid **glycine** may play a role as an inhibitory transmitter in the mammalian spinal cord, and both **glutamic acid** and **aspartic acid** are excitant amino acids found in significant quantities in mammalian CNS. Glutamic acid is the excitatory transmitter of crustacean and insect neuromuscular junctions, rather than acetylcholine. Another amino acid, **histamine,** is widely distributed in nervous tissue and is conjectured to function in the nervous system in a "neuromodulatory role," potentiating the response of neurons to acetylcholine, for example. However, the histamine of neural tissues may not be in the neurons but instead in nonneuronal cells, and, hence, would not qualify as a transmitter.

10-4 THE SYNAPSE

Transfer of information from neuron to neuron or from neuron to effector cell (muscle, gland, or electroplaque cell) requires both *intra-* and *inter*cellular transmission of messages. Intracellular transfer occurs between the conductile, electrogenic portions of the neuron (the subject of the next chapter) and the output membrane component. Intercellular transfer occurs between the output of one neuron and the input of the next in a chain.

It would be logical to expect that the electrical excitation in a cell could induce excitation in another cell, provided the two are in intimate contact. Indeed, cell-to-cell transmission of excitation does occur. Among smooth-muscle cells and in certain other tissues, direct cell-to-cell spread of electrical excitation occurs, but this does not seem to be a very common mode of message transmission. A very small number of direct neuron-to-neuron transmissional junctions has been found to occur in certain nervous systems—crustacean, annelid, and mollusk, specifically. The region of contact is characterized by a relatively low electrical resistance, and leakage of currents from one component cell may affect the other. Such interfaces between neurons have been called **electrotonic** or **ephaptic junctions.**

In most cases, however, where neuron-to-neuron junctions have been carefully examined, electrical spread of excitation does not appear to be the essential mode of transmission. Instead, transmission is effected by the release of specific chemical substances, i.e., neurohumors, that trigger changes in the membrane permeability of the next neuron in the chain. Morphological and physiological evidence points to the conclusion that the vast majority of synapses (certainly in vertebrates) depend upon chemical transmission. (The term **"synapse,"** as first used by Sir Charles Sherrington, can be applied to all junctions between neurons—whether they rely upon electrical or chemical transmission.)

Neurons are interconnected in very characteristic ways. Typically, the axon of one nerve cell forms a synapse with either the cell body (soma) or dendrites of other neurons. Thus, there are two principal types of synapses: axodendritic (axon with dendrite) and axosomatic (axon with cell body) (Fig. 10-8).

As a rule, the dendrites and somata of neurons are the receptive (postsynaptic) components and the axons are the output (presynaptic) components, but this is not an invariant pattern in neural transmission. Though the cell body is necessary for the maintenance of the neuron, it may be bypassed functionally. In many invertebrates, for example, axon-to-axon junctions are formed. Through studies of the fine structure of vertebrate nervous systems, it has become increasingly apparent that while converging axodendritic and axosomatic synapses are very common, parallel interneuronal junctions also occur, such as somatosomatic, dendrodendritic, and axoaxonic junctions. Even somatodendritic and somatoaxonic junctions with the characteristics of chemical synapses have been described. Thus, soma or dendrites may be presynaptic as well as postsynaptic, and axons may be postsynaptic as well as presynaptic (see again Fig. 10-6).

Morphological features of synapses have been well characterized,

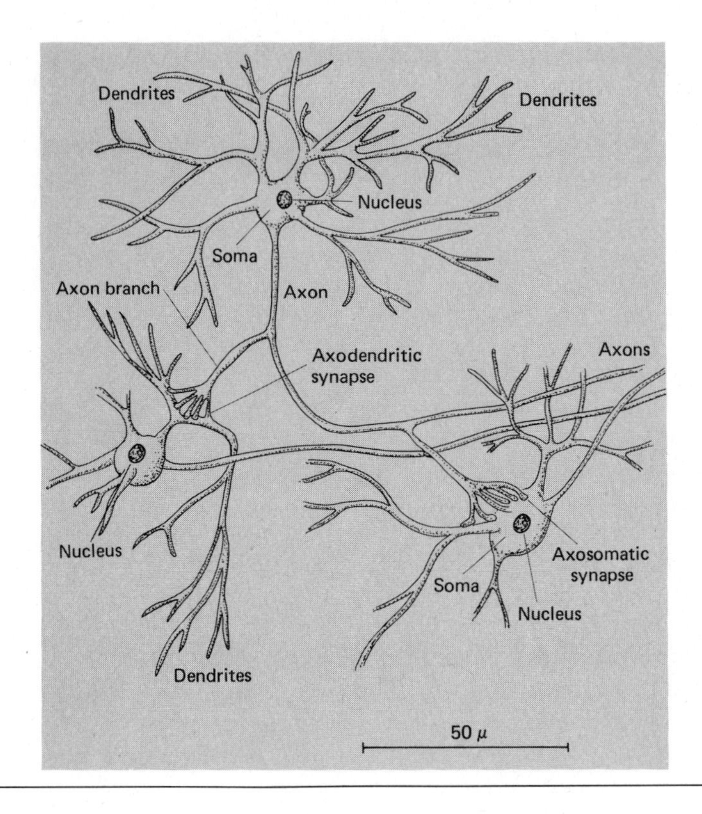

FIGURE 10-8
Schematic representation of three interconnected neurons. Note the branching patterns of dendrites and axons. Axodendritic and axosomatic synapses are illustrated. (*After C. F. Stevens, 1966, Neurophysiology, John Wiley & Sons, Inc., New York, p. 2.*)

especially since electron microscopy was employed in their examination, beginning in 1953. The preterminal portion of an axon expands into a bouton or bulbous ending containing mitochondria, that commonly are arrayed in a palisade along the junctional interface, as well as numerous small vesicles. The interface is, in most instances, a simple appo-

FIGURE 10-9
Diagram of a synapse and terminology of its functional components.

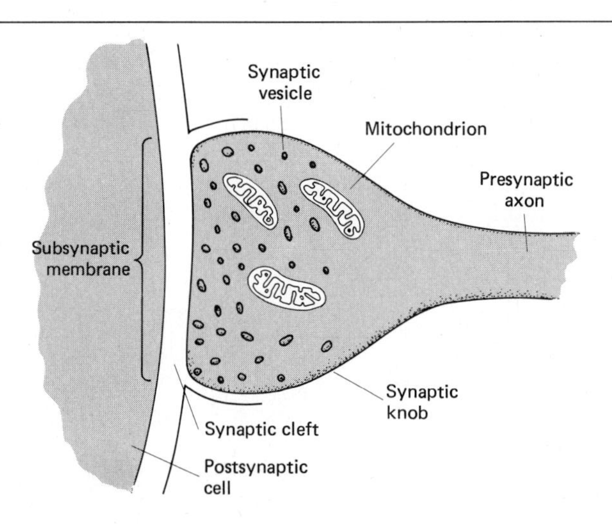

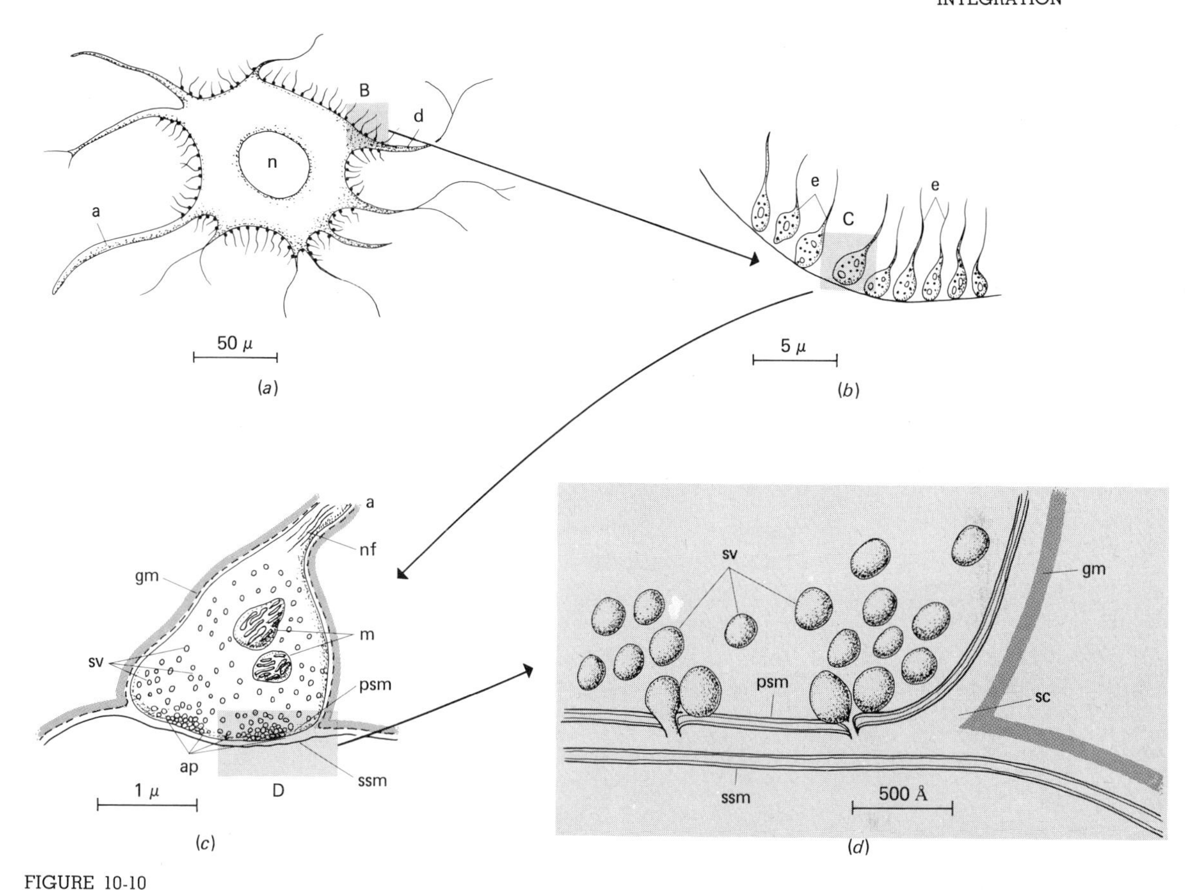

FIGURE 10-10
Diagram showing synapses between bouton nerve endings and a mammalian spinal motor neuron at different magnifications with the light microscope and the electron microscope. (*a*) A motor neuron as seen at medium power with the light microscope: *a*, axon; *d*, dendrites; *n*, nucleus. Note the numerous nerve endings on the cell body which, having a surface area of about 10,000 μm^2, is covered by no less than 2000 such boutons. The enclosure B is magnified about 10 times more in (*b*); e, nerve endings. Enclosure C is magnified 6 times further in (*c*); *a*, axon; *ap*, active areas of synapse; *gm*, glial membrane *m*, mitochondria; *nf*, neurofilaments; *psm*, presynaptic membrane: *ssm*, subsynaptic membrane; *sv*, synaptic vesicles. Enclosure D is magnified 20 times more in (*d*), which is an imaginative interpretation; *sc*, synaptic cleft. (*After E. De Robertis*, 1939, *Submicroscopic morphology of the synapse, Int. Rev. Cytol., 8, p.* 65.)

sition of two cell membranes with an interstitial space about 10 to 20 nm wide, the **synaptic cleft** (Figs. 10-9, 10-10).

The terminals of neurons contain numerous vesicles that vary in shape and size and which often are clustered against the presynaptic membrane. A common type is 20 to 40 nm in diameter, approximately spherical, and clear in the center. Based on the distribution of acetylcholine, this kind of vesicle appears to be associated with cholinergic endings, and thus is thought to be the site of storage of acetylcholine. Slightly larger vesicles, containing a dense particle, are believed to contain norepinephrine. Still larger vesicles, 80 to 90 nm in diameter, and containing a dense spheroidal droplet, occur in a variety of places and sometimes are mixed with small, clear vesicles. What they contain is unknown.

10-5
SYNAPTIC POTENTIALS

In 1952, Fatt and Katz, working at University College in London, England, discovered that acetylcholine appears to be released in multimolecular packets, or quanta, of rather uniform size. This was evidenced by the appearance of minute, transient voltage fluctuations (about 0.5 mV) that were detected in myoneural junctions; these were called *miniature end-plate potentials*. They seem to result from the spontaneous "leakage" of packets (or quantal units) of several thousand transmitter molecules into the synaptic cleft. The arrival of a nerve impulse (action potential) at a neuron terminal increases the probability of a quantum of transmitter being released. A synchronous discharge of a certain number of quantal units (estimated to be about 300) gives rise to the much larger *synaptic potentials*. (Synaptic and other kinds of potentials are described in the next chapter.)

At the time of the discovery of the miniature potentials there was no morphological equivalent of a transmitter quantum, but through studies mainly by De Robertis with the electron microscope, the synaptic vesicles described above were visualized. By electrically stimulating a peripheral synapse, he was able to demonstrate a significant change in the number of vesicles which is related to the frequency of the stimulus (Fig. 10-11).

A widely held, current view, largely formulated by Katz, of what happens during synaptic transmission at cholinergic junctions, is that a nerve impulse (action potential) arriving at the axon terminus results in a depolarization of the membrane, which in turn raises the probability of quantal units of acetylcholine being released from their vesicles. Calcium ions in the external medium are required, for depolarization is

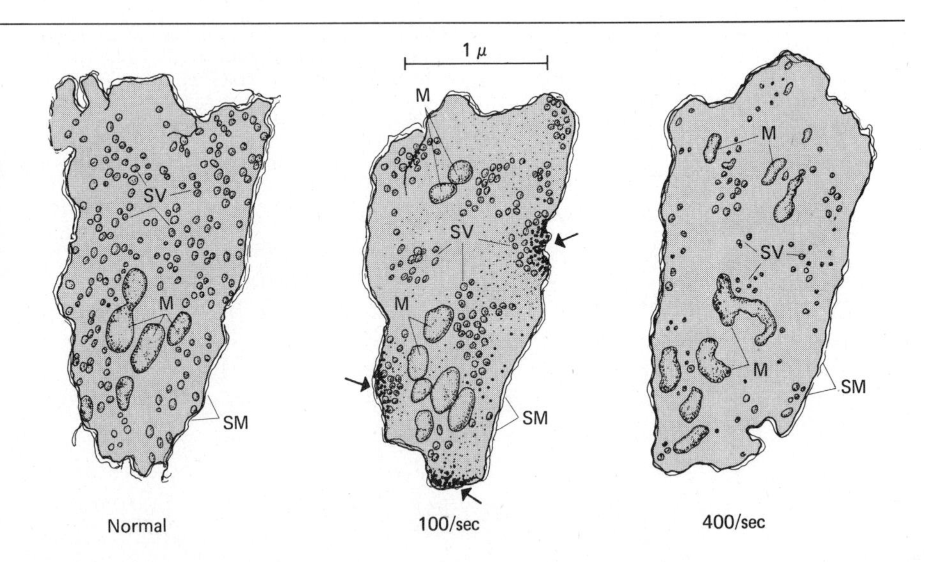

FIGURE 10-11
Diagram showing neuron endings of the adrenal medulla of the normal rabbit, and after stimulation for 10 minutes with supramaximal pulses of 100 and 400 per second. Note the depletion of vesicles (*SV*) and altered mitochondria (*M*). Marked with arrows are denser regions of the synaptic membrane (*SM*) where there is an apparent discharge of material from the vesicles. (*After E. De Robertis,* 1958, *Exp. Cell Res. Suppl.* 5, 360.)

thought to open specific "calcium channels" in the terminal axon membrane; this leads to an influx of calcium ions, which in turn initiates the "quantal release reaction." The latter is envisaged as transient collisions of synaptic vesicles with the axon membrane, with calcium serving to bring about attachment and local fusion between the vesicular and axonal membranes, and then there is an all-or-none discharge of the vesicular contents into the synaptic cleft (Fig. 10-10*d*).

A minority of investigators do not agree with this hypothesis. Nachmansohn, for example, denies that acetylcholine is a chemical mediator between cells. He views the action of this substance as intracellular, serving to trigger permeability changes in the presynaptic neuron and permitting ion movements that, in turn, excite the postsynaptic cell. The majority opinion, however, is that acetylcholine traverses the synaptic cleft, attaches to receptor sites on the postsynaptic membrane, and there induces translocation of ions that can lead to generation of action potentials.

Synaptic transmission at adrenergic junctions is quite a different story. The vesicles, or granules, contain high concentrations of catecholamines and ATP, as well as a specific protein. This intravesicular catecholamine-nucleotide-protein complex constitutes the *reserve pool*, the major storage depot of epinephrine in the adrenal medulla and of norepinephrine in the terminals of adrenergic fibers. It exists in equilibrium with two considerably smaller pools—the *mobile pools* within the vesicles and the cytoplasm. The first stages of synthesis of norepinephrine take place in the cytoplasm: the hydroxylation of tyrosine to dopa and the decarboxylation of dopa to dopamine. The latter then enters the vesicles, where it is converted to norepinephrine and added to the reserve pool.

In addition to synthesis, there is a second major source of norepinephrine in adrenergic terminals, viz., *re-uptake* by active transport of norepinephrine previously released into the extracellular space. In fact, re-uptake is probably the major process responsible for termination of the effects of norepinephrine release, in contrast to the rapid enzymatic inactivation of acetylcholine (Fig. 10-12).

It seems quite likely that two active transport systems are involved in re-uptake: one across the axon surface from the extracellular space to the cytoplasmic mobile pool; the other from the cytoplasmic to the intravesicular mobile pool. The latter system has been rather fully characterized, since it is relatively easy to isolate vesicles, especially from the adrenal medulla. The residues can concentrate catecholamines against a 200-fold gradient across their membranes into the intravesicular mobile pool, which then is in equilibrium with the catecholamine-ATP-protein complex that constitutes the reserve pool.

Adrenergic neurons can sustain the output of norepinephrine during prolonged periods of stimulation without exhausting their reserve supply, provided the mechanisms of synthesis and reuptake of the trans-

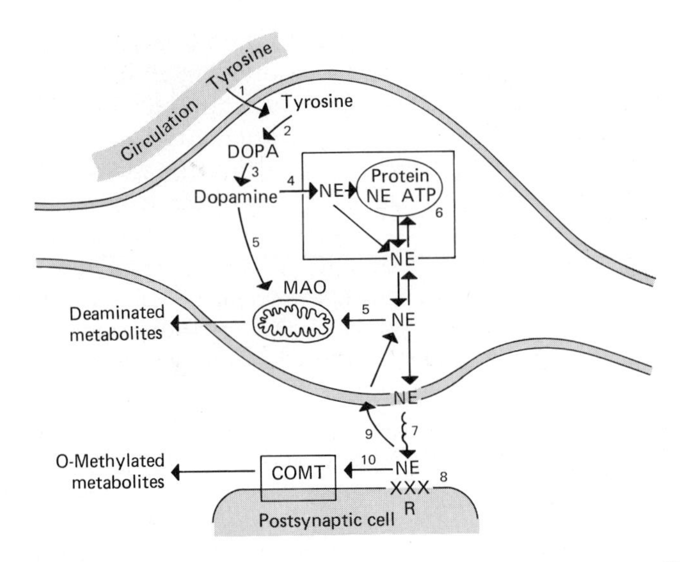

FIGURE 10-12
Schematic model of noradrenergic neuron. The probable sequence of events in neurotransmission by such neurons is: 1, active uptake of tyrosine precursor; 2, conversion of tyrosine to dopa by tyrosine hydroxylase; 3 and 4, decarboxylation and oxidation to form norepinephrine (NE) (cf. Fig. 10-7); 5, sites of metabolic degradations by monoamine oxidase; 6, storage of NE in particulate form; 7, release of "free" NE (Ca^{2+}-dependent) into cleft, effected by action potential; 8, interaction with receptor sites on postsynaptic neuron; 9, inactivation of NE by re-uptake; 10, metabolism by catechol-O-methyl transferase of NE overflow into circulation. (*After J. R. Cooper et al.*, 1974, *The Biochemical Basis of Neuropharmacology*, Oxford University Press, New York, p. 140.)

mitter are unimpaired. The sequence of events in which a nerve impulse effects the release of norepinephrine from adrenergic fibers is not fully known. Entrance of calcium ions followed by extrusion of vesicular contents, much like what occurs in cholinergic endings, is thought to be a key step. There is some evidence to support the opinion, embodied in the so-called *Burn-Rand hypothesis*, that acetylcholine release is the first step in the liberation of norepinephrine. In this view acetylcholine is involved in the mobilization of calcium ions.

There is a pause between the arrival of an action potential at an axon terminal and the development of a *postsynaptic potential* (PSP) in the postsynaptic neuron. Recent studies have shown that this *synaptic delay*, as it is called, is primarily in the release mechanism, not in diffusion time or postsynaptic chemistry.

Synapses may be classified, according to their mode of behavior, as either *excitatory* or *inhibitory*. If a volley of impulses results in the depolarization of a postsynaptic neuron and gives rise to action potentials it is excitatory; the depolarization is called the *excitatory postsynaptic potential*, or EPSP. Excitatory synapses cause an increase in the conductance of the membrane of the postsynaptic cell to specific ions as described in the following chapter. This is a consequence of the association of transmitter substance with receptor sites on the subsynaptic membrane. "Channels" are briefly more "open," allowing increased ion conductances. This condition prevails in many synapses for about a millisecond only and then the transmitter is removed (by diffusion, re-uptake, or inactivation by enzyme). The entire process can be repeated up to several hundred times a second for brief intervals.

The second type of synapse that occurs in the nervous system, the *inhibitory*, can reduce or stop the firing of a nerve cell even though it is receiving synaptic activation from excitatory neurons. Morphologically, inhibitory and excitatory synapses cannot be distinguished; but obviously they are functionally distinct. Apparently, ionic conductance is different in excitatory and inhibitory synapses.

Inhibitory synapses cause an *inhibitory postsynaptic potential* (IPSP), which is almost a mirror image of the EPSP (as seen on a cathode-ray tube, for example). An IPSP makes the cell more negative internally than it is at its usual resting potential. Thus, a more negative potential is more difficult to drive toward the threshold level for spike generation. (A spike is a type of action potential that commonly traverses the axon—the ordinary "nerve impulse.") In simple outline, then, a neuron responds to the algebraic sum of the internal potential changes produced by excitatory and inhibitory synapses.

To summarize transmission at a synapse: An action potential invades the terminal portion of an axon and liberates a minute amount of neurohumor. This agent diffuses to the postsynaptic membrane, where it combines with specific receptor sites and initiates changes in ionic conductance. Specific ions are involved with different transmitter substances and receptor sites at a particular junction. The transmitter is then removed from the synaptic region by a specific mechanism, and recovery to the resting state ensues.

READINGS

Cooper, J. R., F. E. Bloom, and R. H. Roth, 1974, *The Biochemical Basis of Neuropharmacology*, Oxford University Press, New York. In this excellent monograph can be found many details about neurohumors that are not covered in this chapter. It will be particularly interesting to those wishing to know how information about the chemistry of nervous systems is obtained.

Martini, L., and W. F. Ganong (eds.), 1966–1967, *Neuroendocrinology*, Academic Press, Inc., New York, 2 vols. Volume 1 will be found particularly useful as a source of basic information. Chapter 5 of this volume is a very well done treatment of neurosecretion.

McLennan, H., 1970, *Synaptic Transmission*, W. B. Saunders Company, Philadelphia. This is a detailed treatment of the morphology, electrophysiology, and pharmacology of synapses. It includes a chapter on electrical synapses (ephapses) and one on invertebrate synapses.

NEURAL INTEGRATION 11

A **nervous system** is an organized constellation of electrogenic cells[1] specialized for receiving and integrating converging signals, generating new signals, and commanding adaptive behavior. Through communication channels, information is utilized to achieve order and integration of the various parts of an organism.

Neurons receive information in the form of impulses from many other neurons—often hundreds—and may, in turn, transmit information to a like number. Incoming impulses may cause the neuron to fire an impulse or inhibit it from firing, depending upon the nature of the impinging neurons and the recipient neuron's threshold of response, as determined by various factors.

Nervous systems can be extremely complex and require much input of energy to support their activities. The human central nervous system (spinal cord and brain) is estimated to consist of about 10 billion neurons. In a human at rest, the brain, although comprising only about 2 percent of the body weight, consumes 20 percent of the total oxygen intake.

In neuronal systems the information transferred between cells is coded in specific ways and is usually transmitted in one direction only. Two operational terms that will be useful in what follows are *afferent*, which refers to signals going into a neural center, and *efferent*, which refers to signals emanating *from* a neural center.

11-1 THE ORGANIZATION OF NERVOUS SYSTEMS

The *neuron theory*, as originally formulated by Ramón y Cajal in the late 1800s, and subsequently substantiated and amplified by modern studies, asserts that the nerve-cell body and its processes, together called the **neuron,** form the cellular units of the nervous system which are directly involved in nervous function; that all nerve fibers are neuronal processes; that the neuron and all its extensions develop embryologically from a single neuroblast; and that the neuron is a trophic unit, all its processes being dependent upon the nucleated cell body for their maintenance and regeneration.

[1] Electrogenic cells are capable of changing their membrane potential in characteristic ways in response to appropriate stimuli. The basis for this property of nerve cells and other types of cells will be described in Chap. 13.

The neuron doctrine was opposed by a formidable group of investigators, including Golgi, Meynert, Gerlach, and Nissl, who basically adhered to the *reticular theory*. This group maintained that anatomical continuity of fibers and branches was the prevalent condition in the nervous system. We now know that this view is invalid and that Ramón y Cajal was correct in viewing the nervous system as composed of discontinuous functional units. Furthermore, he postulated from the anatomical arrangement of sensory, motor, and internuncial neurons that they are all dynamically polarized; i.e., excitation can be transmitted only from the axon of one neuron to dendrites or soma of the next and, within a neuron, must normally spread from dendritic to axonal poles.

In the classic period of the 1920s through the 1930s, complementary studies of neuroanatomy and neurophysiology led to the concept of the nervous system as a kind of digital computer, with its functional units arranged end-to-end in straight and branched chains, and each unit having an *all-or-none* (binary) response mode. Evidence collected in recent years has forced a number of modifications of this simplistic view.

Contrary to the earlier concept of the all-or-none impulse being propagated along the entire length of the neuron as synonymous with the neuron in action, we now must accommodate such facts as the following: (1) Many parts of neurons cannot respond in an all-or-none fashion and therefore cannot propagate an impulse without a decrement. (2) Input to a neuron can occur at several different kinds of loci in various parts of the neuron, in addition to the dendritic synapses. (3) Responses to afferent excitation do not automatically give rise to its firing, but instead often only help determine the rate of firing of impulses which are initiated at a more distal site—commonly near the base of the axon. (4) Neurons sometimes have two axons and therefore can deliver two nonidentical, pulse-coded outputs at the same time in different directions (Fig. 11-1).

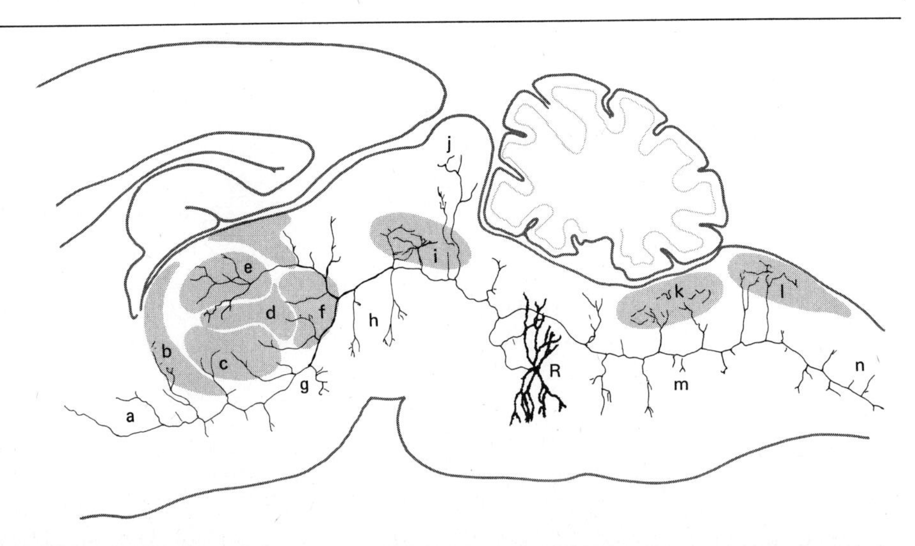

FIGURE 11-1
Sagittal section through the brain of a young rat showing the axonal processes of a single neuron of the reticular formation (*R*). The rostrally-coursing axon supplies collaterals (*a* to *h*) to a number of forebrain, diencephalic, and mesencephalic regions. The posteriorly-directed axon sends collaterals (*k* to *n*) into the brainstem and gray matter of the spinal cord. (*After M. A. Brazier,* 1968, *The Electrical Activity of the Nervous System,* 3d ed., *The Williams & Wilkins Company, Baltimore,* p. 241.)

These insights suggest that the complex behavior of organisms resides in part in the diversity of neuron types and in part in the degrees of freedom of the single neuron. The permutations of the several integrative processes now known to be inherent in neurons permit great complexity.

11-2 MEASUREMENT AND ANALYSIS OF POTENTIALS

The history of electrophysiology has been decided by the development of recording and stimulating instruments. Through the eighteenth century, discoveries in the study of electricity were closely interwoven with advances in the study of muscles and nerves. In fact, for many years the twitch that results when a brief current is discharged through an isolated nerve-muscle preparation of a frog was the most sensitive detector then available for short-lasting electric pulses. By the middle of the nineteenth century it had been demonstrated, by DuBois-Reymond principally, that nerves and muscles are capable of generating electromotive forces themselves. The recording device employed was the slow but sensitive galvanometer.

One of the significant demonstrations of this era was that if one crushes part of a nerve or muscle, the injured region becomes electronegative with respect to undamaged regions. With a galvanometer it can be shown that current—the so-called "injury current"—will flow from the uninjured to the injured region. DuBois-Reymond found that when a muscle is stimulated by a series of shocks applied through its nerve, there is a transient reduction of the "injury current." However, despite ingenious experimental designs, the instruments then available were too slow to reveal accurately the time courses of most of the flickering electrical changes in electrogenic cells.

Just before World War I the vacuum-tube amplifier became available to the electrophysiologist to magnify the small potential changes of electrogenic membranes and to drive recording instruments, such as pen-writing devices. But still there was the problem of the slowness of the recorder. The real breakthrough came with the invention of the cathode-ray tube. Coupled with an amplifier, it provides the most sensitive, inertia-free system available for detecting minute and rapid electrical changes. This combination, called an *oscilloscope*, was introduced into electrophysiology by Gasser and Erlanger in 1922 (Fig. 11-2).

A nerve is a "cable" containing a few or many individual "wires"—neuron processes or fibers (axons and dendrites). As the magnitude (strength) of the stimulus is gradually increased, the size (amplitude) of the electrical response in a nerve increases until all the contained neuron processes are active. The threshold stimulus strength for the response is variable among the different fibers. Such activity recorded from a bundle of neuron fibers excited in concert represents *summed* potentials and is called the **compound action potential** (Fig. 11-3).

Study of the electrical behavior of single neurons was accomplished in the 1930s. Two laboratories, those of Adrian and Matthews in England

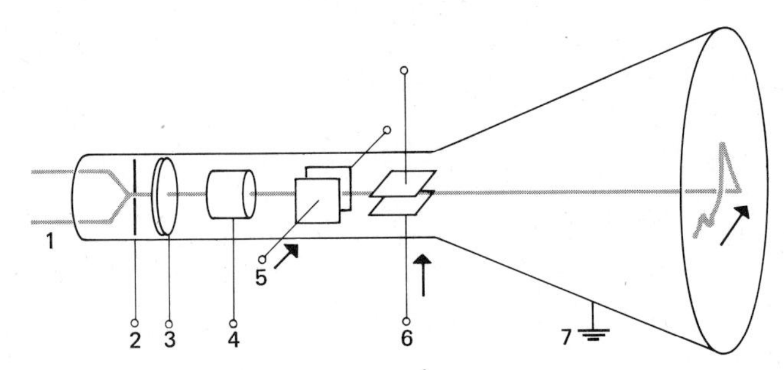

FIGURE 11-2
Cathode-ray oscilloscope. The cathode (1) emits electrons. The control grid, or "brightness control," (2) governs the strength of the electron current. The main anode, or "gun," (3) accelerates electrons. The focusing anode (4) forms a sharp electron beam. A pair of horizontal deflection plates (5) produces the "time-base" deflection. A pair of vertical deflection plates (6) receives the nerve signal from the amplifier. A metallized screen (7) discharges the electron beam after it has produced a bright spot by collision with the fluorescent screen on the front of the tube.

and Bronk in the United States, succeeded in teasing away all neuron fibers until only a single one was left intact. Upon recording from it, they observed, in miniature, the same general characteristics of the whole nerve of which it was a part, but it showed a specific threshold for excitation, was all-or-none in its response, and had a characteristic conduction velocity.

Up to this time potentials were recorded from externally applied electrodes; i.e., both electrodes were in contact with the external surface of the membrane. Two significant events made it possible to do transmembrane recording. One was that Gerard and Ling of the University of Chicago developed a glass micropipette with a tip so fine that it could penetrate into a neuron. When filled with 3 *M* KCl and placed on a silver–silver chloride wire, it could serve as an electrode for penetrating through the cell membrane. It was not necessary to tediously dissect nerves; simply driving the electrode into a nerve made it possible to impale an individual neuron in situ. Thus, by using this as an internal electrode, with a conventional electrode on the outside in contact with the bathing medium, recording of the potential *across* the membrane became possible. (Such microelectrodes cannot be used with all nerves; e.g., little success has been realized with myelinated nerves.)

FIGURE 11-3
Diphasic recording of the action potential of a nerve. (1) A pinch of the forceps stimulates the contained neuron fibers, and a compound action potential moves along the nerve. (2) The A electrode is temporarily negative to the B electrode. (3) Electrodes are equipotential. (4) The B electrode is now negative to A. (5) Electrodes are again equipotential as the impulse passes on. These events, as they might be seen on a rapidly responding "galvanometer," are shown by the position of the indicator on the dial. If the movements of the indicator were plotted as a continuous function, graphs like those to the right of the dial would result. A trace similar to this would be produced by an oscilloscope recording a nerve impulse.

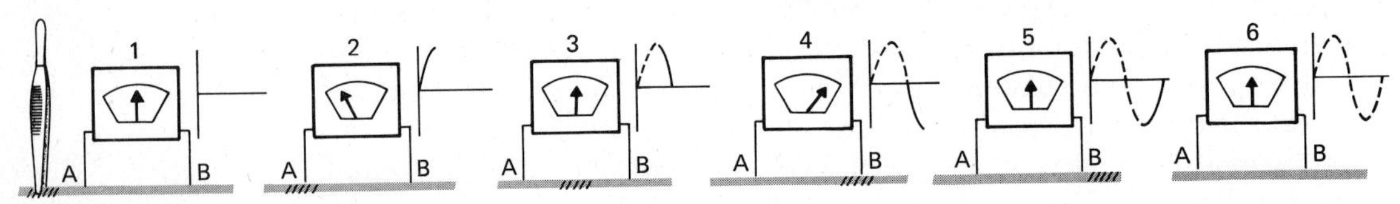

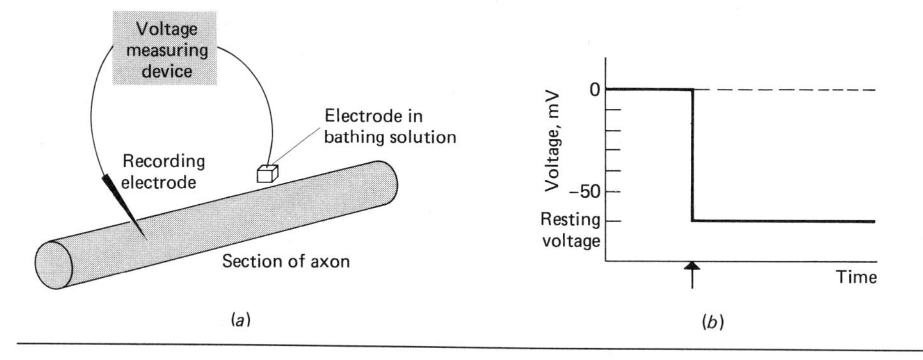

FIGURE 11-4
(*a*) Experimental arrangement for recording axon membrane potentials using a penetrating microelectrode. The reference electrode is in the salt solution surrounding the axon. (*b*) Potential change (voltage) detected as a recording microelectrode enters the axon (↑). The inside is negative relative to the outside of the axon's membrane.

When penetration is effected with such an arrangement, and the potential difference is displayed on an oscilloscope, there will be seen an abrupt change in position of the beam, indicating a voltage difference between the electrodes. This is interpreted to be the membrane **resting potential** (Fig. 11-4).

The second important development that permitted transmembrane recording began when J. Z. Young in 1936 pointed out that the neurons of certain invertebrates, especially cephalopods (squid and cuttlefish), have giant neuron fibers with diameters of 0.5 mm or more. Within the next few years these so-called *giant axons* were established as favorite preparations of neurophysiologists. Hodgkin and Huxley in England and Curtis and Cole in the United States were among the first to exploit these fibers. A standard preparation involves inserting an electrode into the cut end of an axon and recording from it with reference to an external electrode. Most of the basic information underlying modern neuronal physiology has come from experiments with the giant axons of squid. (These will be described in Sec. 11-4.)

11-3 THE REPERTOIRE OF NEURONS

The all-or-none impulse, or **spike,** was the only known form of activity of nerve cells up to 1938. In that year Hodgkin, using giant axons of squid, and very soon thereafter Katz, using the sciatic nerves of frogs, discovered the **local potential,** a graded, essentially passively spreading potential which declines to half-amplitude about every millimeter (i.e., it is decremental). Subsequent work has revealed several more kinds of potentials, which are classified according to the scheme of Bullock in Fig. 11-5.

Resting potential is the potential difference *across* the surface of an inactive cell (Fig. 11-4). Its magnitude can be altered by several conditions; this is probably a normal means of varying responsiveness.

Action potentials are characteristic sequences of potential changes that move *along* membranes of electrogenic cells, such as neurons. **Internal response potentials** are responses to physiologically activated,

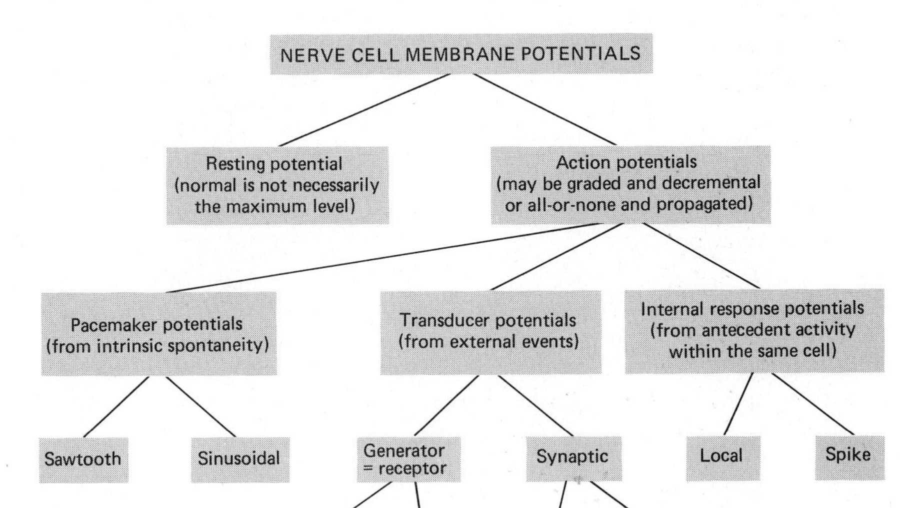

FIGURE 11-5
A classification of the types of membrane potentials that can be recorded from nerve cells collectively. (*After T. H. Bullock,* 1959, *Science, 129,* p. 997–1001.)

antecedent activity *within the same cell.* The **spike potential,** which is characteristic of axons, is regenerative, all-or-none, and propagated. The **local potential** (also called **electrotonic potential**), mentioned above, is graded and decremental. If of sufficient magnitude, it can be a prepotential for spike initiation.

Generator potentials (and **receptor potentials** of receptor cells) and **synaptic potentials** (or **postsynaptic potentials**) are active responses of cells to impinging *external* events of specific kinds, and generally lead to a spike. Thus, collectively they can be called **transducer potentials**—in a rather special way; i.e., transduction here refers to "information" transduction—energy forms in the case of generator potentials and chemical transmitter in the case of synaptic potentials. Incidentally, both generator and synaptic potentials may be *polarizing* (or *hyperpolarizing*) and *depolarizing,* as suggested in Fig. 11-5, the direction of the response depending on the conditions prevailing at the time.

The potentials induced in the postsynaptic neuron by the action of a transmitter chemical, as noted in the preceding chapter, may be of two kinds: excitatory and inhibitory. An **excitatory postsynaptic potential** (EPSP) is a depolarization, usually of a few millivolts, presumably caused by a change in permeability of the membrane of the postsynaptic cell under the brief influence of the chemical transmitter released by the presynaptic neuron. In most synapses, as noted in the previous chapter, the chemical is not known, and in some synapses (electrical ones) the postsynaptic membrane responds to extracellular current caused by the impulse in the presynaptic membrane. Thus, potentials that lead to excitation in postsynaptic neurons, although similar in appearance, may be of different origins.

An **inhibitory postsynaptic potential** (IPSP) may be a depolarization or a hyperpolarization, or it may evoke no change. In contrast to an EPSP, where depolarization is accompanied by an increase in membrane conductance to Na^+ and to either or both K^+ and Cl^-, in inhibition the effect at the synapse is to increase the conductance primarily of Cl^- and/or K^+ but not Na^+. The new *equilibrium potential* (the electrical potential across a membrane that would balance the tendency of ions to diffuse in the direction of their concentration gradient) during the inhibition is then nearer to the resting potential, since the equilibrium potential of either of these ions is close to the usual resting potential. Second, and more important, the increased permeability clamps the membrane at an inhibitory equilibrium potential that is near its resting potential, and, therefore, a much larger flow of Na^+ would be required to depolarize it by a given amount. (The theory of the ionic basis for membrane potentials will be discussed more fully in Chap. 13, and then these relations will become more meaningful.)

Pacemaker potentials are *spontaneous* prepotentials resulting not from an impinging environmental change but rather occurring under normal steady-state conditions. These can be manifested in more than one form: more or less *sinusoidal* (time course is relatively independent of the occurrence of the spike) as occurs in the vertebrate (myogenic) heart, or *sawtooth* (time course dependent on intervention of a spike or local potential to reset the starting condition) as occurs in the neurogenic hearts of crustaceans.

It is probable that some, maybe all, of the different kinds of electrical activity arise from specific varieties of cell membranes; i.e., the surface of an axon probably can be thought of as a *mosaic of membrane types*.

A very important result of membrane specialization is that spikes are (as far as we know) a constant feature only of axons. It is common to find a restricted locus—usually near the origin of the axon—where any of several kinds of arriving prepotentials can cause spike initiation. From this site the spike potential will be propagated without decrement to the terminus of the axon, but throughout the dendritic and somal membranes only graded, decremental potentials are found. Several forms of excitation are integrated in the dendritic zone and soma prior to initiation of the spike (Fig. 11-6).

Evidence is accumulating that in dendritic networks small graded changes in potential in one neuron can synaptically influence electrical activity in other neurons. Some neurons characteristically interact without the benefit of spikes. Thus, a neuron has the capability of processing diverse inputs and determining what information is transferred via its axon to the next neuron.

Other kinds of potentials, known as **afterpotentials**, are sometimes seen following spike potentials; they influence excitation states of neurons for varying durations of time. These will be more fully described in Chap. 13.

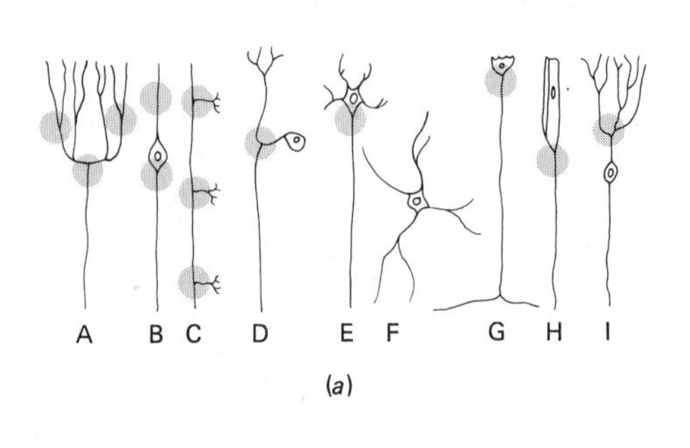

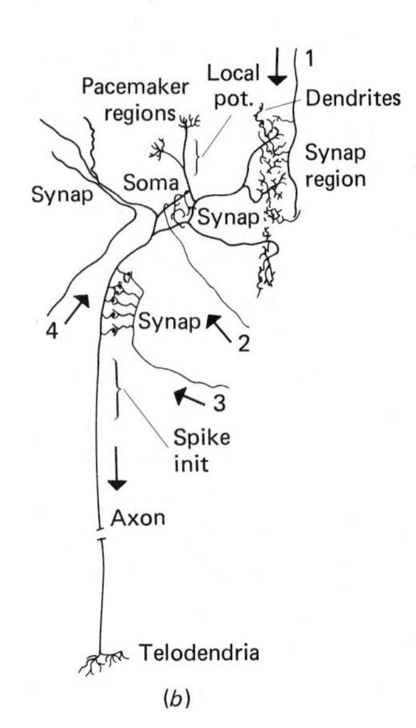

FIGURE 11-6
(a) The anatomical situations in which impulses are initiated. The shaded discs indicate the probable region where neurons of these varied forms have electrically excitable regions of lowest threshold. Neurons with sheaths that have nodes of Ranvier can initiate impulses at the first or second node from the sensitive terminals. *A*, Vertebrate sensory arborization; *B*, typical invertebrate sensory neuron; *C*, interneuron from a ladderlike cord with input and output in every segment; *D*, typical invertebrate motor neuron; *E*, typical vertebrate motor neuron; *F*, amacrine cell (of retina), or local interneuron, perhaps with no axon (or spikes); *G*, stretch receptor cell of arthropod abdomen; *H*, arthropod primary visual cell, with rhabdom; *I*, invertebrate sensory cell as found in some annelids, mollusks, and other groups. (*After G. A. Horridge*, 1968, *Interneurons, W. H. Freeman and Company San Francisco*, p. 70.) (*b*) Schematic representation of a neuron from the cardiac ganglion of a crab. There are several presynaptic pathways converging from diverse sources—inhibiting, exciting, accelerating (1,2,3,4). These produce synaptic potentials in their several special loci. Restricted regions also initiate spontaneous activity ("pacemaker" regions, shown purely diagrammatically), local potentials (labeled in only one place but perhaps repeated elsewhere), and propagated impulses ("spike init," also located arbitrarily). Only the axon supports all-or-none activity. Terminal branches (telodendria) are presumed to carry graded, local potentials. Integration occurs at each of the sites of confluence or transition from one event to the next. (*After T. H. Bullock*, 1959, *Science*, *129*, p. 1000.)

11-4 SOME NEURON SPECIALIZATIONS

Saltatory Conduction

Some axons of vertebrate neurons are surrounded by a sheath of fat-like substance called *myelin* (which is discussed in the next section) and, hence, are called *myelinated fibers*. The sheath is discontinuous, forming interspaces or short segments of exposed axon known as *nodes of Ranvier*. In myelinated axons excitation spreads passively (electrotonically) along the axon with decrement. However, attentuation during passive spread is much less in myelinated than in nonmyelinated axons.

The mode of conduction in myelinated axons involves firing of an action potential (spike) at each node which spreads through the internode segment as a passive, decremental potential which, if greater than threshold, will depolarize the next node and trigger development of another full-sized spike. This is referred to as **saltatory conduction**—because the excitation appears to "jump" from node to node. Because conduction between nodes does not depend upon membrane conductance changes, saltatory conduction is faster than nonsaltatory conduction in comparable neurons (Fig. 11-7).

Giant Fibers

Though it appears quite certain that nervous systems are constructed from individual cellular units, as envisaged by Ramón y Cajal in his neuron theory, there are a few recognized exceptions. For example, there are among annelids and cephalopods nerve-cell syncytia forming the so-called *giant fibers*. The famous axons of decapod cephalopods are the best known of these (Fig. 11-8).

The giant axons of squid and cuttlefish are part of a system of fibers that innervate the mantle muscles, retractor muscles of the head and muscles of the funnel. This system, presumably important in escape behavior, operates the quick contractions by means of which the animal expels a jet of water forcefully from the mantle and darts forward or backward suddenly and quickly.

The axons of the two large cells in the central nervous system, from which the giant-fiber system may be considered to begin (1 in Fig. 11-8), are joined completely across the midline by a bridge of axoplasm (2 in Fig. 11-8) and surrounded by a sheath that is continuous with the sheath of the two neurons. Such a complete fusion implies that impulses generated

FIGURE 11-7
Conduction of an action potential along a myelinated axon. Voltage difference between the inside and outside of the axon's membrane is measured by inserting a microelectrode (cone-shaped) through the membrane to the axon interior. It is compared with an electrode (sphere) in the external bathing solution. At rest, the potential difference across the axon membrane is about one-tenth of a volt, inside negative (horizontal portions of record). When a stimulus initiates an action potential, the potential falls through zero as it reaches the immediate vicinity of the internal electrode, and then returns to resting state as it passes on. In myelinated regions conduction is rapid, passively propagated, and decremental. At the nodes the action potential is regenerated. The dotted curve in the right-hand graph indicates the voltage that would have been recorded at that node if the membrane there had not been able to produce an action potential. (*After C. F. Stevens, 1966, Neurophysiology, John Wiley & Sons., Inc., New York, p. 30.*)

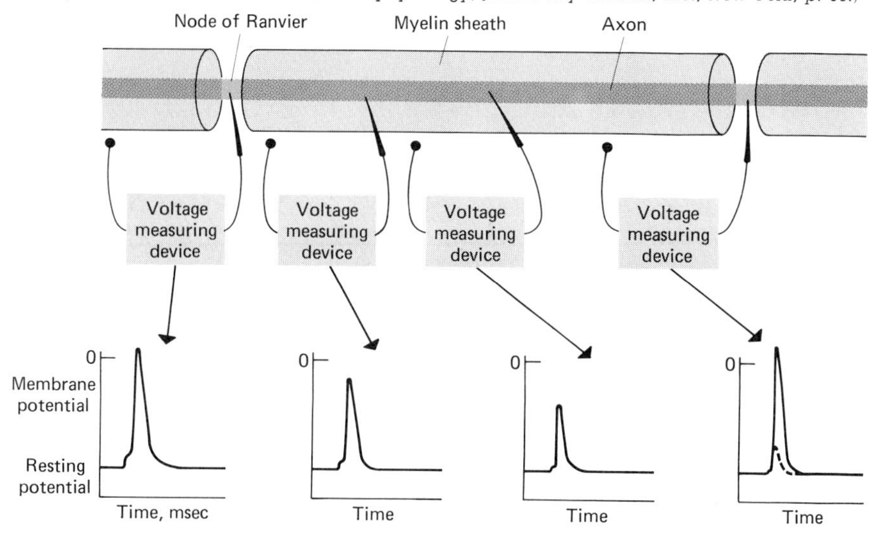

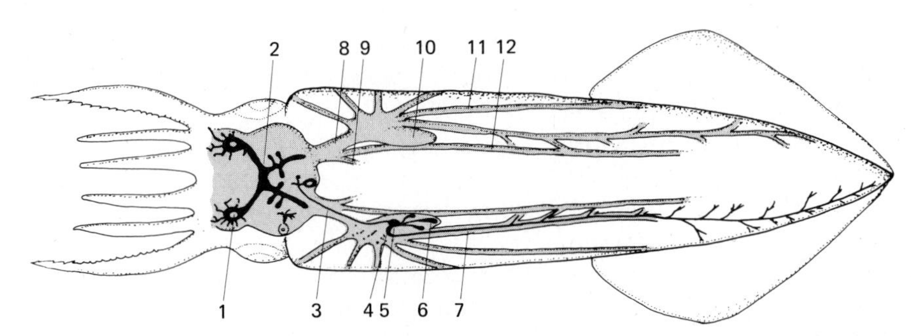

FIGURE 11-8
Diagram of giant fiber system of *Loligo*, as seen from above, the ganglia shown disproportionately large. The nerves are shown in outline on the right, and the giant fibers filled in on the left. 1, Giant cell, with end feet attached to its dendrites; 2, interaxonic bridge; 3, second-order giant axon; 4, distal synapse between second- and third-order axons; 5, proximal synapse between fiber arising independently in the CNS and third-order axon; 6, cell bodies of third-order axons; 7, third-order axon; 8, mantle connective (joining CNS to stellate ganglion); 9, nerve to retractor muscle of head; 10, giant-fiber lobe of stellate ganglion; 11, stellar nerve; 12, fin nerve. (*After J. Z. Young*, 1936, *Cold Spring Harbor Symp. Quant. Biol.*, *4*, p. 4.)

in one of the neurons would always be conducted to the other. This perhaps ensures that both sides of the animal will contract simultaneously in the escape response.

The third-order axons (7 and 11 in Fig. 11-8) are formed by fusion of the processes of numerous small neurons whose cell bodies are enclosed in the stellate ganglion (Fig. 11-9).

The fusion of the axons is complete, so that the resulting giant axon, which may be as much as 1 mm in diameter, is a single neuronal unit, not a number of separate axons enclosed in a common sheath. Giant fibers behave functionally as a single unit, conducting in an all-or-none manner. The great majority of what we know concerning excitation con-

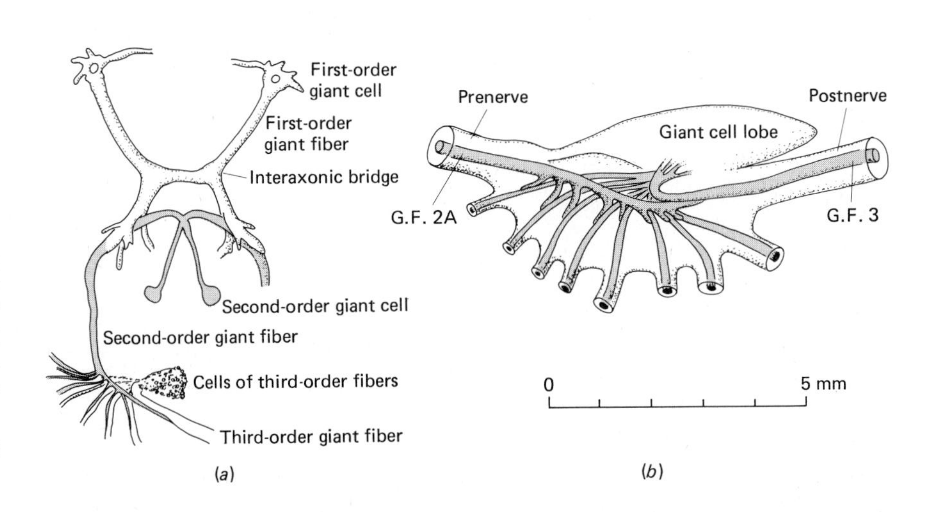

FIGURE 11-9
(*a*) Drawing based on figures originally published by J. Z. Young (1939), showing the sequence of first-, second-, and third-order giant neurons in the squid (*Loligo pealii*). The synapse between the second- and third-order giant fibers has been the one particularly studied. Note that the axons of the third-order giant fibers arise by fusion of cell processes of several neurons. (*b*) Diagrammatic representation of the stellate ganglion with the "giant synapse" between the second-order giant fiber (G.F. 2A) and the third-order giant fibers (G. F. 3) of *Loligo pealii*. (*After T. H. Bullock, and S. Hagiwara*, 1957, *J. Gen. Physiol.*, *40*, p. 1566.)

duction has been gleaned from exhaustive studies of the giant axons of cephalopods. These will be further detailed in Chap. 13.

So that synchronous contractions may occur in muscles located throughout the mantle, there is a variation in size of the axons radiating from the stellate ganglia—the most caudal and longest stellar nerve being the largest. The diameters decrease regularly passing foward, so that the shorter nerves contain the smaller fibers. Hence, the diameter of the axon is adjusted to the distance to be covered. The longer, thicker fibers carry impulses faster than thinner, shorter ones. Thus, excitation will reach all muscles at the same time.

The ventral nerve cord of many annelid worms contains giant nerve fibers. There is considerable variation in their form and arrangement. Some have a single cell body; others have many cell bodies—sometimes one or more in each segment—and are formed by fusion of corresponding interneurons in different segments along the worm. There is also a wide variety in topological arrangement (Fig. 11-10).

In earthworms (*Lumbricus*) there are three of these fibers and they are segmented; i.e., they are divided transversely into units, each sep-

FIGURE 11-10
Interneurons have evolved as giant fibers in a wide range of worms, usually acting as the fast pathways in general shortening of the body. 1 and 2, *Euthalenessa;* 3, *Sigalion;* 4, *Lepidasthenia* and *Euthalenessa;* 5 and 6, *Lumbricus;* 7, *Euthalenessa;* 8, *Eunice;* 9 and 10 *Nereis* and *Neanthes;* 11 and 12, *Arenicola;* 13, *Nereis* and *Neanthes;* 14 and 15, *Halla* and *Aglaurides;* 16, *Mastobranchus;* 17, *Sabella* and *Spirographis;* 18, *Myxicola.* (After G. A. Horridge, 1968, *Interneurons, W. H. Freeman and Company, San Francisco,* p. 99.)

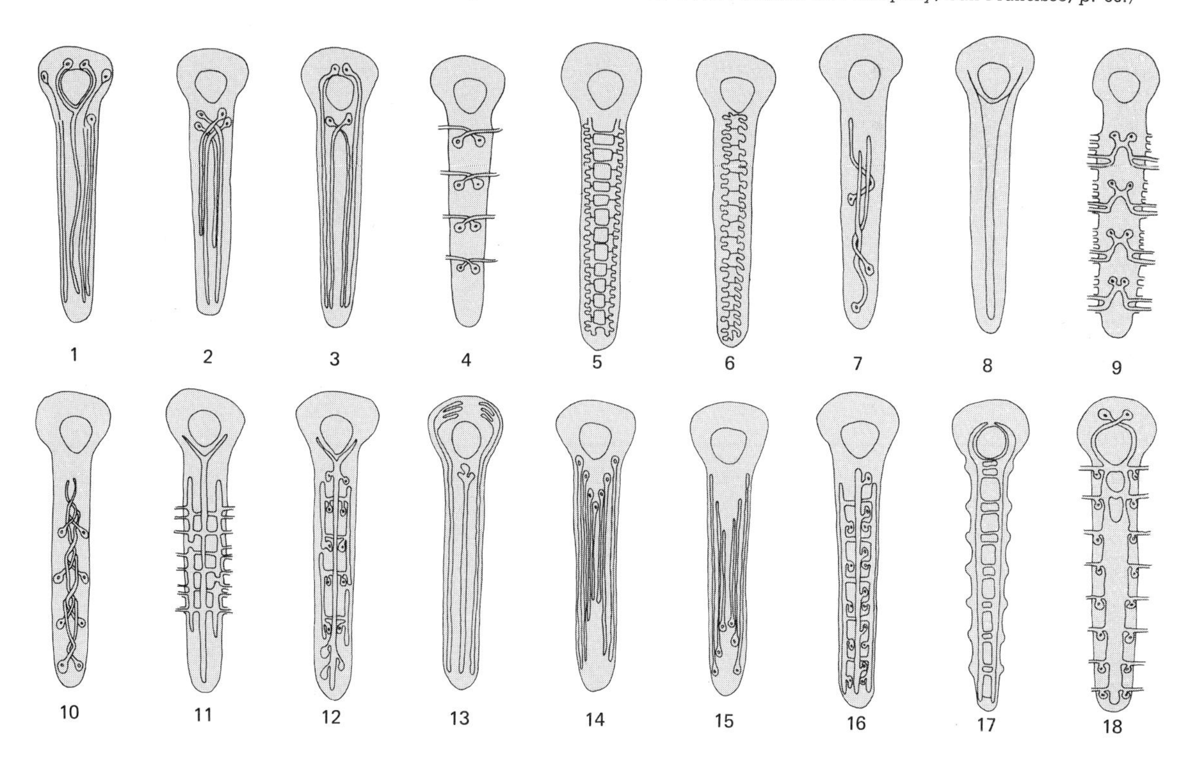

arated from its neighbors by complete membranes (Fig. 11-10, numbers 5 and 6). The two lateral fibers are connected together in every segment by a pathway of low resistance (Fig. 11-10, number 5) and are never active separately except when fatigued.

Electrical activity in the median giant fiber of *Lumbricus* is initiated by touch to the body surface of the anterior 40 segments, whereas the two laterals are activated by touch behind this point. All three may be stimulated electrically at any point, and all three carry impulses in both directions from the stimulated point. Impulses in all three giant fibers cause a rapid twitch of the muscles which shorten the worm, but the median and lateral ones separately cause opposite effects on the direction of pointing of the setae, which project from the sides of the worm. When touched on the head end, median-giant-fiber impulses cause the worm to pull back and the setae to point forward, so serving to anchor the tail. When the tail end is touched the lateral fibers fire, the setae point backward, anchoring the head end, and the tail is drawn up (Fig. 11-11).

Synapses between the giant fibers and segmental motor nerves quickly become fatigued by repeated stimulation, so that twitches are soon replaced by normal locomotion, even though the stimulus may continue. Thus, the animal is not dominated too long by its giant-fiber reflexes.

Multicellular or functional syncytium-type giant fibers are found also

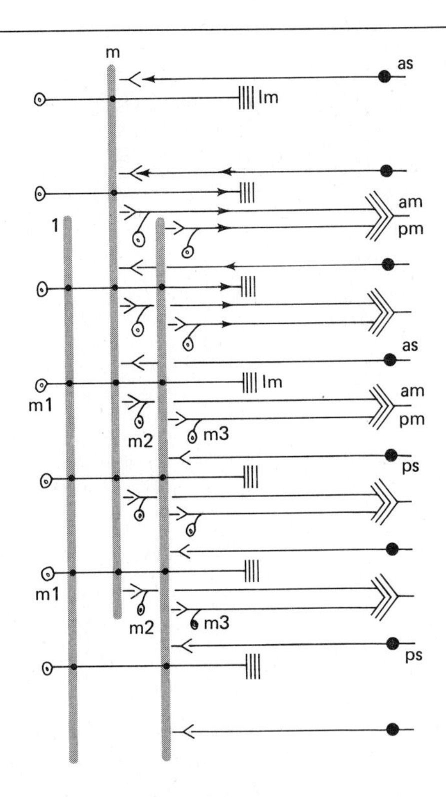

FIGURE 11-11
Connections of the giant fibers of earthworms and some polychaetes such as *Nereis* and *Harmothoë*. Anterior sensory fibers (*as*) connect with the median fiber (*m*), which directs all parapodia or setae forward (*am*). Posterior sensory fibers connect with lateral or mediolateral fibers and direct them backwards (*pm*) all along the animal. All giant fibers connect with motor neurons (*m*1) to the longitudinal muscle, but are selective in their connections to other motor neurons (*m*2,*m*3). Although consistent with the known anatomy, this type of diagram is derived from physiological results. (*After G. A. Horridge, 1968, Interneurons, W. H. Freeman and Company, San Francisco, p. 97.*)

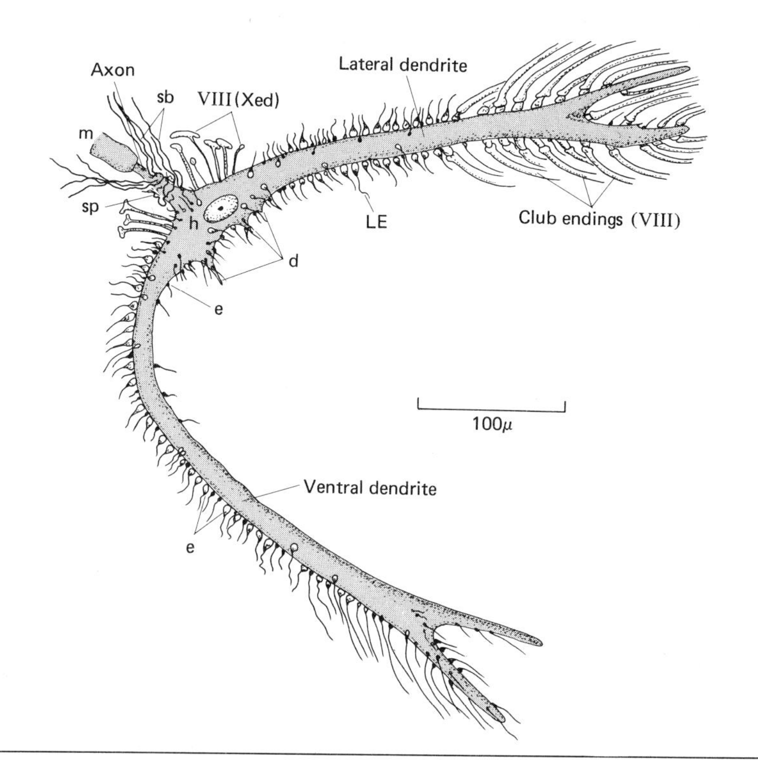

FIGURE 11-12
An example of mosaic segregation of different types of synaptic endings from different sources on specific parts of the dendritic zone of a neuron (Mauthner cell): e, small end bulbs of axons forming synapses with the dendrite; *d*, small dendrites; *m*, myelin sheath of the axon of the Mauthner cell; *sb*, bundle giving origin to spiral fibers (*sp*) in region of "axon cap"; *h*, axon hillock; *LE*, large end bulbs of axons; VIII (*Xed*), small club-shaped terminals of crossed vestibular fibers. (*After D. Bodian, 1952, Cold Spring Harbor Symp. Quant. Biol., 17, p. 3.*)

in crustaceans. They have median, lateral, and motor giant fibers. The quick flipping of the abdomen is presumably mediated by this system of fibers.

There are, it should be mentioned, giant fibers which are not syncytial. Unicellular giant fibers are found in nemerteans, cestodes, numerous polychaetes, balanoglossids, and lower vertebrates. In the base of the medulla of bony fishes and amphibians occur the large *Mauthner cells* of the vestibular reflex system, with fibers up to about 40 μm in diameter that conduct at 50 to 60 m/second at 10 to 15°C (Fig. 11-12).

11-5 NEUROGLIA

In the central nervous system there is a conspicuous class of cells called **neuroglia** (also called **glia**). In fact, in the mammalian cortex, glial cells greatly outnumber the neurons and may make up about half the total volume. Surely, cells as numerous as these need to be given some attention when considering the functioning of the nervous system. An excellent review of what is known about neuroglia, on which the next few paragraphs are based, is that by Kuffler and Nicholls.[1]

Unlike neurons, glial cells retain the capacity to divide throughout

[1] S. W. Kuffler and J. G. Nicholls, 1966, *Ergeb. Physiol.* **57**, 1–90.

life. Two main classes of glial cells have been defined in mammalian brains: *astrocytes* and *oligodendroglia*.

Astrocytes, which exhibit diverse morphologic characteristics, are generally divided into two groups: (1) *fibrous astrocytes*, which contain large bundles of filaments in their cytoplasm and are most prevalent in the white matter (fiber tracts; white because of myelin sheaths), and (2) *protoplasmic astrocytes*, which contain less fibrous material and are found more frequently in gray matter (nerve-cell bodies).

Oligodendroglia have smaller nuclei than astrocytes, their perikarya are sometimes associated with capillaries, they are predominantly located in the white matter of brains, and they are responsible for formation of *myelin*, a lipoid material. Their processes usually wrap around more than one axon (Fig. 11-13).

In the peripheral nervous system another kind of cell, called a *Schwann cell*, forms myelin around axons. If the layer is thick, the axons are said to be *myelinated*. Schwann cells also form a thin envelope or covering around the so-called *nonmyelinated* axons, around nerve endings, and also around the cell bodies in peripheral ganglia (Fig. 11-14).

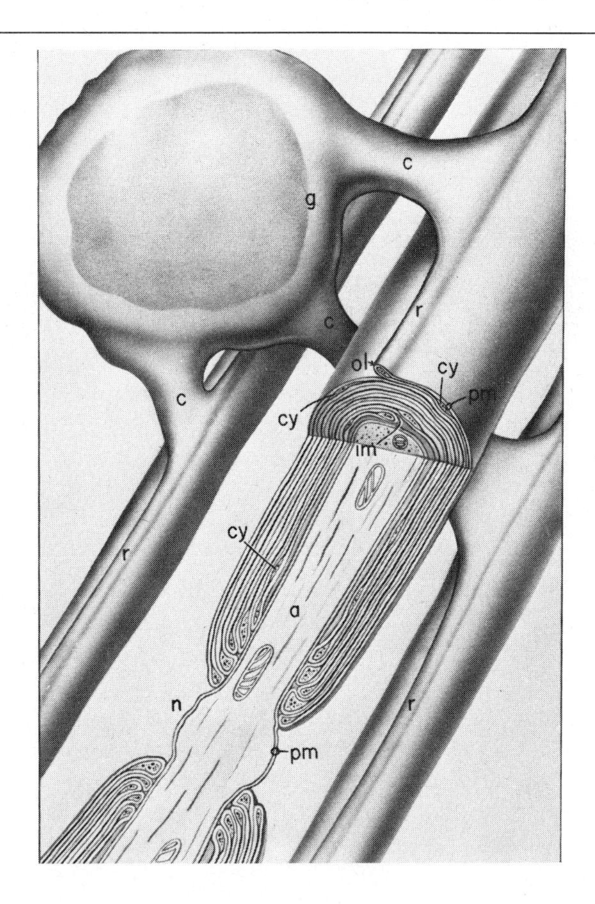

FIGURE 11-13
Diagram of myelinated axons, showing how processes (c) of a neuroglia cell (*g*) enwrap axons (*a*) by means of concentric membranous folds, which ultimately become packed at intervals of 13 to 18 nm units. *cy*, "Trapped" cytoplasm of glial cell; *im*, inner mesaxon and inner loop of plasma membrane; *n*, bare portion of axon or node of Ranvier; *ol*, outer loop of plasma membrane; *pm*, plasma membrane; *r*, external ridge of myelin sheath. (*After M. B. Bunge, et al.*, 1961, *J. Biophys. Biochem. Cytol., 10, p.* 67.)

Schwann cells arise embryologically from the neural crest, but astrocytes and oligodendroglia are derived from *ependymal cells* which line the inner surface of the brain. All types are of ectodermal origin.

Largely on the basis of structural studies, a variety of functions have been ascribed to the glial elements, but supporting experimental evidence is lacking for most of these speculations. Only a few definite statements can be made about neuroglia. They are generally interposed between neurons and other structures, such as blood vessels, the brain surface, and the ventricular system (the chambers within the brain and spinal cord containing cerebrospinal fluid). They resemble neurons in being surrounded by a high-impedance membrane, containing K^+ as their principal intracellular cation, and in having a negative resting potential. Unlike neurons, they do not generate propagated impulses, and, unlike most neurons, they are linked to one another by low-resistance connections.

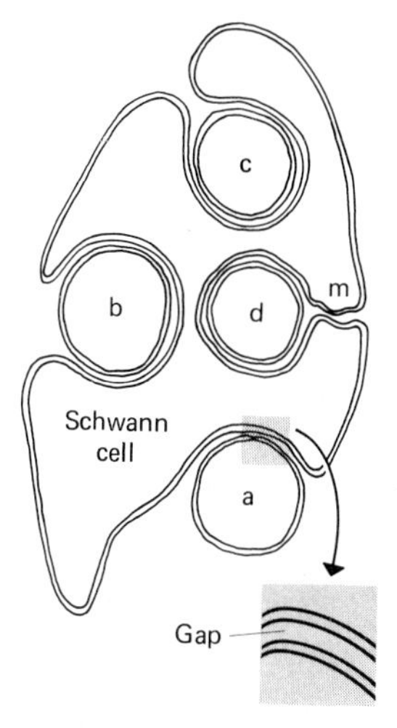

FIGURE 11-14
Diagram of a cross section through a "satellite" Schwann cell in which four "nonmyelinated" axons (*a* to *d*) are embedded. Where the two lips of the Schwann cell meet around axon *d* they form a paired membrane structure known as the *mesaxon* (*m*). A segment of the axon-Schwann membrane is enlarged at the lower right to show the gap between the two unit membranes. (*After J. D. Robertson, 1961, New unit membrane organelle of Schwann cells, in Biophysics of Physiological and Pharmacological Actions, edited by A. M. Shanes, American Association for the Advancement of Science, Washington, p. 68.*)

Neurons and glia are separated by narrow clefts (about 15 nm wide) which serve as channels through which certain ions and metabolites move rapidly through the nervous system. This fluid space prevents the electrical signals that are generated by neurons from spreading to adjacent neurons or to glial cells. However, K^+ liberated from neurons into the intercellular clefts by action potentials depolarizes glial cells, and this in turn is assumed to contribute in some way to the slow potentials recorded with surface electrodes ("brain waves").

Neuroglia are involved in other roles also. Myelin formation has been mentioned. Glial cells also play roles during degeneration, regeneration, and growth of neurons, but these are presently very poorly defined. As noted in Sec. 2-4, neuroglia may be involved in learning.

Phagocytic cells of mesodermal origin, the *macrophages* (sometimes called *microglial cells*), which invade the nervous system from the blood, are another but minor nonneuronal cell found in the nervous system.

11-6 FUNCTIONAL GROUPS OF NEURONS

Neurons in adult animals do not exist in isolation from other neurons but instead always are interconnected in characteristic ways through synapses to form a network called the *nervous system*. Although the details of operation are only dimly perceived, it is quite certain that nervous functions—especially higher-order functions—reside in segregated sets of neurons. Such functional sets, or pools of neurons, often consist of "centers" plus diffuse systems of neurons that interpenetrate with other "centers" to form "circuits."

Complex nervous systems, such as those of vertebrates, are arranged in a hierarchy of sets of neurons, each of which is associated with a single functional role. Functional sets can be categorized as input (e.g., primary sensory pools); output, or motor neuron; intermediate; and cortical, or highest, level. Each set has a unique collection of neuron types (Fig. 11-15).

Generally, the smaller neurons with short axons ramify only within the set and are known as *internuncial neurons* or *interneurons*. Interneurons carry on the principal activities of central nervous systems. From the interneuronal pool, large neurons extend their axons to effectors, such as muscles. These axons, closely packed, form the so-called *fiber tracts* (when inside the brain) or *nerve trunks* (when extending outside the brain). It is these axons that are commonly ensheathed in myelin.

The neurons of the vertebrate spinal cord, which have been extensively analyzed, exemplify the manner in which input neurons, interneurons, and output neurons may be spatially and functionally organized (Fig. 11-16).

Sherrington, an English physiologist, after analyzing and explaining his observations on spinal reflexes, evolved the concept of the motor neuron "pool." Increasing the strength of stimulation (through the sensory inputs) to a pool of responding spinal motor neurons causes a proportionately larger multisynaptic response output. Recent work on spinal neurons has forced a redefinition of the motor neuron pool and also has provided superb evidence that motor neurons of different sizes have different functional properties. The largest neurons have surface areas that are hundreds of times greater than those of the smallest cells, and the diameters of their axons range from less than 0.25 μm to more than 25 μm. The diameters of the axons are directly related to the size of their cell bodies.

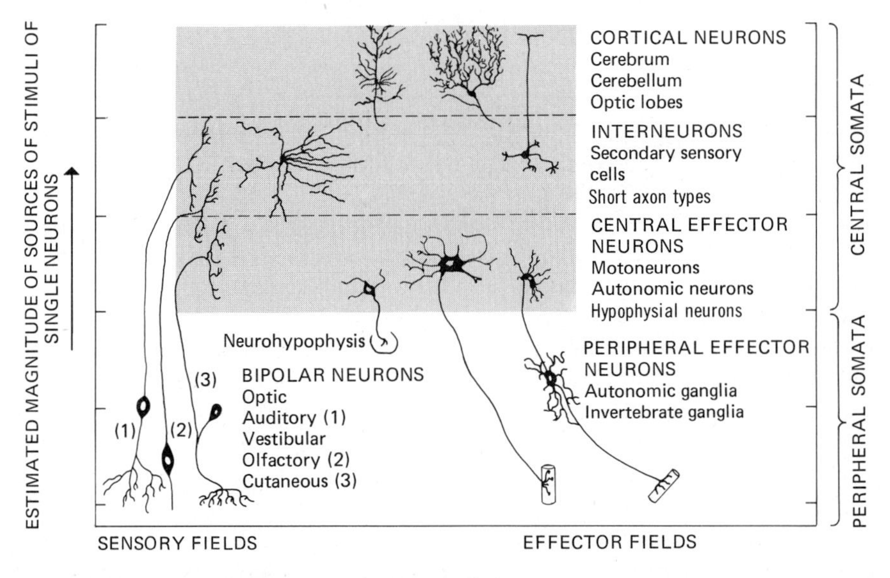

FIGURE 11-15 Major neuronal types in the mammalian central nervous system, arranged according to general role, hierarchical level, and probable magnitude and diversity of sources of synaptic connections. (*After D. Bodian, 1952, Cold Spring Harbor Symp. Quant. Biol., 17, p. 3.*)

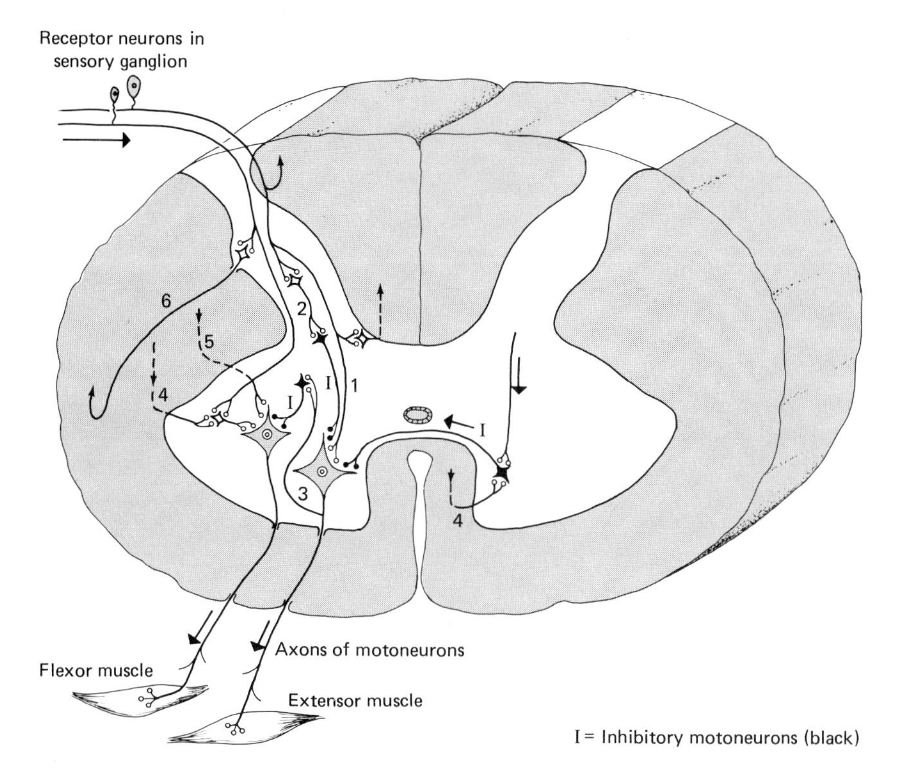

FIGURE 11-16
Diagram illustrating a few simple neuron circuits affecting the motor neuron pool and exemplifying also the major aspects of neuronal circuitry in general (divergence, convergence, delay paths, recurrent loops). A flexor motor neuron, representing a flexor motor neuron "pool," and an extensor motor neuron are shown, since these two pools have interrelated activity (contraction of extensor muscle is concurrent with inhibition of flexor muscle). Small neurons (interneurons) are part of the motor neuron pool or center, and intervene synaptically between receptor neurons and motor neurons, or between descending neurons from brain or upper spinal cord and motor neurons. Receptor neurons have *divergent* telodendria that make synapses with many neurons; motor neurons receive *convergent* synapses from many neurons. (1) Direct connection from muscle receptor neuron to extensor neuron, as in "stretch" reflex such as knee jerk. The "stretch" reflexes are the only known monosynaptic reflexes in vertebrates. (2) Delay pathway, involving two interneurons, and therefore a total of three synaptic delays. One of the interneurons (*I*) is inhibitory in its action on the motor neuron. (3) Recurrent loop from a motor neuron collateral, producing inhibitory effect on neighboring motor neuron through an inhibitory interneuron (*I*), in this case known as a Renshaw cell. (4) Crossed and uncrossed pathways through interneurons, entering the motor neuron pool from extrinsic sources, such as brain or distant levels of spinal cord. (5) Direct extrinsic inputs to motor neurons from cerebral cortex, in primates. (6) Indirect ascending pathway conveying coded receptor signals to brain. (*After D. Bodian, 1967, in The Neurosciences, edited by Quarton et al., Rockefeller University Press, New York, p.* 18.)

Fortunately, the amplitudes of the nerve impulses that are recorded externally from peripheral nerves are directly related to the diameters of the fibers, making it possible to distinguish the signals of large motor neurons from those of small ones. Also, there is a highly significant correlation between the threshold of excitability of individual neurons and the size of the impulses recorded from their axons, the smaller ones being more readily excited than the larger ones. This generalization applies specifically to motor neurons, in particular to the motor fibers supplying muscles in the limbs of cats, which have been extensively studied, but perhaps generally to all neurons.

There are two size classes of motor fibers in the cat. Those from about 1 to 8 μm are called *gamma* motor neurons. They innervate muscle spindles (to be discussed in Chap. 12) and tend to fire impulses spontaneously and continually. The afferent inflow from receptors in resting muscles and from other sources is evidently sufficient to maintain activity in a number of gamma neurons at all times. However, their discharge does not cause development of tetanic (i.e., sustained) tension.

Those cells with fibers ranging from about 9 to 20 μm in diameter are called *alpha* motor neurons. They have high thresholds, do not discharge when the muscles are at rest, and are recruited into a stretch reflex in an order dictated by their relative sizes. The distribution of sizes and thresholds in the pool makes it an anatomically "built-in" grading mechanism which responds autonomously to any input by activating through its output an appropriate number of muscle units. The overall frequency of discharge, or "usage," of a cell is in inverse relation to its size. The tetanic discharge of a large motor neuron may release 200 times as much contractile energy as that of a small one. The morphological basis of this is that large cells have large axons with many terminals, whereas small cells support fewer terminals.

It has been demonstrated that the rate of protein synthesis in neurons is directly related to their firing rate, and the rate of protein synthesis in motor neurons of the mammalian spinal cord is inversely related to cell size. Also, it is of interest that the rate of RNA synthesis in certain neurons is directly proportional to the number of action potentials produced by those neurons, a fact that may have relevance to learning.

All regions of the central nervous system exhibit unique spatial and functional organization. In a few instances groups of neuron cell bodies and dendritic zones are arranged in layers, with input and output components more or less perpendicular to the plane of lamination. This is the case in the vertebrate retina (Fig. 12-24), cerebellar cortex, and cerebral cortex (Fig. 11-17).

In the mammalian cerebral cortex there is a vertical (radial) organization into columns of cells, each of which is related to a specific sensory input. For example, in the somatic sensory cortex the cells in each column are involved with one of the mechanosenses: touch, pressure, joint movement, or hair movement. Furthermore, there is a topographic layout: an adjacent area of the body projects to an adjacent group of columns. The same principle applies to the visual cortex but not so well to the auditory cortex.

Another, and very important, organizational pattern encountered in many parts to central nervous systems is the clustering together of cell bodies to form somewhat distinctive groups called **nuclei.** Such groups are seen, for example, in the integrative centers of brains, such as the hypothalamus, a very ancient part of the basal brain that is present in all vertebrates from fishes to mammals. Specific functions have been ascribed to many nuclei, although in fact they are more like nodal points in complex neural circuits. Behavioral studies reveal that specific brain

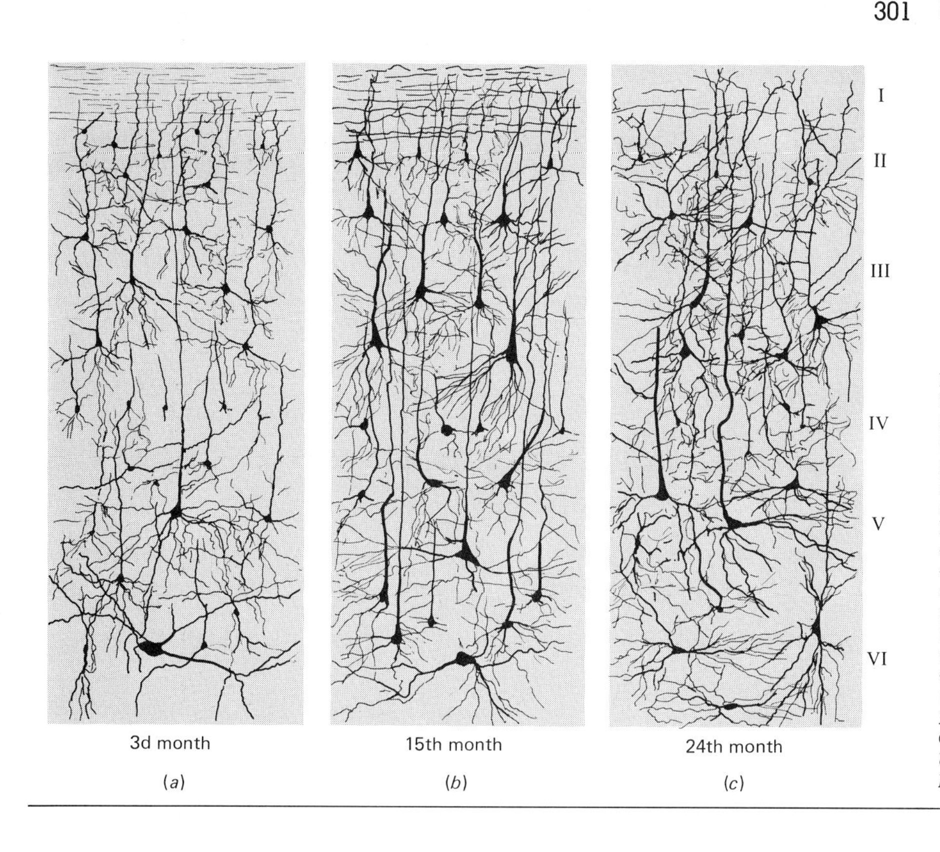

FIGURE 11-17
Tracings from Golgi-impregnated sections from the region of the gyrus temporalis superior in the cerebrum of human children aged 3, 15, and 24 months. Note the vertical layering of cell types and the increased apparent arborization of neurons, and density of dendritic spines, with increasing age. [*After the works of J. L. Conel;* (*a*) *Conel,* 1947; (*b*) *Conel,* 1955; (*c*) *Conel,* 1959, *Postnatal Development of the Human Cerebral Cortex, Harvard University Press, Cambridge, vols. 1 to 6.*)

functions generally involve several nuclei, as switching sites, in diverse parts of the brain that are linked together by neuronal processes to form circuits (Fig. 11-18).

Two standard methods for studying how the vertebrate brain controls behavior are (1) to observe the effects of destroying specified areas or (2) to observe the effects of electrical or chemical stimulation of specific sites. These methods can be illustrated by reference to some studies on the hypothalamus.

If a certain small region of the hypothalamus is destroyed (generally accomplished by passing a cauterizing current through a small wire electrode inserted into the brain) on both sides of the hypothalamus (bilaterally), the animal will cease to eat and will ultimately die of starvation. Electrical stimulation delivered to the same region causes, on the other hand, an animal which has just eaten to satiation to eat still more. It is well documented that this "center" is somehow related to food-seeking and food-consuming behavior. However, in addition, lesions in this region also stop drinking behavior, and stimulation will sometimes evoke sexual responses. Furthermore, an animal with an electrode implanted at or near this site will "work" for electrical stimulation; i.e., if given a switch or lever to close an electrical circuit that delivers brief pulses of current at appropriate levels, through its implanted electrode, the animal will commonly continue to close the circuit for hours at a time, as if the site were a "pleasure" center also.

Such results as these make it difficult to relate specific neuron sets to specific behavior. However, a part of the explanation for such a diversity of effects from stimulation of rather small loci is that the neurons dealing with any one function are not entirely grouped in one area; instead they seem to be built into networks or circuits, so that stimulation at any one point may tap into several different kinds of circuits. Likewise, any one circuit can be encountered at several different loci in the brain.

As will now be apparent, there is great morphological diversity among neurons, and neurons collectively have an impressive repertoire of actions. By capitalizing on the unique properties of diverse types of neurons, the level of complexity afforded by connecting together sensory, motor, and integrative units becomes tremendously large. Correlated neuroanatomical and electrophysiological studies reveal circuits that show *convergence* of input, *divergence* and *amplification* of output through arborized axon collaterals and synaptic telodendria, *feedback*

FIGURE 11-18

I. Three-dimensional reconstruction of nuclei of the rat hypothalamus. (*a*) Major nuclei without interconnecting fiber tracts; (*b*) nuclei with some of the major arriving and departing fiber tracts included. *II.* Cross section through the rat hypothalamus at level indicated by dotted line in (*b*), showing some of the nuclei (circled). The black dots within the circles are neuron cell bodies. (*After W. J. S. Keig,* 1932, *J. Comp. Neurol., 55, p.* 19.)

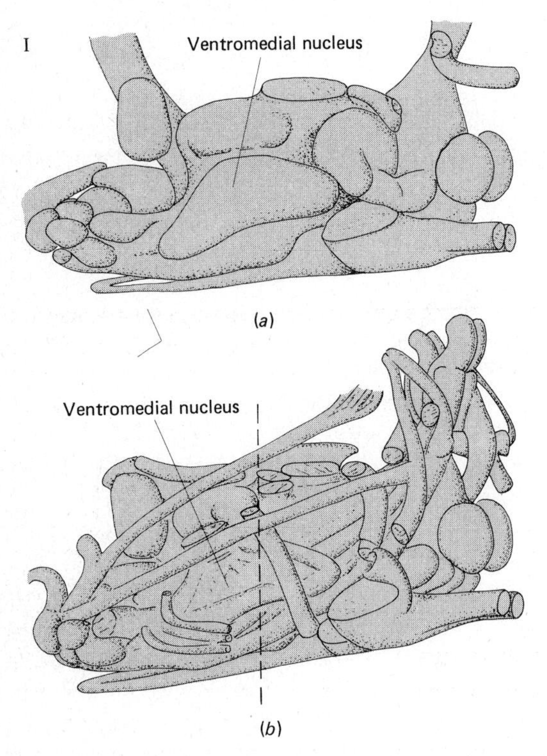

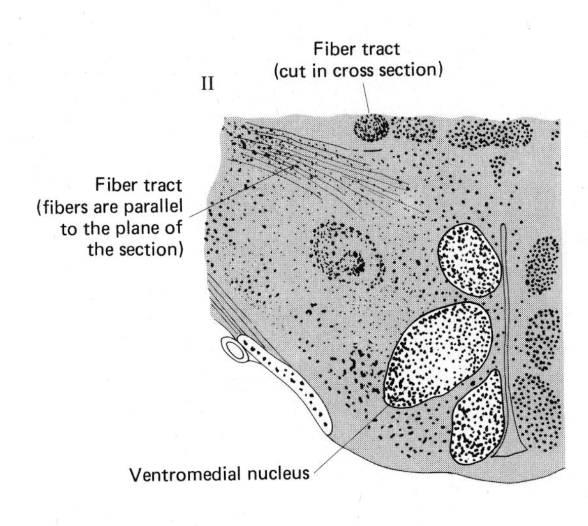

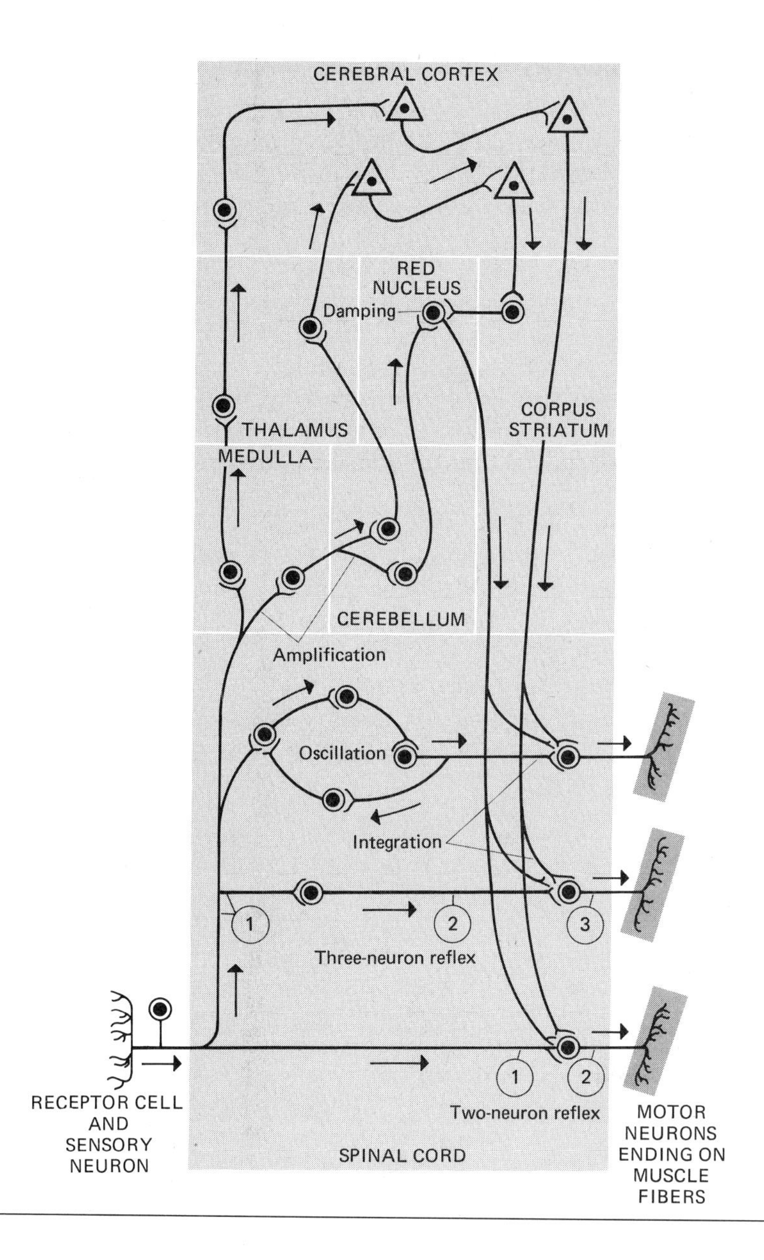

FIGURE 11-19
Simplified diagram of vertebrate nervous system showing sensory input and motor outputs with some possible (highly simplified) interconnecting circuits. Two-neuron reflex pathways probably are rare (see pathway 1 in Fig. 11-16). Spinal circuits involving at least one interneuron are more common. Typically, incoming signals travel upward through the spinal cord (see also 6 in Fig. 11-16) and through several relay centers before arriving at the cerebral cortex. There (or at some lower level) command signals may be generated (or withheld) that will be sent back down the spinal cord to activate a motor neuron (see also 4 and 5 in Fig. 11-16). Amplification of signals through divergence and integration through convergence are illustrated. Self-sustaining, or oscillating, circuits support repetitive and rhythmic actions, such as walking. Damping or suppression of signals can be caused by an inhibitory input modifying or canceling an excitatory input to the same neuron (see 3 in Fig. 11-16).

via axon branches to interneurons (both inhibitory and excitatory), *oscillation* and *reverberation* in loops to sustain for a prolonged time an action once initiated, and *differential rates* of transmission over multiple pathways (Fig. 11-19).

The human cerebral cortex, with its estimated 10 billion neurons each receiving information through several hundred to several thousand synaptic bulbs, boggles in attempting to understand its own complexity.

11-7
LEARNING AND MEMORY CIRCUITS

In Chap. 2 several theories of learning and memory were described. At this time, with a better understanding of neurophysiology, it would be appropriate to read again Sec. 2-4. The reader should now also be adequately prepared to understand the postulated memory system that is described next.

In 1965, Young[1] proposed a unit called the *mnemon*. A mnemon is a module or unit of the memory system of an animal that stores one bit of information. Its components are "a classifying neuron that responds to the occurrence of some particular type of external event that is likely to be relevant to the life of the species . . . and other cells whose metabolism is so triggered as to alter the probable future use of the channels on receipt of signals indicating the consequences of the actions that were taken after the classifying cell had first been stimulated." Young's concept derives from a simplification of a hypothetical interneuron pathway involved with either advance or withdrawal behavior in the octopus. The mnemon is a minimum circuit that learns a "yes/no" decision, based upon interneurons that classify sensory excitation (Fig. 11-20).

In Young's scheme, the classifying cell must be able to operate the switching for a negative or for a positive response and would have to be biased in favor of the output which promotes advance toward unfamiliar things, since we recognize that animals tend to be exploratory. Classifi-

[1] J. Z. Young, 1965, *Proc. R. Soc. London B*, **163**, 285–320.

FIGURE 11-20
(*a*) The components of a single memory unit, or mnemon. The classifying cell records the occurrence of a particular type of event. It has two outputs, producing alternative possible motor actions. The system is biased to one of these (say, "attack"). Following this action, signals indicating its results arrive and either reinforce what was done or produce the opposite action. Collaterals of the higher motor cells then activate the small cells, which produce inhibitory transmitter and close the unused pathway. These may be called "memory cells," because their synapses can be changed. (*b*) Plan of the connections involved in "maintaining the address" of a mnemon in the optic lobe. The question marks show where evidence about connections is lacking. Signals for good taste or pain are ensured of coming back to the appropriate mnemon. This presupposes that the nervous system can differentiate purely neural loops, which itself implies a peculiar form of recognition between growing neurons. *VU* and *VL*, upper and lower vertical lobes of the brain. (*After J. Z. Young*, 1965, *Proc. R. Soc. London B*, 163, 290.)

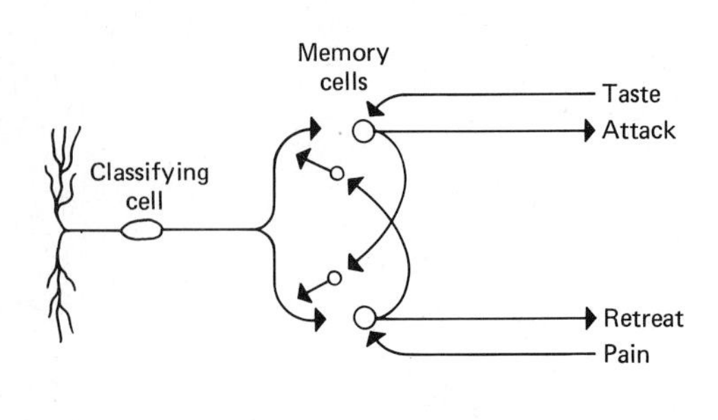

(*a*)

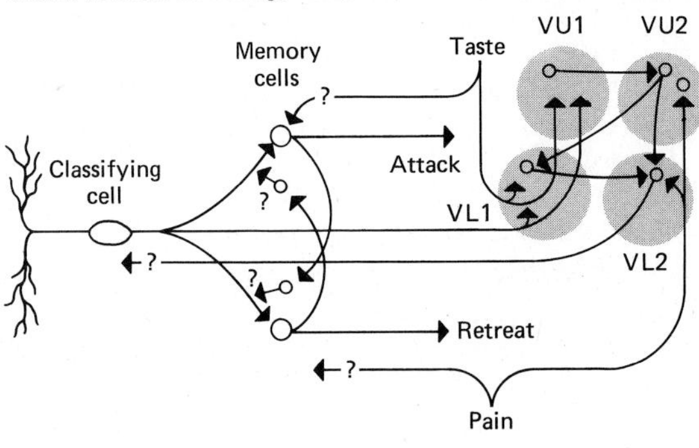

(*b*)

catory systems of interneurons have been encountered in many sensory input pathways. Learning is presumed to modify the probability of action of the interneuron by positive and negative feedback, so that switching through one of the outputs is favored. However, this model may be criticized in that there is no evidence that the minimum memory circuit has so few cells or that classifying cells have as few as two effectively distinct axon terminals. Interneurons probably have more than one output ("yes" and "no" branches), and memory circuits appear to involve clusters of cells. Furthermore, the time scale for reinforcements for most behaviors is several magnitudes greater than that of the short-lived impulses of neurons.

Those who seek a basis for memory in changes in synaptic "connections" may find encouragement in the discovery of a rise in neurotransmitter activity in the visual cortex of rats after prolonged exposure to a "rich" perceptual environment as compared with deprived animals. In the retina, changes in the electroretinogram (a recording of the electrical events associated with and following the photochemical process) have been noted in animals deprived of light for several days. Also, synaptic function at neuromuscular junctions and in spinal reflexes has been shown to be enhanced by use.

Various experiments, with many different species of vertebrates, in which extensive ablations are effected (such as those mentioned in Sec. 2-4), suggest a remarkable plasticity with respect to the ability of the brain to process memory. Opponents of this view argue for the possibility that some tracts may be spared by the ablations and remain intact and functionally undisturbed. These remnants, even if very few, may subsume the originally specified function. As difficult as it is to disprove such arguments, it seems highly unlikely that a small percentage of tracts surviving extensive brain damage would be capable of activating circuits appropriate to the needs of the organism, unless memory "engrams" are widely delocalized throughout the brain. On the other hand, an "engram" might possibly be stored molecularly as a complete entity within a single cell.

READINGS

Horridge, G. A., 1968, *Interneurons*, W. H. Freeman and Company, San Francisco. The individuality of different kinds of interneurons and the complexity of their connectivity patterns are surveyed in many kinds of nervous systems.

Quarton, G. C., T. Melnechuk, and F. O. Schmitt (eds.), 1967, *The Neurosciences*, Rockefeller University Press, New York. A highly recommendable source for information on neurophysiology from molecular aspects to behavior and learning. A second and a third volume with the same title appearing in 1970 and 1974, respectively, have continued this series.

Schadé, J. P., and D. H. Ford, 1973, *Basic Neurology*, Elsevier Publishing Company, Amsterdam. The anatomy and physiology of the mammalian nervous system is described; there are many excellent illustrations.

RECEPTION AND SENSORY INPUT 12

Animals perceive only a small part of their environment. Their sensory organs serve as highly selective filters for (or absorbers of) the great spectrum of physical and chemical energies available in most habitats. The filtered input is somehow "transduced" by the sense organ into a coded electrical "language," ultimately consisting of repetitive impulse discharges (spikes) in nerve-cell axons. The coding process is not simple, as there may be additional filtering, amplification, and modulation of certain components before the pattern is finally transmitted to the analyzing sites in the central nervous system. There is a great deal of variation in how these processes occur in different receptors and in different animals. We shall examine only a few selected examples of representative receptors, and focus mainly on some of the common features shared by many of them.

12-1 THE DIFFERENTIATED ELECTROGENIC CELL

Parker in 1919 set forth in his classic work, *The Elementary Nervous System*, the suggestion that the nervous system of Metazoa evolved from a cell which was a primitive *receptor-effector*, or an *independent effector*. Asconoid sponges can respond to unfavorable situations by closure of their ostia through shortening of contractile elements in cells called *porocytes*, which are perhaps surviving examples of such cells. The epitheliomuscular cells of Cnidaria, another type of independent effector, have a receptive surface exposed to the environment and a deeper-lying portion that is specialized for contraction. Cnidarians also have independent effectors called *nematocysts*, specialized cells which, in response to appropriate chemical or mechanical stimuli, can explosively evert a coiled hollow thread to penetrate or entangle potential prey. Beyond these three examples, it is difficult to find independent effectors. This is understandable because they offer little opportunity for variety in either what is perceived or the response effected. Through extensive cellular specialization and development of functional interconnections between receptors and effectors, integrated responses to a wide spectrum of stimuli have become possible.

Grundfest extended Parker's idea and formulated a hypothesis for evolution and organization of electrogenic cells. He proposes that the receptor and effector components became separated spatially and their separate functions specialized in the course of evolution. Communication between the two was maintained by the development of a conductile axonal component. Some of the cells became specialized as sensory units, while others became distinct neuronal cells with integrative functions. Nevertheless, although much evolutionary specialization has occurred, each electrogenic cell has retained its double role of receptor and effector and has, consequently, three components of action: *input, conductile,* and *output* (Fig. 12-1).

Most neurons are highly elongated and exhibit a functional polarity, in that one pole (input) normally receives information and the opposite pole (output) transmits information to other neurons or to muscle or gland cells. The receptive pole is called the *dendritic zone*, because in many neurons it is highly branched like a tree. The transmitting pole may have terminal branches, called *telodendria*, or may be an unbranched process. The branched or unbranched conductile process that extends between the receptive and transmitting poles, the *axon*, is commonly long—from a few microns to a few meters—and fiberlike. Axons are often sheathed in insulative layers of a lipoprotein, called *myelin* (cf.

FIGURE 12-1
Generalized diagram of the three functionally different membrane structures of electrogenic cells. The input portion, which is not electrically excitable, responds to specific stimuli with a characteristic electrogenesis that may be depolarizing (excitatory) or hyperpolarizing (inhibitory). When the net electrogenesis of the input is depolarizing, it can initiate spikes in the conductile, spike-generating portion. These signals, upon reaching the terminal, output portion of the cell, cause secretory activity which may excite the next postjunctional cell. Though diagramatically represented as distinct, the various types of excitable membrane may be intermingled. (*After H. Grundfest,* 1957, *Physiol. Rev., 37,* p. 343.)

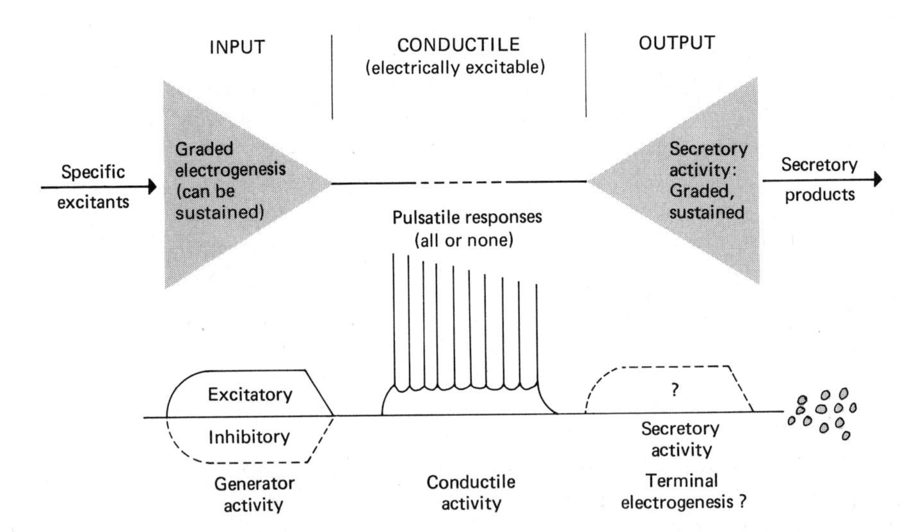

Sec. 11-5). Although under experimental conditions the axon can be shown to conduct in both directions, in normal function it usually conducts unidirectionally.

The dendritic and axonal processes of a neuron arise from an enlarged region known as the *soma, cell body,* or *perikaryon.* Within it are the nucleus, ribosomal aggregates (Nissl substance), the major metabolic machinery, and the center of a cytoplasmic tubular system believed to be involved in transport of secretory and metabolic products toward the axon terminal.

Neurons may be divided into two general classes on the basis of type of input: *Synaptic* and *receptor* neurons. Neurons are in intimate contact with other neurons through synapses. As noted in Chap. 10, most synapses appear to be chemical junctions, but some are electrical. Into the cleft of a chemical synapse the output component of one neuron, the presynaptic neuron, releases a secretory product, variously called a *neurotransmitter, neurohumor,* or *chemical transmitter,* in response to arrival of spikes along the axon. The membrane of the postsynaptic neuron has receptor sites for the neurotransmitter, and thus receives coded information from the presynaptic neuron. Neurons that receive coded chemical information from other neurons we shall call *synaptic neurons.* In the following chapter the actions of these neurons will be the central topic.

Neurons that receive and transduce environmental energy, such as heat, mechanical or electrical energy, and chemical stimuli, are *receptor neurons.* The input components of receptor neurons, in contrast to most synaptic neurons, usually are not highly branched. These portions are differentiated to respond to specific stimulus inputs (Fig. 12-2).

From their geometry, neurons can ordinarily be recognized as a member of a class, but each has unique morphological and functional properties. From the diversity of types illustrated in Fig. 12-2 it is apparent that the textbook pictures of "typical" neurons are a fiction.

The conductile component, the axon, is *electrically* excitable and has a definite *threshold* for stimulation. Any electrical stimulus which exceeds the threshold results in an *all-or-none, propagated* impulse or action potential of brief duration and constant size and shape—a *spike potential.* The spike continues to the terminus of the neuron. Spike potentials will be better described in the next chapter. Our present concern is with the manner in which the impulses that are transmitted along the conductile component are generated at the input of receptor neurons.

Input components serve as "transducers," meaning, in the present context, stimulus energy (chemical, light, heat, pressure) produces an electrogenic response which is an approximate analog of the physical stimulus. This response is a *localized, nonpropagated* potential change that is graded in proportion to its specific stimulus and may be sustained as long as the stimulus is applied (except where there is rapid adaptation; see later). This potential is quite distinct from the phasic, spike action potential characteristic of the conductile component. It is usually called a

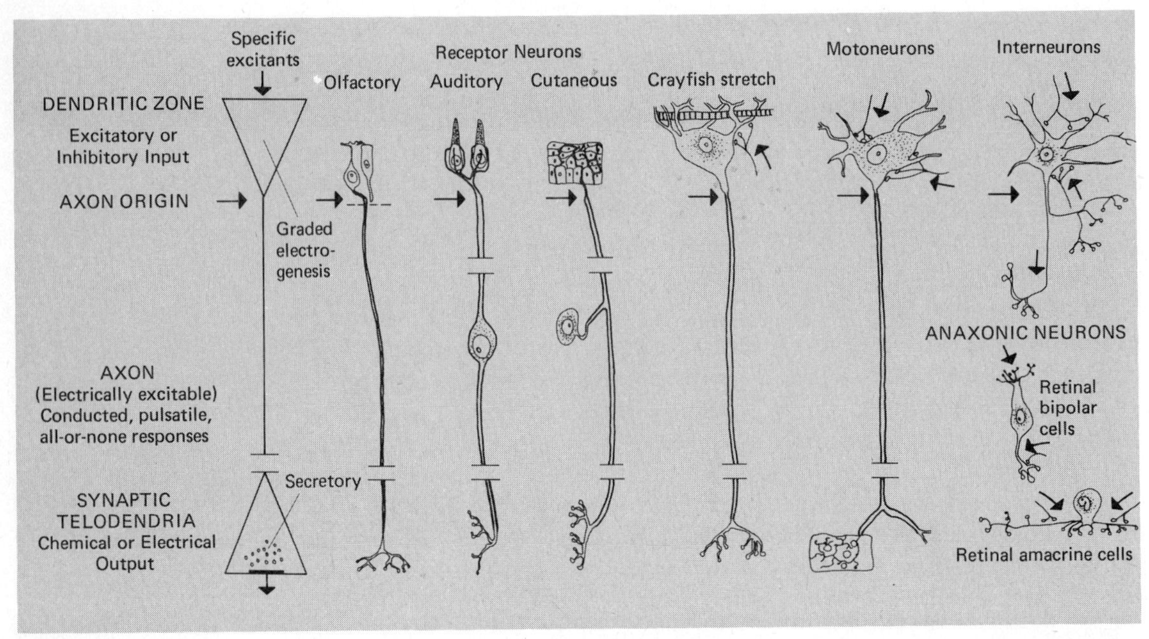

FIGURE 12-2
Diagram of a variety of receptor and effector neurons, arranged to illustrate the idea that impulse origin, rather than cell-body position, is the most reasonable focal point for the analysis of neuron structure in functional terms in neurons with an axon process. In all axon-bearing neurons the four major points of interest (receptor zone, axon origin, conducting zone or axon, and transmitting or synaptic zone) conform to the functional diagram of the generalized neuron proposed by Grundfest. The location of the nucleated portion of the cytoplasm, or perikaryon, however, does not have a constant relationship to the functional geometry of neurons in general, although in any specified type the location is fixed. Anaxonic neurons, in which the spike-conducting region is absent, may be considered as having processes of dendritic or telodendritic (presynaptic) character. (*After D. Bodian,* 1967, in *The Neurosciences, edited by Quarton et al., Rockefeller University Press, New York, p.* 11.)

generator potential (or, if recorded from a receptor cell, it may be called a **receptor potential**) (cf. Fig. 11-5). We shall examine its features in more detail shortly.

12-2
RECEPTORS AND THE RECEPTOR PROCESS

Receptors that respond to stimuli from the external environment are called *exteroceptors*. They are located at or near the external surface of the animal. Receptors that respond to stimuli arising within the organism are called *interoceptors*.

Receptors are responsible for initiating much of the impulse traffic in the central nervous system. Many receptor units autonomously generate impulses in the absence of an apparent stimulus; when stimulated they may increase or diminish activity, depending upon their specific nature. Other units produce outputs only when an appropriate stimulus is impinging on them or when the amplitude of the stimulus is altered.

Receptors can signal not only the existence of a stimulus but its intensity as well. It was Adrian (1928) who pointed out that stimulation of a receptor may result in a train of nerve impulses whose *frequency* is some function of *intensity*. The increase in impulse frequency is commonly found to be approximately linear with respect to the logarithm of the stimulus intensity over much of the dynamic range. How this is accomplished is not known, but the adaptive advantages are obvious. A neuron can conduct only up to several hundred impulses per second, and any sensory system which responded linearly with increasing intensity would soon reach a maximal rate of discharge. Since the range of intensity over which most receptors must respond is enormous (10^{12} for human vision and hearing), a logarithmic-response contour is a practical necessity, even though the whole range may not be covered by a single element (Fig. 12-3).

At this point it will be useful to introduce a generalized diagram of the sensory process, as conceived by Davis. It should be studied briefly and then referred to from time to time as we proceed through the events of reception (Fig. 12-4).

It will be noted in the diagram that in some cases an accessory *sensory cell* receives the input, and in others it is the input component of a *sensory neuron* which directly detects the stimulus. In either case, the same basic processes are presumed to occur. At least the result is the same in terms of generating a coded message that transmits meaningful information to the central nervous system. Sensory cells are not nerve cells but modified epithelial cells that are innervated by sensory endings

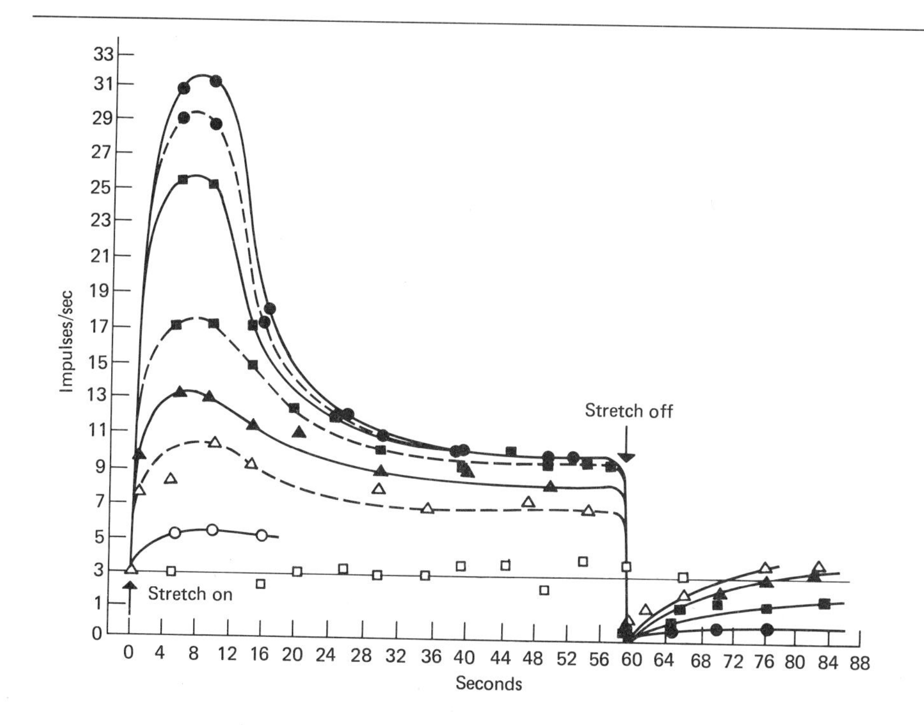

FIGURE 12-3
Rates of discharge of a single stretch fiber of frog skin with different degrees of stretch. Impulse rate of successive experiments at 8-minute intervals in which the preparation is abruptly stretched by:

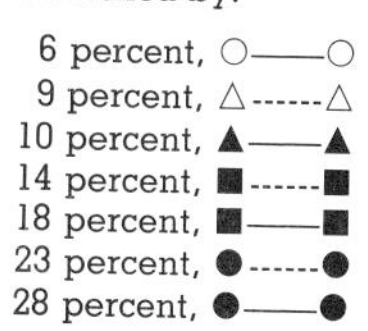

starting from its initial minimal length (100 percent), □, frequency at minimal length. (*After W. R. Loewenstein, 1956, J. Physiol., 133, p. 588.*)

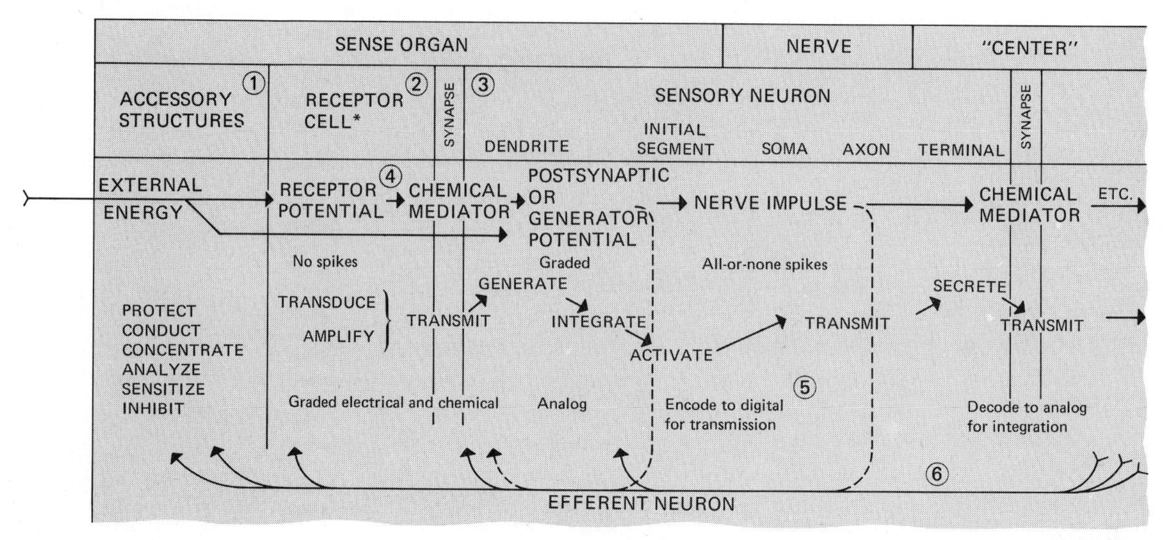

FIGURE 12-4
Generalized plan of sensory action. Not all the features shown in the diagram are present in all sense organs. Thus, (1) accessory structures are absent in some cutaneous sensory systems. (2) There may be no specialized receptor cell, but the receptor response may be in the sensory neuron. (3) There may be no synapse with chemical mediator, but the potential from the receptor cell may stimulate the initial segment directly. In cases 2 and 3 the "receptor" potential is also the "generator" potential. (4) There may be no receptor potential, but the external energy may directly and nonelectrically stimulate the liberation of a chemical mediator. (5) Since an impulse is quantal and is followed by a refractory period during which another impulse cannot be elicited, to call impulses "digital" is not strictly correct, because the intervals between pulses are significant and probably are continuously graded, rather than being a series of fixed intervals. (6) Efferent neurons have been demonstrated in some but not all sense organs. (*After H. Davis,* 1961, *Physiol. Rev., 41,* p. 404.)

of sensory neurons. Responses of sensory cells to environmental stimuli determine the electrical behavior of the associated sensory neurons.

Considerable evidence suggests that the same basic events occur in all the various kinds of receptors, regardless of the stimulus modality. These are (1) absorption of energy, (2) transduction, (3) generation of a receptor (generator) potential, and (4) initiation of a conducted action potential (spike) in the sensory neuron. This may turn out not to be so, but it presently is a convenient conceptual framework that allows one to deal with the complexities of reception as a series of sequential, analyzable steps. As will be seen, not all receptor processes lend themselves equally well to this kind of scheme, indicating either that our information is incomplete or that there are some mechanisms that are not common to all receptors.

Absorption of Energy

In order for an effect to be produced in a receptor, energy must first be absorbed. Receptor cells typically have a low threshold for one form of energy. Thus, we speak of *mechanoreceptors* and *chemoreceptors*. Among mechanoreceptors are those that respond to stretch, rotation of joints, compression, or bending. Light and radiant heat are two portions of the electromagnetic spectrum that are perceived by many organisms, and the receptors are called *photoreceptors* and *thermoreceptors*, respectively. Selective sensitivity determines, then, what form of energy and what segment of an energy spectrum will cause a response in a receptor. These parameters vary widely among organisms. For example, humans do not detect ultraviolet light, but some insects do. Neither do humans have receptors for resolving planes of light polarity, but some arthropods do. Receptors act like "band-pass filters," admitting to the central nervous system certain kinds of information while rejecting other kinds.

It is an oversimplification, however, to state that each receptor is responsive to only one kind of stimulus input. Even the most specialized receptor may be activated by more than one kind of energy, provided the stimulus is sufficiently intense. Thus, receptors are only *relatively* more sensitive to one kind of energy and less sensitive to others. The energy form to which the receptor responds most readily is called the *adequate stimulus* of that receptor.

In the case of chemoreceptors, such as those involved in taste (gustation) or smell (olfaction), there is no apparent absorption of energy. Thus, the first stage of our simplistic scheme is violated. Indeed, as will be mentioned later, we do not know the exact nature of the stimulus for chemoreception.

Transduction

As noted above, it appears that graded electrical change in the membrane of the input component of a receptor is induced by an adequate stimulus for that receptor. How this occurs is unknown. It could be a similar or a different process in the various kinds of sensory units. In crustacean stretch receptors and in visual receptors, stimulation produces a change in permeability of the membrane to various ions. In stretch receptors Na^+ conductance is increased.

The transduction process in photoreceptors involves a chemical event as an essential step in the development of the electrical event: photic breakdown of a visual pigment. In mechanoreceptors it has been postulated that a change of pressure might cause certain changes in molecular organization, either of the plasma membrane or of hydrated ions on either side of it. For other receptors the transduction process is even more obscure.

It needs to be stressed that the response which ensues following the stimulus input generally involves energy changes far in excess of that which initially excites the receptive cell. Therefore, this is not a simple energy transduction in which, for example, mechanical energy or the energy of a given number of photons is transformed into an energetically equivalent change in electrical potential. Instead the stimulus controls or modulates the flow of energy from the electrochemical reservoir maintained by ion pumps (to be described in the next chapter). It is probably satisfactory to assume that what is actually "transduced" is *information.* The stimulating energy initiates a quantified electrical response which is an analog representation of the information content of the stimulus.

Generator Potential

The analog of the stimulus appears to be, at least in most receptors studied, a depolarization, the duration and magnitude of which are proportional to the duration and magnitude of the stimulus. (Receptors can show hyperpolarization also; see "Chemoreceptors," further on.) Such a graded, localized, nonpropagated change of polarization is either a **receptor potential** or a **generator potential.**

Some investigators use the terms "generator" and "receptor" potential interchangeably. However, it is appropriate to use *receptor potential* to designate a relatively slow, graded electrical potential recorded from (or near) *receptor cells.* It may be correlated with either excitation or inhibition of receptor cells. A *generator potential* is also a relatively slow, graded electrical change, but it precedes and evokes the nongraded, all-or-none spike potentials. Receptor cells, it should be noted, do not conduct spikes, whereas sensory neurons do. Nevertheless, through synaptic transduction, receptor cells induce generator potentials in their associated sensory neurons, which are conductors of all-or-none-type (spike) potentials (Fig. 12-4).

Generator potentials have been unambiguously demonstrated in rather few cases. They commonly must be detected by intracellular recording (requiring microelectrode penetration), and it is extremely difficult in certain receptor structures to impale the appropriate site. A generator potential was first detected in muscle spindles by Katz (1950). Subsequently it has been recorded from a variety of sense organs (Fig. 12-5).

The nature of the permeability changes underlying the generator potentials is much less understood than that of the action potentials of axons. However, the amplitude and rate of rise of the potential is related to the sodium concentration.

FIGURE 12-5
Generator and spike potentials in a stretch receptor in the abdomen of a lobster. Stretch was applied at the upward-pointing arrow, and released at the downward-pointing arrow. Note the upward-shifted oscilloscope tracing on which are superimposed spike action potentials. This is the generator potential. Stretching causes the generator potential (12 mV depolarization), which leads to impulse (spike) generation. (*After C. Eyzaguirre and S. W. Kuffler,* 1955, *J. Gen. Physiol.,* 39, *p.* 98.)

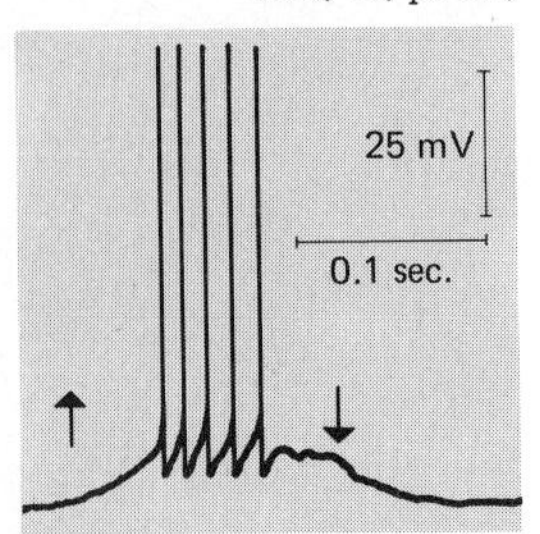

Spike Initiation

The axon of a sensory neuron, like the axon of a synaptic neuron, has its origin in a process from the soma (or sometimes from a dendritic process; see the cutaneous receptor neuron in Fig. 12-2) called the *initial segment.* This portion of the neuron is free of ribosomes and is often constricted to form the so-called *axon neck* of myelinated neurons. Somewhere in the initial segment arise spike potentials, the *frequency* of their discharge being directly related to the *amplitude* of the depolarizing generator potential. Thus, in the receptor, the spike discharge frequency is the final coded indicator of stimulus intensity.

The sequence of events, then, can be summarized as follows: stimulus of intensity $I \rightarrow$ generator potential of amplitude $G = f(I) \rightarrow$ spike discharge of frequency $S = f(G)$. Thus, the stimulus intensity is first coded by *amplitude* modulation; then it is coded by *frequency* modulation (Fig. 12-4, note 5).

It needs to be pointed out that though electrical signs in the receptor cell or in the neurons transmitting information from the receptor are the most elegant evidence that an animal perceives a stimulus, other indications, such as innate behavior patterns or conditioned reflexes, also may be used. For example, the selection of a "preferred" temperature in a thermal gradient indicates that the animal is capable of detecting variations in the ambient temperature. Acceptable chemicals in solution cause extension of the proboscis in dipterans and lepidopterans; extension has been used to show that the insect detects the chemical and also to locate the sites of reception—some of which are on the tarsi (terminal segments of the legs). Through training of bees to come to certain colors for food, it has been learned that the visible spectrum for bees is shortened at the red end (they are red-blind) but is extended into the ultraviolet end.

12-3 MECHANORECEPTORS

The *Pacinian corpuscle* is one of the best-known mechanoreceptors. It consists of an onionlike layered, fluid-filled capsule enclosing a nonmyelinated sensory neuron terminal. The neuron process leading from the capsule is encased in myelin except at the *nodes of Ranvier*, where it is uncovered for a short segment (Fig. 12-6).

The myelin-free, distal end of the neuron is the transducer, being so arranged as to be compressed, and possibly stretched, by rapid compression of the capsule laminae. The capsule can be stripped off, and the free nerve terminal, when mechanically stimulated, will show a depolarization. Stimulation of a very small area of the delamellated terminal fiber will produce a localized depolarization which decreases exponentially with the distance from the point of stimulation. The response is confined to the region of the membrane that is distorted. Increasing the stimulus strength causes a progressively greater degree of deformation and a concomitant increase in the magnitude of the generator potential (Fig. 12-7).

Only generator potentials normally are found in the nonmyelinated Pacinian fiber. The point at which spikes first appear is at the first node of the myelinated portion of the afferent fiber (Fig. 12-6).

The *muscle spindle* of the frog is a biological transducer which measures the relative length and the rate of change of length of a muscle. Typically, during loading of the muscle there is a rapid discharge of nerve impulses, followed by a decline in frequency to a steady level during maintained stretch (Fig. 12-8). (See also Fig. 12-3.)

There is evidence for two distinct components of the spindle potential: an initial dynamic component associated with lengthening of the

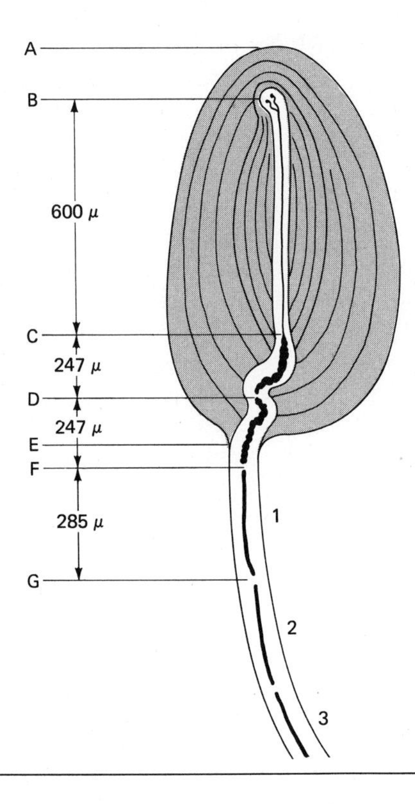

FIGURE 12-6
The Pacinian corpuscle. The axon ending BC is enclosed in a capsule AE; only the region BC is sensitive to pressure. The action potential spikes arise first at F, the first node of Ranvier; the numerals indicate the successive internodes of the myelinated portion of the axon. (*After T. A. Quilliam and M. Sato*, 1955, *J. Physiol.*, *129*, p. 173.)

spindle, and an ensuing static component associated with maintained extension. When the spindle is given a rapid stretch, only one spike with a very short latency occurs during the period of extension. As the velocity of stretching is decreased, a proportionate increase of spikes occurs during the period of extension. The onset of spike activity has a longer latency and a lower peak frequency at slower velocities of stretch. Therefore, for a stretch to a given amplitude, the number of spikes occurring

FIGURE 12-7
Local activity of the nonmyelinated sensory ending of a Pacinian corpuscle after the nonnervous (capsular) tissue has been removed. Two microelectrodes (E_1, E_2), connected to separate amplifier channels, are placed about 350 μm apart. E_1 is about 20 μm from the stylus (*st*) which delivers mechanical stimuli. The preparation is pulled into a device which insulates the first node of Ranvier in contact with a common-ground electrode from the ending. The response to a single mechanical stimulus, as detected by electrodes E_1 and E_2 is shown at right. Note lower amplitude and slower rate of rise of the potential at the more distal electrode, E_2. (*After W. R. Loewenstein*, 1959, *Nature*, *183*, p. 1724.)

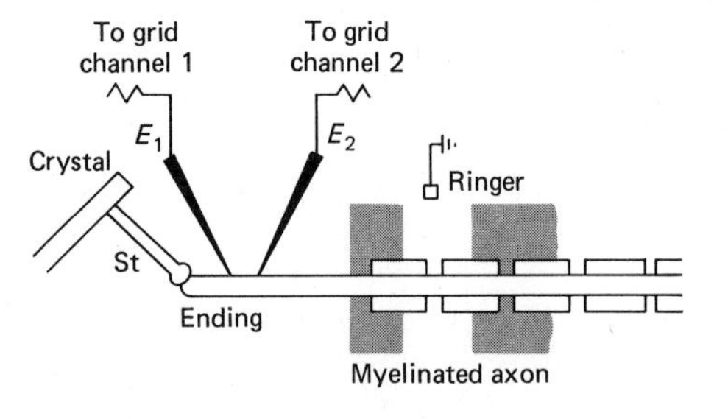

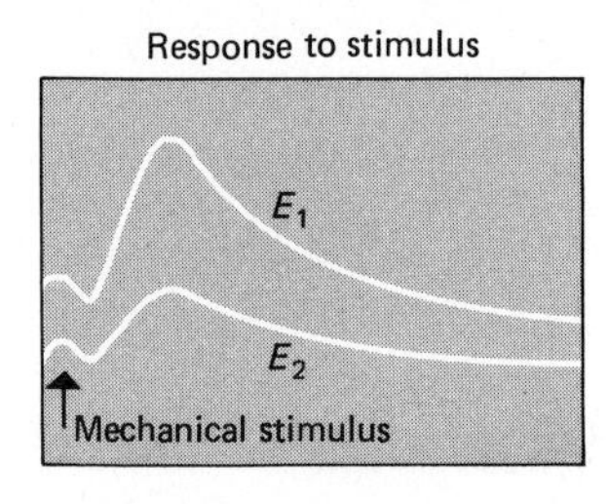

during the dynamic period is inversely related to the velocity of lengthening of the spindle. The spike frequency associated with maintained stretch increases linearly with increasing amplitude of stretch. Except during small or very slow stretches, there is a gap between the spikes associated with the dynamic stretch and those of the static stretch (Fig. 12-9).

An isolated generator (or receptor) potential can be obtained from the muscle spindle by applying a local anesthetic. The pure generator potential obtained has two distinct phases: an initial dynamic phase appearing during the period of lengthening, and a later static phase during maintained stretch. Thus, the two-phased pattern of spikes seen when the spindle is stretched is generated by the two phases of the generator potential (Fig. 12-10).

A spike appears within 0.5 msecond or less after a fast stretch begins. Such a short latency suggests that whatever intermediate processes may be involved in the transduction of the mechanical deformation into spike output, they must be very rapid. The participation of a chemical transmitter, relying on diffusion, seems unlikely.

Decapod crustacea, such as crayfish and lobster, have *muscle stretch receptors* that are functionally analogous to vertebrate spindle receptors. Those in the dorsal abdominal extensor muscles, which extend between abdominal segments, have been studied extensively (Fig. 12-5). In each segment there are two pairs of these stretch receptors (Fig. 12-11).

In simple outline, when the muscle moves, the dendritic membranes are deformed, giving rise to a generator potential with ensuing spike output. However, the stretch-receptor organ is quite sophisticated in its function, and it demonstrates some principles of actions that are found in many other receptors also. There are two kinds of stretch receptor units: *tonic* and *phasic*.

Tonic (or static) and phasic (or dynamic) are terms of convenience for distinguishing two modes of response to stimuli. Actually, these names represent extreme modes of response; i.e., receptors range from the most tonic types through all gradations of tonic-phasic to the most phasic types. *Phasic* receptors are characterized by:

1. **A burst of spike potentials when the stimulus begins (or ends, in some cases)**

2. **Rapid adaptation: while the stimulus persists unaltered, the frequency of spikes rapidly diminishes and ceases (or returns to the rate prevailing before the stimulus was applied)**

Tonic receptors are characterized by:

1. **A burst of spikes at the onset of stimulation, followed by a slow, gradual decrease in spike frequency**

FIGURE 12-8
Response of a frog muscle spindle to varying amounts of stretch. The rate of impulse firing is proportional to the degree and speed of stretching (see text). Note the generator potential, on which are superimposed spike action potentials. Note also that the larger the generator potential, the greater the number of spikes. (*After W. H. Miller et al.*, 1961, *Sci. Am.*, *September*, p. 226.)

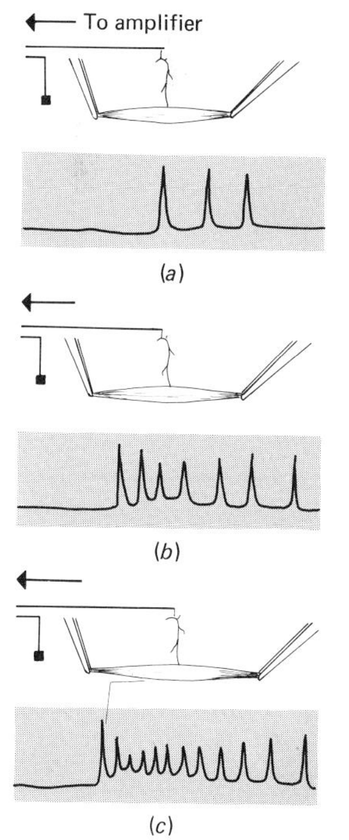

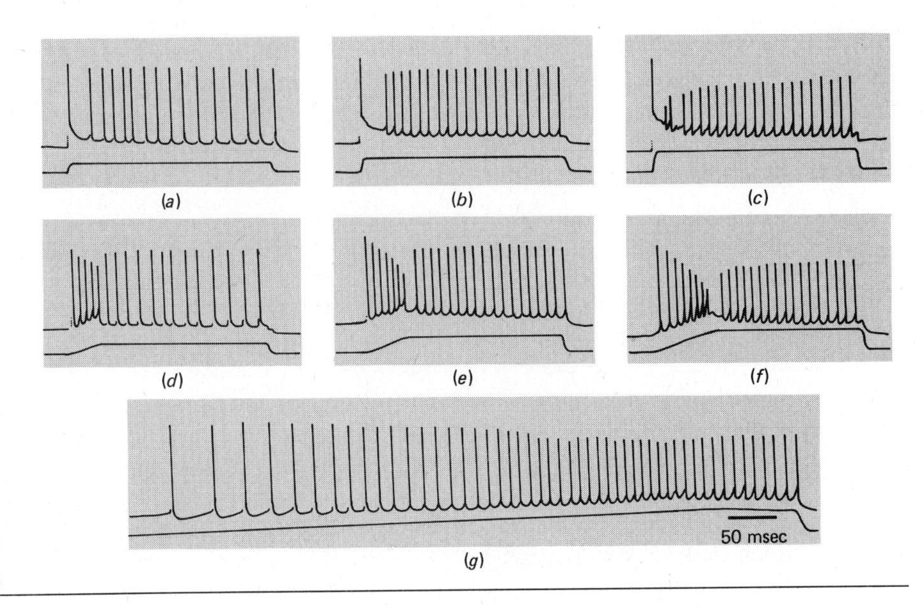

FIGURE 12-9
Evidence for a different mechanism underlying dynamic and static responses of the muscle spindle of a frog, as revealed by responses of the isolated spindle to increasing amplitudes of extension at different velocities. (*a*)–(*c*) At 130 mm/second; (*d*)–(*f*) at 13 mm/second; (*g*) at 1.3 mm/second. Amplitude of extension is the same in (*a*) and (*d*); (*b*) and (*e*); and (*c*), (*f*), and (*g*). Note development of "pause" between dynamic and static discharges with higher velocities and levels of stretching. Time calibration: 50 mseconds. (*After G. M. Shepherd and D. Ottoson, 1965, Cold Spring Harbor Symp. Quant. Biol., 30, p. 101.*)

2. Slow adaptation: while the stimulus persists unaltered the rate of spike firing tends to be maintained without much diminution

3. Spontaneous activity: spikes appear at regular intervals without a change in stimulus

4. A monentary cessation of firing when the stimulus ceases

FIGURE 12-10
Isolation of a receptor potential by application of a solution of local anesthetic to the frog's muscle spindle. Response before (*a*) and after (*b*) blocking of conducted activity (spikes) with 0.2% lidocaine. The isolated spindle is subjected to a stretch of 400 msecond's duration. Note two distinct phases; an initial dynamic phase appearing during the period of lengthening, and a later static phase during maintained stretch. (*After D. Ottoson and G. M. Shepherd, 1965, Cold Spring Harbor Symp. Quant. Biol., 30, p. 107.*)

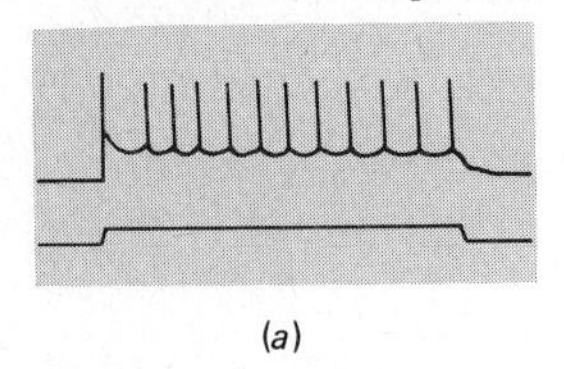

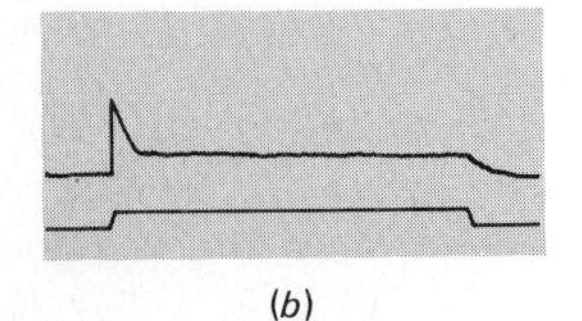

The response of the stretch fiber shown in Fig. 12-3 is a good example of a receptor with tonic features. These response types are further explained in Fig. 12-12.

Spontaneous activity is characteristic not only of tonic stretch receptors but of many other mechanoreceptors and several other kinds of receptors as well. This will receive more comment later.

The crustacean stretch receptor has a means of directly controlling the receptor activity—something not found in vertebrate muscle spindles. An inhibitory neuron extending from the ventral nerve cord synapses on the dendritic branches of the receptor neuron (Fig. 12-11). Stimulation of the inhibitory neuron reduces or abolishes stretch-induced activity in the receptor neuron. Normally a stretch induces a depolarization in the receptor, but if the inhibitory neuron is activated the membrane potential tends to be restored to the resting level. In cases where the receptor membrane is hyperpolarized the inhibitor also tends to return the potential to the resting level (Fig. 12-13).

In contrast to the three representative kinds of mechanoreceptors just discussed, all of which rely upon membrane distortion, or possibly stretch, for stimulation, there is a large second group of peculiar mechanoreceptors incorporating sensory *hair cells*, which typically have struc-

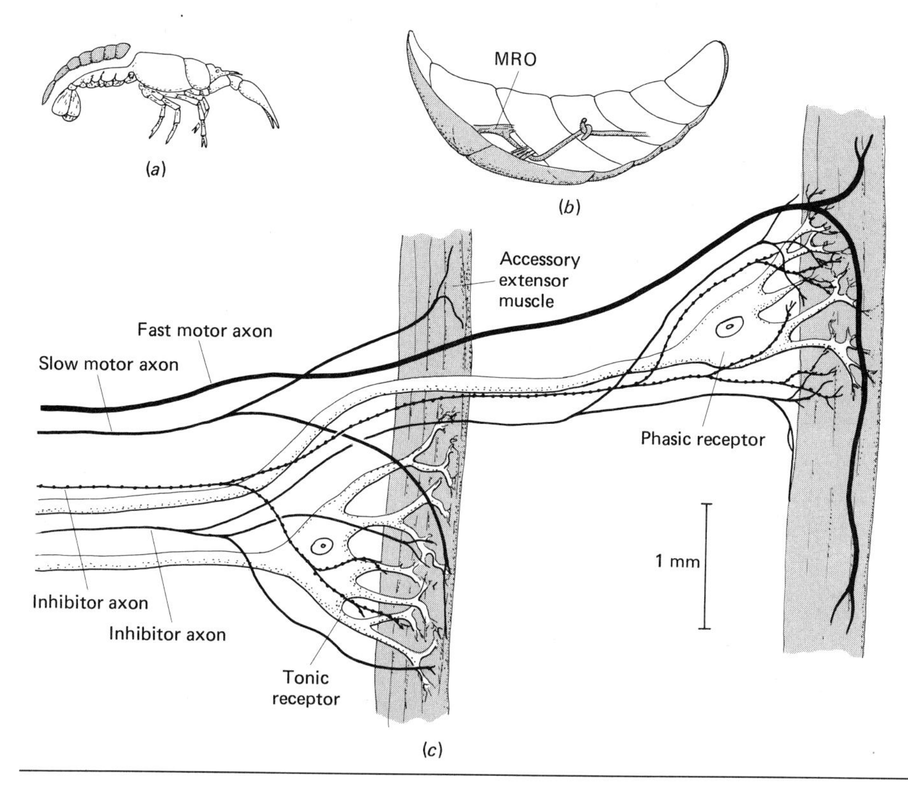

FIGURE 12-11
Diagrams of abdominal-muscle receptor organ (stretch receptor) in a decapod crustacean. (a) and (b) Location of muscle receptor organ (*MRO*) between segments on dorsal abdomen. (c) Phasic (fast) and tonic (slow) receptors attached to their respective muscles. Motor, inhibitor, and sensory nerve fibers are shown. [*After C. A. G. Wiersma* (ed.), 1967, *Physiology of Invertebrate Nervous Systems. The University of Chicago Press, Chicago*]

tural elements that are cilia or cilialike. In this group are the bipolar receptors of arthropods, sensory cells of the acousticolateralis system of

FIGURE 12-12
Diagramatic illustrations of the differences between "on" and "off" effects in tonic receptors (a) and phasic receptors (b). (*After E. Florey,* 1966, *An Introduction to General and Comparative Physiology, W. B. Saunders Company, Philadelphia,* p. 471.)

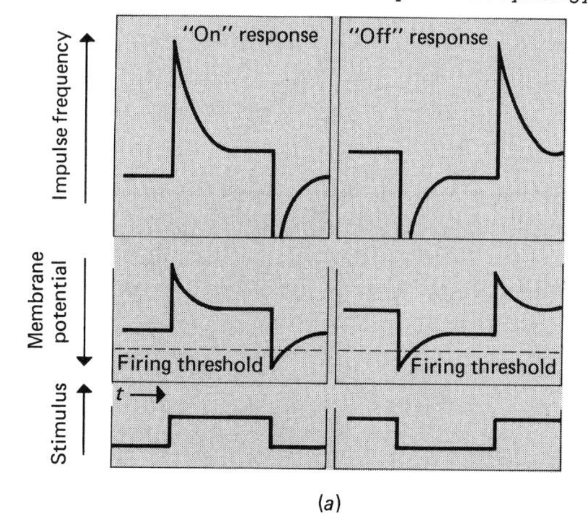

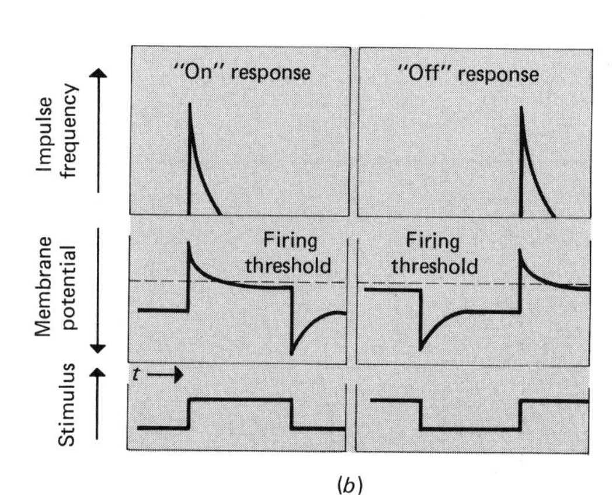

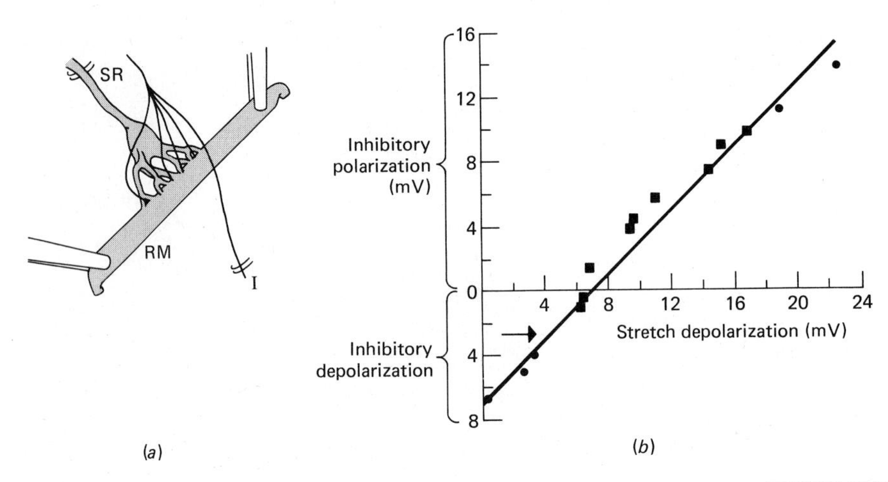

FIGURE 12-13
Action of a stretch-receptor inhibitory neuron. (*a*) Stretch receptor arranged for mechanical stimulation by stretching muscle (*RM*), held by micromanipulator forceps. Receptor activity is recorded by electrodes (*SR*) while inhibitory neuron (*I*) is stimulated via a branch to the second receptor (see Fig. 12-11). (*b*) Plot of *I* effect versus receptor resting potential. *Abscissa*, different amounts of depolarization up to 22.5 mV were produced by stretch without discharging the cell. Zero indicates membrane potential when the cell was completely relaxed. *Ordinate*, inhibitory potential amplitudes. The points lie approximately along a diagonal line with a slope of 45°, crossing the abscissa at the "inhibitory equilibrium level." Note that *I* potential compensates for displacement of membrane potential in either direction from this level at about 6.4 mV. (*A, After J. F. Case*, 1966, *Sensory Mechanisms, The Macmillan Company, New York*, p. 31; *B, after S. W. Kuffler and C. Eyzaguirre*, 1955, *J. Gen. Physiol.*, *39*, p. 155.)

vertebrates, and hair cells in the ampullae of the semicircular canals, saccule, utricule, and cochlea of the inner ear of mammals (Fig. 12-14).

In the hair cells of the *cochlea* there are no spike, or all-or-none, impulses. This is typical of secondary sensory cells which, as noted earlier, are not nerve cells but modified epithelial cells. The hair cell has no axon, and it is assumed that the receptor potential acts directly on the synaptic junction at its base to cause release of a chemical transmitter. (See again Fig. 12-4 and note the position of "receptor cell" in the scheme.)

Considerable difficulty has been experienced in attempting to decide which, if any, of the potential changes that accompany mechanical responses in hair cells are generator potentials. One of the effects of vibrational stimulation of hair cells of the inner ear, labyrinthine organs, and lateral-line organs is to cause comparatively low-level, alternating, potential changes—on the order of 100 μV. These are called **microphonic potentials.**

When sinusoidal or complex sound patterns are applied to the eardrum, microphonic potentials which closely mirror the applied wave form can be recorded from various sites in the inner ear. They arise in the cochlea (probably in the external hair cells), have no apparent threshold and negligible latency, and spread decrementally along the auditory nerve and throughout the tissue surrounding the cochlea.

An additional kind of potential change is believed to be generated by the internal hair cells. Unlike the cochlear microphonic, it continues to increase in amplitude with increasing sound pressure. This type is on the order of a few millivolts. However, this potential is usually not accepted as a receptor potential, and it remains to be determined just which kind of potential generates the action potentials in the auditory nerve.

The *labyrinth*, or vestibular organ, is a complex of vertebrate inner-ear structures located near and connected to the cochlea. It is made up of two principal parts: the *semicircular canals* and the *otolith organs* (saccule and utricle). The primary function of the semicircular canals is to register movement of the body in space, or more precisely, to respond to any change in the rate of movement (i.e., to acceleration or deceleration). Movements of the fluid contained in the semicircular canals stimulate hair cells, which are the mechanoreceptors that detect changes in rate of movement.

The *utricle* of the otolith organ apparently responds to both linear acceleration and tilting. The *saccule* is believed to be associated with reception of slow vibrational stimuli rather than being a part of the vestibular mechanism, but the evidence is conflicting.

The only place that receptor potentials from hair cells seem to have been demonstrated with any degree of certainty is in the *neuromasts* in the *lateral-line organs* of the mudpuppy, *Necturus*. Here intracellular recordings reveal potential changes of up to 800 μV (peak to peak) that are phase-locked with vibratory stimulation (Fig. 12-15).

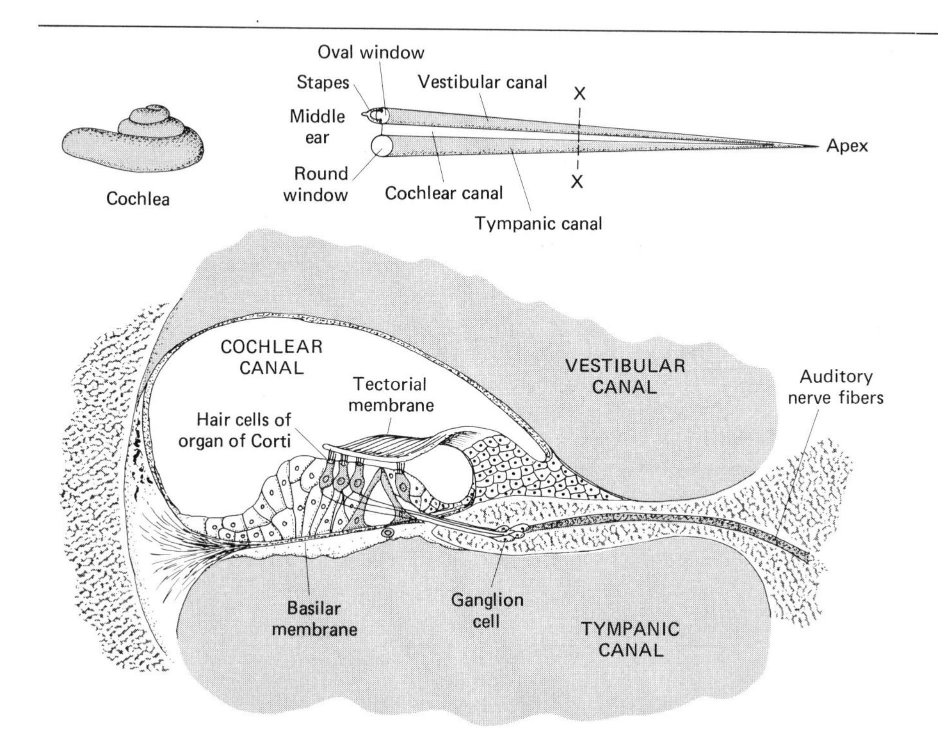

FIGURE 12-14
Diagram of the mammalian cochlea. *Above,* the spiraled cochlea (left) has been uncoiled and opened (right) to show its three canals. *Below,* a section at X-X to show the organ of Corti. Note the position of the hair cells. They are innervated by the sensory fibers of the acoustic nerve. When sound waves vibrate the tympanic membrane (eardrum), the bony ossicles (only the terminal of the three, the stapes, is shown above) transmit the movement to the oval window of the cochlea. The hair cells are stimulated by the shearing force exerted by the tectorial membrane when it and the basilar membrane are set in motion by pressure waves transmitted through the fluid that fills the inner ear.

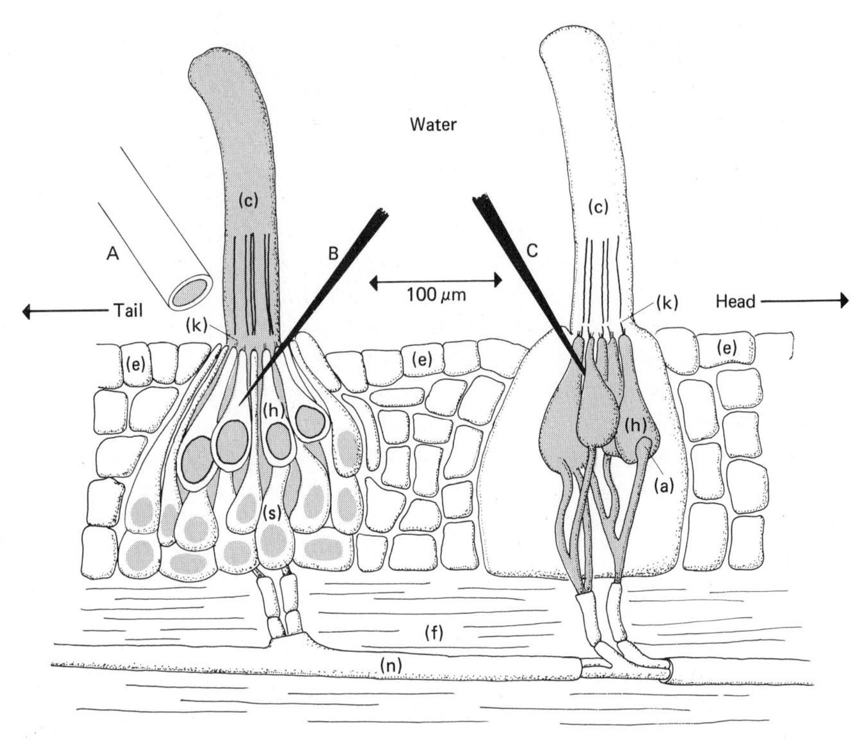

FIGURE 12-15
Two neuromasts in the lateral line on the tail of *Necturus*. Two to six neuromasts, forming a group called a *stitch*, are lined up approximately along the rostrocaudal axis, which is also the axis of motion sensitivity. The neuromast contains about eight hair cells (*h*) whose basal ends are roughly lined up in two rows along the rostrocaudal axis, and whose apical ends converge on a narrow region along the same axis. Each hair cell has a single kinocilium and a number of stereocilia protruding from its apical end (*k*). The hair cells are surrounded and separated by a nest of supporting cells (*s*) that is embedded in the epithelial layer of the skin (e). A cupula (c) extends from the region of hair cells upward into the water. The hair cells are innervated by a nerve bundle running through a tough fibrous layer (*f*) just underneath the epithelial layer of cells. Two large myelinated afferent fibers (*n*) innervate each stitch. Each fiber sends branches to each neuromast. The two fibers lose their myelination just before entering the neuromast and form afferent synapses (*a*) on the basal ends of the hair cells. Pipette A is used as a vibratory stimulator and a source of externally applied chemicals. Pipette B is used for intracellular penetration of hair cells and recording. Pipette C penetrates one neuromast of a stitch and is used to monitor nerve spikes from the two afferent nerves which can be excited from any neuromast in the same stitch (*After G. G. Harris et al.*, 1970, *Science*, *167*, p. 77.)

12-4 CHEMORECEPTORS

Chemoreception is conventionally divided into two categories—*olfaction*, or smell, and *gustation*, or taste. **Olfactory receptors** refer to those cells generally excited by chemical stimuli in relatively low concentration and in a gaseous state. **Gustatory** or **contact chemoreceptors** or **taste receptors** are terms used interchangeably to refer to cells excited by chemical stimuli which contact the receptor as liquids or solutions and act in relatively high concentrations. However, there are no consistent cellular features that permit such distinctions to be drawn, and in a few cases the same receptors mediate both "gustatory" and "olfactory" responses. These are mainly terms of convenience; until better distinctions are formulated they will remain in use.

A very major contribution to our understanding of chemoreception stems from the study of insect chemical sensors. The chemoreceptors of insects are modified epithelial cells located on or near the outside of the body. They are usually associated with easily recognizable specializations of the exoskeleton. It is their easy identification and accessibility that has made them favorable subjects for experimental study.

The dendritic ends of the insect primary chemoreceptor cell are generally highly branched, with each branch redividing into several parallel fingerlike processes, or *microvilli*. It is assumed that the increased surface area of the distal tip facilitates amplification by summation of several minute excitations of different microvilli (Fig. 12-16).

The dendrites of insect chemoreceptors resemble modified cilia in that near the cell body are nine pairs of peripheral filaments, as in cilia or flagella. A central pair of filaments, however, as found in typical motile cilia, is not present.

Mammalian chemoreceptor cells also have hairlike processes at their distal tips. Olfactory receptor cells of the rabbit have 6 to 12 cilia at the ends of the dendrites. A peculiarity is that the axons leading from dozens of these receptor cells are gathered together in bundles and covered by a single sheath. Perhaps some functional integration is indicated by this arrangement, but what it could be remains unknown.

The very small size and inaccessibility of most chemoreceptor cells of

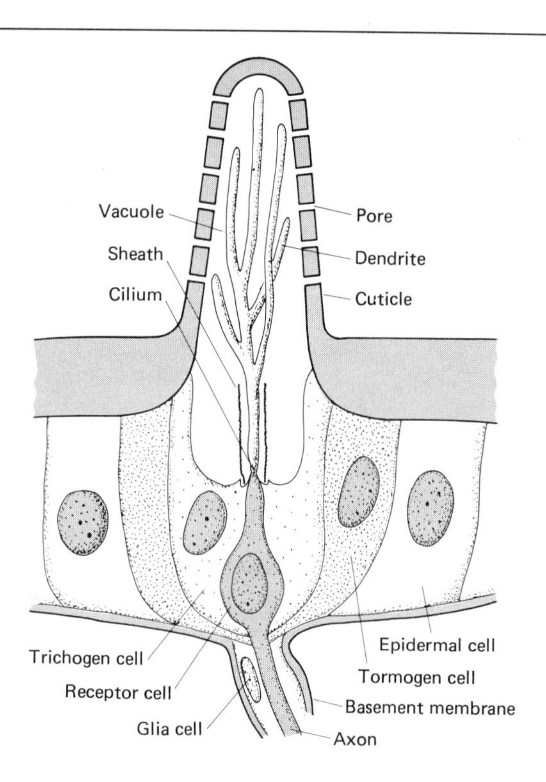

FIGURE 12-16
Diagram of a peg-shaped insect olfactory sensillum in longitudinal section. Note the pores which permit contact between the outside medium and the branched dendritic ending of the receptor cell. (*After D. Schneider and R. A. Steinbrecht, 1968, Symp. Zool. Soc. London, No. 23, p. 281.*)

vertebrates makes penetration by microelectrodes very difficult or impossible. Their minute size also means that their electrical potentials are so small that they are often below the amplitude of the electronic "noise" levels of an experimenter's amplifying system. However, what are said to be pure generator potentials from taste receptors in rats and hamsters have been detected.

Generator potentials also have been recorded from taste receptors of insects. However, insect chemoreceptors exhibit paradoxical features, such as depolarizing when stimulated by some substances and hyperpolarizing in response to others. Moreover, a chemical may depolarize one receptor cell and at the same time hyperpolarize a different cell.

It is commonly more convenient to record from chemoreceptors in such a way that one actually observes compounded generator, spike, or other potentials from several receptor cells. For example, a monophasic negative potential can be recorded from the nasal mucosa when odorous air is passed over it. The assumption can be made that this is the sum of potentials originating somewhere in the dendrites of olfactory cells. However, such an "electroolfactogram" cannot be regarded as a generator potential in the strict sense. Similarly, the overall slow responses of whole antennae of insects to odors are believed to be composed of the generator potentials elicited simultaneously in many olfactory sense cells. Such "electroantennograms," though very useful in studying chemoreception, are probably compounded potentials of more than one type, and may even include responses of mechanoreceptors as well (Fig. 12-17).

The exact properties of the stimulus required for chemoreception are not known. There have been many ingenious ideas about the manner by which stimulating substances interact with chemoreceptor surfaces, but none has as yet been experimentally verified. A technique called *ultravi-*

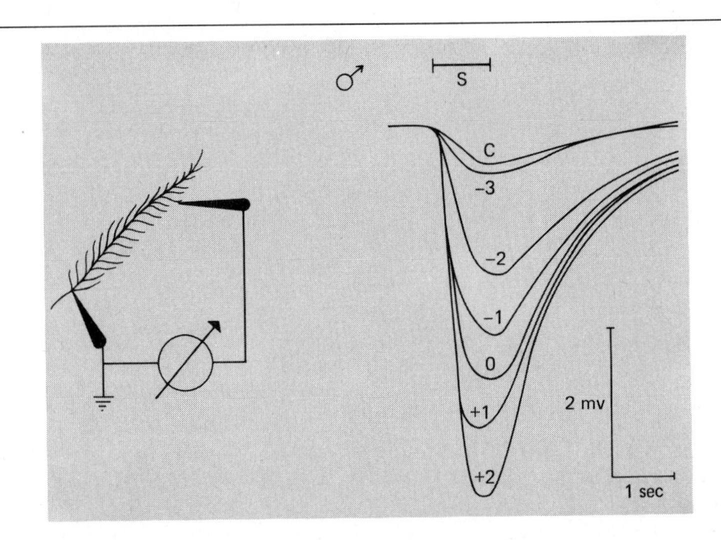

FIGURE 12-17
Recordings of slow, electrical odor responses from the isolated antenna of the male silk moth (*Bombyx*). Recording scheme is shown at the left. The antenna is stimulated by air puffs passing over the odor source, which is filter paper contaminated with the sex attractant (bombykol) of the female moth. The amount of bombykol on the odor sources is indicated for each recording (log values in μg). *C*, control stimulus with air alone; *S*, stimulus duration. (*After D. Schneider*, 1969, *Science*, *163*, p. 1034.)

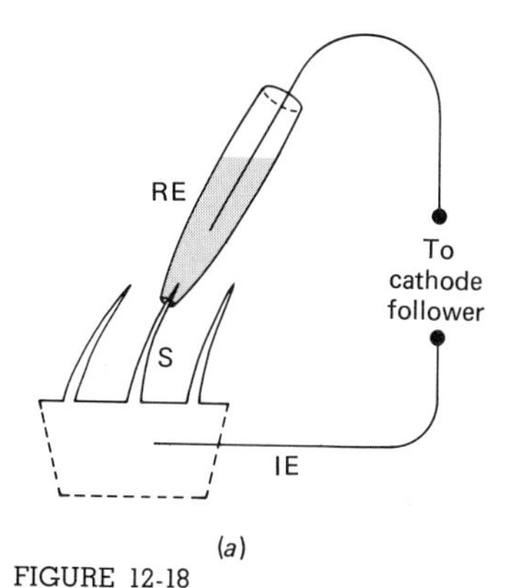

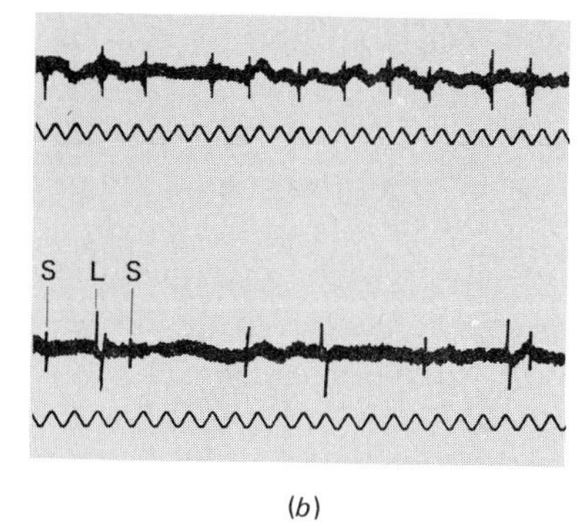

FIGURE 12-18
(a) Technique for recording action potentials from an individual chemoreceptor cell contained in a chemosensory hair (sensillum) (S) on the labellum of a fly. The tube containing the test solution serves as both stimulator and recording electrode (*RE*). It is connected to a cathode followed by a silver–silver chloride wire inserted into the large end of the tube. An indifferent electrode (*IE*) of silver–silver chloride is inserted into the head of the fly. (b) Spike potentials from labellar chemoreceptor cells of a fly. *Upper record:* large spikes from an electrolyte receptor stimulated with 0.5 *M* NaCl. *Lower record:* small spikes (S) and large spikes (L) from the same hair during stimulation with a mixture of 0.1 *M* sucrose and 0.1 *M* NaCl. Time base in both records in 100 cycles/second. (*After E. S. Hodgson et al.*, 1955, *Science, 122*, p. 418.)

olet difference spectroscopy has been used with an "olfactory preparation" (a supernatant fluid from a centrifugation of suspended scrapings from rabbit olfactory epithelium) to observe absorbance changes in the presence of odoriferous substances. The specific absorbance changes observed are interpreted to mean that specific components in the olfactory preparation interact with the stimulating chemical to form a complex. The changes of configuration associated with the formation of this complex might serve as the trigger to initiate the changes in membrane permeability which lead to potential changes. However, in normal receptors stimulated during fractions of a second, the generator potential is very ephemeral (usually disappearing within 20 to 30 mseconds), which suggests that receptor membranes may rapidly rid themselves of or inactivate stimulating molecules. It is a well-established fact that during constant stimulation, chemoreceptors adapt within a few seconds after stimulus application.

Action potentials from individual chemoreceptor cells were first recorded from insects. This was accomplished by a clever technique wherein the stimulating chemical was added to the solution filling the recording electrode (Fig. 12-18).

The best-known taste receptors are the labellar chemoreceptors of flies. Prominent hairs on the labellum (one of the pair of lobes at the end of the proboscis of the mouthparts) contain dendritic endings of two, three, or four bipolar neurons, the tips of which are sensitive to chemical stimulation. One of the labellar receptors responds to sugars, another to cations, and a third to water. A fourth chemoreceptor cell, present in some labellar hairs, has been shown to be an anion receptor. No specific receptors for amino acids have been discovered in flies, but such are found in crustacea.

The frequency of impulses in chemoreceptors is a logarithmic relation to the stimulus concentration over much of the dynamic range, as in other kinds of receptors. Also, as with mechanoreceptors, it is common to find in all types of chemoreceptors low-frequency, spontaneous activity occurring without any application of stimuli by the experimenter. One therefore looks for changes in patterns of impulses when studying such receptors, rather than just for presence or absence of impulses. There are, indeed, cases where the frequency of discharge decreases in the presence of certain stimulants.

An intriguing speculation is that the utility of spontaneous activity in receptors is to permit them to have *fluctuating thresholds* for stimulation. If this is true, then a single, spontaneously active receptor could "scan" a variety of intensities among stimuli!

12-5 THERMORECEPTORS

Although all animals are known to be sensitive to temperature, for most of them study of the mechanisms involved in thermal detection has not begun. This is in large part because of the technical difficulties of the electrophysiological approach. Consequently, behavioral responses have been exploited to a great extent in the study of heat and cold perception.

Arthropods that are ectoparasites of warm-blooded hosts have been shown to have a temperature sensitivity considerably greater than that of nonparasitic arthropods. The warmth of the host is the most important —perhaps the only—sign stimulus releasing and directing the host-finding behavior of the parasite. The sheep tick, *Ixodes ricinus*, can detect air-temperature differences of about 0.5°C with receptors located on its front legs. It will turn toward an object 12°C warmer than the surrounding air at a distance of about 10 mm. *Rhodnius prolixus*, a blood-sucking bug, responds at a range of 4 cm to an object warmed 15°C above air temperature. Its temperature sensors are in the antennae. Mosquitoes orient to a warm odorless surface as readily as to human skin.

Circumstantial evidence has localized the site of reception in a few insects to certain hair sensilla on the cuticle, but definitive proof that they are the thermoreceptors remains to be assembled. Electrophysiological experiments so far have not provided unequivocal evidence for specific temperature receptors in invertebrates.

Among vertebrates, thermal signals appear to be detected by simple neuron processes, there being no identifiable "receptor" structure. For this reason, electrophysiological evidence of the type obtainable with certain mechanoreceptors and chemoreceptors is virtually impossible to procure. Impaling the minute free endings of thermosensitive neurons with microelectrodes is simply not possible.

Fish are extremely temperature-sensitive, detecting changes of water temperature as small as 0.03°C, as demonstrated by conditioned behavior (Fig. 4-9). Skates (Hypotremata) have a system of mucus-filled canals, called the *ampullae of Lorenzini*, which have been considered to be a component of a thermoreceptive system. The frequency of action potentials in the nerves from the ampullae is increased by very small decreases in temperature. The ampullae are often stated to be the only thermoreceptors whose electrophysiology is known. However, Murray has shown that ampullae are also sensitive to weak tactile stimuli, to small voltage gradients in the water, and to slight changes in salinity. Response to such a broad spectrum of modalities suggests that not too much significance should be attached to electrophysiological demonstrations of thermal sensitivity in this, or any other, sense organ. In fact, the anatomical arrangement of the ampullae—deeply embedded in the body and therefore insulated from temperature changes at the surface—makes a thermal function least likely.

Electrical recordings of thermal responses also have been made from cutaneous fibers in rays (Hypotremata), but it is probable that these units were mechanically sensitive as well.

Pit vipers (Crotalinae), such as rattlesnakes, copperheads, and cottonmouths, have a pair of sense organs on the head, below and in front of the eyes, called *facial pits*, which are the most thermosensitive receptors known. They serve as directional distance-receptors of infrared radiation, making it possible for the snake to strike at warm prey in complete darkness.

A pit (in a snake of 1 m length) consists of a cavity about 5 mm deep and about the same dimension wide at the surface. The pit narrows toward the bottom, where there is a thin, free membrane backed by an air space and invaded by numerous nerve endings. The pits have a "field of view" outward and forward through the opening, with a narrow zone of overlap in front of the head, allowing focusing with both pits (Fig. 12-19).

The pit organ is stimulated by radiant heat, of wavelengths between 1.5 and 15 nm. The sensory neurons continually fire nonrhythmic dis-

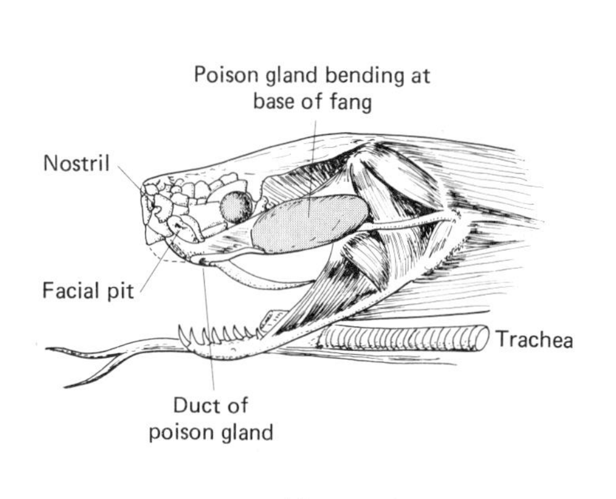

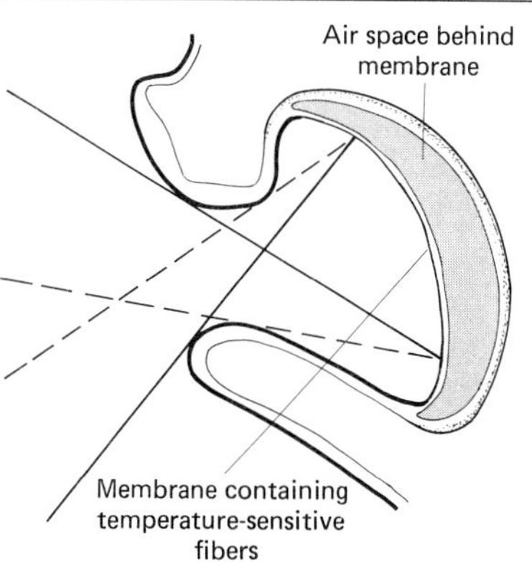

FIGURE 12-19
(*a*) Head of a pit viper after removal of the skin. (*b*) Diagram of longitudinal section of facial pit. The field of view of the pit is indicated by the projecting lines.

charges. Warming increases and cooling decreases the frequency of action potentials. Adaptation is complete in a few seconds. The threshold change of temperature is about 0.002°C, which would be adequate to detect a mouse 10°C warmer than the background that appears in the field of view for less than 0.5 second and about 15 cm away, or a rat at a distance of about 40 cm!

Reptiles regulate their body temperature to a considerable extent behaviorally—i.e., basking in the sun or seeking shade—and by taking advantage of the thermal properties of water, such as buffering and evaporative cooling. Responses of this sort to temperature changes indicate good temperature-detecting receptors, but how they operate and where they are located remains almost entirely undetermined.

Birds and mammals can maintain their body temperatures with less than 1°C variation. (However, this is not the case in all species at all times, as detailed in Sec. 4-6.) The receptors involved are of two kinds: skin and hypothalamic. Among skin receptors, those of tongues have been studied extensively. The tongues of birds have been shown to contain cold and warm receptors. Again, reception is effected by unspecialized-appearing neuron fibers. Among mammals, the best-studied temperature-sensitive fibers are the so-called "cool" and "warm" fibers of the cat's tongue. These fibers have spontaneous, resting discharges at a frequency dependent upon the temperature. On either side of a maximum the rate falls with change of temperature. In addition to this standard response, these fibers show such "paradoxical" features as secondary maxima (Fig. 12-20).

The principle of specific nerve energies becomes extremely strained when attempts are made to apply it to temperature reception. There are cold fibers with a phasic response to a fall in temperature that are also sensitive to mechanical stimuli. There are others that are primarily touch-sensitive but also show a tonic hot response, and, in addition, some of these fibers respond to a fall in temperature (Fig. 12-21).

Regulation of body temperature in mammals depends upon certain "centers" in the hypothalamus. If discrete lesions (made by passing a

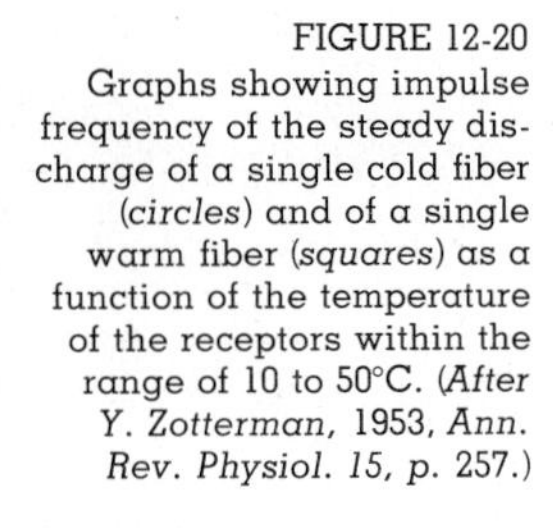

FIGURE 12-20
Graphs showing impulse frequency of the steady discharge of a single cold fiber (*circles*) and of a single warm fiber (*squares*) as a function of the temperature of the receptors within the range of 10 to 50°C. (*After* Y. *Zotterman*, 1953, *Ann. Rev. Physiol.* *15*, p. 257.)

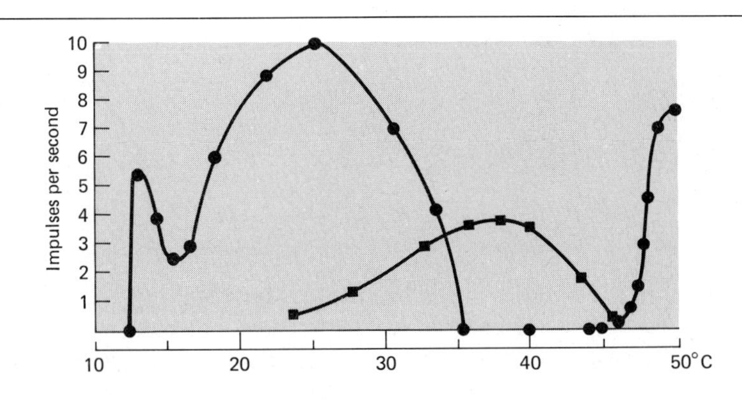

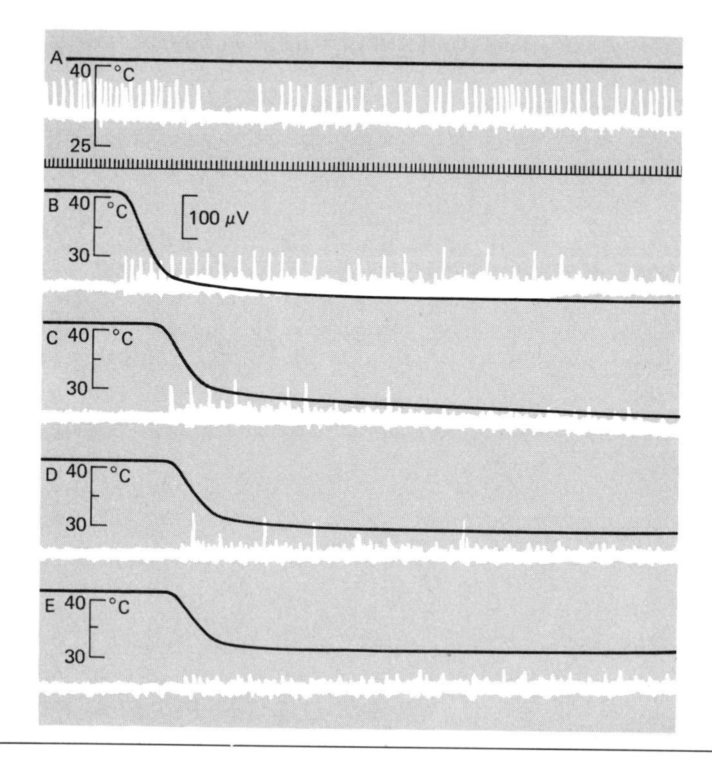

FIGURE 12-21
Oscilloscope recordings from a thin strand of the cat lingual (tongue) nerve obtained on applying mechanical and thermal stimuli to the tongue. The *thin line* shows the temperature of the surface of the tongue. *A*, pressure; *B*, cooling from 41 to 22°C; *C*, cooling from 41 to 26°C; *D*, cooling from 41 to 29°C; *E*, cooling from 41 to 32°C. Time marks, 50 cycles/second. (*After H. Hensel and Y. Zotterman,* 1951, *J. Physiol. 115,* p. 18.)

direct current through the tip of an appropriately placed wire electrode) are made in the rostral hypothalamus, an animal is incapable of regulating its temperature in a warm environment, and if lesions are placed in a specific region of the caudal hypothalamus the animal loses ability to maintain normal body temperature in either a warm or a cold environment. These results, and other related pieces of evidence, indicate that the hypothalamus is the location of a "thermostat" which relies on information collected by cutaneous receptors as well as thermosensors in the hypothalamus itself. Studies with microelectrodes have detected changes of impulse discharge when the temperature of the hypothalamus is locally altered.

Some receptors respond only to stimuli which are sufficiently intense to cause tissue damage, and the energy form (burning, chilling, crushing, cutting) which produces the damage is irrelevant. What is detected is tissue damage, not the nature of the damaging stimulus. Such high-threshold receptors are called **nociceptors,** and their output gives the sensation of pain. Hot and cold nociceptors occur in the skin, and hot nociceptors are found in the tongue. Additionally, many kinds of nonsensory nerve fibers can be stimulated by temperature changes. Mechanoreceptors are often found to be thermally responsive.

Various hypotheses have been proposed to explain the transducing mechanism of temperature reception—i.e., the way in which absolute levels or changes of temperature initiate or alter the existing frequency of nerve impulses. A simple explanation is that temperature change leads to chemical effects—potentiating or inhibiting some dynamic process, for example. There is no evidence to support such a view, however. Another explanation is that radiant heat can initiate processes similar to the photochemical effects involved in reception of visible light; changes of temperature of the receptor itself are not important. Still another hypothesis favors the view that the adequate thermal stimulus is the existence of a gradient of temperature along the free nerve endings, the difference of temperature acting on the membrane potential in a manner analogous to current induction in a thermocouple. Other investigators take the position that the activity of thermoreceptors can be explained without involving special mechanisms; they argue that impulse frequency will tend to be higher at higher temperatures because the afterpotential is shortened and the membrane is more readily depolarized. Obviously, we simply cannot satisfactorily explain thermal sensitivity in terms of present information about neuron function.

12-6 PHOTORECEPTORS

The most specialized and obvious light receptors are eyes, but the general body surface of many animals is also photoreceptive. This diffuse photic sense, called the **dermal light sense,** is not mediated by eyes or eyespots, and the light does not act directly on the effector. In many cases there is no structural sign of cutaneous photoreception; in others, certain cell types are suspected of being photoreceptors, but proof that they are light-receptive is nonexistent or inconclusive. In any case, there are no evident accessory structures associated with these photoreceptors, as there are with eyes. For this reason they are sometimes called *extraoptic photoreceptors.*

Dermal photosensitivity often accompanies a significant concentration of nerve elements within, or just beneath, exposed translucent skin. The existence of photosensitive nerves has been shown by behavioral and electrophysiological means in a variety of animals. Such nerves possess light-absorbing pigments but usually exhibit little other structural specialization to even suggest receptor properties. The identity of the photoreceptive pigments involved is not yet established, although carotenoid-protein and hemeprotein pigments are implicated in a few cases.

Dermal photoreception implies that the responsive elements are at the surface, as many of them are, but deeper-lying extraoptic sites also are common. Certain parts of the nerve cord of arthropods and the diencephalon of vertebrate brains (fish, amphibians, reptiles, birds) are photoreceptive.

The dermal light sense is most widely and characteristically involved in *defensive shadow reactions*—withdrawal reactions of coelenterates and tubiculous worms, for instance, when suddenly shaded. In addition, the dermatoptic sense may inform its possessor of day and night or seasonal rhythms. Its function could be more subtle than we can now deduce. It may be quite wrong to regard extraoptic photoreception as primitive and an evolutionary vestige. The nervous organization behind it, which shows some sophistication, may be importantly involved in information processing. For example, lizards and sparrows that are blinded by eye removal will still respond to long light periods and be stimulated to reproductive activity. The brain—presumably the diencephalon—contains the effective photoreceptors. This suggests that the eyes may not be at all involved in the dramatic changes in state of the gonads of seasonally breeding reptiles and birds.

An extraoptic light sense is not effective in point localization of a stimulus. However, one of the types of activity initiated through this sense is a **kinesis,** a nondirected velocity change of an organism in response to change in intensity of environmental stimulation factors. Most of the photo-stimulated kineses are *orthokineses,* meaning the speed or frequency of locomotion is altered. A few are *klinokineses,* in which there is a change in frequency of turning. (See Sec. 2-1 for additional information on kineses.)

When receptor cells are so arranged that light will strike them only from certain angles, movement with respect to it can become directional. A directed orientation of animals which move in their environment is called a **taxis.** Phototaxis is a feature of animals with eyes and eyespots. (See Sec. 2-1.)

Photoreceptor-containing cups and vessels with varying degrees of aperture constriction are found in each of the major phyletic groups. At the base of the cup is a *retina,* a pigmented layer in which light absorption occurs. Many such structures have light-collecting lenses.

A rather different structural pattern is found in many of the annelids, mollusks, and arthropods. Their eyes, either a simple or a compound type or both, commonly protrude from the body surface (Fig. 12-22).

Epiphysial (or *pineal*) *structures,* arising from the dorsal surface of the brain's diencephalon, are photoreceptive in lampreys (Cyclostomata), bony fishes (Osteichthyes), a few families of frogs (Anura, Amphibia), and lizards (Sauria, Reptilia). In frogs there is an intracranial pineal body that contains photoreceptive structures which resemble rod cells (see below). An extension of the pineal organ, the small, eyelike *frontal organ* on the dorsal surface of the head, also contains rods. A homologous structure, the parietal eye, found on the top of the head in some lizards, is highly developed, with a cornea, a lens, and a retina containing rodlike photoreceptive cells. There is unequivocal proof of the photoreceptive capabilities of the parietal eyes and frontal organs, and of the intracranial pineal components as well in most of the groups mentioned. Pineal photoreceptors show typical light and dark adaptations and have spectral sensitivity maxima very much like conventional lateral

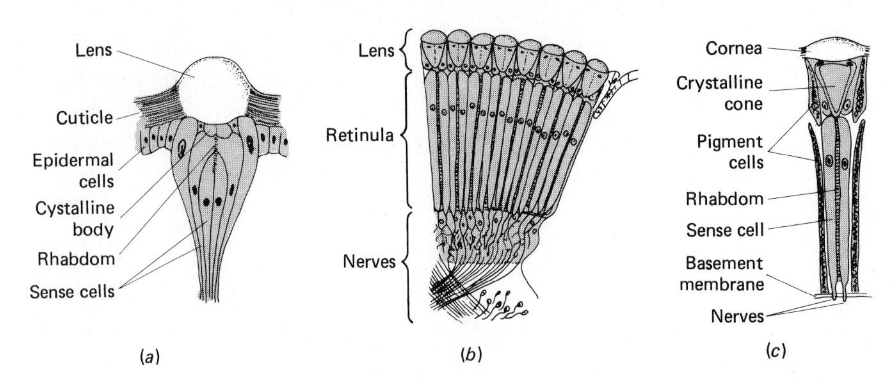

FIGURE 12-22
(a) The simple eye of a caterpillar in vertical section. The cuticle of the body forms the *lens*. The *sense cells* (*retinula*) secrete a vertical rod, the *rhabdom*. (b) An insect compound eye in vertical section. The eye is composed of a number of bundles of cells; each bundle is called an *ommatidium*. (c) Typical structure of an ommatidium. The thickened and transparent cuticle forms the biconvex lens (*cornea*). The *crystalline cone* is a refractive body. The *sense cells*, four to eight in number, together form the *retinula*, which secretes in the axis of the ommatidium a refractive body, the *rhabdom*. Each retinular cell passes at its base into a nerve fiber which pierces the basement membrane of the eye and enters the optic ganglia. Around each ommatidium, separating it from its neighbors, are *pigment cells* (or *iris cells*).

eyes. Just what use is made of pineal-collected photoic information remains an intriguing problem.

The vertebrate eye is the most sophisticated receptor known. This no doubt stems in large part from the fact that the retina, the sensory portion of the eye, is embryologically part of the brain and is, consequently, sufficiently complex to permit extensive analysis of photic information before relaying it to higher centers of the brain. Accessory structures of the vertebrate eye contribute to the focusing of images on the retina, control of brightness, and blocking of extraneous light (Fig. 12-23).

A curious feature of the vertebrate retina is that the photoreceptors lie in the deepest layer and point toward the dark-pigmented choroid layer. Thus, light must pass through neurons, connective tissue, and blood vessels before reaching the sensory cells, except at the fovea, the region of most acute vision of the retina, where there are no blood vessels and no overlying retinal neurons (Fig. 12-24).

The outermost neutral elements, the ganglion cells, whose axons

FIGURE 12-23
Human eye in horizontal section.

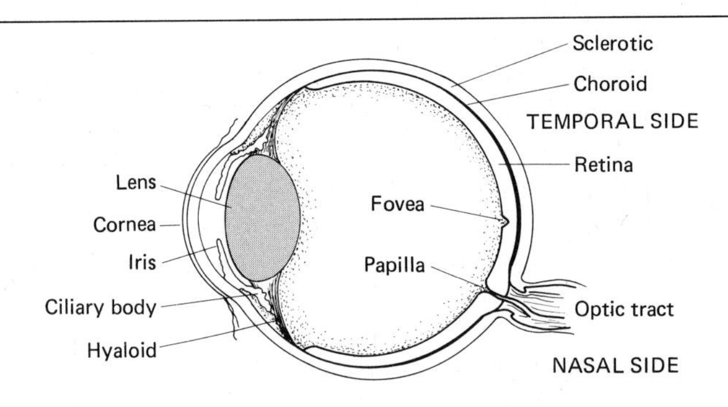

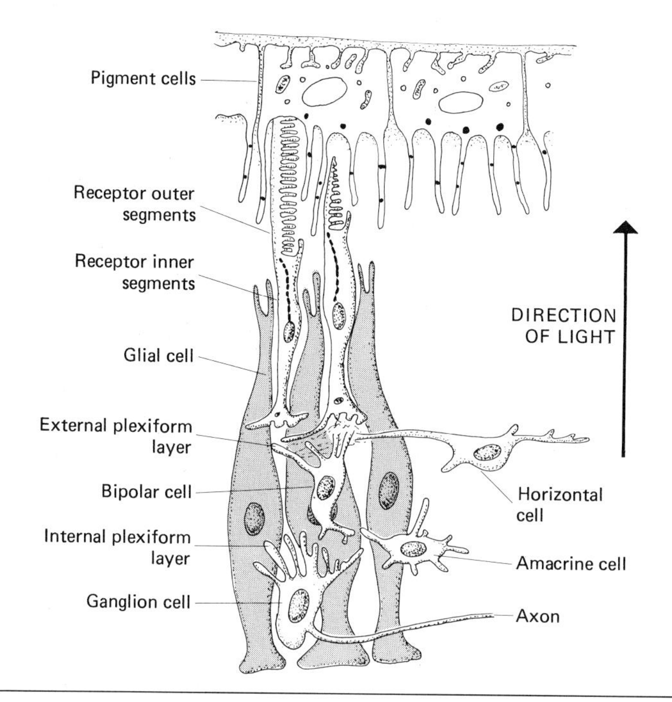

FIGURE 12-24
A section through the vertebrate retina. Synaptic contacts interconnecting the photoreceptors and ganglion cells occur in the external and internal plexiform layers. Lateral connections are formed by horizontal and amacrine cells. Note the close association between the receptor outer segments and pigment cells. Only the ganglion cells possess spike-conducting axons. (*After A. I. Cohen,* 1963, *Biol. Rev., 38, p.* 428.)

make up the greater part of the optic nerve, are not nearly so numerous as the receptor cells. Therefore, there must be considerable convergence in the neural pathway. Convergence is quite great at the periphery of the retina and decreases to a minimum at the fovea. The greater the convergence, the less the visual acuity. Thus, acuity is highest at the central focus, and images become more blurred peripherally.

Two kinds of photoreceptor cells are found in the human retina: *rods,* which respond to dim light (*scotopic* vision), and *cones,* which mediate color vision in brighter light (*photopic* vision). Both cell types have an inner segment, where the major part of the metabolic machinery resides, and an outer segment containing numerous lamellae in which photic excitement is presumed to occur. A neck connecting the two segments has filaments arranged in the conventional pattern of a cilium, suggesting again the common origin of differentiated sensory units from ciliated cells (Fig. 12-25).

In the outer segments of rods and cones there is a high concentration of pigment capable of absorbing photons. The reception process in all eyes and eyespots is built around pigments. All the visual pigments known consist of *retinal,* an aldehyde (oxidation product) or *retinol,* bound as the chromophore to a protein molecule called *opsin*[1] (Fig. 12-26).

The retinals have four carbon-to-carbon double bonds, each of which might potentially exist in either *cis* or *trans* configuration. The forms

[1] Retinal was formerly called *retinine,* and retinol is vitamin A.

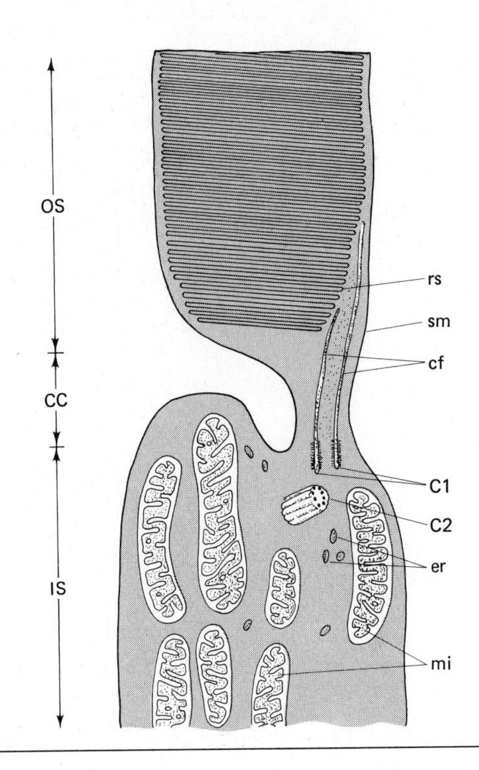

FIGURE 12-25
Ultrastructure of a mammalian rod cell. Diagramatic longitudinal section in region of the connecting cilium (CC) between photoreceptive outer segment (OS) and inner segment (IS). Connecting cilium contains ciliary filaments (*cf*) and two centrioles (C1 and C2). Lamellar organization of outer segment is indicated by rod sacs (*rs*), surface membrane (*sm*), endoplasmic reticulum (*er*), and mitochondria (*mi*).

that are of significance in vision—i.e., the ones with the right shape—are all-*trans* and 11-*cis*. A *cis* linkage always means a bend in the chain. In the case of the 11-*cis* configuration there is a large overlap between the methyl group on C-13 and the hydrogen on C-10, which makes the molecule not only bent but twisted and, therefore, somewhat unstable. However, if kept in the dark, 11-*cis* retinal is fairly stable, but exposure to light changes it to all-*trans*. This is apparently what takes place in the photochemical reaction in the eye. *Rhodopsin*, the pigment of mammalian rod cells and the best-studied of the visual pigments, when exposed in vitro to

FIGURE 12-26
Components of the visual pigments of vertebrates and their usual absorption maxima. The two retinals, 1 and 2, combine with two families of opsins, those of rods and those of cones, to form the four major pigments of vertebrate vision.

						Usual λ_{max} ($m\mu$)
$Retinol_1$	$\underset{NAD-H}{\overset{NAD^+}{\rightleftharpoons}}$	$Retinal_1$	+ Rod Opsin	$\overset{light}{\rightleftharpoons}$	Rhodopsin	500
			+ Cone Opsin	$\overset{light}{\rightleftharpoons}$	Iodopsin	562
(Alcohol dehydrogenase)						
$Retinol_2$	$\underset{NAD-H}{\overset{NAD^+}{\rightleftharpoons}}$	$Retinal_2$	+ Rod Opsin	$\overset{light}{\rightleftharpoons}$	Porphyropsin	522
			+ Cone Opsin	$\overset{light}{\rightleftharpoons}$	Cyanopsin	620

light, breaks down into all-*trans* retinal and opsin. This is a three-step reaction, the first step of which (an isomerization) is a photochemical reaction. The all-*trans* form must be reisomerized to 11-*cis* configuration to regenerate visual pigment. A *cis-trans* isomerization cycle is an intrinsic part of all visual systems known (Fig. 12-27).

The reactions just described can be initiated by the absorption of a single photon by a pigment complex. The result is that the prosthetic group, retinal, progressively assumes the all-*trans* configuration, a shape which will not fit with opsin. It is the freeing of the retinal moiety which in some obscure way leads to an excitation in a retinal neuron that eventually registers in the brain as the sensation of light.

There is good evidence that rhodopsin is indeed an essential molecule in scotopic vision. For example, vitamin A deficiency produces abnormalities in dark vision. Furthermore, the absorption spectrum of rhodopsin corresponds to the spectral sensitivity for scotopic vision of the eye from which it is taken (Fig. 12-28).

It is what transpires between the chemistry of pigment isomerization

FIGURE 12-27
(*a*) Structures of the all-*trans*, 9-*cis* isomers of retinol and retinal. (*b*) Diagram of the rhodopsin system, showing the isomerization cycle. The bleaching of rhodopsin by light ends in a mixture of opsin and all-*trans* retinal. The latter must be isomerized to 11-*cis* before it can regenerate rhodopsin. While that is happening, much of it is reduced to all-*trans* vitamin A, most of which in turn is esterified. These products must be isomerized to or exchanged for their 11-*cis* configuration before engaging in the resynthesis of visual pigments.

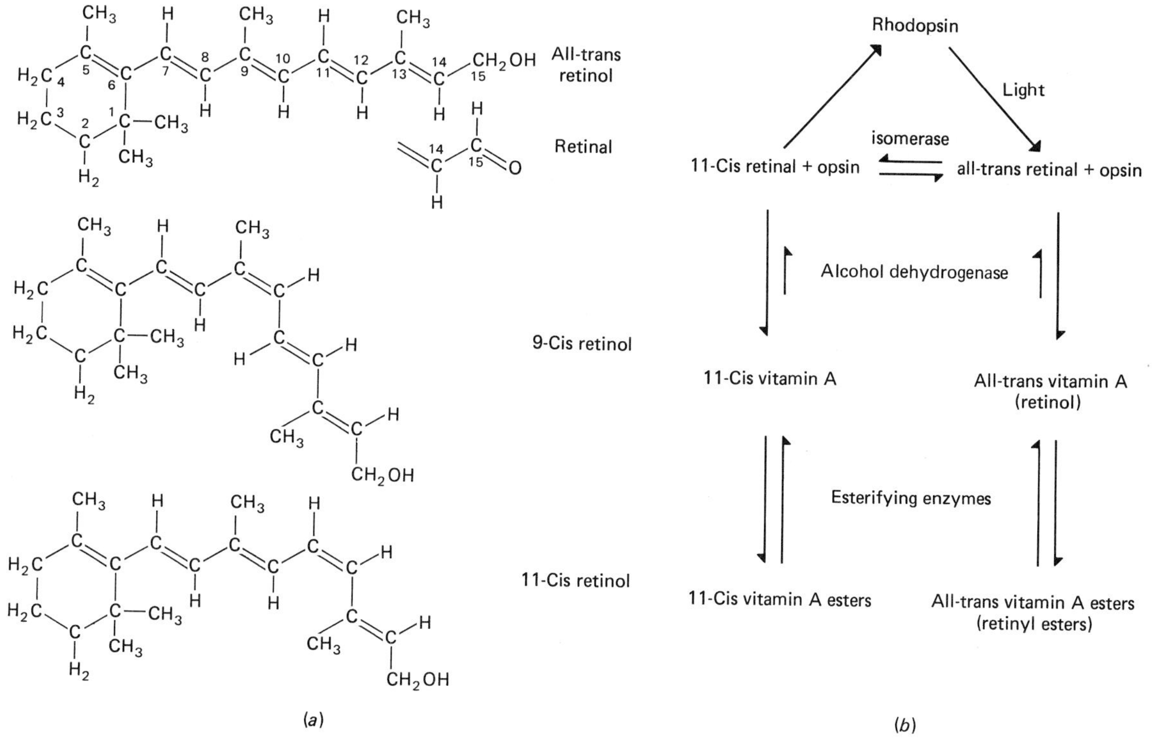

All-trans retinol
Retinal
9-Cis retinol
11-Cis retinol
(*a*)

Rhodopsin
Light
isomerase
11-Cis retinal + opsin
all-trans retinal + opsin
Alcohol dehydrogenase
11-Cis vitamin A
All-trans vitamin A (retinol)
Esterifying enzymes
11-Cis vitamin A esters
All-trans vitamin A esters (retinyl esters)
(*b*)

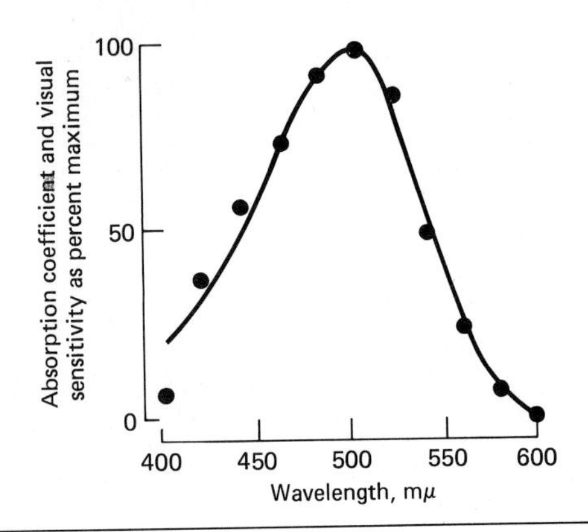

FIGURE 12-28
Coincidence of rhodopsin absorption spectrum (*solid curve*) with human dark-adapted visual sensitivity curve (*circles*).

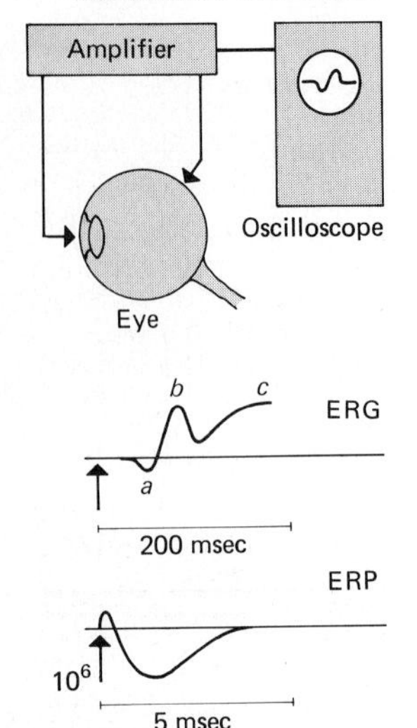

FIGURE 12-29
Diagram to show the essential hookup for observing an electroretinogram (ERG) or early receptor potential (ERP) in a dark-adapted vertebrate eye. Each of these responses is a fluctuation of potential, involving cornea-positive (upward) and cornea-negative (downward) components. The a wave is especially marked in cone-rich retinae. The c wave probably originates in the cells of the pigmented layer. Unlike the ERG, the ERP has no measurable latency. For both types of response to be comparable in amplitude, the flash (↑), that stimulates the ERP must be of the order of 1 million times more intense. Note difference in time bases.

and neural excitation that remains largely unknown in the process of vision. If one places an active electrode on the cornea of an eye and an indifferent electrode somewhere else on the eye or on the body and connects these through an amplifier and oscilloscope, there is seen a characteristic biphasic electrical response, called the *electroretinogram* (ERG), produced by a flash of light. However, following the flash there is a brief period, lasting 1.5 mseconds or more, before the ERG appears. It has been presumed that this latent interval indicates that time is needed to "bleach" the visual pigment to the critical stage and for buildup of "amplification." It has been discovered recently that there is another electrical event with no measureable latency that precedes the ERG—the *early receptor potential* (ERP) (Fig. 12-29).

It seems rather certain that the ERP has its origin in the direct action of photons on visual pigments. An ERP can be obtained in retinas cooled to −3°C or heated to 48°C. Even retinas fixed in glutaraldehyde show ERP-like responses. All that is required is intact—and oriented—rhodopsin. A *photochemical* reaction is certainly indicated. Presumably charge displacements occur in the normal visual process, generating changes of potential between the front and back of the eye. However, there is as yet no evidence that the ERP generated by rhodopsin itself is part of the mechanism of visual excitation.

Rods and cones are too small and too deep in the retina to be easily impaled with recording microelectrodes. However, because photoreception is fundamentally similar wherever found, one can employ more suitable tissues than vertebrate retinae for electrophysiological analysis. The lateral compound eyes of the xiphosuran arachnoid, *Limulus polyphemus*, commonly called the "horseshoe" crab, offer many advantages for study of vision. The sensory structures of its eyes are clusters of receptor cells arranged radially around the dendritic process of a bipolar neuron, called the "eccentric" cell (Fig. 12-30).

Limulus eyes can be bisected, allowing probing with microelectrodes. Illumination of an ommatidium of such a preparation initiates a

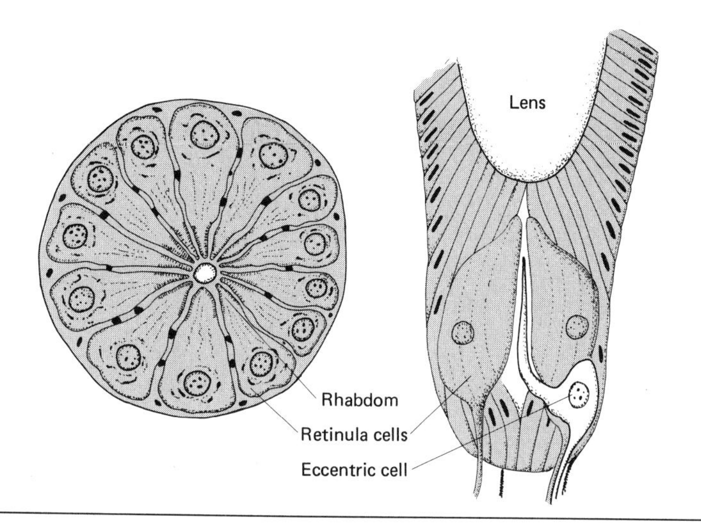

FIGURE 12-30
Ommatidium from the *Limulus* lateral eye in cross section and vertical section. (*After J. F. Case,* 1966, *Sensory Mechanisms, The Macmillan Company, New York,* p. 96; *From R. Demoll,* 1914, *Zool. Jahrb. Abt. Anat., 38,* p. 443.)

receptor potential in the eccentric cell, the magnitude of which is proportional to stimulus intensity (Fig. 12-31).

It is thought that the primary photoreceptor site is the rhabdom surrounding the distal process of the eccentric cell. Light-stimulated permeability changes perhaps cause a current flow between retinula cells and the eccentric cell, which is sufficient to establish a receptor potential. A similar process may occur in the outer segment of the vertebrate rod cell, but it is difficult to make an analogy between the multicellular functional unit of the arthropod eye and the rod cell. [However, it is interesting that the c wave of the ERG of the vertebrate probably originates in the pigment eye layer of the retina, because it disappears when that layer is removed (Fig. 12-29). Thus, pigment cells may play an active role in the visual process and be more than merely a light shield. Metabolically the pigment cells are quite significant, moving vitamin A in and out of rod distal segments during dark and light adaptation.]

12-7 ELECTRORECEPTORS

For some fish the most important sense may be an electric one. Three freshwater families, the Mormyridae and Gymnarchidae of Africa and the Gymnotidae of South America, have many species with weak electric organs which, in conjunction with electroreceptors, apparently are used to discern conductivity features of objects in their environment or possibly for communication between members of the species. The organs, derived from muscle tissue, are the source of low-voltage electrical pulses (Fig. 13-16).

There is considerable diversity in the form of the discharges of electric organs. Some gymnotids, and perhaps all mormyrids, discharge low (about 0 to 5 per second), irregular frequencies, rising to 30 to 100 per sec-

FIGURE 12-31
Response of *Limulus* ommatidium to light. The upper trace is from a retinula cell, and the middle one from an eccentric cell. The lower trace signals the duration of illumination. (*After M. E. Behrens and V. J. Wulff,* 1965, *J. Gen. Physiol., 48,* p. 1088.)

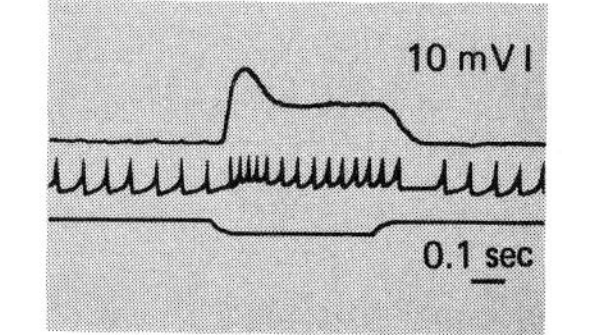

ond when excited. Other gymnotids and the only known gymnarchid (*Gymnarchus*) discharge at high (300 to more than 1000 per second), regular, species-characteristic, and nearly invariant frequencies. Several gymnotids are intermediate in pulse form.

Circumstantial evidence strongly suggests that the receptors involved in electric field detection are on and around the head. Here the skin is perforated in numerous places. The pores lead into tubes often filled with a jellylike substance or a loose aggregation of cells. At the base of each tube is a small, round capsule containing a group of cells. These might be the electric sense organs. Between the cells of the capsule base and the brain run the thickest of the cranial nerves. The brain centers into which these nerves run are quite large and complex; in some of the mormyrids they completely cover the rest of the brain.

As mentioned previously, the ampullae of Lorenzini found in skates are sensitive to low-level electrical stimulation. However, it remains to be proved that these structures are truly electric sense organs.

All the mormyrids, gymnotids (except the electric eel), and *Gymnarchus* continually emit electric pulses. They also all have a rigid swimming posture, propelling themselves with fin movements only. This may be an essential for the operation of the electric sense; to detect distortion of the electric field it may be necessary to keep the electrode system more or less constantly aligned.

In an undisturbed state *Gymnarchus* discharges at the rate of about 300 pulses per second. An object entering the field of perception will change the amplitude of the pulses; this is the effective stimulus for the sense organ. A decrease in the strength of the stimulus causes a lower frequency of impulses in the sensory nerve. On the other hand, an increase in stimulus intensity results in an increase in impulse frequency up to a certain limit.

It seems unlikely in *Gymnarchus* and several gymnotids that each individual discharge is communicated to the sense organ as a discrete stimulus. In fact, in experimental tests the fish respond equally to either high-frequency pulses of short duration or low-frequency pulses of longer duration at the same voltage. Reduction in voltage can be compensated either by an increase of frequency or duration of pulse and, conversely, an increase in voltage can be compensated by a decrease in frequency or duration. Thus, what the fish senses apparently is a product of voltage × duration × frequency—an averaging out of information. The pulse nature of the discharge then carries no information. This may not be true in other electric fishes, especially those in which the discharge frequency changes with the state of excitation of the fish.

Gymnarchus can respond to a continuous direct-current electric stimulus of about 0.15 μV/cm. According to Lissmann's calculations, this means that an individual sense organ should be able to convey information about a current change as small as 0.003 $\mu\mu$A. Extended over the integration time of 25 mseconds, this current corresponds to a movement of only about 1000 univalent ions!

Electrophysiological studies have revealed a remarkable variety of electroreceptor units. For example, one kind spontaneously fires rhythmic pulses, another responds on a 1:1 basis with the electric organ, while still another can fire more than once per electric organ discharge. Some units adapt rapidly and probably detect motion; others adapt slowly and probably sense stationary objects. Such diversity suggests a sophisticated differentiation of function. A social communication role is suggested by observations of changes of frequency when electric fishes approach one another.

It is obvious that not all organisms perceive the world as we humans do. Energy modalities that are useful to us are not universally appreciated; on the other hand, stimuli sensed by many other organisms are completely undetected by us. Thus, the bee, the horseshoe crab, the mudpuppy, the electric fish, and we do not exist in the same world, sensorially speaking.

12-8 CODING IN NEURONS

Our brains know not the real world, but rather an abstraction of certain emphasized aspects of it mediated through sensory receptors and neural circuitry and slightly delayed in time. Following receptor transformation, several different kinds of information of varying precision and detail can be conveyed in neuronal systems:

1. *Quality:* This is signaled by being conveyed along certain transmission routes (the principle of labeled lines).

2. *Locus and form:* The spatial position of a point stimulus, and the spatial extent and form of more complex ones.

3. *Intensity:* A precise estimate of the intensity of a stimulus can be made from the transformed input signal, especially when it can be compared directly with another.

4. *Temporal patterns:* The duration and frequency, as well as spatial translation, of stimuli.

As far as is now known, all the information transmitted between receptors and brains (or equivalent neural "centers") is dependent on nerve impulses—the individual items of the neural code. Because they are brief, discrete, all-or-none electrical events, impulses occupy only small durations of time compared to the intervals of time between them. However, rather than being a digital communication device, a nervous system appears to be in fact a pulse-coded, analog system, as we have seen. But knowing this does not tell us what is the language or code of neurons. That is, we have yet to discover which features of the train of impulses represent the coded information for the myriad discriminations, perceptions, and commands that nervous systems transmit.

An important hypothesis, based on comparison of detected signals and psychophysical observations, is that from the level of receptor transduction and first-order neural coding to the brain, the activity of the transmitting neuron is linear; i.e., any nonlinearity between a brain and the external world is in the initial stage at which impinging stimuli are transduced into nerve impulses. The central nervous system follows in linear fidelity whatever image of the world is reflected to it over primary afferent fibers.

The coding discussed here is mainly in the periphery of central nervous systems. Here the encoding appears to be relatively simple to decipher, using current concepts and methodologies. Central decoding and recoding are yet to be analyzed in any detail, and therein lie the great problems of understanding brain function.

READINGS

Carthy, J. D., and G. E. Newell (eds.), 1968, *Invertebrate Receptors,* Symposia of the Zoological Society of London, No. 23, Academic Press, Inc., New York. A collection of articles on research in some of the most active and promising areas of receptor investigation.

Case, J. F., 1966, *Sensory Mechanisms,* The Macmillan Company, New York. A most excellent small book, especially recommended as an introduction to sensory physiology.

Cold Spring Harbor Symposia on Quantitative Biology, vol. 20, 1965, *Sensory Receptors,* Cold Spring Harbor Biological Laboratory, New York. A valuable collection of reports on the functioning of various types of receptors.

Eyzaguirre, C., and S. J. Fidone, 1975, *Physiology of the Nervous System,* 2d ed., Year Book Medical Publishers, Inc., Chicago. Chapters 6 through 11 deal with sensory systems.

Lissmann, H. W., 1961, Ecological studies of gymnotids, in *Bioelectrogenesis: A Comparative Survey of Its Mechanisms with Particular Emphasis on Electric Fishes,* American Elsevier Publishing Company, Inc., New York. The article cited and others in this volume will introduce the reader to the fascinating world of fish electricity.

Murray, R. W., 1962, Temperature receptors, in *Advances in Comparative Physiology and Biochemistry,* edited by O. Lowenstein, Academic Press, Inc., New York. An account of thermoreception in various animals.

IV

BIOENERGETICS: Cellular Uses of Energy

Mechanisms which link together the several parts of animals into functional totalities that allow the strategies for individual survival and species perpetuation have been outlined in Parts II and III. There the conceptual abstractions were at what is called *systems level*—somewhat arbitrarily defined assemblages of organs and tissues which cooperatively support some particular functions. Thus, we speak of the digestive system, the excretory system, the immune system, the nervous system, and many other systems.

A further abstraction leads to conceptualization of functions in parts of organs and tissues—cells, collectively and individually, carrying on vital processes. So now we turn to consideration of the unique machinery of life at the cellular level. At this level one can best perceive the energetic processes which characterize living things. In the following four chapters we will examine the manner in which energy is invested in mechanisms that move molecules and ions against chemical and electrical gradients, do mechanical work, and drive synthesis. It is the collective and overt manifestations of these mechanisms—excitation, movement, growth, differentiation—which commonly are regarded as the distinguishing features of the living state.

ELECTRO-GENESIS 13

In 1949, K. S. Cole[1] wrote:

It is well established not only that the interior of the living cell is very different from the external inanimate environment in composition, structure and electrical potential, but also that these differences are maintained by a barrier at the surface which is necessary for the life of the cell. Although this barrier may not be positively identified under the microscope—as for example a part of the plasma membrane—it is definitely recognized by other characteristics—such as a rather small permeability to water and to many solutes. The structure which constitutes the barrier to the free flow of ions in and out of the cell is most probably the seat of the immediate source of electrical energy and the origin of the principal bioelectric effects. This structure is our primary concern and for convenience we shall refer to it as the membrane of the cell.

Since this was written a few discoveries have been made about the "morphological" membrane, but its "functional" features have been much better discerned. It is the electrical features of this "barrier" which we now in particular examine, describing its manifestations largely in terms borrowed from the disciplines that deal with electricity, since it has characteristics analogous to those of batteries, resistors, and capacitors. For example, all cells examined have a capacitance of about 1 μfarad/cm^2 at their surface. This is a reflection of the fact that much of the membrane is lipid, which is a dielectric.

The plasma membrane is electrically polarized, largely because of a disparity in distribution of certain ion species across it. Therefore, plasma membranes have a **membrane potential,** or more precisely, a **resting membrane potential,** as defined in Chap. 11 (Sec. 11-2).

13-1 IONIC POTENTIALS

From the quantitative distribution of ions outside (in plasma or other bathing medium) and inside the cell, the potential difference across its membrane can be predicted. In red blood cells, for instance, the interior of the cell should be slightly negative—about 10 mV—relative to the external medium.

[1] K. S. Cole, 1949, *Proc. Natl. Acad. Sci.*, **35**, 559.

This prediction is based upon some relatively simple equations which relate electrical potential to concentration differences of ions. Potentials in biological systems generally are thought to arise from either (1) differences in ion concentration or (2) differences in ion mobilities.

The first situation, the *concentration potential*, can be represented as:

$$E_c = \frac{RT}{nF} \ln \frac{C_1}{C_2} \qquad (13\text{-}1)$$

where R = the gas constant (8.314 V/coulomb mol^{-1} deg Kelvin^{-1})
T = the absolute temperature (t + 273°C)
n = the number of elementary charges on the ion
F = the Faraday constant (96,500 coulombs)
C = the molar concentration of the ion

Thus, E_c, the number of volts (V), depends mainly on the relative concentration of corresponding ions.

The above equation is the *Nernst equation*. For 20°C and monovalent ions (n = 1), and following insertion of constants and conversion to log base 10, the equation becomes the useful form:

$$E_c = 0.058 \log \frac{C_1}{C_2} \qquad (13\text{-}2)$$

For 37°C,

$$E_c = 0.061 \log \frac{C_1}{C_2}; \text{ or in mV, } E_c = 61 \log \frac{C_1}{C_2} \qquad (13\text{-}3)$$

The Nernst equation can be written for every ion in a system. For example, the membrane potential, in millivolts, due to relative K^+ concentration (K_i = potassium inside the cell; K_o = potassium outside the cell; n = 1) would be

$$E_k = 58 \log \frac{[K_i]}{[K_o]} \qquad (13\text{-}4)$$

Each ion in biological systems has a characteristic electromotive force (emf) which can be calculated from the Nernst equation. It is the electrical potential across the membrane that would balance the tendency of the ion to diffuse in the direction of its chemical concentration gradient. Table 13-1 shows the Nernst potentials for the most common ions in a neuron.

Table 13-1
Ionic Concentrations and Equilibrium Potentials across the Cell Membrane of a Cat's Motor Neuron

	Outside, mM	Inside, mM	Equilibrium potential, mV
Na^+	150	ca. 15	ca. +60
K^+	5.5	150	−90
Cl^-	125	9	−70

Note: The resting membrane potential is −70 mV, inside negative.

From Table 13-1 it can be seen that intracellular and extracellular concentrations of the prominent cations, Na^+, K^+, and Cl^-, are markedly different. The exact concentrations vary somewhat among cells and species, but the concentration ratios shown here are typical. Also, these are the major ions that contribute the electrochemical features of most cells so far examined. (The smaller influences of bicarbonate, phosphate, and the divalent cations can be ignored for the moment.) How differences in ionic distributions are established and maintained will be explained later in this chapter and in the following chapter.

If biological membranes were equally permeable to all the common ions found in the body fluids, the membrane potential would average out to the Nernst potentials of those ions and none of the ions would be in equilibrium. Membranes typically have a relatively high permeability to K^+ and a smaller permeability to Na^+. Thus, the resting membrane potential of cells tends to be close to the K^+ potential and is rather far from the Na^+ potential.

The *Goldman equation* takes these considerations into account. It is:

$$E = 58 \log \frac{P_K[K_i] + P_{NA}[Na_i] + P_{Cl}[Cl_o]}{P_K[K_o] + P_{Na}[Na_o] + P_{Cl}[Cl_i]} \qquad (13\text{-}5)$$

where P_K, P_{Na}, and P_{Cl} are the permeabilities of K^+, Na^+, and Cl^- ions, respectively. The Cl^- terms are often omitted from the equation because in the steady state chloride does not contribute to the potential, but rather adjusts itself passively to it (as will become apparent later).

Most of the phenomena so far described in this section can be demonstrated with model systems using artificial membranes or some kind of barrier to establish differences in ionic concentrations or to allow mobility differences of ions to be expressed. Such demonstrations have been extensively used and have been crucial in explaining the origin of concentration and potential differences for biologically significant ions. However, imposed on the physiocochemical properties of ions and the barriers to their free diffusion are some unique features of cell membranes. Through passive and active processes, cells establish chemical gradients, including electrochemical gradients, across their membranes.

13-2 MEMBRANE THEORY OF BIOELECTRIC PHENOMENA

Ostwald in 1890 proposed that bioelectric potentials could be caused by a membrane permeable to cations but impermeable to anions. Nernst's equations relating chemical and electrical potentials had been derived about a year earlier. Drawing especially upon these developments, Julius Bernstein formulated in 1902 a permeability theory of excitation. He proposed that the membrane of a resting cell is permeable to potassium only. Potassium concentrated inside the cell tends to diffuse out but is retarded by electrostatic attraction of *organic anions* not free to diffuse, according to his theory. Sodium cannot diffuse in to make up the K^+ deficit.

Thus, K^+ is poised in an electrochemical gradient, and the inside of the cell is more negative than the outside. During excitation, numerous "pores" open up and other ions, Na^+ and Cl^-, for example, rush in and cause the electric field to suddenly decrease, or *depolarize*.

Bernstein's theory successfully explained the existence of the resting potential, its electric sign, and its approximate magnitude. Furthermore, it was demonstrated subsequently that during passage of a conducted impulse the resistance across the neuron membrane does drop—to one-fortieth of its resting state, indicating indeed that ions have increased mobilities.

Not until almost 40 years after its publication did this theory require serious revision. In 1939 Hodgkin and Huxley in England and Cole and Curtis in the United States, using the new technique of transmembrane recording with squid giant axons, found that during the passage of impulses the potential is not simply abolished, as predicted by Bernstein, but, in fact, for a brief moment it is reversed. Other blows to the Bernstein theory came in the next 2 years when Boyle and Conway demonstrated that muscle fibers were as freely permeable to Cl^- as to K^+ and Fenn showed that radioactive Na^+ became distributed across resting neural membranes in the same relative concentration as nonradioactive Na^+. The conclusion, therefore, was that the membrane is permeable to all common ion species. (However, the resting cell membrane is about 75 times more permeable to K^+ than to Na^+.)

In 1941 Dean hypothesized that a special "pump" within the cell might eject Na^+ at the same rate that it entered. Such a mechanism would keep the internal sodium concentration low and allow potassium and chloride to control the resting potential.

In 1955, Hodgkin, studying cephalopod axons, provided experimental proof that the efflux of Na^+ is an *active* transport process. Furthermore, measurements of K^+ flux were inconsistent with the theory of independent ionic migration. They proposed that part of the *inward* movement of K^+ was linked with the active *outward* movement of Na^+. The remainder of the K^+ flux was passive and directly related to the electric field and K^+-concentration gradient (Fig. 13-1).

By use of tracers, Hodgkin and Keynes revealed that the outward movement of Na^+ (against the gradient) in isolated cephalopod axons could be reversibly stopped by application of metabolic inhibitors (dinitrophenol, azide, cyanide). Transport could then be reinitiated by intracellular injection of adenosine triphosphate (ATP) and arginine phosphate. The inward movement of Na^+ across the membrane was little affected by inhibitors. Their conclusion was that the electrical steady state of the resting cell and its large electrochemical potential gradients are maintained by free energy made available from conventional metabolic reactions. In some unknown way sodium ions are "pumped" back out of the cell as fast as they enter through the slightly "leaky" cell membrane. Recently a Na^+, K^+–activated, ouabain-sensitive ATPase has been implicated in the functioning of the pump.

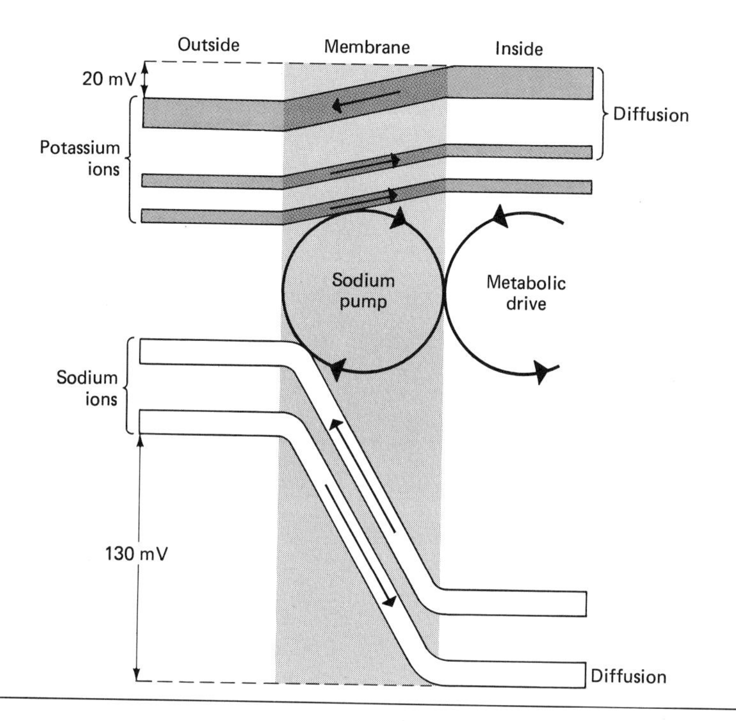

FIGURE 13-1
Active and passive Na^+ and K^+ fluxes of axon membrane creating the steady state or resting potential, as conceived in the sodium-pump theory. Sodium ions are expelled from the interior so that concentration is reduced to about 10 percent that of the exterior fluid. A coupled process drives potassium inward to make a concentration 30 times greater than that outside. Pumping rate must keep up with the passive leakage of these two ion species. Sodium has a higher leakage rate because, being positive, it is under the strong driving force created by the negative interior. There is a net outward leakage of potassium, despite the tendency for it to be retained. (*After B. Katz,* 1961, *Sci. Am., September, p.* 216.)

It is unclear whether the pump acts as a direct source of current by charge separation—i.e., by separating Na^+ from other ions and transferring positive charges outward, thereby causing other diffusible ions (K^+ and Cl^-) to assume an equilibrium distribution across the membrane in accord with the prevailing potential—or whether movement of sodium is an electrically neutral process wherein Na^+ is transferred simultaneously with a negative ion or, alternatively, an outward Na^+ is exchanged for an inward-moving K^+. In the last case the potential would indirectly result from accumulation of K^+ inside the cell and from the differential permeability of the membrane to Na^+ and K^+. Probably the pump is partly neutral and partly electrogenic. In any case, most of the *accumulation* of K^+ in the cell is due to impermeable anions inside the cell.

Support for a specific Na^+-K^+ exchange pump comes from the experiments with metabolic inhibitors, in which it appears that when the influx of K^+ is reduced the efflux of Na^+ also is reduced. Furthermore, active transport of Na^+ is greatly diminished when K^+ is removed from the external medium. Thus, a single, coupled transport of Na^+ outward and K^+ inward is indicated (Fig. 13-1).

Large ionic concentration gradients across the surface membrane result from the action of the pump. According to the ionic theory, the *resting potential* is a secondary consequence of these gradients; it results from the fact that the resting membrane has a much higher conductance for K^+ than for Na^+. In cephalopod axons it has been demonstrated that when extrusion of Na^+ is stopped by inhibitors, the resting potential is not immediately affected. There is a slow decline of potential difference, presumably because of the gradual decline of sodium and potassium gradients as the ions move through the membrane. Therefore, *pumping* of ions makes no significant contribution to the resting potential in these axons. Instead, it is ionic "leakage" which is thought to be of foremost importance in determining the magnitude of the membrane potential and its rapid changes. Each ion has a characteristic electromotive force (*emf*), as was defined earlier.

An illustration of this concept may help to clarify it. In the resting or steady-state condition, there is a potential of about 100 mV across the sarcolemma of vertebrate striated muscle cells. The ionic composition of such cells and their bathing medium can be determined. A set of such data for mammalian muscle cells is shown in Table 13-2.

The more abundant ions in the fluid (interstitial fluid) bathing these cells are Na^+ and Cl^-, whereas K^+ is in low relative concentration. In contrast, the concentration of ions inside the cells (as estimated from chemical analysis of a known weight of tissue and corrected for ions in interstitial spaces) is essentially the reverse, K^+ being high, there being a large amount of organic anions (A^-). The potential difference between these two compartments, as measured by an intracellular electrode, is −90 mV.

The last column on the right in Table 13-2 shows the calculated ionic equilibrium potentials for 37°C. These values can be derived using Eq. 13-3. Thus, as shown, $(K^+)_o = 4\ \mu M/cm^3$ and $(K^+)_i = 155\ \mu M/cm^3$; then $E_K = 61 \log (4/155) = -97$ mV. Therefore, since the membrane potential is −90 mV, little energy would be needed to move K^+ across the membrane, as it appears to be close to equilibrium potential. Chloride, likewise, is

Table 13-2

Approximate Steady-state Ion Concentrations and Potentials in Mammalian Muscle Cells and Interstitial Fluid

Interstitial Fluid		Intracellular Fluid			
	[Ion], $\mu M/cm^3$		[Ion], $\mu M/cm^3$	$\frac{[Ion]_o}{[Ion]_i}$	E_e, mV
Cations		Cations			
Na^+	145	Na^+	12	12.1	66
K^+	4	K^+	155	1/39	−97
H^+	3.8×10^{-5}	H^+	13×10^{-5}	1/3.4	−32
pH	7.43	pH	6.9		
Others	5				
Anions		Anions			
Cl^-	120	Cl^-	4*	30	−90
HCO_3^-	27	HCO_3^-	8	3.4	−32
Others	7	A^-	155		
Potential			−90 mV	1/30*	−90

* Calculated from membrane potential using the Nernst equation for a univalent anion; $t = 37°C$.

Source: After T. C. Ruch et al., 1965, *Neurophysiology*, W. B. Saunders Company, Philadelphia, p. 5.

distributed in a concentration gradient across the membrane so that its electrical potential gradient is in equilibrium with the membrane voltage. Sodium, on the other hand, at a calculated E_{Na} of +66 mV in the steady state, is far away from equilibrium with the membrane voltage.

A very similar set of ion distribution and equilibrium potential data exists for neurons (see again Table 13-1). The resting potential is much closer to E_K than to E_{Na} because the permeability and ionic conductance to Na^+ are much lower than to K^+ (see again Fig. 13-1).

13-3 PROPAGATION OF MEMBRANE EXCITATION

Electrogenic cells are capable of changing their membrane potential in characteristic ways in response to appropriate stimuli. They are commonly long, fibrous cells; such is the usual form of nerve and muscle cells. (Several kinds of electrogenic cells are discussed in the next section.) Such cells are analogous to cables and, indeed, have characteristic "cable constants" of length, time, capacity, and resistivity (Fig. 13-2).

As cables, cells are inferior. Because the conductance of the cell core is millions of times lower than that of a metallic core, and the thin insulating sheath has capacitive and resistive leakages, a small and brief electrical signal will be grossly distorted and attenuated after traveling only a millimeter or two along the cell. However, the continuous cable linkages along the fiber are an essential feature for the passage of excitation, as we shall see. To be a proper conductor of excitation, a cell must compensate for its defective cable properties. The responses of the squid giant axon to brief pulses of electric current have provided much insight into how this is accomplished (Fig. 13-3).

Using the experimental situation shown in Fig. 13-3, either inward or outward rectangular pulses of current can be delivered to an axon through one penetrating electrode and the effect recorded through a second electrode close by (about 50 μm away in the example shown). With the recording electrode in the axon and the stimulating electrode outside it, low amplitude pulses are detected only as momentary deflections—called *capacitive artifacts*—at the beginning and end of the pulse. After

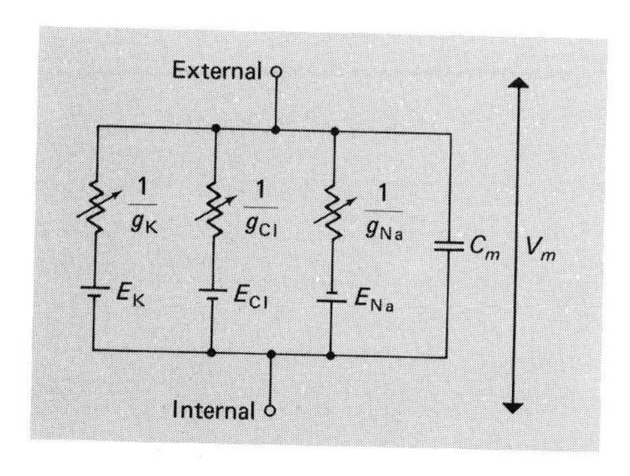

FIGURE 13-2
Equivalent electric-circuit diagram of nerve or muscle membrane. Channels for potassium, chloride, and sodium are shown. Decreases in resistivity (as in depolarization) are effected by increases in ion conductance (hence, the reciprocals of conductance, $1/g_K$, etc., are indicated). C_m = concentration potential; V_m = electrical potential difference across membrane.

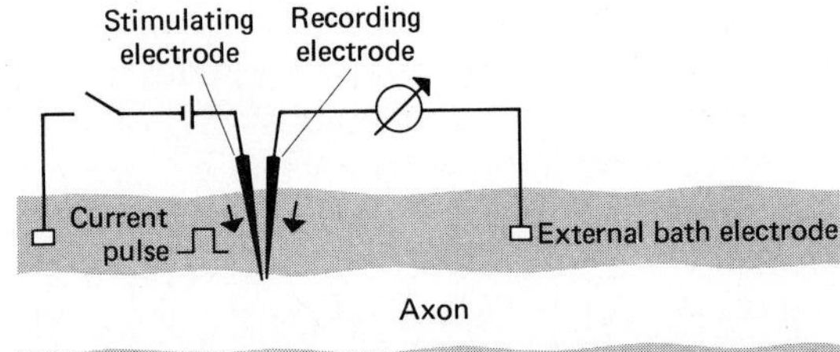

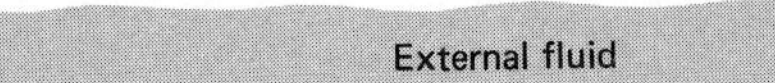

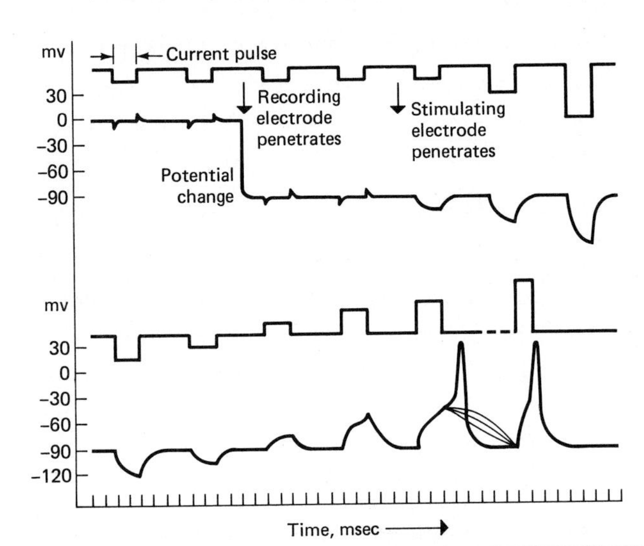

FIGURE 13-3
Recording the membrane potential at the site of stimulation. Square current pulses are applied, producing electrotonic potentials. If the current flows outward (last four pulses, lower set) and is strong enough (last two pulses, lower set) it will give rise to action potentials (spikes). (*After B. Katz, 1966, Nerve Muscle, and Synapse, McGraw-Hill Book Company, New York, p. 34.*)

the stimulating electrode penetrates the cell, inward pulses (downward deflection on the oscilloscope) can induce transient *hyperpolarization* (right part of upper set and left part of lower set of recordings), with a characteristic rounded time course, due to the membrane's having the properties of a "leaky condenser." These responses (which are useful in revealing the cable characteristics of the axon) can be shown to be *subthreshold* and are *not conducted;* 50 percent attenuation occurs every 1 or 2 mm. Such a change in potential is often called an **electrotonic potential.**

If the polarity of the current pulse is reversed so that current flows outward, the same local response is produced, but now it is a *depolarization* (right side of lower set of recordings). As the current strength is increased one approaches a point of electrical instability known as *threshold.* When such a depolarizing pulse ceases, one of two things may happen: the potential either falls back to resting level after a small, variable delay (note that in this case there is a small depolarizing tail on the potential which is called the **local response** or **local potential**), or it may spontaneously continue depolarization to produce a large *action-potential* change known as a **spike** (Figs. 13-3 and 13-4). (Refer again to the classification of potentials in Fig. 11-5).

The spike is a transient, all-or-none event. At its peak the potential across the membrane is reversed, reaching a level of 40 to 50 mV, *inside positive*. The potential then rapidly reverses and returns quickly to resting level. Once the spike potential is elicited, it is propagated along the axon at a constant velocity and without a decrement of signal strength. At its trailing edge there is a *refractory period* of one to a few milliseconds during which another signal cannot be imposed. At the end of this phase the membrane is again ready to be reexcited and conduct another spike (Fig. 13-5).

The rapid changes of membrane potential that occur during propagated excitation in neuron and muscle membranes are believed to be due to changes in *ionic conductance*, not to alterations of the ionic emf, which could result only from the very much slower changes of ion concentrations.

FIGURE 13-4
Analysis of a neuron fiber by stimulating and recording very close together, as shown in the top diagram. In the resting state the membrane potential is about 80 mV negative. Subthreshold stimulating pulses, (*a*) and (*b*), shift the potential momentarily upward. Larger pulses push the potential to its threshold, where it becomes unstable, either subsiding after variable delay (parallel solid and dashed lines) (c) or flaring up into a spike (*d*). The phase of reversed potential sign at the peak of the spike is sometimes called the "overshoot."

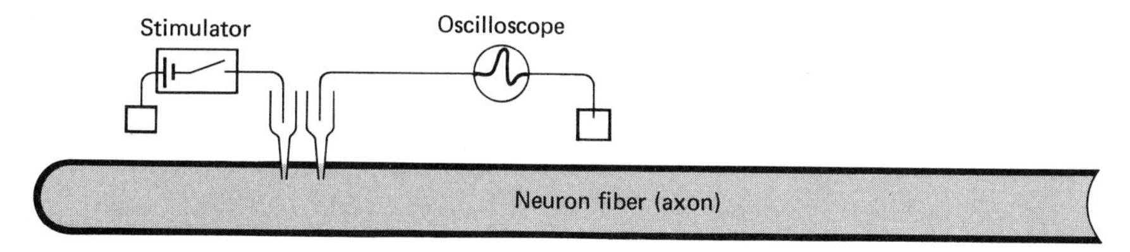

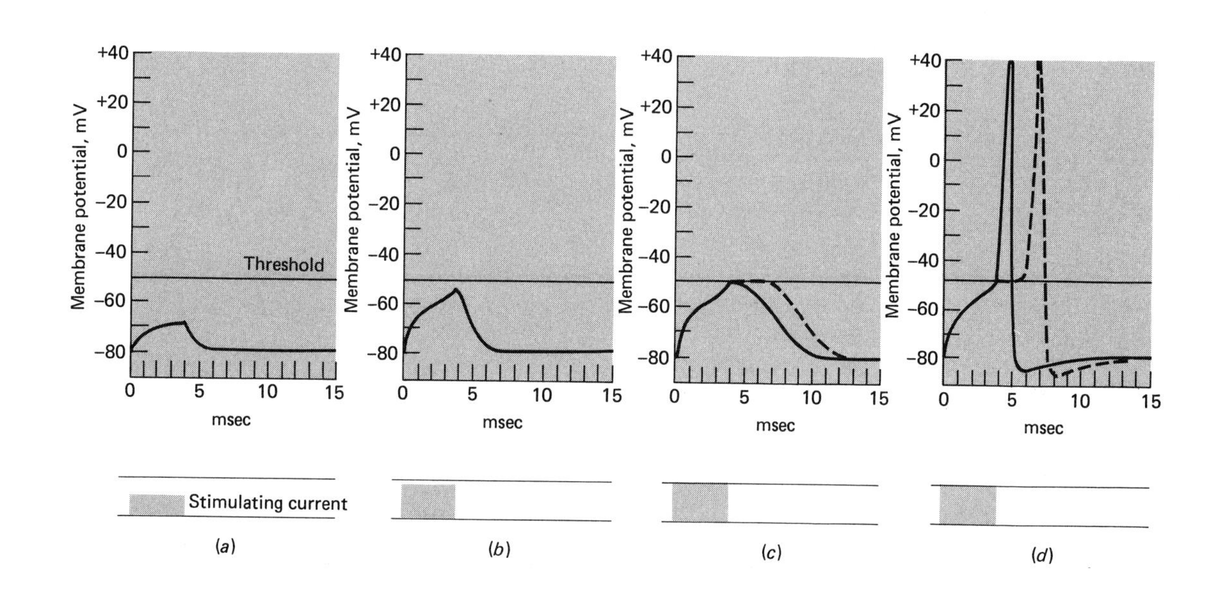

Sodium conductance in the squid axon is a function of membrane potential. At resting potential sodium conductance is very low, but it increases when the membrane is depolarized. It is not known why this occurs, but its significance is evident. As Na^+ enters, carrying positive charge across the membrane, the initial lowering of resting potential difference caused by local depolarizing current will be reinforced, which in turn increases Na^+ permeability, and so on. The membrane potential under these conditions is rapidly displaced toward the sodium potential. Thus, a kind of regenerative autoamplifier is unleashed to produce the explosive depolarization.

Unlike resting potentials, action potentials are directly related to external sodium concentration. Withdrawal of sodium from the bathing medium quickly (but reversibly) abolishes excitability.

If during excitation, Na^+ entry were permitted to go unchecked, theoretically it would continue until the cell interior became sufficiently positive to balance the chemical potential gradient of sodium. This does not occur, according to the ionic theory of action potentials, because the opening of the Na^+ "gates" is only transient (there is "inactivation" of the Na^+-transfer mechanism shortly after its "activation") and an increase in outward conductance of K^+ (down its concentration gradients) begins near the peak of the spike, restoring the membrane potential to its initial state (i.e., inside negative) (Fig. 13-5).

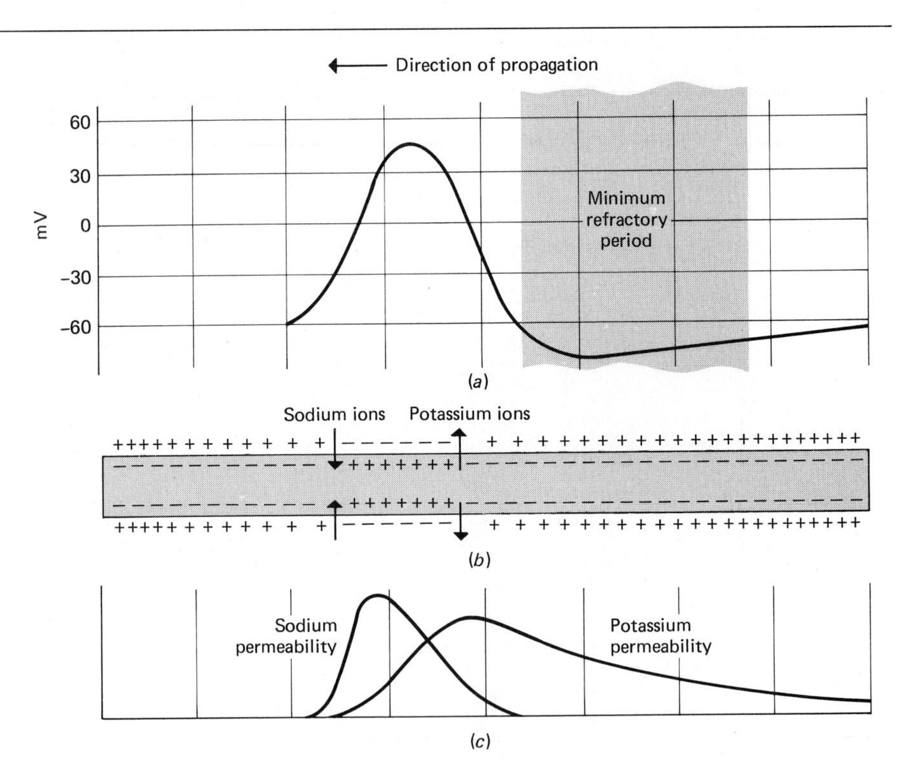

FIGURE 13-5
(a) Potential changes, (b) ion fluxes, and (c) conductance changes during propagation of an action potential.

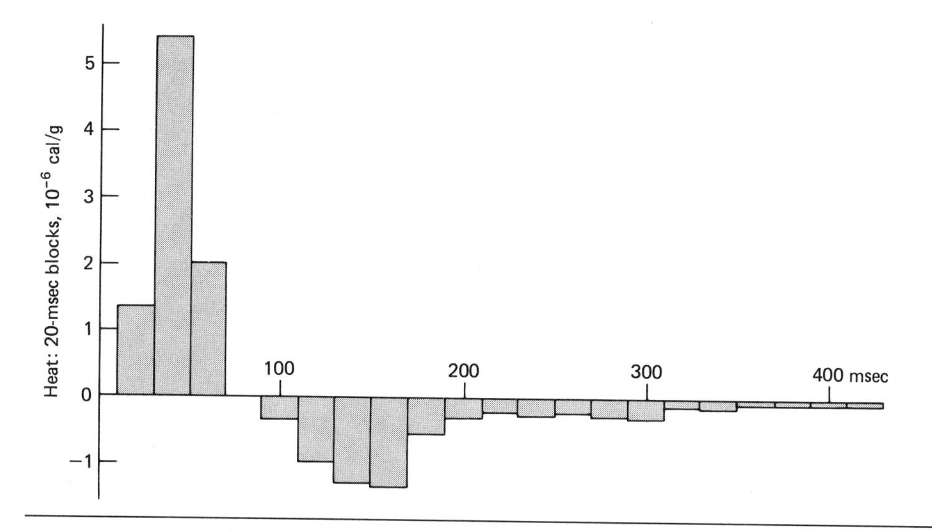

FIGURE 13-6
Heat production and absorption in crab (*Maia*) nerves due to a stimulus. Recorded at 0°C, in 20-msecond blocks (mean of five experiments). The positive heat averaged 8.8×10^{-6} cal/g, the negative heat 6.8×10^{-6} cal/g, and the net heat 2.0×10^{-6} cal/g. (*After B. C. Abbot et al.*, 1958, *Proc. R. Soc. B, 148*, p. 157.)

The refractory period may reflect the time course of the change of the sodium conductance from the inactivated state to the resting state. The membrane at this stage is at a potential near the Nernst potential of potassium—at or even below the resting level, from which it is not readily displaced. The condition for regenerative changes soon returns, and again a spike can be triggered by threshold excitation.

Results from several techniques, such as the use of radioactive ions or microelectrodes that specifically detect Na^+ or K^+, indicate that in the squid giant axon about 3 to 4×10^{-12} mol of sodium per square centimeter of membrane are taken in per impulse and an equal amount of potassium is lost. Calculations indicate that sites totaling much less than 1 percent of the axon surface are involved in the conductance changes. The small changes in ion concentration resulting from the conducting of several impulses are reversed in a few seconds by the Na/K pump, presumably working at a slightly accelerated rate. Neurons are apparently able to conduct thousands of impulses without depleting their accumulated ion reservoirs. However, after periods of intense activity they produce extra heat and require increased amounts of oxygen (Fig. 13-6).

The velocity of an action potential depends in part on the rate at which the membrane capacity ahead of the impulse is discharged beyond the threshold level. Since Na^+ permeability is high in the local region where a spike depolarization has occurred, the membrane potential being near the Na^+ equilibrium potential, and the potential of the adjacent inactive membrane is near the K^+ potential, there is a potential difference between these regions; consequently, current flows from the active region through the intracellular fluid to the inactive region and discharges the membrane capacitor. The return current flows inward through the membrane as a Na^+ current. This *local-circuit* current flow reduces the membrane potential difference in the inactive region, and

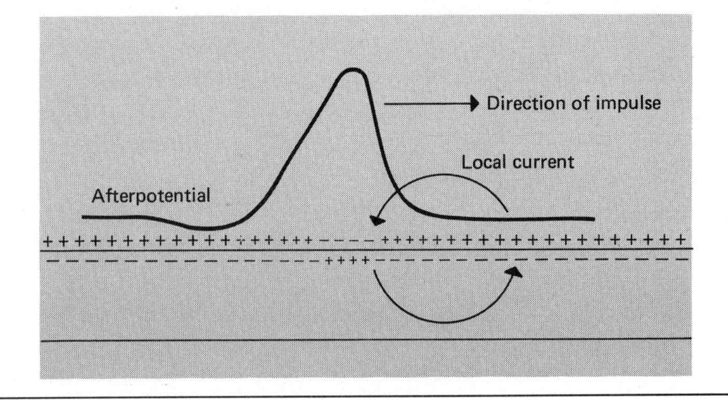

FIGURE 13-7
The propagating action potential. Distribution of the wave of electrical disturbance along a neuron fiber, suggesting the way in which electrotonic current flows from an undisturbed region, and in so doing depolarizes in advance of the main wave.

when threshold is reached, there is a rapid increase in Na^+ conduction, causing the inactive region to approach the Na^+ potential. Thus, it is the cablelike conduction of local-circuit current which links the scattered sites of ion conductance and initiates the ion fluxes that in turn generate more current (Fig. 13-7).

Often the major potential changes of a spike are followed by potential variations of low amplitude and long duration called **afterpotentials.** Sometimes there is a prolonged tail on the falling phase of the spike as the potential slowly returns to resting level; this is a *negative* afterpotential. A prolonged hyperpolarization is called a *positive* afterpotential (Figs. 13-7 and 13-8).

A great amount of what has been said above about neurons is based on the study of a very small number of different preparations. Most of what we know of the excitation of neurons is from experiments using squid giant axons. It is accepted, partly on faith, that the behavior of the isolated squid axon is acting in a "normal" manner, that penetrating microelectrodes actually "see" what is happening ionically, and that electrogenic membranes of other types have properties similar or analogous

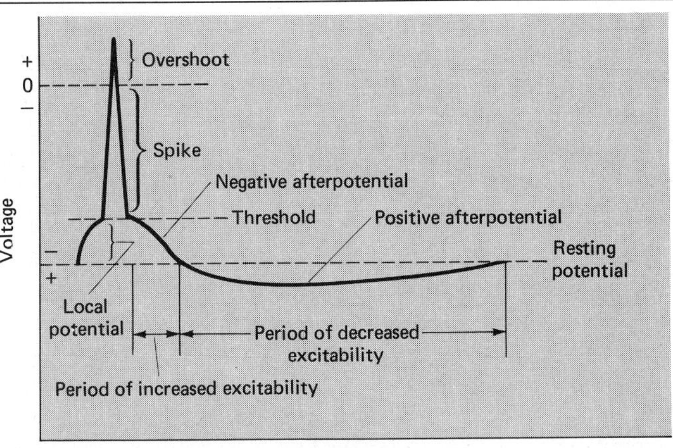

FIGURE 13-8
The excitability cycle of a neuron. A local potential (local process), if greater than threshold, can initiate a spike. The spike may be followed by a negative afterpotential and then by a positive afterpotential. The phases are not proportionately represented; potentials preceding the spike are not adequately shown, and the afterpotentials are very much longer than shown in relation to the duration of the spike.

to those of squid axons. Good evidence can be mustered to support the acceptance of such views and, right or wrong, there is no better preparation than the squid axon presently at hand for a comparably thorough analysis of electrogenic behavior.

The role of Na^+ in the excitation of skeletal muscle appears to be very similar to that in the axon, but some variations in details of electrogenic phenomena are known from other cells. For example, in mammalian heart muscle the secondary increase of K^+ permeability does not take place and the Na^+ conductance does not become completely inactivated, resulting in the appearance of a prolonged plateau of depolarization that provides sufficient time for the systolic contraction of cardiac muscle fibers. During the plateau phase Ca^{2+} enters the fibers. Other kinds of cells behave analogously to the squid axon, but instead of Na^+ entry some may have anion exit (Cl^- in *Nitella*) or entry of a divalent cation (e.g., Ca^{2+}) in regenerative phases. In crustacean muscle, in some mollusk nerve cells, and in frog heart muscle, a substantial inflow of Ca^{2+} causes the rising phase in spike activity. The range of variations on basic ionic themes may be greater than we now suspect, for very few electrogenic cells have been examined in any significant detail.

13-4 ELECTROGENIC CELLS

Electrogenic cells are found in all Metazoa, at least above Porifera, and in a few plants as well. There are several types of electrogenic cells, muscle and nerve cells (neurons) being the best-known ones. A brief survey indicates that electrogenesis is a more widely distributed phenomenon than is commonly appreciated.

Algal Cells

The use of large, multinucleate plant cells (e.g., *Valonia, Nitella, Chara*) for the study of bioelectric potentials was originated by W. J. V. Osterhout in the 1920s. These cells offer several advantages for such study: they either occur singly or are easily separable from adjoining cells, they survive for days or weeks in the laboratory, their large size permits measurement of surface area for expressing resistance and capacitance, and cell sap is in sufficient quantity to be sampled and analyzed.

The cells of the alga, *Chara*, can be as much as 15 cm long and 1.5 mm in diameter. Its large central vacuole contains salts (mainly potassium and sodium chlorides) in a concentration higher than that of freshwater. Consequently, there is a strong osmotic pressure which tends to force against the cell wall the thin layer of cytoplasm which surrounds the vacuole (Fig. 13-9).

Large plant cells have a potential difference across the protoplast, the outer surface being positive to the inner surface in the majority of them. By the use of radioisotopes and chemical analysis of extruded vacuolar sap and cytoplasm, the specific distribution of ions in *Chara*

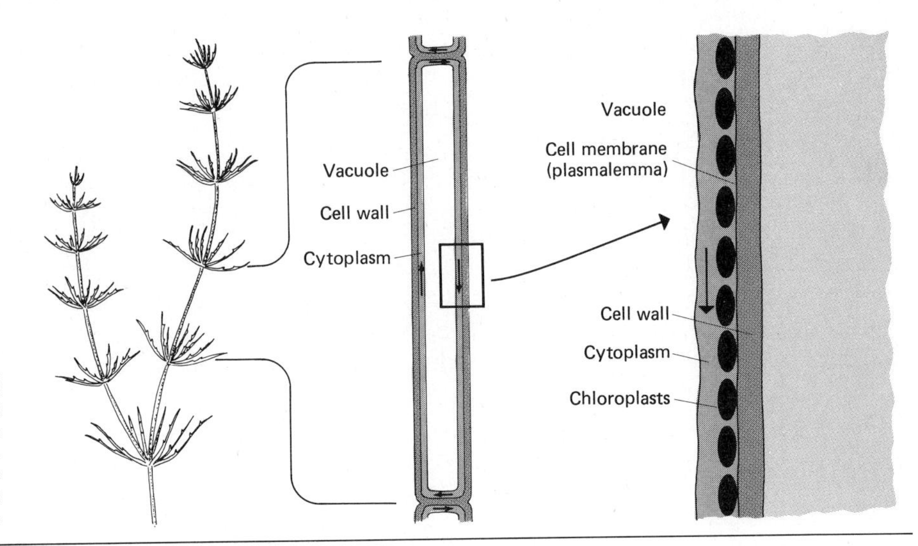

FIGURE 13-9
The large freshwater alga, *Chara australis*, used in experiments on plant electricity. The plant, illustrated at left, has a long stem composed of a succession of single giant cells, one of which is enlarged in the center drawing. A small segment of one side of the cell is in turn enlarged in the drawing on the right.

has been determined. The high salt concentration in the fluids of the cell appears to result from an ion pump which transports Cl^- from the external medium into the vacuole. This tends to make the inside of the cell negative and, as a result, K^+, Na^+, and Ca^{2+}, are pulled in passively. The chloride pump seems to be coupled to a weak outward transport of sodium and calcium. All these ions tend to leak across the membrane opposite to the direction of pumping. Thus, a steady-state condition prevails. Potassium, which is not actively pumped and has the greatest mobility in this cell, is the ion which has most control over the steady-state potential.

When the membrane is "disturbed" by passage of a pulse of appropriate electric current, the inside of the cell becomes momentarily less negative. If the stimulus is of sufficient magnitude (above *threshold*), the stimulated region shows a prolonged response; specifically, membrane potential declines rapidly toward zero and then slowly returns to resting potential. This disturbance may move along the surface of the cell; i.e., it may be *propagated*.

The undisturbed algal cell is relatively impermeable to calcium, but within the stimulated region there appears to be an inward rush of Ca^{2+}. It is the sudden influx of positive ions which lowers the resting potential, resulting in depolarization. This local flow of ions reduces the potential to threshold level in adjacent regions of the membrane, triggering the same electrochemical changes there. Thus, the depolarization moves along the surface as a wave or impulse. Depolarization is quickly followed by a repolarization, effected by exit of K^+ from the cell (Fig. 13-10).

Localized application of certain chemicals or mechanical deformation of the membrane likewise can lead to the propagated disturbance of the resting potential. Regardless of the form of energy input, if it is of threshold level, the response of the membrane spreads along the surface in the same self-sustaining manner. The observed depolarization is not just an electrical event, since it travels at a maximum rate of only a few centimeters per second. In *Chara* it takes about 20 seconds for an action potential to sweep the length of the cell!

The action potential usually travels to the ends of the algal cell; however, it may die out after passing along only part of the cell. If a *Nitella* cell that has been stimulated to conduct an action potential is again stimulated before recovery to the preexisting steady state is attained, there may be a loss of whatever potential has been recovered at that time, and the magnitude of the change will be less than the normal response. For a short time after stimulation no response is possible. *Nitella* cells may be *refractory* to stimulation for a few seconds or even a few hours, as after strong stimulation or rough handling.

Cytoplasmic streaming in plant cells is stopped when an action potential is being conducted. Interestingly, streaming does not stop everywhere at once, but instead ceases progressively along the cell as the depolarization moves along it.

The action potential of large algal cells has no known biological function, in contrast to the action potentials of most other electrogenic cells.

Mimosa

The sensitive plant, *Mimosa pudica*, upon being mechanically disturbed exhibits a movement response that is remarkably rapid for a plant. Its

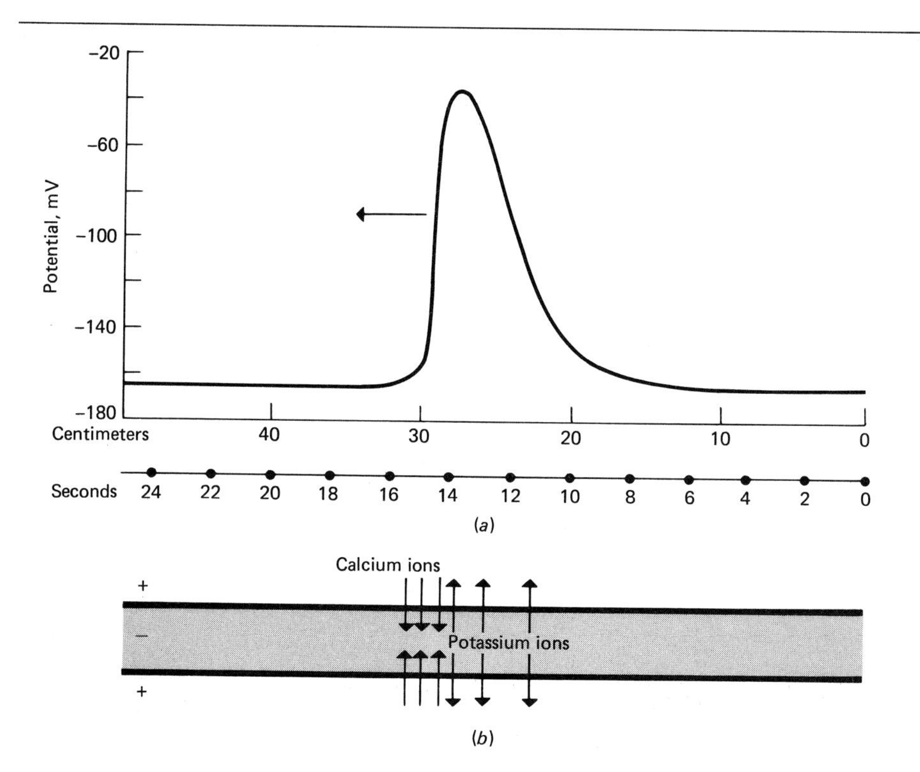

FIGURE 13-10
(*a*) A conducted action potential of a *Chara* cell. Note time and distance scales. The rate of propagation is 1 to 2 cm/second—very slow as compared to that of most other electrogenic cell types. (*b*) Ion fluxes accompanying potential changes; a specific change in membrane permeability to Ca^{2+} is triggered by the action potential, causing depolarization of the membrane; recovery to the polarized condition appears to be caused by the subsequent efflux of K^{+}.

leaves consist of a main stalk or petiole, at the end of which are two to four secondary stalks, each bearing many pairs of leaflets, arranged opposite one another. If a leaflet of *Mimosa* is gently stimulated, that leaflet, and often the one opposite it, will fold up shortly afterward. A more vigorous stimulus is transmitted to a greater distance and causes more leaflets to fold; sufficient stimulation will cause the entire compound leaf to droop (Fig. 13-11).

The mechanism of response was explained by Pfeffer in 1873 as due to the loss of water by certain of the parenchymal cells in swollen regions (called *pulvini*) at the bases of the leaflets, secondary petioles, and the main leaf stem. The leaf and leaflets are normally maintained erect by hydrostatic pressure (or turgor pressure) within the cells of the pulvini. When electrical excitation spreads to a pulvinus, water quickly leaves some of the cells, resulting in a loss of turgor pressure and the characteristic movement of the associated leaf part. Turgor of the cells affected is slowly regained, and the petiole and leaflets eventually assume their former position. The mechanism for effecting the changes in turgor in the cells of pulvini is not known.

FIGURE 13-11
The sensitive plant (*Mimosa pudica*). Touching the leaflets will cause them to fold up. Vigorous mechanical disturbance will start a spreading excitation which will cause the leaflets to fold and the petiole to droop, as in the lower leaf.

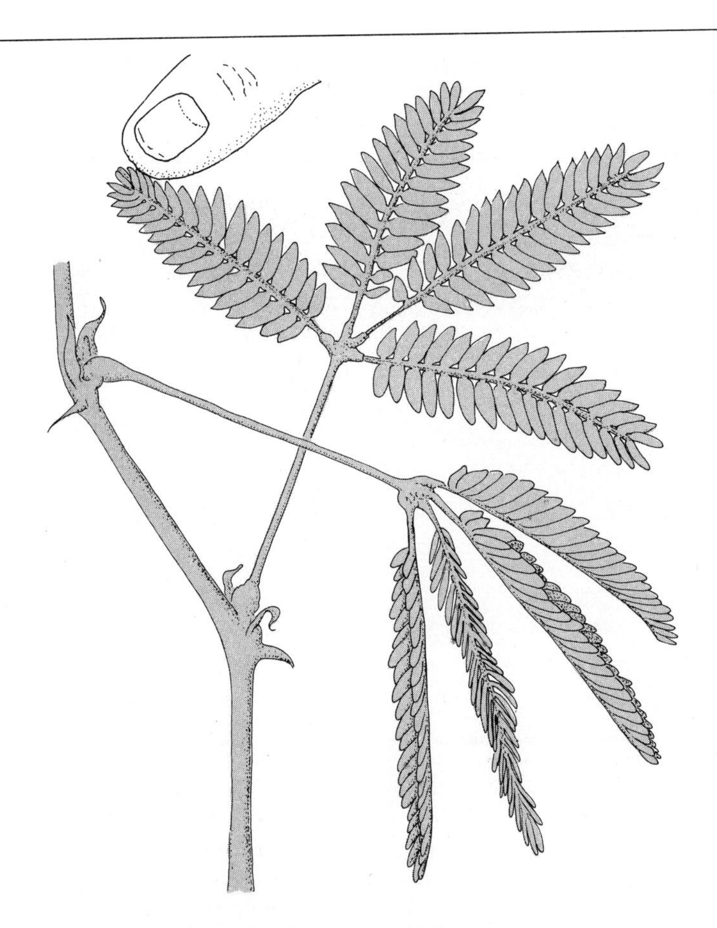

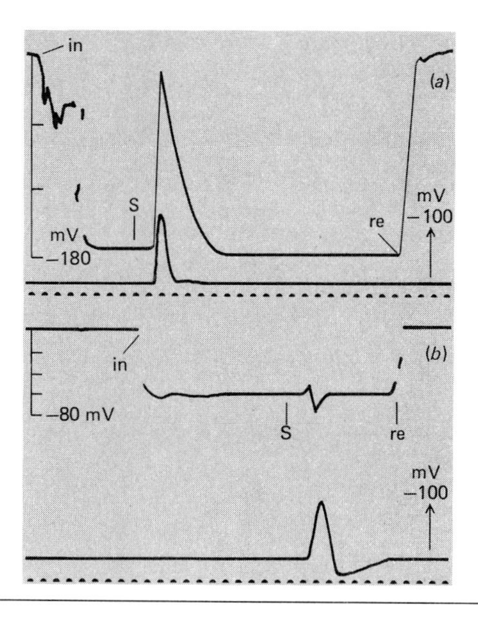

FIGURE 13-12
Membrane potentials (upper trace) in cell of protoxylem (a) and pitch (b) of *Mimosa*. Microelectrodes inserted into cells at *in* and removed from cells at *re*. Petiole is stimulated at *S*, Simultaneously recorded diphasic action potentials (lower traces) in both (a) and (b) are detected by external electrodes. Time marks, 1 second (*From T. Sibaoka*, (1962, *Science, 137*, p. 226.)

Excitation in *Mimosa* appears to be conducted in the petiole by phloem (vascular tissue serving in nutrient conduction and consisting of several kinds of cells) and by protoxylem (elongated parenchymal cells just inside the water-conducting xylem vessels). The membrane of the excitable cells is polarized with the cell interior about 160 mV negative to the exterior. During activity a typical depolarization occurs (Fig. 13-12).

Venus's Flytrap

The Venus's flytrap, *Dionaea muscipula*, is one of a small group of carnivorous plants, capable of trapping insects by leaf movements and digesting them. The plant consists of a rosette of leaves, each of which terminates in a two-lobed blade, hinged down the middle. From the blade margins grow long, stiff, hairlike teeth. On the upper surface of the blade are sensitive hairs and glands. Some of the latter produce digestive enzymes, and others a sugary secretion which attracts insects (Fig. 13-13).

The leaf almost always contracts after two touches of the sensitive hairs. Occasionally contraction will require three stimuli at 2-second intervals. Rarely will the leaf contract on first stimulus. After stimulation of the sensitive hairs, and prior to leaf closure, action potentials similar in contour to those of other electrogenic cells can be recorded (Fig. 13-14).

An adequate stimulus causes the leaf lobes to come quickly together (sometimes in less than a second), the marginal teeth forming a cage and preventing escape of the insect. If an insect is caught, the leaf remains closed for at least 10 days, but if the leaf closes and there is no insect present, it will reopen in 5 to 10 hours.

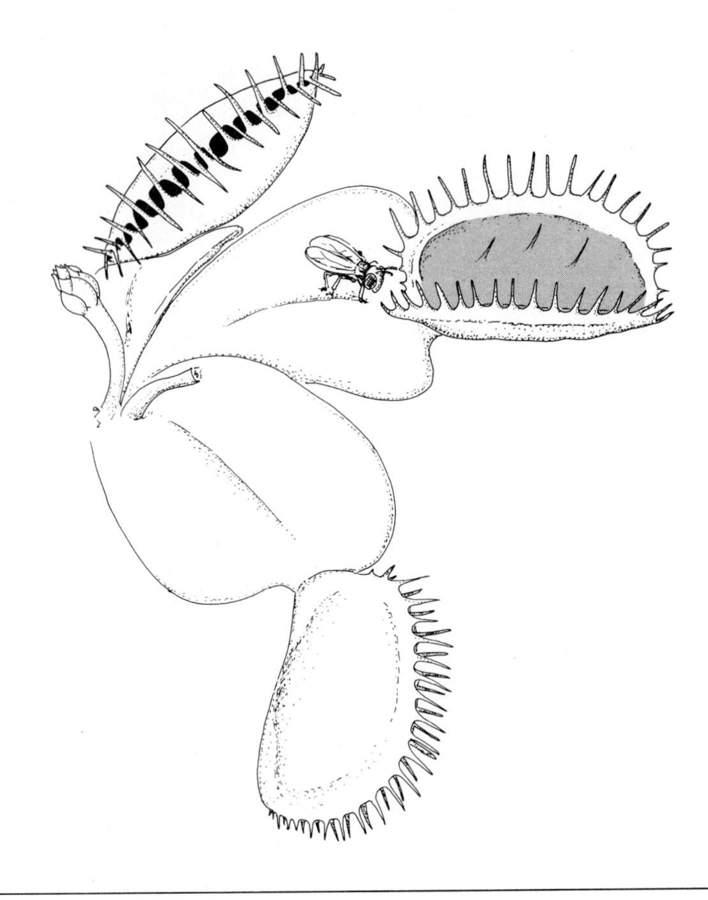

FIGURE 13-13
Venus's flytrap (*Dionaea muscipula*). The leaf lobes at the right are open. An insect crawling onto the trap will trigger closure (sometimes in less than a second) if the sensitive hairs on its surface are touched. The upper one has closed, the marginal teeth preventing escape of a captured insect.

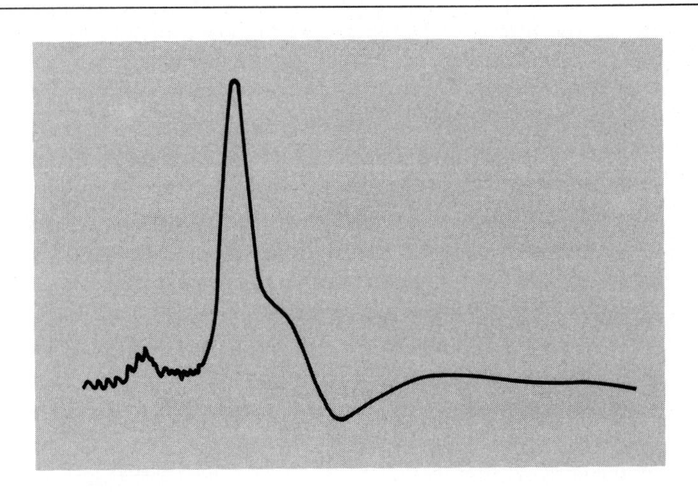

FIGURE 13-14
Oscillograph record of an action potential of Venus's flytrap. Electrodes on underside of trap. Bending of trigger hair used as stimulus. Time scale logarithmic. (*After O. Stuhlman and E. B. Darden,* 1950, *Science,* *111,* p. 491.)

Muscle Cells

Among animals, muscle cells and neurons and the electroplaque cells of the electric organs of fishes are the most obviously electrogenic.

We know that contraction of muscles is associated with membrane depolarization. In vertebrate skeletal muscle the action potential spreads from its point of initiation—normally the neuromuscular junction—at a speed of about 5 m/second (Fig. 13-15).

As will be discussed in Chap. 15, excitation spreads inward from the sarcolemma, presumably on the transverse tubular system, causing the release of Ca^{2+} and the ensuing chemical events of contraction. It is presumed that sequestering of Ca^{2+} by the tubules accompanies repolarization of the membranes.

Electroplaques

For many years the generation of electricity by certain fishes has interested physicists and physiologists. Franklin, Cavendish, Volta, Davy, DuBois-Reymond, Faraday, and Bernstein are among the many eminent scientists who have investigated electric fishes. They especially interested scientists of the eighteenth and nineteenth centuries because the discharge of electric organs was easy to detect with the relatively insensitive electrical measuring devices then in use (Fig. 1-3).

Electric organs are found in several species of fish, including an African catfish (*Malapterurus*), the Mediterranean electric ray (*Torpedo*), the North American stargazer (*Astroscopus*), a family of African fishes

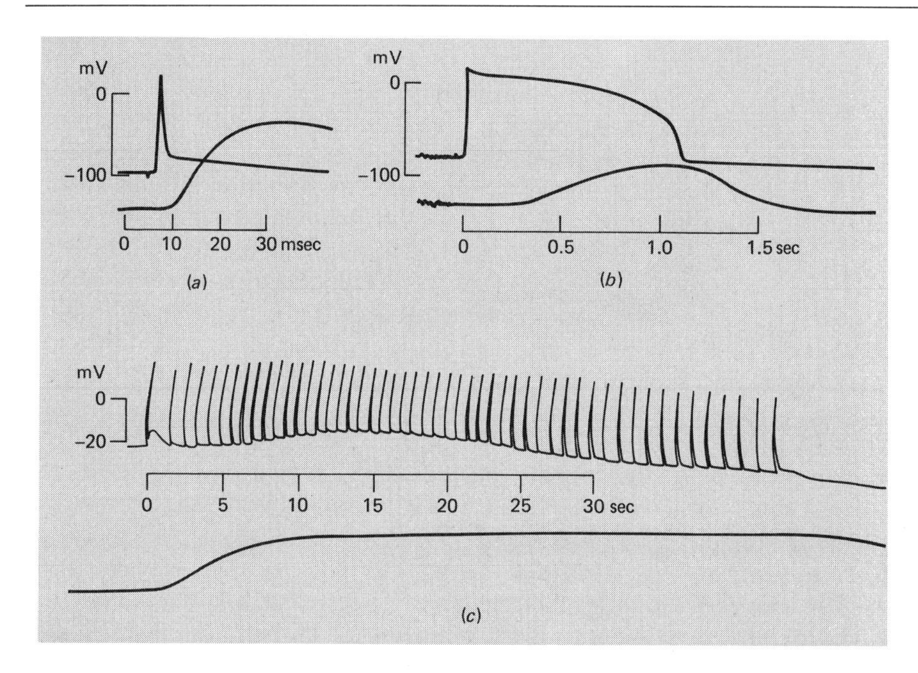

FIGURE 13-15
Simultaneously recorded transmembrane potentials (*upper trace*) and contraction (*lower trace*) in three types of muscle. (*a*) Isolated frog skeletal muscle fiber. (*b*) Whole frog ventricle; action potential recorded from one "cell." (c) Strip of pregnant rat uterus (smooth muscle); action potential recorded from one cell; spontaneous activity. (*After T. C. Ruch et al.*, 1955, *Neurophysiology, W. B. Saunders Company, Philadelphia,* p. 139.)

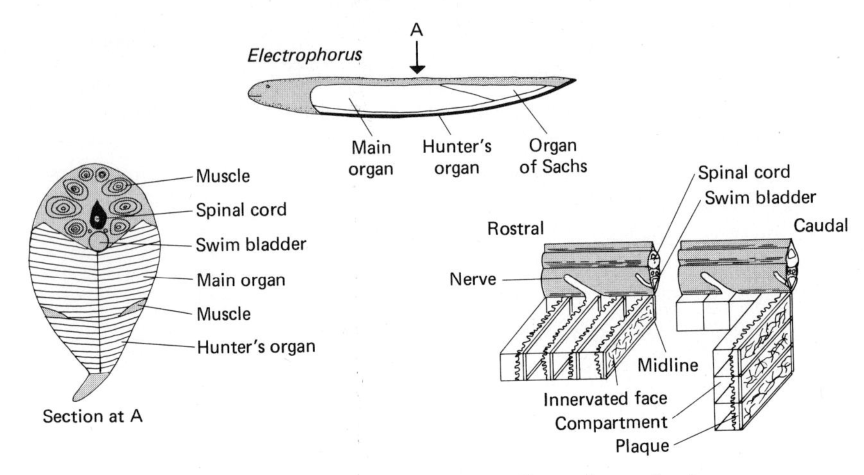

FIGURE 13-16
Structure of the electric organ of the electric eel (*Electrophorus electricus*). The main organ has about 40 columns of electroplaques on each side of the midline at the widest plane of the fish. Its discharge can stun small prey fish. Hunter's organ, separated by a strip of muscle from the main organ, contains up to 15 columns on each side. Its function is not known with certainty. The organ of Sachs gives off low-amplitude periodic discharges and apparently serves in the detection of changes in the electric field which surrounds it. (*Partly after H. Grundfest, Electric Organ (Biology), McGraw-Hill Encyclopedia of Science and Technology, 4, p. 479,* 1971.)

called *mormyrids*, the South American *gymnotids* and the closely related "eel," *Electrophorus*, as well as some skates and rays.

Electric organs consist of modified muscle fibers called **electroplaques** or **electroplates.** They are derived from branchial or skeletal muscle in elasmobranch fishes and from ocular muscles in stargazers. Typically, a number of disklike, noncontractile electroplaques are stacked in columns. Each electroplaque has a nerve connected to one face but not to the other; the innervating fibers usually branch extensively and make numerous synaptic contacts with the electroplaques. In *Electrophorus* the greater part of the lateral four-fifths of the fish is occupied by the electric organ, there being at least 6000 cells in series (Fig. 13-16).

The columns of electroplaques are synchronously activated by neural input and then discharge as in a voltaic pile. On open circuit (i.e., in air) a large *Electrophorus* can produce brief (3-msecond) pulses up to 600 V in potential; short-circuited, the peak current is about 1 A. However, in terms of power output, *Torpedo* is the most potent fish known; it generates a current of several amperes at about 50 V. The mormyrids, gymnotids (other than *Electrophorus*), and the rays have smaller electric organs which generate only 1 to 2 V. The output of *Astroscopus* and *Malapterurus* are between these extremes.

It was not until electroplaques were penetrated by microelectrodes that an explanation was possible as to how their additive discharge is achieved and what the source of the energy for current flow is (Figs. 13-17 and 13-18).

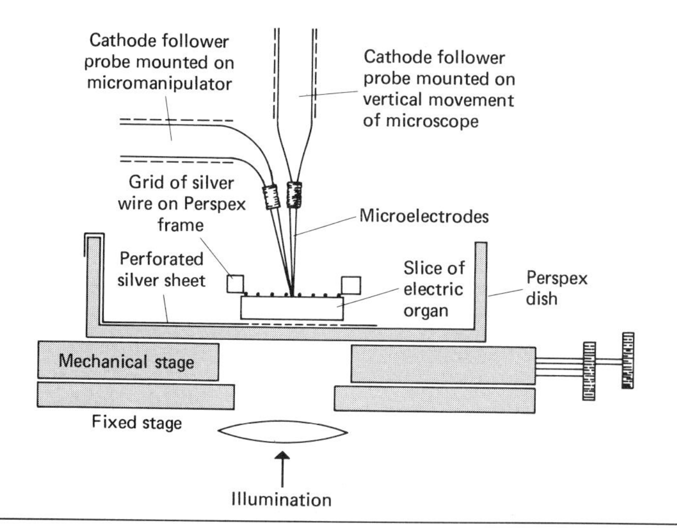

FIGURE 13-17
Simplified diagram of an experimental arrangement for stimulation and recording of membrane potentials from electroplaques of *Electrophorus.* Stimulation is applied through the perforated silver sheet and the grid of silver wire. Response is recorded through penetrating microelectrodes. (*After R. D. Keynes and H. Martins-Ferreira,* 1953, *J. Physiol., 119,* p. 317.)

Because only one side of the electroplaque depolarizes,[1] each becomes temporarily electrically polarized with the innervated side negative to the noninnervated face. Thus, each column of electroplaques behaves as a series of electric elements consisting of alternating positive and negative "plates." Each electroplaque contributes several millivolts—as much as 150 mV in *Electrophorus*—which sum in series for

[1] In freshwater forms both faces of the electroplaque participate in the response; not all electric organs discharge in the same way even among marine forms; however, the net result is similar to that described here.

FIGURE 13-18
Membrane potentials recorded from single electroplaques of *Electrophorus* using the arrangement shown in Fig. 13-17. Inserts show positions of microelectrode tips. In (*a*) and (*b*) the innervated side points upward; in (*c*) it is downward. Note that resting potentials (about −80 mV) are recorded only when one electrode penetrates the membrane (arrangement B), and that spikes occur only at the innervated membrane. (*After R. D. Keynes and H. Martins-Ferreira,* 1953, *J. Physiol., 119,* p. 323–324.)

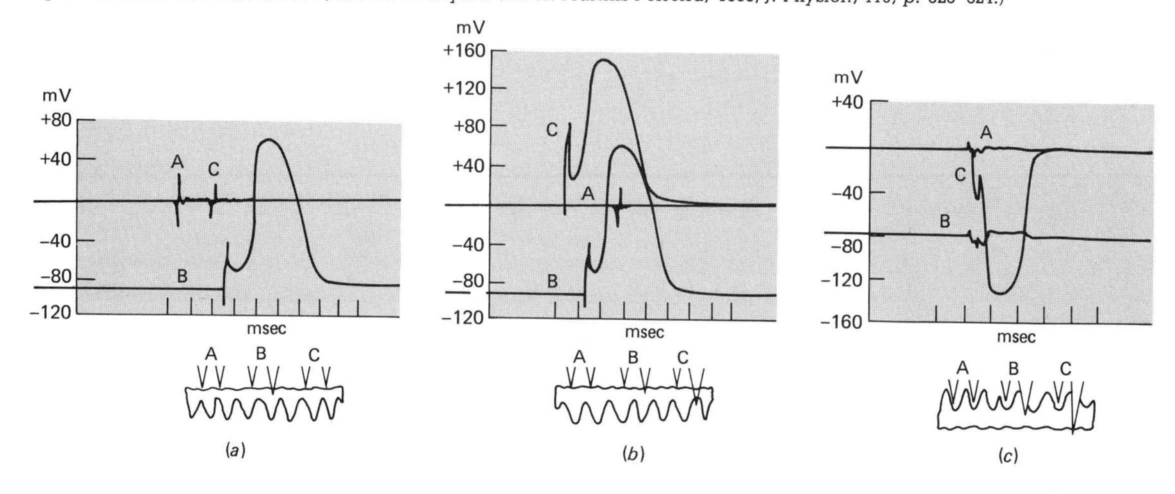

the *voltage* developed by the column. The *current*, however, is proportional to the number of elements in parallel and to the surface area of each electroplaque. This is why *Electrophorus*, which has several thousand plates in columns, but few parallel columns, discharges several hundred volts at low current, and *Torpedo*, which has many short columns in parallel, produces lower voltage but more amperage.

The function of electric organs in those fishes generating large outputs is to stun prey. In those producing only a volt or two it appears that the electric discharge is used as an electrosensory mechanism for detecting nearby objects and predators, as described in Chap. 12 (Sec. 12-7).

Electrogenic cells of all kinds have in common an electrically polarized membrane and a characteristic set of responses of that polarity to perturbations. It seems likely that the asymmetric distribution of charge across their cell membranes is due in part to the activity of ion "pumps" in the membranes. Energy is required to drive the "pumps," and they seem to be universally dependent upon the availability of ATP. The pumping of Na^+ out of cells (or perhaps more accurately, the coupled exchange of Na^+-K^+ across the surface membrane) is not a process which is restricted to electrogenic cells. Instead, it may be a ubiquitous cellular mechanism and one which might be fundamental to the movement of other materials across cell boundaries, as will be detailed in the following chapter.

READINGS

Bures, J., M. Petran, and J. Zachar, 1967, *Electrophysiological Methods in Biological Research*, Academic Press, Inc., New York. A thorough guide to techniques and instrumentation for electrophysiology, with a good introduction to the theory of electrogenic phenomena.

Junge, D., 1976, *Nerve and Muscle Excitation*, Sinauer Associates, Inc., Sunderland, Mass. A concise compilation of information about excitation, supported by extensive mathematical derivations.

Katz, B., 1966, *Nerve, Muscle, and Synapse*, McGraw-Hill Book Company, New York. One of those rare little volumes that is both authoritative and enjoyable to read.

Ruch, T. C., et al., 1965, *Neurophysiology*, 2d ed., W. B. Saunders Company, Philadelphia. A comprehensive text. The first two chapters deal with the kinds of material covered in this chapter.

Stevens, C. F., 1966, *Neurophysiology: A Primer*, John Wiley & Sons, Inc., New York. A basic book for those who want a nonrigorous and brief summary of neuron physiology. The illustrations are excellent for the beginner.

ACTIVE TRANSPORT

14

Matter enters, temporarily constitutes, and then passes from organisms. It is certain that great fluxes of matter—as atoms and molecules, and perhaps as aggregates of molecules, as well—traverse the boundaries of cells and penetrate the various membranes of organelles.

Cells engage in a variety of processes which achieve a relative homeostasis of the internal cellular environment. Through absorption and secretion, the cell is able to maintain a dynamic state of intermediary metabolism. Within the open thermodynamic system of the cell are subcompartments with still more restricted regulation. It is by the general phenomenon of **transport** that these conditions are made possible.

The manner in which these movements are effected and controlled is largely obscure. Nevertheless, the exchange between organisms and their environment has been well studied, and the various discoveries hold forth promise of understanding much about the uniqueness of the living state.

Inevitably, in dealing with movement of substances into and out of cells, we encounter problems of the nature of membranes and their functional role in hindering or facilitating transport. The exchanges that occur at the surfaces of cells are, obviously, of prime importance. The outer boundary is the most accessible for study, so, consequently, almost everything we know about movement of materials through biological membranes derives from investigations of the plasma membrane. A great deal of the exquisite regulation, or biochemical control, practiced by cells undoubtedly is effected by membranes, particularly plasma membranes. Not only do membranes participate in the regulation of cell interiors, they also commonly control the composition of extracellular fluids such as blood and lymph. In this chapter, however, we will be emphasizing the biochemistry of transport and the theoretical mechanisms that underlie it. Only incidentally will structure be discussed; that aspect will be covered in Chap. 20.

14-1 MIGRATIONS THROUGH MEMBRANES

Cell biologists commonly assume that at the surface of all cells there is a special and highly ordered arrangement of specific molecules called the *plasma membrane* or *cell membrane*. For almost four decades the reigning concept of this membrane's anatomy has been that of a bimolecular layer of phospholipids, coated on both surfaces by protein. There have been variants of this basic idea, but fundamentally most hypotheses of membrane structure have accepted the lipoprotein arrangement, as first formalized by Davson and Danielli in 1935.

The chemical and physical nature of the plasma membrane is largely inferred from studies of penetration of various solutes into cells; i.e., the various models of membrane structure are primarily attempts to reconcile a great mass of permeability data with a meager set of information on the chemical nature of the membrane. Thus, *functional* studies have been of first importance in the formulation of *structural* concepts of membranes.

Among the significant studies which have shaped our ideas of membrane structure are those of Overton (published in 1895 to 1900). He found that substances penetrate cells in the same relative order as the order of their oil-water solubility ratios (Fig. 14-1).

The ratio of solubility of a compound in oil or a fat solvent to its solubility in water is called its **partition coefficient.** The more nonpolar a compound is, the more soluble it is in oil or fat solvents, and, therefore, the higher is its partition coefficient. The more polar compounds are more soluble in water. They, therefore, have lower partition coefficients. For example, lower-series alcohols, such as methyl and ethyl, are highly

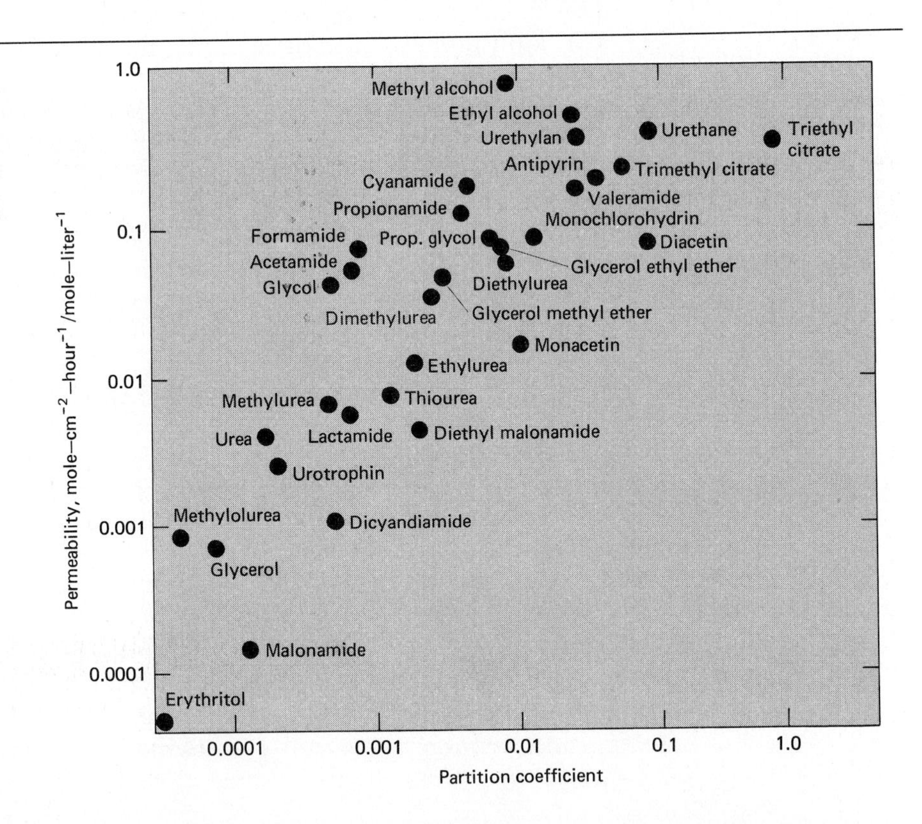

FIGURE 14-1 Permeability of the cells of *Chara* to nonelectrolytes, plotted against olive oil–water partition coefficients. (*After R. Collander*, 1937, *Trans. Faraday Soc.*, 33, p. 985.)

polar and are, consequently, very soluble in water but only slightly soluble in oil or fat solvents. Their partition coefficients are low, and they penetrate rather slowly. Higher alcohols, such as butyl and amyl, are sparingly soluble in water but miscible with fat solvents in all proportions and have high partition coefficients.

Compounds which dissolve readily in oil or fat solvents are commonly called **lipophilic** (meaning "lipid-loving") or **hydrophobic** ("water-hating"). Conversely, compounds which are not soluble in oil but dissolve in water are often called **hydrophilic** ("water-loving"). Obviously, these are relative terms and of limited usefulness.

The direct relationship between partition coefficient and permeability provided the basis for the assumption that the plasma membrane contains a large proportion of lipoidal material.

If the plasma membrane is a reality and if its principal barrier to free diffusion is lipoidal, there appear to be aqueous channels penetrating it. Cells generally are highly permeable to water. If, for example, a sea-urchin egg is placed in diluted seawater, it swells because of entry of water. The diluted seawater is said to be **hypotonic** to the sea-urchin egg; i.e., diluted seawater has a lower osmotic pressure than the egg. If the seawater is sufficiently dilute, the egg will swell until it bursts. On the other hand, a sea-urchin egg placed in salt water more concentrated than seawater will lose water and shrink. The suspending solution is then **hypertonic** to the egg. In seawater the egg neither swells nor shrinks; the seawater is said to be **isotonic.**

Erythrocytes are readily ruptured (*hemolyzed*) when placed in hypotonic solution. Conversely, red cells in hypertonic solutions shrink, and their surfaces become wrinkled. A 0.7% solution of sodium chloride is isotonic with frog erythrocytes, and a 0.9% solution is isotonic with mammalian erythrocytes.

In vacuolated plant cells, the cytoplasm (termed the "protoplast" in this case) normally is held firmly against the rigid cell wall. When the cell is placed in a hypertonic solution there will be a gradual shrinkage in the volume of the entire cell because of the loss of water from the cell sap of the vacuole. If the cell is observed under a microscope it will be seen that eventually the protoplast recedes from the cell wall. This phenomenon of "balling up" into a more or less spherical mass within the chamber formed by the cell wall is called **plasmolysis.** A plasmolyzed cell placed in water will slowly swell and regain a turgid state as a result of movement of water into its vacuole (Fig. 14-2).

This phenomenon of extreme volume change is confined to vacuolated plant cells, and it is quite doubtful that it has any relation to normal transport processes. In fact, it is quite doubtful that the plasma membrane (or the plant cell as a whole) is responsible for this behavior. For one thing, the osmotic pressure of the vacuolar fluid is two to five times less than the pressure in the protoplast. Using osmotic principles, it is impossible, then, to explain how water moves from the protoplast to the vacuole or why water does not tend to come out of the vacuole. The best

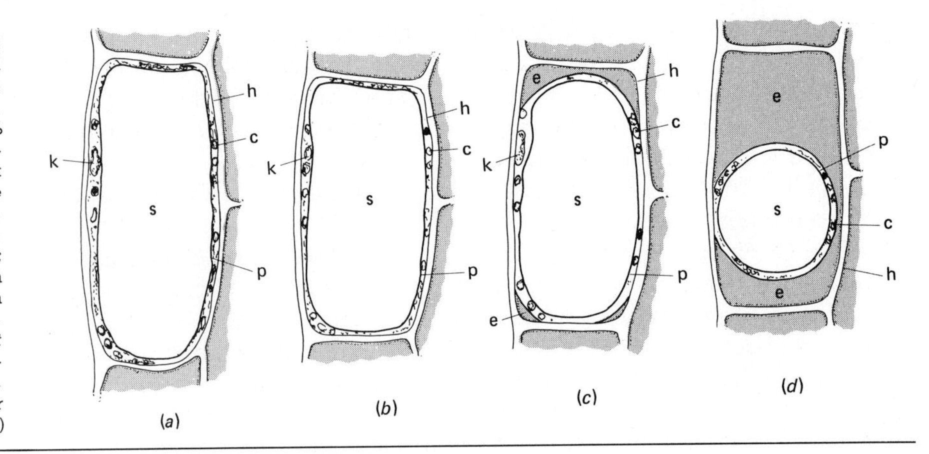

FIGURE 14-2
Plasmolysis of vacuolated plant cells. (a) Normal cell from cortical parenchyma. (b) The same cell in a 4% solution of potassium nitrate. (c) The same cell in a 6% solution. (d) The same cell in a 10% solution. *h*, cell wall; *p*, protoplast; *k*, cell nucleus; *c*, chloroplasts; *s*, cell sap; *e*, salt solution which has passed through the cell wall. (*After F. C. Steward*, 1964, *Plants at Work*, *Addison-Wesley Publishing Company, Inc., Reading, Mass., p. 105; after* Hugo deVries.)

guess at present as to why vacuolated plant cells act as *osmometers* (in truth, only the vacuole obeys the law of osmosis) is that the *vacuolar membrane* is a kind of semipermeable membrane and water can move freely through the cell between the outside medium and the vacuole. Nevertheless, such cells point to the relative ease with which water moves into and out of cells.

Returning to the familiar problem of the nature of the plasma membrane, how can lipid-soluble substances and water both penetrate the same membrane with ease? This has long puzzled biologists and led them to make membrane models by the dozens. However, despite all the various cells and preparations studied, the many hours invested, and the numerous analyses and calculations, the true nature of membranes still eludes us. Some contemporary cell biologists even make convincing arguments that the idea of an anatomical entity at the boundary of cells is erroneous, there being instead a highly labile arrangement of molecules which are special only in that at that particular moment they happen to form the outer limit of the living unit.

It is quite possible to deal with the problem of how things enter and leave cells without giving much attention to what kind of special total structure, if any, surrounds the cell. In fact, we will now proceed in this way, returning only at the end of this chapter to certain aspects of membrane structure. This kind of presentation is an example of the *black-box* approach, and it is exactly the manner in which, until quite recently, studies of cell permeability were carried out.

14-2
BASIC PHENOMENA

The phenomena which we are about to consider can be lumped together under the term **"transport,"** meaning the modes by which a substance passes from one phase to another, appearing in the same state in both phases. By phase we usually mean, in biological systems, compartments separated by membranes. The passage of glucose from blood into a cell and the movement of an acetyl group into a mitochondrion are examples of movement of solutes between phases. Nothing is demanded by the term "transport" concerning energetics or mechanisms. All it indicates is that there is discernible flux of a substance which retains its identity.

Underlying all kinds of biological transport is the phenomenon of **diffusion,** the random movement of particles. Because of diffusion, there is a tendency for a substance to move by its own kinetic energy from higher to lower concentration, or in the case of charged particles, from higher to lower electrical potential. The student should have the fact indelibly stored in his mind that in order for any transport mechanism to operate, there must be movement over short distances by diffusion.

The net rate of diffusion of molecules bearing no net charge (and thus uninfluenced by an electrochemical gradient) depends on the concentration gradient, $S_1 - S_2$. There is a linear relationship between *concentration* and *flux* (the latter meaning rate of one-way movement). The ratio of the fluxes across a boundary ($1 \rightarrow 2$, and $2 \rightarrow 1$) equals the ratio of their concentrations, S_1/S_2. At equilibrium for two similar aqueous phases, this ratio becomes 1.

If a membrane is involved, the rate of diffusion also depends upon the permeability of the material traversed, which we can represent as K_D. Thus, the velocity of migration through a unit of membrane is

$$V = K_D(S_1 - S_2)$$

If a solute is diffusing through pores that are small in relation to its size, the migration rate may not increase linearly with concentration, since a greater number of collisions is not favorable for passage through the pores. This is called **restricted diffusion.**

Diffusion is influenced in some cases by **solvent drag.** This occurs if bulk flow of a solution is taking place through a porous membrane. Electroosmosis (see Fig. 15-22) or a difference in hydrostatic or osmotic pressures between the two solutions bathing a membrane, will cause net flow of solvent. Consequently, solute particles diffusing in the direction of flow will be speeded up, but particles diffusing in the opposite direction will be slowed.

About 30 years ago it was thought that diffusion as a driving force and a plasma membrane that was "differentially" or "selectively" permeable could eventually explain the whole spectrum of disparities in distribution of solutes across cell surfaces. Since then it has been discovered that such passive processes do not provide adequate explanations, and terms like "permeability" and "diffusion barriers" are now of little utility.

The diffusion phenomena just described apply to physical membranes, or so-called *artificial membranes*. But when we turn to the study of membranes of cells, many kinds of results are obtained which indicate that a variety of modes of transport occur, many of which involve active participation of the membranes.

14-3 MODES OF TRANSPORT

Many molecules, especially hydrophilic ones, show unusual features in their migrations through membranes, even when they are moving along a normal concentration gradient. For example, certain solutes that would not be expected to enter a lipid phase readily are found to penetrate cells with ease. In other cases, the migration is strongly influenced by temperature, contrary to simple diffusion situations. A case in point is the observation that the time for a 5.5% solution of glucose to come to equilibrium across human red blood cells increased from about 20 minutes to 3 to 4 hours when the temperature was lowered from 40 to 30°C. This corresponds to a Q_{10} of about 10! The Q_{10} is the factor by which a reaction is increased for a rise of 10°C:

$$Q_{10} = \left(\frac{K_1}{K_2}\right)^{10/(t_1-t_2)}$$

where K_1 and K_2 are velocity constants corresponding to temperatures t_1 and t_2. Although some functions are linear, most are logarithmic with respect to temperature. The temperature range for which a Q_{10} is calculated must be specified because it is higher in low ranges than in high ranges. Physical events have lower Q_{10} values than thermochemical events. The latter typically have a Q_{10} of 2 to 3. Diffusion has a Q_{10} of less than 1.5.

Another phenomenon characteristic of some kinds of cells is that as the level of a solute increases, the flux eventually reaches a maximum. Further increases in concentration do not result in increased flux, suggesting that a reaction involving a limited number of structures or sites is involved. Such a hypothetical structure or group is said to *mediate* transport, and thus we have situations called **mediated transport.** If the transport can occur only *down* the concentration gradient, the mediation is *passive* and we have what Danielli has called **facilitated diffusion.** (Transport *up* a gradient will be considered shortly.)

In certain instances it has been observed that if cells are allowed to come to equilibrium with a solute, and then a second, competing solute is added to the suspending solution, the first solute will move out of the cells. Consider for example the data on efflux of glucose from human erythrocytes shown in Fig. 14-3.

The system described in Fig. 14-3 is always effectively saturated with glucose at the external face. When certain sugars (galactose, mannose, or xylose) are added, they appear to pull out a certain amount of glucose from the cells.

What happens is that this effect, for the most part, is due to the blocking of glucose *uptake* that would otherwise occur in the absence of those sugars. It has been shown, for instance, that galactose uptake is equivalent to the extra loss of glucose caused by the presence of galactose. That is, there is a mole-for-mole exchange of galactose for glucose.

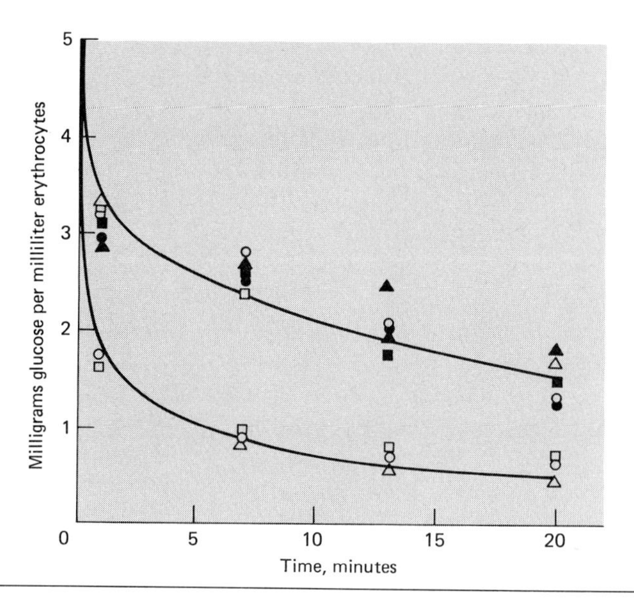

FIGURE 14-3
Efflux of glucose at 0°C from human erythroctes (pre-loaded with glucose to some 160 m*M* at 37°C) in the presence of various sugars (at 28 m*M*) added to the efflux medium. *Upper curve*—nonexchanging sugars: ○, □, saline control; △, L-sorbose; ●, D-fructose; ■, D-arabinose; ▲, D-ribose. *Lower curve*—exchanging sugars: ○, D-galactose; □, D-mannose; △, D-xylose. (*After L. Lacko and M. Burger*, 1961, *Nature, 191*, p. 881.)

The term "**competitive exchange**" can be used to describe such results. Its chief characteristic is that the major part of the apparent increase in efflux results from a decrease in the *parallel* influx, this decrease being brought about by the presence of a competing substrate.

In certain other situations, it has been observed that if cells are allowed to come to equilibrium with a solute and then a second, competing solute is added to the suspending solution, the first solute will move *out* of the cells to such an extent that it will create a *concentration gradient* while the second sugar is moving *in* along its own concentration gradient! The term **flow driven by counterflow** is used to describe this phenomenon.

Counterflow and competitive exchange diffusion both result from the same mechanism, the difference between the two phenomena involving only a matter of differing experimental conditions. To demonstrate counterflow there initially must be an absence of a net flux of the driven substrate. Then, the driving substrate is added. The resulting competition between the two substrates will hinder inward movement of the driven substrate, but as long as the driving substrate remains in low internal concentration, outward movement of the driven substrate will be largely unaffected.

In contrast, to demonstrate competitive exchange diffusion, the driven substrate must be at a high concentration at both sides of the membrane; then the addition to one face of a still higher concentration of the driving substrate prevents the uptake of the driven sugar at the face, and thus a one-to-one exchange of driving and driven substrates occurs.

Both phenomena require that the pathways of entry and exit be physically separate (i.e., inward and outward fluxes do not interfere with

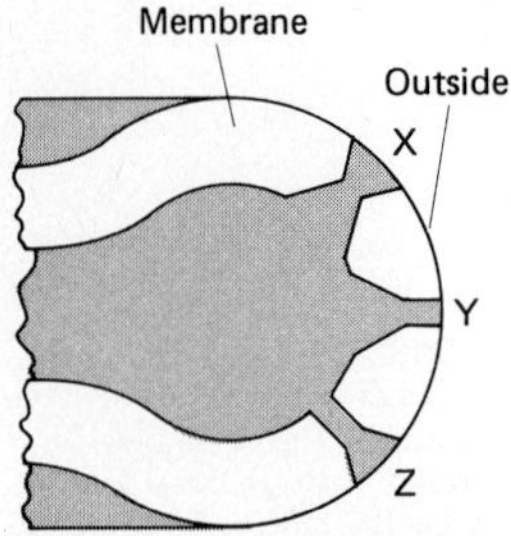

FIGURE 14-4
Diagrams to represent simple, fixed transport sites: *X*, visualization of a chemical site located so that it can receive an appropriate solute molecule from either phase; once received, the solute molecule can be dissociated equally well into either phase; *Y* and *Z*, sites constructed so that they can receive solute molecules only from one side and pass them to the other, i.e., site *Y* is supposed to be subject to mass-action effects only from the inside; site *Z* only from the outside. (*After N. H. Christensen, 1962, Biological Transport, Addison Wesley, W. A. Benjamin Inc., Advanced Book Program, Reading, Mass., p. 19.*)

each other). This condition can be achieved either by a system of unidirectional pores reserved for entry or exit or by a mobile-site (or mobile-carrier) mechanism (Figs. 14-4 and 14-5).

The evidence strongly favors the concept of mobile sites over fixed sites for mediated transport. For example, exit as well as entrance of glucose can be blocked by action confined to the external face of membranes, as when nonpenetrating competitive inhibitors (such as polyphloretin phosphate) are applied. If fixed sites existed, it should be possible to demonstrate the preferential inhibition of entry over exit or vice versa, but in no case of mediated transport has this been done. A physical translocation of some part of the transport system definitely is indicated.

Another phenomenon, called **accelerative exchange diffusion,** tends to support the mobile-site hypothesis. Red blood cells loaded with radioactive glucose to a low level internally will exchange glucose at a faster rate as the external concentration of glucose is increased. That is, the transported molecule tends to facilitate the mobility of the carrier (as the substrate-carrier complex, probably).

A similar kind of acceleration of exchange has been demonstrated for glycine transport by ascites tumor cells.

Stein has summarized the carrier hypothesis as follows: A **carrier** is a hypothetical element present in the cell membrane, which is endowed with the following properties: (1) it can combine with the molecule (or ion or group) to be transferred to form a more or less transient complex; (2) the complex has the property of being able to cross the membrane—to translocate; (3) the carrier may or may not be able to translocate when not combined with the substrate; and (4) the substrate permeates at a negligible rate when not combined with carrier.

The concept of a "carrier" is, as Stein points out, a convenient fiction, and one need not be overly concerned about its reality. Its virtue is that it effectively accounts for the available experimental data. More will be said about carriers in Sec. 14-6.

For convenience, cellular transport may be classified as occurring at three levels of organization:

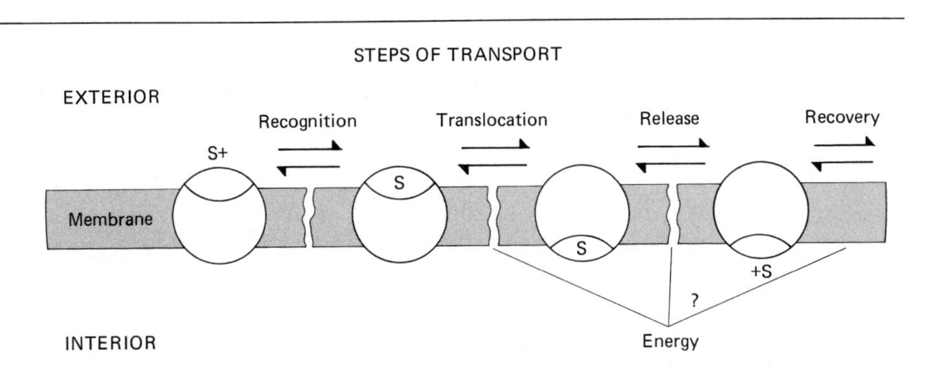

FIGURE 14-5
A generalized concept of transport involving a mobile carrier. The substrate is bound to a specific site on the outer membrane surface; translocation of the substrate through the membrane occurs; the substrate is released inside the cell; the system returns to its original state. Perhaps this last step is the energized one. (*After A. B. Pardee, 1968, Science, 162, p. 633.*)

1. **Primary order: the translocation of a substance across a single, integrated biological membrane**

2. **Secondary order: transcellular transport**

3. **Tertiary order: transport resulting from integrated organ function**

Primary transport, in which only one membrane is crossed, generally involves only one mechanism for the transport of a substance. The two commonest forms of primary transport are diffusion and active transport (see below). Facilitated diffusion is a frequently employed process of primary transport, also.

These unitary processes are mechanisms for movement of substances *through* membranes. There are additional mechanisms for movement of materials *across* membranes. Included in the latter is the translocation of a solution or suspension enveloped in a membrane—so-called **vesicular transport.** What is known of this process comes from interpretation of electron micrographs. It appears that to effect absorption, a vesicle forms from an inpocketing of the outer membrane of a cell through conformational changes, eventually disengages from it, and forms a spherical vesicle. Secretion is essentially the reverse process.

Secondary transport—the translocation of a substance across cells—generally involves more than one unitary process operating in sequence. These processes may be referred to as either absorption or excretion, depending upon the polarity of the cell and its position in the organ of which it is a part.

Tertiary transport may involve several unitary mechanisms, in addition to several forms of secondary transport. For example, the net excretion of a substance by a complex kidney represents the summation of sequential processes of secondary transport.

14-4 ENERGY-DRIVEN TRANSPORT

An intriguing hypothesis about how membrane transport works is that there possibly is an interrelationship between metabolite and cation transport. The former appears to be a passive process in certain cases, a facilitated diffusion, in fact, but the latter has thermodynamic characteristics that would demand expenditure of free energy. Because both processes can be considered in the same conceptual framework, the active component becomes much more amenable to analysis.

Christensen and colleagues found, while studying transport of amino acids by duck erythrocytes, that if half the sodium in the medium was replaced by potassium, the amount of glycine and alanine accumulated was reduced. Also ascites tumor cells were observed to lose potassium ions as they accumulated amino acids.

Subsequently, it was shown that the presence of sodium is required for sugar transport by the intestinal wall. In fact, an intact, energy-requiring transport system for sodium has been demonstrated to be essential for the accumulation of sugar. Thus, facilitated diffusion of sugars and amino acids may depend upon the *active* transport of sodium. Crane[1] has proposed a model to explain such observations (Fig. 14-6).

In this model, a facilitated diffusion system, E_1, transports glucose (G) only in combination with a sodium ion (Na). E_1 is located at the mucosal surface of the intestinal epithelial cells. A complex, formed by sugar (or amino acid in amino acid–accumulating systems), sodium ion, and the carrier, diffuses across the membrane. Within the cell the complex dissociates. At the opposite side of the cell is the system, E_2, which can actively extrude sodium from the cell—a sodium "pump." The maintenance of sodium at a low level within the cell establishes an electrochemical gradient of sodium ions directed inward from the cell's exterior. This gradient serves to drive sodium, together with sugar or amino acid, into the cell by the facilitated diffusion system.

It should be noted that sodium in this model undergoes transcellular (secondary order) transport; it is moved completely through the cell by two sequential transmembrane systems.

We have introduced the term **"active transport"** without defining it. It is a surprisingly difficult term to define. The definition proposed by Rosenberg in 1948 is the most widely accepted: An active-transport process is one that results in the net movement of a substance from a region of lower electrochemical potential to a region of higher electrochemical potential. However, this definition does not exclude many flows that are not *directly* dependent upon metabolic processes, as in solvent drag and

[1] R. K. Crane, 1965, *Fed. Proc.*, **24**, 1000–1006.

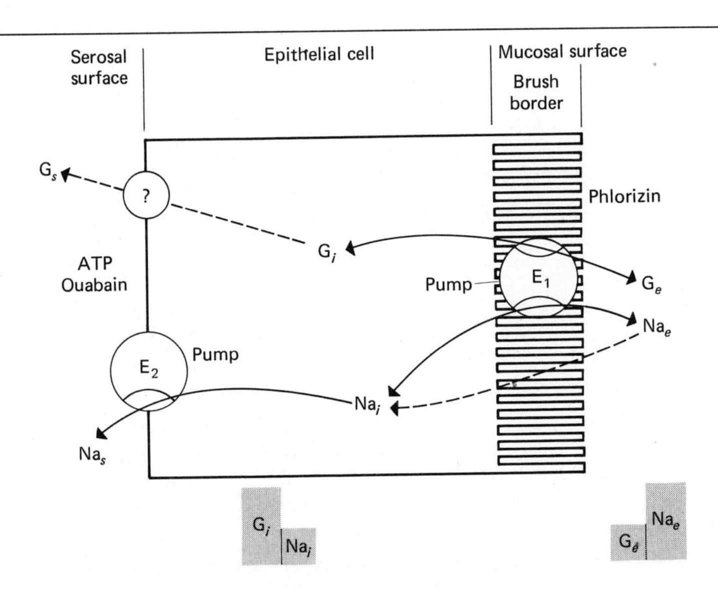

FIGURE 14-6
Model for the interrelations of sodium and glucose transport in intestine. E_1 is a cotransporting sodium- and glucose-requiring facilitated diffusion system, sensitive to phlorizin, and E_2 is the sodium pump, sensitive to ouabain and requiring ATP. The height of the blocks at the foot of the figure represent the prevailing levels of glucose (G) and sodium (Na) within (subscript *i*) and at the external surface (subscript e) of the epithelial cell.

the passive uphill movement of ions paired with other ions that are being actively transported.

We actually can do no better than fall back on the negative definition of Ussing: A substance may be regarded as actively transported only if transfer of the substance across a membrane cannot be accounted for by the action of the forces of diffusion, electrical potential gradient, solvent drag, or these forces in any combination.

Active transport includes both cases where a substance is moved against its electrochemical gradient (uphill) and those in which it is transferred along its potential gradient (downhill) at a rate faster than would be expected from the magnitude of the above forces.

The movement of a substance by active-transport processes results in increasing its free energy. The source of the energy which drives the process is, of course, metabolism, but rather little is known about the manner in which transport is linked to metabolism. It has been established that active transport of sodium and potassium by mammalian erythrocytes is dependent upon the anaerobic breakdown of glucose. Tortoise red cells use the energy from the aerobic pathway, while duck red cells and ascites tumor cells can use either glycolysis or aerobic oxidation.

An impressive portion of a cell's available energy goes to support transport. It is estimated to be 30 percent in red blood cells and 15 percent in ascites tumor cells. Such amounts suggest the biological importance to the cell of maintaining the active transport systems.

Many hypotheses have been proposed to account for mechanisms of active transport. Some are designed to explain only specific situations, while others are intended to be more comprehensive. Following are some of the various theories:

1. There is a simple membrane carrier which combines specifically with its substrate and diffuses across the membrane. Unidirectional transport is brought about by maintaining a concentration gradient for the complex in the direction of transport by *synthesis* of the carrier at one boundary and *consumption* of the carrier at the other boundary.

2. A membrane carrier is *converted* at one boundary into an active form which combines with its substrate. At the other boundary it is reconverted to another form which cannot combine with that substrate (Fig. 14-7).

3. For anion and cation transport, specifically, a *redox pump* mechanism has been proposed. Transference of electrons along the iron of heme groups in respiratory enzyme chains creates the driving force. The transfer of an electron from Fe^{2+} to create Fe^{3+} would cause an additional anion to move toward the iron. Cation transport would be accomplished by an exchange of the hydrogen ions, which are also being moved in the respiratory chain, for the cations. This mechanism has the flaw that it cannot explain the high degree of specificity commonly displayed by transport mechanisms. To overcome this shortcoming, a carrier, driven by the transfer of electrons, is postulated.

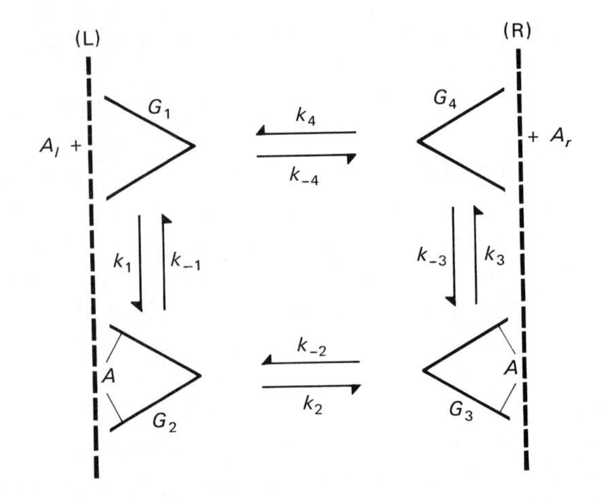

FIGURE 14-7
A model for membrane transport. The gate G_1 receives the solute A_1 from the left. A reorganization is triggered, orienting the gate toward the other phase (G_3), where the unchanged solute is released to the right. The gate can then be returned to its original orientation. If uphill transport is to be achieved, energy must be supplied to drive k_1, k_2, k_3, or k_4. For example, energy may be supplied to k_4 to prepare the site so that the entry of A_1 from the left will trigger a relaxation to the condition G_3. (*After C. Patlak*, 1957, *Bull. Math. Biol.*, *19*, p. 209.)

4. The *fluid-circuit* hypothesis has been proposed to account for salt uptake by the small intestine of the dog. It assumes that the membrane (intestinal wall) contains two sets of pores. Through one set a salt solution passes and through the other set water, or at least a less concentrated solution, is returned. The driving force is thought to be electroosmosis. This mechanism, like the redox pump, lacks specificity.

5. A mechanism involving *configurational changes* in a contractile protein has been proposed. The macromolecule uncoils or extends, breaking intramolecular bonds; ATP presumably provides the energy for thus "cocking" the system. When an appropriate substrate encounters the extended protein, it is adsorbed, causing contraction or coiling. The substrate is thereby whisked across the membrane (Fig. 14-8).

6. Another mechanical model envisages a lattice of macromolecules which can expand and contract. When it is expanded at one boundary, molecules or ions of the right species can move into the lattice, and when it contracts in a peristaltic manner they are pumped across the membrane.

Unfortunately, it is too soon yet to embrace confidently any particu-

FIGURE 14-8
A hypothetical, propelled shuttle or carrier mechanism of transport. An enzymatic center, such as phosphatase or choline esterase, supplies energy to actuate a contractile protein. The latter has an adsorption site which will be on one side of the membrane when contracted and on the other side when extended. Contraction is triggered by adsorption of the substrate. (*After J. F. Danielli*, 1954, *Symp. Soc. Exp. Biol.*, *8*, p. 509.)

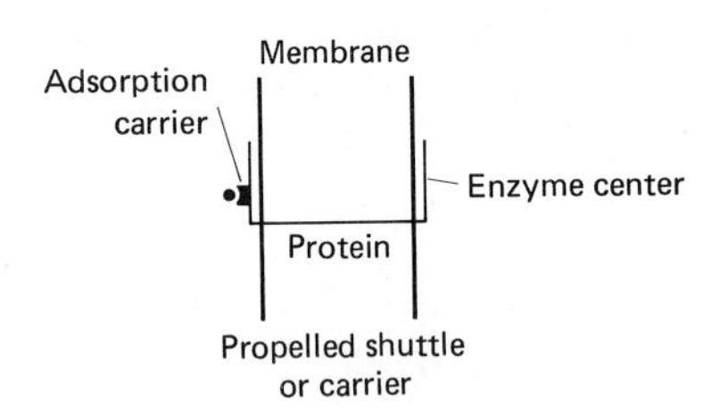

lar hypothesis and declare it superior to any other explanation of active transport.

14-5 TYPES AND SPECIFICITY OF ACTIVE TRANSPORT

Chemical sites mediating transport are analogous to enzymes in that they are highly, though not absolutely, specific. For example, a system that transports glucose into animal cells also transports the closely related sugars, 3-O-methyglucose and 2-deoxyglucose. Closely related substances often seem to be transported by the same mechanism, but not necessarily with the same efficiency. On the other hand, substances that are quite different sometimes are found to compete for the same site.

Because active-transport processes are interwoven with passive ones, it is extremely difficult in most situations to demonstrate rigorously that a substance is being transported actively. Only in a few cases has unequivocal proof been provided for the active nature of a transport process. In others, although the proof is not conclusive, active transport is strongly indicated.

The best characterized of the active-transport mechanisms are those for ions. Many cells have been shown to transport ions. These include algal cells, roots of plants, red blood cells, muscle and nerve fibers, and epithelial cells. The latter have been favorite subjects for study.

Frog-skin epithelial cells transport sodium chloride from the outside medium to the inside of the skin. The classic work of Ussing has demonstrated that if the skin is placed as a barrier separating two identical, balanced salt solutions, sodium chloride moves from one compartment to the other (in the direction from outside to inside of the skin), creating an electrical potential. Even when an opposing voltage is applied to eliminate any electrical gradient, sodium will still be transported (Fig. 14-9).

Unfortunately, the frog skin is a complex, multilayered tissue, and not a homogeneous cell membrane. It is probable that only the basal layer of cells participates in the transcellular transport of sodium, the outer, more progressively keratinized layers being nonfunctional metabolically.

The frog skin is a very effective transport system. Krogh found that an intact frog could take up Na^+ when its concentration in the frog's body fluids are 100 meq/l and in the surrounding medium was only 0.01 meq/l—i.e., a concentration ratio of 10,000:1. Furthermore, the mechanism is completely specific for sodium ions.

Active transport of sodium also has been demonstrated in rat intestine, vertebrate kidney, and various muscles and neurons. As noted in the preceding chapter, transport of sodium and potassium seems to be linked in neuron membranes. This appears to be the case also for frog skin and for membranes of red blood cells and muscle cells as well.

Potassium appears to be actively taken up by yeast at the outset of fermentation in exchange for hydrogen ions. The transport mechanism is quite specific, accepting other ions only when sufficient potassium is lacking.

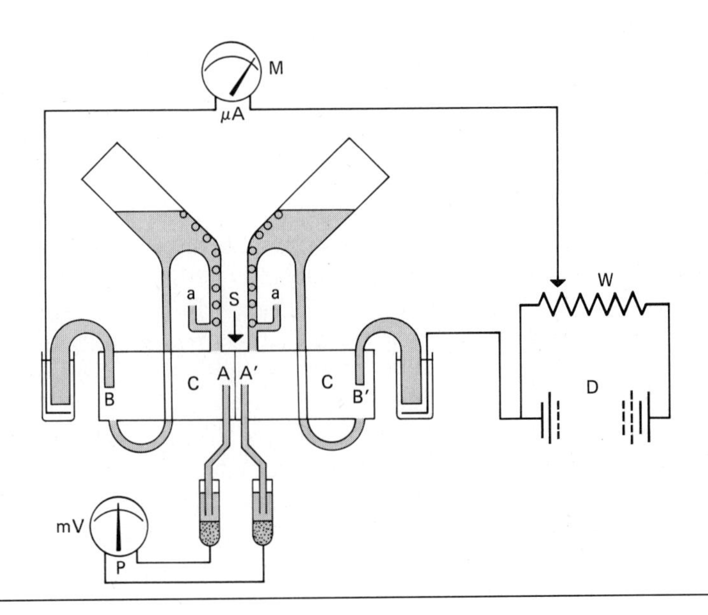

FIGURE 14-9
Diagram of apparatus used for determining Na flux and short-circuit current across a piece of frog skin. *C*, celluloid chamber containing, on each side of the skin, 40 ml Ringer's fluid; *S*, skin; *a*, inlets for air; *A*, *A′*, agar-Ringer bridges connecting outside and inside solutions, respectively, with calomel electrodes; *B*, *B′*, agar-Ringer bridges used for applying outside emf; *D*, battery; *W*, potential divider; *M*, microammeter; *P*, tube potentiometer. (*After H. H. Ussing*, 1954, *Symp. Soc. Exp. Biol.*, *8*, p. 415.)

Active transport of both chloride and sodium ions has been demonstrated in two algal species, *Nitellopsis obtusa* and *Halicystis ovalis*. Their cells are large, and the vacuole contains enough sap to permit chemical analysis. Chloride ions are transported into the vacuole, while sodium is extruded through the plasma membrane. Active transport of chloride by a portion of the nephron of kidneys was noted in an earlier chapter. Active chloride transport also has been declared to occur in frog skin, gastric mucosa (several species), mammalian intestinal mucosa, turtle and toad bladder, teleost gill, rabbit cornea, and the salt-secreting rectal gland of marine elasmobranchs.

Excised roots of plants have been extensively used for study of transcellular transfer of salts. Salts can be taken up from very dilute solutions by certain plants. It is not certain whether it is anions or cations that are being actively transported, however. The transport mechanism is not highly specific but does show varying degrees of efficiency in transport of the different ions.

The gastric mucosa of animals likewise shows considerable specificity in ion transport, moving not only chloride but bromide and iodide as well. The thyroid transports the same three anions, but iodide is accumulated more effectively. If rats are fed excess chloride or bromide, the iodide content of their thyroids will decrease and the animals will develop goiter. This indicates transport of all three ions by the same mechanism.

Transport of phosphate ions by the bacterium *Staphylococcus aureus* has been rather convincingly demonstrated.

Active transport of several organic substances takes place in the vertebrate kidney. For example, active reabsorption of amino acids occurs in the proximal tubules. On the basis of competitive studies, there seem to be three kinds of mechanisms for transport of amino acid groups. The amino acids will compete within groups but not between groups. Histidine, lysine, and arginine are one group; leucine and isoleucine are a second; and glycine, alanine, glutamic acid, and creatine (which is not an amino acid) form the third.

Transport of amino acids by erythrocytes and tumor cells was mentioned earlier. Glucose is actively transported by the proximal tubules of kidney, as well as by intestine.

Analysis of the gas in swim bladders of fishes has revealed some intriguing indications of active transport. For instance, it has been known for some time that in certain fishes the oxygen in their swim bladders may be at a pressure of more than 100 atm. In others, the gas in the swim bladder is mainly a mixture of nitrogen and argon. It has been proved that the gas composition is not derived as a by-product of metabolic events. Rather, molecular gas is deposited against steep gradients. How active transport of gas occurs is quite unknown.

14-6 MECHANISMS AND MEMBRANES

The great goal of transport study is a description in three dimensions of the structural components of active-transport mechanisms and their modes of operation. Such an achievement does not seem very near at hand, and in the meantime more attainable ends are being sought, such as the identity of carriers and the nature of energy-coupling mechanisms that permit active transport.

It is assumed that transport molecules are proteins, because proteins are the only molecules which could have the degree of specificity displayed by transport processes. Indeed, that proteins are required is shown by the fact that phenylisothiocyanate, which reacts with proteins, inhibits transport. Also, inhibitors of protein synthesis, such as the antibiotic chloramphenicol, prevent production of transport systems in bacteria.

Because drugs and inhibitors have been used so successfully in attacks on the mode of action of enzymes, they also are being used in investigations of transport systems, since the latter act so much like enzyme systems. (However, there may be some differences between metabolic enzymes and transport proteins; covalent bonds may not be formed between substrate and carrier, for example.)

The drug phlorizin, a glycoside, inhibits glucose transport in kidney tubules (causing renal glycosuria) and in the intestine (Fig. 14-6). This drug may combine with and block membrane carrier sites. By use of phlorizin fragments and analogs, it has been possible to obtain data about which reactive groups of the carrier are involved in transport.

Several inhibitors of the facilitated diffusion systems for glucose and glycerol of erythrocytes are known. Their actions indicate the presence of a thiol residue at the active center of the glucose system and at least two groupings (amino and imidazole groups) at the active center of the glycerol system. In addition, companion studies have provided information on the number of transport sites per cell for certain substrates.

As will be discussed in Chap. 20 (Sec. 20-3), metabolism of lactose (α β-galactoside) in *Escherichia coli* is under the control of inducible structural and regulator genes. The *z* gene controls the synthesis of β-galactosidase, and the *y* gene controls the synthesis of a component essential for galactoside accumulation. Kinetic analysis indicates that the maximum rate of hydrolysis (V_m) and the apparent Michaelis-Menten constant (K_m) of the β-galactosidase isolated from the bacterial cell are different from those shown by the enzyme in vivo. The maximum velocity of galactoside hydrolysis of intact bacteria is slower and corresponds to an initial, rate-limiting step: penetration into the cell.

A considerable amount is known about the specificity of the galactoside transport system, as well as the effects of competitive inhibitors and sulfhydryl-inactivating agents. Lactose accumulation is a carrier-mediated process and appears to involve a true, active-transport mechanism, capable of net transfer against a chemical potential.

If *E. coli*, induced for the lactose operon, are treated with a metabolic inhibitor, such as cyanide or 2,4-dinitrophenol (DNP), it can be demonstrated that the kinetics of hydrolysis will have the characteristics of a facilitated or carrier-mediated diffusion. Then, if the same suspension is washed to remove the metabolic inhibitor, they will revert to transporting actively. The conclusion is that by addition of an inhibitor to deenergize it, the same transport system can be changed from an active transport to a facilitated diffusion mechanism. Active transport, therefore, can result from facilitated diffusion coupled to energy-yielding, biochemical reactions.

The hypothetical carrier of this system is one of a class of postulated substances called by microbiologists **permeases.** This is a bad term on at least two counts. First, the "-ase" ending indicates that the substances are enzymes. The Commission on Enzymes of the International Union of Biochemistry has ruled (in 1961) that a term containing the "-ase" suffix should be used only for a single enzyme catalyzing a known reaction. The postulated carriers thus are not qualified for such naming. Second, the term has been uniquely applied to bacteria, and it is by no means certain that their transport mechanisms are unique, since all such mechanisms remain still to be fully characterized.

Attempts are being made to demonstrate the reality of carriers. Several different membrane-associated proteins have been discovered which "recognize" (i.e., bind) specific substrates. Among these is the *y* gene product of *E. coli*. Through a chemical method of labeling with radioisotopes, a highly purified, membrane-bound lipoprotein has been isolated from induced *E. coli* that is not present in induced cells. The specifically labeled protein, which is called *M protein* because of its association with the membrane, is postulated to be either the carrier itself or an essential part of the carrier. Its molecular weight is 31,000.

Two mechanisms have been proposed to explain the working of this carrier. One hypothesis proposes that energy could decrease the affinity of the carrier for galactoside at the inner border of the membrane, and thus the fraction of carrier bound to galactoside would be smaller on the inner membrane surface as compared with the outer surface. This would be a "push" mechanism. A second proposal envisages a "pull" mechanism; viz., energy acts at the outside of the cell membrane, increasing affinity for galactoside. It has not as yet been possible to determine which of these two alternatives is correct, or if either is correct.

A protein which may be the carrier that transports calcium in intestine has been isolated. It has a vitamin D dependency, and it corresponds in distribution within the intestine to the regions that transport calcium.

Another potential transport protein, one that binds sulfate, has been isolated from *Salmonella typhimurium*. Also, a protein from *E. coli* that binds neutral amino acids has been isolated and crystallized. In addition, proteins that bind specific sugars are known and are being investigated.

Most of the "transport proteins" have molecular weights of about 30,000 and have one specific binding site with a dissociation constant of 10^{-5} to 10^{-6} mol/l.

There is no conclusive evidence on hand yet that these truly are transport proteins. It is indirect information that links proteins and transport. The so-called *transport proteins* do appear to be membrane-bound and are released by osmotic shock—a relatively gentle treatment in which cells are rapidly transferred from a sucrose solution of high osmotic pressure to a dilute salt solution—suggesting that they are located on the surface. However, other proteins also are released by this treatment.

Evidence is accumulating that carrier-mediated sodium transport in a large variety of tissues involves an enzyme which can function as an ATPase. The enzyme, a phospholipoprotein complex, requires the presence of both Na^+ and K^+ and, like other ATPases, is stimulated by Mg^{2+}. It is ubiquitously associated with animal tissues transporting Na^+ at high rates, appears to be localized in cell membranes, and is inhibited by ouabain (see Fig. 14-6) in a quantitative relationship that corresponds to the effects of ouabain on sodium transport. This enzyme, known as *Na^+, K^+–activated ATPase* or *Na^+, K^+–translocating ATPase*, is a component of the best-studied ion-transport system to date. In human red blood cells the hydrolysis of ATP drives the extrusion of three sodium ions in exchange for two potassium ions. This system establishes an electrochemical gradient for the sodium ion that may be used to drive the active transport of sugars, amino acids, and other substances (Fig. 14-10).

Na^+,K^+–activated ATPase systems show allosteric features, responding either to Na^+ or K^+ by increasing transport efficacy. The systems may be composed of two different kinds of subunits: catalytic ones interacting with ATP, and regulatory subunits interacting with Na^+ and K^+. The latter subunits could be the carriers for sodium and potas-

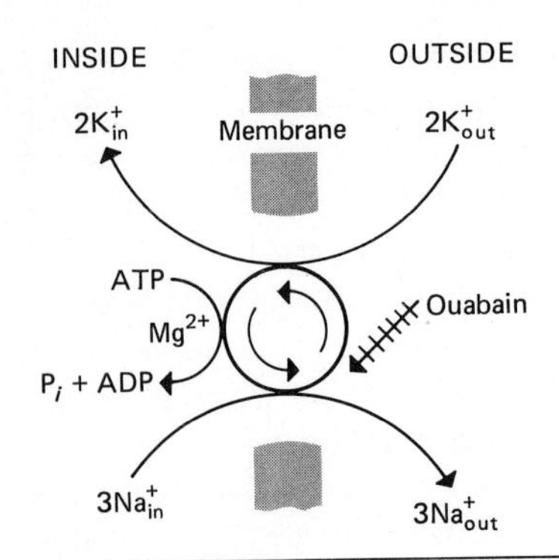

FIGURE 14-10
Simplified model of the Na^+,K^+–activated ATPase of human erythrocytes. ATP reacts on the inner surface of the plasma membrane. The inhibitory effect of ouabain is exerted on the outside. (*After R. L. Post et al.*, 1969, *J. Gen. Physiol.*, *54*, p. 207s.)

sium. Other specific subunits might be incorporated into the complex for transport of other solutes.

There seems to be a Ca^{2+}-activated ATPase system in plasma and mitochondrial membranes that is similar to the Na^+,K^+–stimulated system. Ca^{2+} appears to be transported out of the cell against an electrochemical gradient, corresponding to a concentration gradient of 1:20. The mechanism is like the Na^+ transport system in that ATP hydrolysis provides the driving energy, but unlike Na^+ transport it does not require the movement of a cation in the opposite direction. The activating effect of Ca^{2+} on the ATPase system is restricted to the internal side of the membrane. Unlike the Na^+ and K^+ transport system, ouabain does not inhibit Ca^{2+} transport.

Mitchell has championed the view that mediated transport is effected by enzymes which are constituents of membranes. Part of the plasma membrane, in this concept, consists of plugs of protein. Included in the protein may be normal enzymes of cellular metabolism, as well as Na^+,K^+–activated ATPase and specialized "enzymes" concerned only

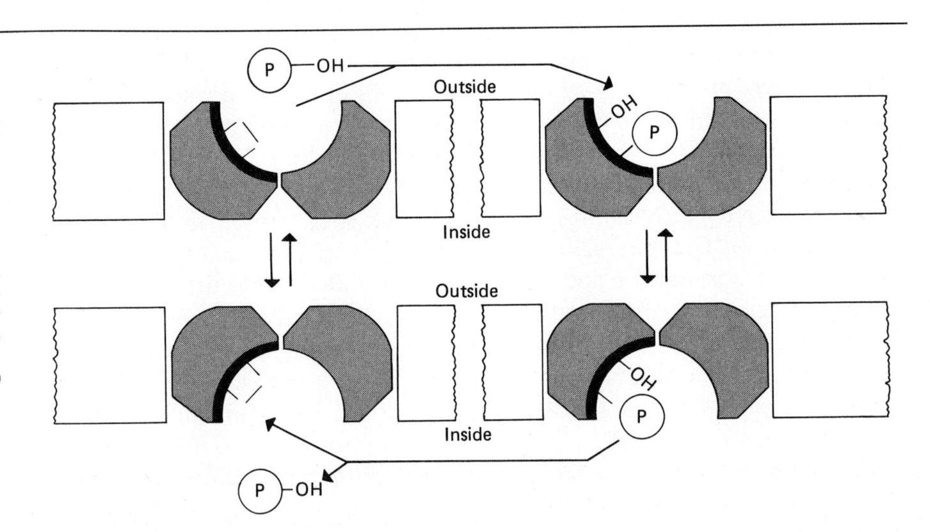

FIGURE 14-11
The "translocase" model of Mitchell, shown here as catalyzing the translocation of a phosphoryl anion, the ringed P, from the outside to the inside of the cell. The hatched areas represent the "translocase," a membrane-bound "enzyme" which is available alternately to the inner and outer faces of the membrane, whether free or in combination with the permeant. (*After P. Mitchell*, 1957, *Nature*, *180*, p. 135.)

with transport. Mitchell also has suggested that in some cases the normal enzymes of metabolism, if located in an appropriate fashion in the membrane and if endowed with certain specialized active-center properties, may be used for transport across membranes. Such a "translocase" mechanism has been hypothesized for phosphate transport (Fig. 14-11).

The specialized enzymes of transport, in order that they be anchored in the membrane, probably make firm bonds with membrane lipids. Transport enzymes, in fact, may well be lipoproteins, as suggested by some of the information already presented. Such enzymes would be expected to have a high proportion of nonpolar side chains, which could form a shell enclosing a hydrophilic interior through which the substrate would pass.

There is evidence, indeed, that the membrane lipids themselves may play an active role in transport. Experiments, especially with the salt-secreting glands of birds (nasal glands), have shown that when the cells start to secret sodium there is a cycle of energy-requiring events that involves certain of the phospholipids. When sodium transport stops, the phospholipid is converted to its former condition. These reactions are interpreted as having something to do with turning on and off the sodium transport machine.

Cells in tissue culture were observed by Lewis in 1931 to form invaginations along their periphery which were frequently pinched off to become vacuoles. He called this process **pinocytosis** and likened it to a drinking of extracellular fluid by cells. The phenomenon has been reported to take place in a variety of cell types. As a possible mode of transport, pinocytosis has been a subject of increasing interest. However, such a nonspecific inclusion of extracellular fluid would appear to produce more transport problems than it would solve. If it is a device for capture of specific solutes, large volumes of water would have to be voided and other solutes would have to be expelled from the cell. The observed specificity and kinetics of transport cannot be reconciled with such a mechanism.

The plasma membrane often appears in electron micrographs to be highly folded, a modification that may be of significance in transport. Infoldings are frequently seen to be in close relationships with mitochondria. An elaborate set of membranes is found in striated muscle cells. Certain of these membranes probably transport calcium during a phase of the contraction cycle. Isolated portions of the membrane can concentrate Ca^{2+} as much as 1000-fold. This active transport of Ca^{2+} is thought to be mediated by a Mg^{2+}-dependent and Ca^{2+}-activated ATPase. The sequestering of Ca^{2+} by membrane systems is thought to be necessary for muscle relaxations, as explained in the following chapter.

Many significant aspects of transport have not been touched upon here. It should be apparent, however, that functional membranes dividing compartments constitute a unique biological development and that transport processes participate centrally in virtually all physiological and biochemical control systems.

READINGS

Christensen, H. N., 1975, *Biological Transport*, 2d ed., W. A. Benjamin, Inc., Reading, Mass. This is a fairly detailed treatment of transport from a largely biophysical viewpoint.

Davson, H., and J. F. Danielli, 1952, *The Permeability of Natural Membranes*, 2d ed., University Press, Cambridge. A classic work which first appeared in 1943, now more important as a historical account of early permeability and membrane studies than as a source of information on transport processes.

Dowben, R. M. (ed.), 1969, *Biological Membranes*, Little, Brown and Company, Boston. A collection of articles by nine authors emphasizing various features of membranes. Some of the material is covered by more than one author, leading to redundancy, and much of it is esoteric, but the chapters by Davis (4) on metabolic aspects and by Schacter (5) on molecular mechanisms are lucid and valuable.

Nystrom, R. A., 1973, *Membrane Physiology*, Prentice-Hall, Inc., Englewood Cliffs, N.J. This is a very readable and integrated presentation of cytological, physiological, and molecular information on plasma membranes of animal cells.

Stein, W. D., 1967, *The Movement of Molecules across Cell Membranes*, Academic Press, Inc., New York. A very good account of the various types of transport and theories of mechanisms. Much of the material is devoted to the author's own work and interpretations.

MOTILITY AND CONTRACTILITY 15

"We know life only by its symptoms, and what we call life is to a great extent the orderly interplay of the various forms of work. Since the dawn of mankind, death has been diagnosed mostly by the absence of one of these forms, the one that expresses itself in *motion*."[1]

Motion is best exemplified by muscles, but motion in a living organism is not restricted to contraction and relaxation of muscle cells. The light microscope reveals a marvelous diversity of movements in a variety of forms—diatoms glide, cilia and flagella beat, chromosomes divide and migrate, particles spurt about within cells, pseudopods bulge out and shrink, cytoplasm streams around and about, mitochondria bend and contort, and so on. In fact, it seems quite likely that all cells are capable of utilizing metabolically released energy to do mechanical work. Some motions occur within certain specialized structures, but others appear to be a feature of undifferentiated or nonspecialized portions of cells.

When viewed properly with a microscope, the plastids, mitochondria, and various granular bodies of cells are seen to have independent movement. This is remarkable on at least two counts: (1) the forces which cause movement seem to be generated in various places and in any direction in the cell, and (2) electron micrographs reveal elaborate internal structure which would appear to preclude the kinds of freedom of movement that cellular inclusions are known to show. There is obviously a plasticity and a dynamism inherent in cell structure that is not revealed in the shadowy images of the electron microscope.

Knowledge of the true nature of movements of cells and parts of cells is not well advanced. This is largely because one generally has to work with intact cells where the contractile machinery exists among various other kinds of machinery. If any part of a cell is damaged, normal motility in that cell commonly is lost. Though studies which employ molecules or functional pieces isolated from cells can be instructive, investigations of motility must deal ultimately with the complex cell as the smallest functional unit.

[1] A. Szent-Györgyi, 1960, *Introduction to a Submolecular Biology*, p. 10, Academic Press, Inc., New York.

Most of what we know about cell movements comes from studies of muscle cells, especially the striated types; rather little is known about motility in other kinds of cells. Nevertheless, there have been some interesting recent developments in the study of primitive systems. These will be summarized after some of the better-characterized features of muscles are presented.

15-1 STRUCTURE AND FUNCTION OF MUSCLES

The larger organisms have special contractile cells. These are commonly grouped to form muscle tissues which may provide locomotion for the entire organism, pump body fluids, propel food through the gut, or form special sacs for storage or expulsion of various substances. When combined with a rigid framework on which to do work, muscles can perform very complex movements.

The skeletal muscles of vertebrates have been better characterized than any other motile structure. They are large, readily accessible, and perform well in experimental situations. (They also are a preferred diet for many predators, including gourmands.)

Skeletal muscles provide one of the means by which organisms respond to inputs of stimuli. They often constitute a major part of an animal's structure. For example, those of the human comprise about 40 percent of the weight of the body. Such muscles, designed for rapid and brief, or phasic, movements, are controlled by a nervous system. They usually function as part of a lever system, being attached at their ends by tough connective tissue (tendons) to portions of the skeleton. Often they occur as antagonistic pairs, one causing movement in a certain direction and the other in the opposite direction. Sometimes the contraction of one reflexly inhibits contraction of the other (Fig. 15-1).

Surrounding skeletal muscles is a connective-tissue sheath (the epimysium) from which septa (the perimysium) extend at irregular intervals

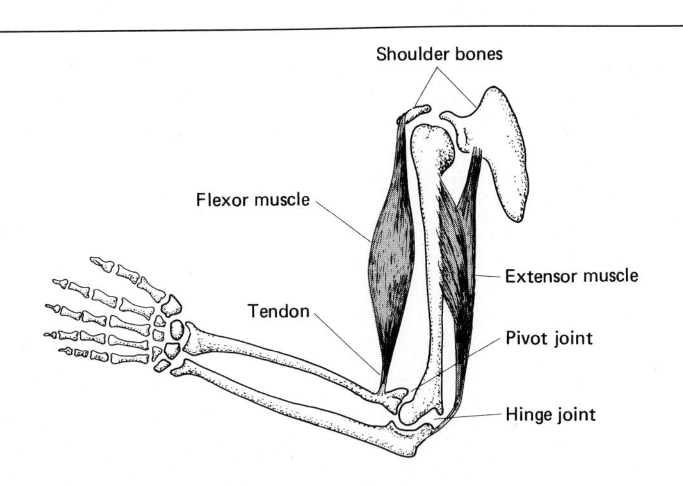

FIGURE 15-1
Antagonistic muscles of the human arm. Contraction of the flexor muscle (biceps) bends the arm at the elbow. Contraction of the extensor muscle (triceps) straightens the arm. The pivot joint at the elbow is also shown. Rotation of the lower arm on its long axis involves movement at this joint.

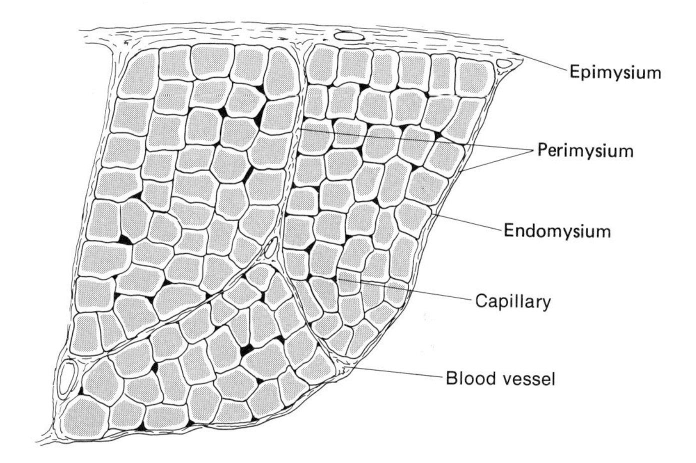

FIGURE 15-2
Diagram of the connective-tissue framework of a voluntary muscle. From the epimysium, which surrounds the whole muscle, septa pass inward to surround variously sized bundles of fibers. The connective-tissue sheaths of the bundles constitute the perimysium and from it delicate strands continue between the individual muscle fibers as the endomysium. Larger blood vessels are present within the perimysium, and capillaries lie between the individual muscle fibers. (*After E. H. Walls, in G. H. Bourne, 1960, The Structure and Function of Muscle, vol 1, Academic Press, Inc., New York, p. 23.*)

into the muscle and surround bundles (fasciculi) of muscle fibers. Continuous with the perimysium are connective-tissue strands (the endomysium) which surround each muscle fiber. All these connective tissues blend with the tendons at the terminals of muscles, which in turn are attached to bone (Fig. 15-2).

Muscle fibers of vertebrates may extend from the tendon of origin to the tendon of insertion, often being several centimeters in length. Others may attach to only one tendon, and some may reach neither end of the muscle. The diameter of such fibers varies between 10 and 100 μm. They also are multinucleated, a characteristic which has inspired the opinion that the fiber is a syncytium, arising by the fusion of separate cells, but it is not certain that this is so. Thus, the fibers of skeletal muscles are elongated, atypical cells (Fig. 15-3).

Skeletal-muscle fibers are encased by a plasma membrane-basement complex, the **sarcolemma,** and contain a fluid matrix in which lie parallel arrays of **myofibrils.** These are the contractile components. As seen in the light or electron microscope, they are transversely banded in a regular pattern, due to variations in density along their length. They are commonly about 1 μm in diameter and appear to extend the whole length of the fiber. The banded or striated appearance of the muscle cell, as seen in the light microscope, is due to the fact that the equivalent bands of its constituent fibrils are aligned in register (Fig. 15-3).

With a polarizing microscope it can be shown that the darker bands of myofibrils are **anisotropic,** which means that there are variations in some characteristic along different axes. (This is in contrast to **isotropic** substances in which there is random arrangement and therefore all

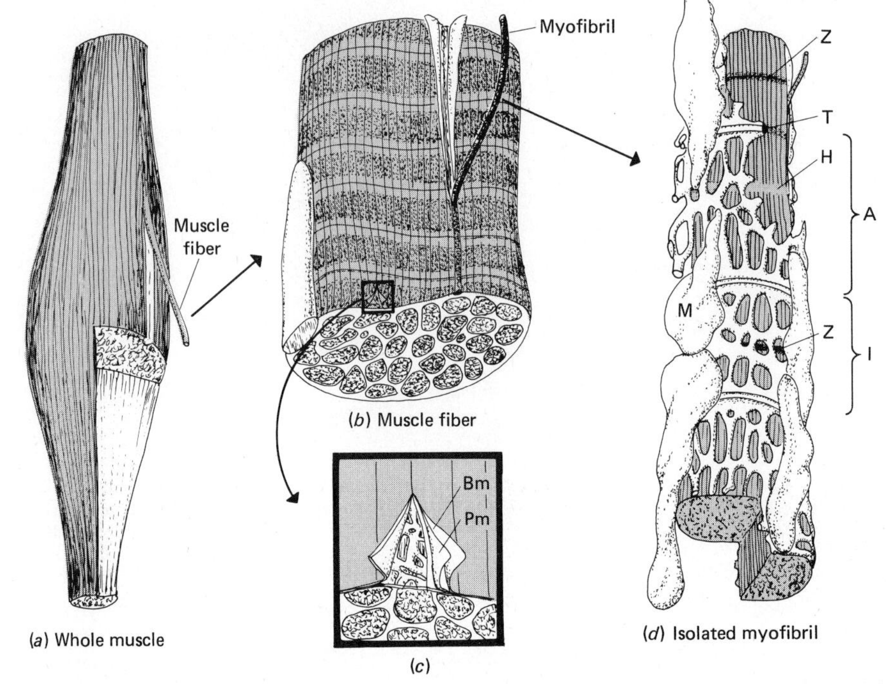

FIGURE 15-3
The structural organization of mammalian skeletal muscle at different levels of magnification. (*a*) The whole muscle. (*b*) The resting muscle fiber whose striated appearance is the net result of many contractile units, the myofibrils, being regularly aligned so that their crossbands are in register. (Magnification about × 1200.) (*c*) The components of the sarcolemma complex (*Bm*, basement membrane; *Pm*, plasma membrane) which encases the individual muscle fiber. (Magnification about × 25,000.) (*d*) The isolated myofibril whose crossbands are a result of the disposition of the thick and thin myofilaments. Embracing the myofibril is a complex network of membrane-limited tubules, the *sacroplasmic reticulum*, as it may appear in the "typical" mammalian muscle fiber. It has a repetitious pattern that bears a specific relationship to the crossbands of the myofibril. At the junction between an A band and I band the tubules are organized into a special complex called a *triad* (T) which is composed of three transverse longitudinal elements: a separate intermediate element (or T system) and two lateral transverse tubular elements which are respectively confluent with separate longitudinal anastomosing tubular systems overlying the adjacent A and I bands. (Magnification about × 50,000.) (*After H. M. Price,* 1963, *Am. J. Med., 35, p.* 590.)

physical properties are the same along all axes.) Anisotropy is generally characteristic of biological fibers because they are essentially ordered bundles of long crystals, each of which is anisotropic. Optical anisotropy can be demonstrated by passing a beam of light through the fibers at various angles of incidence. Two beams will emerge which can be shown to be plane-polarized, with their planes of polarization at right angles to each other. This phenomenon of *double refraction* is called **birefringence** (Fig. 15-4).

One of the emergent beams, the *ordinary*, obeys the usual laws of refraction at crystal surfaces, but the second beam, the *extraordinary*, does not. In contrast to the ordinary wave, which travels through the crystal with the same speed in all directions and therefore has a single index of ordinary refraction, η_o, the extraordinary wave travels with a speed that varies with the direction and therefore with a variable index of refraction, η_e. The degree of birefringence is given by the equation

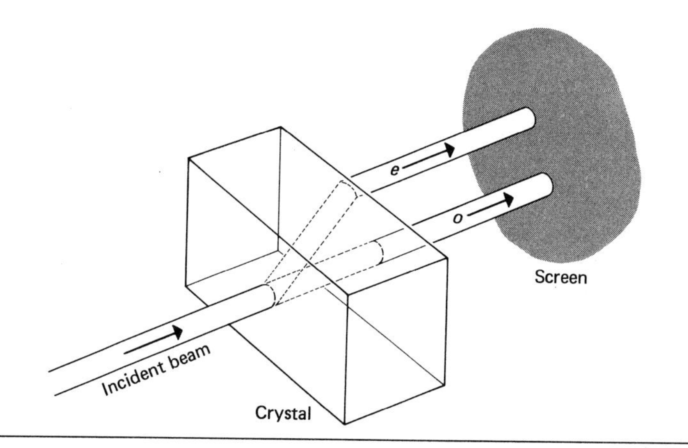

FIGURE 15-4
Double refraction, or birefringence, by a calcite crystal. A beam of unpolarized light is split into two beams, o, the ordinary and e, the extraordinary, which are polarized at right angles to each other. (*After D. Halliday and B. Resnick, 1962, Physics, John Wiley & Sons, Inc., New York, p. 1157.*)

$$\Gamma = \eta_e - \eta_o$$

An explanation of this phenomenon in terms of atomic structure is beyond the scope of this book. Fundamentally it results from the fact that the structural asymmetry of anisotropic materials manifests itself in the strength of restoring forces that act upon electrons which are caused to oscillate as they are propagating light. As these vary with direction, the direction of the transverse vectors of the wave will determine the speed of travel.[1]

Structures which produce contractile force generally are anisotropic. The marked birefringence of certain regions in muscle fibrils has served as a useful indicator of molecular arrangements. Many other biological systems are anisotropic, and quantitative investigations using polarized light have provided some understanding of their ultrastructure.

The darker bands of myofibrils are strongly birefringent and are therefore called *anisotropic bands*, or simply *A bands*. The alternating lighter bands are weakly birefringent, or nearly isotropic, and hence are called *I bands*. In addition to these bands, the living fiber can be seen to have a structure, called the *Z line* (from the German *Zwischenscheibe*, meaning "intervening disc"), bisecting the I bands. The interval from one Z line or disc to the next is called a **sarcomere.** It is regaded as a structural and functional unit of contraction (Fig. 15-5).

[1] An excellent analysis of polarization and isotrophy can be found in D. Halliday and R. Resnick, 1966, *Physics*, Chap. 46, John Wiley & Sons, Inc., New York.

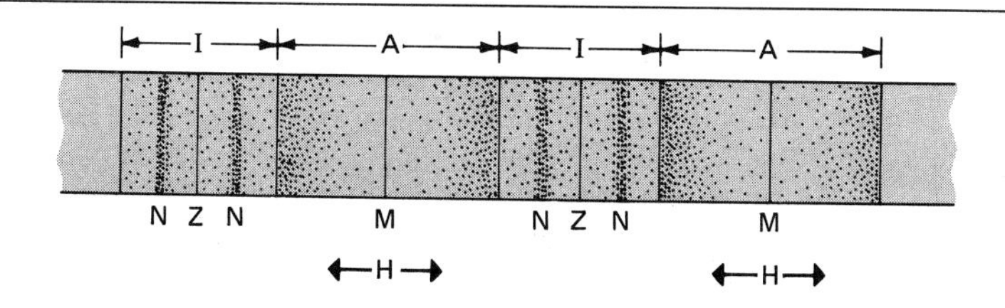

FIGURE 15-5
Diagram in which the main striations seen in a myofibril are named.

Additional bands (zones and lines) can be detected in fixed and stained material. Which particular features are seen depends upon the nature of the muscle and its condition (contracted, relaxed, or stretched) at the time of fixation, as well as on the techniques used to prepare and observe the structure. In the center of the A band of contracted fibrils the optical density is sometimes lowest. This band is called the *H zone* (from the German *heller*, meaning "clearer" or "brighter"). In the middle of the H zone, and bisecting the A band, is an ill-defined band called the *M line* (from *Mittelscheibe*, meaning "intermediate disc"). Occasionally a narrow dark striation known as the *N line* (from *Nebenscheibe*, for "disc next to or beside") is seen between the Z line and A band. None of the various bands is present during all phases of contraction, relaxation, or stretch (Figs. 15-5 and 15-6).

Observations made with phase-contrast and interference microscopes suggest that striated-muscle fibrils contain parallel arrangements of fibrous elements. Studies of living muscle using x-ray diffraction techniques (which are based on the fact that regularly ordered structures of suitable dimensions diffract x-rays in characteristic ways) support the concept of parallel, ordered filaments. Electron micrographs often reveal two kinds of filaments, differing in length, diameter, and position, each forming sets of fibers which interdigitate with one another. Electron micrographs of transverse sections of fibrils show that near the Z line the thin filaments are arranged in an approximately square lattice, but where the arrays of thick and thin filaments overlap they form a double hexagonal lattice with the thin filaments in trigonal positions (Fig. 15-7).

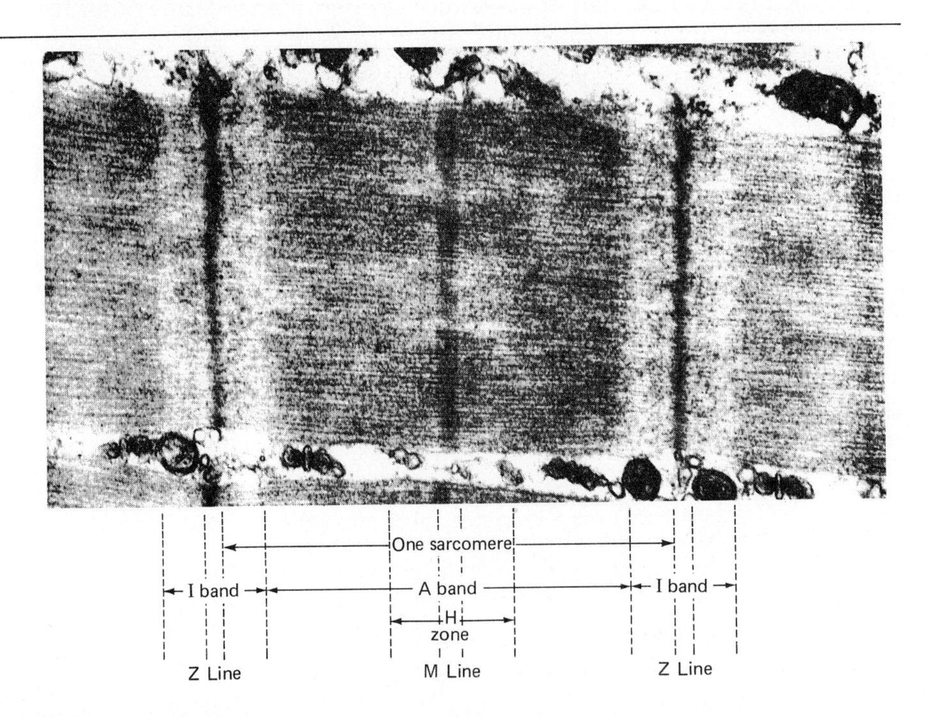

FIGURE 15-6
Electron micrograph of striated muscle. The sarcomere is bounded by Z lines. Each sarcomere contains an A band, of thick filaments, and two halves of I bands, consisting of thin filaments.

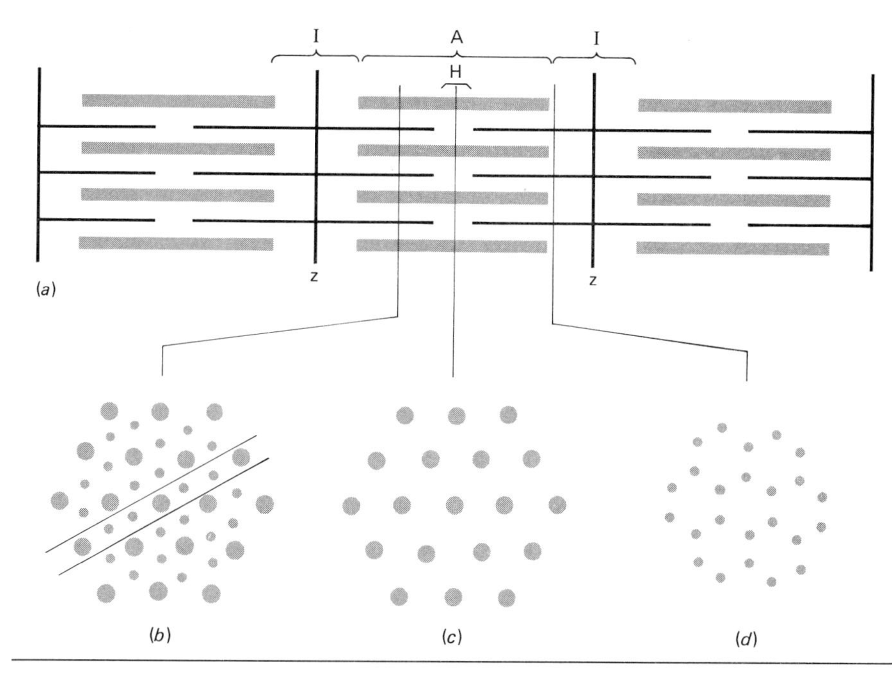

FIGURE 15-7
Arrangement of filaments within the bands of a myofibril. Sarcomere length corresponds to slightly extended length in the body. Transverse distances are grossly exaggerated. (*a*) Longitudinal view showing overlap of thick and thin filaments and their relationship to striation bands. (*b*) Cross section of region of overlap showing double hexagonal array. (*c*) Cross section through H zone showing hexagonal array of thick filaments only. (*d*) Cross section through I band showing thin filaments only. (*After H. E. Huxley,* 1962, *Chap.* 7 *in The Cell, vol.* IV, *edited by J. Brachet and A. E. Mirsky, Academic Press, Inc., New York, p.* 375.)

The thick filaments, which form the A bands, typically are about 16 nm in diameter and 1.5 μm in length. The thin filaments are only about 6 nm in diameter and 1 μm long (from Z line to the tip in the A band). In the I band the thin filaments are present alone. Where the thick and thin filaments interdigitate, at the ends of the A band, the optical density is greatest. The central region of the A band, which has no filament overlap, is the H band. The M line is thought to result from slight bulging in the middle of the thick filaments.

The thick filaments appear in electron micrographs to have short, lateral projections about every 40 nm, except in the central region, where a space of about 0.15 to 0.2 μm has none. The projections seem to connect with the thin filaments in regions of interdigitation.

Vertebrate *cardiac* muscle presents much the same cellular and subcellular appearance as just described for skeletal muscle but, in addition, has special junctions between cells known as **intercalated discs.** At the discs, the cell boundaries are close together (about 10 nm apart) and greatly folded so that their surface areas are increased. It is thought that this configuration provides a low-resistance region for current spread from active to inactive cells during propagation of the excitation which induces contraction of the heart. The discs are located between the sarcomeres, and thus I bands adjoin them (Fig. 15-8).

The thick and thin myofilaments constitute the contractile machinery of the sarcomere. They are polymers of proteins. The *thick* filaments are thought to be composed of a protein called **myosin.** This molecule is a long double α-helix chain with a molecular weight of about 500,000. It

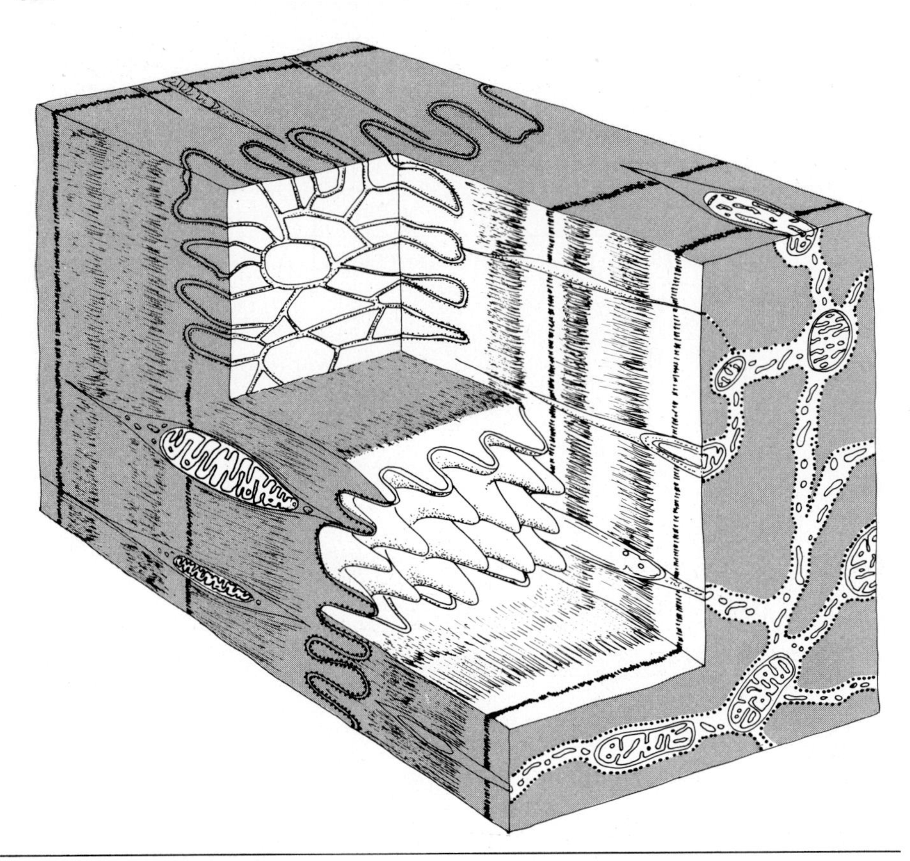

FIGURE 15-8
Three-dimensional diagram of the fine structure of a small segment of the junctional region between two cardiac muscle cells, showing an intercalated disc. The disc consists of highly interdigitated plasma membranes of two closely apposed muscle cells. The dense material resembling the substance of a Z band is concentrated in the cytoplasm adjacent to the cell membranes. Myofilaments of the I band insert into this dense material. The discs transect fibers in a stepwise manner. The disc in the upper portion of the diagram is shown in cross and longitudinal sections; in the lower portion of the diagram, the cell surface of one fiber is shown in three dimensions. Mitochondria containing small, dense granules and a few tubular and vesicular elements of endoplasmic reticulum appear in the sarcoplasm between the myofibrils.

can be split into two kinds of subunits, or *meromyosins*, by trypsin digestion. One is called light meromyosin (LMM) and the other heavy meromyosin (HMM), because of their differences in rates of sedimentation. HMM has ATPase activity and the capacity to combine with other muscle protein, actin; LMM has neither of these properties. Isolated myosin molecules have been examined in the electron microscope, and the HMM fragments are seen to be elongated structures with a globular region at one end. LMM fragments look like simple linear strands.

Precipitates from purified myosin solutions yield filaments with clusters of globular projections at both ends and a projection-free region in the middle. H. E. Huxley has interpreted this configuration to be due to parallel aggregations of myosin molecules oriented with their globular heads in one of two opposite directions and their linear tails overlapping (Fig. 15-9).

When resolution is good, electron micrographs of the *thin* filaments appear to indicate two beaded chains of globular subunits twisted around one another in a double helix. These filaments are thought to contain the protein called **actin.** Filaments prepared from actin solutions have the same beaded appearance. Actin can be isolated as a dispersed

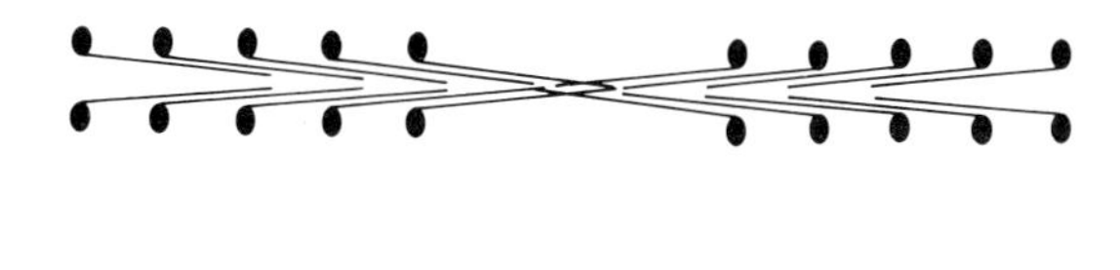

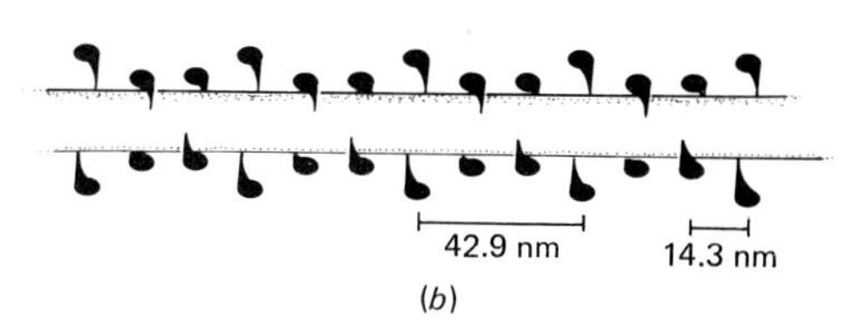

FIGURE 15-9
(a) The arrangement of the individual myosin molecules in the thick filament of skeletal muscle illustrated as a two-dimensional structure. The central core of the filament is formed by the long tails of the molecules, and the heads (globular regions) are at the outer ends. Light meromyosins form most of the tail; the remainder of the tails plus the globular heads are heavy meromyosin. (b) The relative, three-dimensional positions of the helically arranged HMM heads projecting from an end of a thick filament. These apparently are the cross bridges. There are six pairs of bridges per revolution, repeating at 42.9 nm.

monomer, called G-actin (for globular), and, by the addition of neutral salts and ATP, can be transformed to the fibrous polymer, F-actin (Fig. 15-10).

In addition to actin and myosin, other proteins are associated with the contractile apparatus. Although minor constituents, they appear to serve unique functions in muscle movements. **Tropomyosin** is a long, straight molecule, in a double-helix configuration, that is present in the Z disc and also forms the core of the thin-filament structure (Fig. 15-10).

Troponin along with tropomyosin is present in the thin filaments. In fact, what formerly was called tropomyosin turned out to be a mixture of tropomyosin and troponin. The sensitization of purified actin and myosin that was first attributed to tropomyosin now is known to be due to troponin. Unlike tropomyosin, troponin is a globular protein with a molecular weight of about 80,000. Antibody staining has shown that troponin occurs on the thin filaments at intervals of about 40 nm (Fig. 15-10).

Troponin was first considered to be a homogeneous protein, but later was found to be separable into two components, troponin A and troponin B. The Ca^{2+}-sensitizing feature was related to the A subunit, whereas the B subunit acted as an inhibitor of the interaction between myosin and actin. Still later, troponin was found to consist of three components when the B unit was subdivided. The largest of these newly discovered moieties (troponin T; mol. wt. about 37,000) is the subunit involved in the affinity of troponin for tropomyosin. The other (troponin I; mol. wt. about 22,000) is that which inhibits myosin-actin interaction. The original troponin A (now called troponin C; mol. wt. about 17,000) is the Ca^{2+}-binding protein.

FIGURE 15-10
Model for the molecular structure of thin muscle filaments. The arrows indicate the link between neighboring tropomyosin molecules. (*After S. Ebashi,* 1974, *Essays in Biochemistry, vol. 10, Academic Press, New York, p.* 6.)

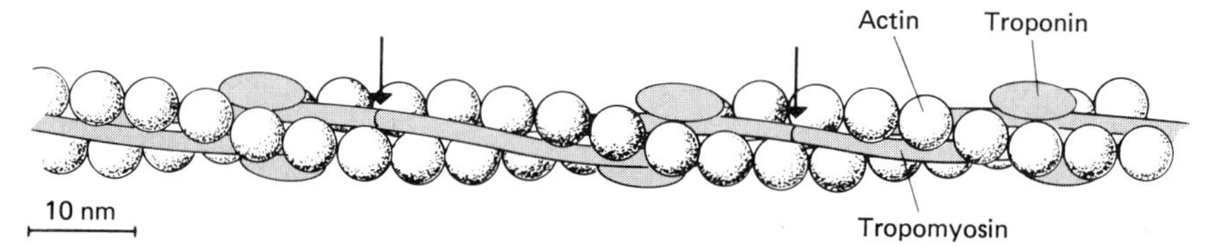

α-**Actinin** is present in the Z disc. This protein specifically binds F-actin and accelerates myosin-actin interaction. β-**Actinin** in another minor protein which has been shown to interact with actin.

The foregoing description of structure is based almost entirely upon intensive study of only a small number of types of vertebrate skeletal muscle, especially the rabbit psoas. The few other muscles that have been studied seem to have in common with vertebrate skeletal muscle certain features, viz., (1) there are two kinds of filaments in overlapping assemblies, (2) actin filaments appear to have the same structure, (3) projections (cross bridges?) are present on a second kind of fiber which may contain myosin, and (4) actin filaments are attached either to Z lines (in transversely striated muscles) or to "dense bodies" (in obliquely striated and unstriated muscles).

There are distinct structural differences among muscles, nevertheless. Even among striated muscles there are variations in fiber size, sarcomere length, development of the reticulum, and ratios of thick to thin filaments. A third kind of filament, in addition to the thick and thin ones, apparently has been detected in a few kinds of muscles. These, the so-called *T filaments* or *superthin filaments*, are extremely thin (about 2.5 nm in diameter) and may extend through the sarcomere from Z disc to Z disc.

A provisional classification of muscles proposed rather recently is based upon the structure of their fibers, especially as revealed by electron microscopy. It recognizes (1) striated muscles, (2) helical smooth muscles, (3) paramyosin smooth muscles, and (4) classic smooth muscles.

Helical smooth muscles are found in the locomotory muscles of annelids and cephalopods, as well as in other mollusks and in other phyla. They are unique in having smooth fibers (i.e., not striated) with helically arranged bundles of filaments. Such fibers give the appearance in phase contrast of a diamond lattice, or "double-oblique striation," when the microscope is appropriately focused to reveal both aspects of the helix.

Paramyosin smooth muscles contain fibers with longitudinal filaments of large size (sometimes more than 0.1 μm in diameter) which appear to be constructed from ribbon-shaped elements. They contain very large amounts of a water insoluble protein called **paramyosin** (or **tropomyosin A**) which forms the core of the thick filaments in various kinds of invertebrate muscles. It is found in muscles which typically serve tonic or "holding" functions, such as adduction of the valves of mollusks, and in some other muscles as well. Paramyosin muscles will be discussed further in the section on mechanics.

A great number of muscles have cells lacking striations of any kind and are generally called *smooth muscles*. They usually contain small, longitudinally or obliquely oriented filaments but are not grouped into well-differentiated fibrils. Though smooth muscles may have two kinds of filaments, it remains to be demonstrated in some cases—in vertebrate gut and uterus muscle, for example—that actin and myosin are located in separate filaments.

15-2 CONTRACTION OF STRIATED-MUSCLE CELLS

The *sliding-filament theory* of H. E. Huxley has gained general acceptance as an explanation of the mechanical events during muscle contraction and extension. It accounts for the various configurations that are seen during shape changes, and it is consistent with data from x-ray diffraction studies of muscles in various states.

Huxley's theory very satisfactorily explains the observation that muscles, over a considerable range of starting lengths, can shorten, exert tension, and perform mechanical work. According to the theory, when a muscle is contracting a relative sliding force is generated between two kinds of filaments by moving *cross bridges,* which are a permanent part of myosin filaments, reacting during contraction with a succession of sites (in a ratchetlike fashion) on the actin filament. In this way force is produced as the filaments slide past one another. When in the resting state, the filaments can slide past one another freely because the bridges are not attached, but as soon as the system is activated it can begin to develop tension because bridges can be linked up within a range of starting lengths.

The projections from myosin molecules, mentioned in the preceding section, are arranged around the longitudinal axis of the filament backbone in a spiral manner, six per revolution, as shown in Fig. 15-9. Again, these projections are the head part of the HMMs, which possess ATPase activity and the actin-combining property of the whole parent molecule. The sliding-filament theory proposes that they constitute the cross bridges which act between myosin and neighboring actin molecules. Because there are six bridges per revolution, it is reasonable to expect each myosin filament to be surrounded by six actin filaments. Such is indeed the case in many vertebrate skeletal muscles (Fig. 15-7*b*), but is by no means the only configuration seen in other vertebrate or invertebrate skeletal muscle.

Observations of living fibrils under the microscope have shown that while the length of a fibril changes the A bands remain constant in length. As the fibril shortens the I band shortens, and as it is stretched the I band lengthens. Also, the width of the H zone changes in proportion to the length of the sarcomere. In the rabbit psoas the I band completely disappears and each Z line touches the ends of two adjacent A bands when the fibril has shortened to about 65 percent of resting length. All these changes can be accounted for by a sliding-filament mechanism (Fig. 15-11).

Further support for the sliding-filament theory has come from measurements of the tension developed by a single muscle fiber (the semitendinosus of frog) as a function of sarcomere length (Fig. 15-12). As the stretched sarcomere shortens, the tension first rises from zero to its maximum. This would be expected as the number of bridges between the thin filaments and thick filaments progressively increases. Then, as shortening proceeds, the tension remains constant because the number of links is

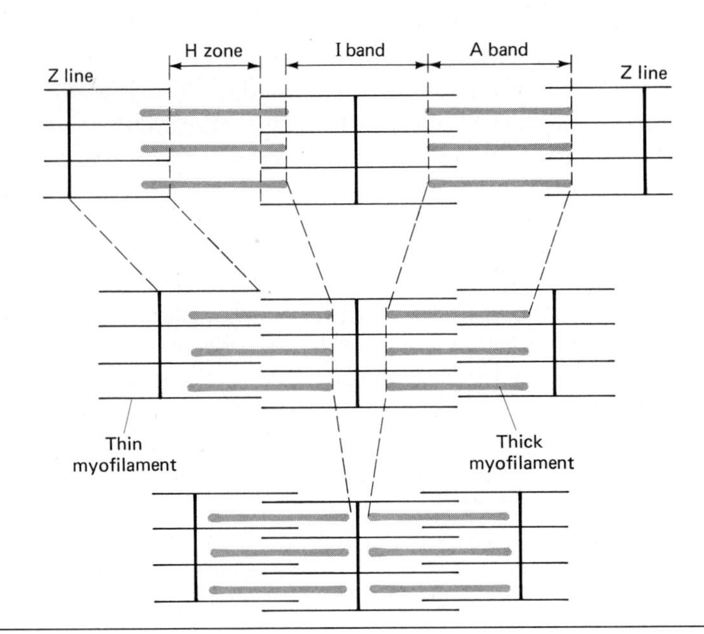

FIGURE 15-11
Changes in banding pattern as thick and thin filaments move past one another during contraction. (*After A. J. Vander, 1970, et al., Human Physiology, McGraw-Hill Book Company, New York, p. 215, 1970.*)

constant. The middle of the thick filaments has no cross bridges. At still greater shortening, the thin filaments cross the center and begin to overlap, entering regions of antiparallel cross bridges. Tension falls progressively, presumably because the orderly attachment of bridges is disturbed by double overlap of filaments. Finally, the ends of the thick filaments butt against the Z discs.

FIGURE 15-12
Changes of tension with shortening as accounted for in the sliding-filament hypothesis. (*a*) Standard filament lengths. $a = 1.60\ \mu m$; $b = 2.05\ \mu m$; $c = 0.15$ to $2\ \mu m$; $z = 0.05\ \mu m$. (*b*) Tension-length curve from part of a single muscle fiber (schematic summary of results). The arrows along the top show the various critical stages of overlap that are portrayed in (c). (c) Critical stages in the increase of overlap between thick and thin filaments as a sarcomere shortens. (*After A. M. Gordon et al., 1966, J. Physiol., 184, pp. 185 and 186.*)

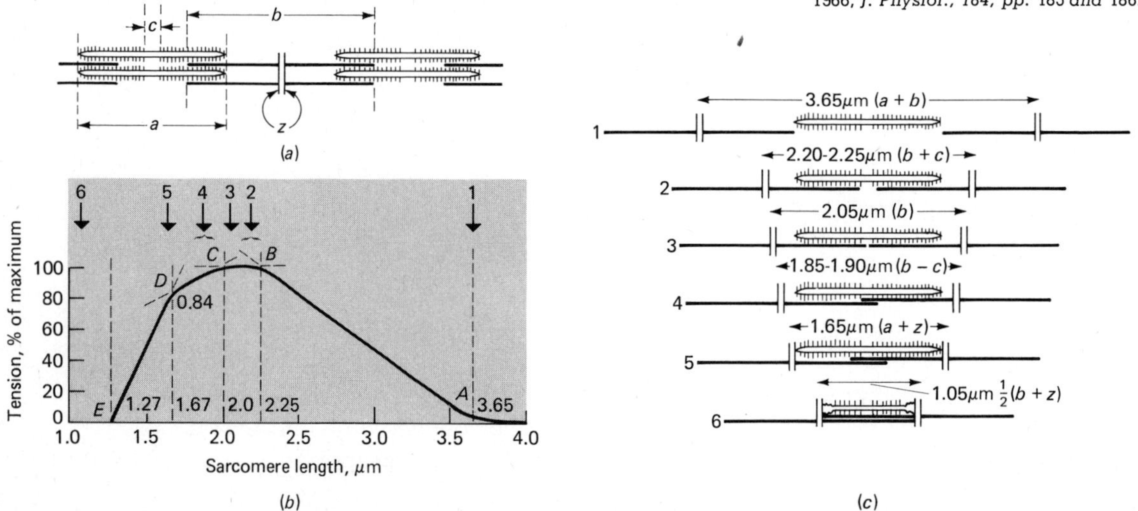

Contraction cannot presently be explained in all muscles in terms of this attractive theory. A "sliding-filament" mechanism probably operates in certain smooth muscles, but direct evidence for this view has not been obtained as yet. Although birefringence measurements and x-ray diffraction patterns of smooth muscles indicate that the length of individual filaments does not change appreciably with changes of muscle length, there is no evidence that they slide past one another.

Even among striated fibers there are some interesting differences in mechanisms. One has been described in the adductor muscle of the barnacle, *Balanus nubilis,* which can shorten down to about one-sixth of its resting length. Instead of the thick filaments simply abutting the Z discs at maximum contraction, they pass through them and overlap filaments in adjacent sarcomeres.

15-3 ACTIVATION OF MUSCLES

The fibers of skeletal muscles are generally organized into functional sets, a few or many of them receiving innervation from one nerve fiber. The motor neuron and the muscle fibers it innervates are known as a **motor unit.** The neuromuscular junction of typical vertebrate skeletal muscle is known as the **motor end plate.** When this type of muscle is stimulated by neural input, a conducted action potential sweeps across the sarcolemma and a rapid, simultaneous contraction of several muscle cells ensues. Muscles exhibiting fast contraction are called *phasic* muscles. Similar action potentials have been recorded from certain insect and crustacean muscles. Other muscles do not conduct excitation in this way but instead develop junction potentials at the end-plate terminals of the same neuron which impinges at many sites upon the membrane.

Some muscle cells have multineuronal innervation, with different ones for fast or slow contractions, and sometimes one for inhibition. Thus, gradations of rate or strength of contraction may be effected by varying the number of active motor units, varying the frequency of motor stimulation of individual units (increasing frequency causing greater contraction), or exciting different nerve fibers where multineuronal innervation exists.

Those muscles which are arranged around hollow structures and have one portion of their cells pulling against another portion are generally referred to as *tonic* or "holding" muscles. They are the main tissue of such organs as the gut, the bladder, some hearts, and the body walls of annelids and holothurians. They generally contract more slowly than phasic muscles and usually are nonstriated. Like the phasic muscles, many tonic muscles are organized into motor units, but others contract autonomously. Furthermore, conduction in some tonic muscles is entirely by nerves, but in others conduction from muscle cell to muscle cell occurs, as is the case for vertebrates visceral muscle.

Vertebrate cardiac muscle exhibits several features which resemble in part phasic muscle and in part tonic muscle. Like skeletal muscle it is striated and moves rapidly, but it contracts and relaxes spontaneously and rhythmically, like many phasic muscles, and it also conducts excitation from cell to cell.

Again we turn to striated-muscle cells for information about how excitation leads to contraction, a process called *excitation-contraction coupling*. Completely surrounding each muscle fiber is the electrically polarized **sarcolemma** (so named by Bowman in 1848, who also demonstrated its semipermeable character). According to electron microscopists, the sarcolemma consists of a plasma membrane and a sturdy, outer basement membrane (Fig. 15-3c). Depolarization of this surface membrane is the usual stimulus for contraction.

How the activating stimulus is conducted from the surface membrane to the protein filaments inside has been a challenging problem. It has been speculated that some chemical agent diffuses inward from the surface, but because such a process would be too slow to account for the rapid responses of muscles, it seems unlikely. More probable is some special conducting apparatus between the surface and the contractile elements. The labyrinthine, continuous system of membranes, known as the **sarcoplasmic reticulum,** which pervades the spaces between myofibrils, has been suggested for such a role. The membranes are associated in specific ways with the myofibrils, and thus certain structural features reoccur at fixed positions in every sarcomere. However, the sarcoplasmic reticulum is considered to be similar to the endoplasmic reticulum of other cells and, therefore, may actually serve several functions (Figs. 15-3*d* and 15-13).

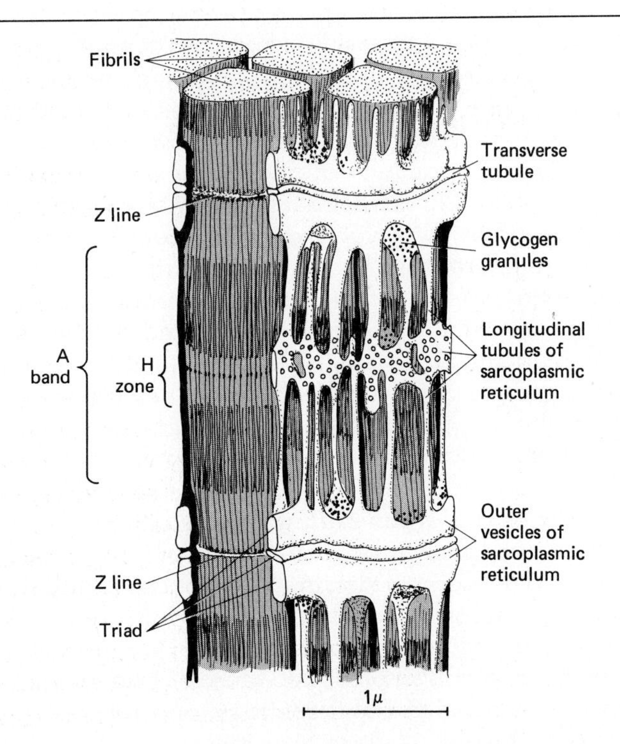

FIGURE 15-13 Sarcoplasmic reticulum of a striated muscle. (*After L. D. Peachey, 1965, Excerpta Medica International Congress Series, No. 87, p. 391.*)

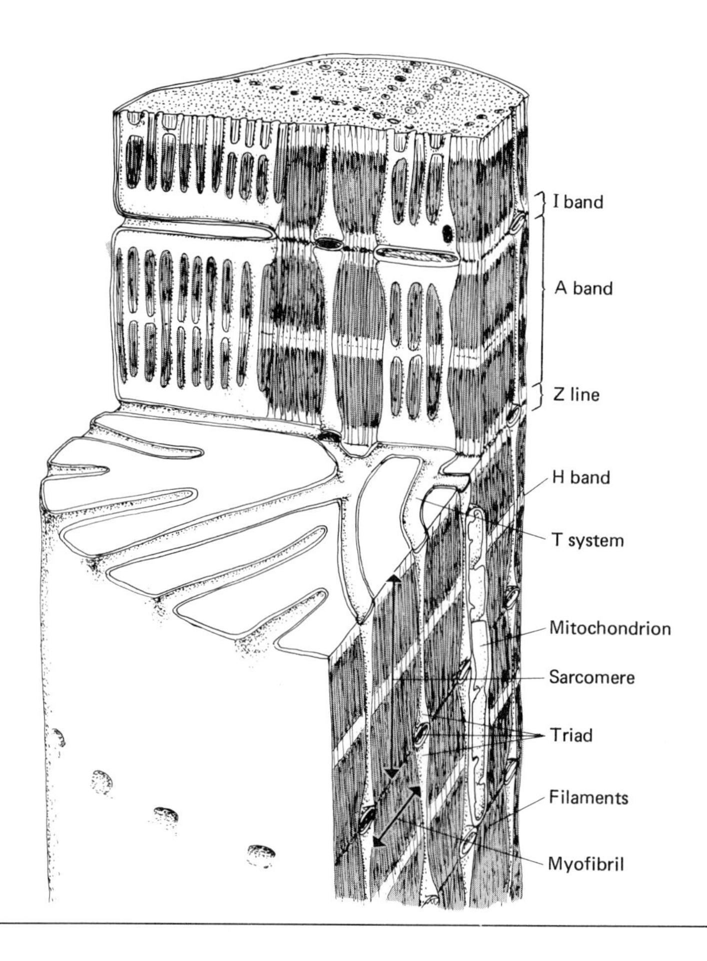

FIGURE 15-14
Three-dimensional representation of part of a skeletal muscle showing the membrane systems. Extensions of the membranes form a network of transverse tubules (the T system) which impinge upon the myofibrils at regular intervals. Tubules of the endoplasmic reticulum form special junctions with the T system to form "triads." (*After K. R. Porter and C. Franzini-Armstrong*, 1965, *Sci. Am. March, p.* 75.)

The existence of a special electric path from the surface to the contractile apparatus has been suggested by a discovery of A. F. Huxley and coworkers. They found that if a microelectrode was used to apply a highly localized stimulus to the surface membrane of a muscle fiber, only when the electrode was over the Z line would the underlying fibril contract. The contraction was confined to the immediately adjacent half I bands and it spread inward toward the center as the amplitude of the surface depolarization was increased. Thus, the response of the contractile apparatus appears to be graded (not all-or-none) and local (not propagated).

The position for effective electrical stimulation of contraction corresponds to where certain tubules of the fibril's membrane system connect with the surface membrane. Electron micrographs have been published which suggest that these tubules are actually a part of the surface membrane, extending deep into the interior of the fiber as transverse channels or the *T system*. Several experiments have indicated that these channels are open to the exterior and they contain extracellular fluid (Figs. 15-13 and 15-14).

There seems to be no continuity between the membranes of the T system and the sarcoplasmic reticulum. However, they do form close associations. Two vesicles of the endoplsmic reticulum with a T-system tubule between them form a configuration that has been named the *triad*. In some muscles the triad is associated with the Z line, but in others triads are opposite the junctions of A bands and I bands.

The physiological role of the T system definitely seems to be to conduct impulses inward from the cell surface. The T system has qualitatively the same electrophysiological properties as the surface membrane. Presumably, the excitation causes *calcium ions* to be released from certain specialized vesicles of the sarcoplasmic reticulum (and, perhaps, T-system tubules). Ca^{2+} is the trigger for activating the contractile process. At least for some muscles, it seems certain that calcium ions inhibit the troponin inhibition of the actin-myosin ATPase, which is Mg^{2+}-activated. Most interestingly, an autoradiographic study of ^{45}Ca distribution in a muscle fiber indicated that calcium is localized either in the interfibrillar space over the middle of the I band (i.e., about the Z line) or within the fibrils in the region of overlap between the thick and thin filaments.

Relaxation appears to depend upon sequestering of calcium by the sarcoplasmic reticulum. It has been proposed that a reticular calcium pump transports the ion against a large concentration gradient and thus reduces sarcoplasmic calcium to a level that allows cross-bridge breakage and thereby terminates mechanical activity. The pump is a *Ca^{2+}-activated ATPase* system, as mentioned in Chap. 14. Exactly what activates this particular reticular pump is not presently clear.

15-4 MUSCLE ENERGETICS

Muscle proteins are distinctive in their ability to transform energy from chemical form directly to mechanical form. (Human-made machines commonly change chemical to heat energy before doing work.) Although the chemical events of contraction and relaxation are not well understood, it is known that only nucleotide triphosphates will evoke contraction (ATP[1] is the most effective of these) and that magnesium ion is essential for contraction to occur. To effect relaxation, calcium ions must be pumped back into the reticulum at the end of activation, and this stops ATPase activity.

Evidence has come from many diverse experiments that the energy of hydrolysis of ATP is in some manner stored in the local interaction between myosin and released upon movement of the filaments. For example, ATP has been shown to cause contraction in simplified, synthetic contractile systems, and even a mixture of reconstituted, purified actin and myosin will simulate contraction on addition of ATP, though very

[1] See Fig. 17-3 for structure of ATP.

much more slowly than natural actomyosin. Evidence obtained from living muscles shows that added ATP is consumed in proportion to the work performed. By poisoning muscles with fluorodinitrobenzene, which inhibits phosphorylation of ADP by creatine phosphokinase, it has been shown that ATP breakdown occurs as an early event in contraction. Normally ATP concentrations fall very little because of rapid transphosphorylation from *creatin phosphate* (or arginine phosphate in most invertebrates).

The chemical energy for muscle work is ultimately derived from carbohydrate and lipid metabolism. A muscle can do work as long as its glycogen store lasts, even in the absence of oxygen. Anaerobic oxidation in muscle leads to the formation of lactate as an end product, and a concomitant fall in pH. Excessive lowering of pH will lead to a sustained contraction known as *rigor*. Even under aerobic conditions lactate can be detected after only a few twitches. For many years it has been believed that the re-formation of glycogen from lactate occurs in muscle cells, but considerable evidence now argues against such a reaction. The lactate escapes into the circulation and is oxidized in other organs, especially in heart muscle.

Shortly after the onset of activity, phosphorylase *b*, a less active form of the enzyme present in the resting muscle, is converted into the highly active form, phosphorlase *a*, which breaks the glycosidic bond of glycogen to free glucose. Electrical stimulation or exposure to epinephrine increases phosphorylase *a* activity in rat and frog skeletal muscle. Epinephrine rapidly activates the enzyme in mammalian cardiac muscle also. The conversion of phosphorylase *b* to phosphorylase *a* is catalyzed by phosphorylase *b* kinase. It has been assumed that stimulation of muscle leads to an increase in cyclic 3′,5′-adenosine monophosphate (cAMP), which in turn activitates the kinase (cf. Fig. 9-22). Other enzymes of the glycolytic sequence also are believed to be activated during contraction. Thus, when the contractile machinery of a muscle is stimulated, the supporting metabolic systems also are set in motion. However, just how much of a contribution glycolysis makes to contraction is uncertain.

Conventional aerobic oxidation leading to oxidative phosphorylation proceeds in mitochondria, which in muscle are usually called **sarcosomes.** (One kind of these particles, those from beef heart, has provided most of the information we have about structure and enzyme function in mitochondria.) Sarcosomes lie very close to myofibrils and perhaps have direct contact with them (Fig. 15-15).

In the substance between myofibrils is a red-pigmented protein, **myoglobin.** It has been speculated that it serves as a local, static store of oxygen because it binds oxygen at lower tensions than does hemoglobin. Presumably it could release oxygen during periods of circulatory arrest or during times of excessive oxygen demand. A more recent suggestion is that it functions to speed up diffusion of oxygen from blood to mitochondria.

FIGURE 15-15
Electron micrograph of insect flight muscle. Notice the densely packed sarcosomes between muscle fibers. (Courtesy of *David S. Smith.*)

15-5 MECHANICS OF MUSCLE

Since not all of a muscle is contractile machinery, a portion of the total work in contraction is invested in overcoming viscosity and elasticity of the total structure. Muscle is commonly regarded as having its contractile machinery in series (and partly in parallel) with elastic elements. Just what these elements are is uncertain. The tendons are a part, but their relative inelasticity makes it likely that a part of the elasticity resides in the contractile machinery itself. The mechanical situation can be represented in a highly simplified form by a model such as that shown in Fig. 15-16.

Much of what has been learned about the nature of contraction has come from experimenting with isolated muscle or muscle fibers in two kinds of recording situations. In one of these, a muscle is rigidly fixed at each of its ends and then stimulated. It will develop tension, though prevented from shortening. The response, known as **isometric contraction,** is something of a misnomer, since if a muscle does not shorten there is no "contraction." In truth, unless prevented from stretching its noncontractile structures, such as connective-tissue sheaths and tendons (series elastic components), the rigidly fixed muscle does shorten its contractile components. Also, to record the event, some slight movement must occur in the tension-measuring device. To some extent this type of contraction resembles the tonic contraction of postural muscles. Theoretically, no work is done in this situation because no load is moved; all the expended energy appears as heat.

In a second kind of experimental situation the muscle is arranged so that it is able to move a load by shortening. The resistance offered by the load will be less than the tension developed, and the muscle will perform mechanical work. A muscle under these conditions is said to engage in **isotonic contraction.** Models representing isometric and isotonic contraction are explained in Fig. 15-17.

Within a few milliseconds after activation (depolarization of the sarcolemma) and before tension starts to develop, there is a decrease of extensibility, or a stiffening of a muscle. This is the onset of a condition in the contractile elements known as the **active state.** As defined operationally by A. V. Hill, active state is the maximum load a muscle can support without further lengthening. If the true isometric response of the contractile elements could be recorded—i.e., free of the distortion contributed by the viscous elastic elements—one would obtain an active-state curve. In fact, this has been done rather well by a combination of mechanical methods, as next explained.

Following stimulation there is generally a latent period of a few milliseconds before tension starts to develop. If a small, quick stretch is applied during this period, the elastic elements are stretched and the muscle can develop tension unlimited by the velocity of shortening. Thus the active state is manifested. However, if the stretch is applied at the

FIGURE 15-16
Diagram of a mechanical model of a muscle fiber to explain the relationship of the elastic and the contractile components.

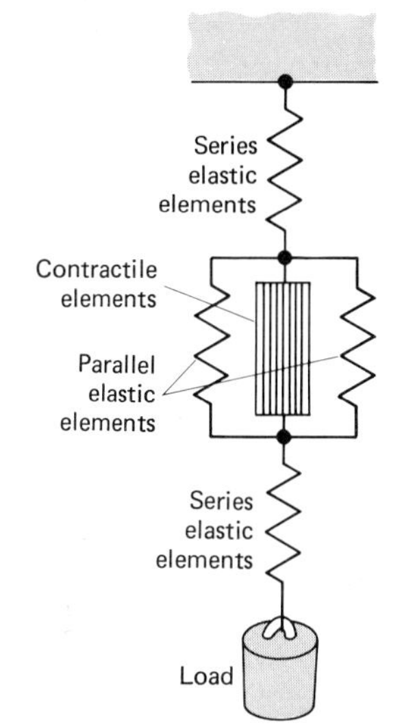

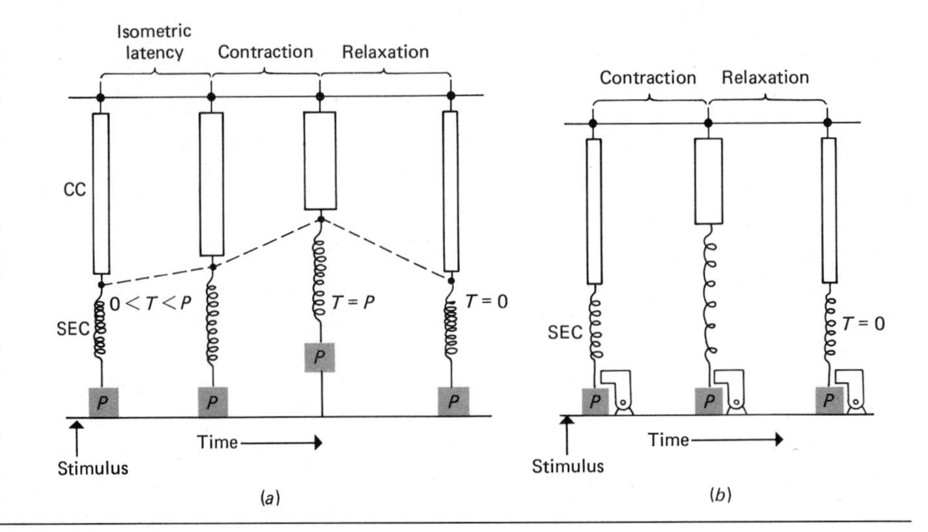

FIGURE 15-17
(a) Mechanical equivalent diagram of isotonic muscle contraction. CC, contractile components; P, load; SEC, series elastic components; T, tension. (b) Mechanical equivalent diagram of isometric contraction. Symbols as in (a). Note that tension is applied to the load even though the muscle as a whole does not shorten. (After A. Sandow, 1961, *Energetics of muscular contraction, in Biophysics of Physiological and Pharmacological Actions, edited by A. M. Shanes, American Association for the Advancement of Science, Washington, pp. 417 and 422.)*

peak of the twitch, no additional tension is developed because the velocity of movement is zero, and the contractile and elastic elements are at the same tension. Thus, quick stretches applied during the development of tension reveal the form of the active state at that time.

The decline of the active state can be detected by quick releases applied at various intervals after the stimulus. Changes of length must be sufficient to permit shortening of the elastic components so that the twitch must start anew from zero tension. A series of twitchlike tension curves is obtained, the peaks of which lie on the decaying active-state curve (Fig. 15-18).

Repeated stimulation of striated muscle will cause a sustained or **tetanic contraction,** because the duration of the active state is sustained by successive activations. The tetanic tension, or tetanus, thus serves as a direct measure of the maximum development of the active state.

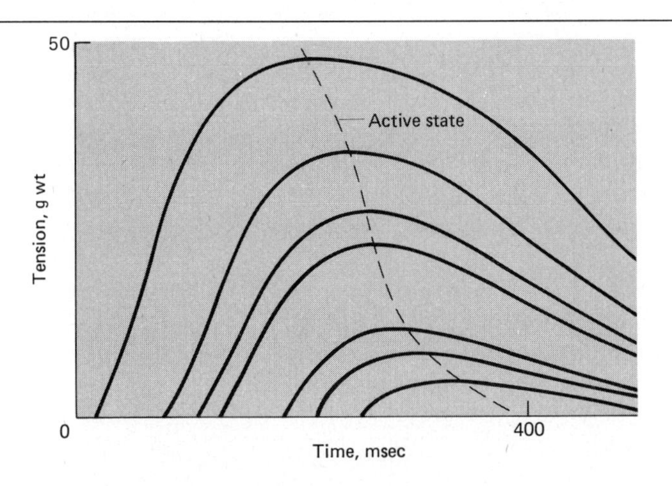

FIGURE 15-18
Tension-redevelopment curves recorded at various intervals after stimulation. The muscle is allowed to develop tension and stretch the series elastic components isometrically, then suddenly is released. The redevelopment of tension is recorded as a truncated twitch, the peak of which represents the active state, since neither the elastic nor the contractile components are moving at that instant.

Thus, the active-state curve gives the velocity of contraction and relaxtion of the contractile elements, whereas the time course for contraction of the muscle as a whole lags behind and is distorted by the viscous elastic elements (Fig. 15-19).

In some types of muscle fibers a single motor impulse can induce a noticeable contraction called a **twitch** (Fig. 15-19). Muscles showing this response are called "twitch" or "fast" muscles. A twitch is almost "all-or-none" and the active state may be almost fully developed following each activation. The classic of the physiology laboratory, the frog gastrocnemius, is such a muscle. Fast muscles are relatively rare. Most muscles require repetitive stimulations to induce a notable contraction and, consequently, are "slow" muscles. Slow fibers only partially develop the active state following each membrane activation. Maximal contraction is achieved only by repetitive stimulation as the active state summates to its maximum. Obviously, the distinction between "fast" and "slow" muscles is relative, there being a continuum in degrees of responsiveness to single inputs (Fig. 15-20).

Some unusual (and poorly understood) properties are shown by certian molluskan muscles. For example, if a pelecypod, such as *Pecten maximus*, is caused to close its valves on some rigid object, such as a block of wood, and then later the object is removed, the valves will remain rigidly set and resist opening or closing. This phenomenon was first described by von Uexküll in 1912, who referred to the adductors as "set" or "catch" muscles, indicating that they operated on the principle of a ratchet. He noted also that contraction and relaxation were functions of separate neural inputs. The adductor muscle of *Pecten* consists of two portions, one part effecting fast or phasic contractions, and the other the sustained or tonic contractions. It is the tonic muscles that contain the water-insoluble paramyosin mentioned earlier (Sec. 15-1).

The anterior byssus retractor muscle (ABRM) of the mussel, *Mytilus*,

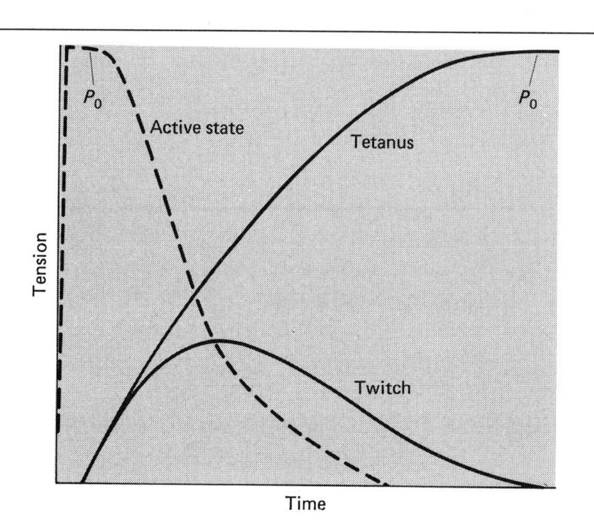

FIGURE 15-19
Diagram showing the relation between tension and time for a twitch, a tetanus, and the active state in a vertebrate striated muscle. P_0 is the maximum tension of which the fiber is capable. Stimulus begins at zero time.

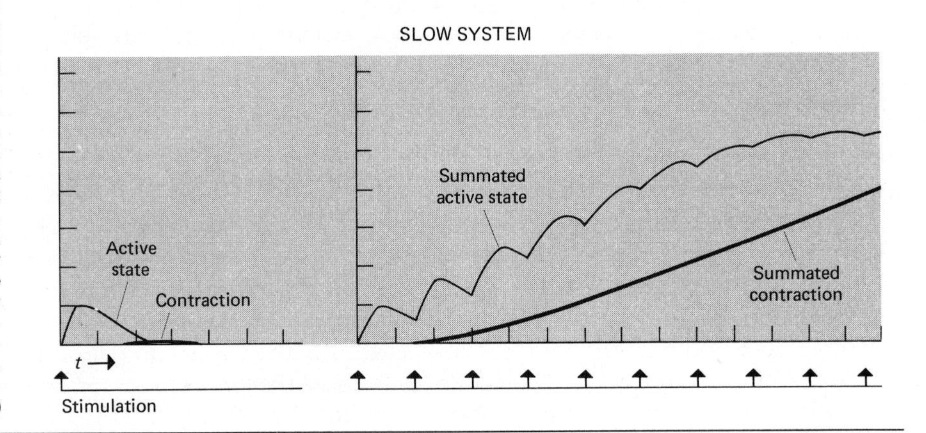

FIGURE 15-20
Active state and tension development of "fast" and "slow" muscle contraction. In both cases the active state can be evoked repeatedly, but in the fast system a single nerve impulse leads to nearly maximal development of the active state, whereas in the slow system the active state develops only partially. Between these extreme cases are all kinds of intermediates. A typical example of the fast system is the frog gastrocnemius muscle. A typical example of the slow system is the opener muscle of the crayfish claw. (*After E. Florey, 1966, An Introduction to General and Comparative Physiology, W. B. Saunders Company, Philadelphia, p. 549.*)

is another catch type muscle and one that has been rather extensively studied. It sustains tension for days on the attached byssus threads, which are tough fibers secreted by a gland in the mussel's foot that serve to hold the animal to the substratum. It is a most curious muscle in that only infrequent activation (an impulse every few seconds) maintains it in sustained passive tension and the active state decays long before the decay of tension. The metabolic cost of supporting this condition is extremely low, and high levels of tension are maintained for long periods without signs of fatigue.

Like the tonic muscles of *Pecten* and other mollusks, the ABRM contains paramyosin, a correlation which has led to the idea that this protein is involved in the catch mechanism, perhaps by being changed to a crystalline state. However, there is little evidence to support such a conclusion. The holding mechanism remains to be elucidated.

The muscles of flight and sound production in certain insects have the highest contraction frequencies known. For example, wings beat at about 240 cycles/second in honeybees and 580 cycles in mosquitoes. A midge is reported to achieve 1046 cycles/second. Muscles in these insects connect with the wings indirectly through a complex system of levers in the thorax. In flight the wings, thorax, and muscles operate as a resonant system, the muscles pulling in synchrony with the resonant frequency of the system. Nerve impulses are required to keep the oscillations going, but the frequency of nerve impulses arriving in the muscles is lower than the rate of wing beats. In flies there may be only one muscle spike per 5 to 20 wing beats. Such muscles are, therefore, said to be *asynchronous*. In butterflies and dragonflies, which have much slower wing beats, the frequencies of nerve impulses and wing beats are the same. Their flight muscles are, therefore, *synchronous* ones.

The asynchronous flight muscles of insects differ in many of their properties from normal striated muscles. Because they are composed of a few large fibrils they are sometimes called *fibrillar muscles*. They are very resistant to stretch, even in the unstimulated resting condition. Their stiffness increases only by a factor of about 2 during maximal excitation. This stiffness, a parallel elasticity, is not attributable to the sarcolemma and must reside in the myofibrils themselves. Interestingly, the apparent modulus of elasticity of active fibrillar muscle varies indirectly with the frequency of its oscillation. Furthermore, the isolated fibrillar muscle behaves like normal striated muscle in that it twitches in response to single stimuli and shows a smooth tetanus with high-frequency, repetitive stimulation. Rhythmic, mechanical activity occurs only when the muscle is connected to a suitable load. Additionally, fibrillar muscle does not behave like normal striated muscle in response to rapid increases or decreases of length. In frog muscle, for instance, there is simply a transient rise or fall before the steady tension is reestablished at the new length. Fibrillar muscle under the same conditions shows not only immediate transient changes of tension, but also a further rise or fall after a slight delay. Such concepts as "active state," as developed in the study of vertebrate muscles, do not explain these features of insect fibrillar muscles.

It should now be apparent that there is great diversity in the nature of movements of the different types of muscle cells. Their specializations afford excellent examples of structure-function interrelationships. Also, the force generated by certain contracting-muscle cells is impressive; no other cells are capable of such singular accomplishments. As noted earlier, their direct use of chemical energy to do work distinguishes muscles from any man-made devices. Muscles are remarkable in yet another way; they can be stopped from moving. Though our attention tends naturally to concentrate on the magnificence of the contracting movements of muscles, it should be appreciated that they also can be controlled as to when and how much they move.

15-6 MOVEMENTS OF CELL PARTICLES AND ORGANELLES

Particles in a large variety of cells have frequently been observed through the light microscope to show movement that is statistically different from Brownian motion and independent of cytoplasmic streaming. Such particles are generally not identified specifically, but are probably certain of the small organelles and granules, such as yolk and melanin. This form of movement in a particular cell type is often restricted to one class of particles. Their motion is characteristically saltatory, i.e., discontinuous, stop-go, or jumplike. A given particle will not move for perhaps minutes and then suddenly traverse distances of up to 30 μm at a velocity as great as 5 μm/second and then suddenly stop. Such particles usually seem to be independent of neighboring particles in their movements, and during motion they do not appear to change form or size.

Some investigators of this phenomenon have proposed that the mechanism which causes these motions is the same in all cells in which they occur and that the motive force arises from a system outside the particles. It is speculated that there are many contractile fibrils attached to the surface of the granules, with their other ends attached to the plasma membrane or to the endoplasmic reticulum. These contractile elements are presumed to be formed and broken down continuously.

Studies of the ultrastructure of cells that exhibit saltatory motion of particles have revealed fibrillar material in the form of *microtubules* or *microfilaments*. However, there seems to be a lack of constancy between the structure of the fibrillar material and the distribution of particle movements among cell types. Thus, we are left with the mechanisms for such motion yet to be explained with certainty.

Certain animals which show physiological color change have cells containing granules of pigment which can migrate into and back from processes of the cell. The granules of melanophores in fish, for example, are aligned in files along straight microtubules, as if they are being oriented and moved along specific channels by the tubules.

Mitochondria have been observed to undergo a variety of shape and size changes. There have been reports of formation of loops by end-to-end fusion, fusion of two or more mitochondria, formation of syncytial complexes of several mitochondria, amoeboidlike extensions and retractions of arms, and local contractions. Mitochondria are found to assume a variety of structural and functional forms in response to a number of agents and conditions. Such substances as Ca^{2+}, thyroxine, heavy metals, and a variety of detergents, as well as hypotonic conditions, all cause mitochondria to swell. Other agents and conditions inhibit swelling and induce some contraction. These include exposure to Mg^{2+}, ATP, and hypertonic sucrose. The response to ATP seems to be a true contraction, causing the extrusion of water molecules from the mitochondria. It seems likely that a protein is responsible for these contractions, and, in fact, a contractile protein which changes shape and size in the presence of ATP and catalyzes the splitting of the terminal phosphate of ATP in the presence of Ca^{2+} can be extracted from mitochondria.

Studies of the ultrastructure of contractile vacuoles of several organisms have shown that they contain fibrillar or tubular components about

20 nm in diameter. It seems likely that these fibrils may be responsible for contracting the vacuole and for controlling their discharge through ducts to the exterior.

When the eukaryotic cell divides, there is formed a fibrillar spindle which is somehow involved in the poleward movement of the chromosomes during anaphase. It is disassembled when the chromosomes are separated into daughter cells. By treating dividing sea-urchin eggs with mild detergent (digitonin) and then gently washing and centrifuging, Mazia was able to isolate intact mitotic apparatuses consisting of asters, spindle, and chromosomes (Fig. 15-21).

Spindle fibers are anisotropic, irregular in size and shape, and composed of filaments or microtubules, about 15 nm in diameter, which in cross section have a dense cortex and light center.

The mechanical integrity of spindle fibers in situ has been demonstrated by micromanipulation studies. They can be pulled and stretched, and when fibers are broken new ones can be re-formed rapidly. In fact, the fibers in living cells are extremely labile and dynamic in nature, in contrast to the stable spindles isolated from cells. It has been speculated that spindle fibers consist of oriented polymers in an equilibrium with dissociated molecules.

The mechanism by which the fibers move chromosomes to opposite poles has not been resolved. They do not become thicker as they shorten, but rather seem to disappear into the centrosphere region. Certain evidence based upon optical properties has been used to hypothesize that instead of folding or unfolding polypeptide chains, the fibers may elongate by adding and aligning additional molecules or shorten by removing molecules. Whether or not these events account for the motive force that separates chromosomes is an unanswered question. The velocity of chromosome movements is on the order of one to a few microns per

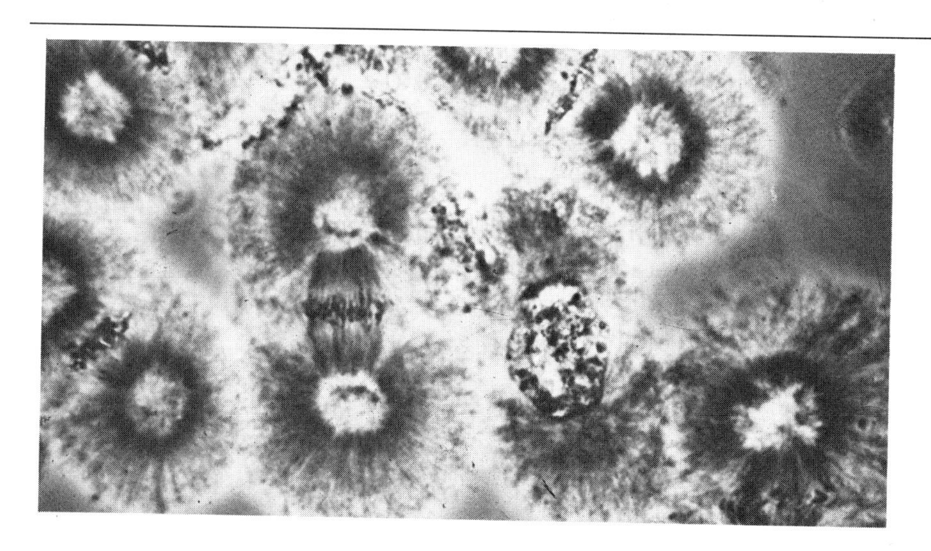

FIGURE 15-21
Isolated mitotic apparatus at anaphase (developing eggs of the sea urchin *Strongylocentrotus purpuratus*). Note mitotic centers, astral rays, chromosomal fibers connecting chromosomes to centers, and interzonal regions of spindles between separated chromosomes. Phase contrast ×2,000. (*Courtesy of Daniel Mazia.*)

second, and perhaps the growth and dissolution of filaments is in accord with these rates.

15-7 CYTOPLASMIC STREAMING AND AMOEBOID LOCOMOTION

Many cells, when examined under high power of the light microscope, can be seen to have continuous currents moving through them. Often, particularly in plant cells, a movement known as **cyclosis** occurs in the endoplasm region, i.e., between the central vacuole and a thin layer of ectoplasm, or cortical gel, associated with the cell membrane. The entire endoplasm streams in a circular manner. Various studies have shown that the velocity of movement, which in algal cells can be as high as 60 μm/second, is greatest in the most peripheral part of the endoplasm. This fact has been used to argue that the driving force for cytoplasmic streaming originates near or at the interface between the endoplasm and the ectoplasm.

In the alga *Chara*, there are tubules which can just be resolved by the light microscope that have been observed to rotate rapidly and to aggregate and form hollow polygons which undulate. In electron micrographs parallel arrays of tubules seen near the surfaces of the cortical gel appear to be bundles of microtubules, the latter being about 20 nm in diameter. They appear to be the same kind of microtubules that are present in mitotic spindles. Thus, microtubules seem to be associated with sites of cytoplasmic movements, but the evidence that they are responsible for streaming is quite inadequate. Calculations based upon estimates of fibrillar tensions of muscles have suggested that the forces involved in cyclosis would be quite small, so that if a fibrillar mechanism is responsible, the number of fibrils per unit area required to produce the movement would be minute.

There is evidence that the motive force for streaming is dependent upon ATP. This suggests, by analogy with muscle contraction, that contractile proteins at or near the gel-sol interface may be responsible for the movement. In some unknown way shape changes, such as winding and unwinding of helical molecules, may move the sol-phase substance past the fixed, gel-phase material.

It also has been proposed that the force for steaming is electrical, rather than mechanical. According to this idea, in addition to the active transport of ions which produces electrical potential across the plasma membrane, there are local potential differences *along* the surface of the membrane of certain cells, due to a mosaic arrangement where ions are pumped out at some places and then leak back through the membrane at others. The local, circular current flow thus induced provides the force to drive the cytoplasm (Fig. 15-22).

This scheme relies in part for support on the fact that neuron fibers produce local currents during the passage of an action potential. Also, it has been noted that cytoplasmic streaming is arrested when cells of the

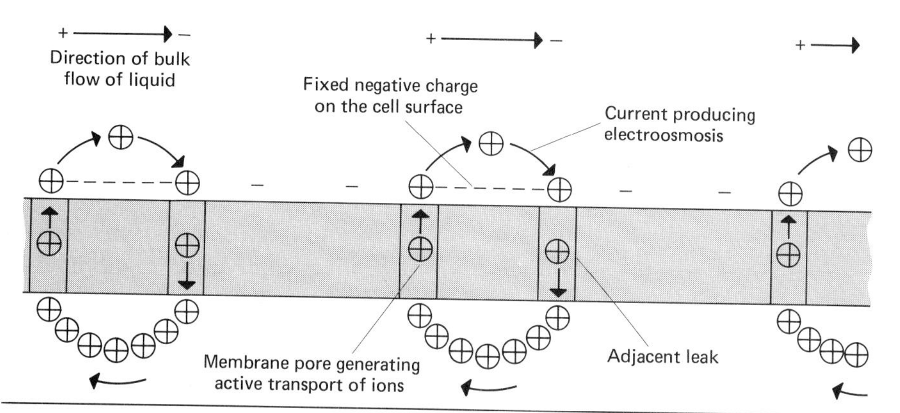

FIGURE 15-22
A mechanism of electroosmosis that might account for streaming. Adjacent to ion-pumping pores are regions of membrane leak. The circular currents produced provide the force to drive cytoplasmic streaming.

alga, *Nitella*, are stimulated by electrical pulses and thereby are caused to generate and to conduct action potentials. Charges of membrane potential induced by altering ion concentrations across the membrane also affect streaming. Therefore, the nature of the electrical potentials of the plasma membrane seems to be importantly involved in the mechanism of streaming.

Some cells and a few primitive plant cells are capable of a kind of locomotion called **amoeboid movement,** a form of cytoplasmic streaming which results in net locomotion by the cell. It has been best studied in the large *Amoeba proteus*, which moves unidirectionally by extending broad, lobose pseudopodia. For almost 150 years theories have been offered to explain how amoeboid movement is accomplished; we will examine only some representative ideas about the process.

One suggestion is that amoeboid movement is rather like cytoplasmic streaming, with long molecules or fibers moving in opposite directions. It requires that some sort of interaction (shearing force) occur between elements in the stationary (gel) zone and the moving (sol) region to produce motive force. The nature of such a mechanism may be like that which is proposed for actin-myosin sliding in muscle cells (Fig. 15-23*a*).

Others propose that the motive force is exerted at the front of the amoeba and "pulls" it forward. They believe that the semiliquid endoplasm steaming up the center of the cell is pulled forward by the contraction resulting from gelation occurring near the head end, or in the "fountain zone" (Fig. 15-23*b*).

On the other hand, a long-standing theory, developed by several investigators, holds that the motor region is near the tail. In this rear-gel contraction theory the tail is considered to be physiologically, as well as morphologically, a distinct region (Fig. 15-23*c*).

One of the supporting pieces of evidence for this last interpretation is that when ATP is injected into various sites in the cell, a vigorous contraction begins immediately and the cell membrane becomes wrinkled at the site of injection. This wrinkling resembles the normal appearance of the

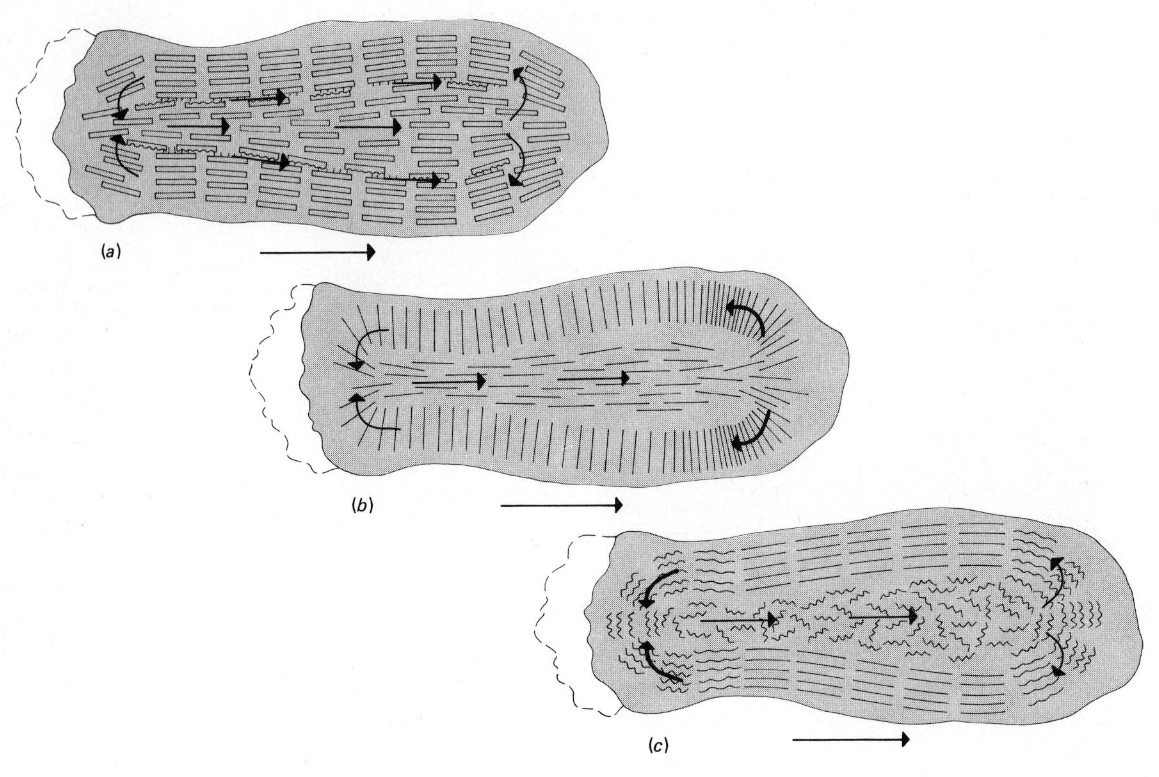

FIGURE 15-23
Theories of amoeboid movement are illustrated by the diagrams. (*a*) The sliding or gel-endoplasm shearing theory. Forces are developed between long molecules, the outermost ones at the edge of the gel phase pushing by some kind of "ratchet" or "bridge" mechanism the more inner molecules of the sol phase. (*b*) The "fountain" theory. The motive force results from contraction of the cortical gel near the head end. (c) The rear-gel contraction theory. The contraction of fibrous molecules at the tail end of the animal causes locomotion.

amoeba's tail and suggests that in the tail region there may be a continuous release of ATP. Furthermore, when ATP is injected into the front end of the amoeba, it reverses direction and the front end becomes the tail. However, because the effects of ATP depend in large part on concentration and ionic conditions, the interpretation of such results is very difficult. On the other hand, a myosinlike ATPase has been isolated from *Amoeba*.

Movement of certain slime molds (myxomycetes) is somewhat like amoeboid movement. However, a slime mold is an irregular, branching plasmodium, which may be moving in various directions simultaneously. *Physarum polycephalum* is the best known of these molds; its specific name derives from its "many-headed" nature. *Physarum* may spread over an area of 30 cm^2 or more and contain hundreds of synchronously dividing nuclei. It also shows an interesting kind of pulsating movement, the branches or filaments advancing for a few moments and then reversing direction, followed by another advance, etc. This motion has been called **shuttle streaming.** Movement is generally greater in one direction than in the other, so that the organism as a whole advances or retreats (Figs. 15-24 and 15-25).

There have been reports of musclelike proteins found in extracts of myxomycetes, and it has been assumed that these provide the motive force for streaming. One has been described as myosinlike (*myxo-*

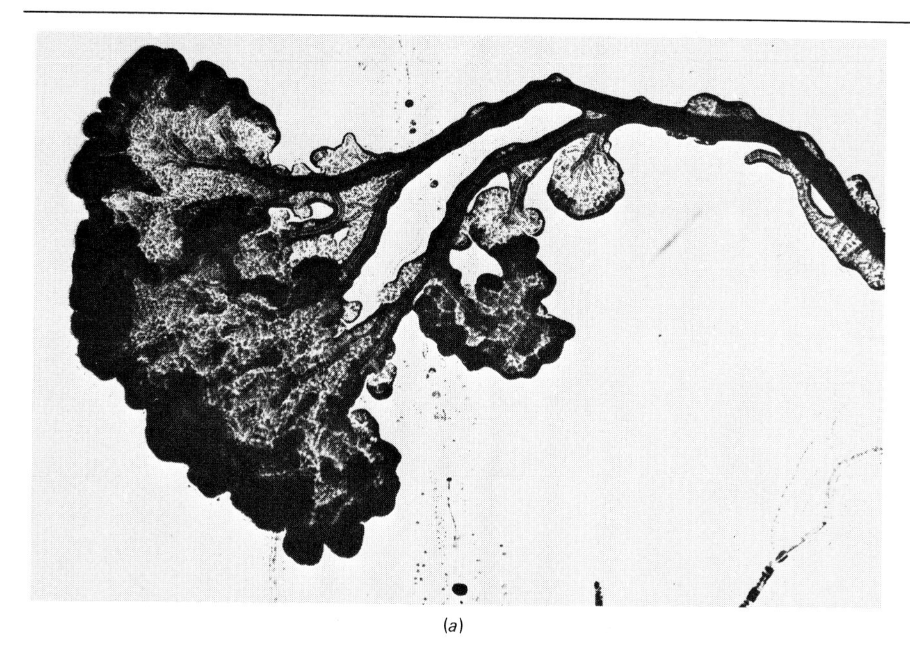

(*a*)

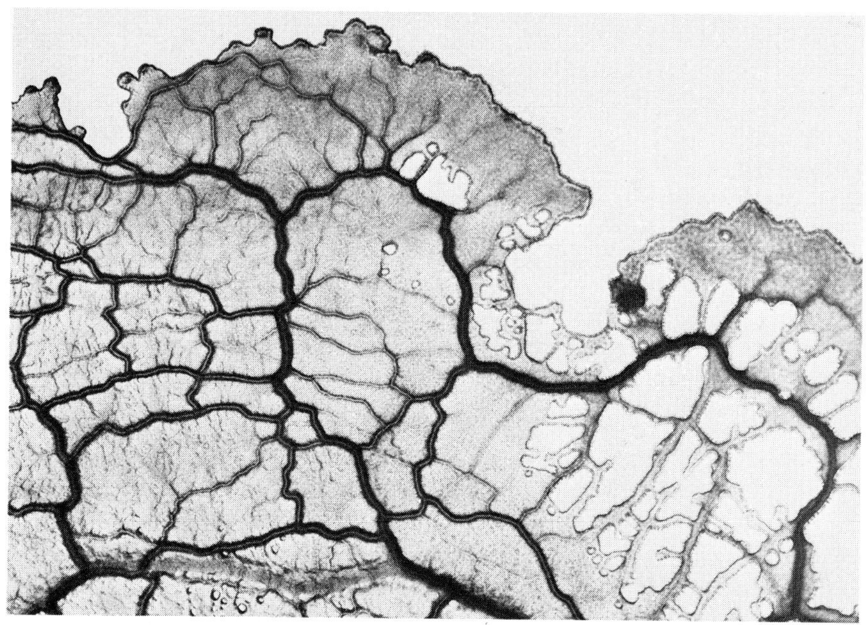

(*b*)

FIGURE 15-24
(*a*) A photograph of the myxomycete slime mold, *Physarum polycephalum*, showing its branched plasmodium. (*b*) A close-up of the plasmodium showing a network of veinlike structures that enclose a flowing protoplasmic stream, which rhythmically reverses its direction of flow. (*Carolina Biological Supply Company.*)

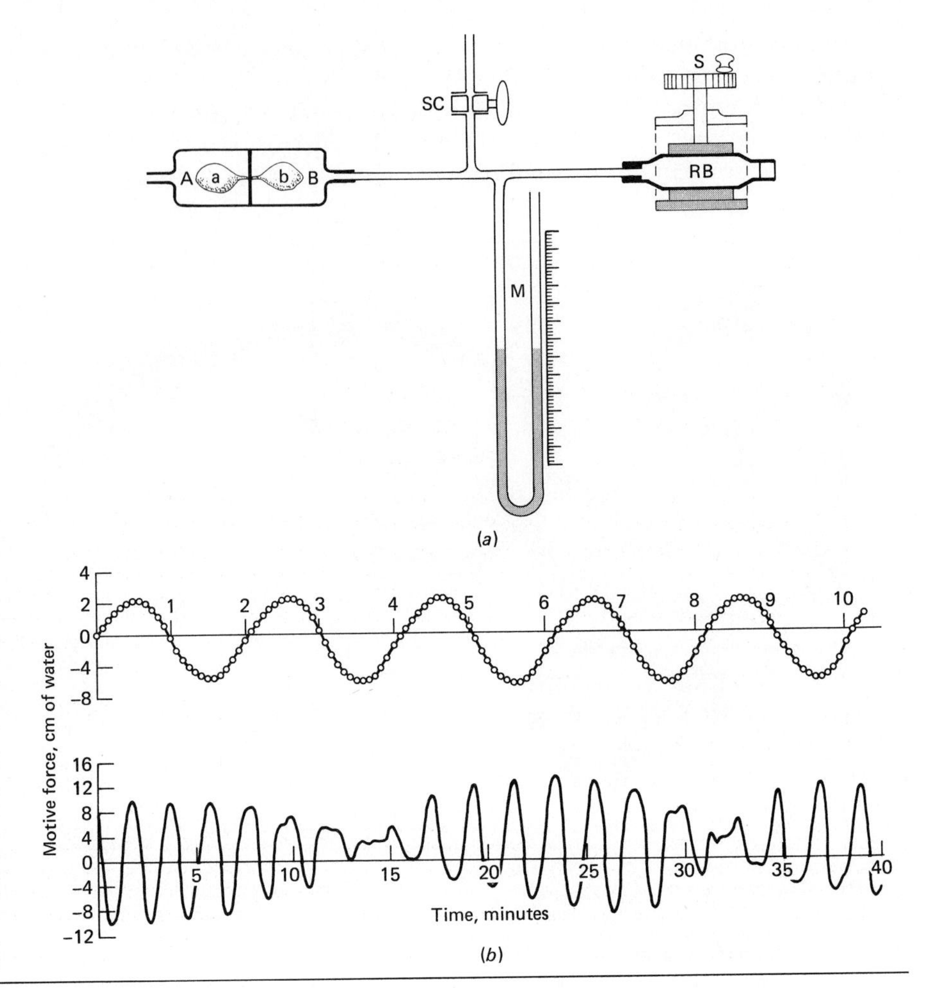

FIGURE 15-25
(*a*) Diagram showing the general arrangement for measuring the motive force of protoplasmic streaming in a myxomycete plasmodium. Note double chamber (*A,B*) with myxomycete (*a,b*) connected through a small hole between the chambers. *M* is a water-filled manometer, *SC* a stopcock, and *S* a screw to control the pressure applied to the bulb (*RB*). (*b*) A recording of the motive force in centimeters of water. Note the beatlike wave pattern under normal conditions. The upper line is a recording showing all the points recorded several times a minute. The lower line shows a record in which the points are left out and the abscissa is compressed to make possible the showing of data for a longer period of time. (*After N. Kamiya,* 1959, *Protoplasmatologia, 8,* sec. 3*a*, pp. 41, 43.)

myosin), since it is an ATPase that is stimulated by Ca^{2+}, is inhibited by Mg^{2+}, and has some physical properties like those of the muscle protein. More recently the presence of an actinlike protein in a slime mold has been described. Its molecular characteristics are said to be like the G-actin of muscle. This protein can be combined with myosin A from muscle, and the complex will act much like the actomyosin from muscle.

15-8 MOVEMENTS OF CILIA AND FLAGELLA

Cilia and flagella are motile, hairlike organelles which project from the free surfaces of certain cells. The term "flagella" is used when a few, long structures are present, and "cilia" when they are short and numerous. They are found in all animal classes except nematodes and crustaceans and in all plants except angiosperms and gymnosperms. Many bacteria have flagella, but their structural organization is different from that of other organisms.

Studies with the electron microscope have provided a very good idea of how cilia and flagella are structured. All of them, whether from plants or animals, are remarkably alike. Typically, they consist of a complex bundle of fibers (or axoneme) with a matrix which is enclosed by a membrane. In most of the organisms examined it is not certain that this membrane is continuous with the plasma membrane. The axoneme is almost always arranged as a bundle of nine pairs of outer fibers and two central fibers which extend its entire length. A few cases have been reported in which the two central fibrils are absent (Fig. 15-26).

Flagellar motion is often more complicated than ciliary motion. Cilia are always stroked like an oar, but some flagella move only at the tip, like a propeller. Flagella of bacteria are thought to rotate in a conical path, causing the bacterium itself to rotate in the opposite direction. Cilia usually beat continually and act as independent effectors, but in some invertebrate embryos they appear to be coordinated by nerves. In protozoans the beat of cilia appears to be coordinated by a system of fibers.

The most common type of ciliary beat consists of an effective stroke and a recovery stroke. During the effective or power stroke the cilium moves as a rigid, slightly curved fiber, bending at its base. In the recovery stroke a bending starts near the base of the cilium and proceeds as a wave toward the tip (Fig. 15-27).

An interesting hypothesis proposes that during the effective stroke five of the nine outer filaments contract (or bend) simultaneously to produce the effective stroke and the other four are idle at such times. During the recovery stroke, these four fibers contract slowly, beginning near the base. The two central fibers may be conductive elements which coordinate these movements. Numerous other hypotheses also have sought to relate flagellar structure to function.

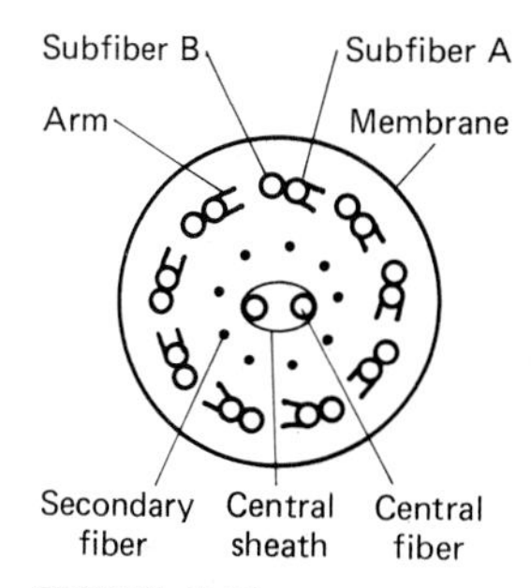

FIGURE 15-26
Diagram of the structure of a typical flagellum as seen in cross section. All cilia and flagella so far studied have nine pairs of longitudinal tubules, usually arranged around two central fibers, but the central fibers are not always present and may not be essential for function. On the peripheral double fibers are short projections, referred to as "arms," which are consistent in direction. A secondary set of fibers is interposed between the central and outer fibers.

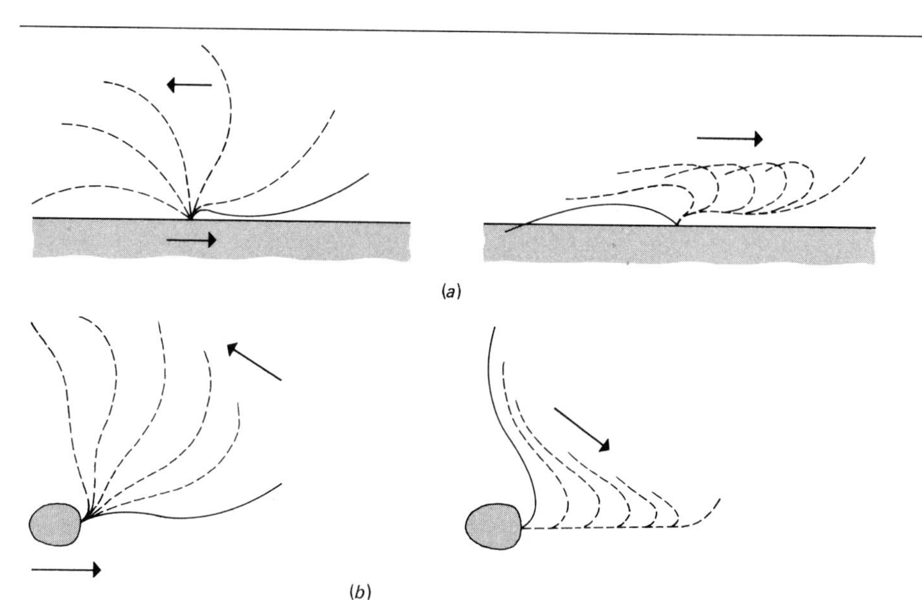

FIGURE 15-27
(a) Motion of a cilium on an organism moving toward the right involves an effective stroke (*left*) followed by a recovery stroke (*right*). Cilia are quite short. (b) Motion of a flagellum propels a protozoan to the right. Effective stroke is at left, recovery at right. Flagella are longer than cilia and can move in several ways.

Some recent studies of the chemistry of cilia and flagella have given further insight into the motility mechanisms of cilia and flagella. Flagella isolated from bacteria contain carbohydrate, RNA, and a large quantity of a protein monomer (mol. wt. 40,000) which has been called *flagellin*. Flagellin monomers will aggregate in acid solution to form fibers that are indistinguishable in electron micrographs from native flagella.

Studies of the isolated cilia of *Tetrahymena pyriformis*, a ciliate protozoan, reveal that after their highly insoluble membranes have been disrupted, a protein, soluble in salt solutions at neutral pH, can be obtained which has ATPase activity. This molecule, called *dynein*, seems to be a major constituent of the fiber structures. What is believed to be the monomer of dynein (called *14S*, from its sedimentation constant in density-gradient centrifugation) has a molecular weight (perhaps 540,000) that is about the same as that of muscle myosin. Its ATP-splitting activity is stimulated by Ca^{2+} and, unlike muscle myosin, also by Mg^{2+}. Polymerized dynein (i.e., a linear polymer of 14S dynein) is strikingly similar to myxomyosin of slime molds. Both have a sedimentation constant of 30S, consist of rodlike particles of variable length (up to 700 nm) and with diameters of 6 to 8 nm, and have ATPase activity.

Detection of acetylcholinesterase in a few ciliate protozoans and sperm has pointed to the possibility of a control system based on the release and hydrolysis of acetylcholine.

15-9
ACTIONS OF ATP IN MOTILE SYSTEMS

An interesting classification of the movements of structures on the basis of the manner in which they are influenced by ATP has been made by Hoffman-Berling (see citation in Readings). He suggests that all movements known at present are either produced by ATP or reversed and inhibited by ATP.

There is general agreement that ATP is the energy source for flagellar and ciliary movement. Isolated flagella beat at a maximum frequency when ATP is at a concentration of about $10^{-3}\,M$ and at lower frequencies with either lesser or greater concentrations. It is not known whether ATP hydrolysis is associated with contraction or relaxation, since flagella show only cyclic movements. There is some evidence that in amoebae (as noted above) and slime molds ATP also has a mechanochemical action. During cytokinesis ATP seems to cause contraction in the middle of the cell and relaxation at the cell poles, but it is generally conceded that there is no evidence that ATP influences the spindle apparatus. Finally, there is little doubt that ATP is intimately involved in muscle movement.

ATP-induced contraction, according to Hoffman-Berling, is an elementary, evolutionary acquisition and is not limited to specialized cells, such as those of muscles. Thrombocytes and fibroblasts (branched or elongated, undifferentiated cells from tissue culture) will actively contract in the presence of ATP. Like muscle cells they also require Mg^{2+} for contraction. In fact, glycerol-extracted fibroblasts will shorten by the same amount (80 percent) as glycerol-extracted muscles, and they will develop tension. Fibroblasts contract much more slowly than muscle cells, however, and have much less tensile strength. The contraction is even reversible; a rounded fibroblast can be caused to reextend by inhibiting contraction.

Quite a different situation seems to exist in certain ciliate protozoans, such as *Vorticella*. They have a motile stalk by which they are attached to a substratum. The stalk contains a contractile spiral element, the *myoneme*, in a flexible sheath. In the myoneme there are microscopically visible fibrils, 5 to 6 nm in diameter, which, like other such fibrous structures, exhibit birefringence. When the myoneme contracts, like a coiled spring, the body is pulled downward, and when it relaxes the stalk straightens out (Fig. 15-28).

Unlike muscles, stalks do not contract when treated with ATP and Mg^{2+}. ATP does appear to have an action, however. It is the only known substance which causes myoneme relaxation. ATP does not cause a permanent relaxation, but instead induces a rhythm of spontaneous relaxations and contractions. Hoffman-Berling proposes that the chemical

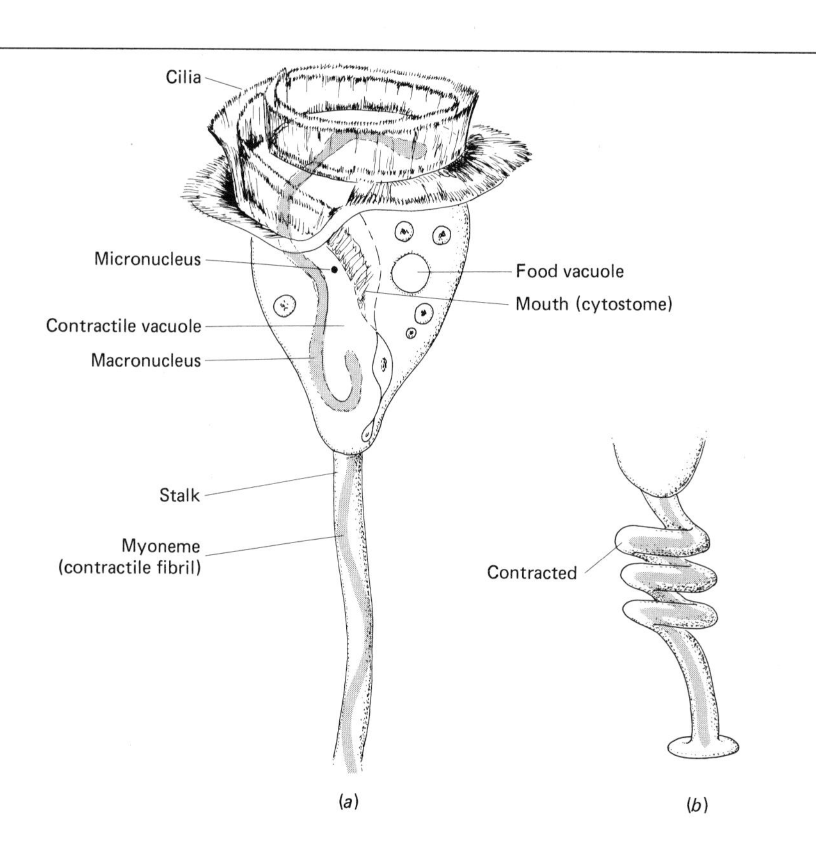

FIGURE 15-28
(a) The sessile peritrich, *Vorticella*. (b) The stalk of *Vorticella*, showing the helically contracted myoneme. (*After P. A. Meglitsch*, 1967, *Invertebrate Zoology*, Oxford University Press, Fair Lawn, N. J., p. 68.)

energy of ATP is utilized by means of a complicated process in the relaxation phase and it is stored in the contractile system.

It is obvious that there are marked similarities between muscle contraction and more primitive mechanisms of motility. The requirement for ATP, the need for metal ions, and the ubiquity of fiber arrays are unifying themes. Nevertheless, there is certainly a great amount of variation in the structure-function of contractile devices, and it is doubtful that one mechanism will explain the whole gamut of their movements. In fact, multiple macromolecular mechanisms probably exist within certain contractile cells.

READINGS

Allen, R. D., and N. Kamiya (eds.), 1964, *Primitive Motile Systems in Cell Biology*, Academic Press, Inc., New York. A collection of articles by various investigators from a symposium on cytoplasmic streaming, amoeboid motion, particle movement, and mitosis.

Bourne, G. H. (ed.), 1972–1973, *The Structure and Function of Muscle*, 2d ed., 4 vols., Academic Press, Inc., New York. A multiauthored, interdisciplinary, thorough assessment of the information available at the end of the sixties.

Hoffman-Berling, H., 1960, Other mechanisms producing movements, in *Comparative Biochemistry*, vol. 2, edited by M. Florkin and H. S. Mason, Academic Press, Inc., New York. Movements of various kinds are classified and discussed, chiefly in terms of the manner in which they are influenced by ATP.

Proceedings of a Symposium, New York Heart Association, *The Contractile Process*, Little, Brown and Company, Boston, 1967. (Published also as a supplement to *The Journal of General Physiology*, **50**, (6), part 2, 1967.) This collection contains contributions on contractile processes in fibrous macromolecules, cilia, microtubules, and muscles.

Prosser, C. L., 1973, *Comparative Animal Physiology*, W. B. Saunders Company, Philadelphia. Chapter 16 presents an analysis of muscle structure and function in a great variety of animals.

Wilkie, D. R., 1968, *Muscle*, St. Martin's Press, Inc., New York. This small volume (64 pages) can be recommended as an authoritative, readable account of the structure and physiology of muscles.

BIOSYNTHESIS 16

The illuminated green plant synthesizes all its organic constituents from inorganic salts, carbon dioxide, and water. Heterotrophs derive many of their organic molecules, such as glucose, fatty acids, amino acids, and vitamins, from green plants, but they also actively modify and synthesize numerous molecules. Smaller molecules are common to all organisms; they are ubiquitous and identical in the biosphere and are freely traded about. It is in their macromolecules that organisms differ. Uniqueness, or identity, as a species and as an individual, is an attribute of an organism's macromolecules.

How organisms go about fashioning some of their various molecules—small and large—is what this chapter is about.

16-1 MAINTENANCE AND GROWTH

In some areas of the human gastrointestinal tract, all the cells of the mucosal surface are replaced in less than 24 hours. Human red blood cells have an average life-span of about 120 days. Since there are about 25 trillion red blood cells in an adult human, that number of cells will be produced in a 120-day period; that works out to approximately 2,400,000 cells produced per second!

Such prodigious accomplishments illustrate the magnitude of *dynamic turnover* and the synthetic capacities of an organism. They also call attention to the great costs in energy and material for just maintaining an organism in statu quo.

In addition to *maintenance*, at certain times *growth* and *regeneration* make demands upon biosynthetic mechanisms. For example, the antlers of deer (Cervidae) are grown in a few months and then shed. These are true bone and can weigh up to 22.5 kg (50 lb). Under favorable conditions maize plants sometimes grow 12 cm in 24 hours. Young bamboo shoots grow as rapidly as 60 cm in a day. The bacterium, *Escherichia coli*, can divide every 20 minutes in an optimum environment, utilizing at least 2,500,000 ATP phosphorylations per second. In their phases of rapid growth, organisms are capable of doubling their mass in remarkably short times (Table 16-1).

When the limb or tail of a salamander is amputated, the wound is closed in a matter of hours, and within several weeks a new appendage will be regenerated. If one or even one-and-a-half kidneys are removed from a rat, the remaining cells will begin to proliferate. Mitotic activity reaches a peak within 48 hours after surgery. These are examples of the regenerative capacities of organisms. Regeneration is not a feature of all organisms or even all parts of organisms. The response to deletion of parts varies widely in organisms, from wound healing to complete regeneration. However, in all cases there is a rapid unleashing of a capacity to change cell structure and make new cells. Developmental biologists have been long puzzled by what it is that regulates the size and numbers of structures, and what happens when the equilibrium is changed. Many pathological conditions, including cancers, cause excessive cell proliferation and, consequently, heightened molecular synthesis. We know very little about the nature of the normal regulation of many aspects of biosynthesis, but some of what is known of the chemistry will now be recounted.

16-2 SYNTHESIS OF AMINO ACIDS

Amino acids are some 20 molecules with both an amino ($—NH_2$) and carboxyl group. The general formula for most amino acids is:

$$H_2N-\overset{\displaystyle R}{\overset{|}{\underset{\underset{\displaystyle H}{|}}{C}}}-COOH$$

where *R*, called the *amino acid side chain* or *residue*, represents a variety of structures. The amino group of most amino acids is in a position α to the carboxyl group.

The simplest amino acid is:

$$H_2N-\overset{\displaystyle H}{\overset{|}{\underset{\underset{\displaystyle H}{|}}{C}}}-COOH$$

Glycine

Table 16-1
Time Needed for Organisms to Double Their Mass

Escherichia coli	20 min
Fly larva	13 hr
Silkworm	68 hr
Rabbit at birth	6 days
Pig at birth	6–7 days
Sheep at birth	10 days
Guinea pig at birth	18 days
Horse at birth	60 days
Human at birth	180 days

Source: From M. Sussman, 1964, *Growth and Development*, Prentice-Hall, Inc., Englewood Cliffs, N.J., p. 21.

The next simplest is:

$$
\begin{array}{c}
CH_3 \\
| \\
H_2N-C-COOH \\
| \\
H
\end{array}
$$

Alanine

All naturally occurring amino acids, with the exceptions of glycine (which can have no asymmetric carbon) and a few that are found in certain compounds (mostly antibiotics), are in the *l* configuration.

Three amino acids contain sulfur:

$$
\begin{array}{l}
CH_3 \\
| \\
S \\
| \\
CH_2 \\
| \\
CH_2 \\
| \\
H_2N-CH-COOH
\end{array}
$$

Methionine

$$
\begin{array}{l}
S-CH_2-CH(NH_2)-COOH \\
| \\
S-CH_2-CH(NH_2)-COOH
\end{array}
$$

Cystine

$$
\begin{array}{c}
NH_2 \\
| \\
HS-CH_2-CH-COOH
\end{array}
$$

Cysteine

Various ring structures form amino acid side chains; e.g., the indole in tryptophan:

$$
\text{(indole ring, N–H)}-CH_2-\underset{}{\overset{NH_2}{\overset{|}{CH}}}-COOH
$$

benzyl in tyrosine:

$$
HO-C_6H_4-CH_2-\overset{NH_2}{\overset{|}{CH}}-COOH
$$

and pyrrole in proline:

$$
\begin{array}{l}
H_2C-CH_2 \\
\;|\qquad\;\, | \\
H_2C\quad CH-COOH \\
\quad\; N \\
\quad\; | \\
\quad\; H
\end{array}
$$

Proline is called an *imino* acid because it has an —NH group instead of —NH_2. It is nevertheless a protein subunit.

Most microorganisms and plants are capable of carrying out the de novo synthesis of amino acids, but most animals lack about half these synthetic capacities. For the latter organisms, amino acids may then be considered as "essential" if they are not synthesized by the organism and "nonessential" if they are. However, the "essentiality" of an amino acid is sometimes a function of many things, including other components of the diet, physiological state, age, nature of the intestinal flora, and even the criteria employed for its determination. In many instances a clear-cut need for the assembled amino acid has been demonstrated. For example, leucine, isoleucine, valine, lysine, methionine, phenylalanine, histidine, tryptophan, and threonine have been found to be essential for all species of higher animals studied (see Table 5-4).

Several methods have been used to reveal the pathways by which organisms synthesize their molecules. We can use amino acid synthesis to illustrate one of the most important of these methods, viz., use of *isotopes*.

Microorganisms, such as the bacterium, *E. coli*, can be grown on specifically labeled carbon sources, such as glucose labeled in the C-1 or C-6 position with ^{14}C. After assimilation of the labeled material, a protein fraction is isolated. The protein can then be hydrolyzed and the isolated amino acids degraded to see where the label is found. Such a procedure is laborious and seldom provides unambiguous information about the reactions by which carbon was incorporated into amino acids.

A more useful isotope technique is to feed the organism uniformly labeled ^{14}C-glucose (i.e., glucose in which the isotope occurs with a uniform frequency in all carbon positions); subsequently all metabolites will be uniformly labeled. (This, by the way, clearly indicates the fluidity and continuity of all metabolic pathways.) Now, if a suspected intermediate is added to the growing culture (provided it enters the cell in quantity), it will compete with and dilute the intracellular pool of labeled intermediate. The result is that there will be a diminution in radioactivity of the product if the added compound truly is an intermediate.

Here is an example of how this method has been used. A culture of bacteria is grown on a medium containing uniformly labeled glucose and an unlabeled, suspected intermediate of amino acid synthesis, homoserine. A second culture is grown in the same way, but without homoserine. When the amino acids are isolated (generally by chromatography), the amino acids threonine, methionine, and isoleucine are found to have far less radioactivity in the culture to which homoserine was added. This suggests that homoserine is an intermediate in the biosynthesis of these three amino acids. Several related and similar kinds of experiments have shown that this conclusion is correct. In fact, the synthetic pathway for these amino acids has been mapped completely. They can be considered as derivatives of participants of the tricarboxylic acid cycle, and belong to what is referred to as the aspartic family of amino acids; viz., aspartic acid, lysine, methionine, isoleucine, and threonine. Their synthetic pathways are shown in Fig. 16-1.

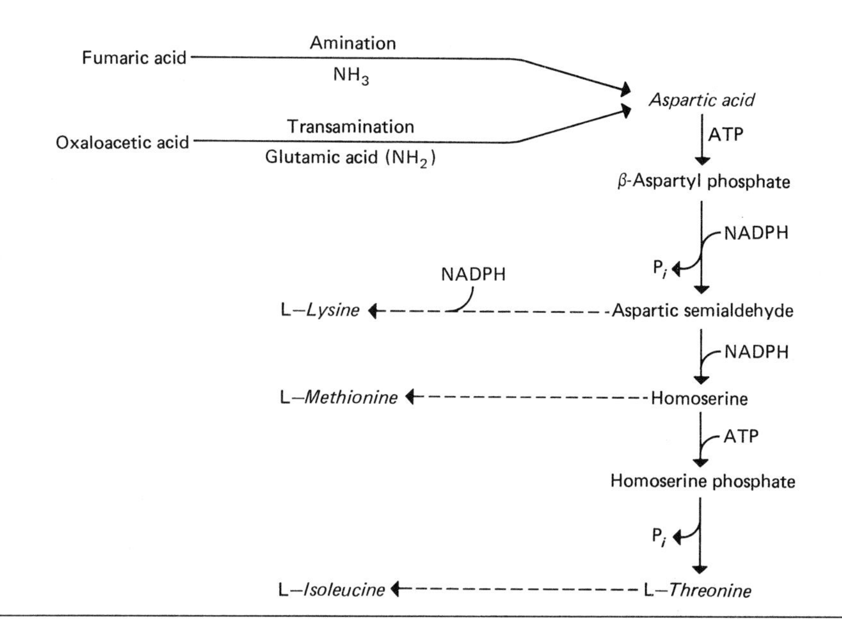

FIGURE 16-1
Synthesis of aspartic family of amino acids. The major points of energy input (phosphorylations and reductions involving coenzymes) are indicated. The dashed arrows mean that a number of intermediate steps are omitted. NADPH is the reduced coenzyme, nicotinamide adenine dinucleotide phosphate.

Note the involvement of reduced coenzyme in the sequences of Fig. 16-1. Coenzymes are to be regarded as sources of chemical energy; the reductive steps shown here make use of the energetic electrons of reduced nicotinamide adenine dinucleotide phosphate (NADPH).

You should now refer to Fig. 19-2 and note that the members of the *aspartic acid family* of amino acids are catabolized on entry into the tricarboxylic acid (TCA) cycle by way of oxaloacetate. This molecule is either the precursor for synthesis or the product of catabolism of the five members of this family.

In Fig. 19-2 the site of contact with the TCA cycle of another family of amino acids also indicated; viz., the *glutamic family*. Glutamate is produced by amination of α-ketoglutarate and then, by reduction and cyclization, gives rise to proline. An alternate pathway from glutamate, involving reduction and then phosphorylation by ATP, produces arginine. A third route, driven by active acetaldehyde (acetyl-CoA) and a reduction, synthesizes lysine.

A third group of amino acids is called the *pyruvic acid family*. Pyruvic acid serves as the precursor for alanine, valine, and leucine. The phosphorylated derivation of pyruvate, glyceraldehyde 3-phosphate, is used to form serine, glycine, and cysteine.

In Fig. 19-2 two classes of amino acids, *glucogenic* and *ketogenic* (arrows *A* and *B*, respectively), are indicated as related to the terminal steps of glycolysis. The distinction between the two classes depends on whether ketone bodies (acetoacetic acid, acetone, etc.) or glucose (or glycogen) results from catabolism of the particular amino acid. On this basis, leucine is ketogenic, whereas isoleucine, lysine, phenylalanine, and tyrosine are both glucogenic and ketogenic, and the remainder of the amino acids are glucogenic.

Synthetic pathways for amino acids containing an aromatic ring (phenylalanine, tyrosine, and tryptophan) have been worked out largely by use of *auxotrophic mutants* (Gr. *auxein*, to add to; *trophe*, food). Bacterial mutants unable to carry out certain syntheses occur spontaneously, but at a low frequency (of the order of 10^{-8}). By use of ultraviolet irradiation or chemical mutagens, their frequency can be increased, and specific techniques exist for the selection of desired mutants. One widely used method is called *penicillin selection*. After exposure to the mutagen, the bacterial culture is allowed to grow for a few generations in a medium which includes the nutrient for which a mutant is required. The culture is then washed several times and resuspended in a medium that is complete, *except* for the nutrient which would allow the required mutant to grow. This medium also contains penicillin. All cells will be killed except the required mutant, because penicillin kills only cells that are in the process of active division. The action of penicillin appears to involve the inhibition of the incorporation of a component into the mucopeptide polymer which is an essential constituent of the normal cell wall. Since the mutant does not grow and divide in the medium lacking its required nutrient, it will escape killing and can be obtained in pure culture.

Such mutants are used in studies which reveal their synthetic limitations. Some that require tryptophan for growth will survive in a medium without tryptophan if anthranilic acid or indole is added instead. Other mutants secrete indole into the growth medium, and still another mutant accumulates anthranilate, conditions not found in the wild type. From this and other such information, a tentative scheme for biosynthesis of tryptophan has been formulated:

$$\text{Glucose} + \text{NH}_4 \xrightarrow[(3)]{} \text{anthranilic acid} \xrightarrow[(2)]{} \text{indole} \xrightarrow[(1)]{} \text{tryptophan}$$

The numbers in parentheses indicate the number of "genetic blocks" of the different classes of mutants and the reactions which they are unable to perform.

Some of the most significant discoveries about the genetic control of biosynthesis were made by Beadle and Tatum, using primarily the red bread mold, *Neurospora*. The techniques they used were very similar to those just described for auxotrophic mutants of bacteria, except that spores were individually selected and cultured to obtain mutant strains (Fig. 16-2).

Wild-type *Neurospora* can be grown in pure culture on a medium containing glucose as a source of carbon and energy, ammonium chloride as a source of nitrogen, a few mineral salts (such as sulfate and phosphate), and two vitamins. From this minimal medium the mold can synthesize about 20 amino acids and a host of other kinds of molecules. Some mutant strains have been found which are unable to synthesize

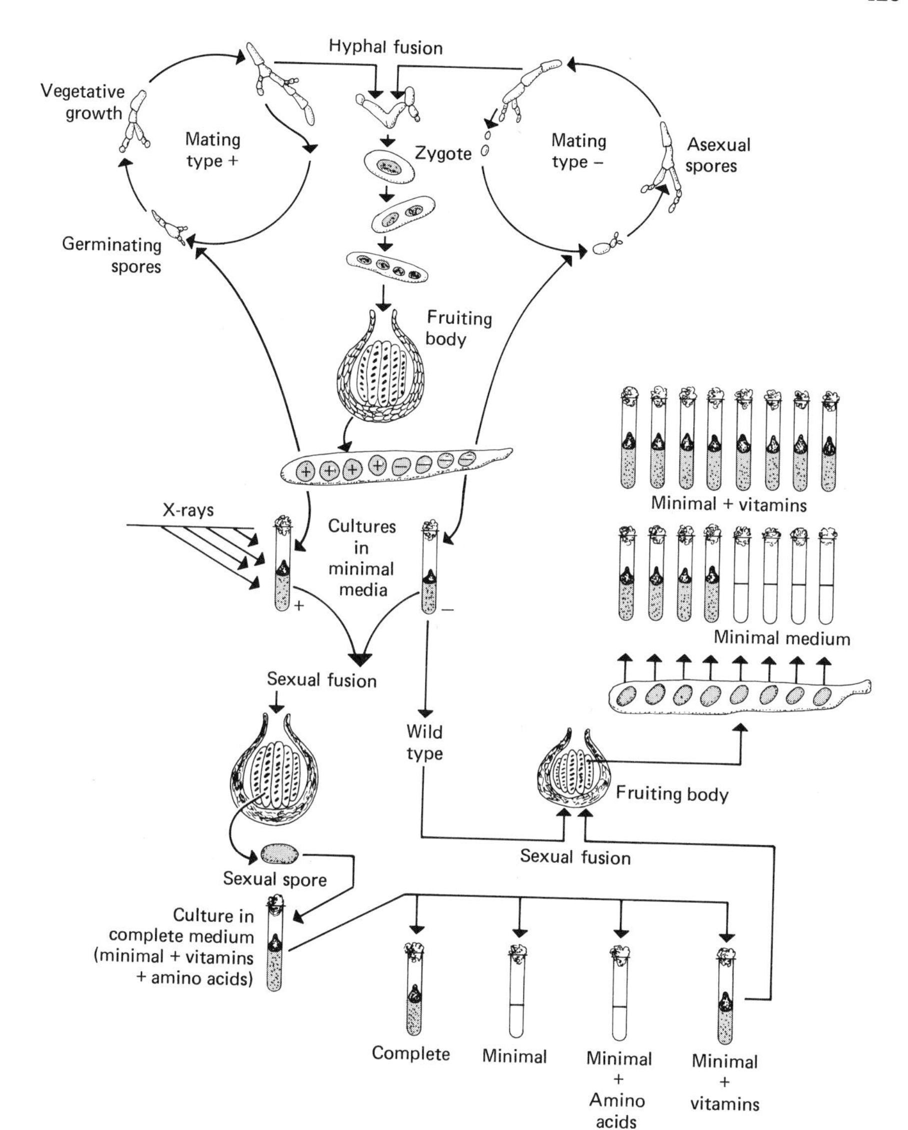

FIGURE 16-2
The life cycle of *Neurospora* and a diagramatic representation of the classical experiments which demonstrated the genetics of its vitamin requirements.

certain of the amino acids. Several mutants, for example, cannot synthesize arginine. Since the steps in the pathway are known, it is possible to ascertain where synthesis is blocked in each mutant (Fig. 16-3).

By determining positions of the blocks in a variety of biosyntheses, it has been shown that the functional distinction between the wild type and mutants is that the mutants have lost the capacity to produce one proper enzyme in a biosynthetic pathway. The fact that each step requires a different enzyme suggested to Beadle and associates that the function of a gene can be equated with the formation of a protein: the famous "one gene–one enzyme" hypothesis. We now know this is not quite correct be-

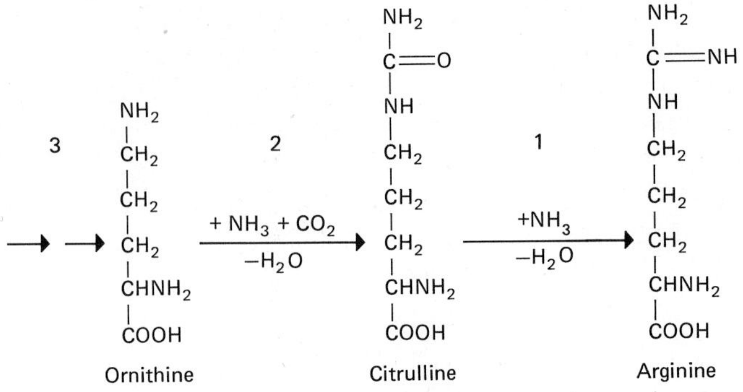

$$\xrightarrow{}\xrightarrow{}\ \underset{\text{Ornithine}}{\overset{3}{NH_2-CH_2-CH_2-CH_2-CHNH_2-COOH}}\ \xrightarrow[-H_2O]{2\quad +NH_3+CO_2}\ \underset{\text{Citrulline}}{NH_2-C(=O)-NH-CH_2-CH_2-CH_2-CHNH_2-COOH}\ \xrightarrow[-H_2O]{1\quad +NH_3}\ \underset{\text{Arginine}}{NH_2-C(=NH)-NH-CH_2-CH_2-CH_2-CHNH_2-COOH}$$

(a)

Arginine-requiring mutant	Growth on				Reaction blocked
	Minimal	Ornithine	Citrulline	Arginine	
1	−	−	−	+	1
2	−	−	+	+	2
3	−	+	+	+	3

(b)

FIGURE 16-3
(a) Part of the pathway of arginine biosynthesis in *Neurospora*. (b) Analysis of the growth requirements of three classes of arginine-requiring mutants of *Neurospora*.

cause many—perhaps most—enzymes are an association of small polypeptides (subunits) which derive from different gene loci.

In several cases the method of enzymatic analysis has been used as a more definitive proof of postulated sequences. Mutants requiring a certain nutrient sometimes can be shown to lack the enzyme for making that particular molecule. When this information is added to data from methods using isotopic competition and auxotrophic mutants, a firm basis is provided for making conclusions about a biosynthetic pathway.

16-3 SYNTHESIS OF FATTY ACIDS

Lipids are greasy or fatty materials formed from combinations of C, H, and O. They include a homologous series of unbranched compounds called *saturated aliphatic* or *fatty acids* with the general formula RCOOH, where *R* represents an aliphatic chain of the type $CH_3(CH_2)_n$. The simplest is formic acid:

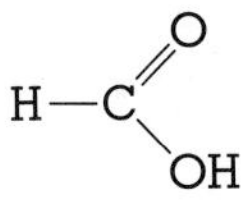

$$H-C(=O)-OH$$

Adding a second carbon creates the complex known as acetic acid:

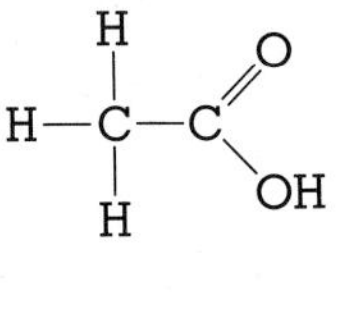

$$H-\underset{H}{\overset{H}{C}}-C(=O)-OH$$

Unsaturated fatty acids have one or more double bonds, which permit the possibility of geometric isomerism (*cis* and *trans* forms). Naturally occurring fatty acids generally are all *cis* forms. Most of the saturated and unsaturated fatty acids share the interesting feature of having an even number of carbon atoms. Some of these are listed in Table 16-2.

Dibasic and polybasic acids have two or more carboxyl groups. Examples of such acids are:

$$\begin{array}{c}COOH\\|\\COOH\\\text{Oxalic acid}\end{array}\qquad\begin{array}{c}COOH\\|\\CH_2\\|\\CH_2\\|\\COOH\\\text{Succinic acid}\end{array}\qquad\begin{array}{c}H-C-COOH\\\|\\HOOC-C-H\\\text{Fumaric acid}\end{array}$$

Succinic and fumaric acids are important participants in the TCA cycle.

The *hydroxy acids* contain one to several —OH and one or more —COOH radicals. Lactic acid is a monohydroxy compound which has two-position isomers; the projection formulas are:

$$\begin{array}{c}COOH\\|\\H-C-OH\\|\\CH_3\\\text{L-Lactic acid}\end{array}\qquad\begin{array}{c}COOH\\|\\HO-C-H\\|\\CH_3\\\text{D-Lactic acid}\end{array}$$

D-Lactic acid is formed by metabolism in muscle, and L-lactic acid is produced by microorganisms from lactose. Glyceric acid, $CH_2OH \cdot COOH \cdot COOH$, is an example of a dihydroxy monoacid, and malic acid, $COOH \cdot COOH \cdot CH_2 \cdot COOH$, is a monohydroxy diacid.

Table 16-2
Some Natural Saturated and Unsaturated Fatty Acids

Common name	Systematic name	Formula
Saturated acids		
Butyric	Butanoic	C_3H_7COOH
Caproic	Hexanoic	$C_5H_{11}COOH$
Palmitic	Hexadecanoic	$C_{15}H_{31}COOH$
Cerotic	Hexacosanoic	$C_{25}H_{51}COOH$
*Unsaturated acids**		
Oleic	9-Octadecenoic (*cis*)	$C_{15}H_{29}COOH$
Linoleic	9,12-Octodecadienoic (*cis, cis*)	$C_{17}H_{29}COOH$
Linolenic	9,12,15-Octodecatrienoic (*cis, cis, cis*)	$C_{17}H_{29}COOH$
Clupanodonic	4,8,12,15,19-Docosapentaenoic	$C_{21}H_{33}COOH$

* Numbers indicate double-bond positions.

A few *aldo acids* are known, and they include glucuronic acid, an intermediate in a major pathway of carbohydrate metabolism. The natural form is the dextrorotatory compound:

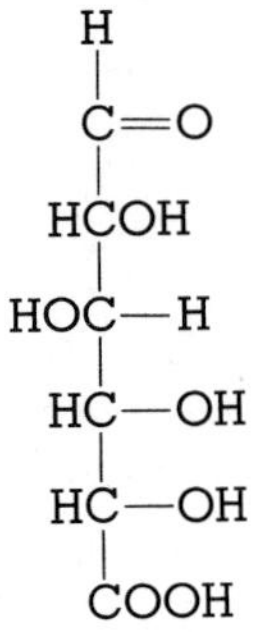

```
     H
     |
     C=O
     |
   HCOH
     |
  HOC—H
     |
    HC—OH
     |
    HC—OH
     |
     COOH
```

D-Glucuronic acid

Keto acids include the following important metabolic intermediates:

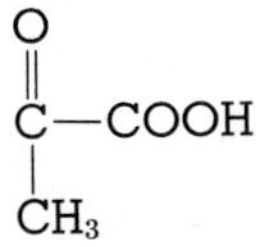

```
                                     O                  O
                                     ||                 ||
                  O                  C—COOH             C—COOH
                  ||                 |                  |
  O               C—COOH             CH2              HC—COOH
  ||              |                  |                  |
  C—COOH          CH2                CH2                CH2
  |               |                  |                  |
  CH3             COOH               COOH               COOH
```

Pyruvic acid 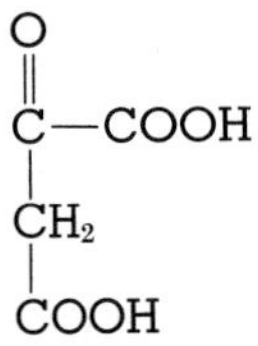   Oxalacetic acid   α-Ketoglutaric acid 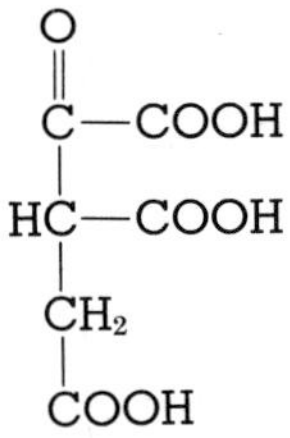   Oxalosuccinic acid

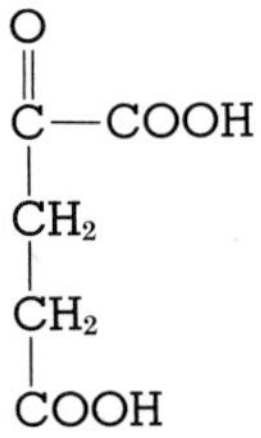

The *fats* are fatty acid esters of glycerol and are commonly called triglycerides. They are the most common natural kind of lipid. The oils of nuts and seeds and the fat of animals are mixtures of triglycerides. *Simple glycerides* have the three glycerol hydroxyls esterified with identical fatty acids, and *mixed glycerides* contain two or three different fatty acids:

Simple Glyceride	Mixed Glyceride
$H_2C—O—CO—C_{17}H_{35}$	$H_2C—O—CO—C_{17}H_{33}$
$HC—O—CO—C_{17}H_{35}$	$HC—O—CO—C_{17}H_{35}$
$H_2C—O—CO—C_{17}H_{35}$	$H_2C—O—CO—C_{15}H_{31}$
Stearin	α-Olea-δ′-β palmitostearin

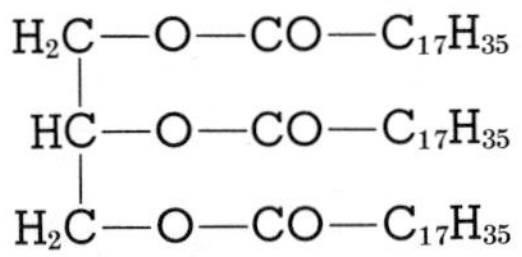

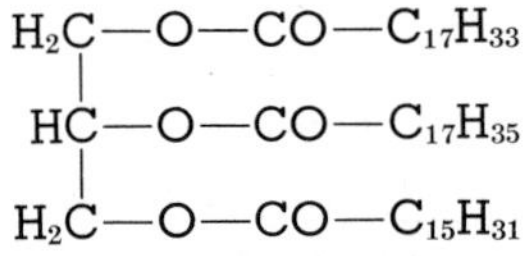

Waxes may be defined as the esters of higher fatty acids and of higher monohydroxy alcohols, but they are chemically quite varied and there are exceptions which do not fit this definition. Waxes occur in the secretions of insects, on the skins and furs of animals, and on the leaves and fruit of plants.

Phospholipids are a diffuse group of compounds that are classed together partially on the basis of solubility and partially on the basis of the ester phosphorus present in the compounds. The major phospholipids are phosphatidyl cholines (lecithins), phosphatidyl ethanolamines (cephalins), and sphingomyelins.

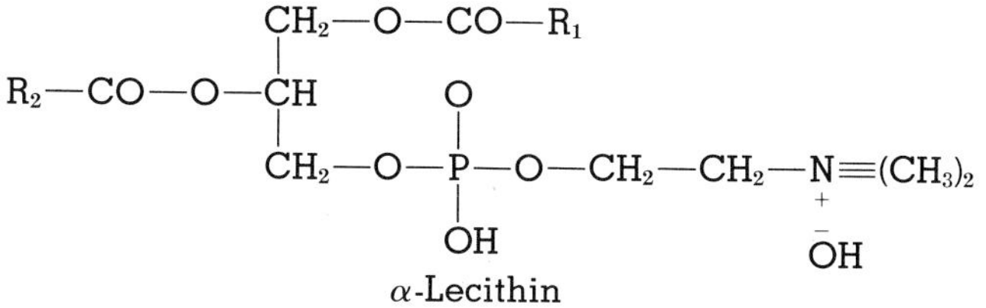

$$\begin{array}{l} \quad\quad\quad\quad\quad\quad\quad CH_2-O-CO-R_1 \\ R_2-CO-O-CH \quad\quad\quad\quad O \\ \quad\quad\quad\quad\quad\quad\quad CH_2-O-\overset{|}{P}-O-CH_2-CH_2-\overset{+}{N}\equiv(CH_3)_2 \\ \quad\quad\quad\quad\quad\quad\quad\quad\quad\quad\quad OH \quad\quad\quad\quad\quad\quad \bar{O}H \end{array}$$

α-Lecithin

Phospholipids, especially lecithins, form moderately stable combinations or complexes with many different substances, such as other lipids, carbohydrates, proteins, and heavy metal salts. This property figures importantly in the role of phospholipids in formation of membrane structures.

Many other kinds of lipids are found in tissues. Only a few more will be mentioned to suggest the variety of chemistry they exhibit.

An interesting class of carbohydrate-rich lipids, called *gangliosides*, has been found principally in neural tissue. Another class, the *cerebrosides*, from brain, contain sphingosine, a fatty acid, and the sugar galactose. There are also sulfur-containing lipids, *sulfolipids*, and ones containing amino acids, *proteolipids*, that are found in a variety of tissues.

Steroids are a class of lipids which contain a nucleus of three six-membered rings and one five-membered ring. For example:

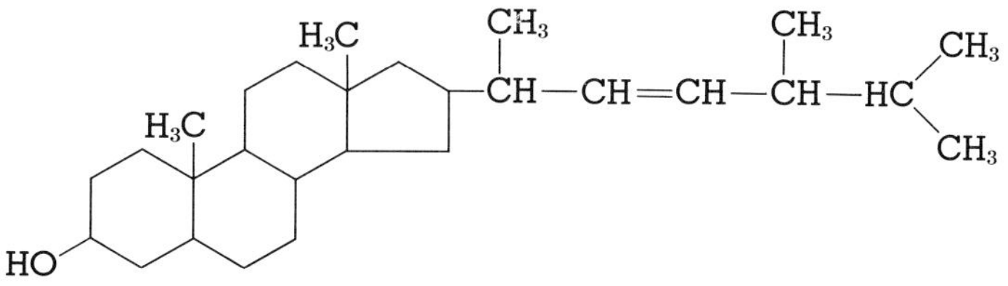

Ergosterol, $C_{28}H_{43}OH$

The sterols, bile acids, sex and adrenocorticoid hormones, and some alkaloids contain the same ring structure.

The basic building block for fatty acid synthesis is acetyl-CoA, arising principally from the intramitochondrial oxidation of pyruvate (Fig. 16-4). Although fatty acid oxidation occurs primarily or exclusively in mitochondria, synthesis of fatty acids occurs among the "soluble" enzymes. There are two ATP-dependent reactions known by which extramitochondrial generation of acetyl-CoA occurs, one using citrate and the other acetate. However, it appears that extramitochondrial acetyl-CoA is largely derived from a carnitine-mediated transport of acetyl groups across the mitochondrial membrane (Fig. 16-4).

A key step in the biosynthesis of fatty acids is the formation of malonyl-CoA from CO_2 and acetyl-CoA:

$$\underset{\text{Acetyl-CoA}}{CH_3-CO-SCoA} + CO_2 + ATP \xrightarrow{Mg^{2+}} \underset{\text{Malonyl-CoA}}{HO_2C-CH_2-CO-SCoA} + ADP + P_i$$

The enzyme for this reaction is acetyl-CoA carboxylase. Note the fixation of CO_2. Biotin (a B-group vitamin) serves as the transcarboxylating coenzyme in this reaction.

The principal product of fatty acid synthesis by soluble, cell-free extracts is palmitic acid:

$$\underset{\text{Acetyl-CoA}}{CH_3—CO—SCoA} + 7\underset{\text{Malonyl-CoA}}{HO_2C—CH_2—CO—SCoA} + 14NADPH + 14H^+ \longrightarrow$$

$$\underset{\text{Palmitic acid}}{CH_3—(CH_2)_{14}—COOH} + 8CoASH + 14NADP^+ + 6H_2O + 7CO_2$$

Other long-chain fatty acids, C_{10} to C_{14}, are formed in lesser amounts.

Fatty acid synthesis has an absolute dependence upon several factors, including ATP, and NADP [rather than nicotinamide adenine dinucleotide (NAD), which is required as the coenzyme in the oxidative pathway] (Fig. 20-7). Systems from three sources that synthesize fatty acids have been investigated in considerable detail: *E. coli*, pigeon liver, and yeast. The *E. coli* complex includes an acetyl transferase, a malonyl transferase, a condensing enzyme, an acyl-carrier protein (ACP), a β-ketoacyl-ACP reductase, and an enoyl-ACP hydrase. The acyl intermediates are bound throughout the course of the synthetic process to the ACP molecule. In fact, free intermediates in fatty acid biosynthesis have never been detected, suggesting that they are tightly bound to the enzyme complex at all times.

The synthesis of various lipids, such as *glycerides*, *steroids*, and *terpenes*, largely consists of the same kinds of reactions employed in fatty acid synthesis. They are mainly a variety of repeated condensations of two-carbon fragments with acetyl-CoA serving as the donor of carbons. Addition of carbons by twos accounts for the even number of carbons in natural fatty acids.

Most of the enzymes of lipid biosynthesis, with the exception of those

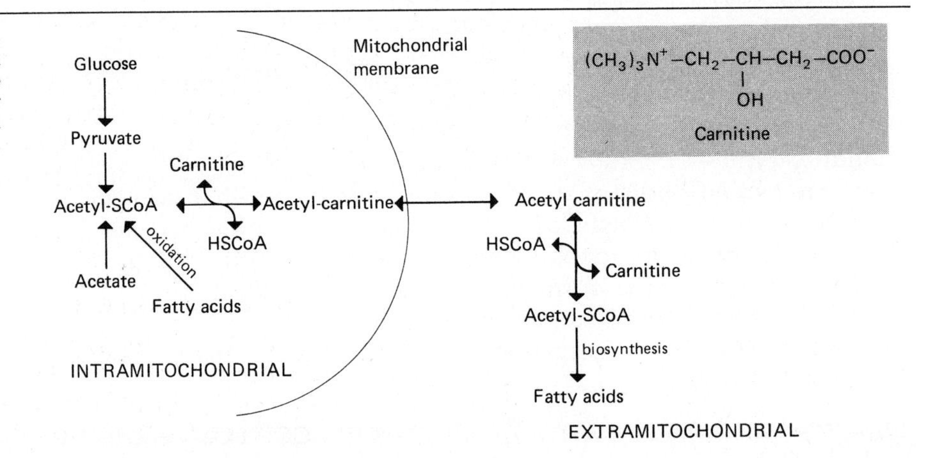

FIGURE 16-4
A schematic representation of the function of carnitine as acetyl-group carrier across the mitochondrial membrane.

for the initial synthesis of fatty acids (which are soluble), are localized in microsomal membranes. That is, these reactions occur in the membranes of the endoplasmic reticulum, rather than in ribosomes like the protein-synthesizing reactions.

16-4 SYNTHESIS OF CARBOHYDRATES

The **sugars**, or "oses," are C, H, and O compounds characterized by having the group

$$
\begin{array}{c}
\diagdown C{=}O \\
| \\
CHOH
\end{array}
$$

and can be classified on the basis of the number of carbons in their structure. Thus, the simplest are two-carbon ones called *bioses;* those with three carbons are *trioses*, etc. There is a second scheme which classifies sugars as *aldoses* and *ketoses*. For example, the natural three-carbon sugars are glyceraldehyde, an aldose, and dihydroxyacetone, a ketose:

D-Glyceraldehyde	Dihydroxyacetone
CHO	CH_2OH
H—C—OH	C=O
CH_2OH	CH_2OH

It will be noted that the middle carbon of glyceraldehyde is asymmetrical. The configuration shown here represents the D form. Most naturally occurring sugars are spatially related to D-glyceraldehyde and are so classed, even though they may actually be levorotatory. Thus, prefixes such as D(−) indicate a compound in the "D" series but are levorotatory. D(+)-glucose is indicated to be in the "D" series and is dextrorotatory. A few naturally occurring sugars belong to the "L" series, including arabinose, a *pentose*, and galactose, a *hexose*.

Speaking of hexose, the simple formula $C_6H_{12}O_6$ represents 16 different simple sugars, all with different spatial arrangements of their constituent groups. All 16 can be synthesized, but only six occur in nature:

L-Galactose	D-Galactose	D-Mannose	D-Glucose	D-Ribulose	D-Fructose
CHO	CHO	CHO	CHO	CH_2OH	CH_2OH
HOCH	HCOH	HOCH	HCOH	C=O	C=O
HCOH	HOCH	HOCH	HOCH	HCOH	HOCH
HCOH	HOCH	HCOH	HCOH	HCOH	HCOH
HOCH	HCOH	HCOH	HCOH	HCOH	HCOH
CH_2OH	CH_2OH	CH_2OH	H_2COH	CH_2OH	CH_2OH

The first four compounds will be recognized as aldoses and the last two as ketoses.

Although the sugars shown above are represented as straight chains, it is believed that those containing four or more carbons usually exist in cyclic form when in solution.

If an oxygen atom is removed from a hydroxyl group there is formed a *deoxy-* or *desoxy-sugar*. Such a compound is the sugar moiety of deoxyribonucleotides. Written in ring form, its structure and that of the related D-ribose appear as follows:

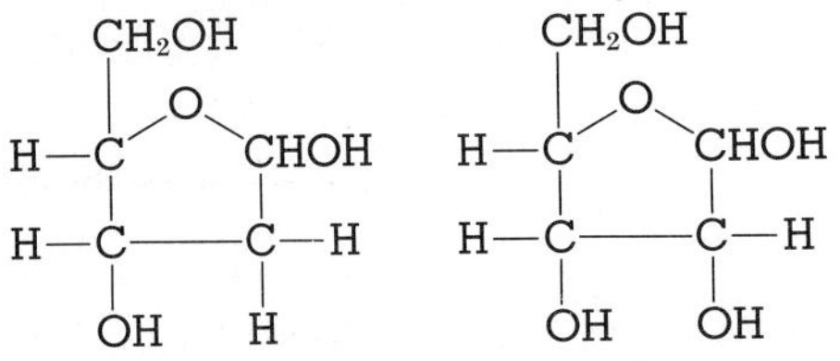

```
    CH2OH                  CH2OH
    |  O                   |  O
   /     \                /     \
H—C       CHOH         H—C       CHOH
  |        |             |        |
H—C————————C—H         H—C————————C—H
  |        |             |        |
  OH       H             OH       OH
D-2-Deoxyribose          D-Ribose
```

Amino groups may be substituted for various hydroxyl groups of sugars to form *amino sugars*. Only a few are known, the best-characterized one being glucosamine.

Many nonsugar substances are convertible to glucose, pentose, or polysaccharides. This can be accomplished through the glycolytic reactions, all of which are directly or indirectly reversible, provided a metabolic energy supply is available. Thus, the so-called glucogenic amino acids, pyruvate, lactate, and members of the TCA cycle, can serve as sugar precursors (Fig. 19-2). The making of sugars (principally glucose) from nonsugar molecules is called **gluconeogenesis.** (Hormonal influences on gluconeogenesis are discussed in Chap. 9, Sec. 9-4, and schematically illustrated in Fig. 9-7.)

There are three stages at which the glycolytic reactions are not readily reversible and, consequently, the glycolytic and gluconeogenic reactions are not the reverse of one another. One is the conversion of glucose to glucose-6-P, which is catalyzed by hexokinase. The reverse (gluconeogenic) reaction is a hydrolysis of glucose-6-P catalyzed by glucose-6-phosphatase. This reaction is important not only in gluconeogenesis, but also in the conversion of liver glycogen to blood sugar in mammals (see Fig. 9-7).

The second block to gluconeogenesis is between fructose-6-P and fructose-1,6-diP (Fig. 9-7). This is a key step in metabolic control. The enzyme of glycolysis, phosphofructo-1-kinase, is inhibited allosterically by ATP and is activated by ADP or AMP.[1] Thus, when ATP levels are high, this step becomes rate-limiting and all subsequent reactions of the TCA cycle and associated electron transfers are proportionately retarded. In this condition glucose synthesis is favored. As a further control, fructose 1,6-diphosphatase, the glucogenic enzyme, is allosterically inhibited by AMP. Thus, high AMP (and low ATP) favors glycolysis.

[1] See Fig. 17-3 for the structures of ATP, ADP, and AMP.

The third block is the step between phosphoenol pyruvic acid and pyruvic acid (Fig. 9-7). This is one of the two sites of ATP generation in glycolysis, and the thermodynamic equilibrium strongly favors pyruvate formation. The block is circumvented in gluconeogenesis by converting pyruvate first to oxaloacetate and then to phosphoenolpyruvate. Both steps require phosphorylation by nucleotide triphosphates. (The glycolytic and glucogenic reactions are shown in more detail in Fig. 19-3.)

A large number of interconversions and syntheses of carbohydrates involves esters of aldoses and nucleoside *di*phosphates. They are formed by specific *pyrophosphorylases*. One of the most widely used of these compounds results from the reaction of glucose-1-P and uridine triphosphate (UTP) (the latter being like ATP except that the base is uridine instead of adenine) to form a nucleoside diphosphate glycoside, as follows:

$$\text{Glucose-1-P} + \text{UTP} \xrightarrow{\text{UDP glucose pyrophosphorylase}} \text{UDP-glucose} + \text{PP}$$

Note that *pyrophosphate* (PP) is a product of this type of reaction. The structure of the resulting UDP-glucose is shown in Fig. 16-5.

The nucleoside diphosphate sugars participate in several kinds of reactions, many of which involve the transfer of the sugar moiety, which is a reactive group and is called a *glycosyl residue*, to an acceptor molecule with the formation of a new glycosidic linkage.

An example of a synthesis using nucleoside diphosphate is provided by the following reaction:

$$\text{UDP-Glucose} + \text{fructose} \rightleftharpoons \text{sucrose} + \text{UDP}$$

Fructose serves as the acceptor of the glycosyl, a glucose residue, and the disaccharide sucrose is formed. The nucleoside diphosphate glycoside serves as a coenzyme in reactions of this kind.

The synthesis of three of the main polysaccharides encountered in nature, viz., *cellulose*, *starch*, and *glycogen*, will illustrate the common mechanism by which carbohydrate *polymers* are formed. In all there is transfer of the active glycosyl residue from a nucleoside diphosphate coenzyme to a primer. The primer is a chain of glycosyl residues, the length of which varies in each case.

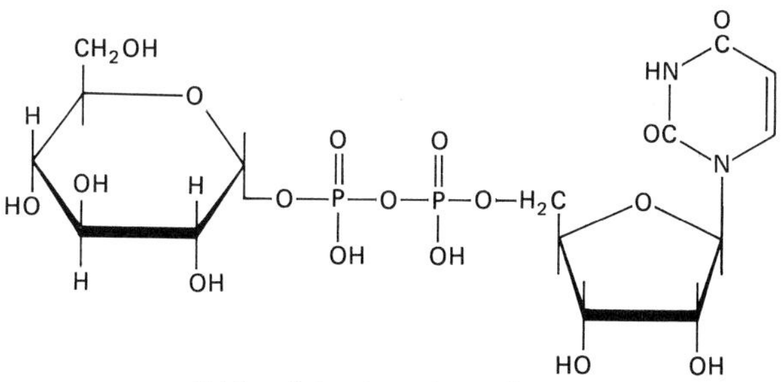

Uridine diphosphate glucose (UDPG)

FIGURE 16-5 Uridine diphosphate glucose.

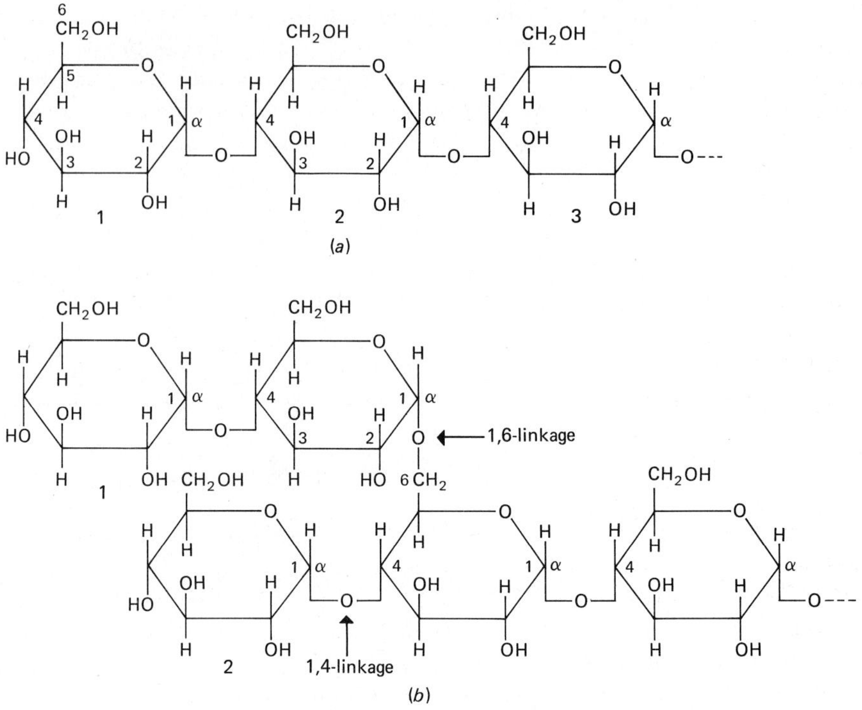

(a)

1,6-linkage

1,4-linkage

(b)

FIGURE 16-6 Starch contains glucose units joined through α-glycosidic linkages of carbons 1 and 4, forming (a) amylose, in which the glucose units are linked in an unbranched chain, and (b) amylopectin, which contains chains of glucose units like those of amylose but also has branches of these chains linked through the 6-OH of glucose. By convention the carbons are numbered in the sequence shown.

When a macromolecule is made up of a number of similar subunits, the subunits are called *monomers* and the macromolecule is called a *polymer*. Generally, polymers of any arbitrary length can be formed because the monomers forming them have two different reactive groups; one of the groups of a monomer forms a linkage with the other kind of group on another monomer, in a way exemplified by the manner in which amino acids form peptides. Macromolecules may be formed from identical monomers, as in the case of starch which is composed of glucose subunits, or they may be different, as in the case of proteins and nucleic acids.

Sugar polymers, called *polysaccharides*, are formed by large numbers of sugar molecules joined by glycosidic bonds. Cellulose, starch, and glycogen are variants of glucose polymerization.

In both liver and muscle, glycogen is synthesized from UDP-glucose and a primer. The primer can be a very small glucose polymer in this case; even the disaccharide maltose will do, but glycogen itself is a much better acceptor. The reaction, catalyzed by UDP-glucose-glycogen transglucosylase (also known as glycogen synthetase), can be represented as:

$$\text{UDP-glucose} + \underset{\substack{\text{Glycogen} \\ \text{(primer)}}}{(C_6H_{10}O_5)n} \longrightarrow \underset{\text{Glycogen}}{(C_6H_{10}O_5)n+1} + \text{UDP}$$

Glycogen synthetase produces a straight-chain polysaccharide because only α-1,4 linkages are made. A second enzyme is present which initiates branching chains (and cross links) by transposing 1-4 to 1-6 linkages. This enzyme is amylo-(1,4 $\rightarrow$ 1,6)-transglucosidase, or branching enzyme for short. The linkages characteristic of polysaccharides are shown in Fig. 16-6.

What is the energy cost of making a typical polysaccharide molecule? About 2000 ATPs.

The nucleoside diphosphatases and other enzymes for polysaccharide synthesis are localized in some cells in "rough" endoplasmic reticulum (that lined with particles, in contrast to smooth-surfaced reticulum). They are contained within a particle (microsomal fraction) that can be isolated by centrifugation, suggesting that the polysaccharide-synthesizing enzymes occur in a complex.

16-5 SYNTHESIS OF NUCLEOTIDES AND NUCLEIC ACIDS

We have just seen how nucleotides participate as coenzymes in carbohydrate synthesis. Their roles as coenzymes in energy transfer and as regulators of enzymes will be discussed in a later chapter.

In this chapter, after a brief summary of nucleotide synthesis, the formation of polymers of nucleotides, the nucleic acids, will be considered.

The purine nitrogen bases, *adenine* and *guanine*, and the pyrimidine nitrogen bases, *cytosine*, *thymine*, and *uracil*, are of foremost biological importance. Their structures are shown in Fig. 16-7.

The *purine* ring is readily synthesized by nearly every organism

FIGURE 16-7
The purine and pyrimidine bases and the pentoses which, together with phosphate, form the nucleotides that are polymerized to form the nucleic acids, deoxyribonucleic acid (DNA) and ribonucleic acid (RNA).

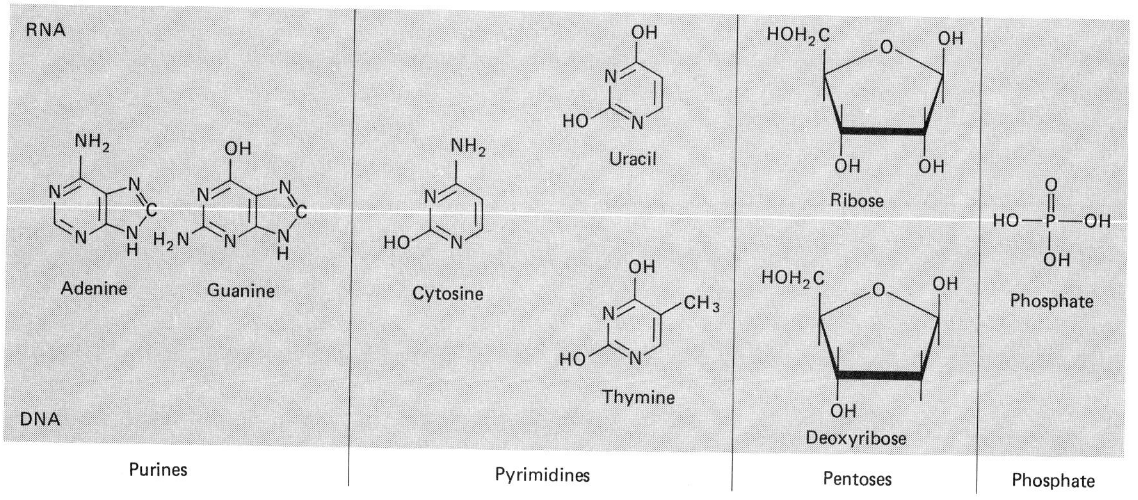

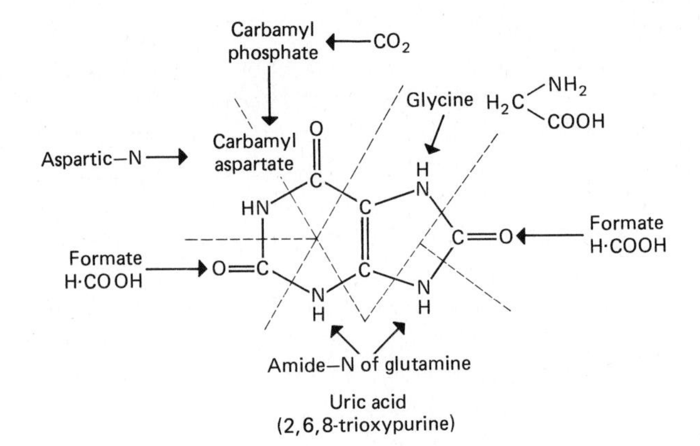

FIGURE 16-8
Sources of the atoms in the purine uric acid. Broken lines divide the molecule into the component parts from which it is synthesized.

known. By isotope labeling it has been determined that the ring is formed by contributors from five different precursors (Fig. 16-8).

The purine nucleus results from an elaborate sequence of reactions which produces an entire nucleotide. Inosinic acid, or inosine monophosphate (IMP), is the nucleotide first formed (Fig. 16-9).

Inosinic acid can be converted to the common nucleotides, adenylic acid (adenosine monophosphate, or AMP) and guanylic acid (guanosine monophosphate, or GMP), as shown in Fig. 16-10.

It is in the form of the diphosphate and triphosphate that we have most often encountered these nucleotides, and in that form they serve as precursors for the nucleic acids. The di- and triphosphate nucleotides are formed by successive kinase reactions:

$$\text{NMP} \xleftrightarrow[\text{Mg}^{2+}]{\text{ATP} \quad \text{ADP}} \text{NDP} \xleftrightarrow[\text{Mg}^{2+}]{\text{ATP} \quad \text{ADP}} \text{NTP (e.g., ATP, GTP)}$$

In contrast to the purine ribonucleotides, the ring forming the nucleus of *pyrimidines* is synthesized prior to its attachment to ribose 5-phosphate. The key intermediate is orotic acid, which is synthesized from ammonia, carbon dioxide, and ATP (Fig. 16-11).

Uridylic acid (or uridine monophosphate, or UMP) appears to be the first pyrimidine nucleotide formed. It can be sequentially phosphorylated to form the di- and triphosphates (UDP and UTP), like the purine mononucleotides. It appears that cytidine nucleotide synthesis involves the amination of UTP to form cytidine triphosphate (CTP). The third pyrimidine precursor of nucleic acids, thymidylic acid (TMP), also arises from uridylic acid.

We have now had a cursory examination of the mechanisms by which nucleoside polyphosphates are synthesized. Within the context of biosynthesis we are interested primarily in these compounds as precursors of nucleic acids. From genetic and biochemical studies we know that specificity of an organism is determined by the "information" of polymerized nucleotides, especially deoxyribonucleic acid (DNA), the genetic

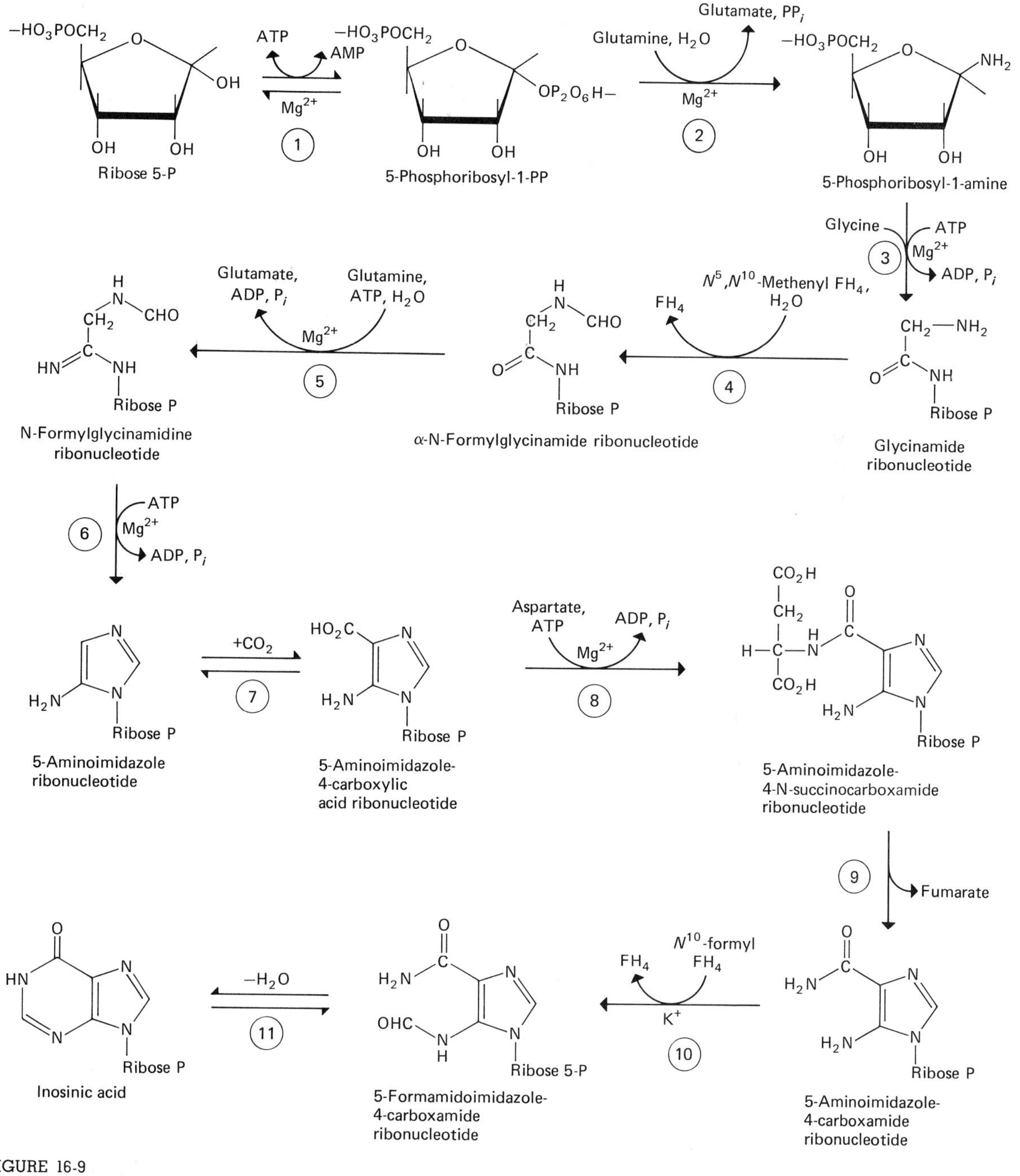

Ribose 5-P

① ATP, AMP, Mg^{2+}

5-Phosphoribosyl-1-PP

② Glutamine, H_2O; Glutamate, PP_i; Mg^{2+}

5-Phosphoribosyl-1-amine

③ Glycine, ATP, Mg^{2+}; ADP, P_i

Glycinamide ribonucleotide

④ N^5,N^{10}-Methenyl FH_4, H_2O; FH_4

α-N-Formylglycinamide ribonucleotide

⑤ Glutamine, ATP, H_2O; Glutamate, ADP, P_i; Mg^{2+}

N-Formylglycinamidine ribonucleotide

⑥ ATP, Mg^{2+}; ADP, P_i

5-Aminoimidazole ribonucleotide

⑦ $+CO_2$

5-Aminoimidazole-4-carboxylic acid ribonucleotide

⑧ Aspartate, ATP; ADP, P_i; Mg^{2+}

5-Aminoimidazole-4-N-succinocarboxamide ribonucleotide

⑨ Fumarate

5-Aminoimidazole-4-carboxamide ribonucleotide

⑩ N^{10}-formyl FH_4; FH_4; K^+

5-Formamidoimidazole-4-carboxamide ribonucleotide

⑪ $-H_2O$

Inosinic acid

FIGURE 16-9
Synthesis of a purine ribonucleotide, inosinic acid (inosine monophosphate or IMP). The reactions are 1, a pyrophosphorylation, an uncommon type of ATP-phosphorylation; 2, the transfer of an amino group (amination) from glutamine to PRPP; 3, an ATP-dependent conjugation with glycine through formation of an amide bond; 4, a formyl is added with folic acid (a vitamin) coenzyme, tetrahydrofolic acid (FH_4), serving as intermediary; 5, a second ATP-dependent amination in which glutamine is the amine donor; 6, ring closure by an ATP-dependent dehydration; 7, a carboxylation using carbon dioxide from bicarbonate; 8, ATP-dependent introduction of aspartic acid; 9, elimination of the carbon chain as fumaric acid; 10, introduction of the last carbon atom as a formyl group to complete the purine ring; FH_4 again is the coenzyme, but the formyl is in a different position than in reaction 4; 11, ring closure with dehydration completes the purine ring.

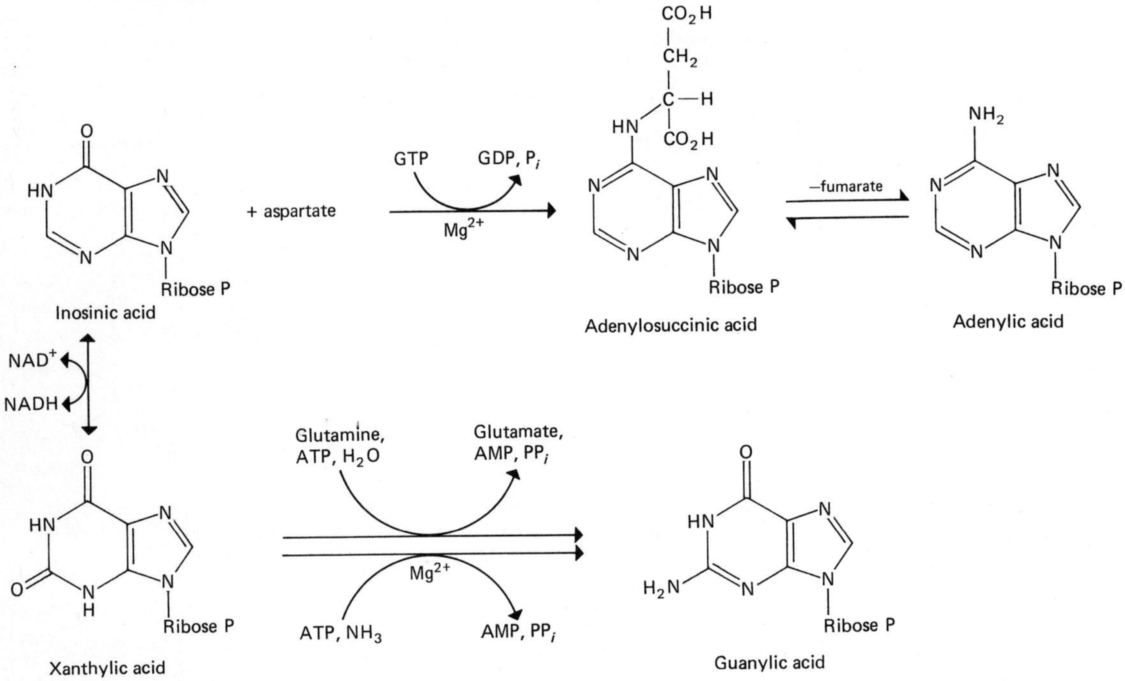

FIGURE 16-10
Conversion of inosinic acid to adenylic and guanylic acids. The pathway to adenylic acid requires guanine triphosphate (GTP) as a coenzyme. In the pathway to guanylic acid, NAD^+ serves as coenzyme, and in the conversion of xanthylic acid either glutamine or ammonia serves as amine donor, depending on the species in which the reaction occurs.

material. The closely related substance, ribonucleic acid (RNA), is involved with translation and transcription of the genetic code of DNA.

Both DNA and RNA are linear polymers of four different nucleotides. The sugar moiety of the nucleotides of RNA is ribose, whereas that of DNA is deoxyribose. They differ further in their base composition. In DNA the nucleotides are *d*-guanylic acid (G), *d*-cytidylic acid (C), *d*-adenylic acid (A), and thymidylic acid (T). The first three of these bases also are found in RNA, but uridylic acid (U) replaces thymidylic acid (Fig. 16-7). The nucleotide bases composing DNA and RNA are joined together through their phosphate and sugar groups to form linear chains (Fig. 16-12).

The number of nucleotide residues in a molecule of nucleic acid varies from about 80 to over a million. However, it has been shown that in DNA the ratios of A to T and G to C are close to 1. From this finding and from x-ray diffraction patterns and titration data, the structure of DNA was deduced by Watson and Crick to be two linear strands of polynucleotides in a double helical form, held together by hydrogen bonds between bases A . . . T and G . . . C, which are opposite one another in the two different strands (Fig. 16-13).

Such a structure does not restrict the sequence of bases in one

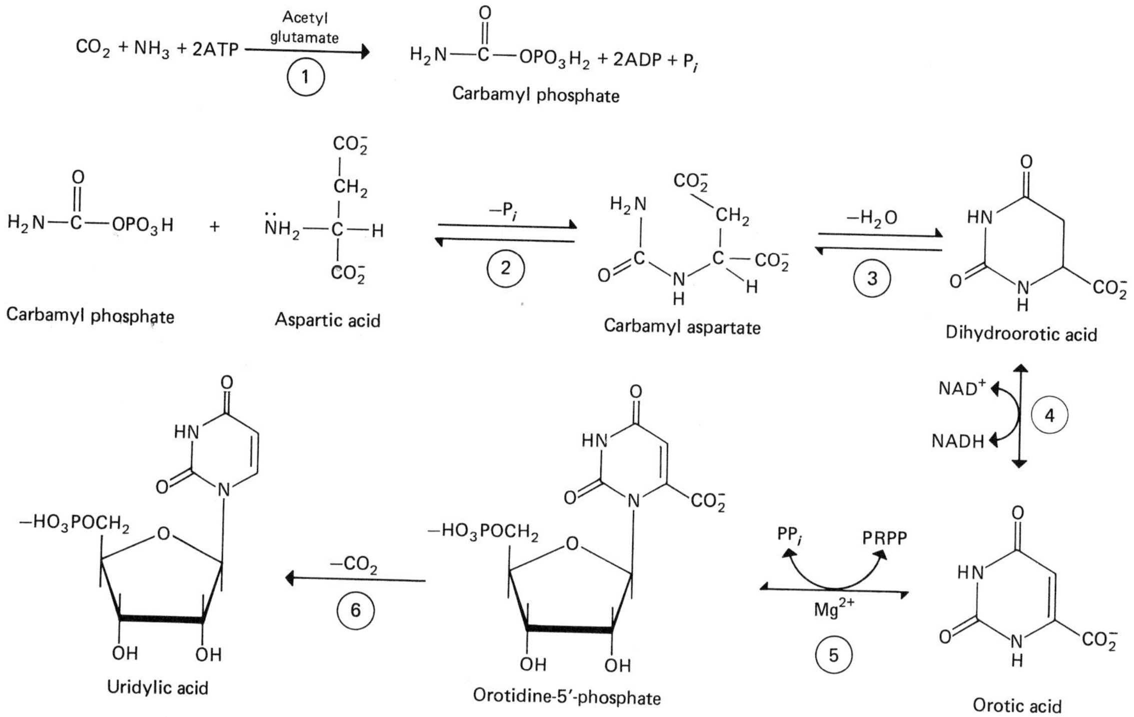

$$CO_2 + NH_3 + 2ATP \xrightarrow[\text{(1)}]{\text{Acetyl glutamate}} H_2N{-}\overset{O}{\overset{\|}{C}}{-}OPO_3H_2 + 2ADP + P_i$$

Carbamyl phosphate

Carbamyl phosphate + Aspartic acid $\underset{(2)}{\overset{-P_i}{\rightleftharpoons}}$ Carbamyl aspartate $\underset{(3)}{\overset{-H_2O}{\rightleftharpoons}}$ Dihydroorotic acid

Dihydroorotic acid $\underset{NAD^+ \to NADH}{\overset{(4)}{\rightleftharpoons}}$ Orotic acid

Orotic acid $\underset{Mg^{2+}\ (5)}{\overset{PRPP \quad PP_i}{\rightleftharpoons}}$ Orotidine-5′-phosphate $\underset{(6)}{\overset{-CO_2}{\longrightarrow}}$ Uridylic acid

FIGURE 16-11
Synthesis of the pyrimidine ribonucleotide, uridylic acid (uridine monophosphate, or UMP). The reactions are 1, synthesis of carbamyl phosphate (as it occurs in mammals); acetyl glutamate serves as coenzyme; 2, the carbamyl moiety is donated to the α-amino group of aspartate; 3, a cyclodehydration; 4, oxidation in a NAD^+-linked reaction to complete the pyrimidine ring; 5, attachment of ribose-5′-phosphate (as PRPP); 6, decarboxylation.

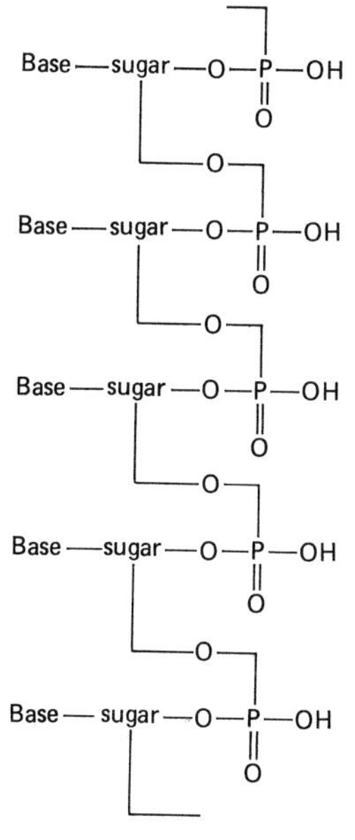

Base—sugar—O—P(=O)—OH
Base—sugar—O—P(=O)—OH
Base—sugar—O—P(=O)—OH
Base—sugar—O—P(=O)—OH
Base—sugar—O—P(=O)—OH

FIGURE 16-12
The manner by which nucleotides are linked to form a linear polymer, as in DNA and RNA.

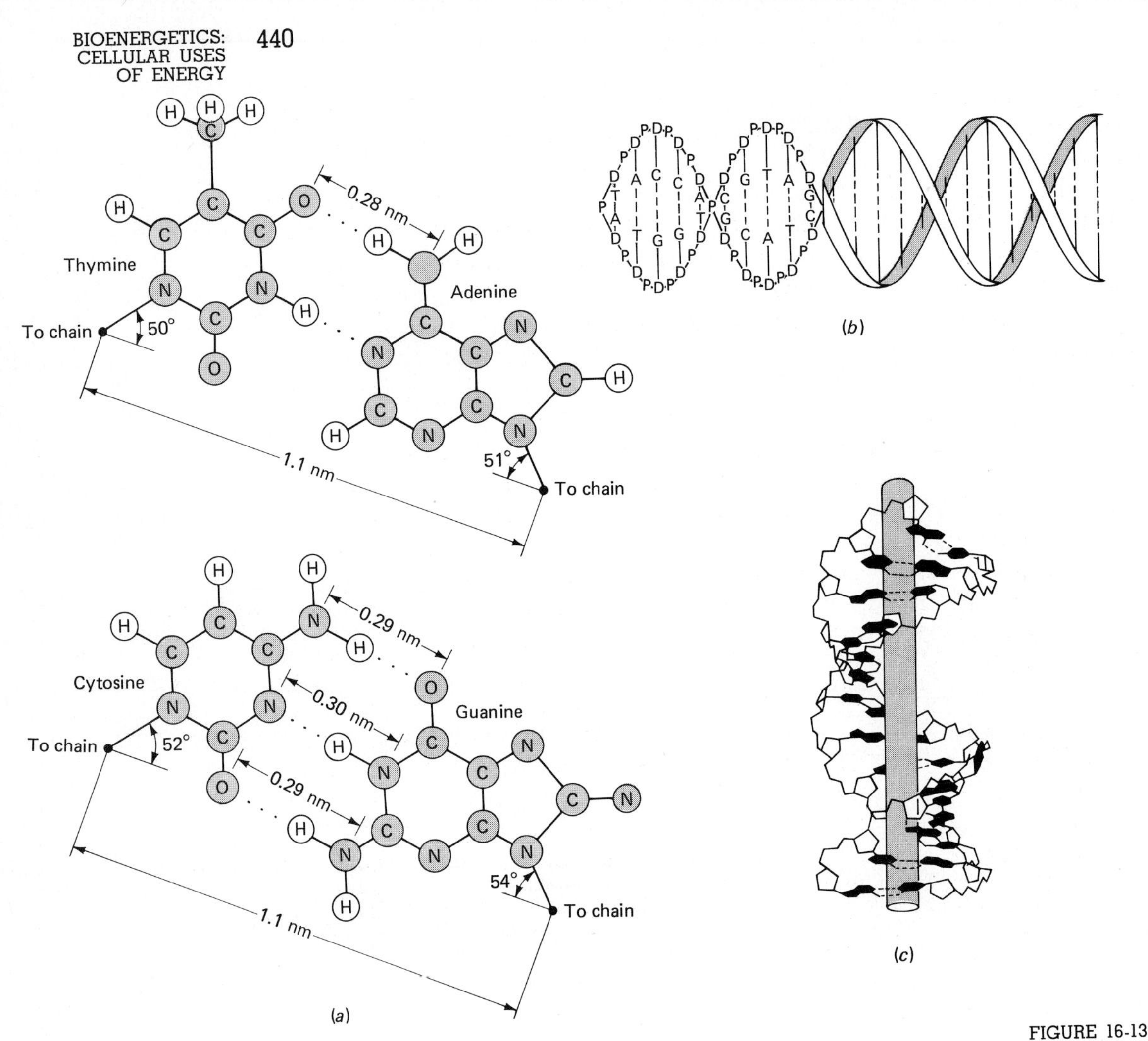

FIGURE 16-13
Structure of DNA. (a) The pairing of adenine with thymine and of cytosine with guanine by means of hydrogen bonding. (b) The double helix of DNA formed by two polynucleotides. The two phosphate (P)-sugar (D) chains are represented as ribbons on the outside of the helix and the bases (A, T, C, G) on the inside. (c) Drawing of DNA model as proposed by Watson and Crick. The dotted lines represent hydrogen bonds of the bases. The rod in the center is a support for the model, not a real structure.

strand, but because of the requirement for *base pairing* the sequence in one strand predetermines the sequence in the other. Thus, DNA, in the model of Watson and Crick, has the special feature of *complementarity*. It is this property which can account for replication of genetic material during cell division. It is assumed that the paired strands separate and each replicates by the formation of its complement. Thus, each strand serves as the template for the synthesis of a new strand (Fig. 16-14).

The requirement of DNA for the synthesis of more DNA introduces a new feature in biosynthesis. Other types of autocatalysis are known; e.g., ATP is required to make more ATP during anaerobic glycolysis. However, in the case of DNA we have a situation in which a molecule separates in half, and then the halves serve as templates for the formation of their complementary halves.

The first in vitro synthesis of a polynucleotide was accomplished in Ochoa's laboratory in 1955 when a high-molecular-weight (70,000; equal to about 30 residues), RNA-like molecule was made from 5′-ribonucleoside diphosphates. His system contained an enzyme, polynucleotide phosphorylase, obtained from the bacterium, *Azotobacter*. Incubation of all four ribonucleotides with this enzyme produced polynucleotides in which the base sequence was random, the composition depending upon the percentage of each nucleotide added in the reaction mixture. If only one nucleotide was present, a polymer of that single nucleotide was obtained. This is not the way real RNAs are made, of course; specific coding by DNA occurs in cells. Additionally, it appears that polynucleotide phosphorylase is not involved directly in the cellular mechanism of RNA coding, but the synthetic polymers made by this system were an important step in understanding functional nucleic acids.

In vitro incorporation of ^{14}C-labeled thymidine into DNA was reported by Kornberg in 1956. Two crude enzyme fractions from *E. coli* mediated this reaction. The following year he reported the incorporation of all four deoxynucleotide triphosphates in a system containing a small amount of "primer" DNA, Mg^{2+}, and a single enzyme. The enzyme, when later purified, was designated *DNA polymerase* (or DNA nucleotidyl transferase).

It appears that in Kornberg's system the enzyme catalyzes replication of the primer molecule, in the manner postulated by Watson and Crick (Fig. 16-14). The base composition of the DNA produced corresponds closely to the base composition of the primer. Other techniques bear this out also, and, in the case of DNA synthesized from a phage template, the in vitro product is biologically active as judged by its infectivity for *E. coli* cells. However, several lines of evidence suggest that these in vitro syntheses are not quite the same as the normal process of synthesis in vivo. For example, some of the polymerases are almost inactive with native, double-stranded primer. Also, although active DNAs are produced, upon extended synthesis abnormal DNA molecules are produced. Thus, replication of a double-stranded DNA molecule in vivo is a more complex event than in vitro synthesis of a single complementary strand on a single-stranded template.

DNA-dependent synthesis of specific RNA was reported in 1960 by three different laboratories, one using bacterial, one using plant, and one using animal cells. All three of the systems required the presence of the four ribose nucleoside triphosphates, Mg^{2+} (or Mn^{2+}), primer DNA, and an RNA polymerase. The enzyme from plant and animal cells occurs as a bound component of a DNA-nucleoprotein aggregate which, perhaps, forms part of the proteinaceous sheath of DNA molecules in the

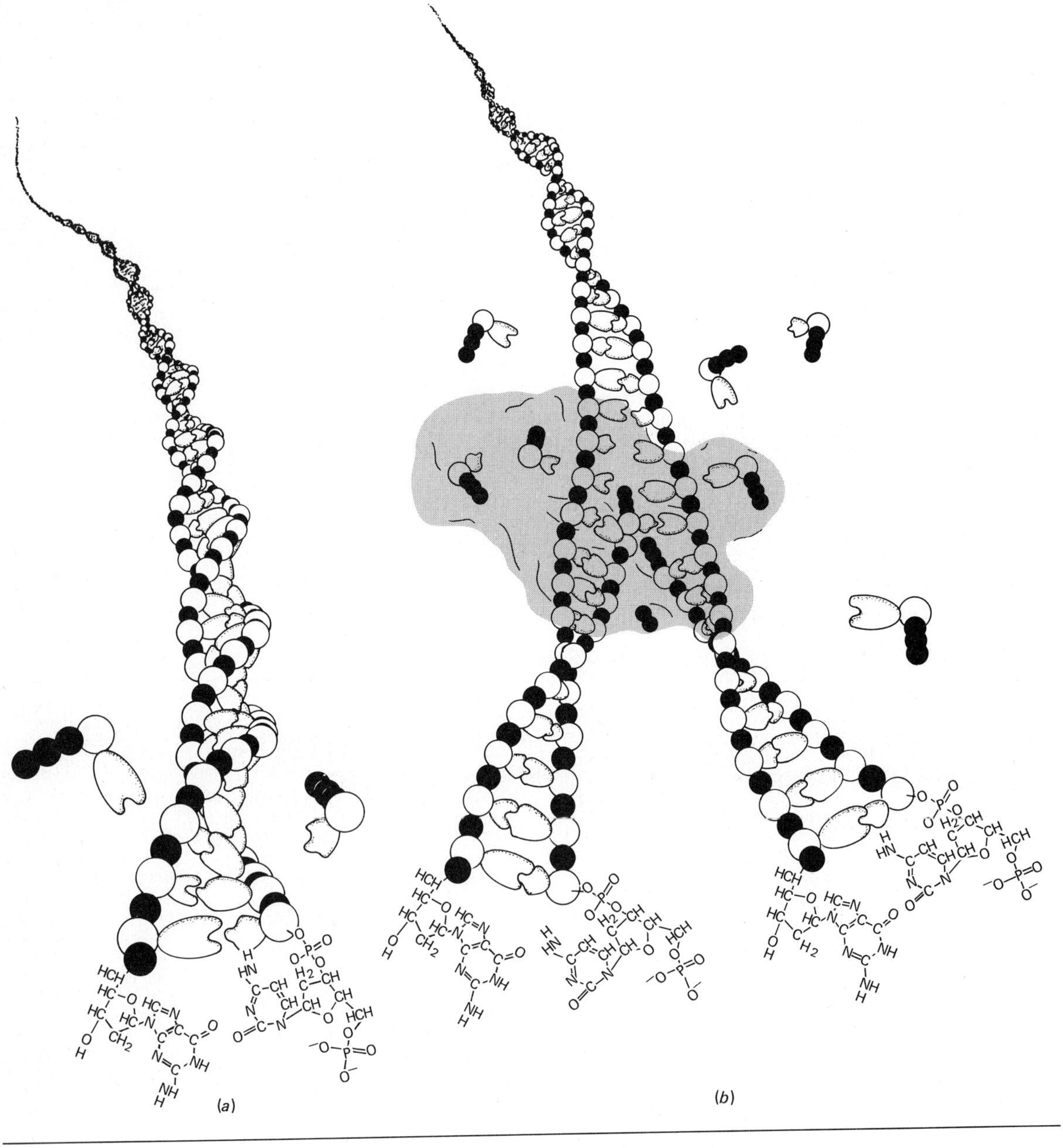

(a) (b)

interphase nucleus. By contrast, bacterial RNA polymerase is a readily soluble enzyme.

An RNA polymerase system in vitro will copy both strands of the DNA template. RNA polymerse in vivo transcribes only one strand of the double helix. Presumably, the DNA strands unwind only locally to permit

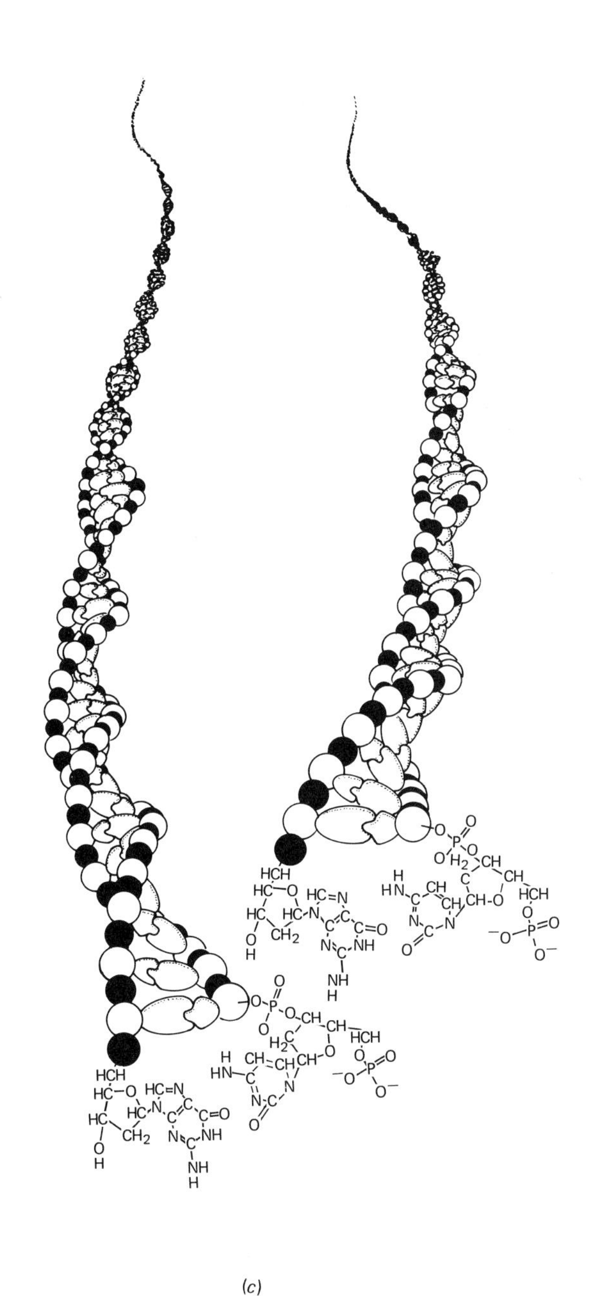

FIGURE 16-14
Interpretation of a replicating DNA helix. The two complementary strands separate (*a*), and each one acts as template for the synthesis of a new complementary strand (*b*). Precursor molecules are 5′-nucleotide triphosphates. The cloud-like form at the replication fork represents a molecule of DNA polymerase, the enzyme which catalyzes the replication. In (*c*) replication has been completed. (*After E. J. DuPraw, pp.* 349–351.)

transcription, then rewinds in a fully conserved form (Fig. 16-15).

To synthesize a molecule of RNA, about 6000 phosphorylations by ATP are required; DNA consumes about 120,000 ATPs per molecule. The synthesis of a protein molecule, our next subject, costs only about 1500 phosphorylations.

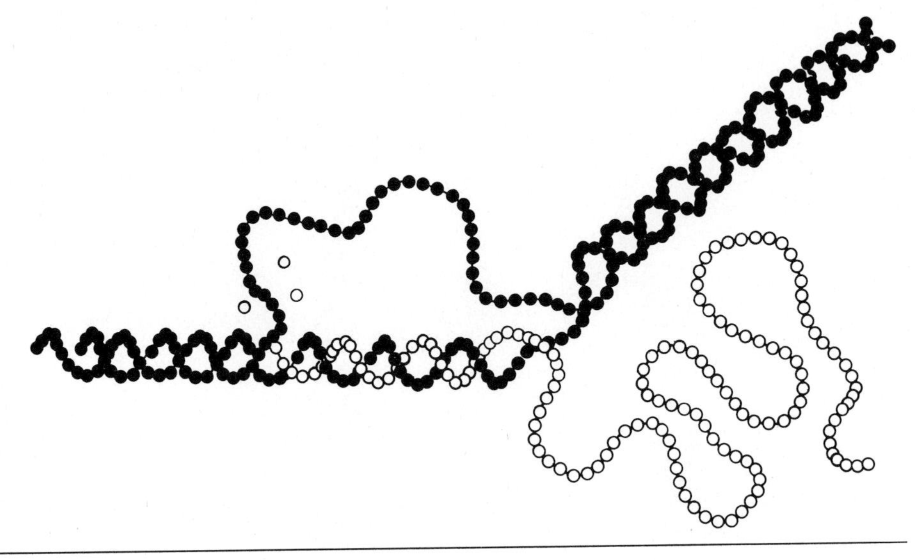

FIGURE 16-15
A model illustrating transcription of an RNA chain from a DNA helix. The original helix (dark beads) unwinds only locally, then rewinds in a fully conserved form. Only one strand of DNA is transcribed. (*After E. J. DuPraw,* p. 357.)

16-6 SYNTHESIS OF PROTEINS

DNA is the central informational molecule of a cell. In a eucaryotic cell, most of the DNA is localized in the nucleus, but it also occurs in chloroplasts and in mitochondria along with protein-synthesizing machinery. The DNA of each species is unique, containing coded information required to specify the amino acid sequence for synthesis of all the different proteins of the organism. Among the proteins synthesized are the all-important enzymes. All biosynthetic sequences require enzymes at most steps. Obviously, certain enzymes are required for synthesis of nucleic acids, and since nucleic acids guide enzyme synthesis, we are presented with a most curious circularity. Which came first, enzymes or nucleic acids? It is baffling to ponder how such a self-sustaining system originated.

Proteins comprise up to 70 percent of the total dry weight of organisms. *E. coli* is estimated to make as many as 1400 protein molecules per second. A protein contains, on the average, about 500 amino acids, and at least two molecules of ATP are required for each peptide bond formed. These figures should suggest the relative importance of protein synthesis in bioenergetics. In fact, it is probable that 90 percent of the energy invested in biosynthesis goes into protein formation, the synthesis of all the other kinds of molecules discussed up to now accounting for the remaining 10 percent.

Historically, the study of protein synthesis emerged more from cytochemical than from purely biochemical studies. It was by fractionation of cells into morphological components that systems for protein synthesis were established. Homogenates of liver cells, for example, may be separated by differential centrifugation into three fractions: (1) free nuclei and cell fragments, (2) granules (mostly mitochondria) visible by light microscopy, and (3) minute granules, called *microsomes*, with high RNA and phospholipid content, that are not visible by light microscopy. The importance of this third fraction in protein synthesis was realized when radioactive amino acids injected into an organism were subsequently found to be mainly concentrated in the microsomal fraction.

The microsomal fraction was examined when electron-microscopic techniques became available. It was then established that in contained membranes from the endoplasmic reticulum and particles which correspond to the granules seen on "rough" endoplasmic reticulum. Most of the RNA of the microsome fraction is associated with these granules, and so they are generally called *ribonucleoprotein* (RNP) *particles* or *ribosomes*.

Although its significance was not recognized at the time, many years ago it was observed that certain parts of cells stained readily with basophilic dyes. Those cells most stainable with such dyes are the most active in protein synthesis. It was subsequently shown that this characteristic staining is largely due to the presence of RNA. This was the first suggestion that RNA is involved with protein synthesis.

The next major discovery in protein synthesis was that radioactive amino acids will be incorporated by cell-free systems. Siekevitz in 1952, using microsomes from rat liver and a soluble factor from mitochondria, incorporated labeled alanine into proteins. The next year Allfrey used a system of microsomes and whole mitochondria from mouse pancreas to incorporate amino acids. The conditions that were found necessary for the in vitro incorporation of amino acids included the conditions that were necessary for the formation of ATP (Table 16-3).

Further refinements of cell-free, protein-synthesizing systems followed, and it was shown that the requirement for mitochondria and

Table 16-3
Ability of Various Isolated Morphological Liver-cell Fractions to Oxidize Substrate, Form ATP, and Incorporate Radioactive Amino Acids into Their Proteins

Fraction	Oxygen consumption	ATP formed	Counts/min mg^{-1} protein
Homogenate	37.4	4.0	10.8
Mitochondria	8.2	4.2	1.3
Microsomes	0.4	0.0	1.1
Supernatant	0.8	0.0	0.4
Mitochondria plus microsomes	14.0	4.1	10.2
Mitochondria plus supernatant	9.7	4.1	1.5
Mitochondria plus microsomes plus supernatant	18.8	3.8	4.3
Mitochondria plus boiled microsomes	9.7	4.4	1.2

Source: From A. G. Loewy and P. Siekevitz, 1963, *Cell Structure and Function*, Holt, Rinehart and Winston, Inc., New York, p. 164.

aerobic conditions could be eliminated by substituting ATP or a triphosphate-regenerating mechanism. It was also found that a soluble-protein fraction (supernatant) was required. This fraction was later shown to contain **activation enzymes** which mediate the adenylation of free amino acids (AA).

$$\text{Enzyme} + \text{AA} + \text{ATP} \rightleftharpoons \text{(adenyl-AA)Enz} + \text{PP}$$

A distinct enzyme is believed to catalyze the activation of each of the different amino acids. Several of these amino acid–specific enzymes have been isolated and identified.

Another cofactor of cell-free systems was discovered to be a low-molecular-weight (about 25,000) RNA. This form, later found in the soluble fraction, was called *soluble RNA* (sRNA) and is quite distinct from bound, *ribosomal RNA*. It appeared that sRNA acted as an acceptor of "activated" amino acids:

$$\text{(Adenyl-AA) Enz} + \text{sRNA} \rightleftharpoons \text{AA-sRNA} + \text{AMP} + \text{Enz}$$

The short sRNA molecules thus appeared to serve as adapter molecules, assembling at specific sites on a nucleic acid template through their complementary nucleotide sequences while transporting specific amino acids bound to their terminal positions. In recognition of these functions, sRNA is sometimes referred to as "adapter" RNA or "transfer" RNA.

More than 20 different sRNAs, corresponding to the different amino acids and amino acid–activating enzymes are known, and sRNA–amino acid complexes have been isolated from a variety of organisms. The amino acid is bound to sRNA by a 3′-ester linkage with a terminal adenosine group. All sRNA molecules end in a 3′-cytosine-cytosine-adenosine sequence, and if this terminal triplet is missing an amino acid will not be bound. Binding of the different amino acids to their specific sRNA is catalyzed by the specific activation enzymes (or aminoacyl RNA synthetases) through ATP activation (Fig. 16-16).

It was first assumed that ribosomal RNA might provide the template on which sRNA–amino acid complexes would orient, but the similarity in base composition among ribosomal RNAs from a variety of species made this appear unlikely. A third RNA, known as *template* or *messenger RNA* (mRNA), was discovered in *E. coli* infected with T_4 phage. This proved to be a single-stranded, high-molecular-weight RNA. Following its discovery, evidence has accumulated to support the concept that ribosomes provide only an unspecific mechanism for assembling polypeptide chains. The information for orienting sRNAs to form specific peptide sequences is contained in mRNA.

The requirements for protein synthesis, therefore, appear to be: ATP, 20 amino acids, ribosomal RNA, transfer RNA, and messenger RNA. As a simplified scheme we can write:

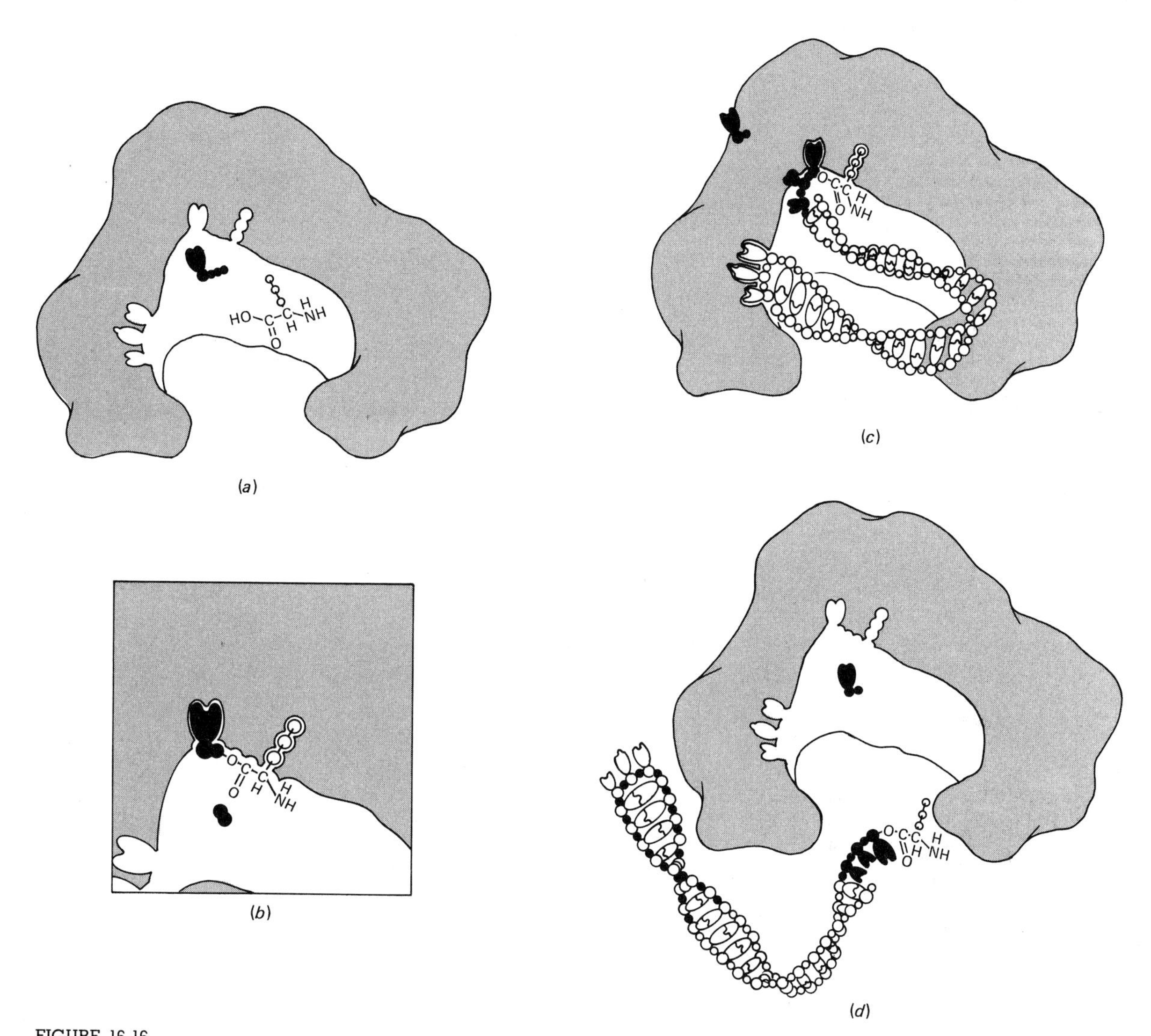

FIGURE 16-16
Diagram illustrating the reactions catalyzed by an aminoacyl RNA synthetase. In (a) and (b) an amino acid is adenylated at the expense of ATP, with pyrophosphate as a by-product; in (c) and (d) the specific adenyl amino acid is coupled to the adenyl end of a specific transfer RNA molecule, with AMP as a by-product. (*After E. J. DuPraw, pp.* 374–375.)

$$n(\text{sRNA-AA}) + \text{ribosome} + \text{mRNA} \xrightarrow[\text{ATP}]{\text{GTP}} \text{protein-mRNA-ribosome} + n(\text{sRNA})$$

GTP is specifically required for the transfer of amino acids from sRNA to protein, but the exact role of GTP is not known. Ribosomes have associated with them a high and specific GTPase activity.

Some investigators consider it likely that mRNA is used as a template by several ribosomes at a time. This is supported by the fact that sedimentation data indicate that ribosomes form aggregates, and some electron micrographs show aggregates of ribosomes which have been interpreted as strands of mRNA connecting ribosomes. Such an aggregate is called a **polyribosome.** Each ribosome is believed to begin "reading" the message at the same place, but several ribosomes can be reading at different places along the message at any one time (Fig. 16-17).

The concept that nucleic acid molecules might contain "information" in the form of specific nucleotide sequences met a formidable obstacle in the fact that proteins contain about 20 different amino acids, but nucleic acids contain only four kinds of nucleotides. Various coding mechanisms were proposed, but it became apparent that the minimum coding ratio must be *three* nucleotides for each amino acid (since two nucleotides provide only 16 different combinations). However, a code of three permits 64 possible combinations, and only 20 are needed. Nevertheless, it does appear that the "message" is read in threes from a fixed point in a polynucleotide sequence, meaning that there may be alternative trinucleotides for any one amino acid. Indeed, alternatives have been found for all the amino acids except methionine and tryptophan. For example, AAG and AAA are both codes for lysine, UUU and UUC both signify phenylalanine, and UCC and UCU are for serine. In a few cases four and even six alternative code words have been found for one amino acid. There are also nonsense triplets identified in some codes;

FIGURE 16-17
Interpretation of two ribosomes "reading" a messenger RNA strand. The mRNA is associated with the smaller (30S) ribosomal subunit, while the developing polypeptide chain is associated with the larger (50S) subunit. New amino acids are added to the carboxyl end of the polypeptide chain one at a time. Each precursor amino acid is brought to the site of synthesis as an adenyl ester on the C-C-A end of a specific transfer RNA molecule. The anticodon of the transfer RNA recognizes a codon triplet in the mRNA, permitting it to insert the appropriate amino acid in the polypeptide chain. (*After E. J. DuPraw, p. 386.*)

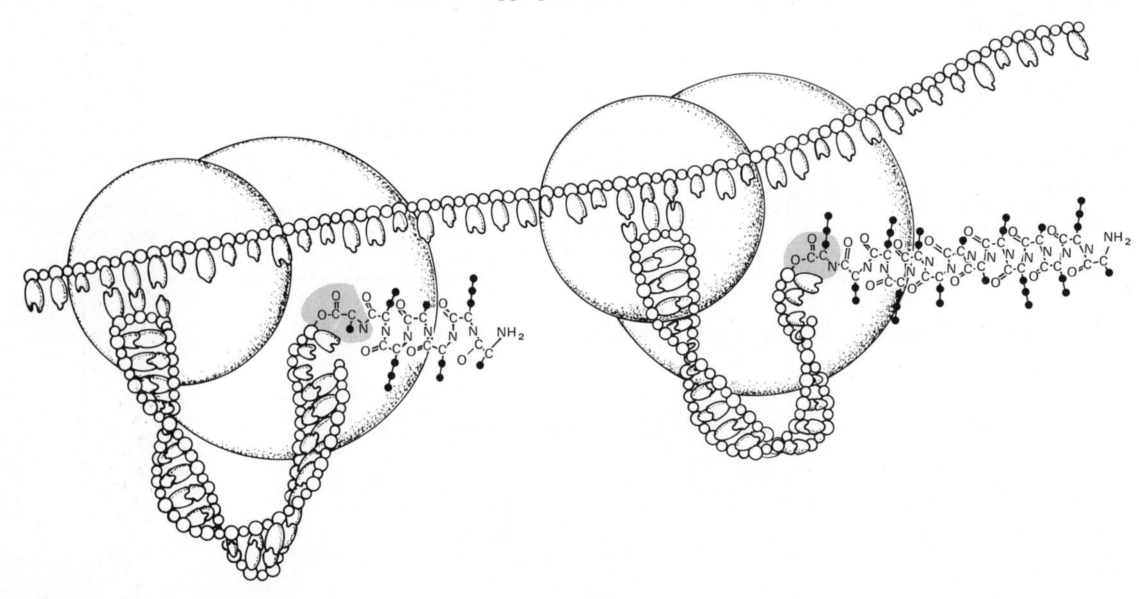

i.e., they do not code any amino acid. However, they may serve as a signal for chain termination in protein synthesis.

A detailed explanation of the genetic code and its various implications for protein synthesis is beyond the scope of this text. This is currently such a popular subject that the student will have no trouble finding literature that treats it in depth; some of the references at the end of the chapter will provide such information in detail.

We will end this section with an example of the sophisticated state of the art of speculation in this burgeoning area. On the assumption that three nucleotides are required to specify one amino acid, the number of ribosomes in a polyribosome group and the theoretical length of the mRNA strand for hemoglobin synthesis have been calculated. It is estimated that the mRNA should be about 150 nm long (146 amino acids, requiring 438 nucleotides of 0.34 nm per nucleotide). Neighboring ribosomes have a center-to-center spacing of 30 to 35 nm; therefore, a 150-nm hemoglobin messenger should link five ribosomes. Polyribosomes seen in electron micrographs appear to contain about that number of ribosomes.

16-7 BIOSYNTHETIC DEFECTS

Sometimes the genetic code of an organism contains one or more sequences of bases in DNA that do not correspond to normal sequences. An altered code can be produced naturally (spontaneously) or by physical or chemical means. Because the alteration is in the DNA, it is potentially inheritable by the offspring of the individual possessing such a genetic *mutation*.

Genetic abnormalities in that part of the DNA molecule which codes for hemoglobin synthesis have been studied in great detail. *Sickle-cell anemia* is a condition in which red blood cells are abnormally shaped, cannot efficiently transport oxygen, and are rapidly destroyed by the spleen. A person afflicted with sickle-cell anemia has hemoglobin (HbS) with properties that differ from those of normal hemoglobin (HbA). The difference stems from an abnormal sequence of amino acids. The HbS mutation occurs in the β chain. (Hemoglobin structure is discussed in Chap. 6, Sec. 6-4; Fig. 6-25 shows assembled α and β chains.) The amino acid sequence for the normal HbA is:

Valine-histidine$^{\oplus}$-leucine-threonine-proline-*glutamate*$^{\ominus}$-glutamate$^{\ominus}$-lysine$^{\oplus}$---

The sequence for the abnormal HbS is:

Valine-histidine$^{\oplus}$-leucine-threonine-proline-*valine*-glutamate$^{\ominus}$-lysine$^{\oplus}$---

Thus, valine has replaced glutamate. Glutamic acid has a negative charge, but valine is a neutral molecule. For this reason HbS moves more slowly in an electric field and can be separated from HbA by electrophoresis.

The code for glutamate is GAA, and that for valine is GUA. Therefore, the mutation to produce sickle-cell hemoglobin may be caused by a single base change in DNA.

Several diseases stemming from metabolic defects have genetic bases. These are called *inborn errors of metabolism*. They are usually caused by failure to synthesize enzymes or to synthesis of enzymes that lack proper catalytic activity. The amino acid, phenylalanine, when excreted normally is converted to carbon dioxide and water. One stage in the conversion produces alkapton (homogentisic acid). Certain individuals lack the enzyme for further conversion, and so they excrete a large amount of alkapton in their urine, a condition known as a *alkaptonuria* (Fig. 16-18).

The urine of a person with alkaptonuria turns dark on exposure to air because the homogentisic acid oxidizes to form a dark-colored compound. As the disease progresses, the cartilage of the ears and the eyes (sclera) darken and finally arthritis develops.

Phenylketonuria is a disease stemming from inadequate activity of the enzyme, phenylalanine hydroxylase, which converts phenylalanine to tyrosine. It is characterized by abnormal amounts of phenylalanine and phenylpyruvate in the blood, cerebrospinal fluid, and urine. Children who have this disease become feebleminded. The hair becomes lighter, because tyrosine is a precursor for the melanin pigments that impart dark colors (Fig. 16-18).

Another pathological condition, due to deficiency of a decarboxylase, results in the accumulation of valine, leucine, and isoleucine, plus their corresponding ketoacids in the urine. Because of the absence of a functional xylulose dehydrogenase, xylulose is excreted by persons with a condition known as *pentosuria*. Usually, there are no clinical abnormalities, however. A more serious abnormality is *galactosuria*, a genetic defect that afflicts newborns because galactose is a component of lac-

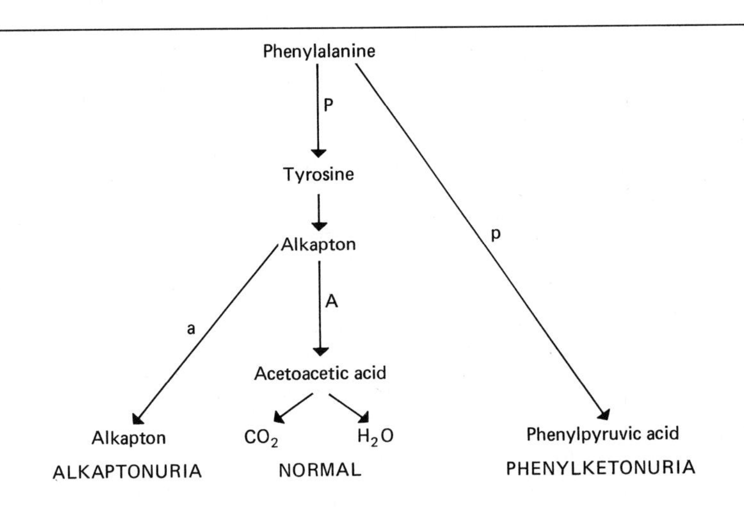

FIGURE 16-18
Metabolism of phenylalanine. *Pp* and *Aa* are two pairs of alleles. Normal persons have the dominant genes, *P* and *A*. An individual afflicted with alkaptonuria has only recessive genes, *aa*, lacks the enzyme to oxidize alkapton (homogentisic acid), and excretes this intermediate product. In the case of phenylketonuria, the individual has only recessive genes, *pp*, and does not convert phenylalamine to tyrosine.

tose, the sugar of milk. There are also *glycogen-storage diseases*, most of which are characterized by the absence of a specific enzyme.

It is speculated that the *aging* process is due to the occurrence and accumulation of a series of errors in cellular enzymes. Occasional mistakes in the synthesis of most proteins are tolerable because there should be enough normal molecules to compensate for the defective ones. However, when the code for any of those enzymes that are involved in the most fundamental operations of a cell becomes defective, the errors can be passed on to the daughter cells, and thus a clone of defective cells could develop over a period of time. Eventually a critical threshold is reached, according to this hypothesis, at which cells or organs become so burdened by errors that they no longer are viable.

READINGS

Cohen, G. N., 1967, *Biosynthesis of Small Molecules*, Harper & Row, Publishers, Incorporated, New York. A useful monograph on synthesis of amino acids, carbohydrates, lipids, vitamins, and coenzymes.

DuPraw, E. J., 1968, *Cell and Molecular Biology*, Academic Press, Inc., New York. Chapters 13 through 18 contain a wealth of information on biosynthesis and genetic coding.

Lehninger, A. L., 1971, *Bioenergics*, 2d ed., W. A. Benjamin, Inc., Menlo Park, Calif. Chapters 7 and 8 present lucid accounts of polysaccharide, lipid, and protein synthesis.

Mahler, H. R., and E. H. Cordes, 1968, *Basic Biological Chemistry*, Harper & Row, Publishers, Incorporated, New York. Chapters 18 through 21 deal with nucleotides, nucleic acids, and protein synthesis.

V

BIOENERGETICS: Energy Capture and Transfer

The biomass—the total living material—is in dynamic interaction with the sun and the earth. Within the world biomass there are no isolated parts, metabolically speaking; instead there are characteristic sequences of energy flow—or "food chains"—through the organisms of an ecosystem, as well as between ecosystems. In fact, the entire biomass can be considered as a single energetically interconnected ecosystem, the "ecosphere."

In the next three chapters our central concern will be with the tactics used by organisms in the strategy to secure energy for support of all their processes and mechanisms—many of which have been described in the preceding chapters of this book. It is fitting that we now examine, therefore, in some detail, the chemistry, from an energetic interest, that is common to most cells. Energy capture and transfer are basic to all other events in organisms. Because living things appear to go about their energetics in quite similar ways, it has been possible to integrate the considerable amount of information from a variety of species into a rather complete outline of the fundamentals of metabolism.

The very essence of the living state is energy flow. So much of biological study deals with the structural aspects of organisms that it is quite easy to forget that it is the continuous expenditure of energy which orders and maintains the "fabric" of life. The emphasis is misplaced too often, for as Needham reminds us: "It is usual to suppose that living organisms are material systems competing for energy, whereas it seems more likely that they are energy-systems competing for materials" (A. E. Needham, 1959, The origination of life, *Q. Rev. Biol.*, **34**, 202).

STEADY-STATE SYSTEMS 17

There is an irreversible flow of energy from the sun to the earth. It impinges on the earth as high-grade, fully convertible electromagnetic energy. In fact, the ultimate source of energy for the biomass is sunlight, captured and transformed to chemical energy by photoautotrophs. Organisms use electromagnetic energy to form medium-grade, partially convertible, organic compounds. Eventually it is transformed to low-grade, nonconvertible energy. Thus, the energy cycle is said to be *open*. In contrast, the matter of the planet is used over and over again, so its cycle is *closed* (Fig. 5-3). Matter, in addition to structuring organisms, is also the vehicle for the passage of energy through the organism. (Since in a relativistic sense, matter (mass) and energy are interconvertible, the distinction implied here is operational.)

A unique feature of life is that the equilibrium state is avoided. Instead of tending toward equilibrium with their environment, living entities constitute *open systems*, maintaining *steady-state* conditions, and continuously exchanging energy with the universe. In this chapter we shall explore the significance of this statement.

17-1 THERMODYNAMICS

Historically, the primary concern of those studying energy transformations was with transformations which accompany material changes, and the name "thermodynamics" was evolved. Subsequently, interest shifted more toward emphasis on energy functions to describe the state of a material system and to formulate rules that govern transitions from one state to another. The energy functions came to be used as a mode of bookkeeping in correlating the behavior of matter. Hence, the name *energetics* is perhaps more appropriate than *thermodynamics* to describe this field of knowledge. *Bioenergetics* describes the particular arm of this field extending through biological systems.

The familiar first law of thermodynamics embodies the concept that in any system, although the values of the work (W) done by the system and the heat (Q) absorbed in going from one state to another vary with the method followed, the difference, $Q - W$, is constant. The change of energy (ΔE) of the system can be written:

State 1

$$\Big\downarrow \Delta E = \Delta Q - \Delta W \qquad (17\text{-}1)$$

State 2

Thus, energy in such a system is defined in two measurable variables: heat and work. [The symbol Δ (delta) means "change of."]

Note that energy is really more than the "ability to do work," as it is often defined. It also should be pointed out that the energy value does not depend upon the previous history of the system, but only on the state of the system.

It is often useful to be able to predict whether a certain bioenergetic change is likely or not likely to occur. The first law does not provide any clues as to which is the case. It is customary to assume that chemical or physical events will occur spontaneously only if the final energy state of the system is lower than the first. There are some surprising exceptions, and in fact, there are spontaneous reactions where ΔE is negative, zero, or positive. A transformation can occur in which the energy of the final state is higher than the first, by absorbing energy from the surroundings during the transformation. How, then, does one distinguish a chemical system that is capable of absorbing energy from its surroundings and going to a higher energy state from one that cannot? The answer lies within the domain of the second law of thermodynamics. The key is the *entropy change* (ΔS). If

(ΔS) system + (ΔS) surroundings = positive number

then the transformation may occur spontaneously. Thus,

$\Sigma\Delta S > 0$; spontaneous change possible
$\Sigma\Delta S = 0$; system at equilibrium

Entropy has been variously defined. It may be considered as essentially a mathematical function to be viewed as an index of condition or character (perhaps somewhat analogous to a cost-of-living index or to pH as an index of acidity) rather than as a measure of some imaginary fluid. It is *an index of the capacity for spontaneous change*. Entropy also can be defined in terms of disorder or randomness—specifically, a quantitative measure of the atomistic disorder of a body or system.

By virtue of the way entropy is defined it *increases* as the capacity of an isolated system for spontaneous change *decreases*. In modern molecular-statistical energetics, where the degree of disorder of a system often is identified as entropy, the positive sign turns out to be a logical one.

The classical laws of thermodynamics were formulated for *closed systems*—ones in which no material enters or leaves. Such a system must eventually attain a time-independent equilibrium state with maximum entropy. Living systems are *open systems*, because they exchange materials with environment and continuously build up and break down components. The open nature of living systems accounts for many of their characteristics that seem contrary to the laws of physics and, consequently, have been considered "vitalistic" features. An open system *may* attain a time-independent state where the composition of the system remains constant as a whole, but there is a continuous flow of the component materials. This is called the *steady state.*

The total change of entropy in an open system can be represented as follows:

$$\Delta S = \Delta_e S + \Delta_i S \qquad (17\text{-}2)$$

where $\Delta_e S$ = the change of entropy by import
$\Delta_i S$ = the production of entropy due to irreversible processes in the system, such as chemical reactions, diffusion, and heat transport.

The term $\Delta_i S$ is always positive, according to the second law, but $\Delta_e S$ may be negative or positive. Therefore, the total entropy change can be negative or it can be positive.

By importing "order" from the environment, a living thing has the marvelous faculty of delaying its decay into thermodynamic equilibrium, or as Schrödinger[1] so cleverly states, it "feeds upon negative entropy, attracting, as it were, a stream of negative entropy upon itself, to compensate the entropy increase it procures by living and thus to maintain itself on a stationary and fairly low entropy level."

So, here is a striking contrast between inanimate and animate nature. The universe approaches entropy death, where all higher forms of energy such as mechanical, chemical, and light energy are converted into heat of low temperature. Living systems, on the other hand, show a trend toward development and evolution of increasing complexity or toward higher order and greater heterogeneity. This tendency toward increasing complication and decrease in entropy was what was previously believed to be "vitalistic" about organisms.

17-2 FREE ENERGY

The use of the entropy term to predict potentially spontaneous reactions is not practical, since it requires that one evaluate the changes in entropy not only of the substances undergoing transformation but of the surroundings as well. The space included in such changes usually can be determined only in special, isolated laboratory situations. Thus, it is desirable to restrict attention to the materials of the reaction and ignore surroundings. These objectives are realized in the concept of the *free-energy function* of Gibbs and Helmholtz. In simple terms, *free energy is that portion of the total energy of a system available to do work under iso-*

[1] E. Schrödinger, 1945, *What Is Life?* The Macmillan Company, New York, p. 74.

thermal conditions. How is it determined? As stated in the first law of thermodynamics:

(17-1) $\Delta E = \Delta Q - \Delta W$

Now, in chemical reactions in which pressure is constant, the heat change (ΔQ) is called *enthalpy* (ΔH). So, we write:

(17-3) $\Delta E = \Delta H - \Delta W$

If the reaction also undergoes no change in volume, and no work is produced, then:

(17-4) $\Delta E = \Delta H$

Thus, measuring the heat change of chemical reactions under conditions of standard temperature and pressure (*STP*)[1] gives information on the total energy change of the reaction. This can be done by direct calorimetric measurements.

Again using simple terms, in any transformation a portion of ΔE will not be available to do work. That portion is entropy, and its magnitude is a function of temperature. Thus, entropy and enthalpy are related to free energy in the following way:

(17-5) $\Delta E =$	ΔH	$=$	ΔG	$+$	$T\Delta S$
	Change in enthalpy (total change in energy between system and surroundings)		Change in free energy (available to do work)		Entropy (not available to do work; function of absolute temperature)

Rearranging:

(17-6) $\Delta G = \Delta H - T\Delta S$

If ΔG decreases, $T\Delta S$ increases, provided no heat is exchanged between the system and its surroundings. If heat is lost to the surroundings, then the decrease in ΔG of the system is greater than the gain of $T\Delta S$, etc. The *maximum* energy available under isothermal conditions to do work in a frictionless system is equal to ΔG. (In some biochemical publications, instead of indicating free energy by ΔG it is called ΔF.)

The value of ΔG may be used to predict whether a reaction will proceed, given the proper conditions, without an uptake of energy. Those which will proceed spontaneously under appropriate conditions occur with a net release of free energy and are called *exergonic* reactions. By

[1] In calculating heat and energy exchanges the initial and final states of all chemical processes are expressed in terms of an arbitrary chosen *standard state*, which is its most stable form at 1 atm pressure at 25°C (298°K). If a reaction occurs in a solution, the standard state of a reactant or product is defined as the *thermodynamic concentration* or *activity:* the molar concentration of a substance multiplied by its activity coefficient. Substances behaving ideally have an activity coefficient of 1; usually, the thermodynamic activity is a quantity less than its molar concentration.

convention they are assigned a negative sign ($-\Delta G$). Reactions which require a net increase of free energy in order to proceed are called *endergonic* reactions and are assigned a positive free-energy sign ($+\Delta G$). It must be clearly understood that although a negative ΔG is necessary for a reaction to occur spontaneously, it does not mean necessarily that one will occur. That is, ΔG and reaction rates are not related. (Also, because a compound has high energy does not mean that it is unstable.) Biological reactions proceed at measurable rates because enzymes are present. Thus, enzymes are the devices that secure the energy status of organisms (Fig. 17-1).

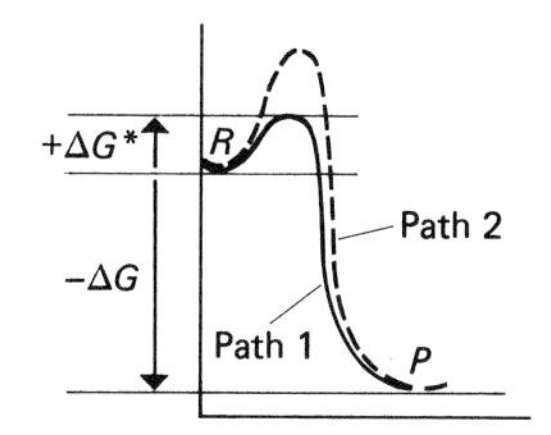

FIGURE 17-1
Hypothetical path for an enzymatic reaction. To pass from reactants (R) to products (P) an amount of energy (ΔG^*) must be invested to overcome the barrier of activation energy. The reaction may take different paths (two are indicated), each requiring different activation energy, but the net difference in free energy between the initial and final states is independent of the path of the reaction. The free energy released by the reaction is $-\Delta G$.

Merely specifying the nature of a chemical reaction does not predict the value for ΔG; its magnitude depends also on the conditions under which the reaction is carried out. We know that any chemical process will reach a certain equilibrium concentration of reactants and products and no further net chemical change takes place. That is to say:

$$\text{Reactants} \rightleftharpoons \text{products}$$

There is a constant which expresses the chemical equilibrium reached by a single isolated reaction, the so-called *thermodynamic equilibrium constant*, which has the form

$$K = \frac{(\text{products})}{(\text{reactants})} \qquad (17\text{-}7)$$

In situations where there are several components in the reaction, the equilibrium constant is the *product* of the active masses of the reaction products divided by the *product* of the active masses of reactants at equilibrium. For the reaction

$$A + B \rightleftharpoons C + D$$

the equilibrium constant is

$$K = \frac{(C)(D)}{(A)(B)} \qquad (17\text{-}8)$$

17-3 COMPUTATIONS OF STANDARD FREE ENERGIES

It is obvious that a standard method of representing free energies is required. One way this is accomplished is by the simple relation:

$$\Delta G^\circ = -RT \ln K \qquad (17\text{-}9)$$

where R = the gas constant (1.987 cal/deg mol^{-1})
T = the absolute temperature
$\ln K$ = the natural logarithm of the equilibrium constant

The symbol ΔG° designates the *standard free-energy change*, the gain or loss of free energy in calories under standard conditions (1 *m* for reactants and products in solution, 1 atm for gases, temperature 25°C).

Since most biological reactions occur at or near pH 7, another symbol, $\Delta G^{\circ\prime}$, is used to indicate the standard free-energy change at pH 7.

Table 17-1 shows the relationship between the equilibrium constant of a reaction and the calculated standard free-energy change. When the equilibrium constant is high (i.e., the reaction tends to go to completion), the standard free-energy change is negative and the reaction proceeds with a decline of free energy. When the equilibrium constant is low, the free-energy change is positive and, thus, energy must be put into the system to transform a mole of reactant to a mole of product under standard conditions. We can summarize as follows:

If $\Delta G^{\circ\prime}$ is + ; $R_1 + R_2 \leftharpoondown\!\!\rightharpoonup P_1 + P_2$
$\Delta G^{\circ\prime}$ is 0; $R_1 + R_2 \rightleftharpoons P_1 + P_2$
$\Delta G^{\circ\prime}$ is − ; $R_1 + R_2 \rightleftharpoons P_1 + P_2$

If $\Delta G^{\circ\prime}$ is known, K can be calculated using Eq. 17-9, or conversely, the equilibrium concentrations of reactants and products can be measured to determine K and then $\Delta G^{\circ\prime}$ can be calculated. The latter method is most useful for reactions with $\Delta G^{\circ\prime}$ between plus and minus 3000 cal (see Table 17-1), because it is difficult to measure the very small amounts of compounds present at equilibrium in reactions with extreme equilibrium constants.

Equilibrium measurements are especially applicable for hydrolyses and rearrangements. Let us consider an equilibrium that has been studied extensively—a step in glycogen metabolism catalyzed by phosphoglucomutase:

$$\text{Glucose-1-phosphate} \rightleftharpoons \text{glucose-6-phosphate}$$

At 25°C and pH 7,

$$K = \frac{(\text{glucose-6-phosphate})}{(\text{glucose-1-phosphate})} = \frac{0.019}{0.001} = 19$$

Table 17-1
The Numerical Relationship between the Equilibrium Constant and $\Delta G^{\circ\prime}$ at 25°C

K	$\Delta G^{\circ\prime}$, cal/mol
0.001	+4089
0.01	+2726
0.1	+1363
1.0	0
10.0	−1363
100.0	−2726
1000.0	−4089

Therefore,

$$\begin{aligned}\Delta G^{\circ\prime} &= -RT \ln K \\ &= -(1.987)(298)(\ln 19) \\ &= -(1.987)(298)(2.303)(\log_{10} 19) \\ &= -(1363)(1.28) \\ &= -1745 \text{ cal/mol}\end{aligned}$$

This tells us that there is a decline in free energy of 1745 cal when 1 mol of glucose-1-phosphate is converted to 1 mol of glucose-6-phosphate at 25°C and pH 7 under conditions whereby the concentration of each is maintained at 1 *M*.

Knowing the standard free-energy change and the concentration of participants, the ΔG of any reaction can be calculated as follows:

$$\Delta G = \Delta G^{\circ\prime} + RT \ln \left[\frac{(P_1)(P_2)}{(R_1)(R_2)} \right] \qquad (17\text{-}10)$$

Thus, it can be seen that the ΔG of a reaction depends upon the initial concentrations of reactants and products and that the concentrations determine the *direction* in which the reaction will proceed. This is called *concentration effect* and is especially important in real biological reactions, especially where the $\Delta G^{\circ\prime}$ is positive. By establishment of proper concentrations, such reactions can be made to have a $-\Delta G$ and, therefore, occur spontaneously.

Table 17-2 shows calculated standard free-energy changes of some of the chemical reactions known to occur in cells. Note the large decline in free energy of oxidation reactions, a major source of energy for biological processes.

Table 17-2
Standard Free-energy Changes at pH 7 and 25°C of Some Chemical Reactions

	$\Delta G^{\circ\prime}$, kcal/mol
Oxidation	
Glucose + $6O_2 \rightarrow 6CO_2 + 6H_2O$	−686
Lactic acid + $3O_2 \rightarrow 3CO_2 + 3H_2O$	−320
Palmitic acid + $23O_2 \rightarrow 16CO_2 + 16H_2O$	−2338
Hydrolysis	
Sucrose + H_2O → glucose + fructose	−7.0
Glucose-6-phosphate + H_2O → glucose + H_3PO_4	−3.3
Glycylglycine + H_2O → 2 glycine	−2.2
Rearrangement	
Glucose-1-phosphate → glucose-6-phosphate	−1.7
Fructose-6-phosphate → glucose-6-phosphate	−0.4
Elimination	
Malate → fumarate + H_2O	+0.75

Source: From A. L. Lehninger, p. 32.

Oxidation-Reduction Potentials

For oxidation-reduction reactions one method is especially useful in calculating ΔG—viz., measurement of the tendency of substances to donate or accept electrons. The value obtained is termed an *oxidation-reduction potential.* It is possible to measure the electromotive force of oxidation-reduction systems because they can serve as part of a battery.

A scale of oxidation-reduction potentials can be established by comparison with a common standard whose potential is arbitrarily set equal to zero. The common standard is the *hydrogen electrode,* $\frac{1}{2}H_2 \rightarrow H^+ + e^-$, the potential of which is considered to be zero for a solution containing hydrogen ions at unit activity in equilibrium with hydrogen gas at 1 atm pressure. The oxidation-reduction potential of any electrode (reaction system) can then be measured relative to the hydrogen electrode, as explained in Fig. 17-2.

The difference between the oxidation-reduction potential of any electrode and the hydrogen electrode is given by the general equation:

$$E = E_0 - \frac{RT}{nF} \ln \frac{\text{(reduced)}}{\text{(oxidized)}} \tag{17-11}$$

where E = the observed potential difference in volts
E_0 = the standard oxidation-reduction potential for the electrode being measured
R = the gas constant
T = the absolute temperature
F = the faraday (23,000 cal/absolute volt equivalent)
n = the number of electrons per gram equivalent transferred in the reaction

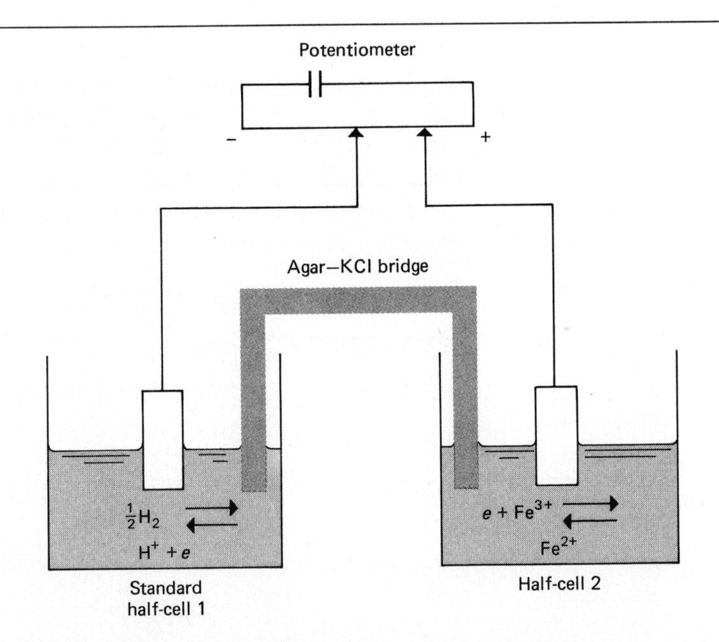

FIGURE 17-2
Design of apparatus used for the determination of oxidation-reduction potentials. One half-cell is a standard hydrogen electrode, and the other is the system to be measured. Inert metal electrodes are connected through a potentiometer, which measures the potential (voltage) difference between the two half-cells. An agar-salt bridge completes the circuit by permitting migration of ions but preventing direct chemical reaction between the components of the half-cells. The imposed voltage just necessary to prevent electron flow is the oxidation-reduction potential.

There are two conventions for indicating the sign of oxidation-reduction potentials: one assigns more negative potentials to systems that have an increasing tendency to donate electrons (reduction potentials), and the other assigns more negative potentials to systems that have an increasing tendency to accept electrons (oxidation potentials). The former is the generally preferred convention and will be used in this book. Table 17-3 shows standard reduction potentials of some biologically important systems. Any system placed lower in this table would have a tendency to donate electrons to any higher system. It will be useful to refer back to this table when we come to Chap. 19, which deals with oxidations.

Work and Electron Transfer

Now that we know how to find potentials of substances, how do we translate them into free energies? Here is one way. By definition, the free-energy change in a reversible process is equal to the maximum work, W_{max}, possible in a frictionless system:

$$\Delta G = W_{max} \qquad (17\text{-}12)$$

Table 17-3

Standard Reduction Potentials for Systems of Biological Importance

System	E'_0 (pH 7), V
$^1/_2O_2/O^{2-}$	0.82
Oxygen/water	0.815
Ferric/ferrous	0.77
Nitrate/nitrite	0.42
Oxygen/hydrogen peroxide	0.30
Cytochrome a; ferric/ferrous	0.29
Cytochrome c; ferric/ferrous	0.26
Crotonyl-SCoA/butyryl-SCoA	0.19
Methemoglobin/hemoglobin	0.17
Cytochrome b_2; ferric/ferrous	0.12
Ubiquinone; ox/red	0.10
Metmyoglobin/myoglobin	0.046
Fumarate/succinate	0.03
Methylene blue, ox/red	0.01
Yellow enzyme; FMN/$FMNH_2$	−0.122
Pyruvate + ammonium/alanine	−0.13
α-Ketoglutarate + ammonium/glutamate	−0.14
Oxaloacetate/malate	−0.17
Pyruvate/lactate	−0.19
Acetaldehyde/ethanol	−0.20
Riboflavin, ox/red	−0.21
Glutathione, ox/red	−0.23
NAD^+/NADH	−0.32
Pyruvate/malate	−0.33
Carbon dioxide/formate	−0.42
H^+/H_2	−0.42
Acetate/acetaldehyde	−0.60
Succinate/α-ketoglutarate	−0.67
Acetate + carbon dioxide/pyruvate	−0.70

The potential (E) existing between two such reaction systems can be expressed as work, since each mole of electrons (6.02×10^{23} electrons = 1 faraday = 96,487 coulombs) flowing through this potential yields work and can be represented as

Work = potential × coulombs = joules

and since

joules/4.18 = calories

it then follows that if E is known, the free-energy change can be calculated. That is to say, the electromotive force is proportional to the maximum work per electron transfer and is related to ΔG by the equation

$$\Delta G = -nFE \tag{17-13}$$

or

$$\Delta G^{\circ\prime} = -nFE_0' \tag{17-14}$$

where terms are as defined above and E_0' is the standard oxidation-reduction potential for the electrode being measured at pH 7. Again, the negative sign represents free-energy decrease.

Here is an example of how reduction potentials can be used to calculate free energy: As shown in Table 17-3, the E_0' for the system (cytochrome c Fe^{3+}/cytochrome c Fe^{2+}) is +0.26 V and for ($\frac{1}{2}O_2/O^{2-}$) it is +0.82. The difference between the half-cells would be +0.56 V. Now if cytochrome c Fe^{3+} donates electrons to oxygen (the acceptor), what would be the standard free-energy change? By using Eq. 17-13, and noting that *two* moles of electrons are involved, we can find:

$$\begin{aligned}\Delta G^{\circ\prime} &= -nFE_0' = -2FE_0' \\ &= -(2 \times 96{,}487 \times 0.56) = -108{,}065 \text{ joules} \\ &= -108{,}065/4.18 = -25{,}853 \text{ cal}\end{aligned}$$

Calorimetry

Still another method for calculating free energy is based upon Eq. 17-6:

$$\Delta G = \Delta H - T\Delta S \tag{17-6}$$

The heat released at constant temperature and pressure, ΔH, can be measured by calorimetry. One way of determining entropy, ΔS, is to measure the heat absorbed per degree by each compound separately at temperatures ranging from the temperature of the reaction to as low a temperature as possible. Such measurements are extremely difficult and must be done with great precision. For most biological systems determination of entropy is not practical.

Here is an example of how this method is used, however. The oxidative conversion of glyceraldehyde-3-phosphate to 3-phosphoglyceric acid under standard conditions can be represented as:

$$\underset{\text{(GP)}}{\text{Glyceraldehyde-3-phosphate}} + H_2O \longrightarrow \underset{\text{(PGA)}}{\text{3-phosphoglyceric acid}} + H_2$$

The change in enthalpy, ΔH, is +4000 cal, as determined by calorimetry. The entropy change, $T\Delta S$, is equal to $T \times$ (entropy of products − entropy of reactants). It is 0.9 cal/deg for the conversion of GP to PGA. The value of S for H_2 is 31.12 and for H_2O is 16.45 cal/deg. Thus the entropy change at 20°C is

$$\begin{aligned} T\Delta S &= 293\,(S_{PGA} - S_{GP} + S_{H_2} - S_{H_2O}) \\ &= 293\,(0.9 + 31.12 - 16.45) = 4560 \text{ cal} \end{aligned}$$

Therefore,

$$\Delta G^\circ = \Delta H - T\Delta S = +4000 - 4560 = -560 \text{ cal}$$

Before leaving the subject of free-energy change a note of caution is warranted. Real biological reactions do not occur as isolated systems and they may not reach equilibria. Within a cell, reactions are generally members of chains of reactions in which the products of one reaction become the reactants of the next:

$$A \xrightarrow{1} B \xrightarrow{2} C \xrightarrow{3} D \xrightarrow{4}$$

The members of such a system are in a "steady state"; their concentrations do not change with time because when a molecule of A is converted into B, a molecule of B is simultaneously converted into C, etc. The amount of any one component in the steady state is determined by the rates of reactions 1, 2, 3, 4, etc., not by their equilibrium concentrations as isolated reactions. It is the activation energy, not the overall difference of free energy between each reactant and its product, that determines the rate of reaction. Therefore, an equilibrium reaction in the test tube does not predict the concentration of a compound within the cell.

17-4 TRANSFER OF CHEMICAL ENERGY

All organisms, regardless of the primary source of their energy, conserve and use energy not as heat but as chemical energy. The compound that occupies the most central role in bioenergetic processes is **adenosine triphosphate** (ATP) (Fig. 17-3).

It is common to encounter the statement that ATP has "high-energy" bonds in the phosphate moieties. Such a name gives a misleading impression as to the nature of the quantity being considered. Since fundamentally what is referred to are changes in potential, ΔG, when certain groups are transferred from one molecule to another, a more appropriate name is *group-transfer potential.*

The term "bond energy" has a very definite meaning in the field of energetics, quite different from that implied by the biochemical term "high-energy bond." What is meant by "high-energy phosphate bond" is that the *difference* in energy content between the reactants and products

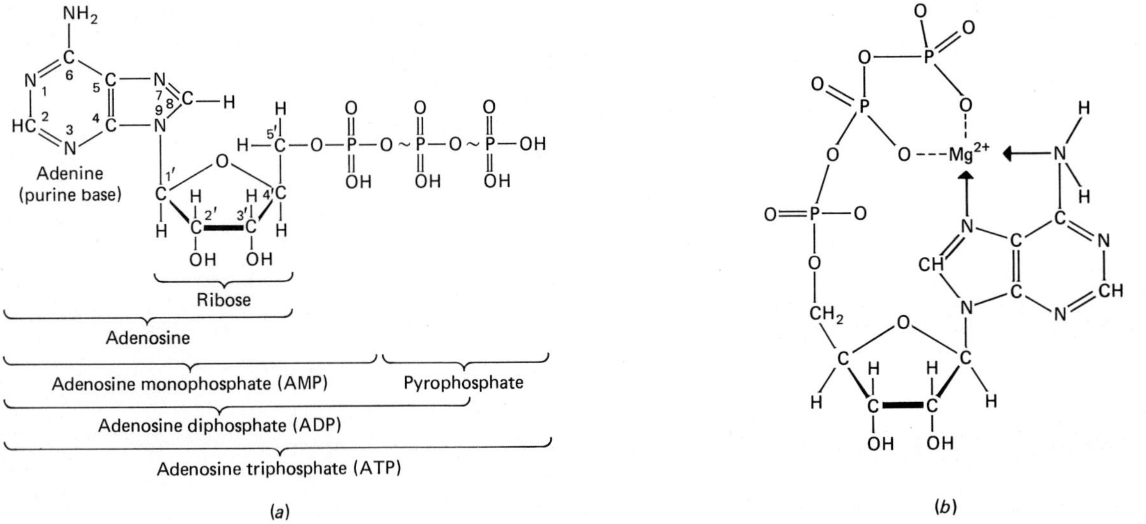

FIGURE 17-3

(*a*) The chemical structure of ATP. The standard numbering of the various atoms is indicated. The molecule contains the purine, adenine, and a pentose, D-ribose, attached to the adenine through a glycosidic linkage. This compound alone would be called *adenosine*, and is classified as a nucleoside. When phosphate is added it forms a nucleotide. If one phosphate group is attached in ester linkage at the 5′ position, it is called *adenosine monophosphate* (AMP). When a second phosphate is added in anhydride linkage with the 5′ phosphate, the complex is *adenosine diphosphate* (ADP), and if there is a third phosphate group it is *adenosine triphosphate* (ATP). The symbol ~ designates the so-called "high-energy" bonds. (*b*) A possible structure of ATP as a Mg^{2+} complex, which is thought to be its common, stable form.

of hydrolysis is relatively high; the free energy of hydrolysis is not localized in the actual chemical bond itself (Fig. 17-4).

The special symbol, ~, used to designate high phosphate-transfer potential groups implies that there is a concentration of energy between P ~ P which tends to make the terminal phosphate fly off whenever possible. It does not do this spontaneously, however, for energy has to be put

FIGURE 17-4

Explanation of the difference between "bond energy" and "phosphate transfer (or group) potential." If one refers to bond energy, the process shown in (*a*) would represent the situation, and the energy change (ΔE) would be 50 to 100 kcal/mol. However, in living systems it is the transfer reaction that is of greater interest, as shown in (*b*) with water as the acceptor. The free-energy change ($\Delta G'$) of this process is −7 kcal/mol.

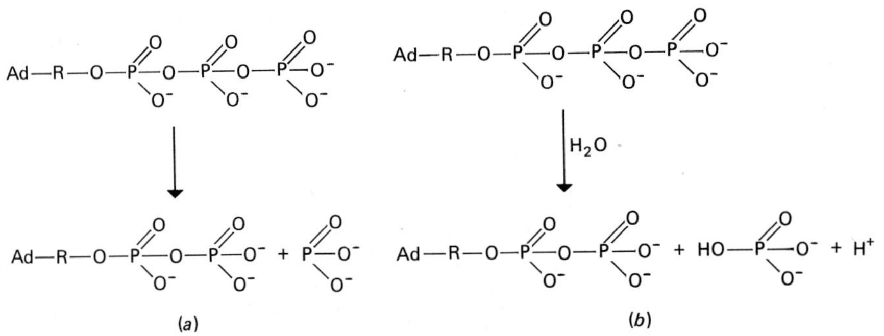

into ATP to break the terminal phosphate bond. It is this energy which must be forced into the molecule to break a bond between two atoms that should be called the bond energy.

However, we are not interested in the change in internal energy necessary to break the bond, but rather in the chemical potential—the free energy—when a compound such as ATP transfers one of its substituent groups to another molecule. The ΔG for such a transfer will depend upon the nature of the donor and acceptor molecules. Since one is mostly interested in comparing group-transfer potentials of several kinds of donor molecules, it is convenient to have some standard conditions for comparative purposes. Water is the standard acceptor molecule and, using standard conditions, *standard transfer potentials* can be determined.

Examples:

$$\text{ATP} + H_2O \longrightarrow \text{ADP} + HPO_4 \qquad \Delta G^\circ = -7\ \text{kcal}$$
$$\text{Glucose-6-phosphate} + H_2O \longrightarrow \text{glucose} + HPO_4 \qquad \Delta G^\circ = -3\ \text{kcal}$$

These reactions are hydrolyses. By establishing the magnitude of the standard free energies of hydrolysis of various compounds, we can predict their relative affinity for transfer groups and decide whether particular transfer reactions are compatible with thermodynamic requirements. Table 17-4 shows the standard free energies of hydrolysis of several phosphate compounds.

Why was ATP selected as the "common currency" for energy exchange? Probably because it occupies an *intermediate* level on the transfer potential scale, as shown in Table 17-4. By occupying such a position it can *couple* biological reactions. The degradation of food by organisms is a series of coupled reactions in which energy is released. That energy is transferred via ATP and used for performing work: electrical, osmotic, chemical, and mechanical.

Table 17-4
Standard Free Energies of Hydrolysis of Phosphate Compounds

Compound	$\Delta G'$, cal/mol	Direction of phosphate group transfer
Phosphoenolpyruvate	−12,800	↓
1,3-Diphosphoglycerate	−11,800	
Phosphocreatine	−10,500	
Acetyl phosphate	−10,100	
ATP	−7,000	
Glucose-1-phosphate	−5,000	
Fructose-6-phosphate	−3,800	
Glucose-6-phosphate	−3,300	
3-Phosphoglycerate	−3,100	
Glycerol-1-phosphate	−2,300	

A carbohydrate can be oxidized directly with oxygen to yield energy. For example, if a mole of glucose were to be oxidized completely:

$$6O_2 + C_6H_{12}O_6 \longrightarrow 6CO_2 + 6H_2O \qquad \Delta G^\circ = -686{,}000 \text{ cal/mol}$$

But this reaction would simply yield heat that an organism could not use. The grand strategy of life is to capture that energy in small "packets" by releasing the energy of compounds in controlled, sequential steps. And it is not heat that is captured, but rather chemical energy. The function of phosphate compounds in bioenergetics is to act as intermediates in the capture and transfer of this energy.

We may write the above equation as it occurs in organisms as follows:

$$C_6H_{12}O_6 + 6O_2 + 38ADP^{3-} + 38HPO_4^{2-} + 38H^+ \longrightarrow 6CO_2 + 44H_2O + 38ATP^{4-} \qquad \Delta G^\circ = -686{,}000 \text{ cal/mole}$$

It is to be expected that reactions which have a negative free-energy change are thermodynamically feasible. However, in biological systems reactions can occur that have positive free-energy changes. One of the ways to make this feasible is to have the reaction with a $+\ \Delta G$ "coupled" to another reaction which has an appropriate $-\Delta G$. Here is an example:

$$\text{Glucose} + \text{ATP} \longrightarrow \text{glucose-6-phosphate} + \text{ADP} \qquad \Delta G^{\circ\prime} = -5.1 \text{ kcal}$$

$$\text{Glucose-6-phosphate} \longrightarrow \text{glucose-1-phosphate} \qquad \Delta G^{\circ\prime} = +1.8 \text{ kcal}$$

$$\text{Glucose-1-phosphate} + (\text{glycogen})_n \longrightarrow (\text{glycogen})_{n+1} + \text{ADP} + \text{PO}_4 \qquad \Delta G^{\circ\prime} = -0.5 \text{ kcal}$$

$$\text{Sum: Glucose} + \text{ATP} + (\text{glycogen})_n \longrightarrow (\text{glycogen})_{n+1} + \text{ADP} + \text{PO}_4 \qquad \Delta G^{\circ\prime} = -3.8 \text{ kcal}$$

The overall reaction has a negative $\Delta G'$ under physiological conditions. In standard conditions energy would be required for the conversion of glucose-6-phosphate to glucose-1-phosphate, but under the conditions that occur in a cell the ΔG must actually be negative. This is accomplished by a steady-state condition in which there is a buildup of glucose-6-phosphate by the previous reaction and a removal of glucose-1-phosphate by the subsequent reaction (recall concentration effect, discussed earlier). By knowing the free-energy change of a reaction one can predict with great accuracy the relative concentrations of the participants required to provide the driving force to make a reaction go in a certain direction. How enzyme molecules are used to make sequences of reactions feasible is the subject of the next chapter.

READINGS

Klotz, I. M., 1967, *Energy Changes in Biochemical Reactions.* Academic Press, Inc., New York. An introduction to thermodynamics and its applications in biology; highly recommended for those interested in bioenergetics.

Lehninger, A. L., 1971, *Bioenergetics*, 2d ed., W. A. Benjamin, Inc., New York.

An excellent, readable book; strongly recommended as a constant companion.

Pardee, A. B., and L. L., Ingraham, 1960, Free energy and entropy in metabolism, in *Metabolic Pathways*, vol. 1, edited by D. M. Greenberg, Academic Press, Inc., New York. A very good discussion of the concepts of free energy and entropy in biochemical reactions.

BIOCATALYSIS 18

One of the intellectually pleasing things about free-energy changes is that the pathway of intermediate conditions of reaction is irrelevant; only the nature of the initial and final states determines the magnitude and sign of the free-energy change. Thus, it follows that the presence of enzymes in these systems is irrelevant to the calculations; the enzyme only accelerates the approach to equilibrium and is generally considered not to influence the equilibrium point attained. With or without an enzyme, the free-energy change will be the same. However, the fact that a reaction will proceed with a decline of free energy (spontaneous reaction) does not mean that the reaction will occur at a perceptible rate and perhaps will not occur at all. Pure glucose in a bottle on a shelf at 25°C may remain unoxidized for centuries. Given proper conditions, it will readily be oxidized. Enzymes are a vital contributor to the "conditions" for biological reactions; they *lower the activation energy of a reaction* and thereby can accelerate the rate of a chemical reaction.

Enzymes are like other catalysts in that they are not consumed in the reaction they influence. The substance acted upon by an enzyme is called its **substrate.**

The amount of enzymes known to be present in tissues comes to a considerable fraction of the total soluble protein present. Over 700 different enzymes are known. It has been suggested that the great majority, if not all, of the protein that can be extracted from tissues is enzymes. This may not be true, but it should be pointed out that several cases are known where protein molecules combine enzymatic activity with other functions. For example, myosin is a contractile macromolecule of muscle and it also has enzyme activity (an ATPase). Thus, large protein molecules may have specific parts (called *active sites*) serving a catalytic function while other portions subsume other biological roles. This subject will be discussed further in Sec. 18-4.

18-1 NATURE OF ENZYMES

Enzymes are remarkable molecules in several respects, as a brief resumé of some of their general properties will indicate.

Protein Structure

The first establishment of the fact that an enzyme is a protein was in 1926, when Sumner crystallized *urease* from jack-bean meal. Of the dozens of enzymes that have been isolated and crystallized, all are protein.

Some, however, have associated parts that are not protein, as will be discussed in Sec. 18-3.

Most soluble enzymes are globular proteins whose molecular weights range from about 10,000 up to millions. These macromolecular catalysts are composed of chains of 20 different residues of amino acids arranged in precise sequences, as determined by the genetic code. The average molecular weight of amino acid residues is 100 (a useful figure to remember), so the smallest enzymes would be composed of approximately 100 residues and the largest ones of a few thousand. The primary (covalently bonded) structure of a few enzymes are known (Figs. 18-1 and 18-2).

These long, linear arrays of amino acids appear to be folded into exactly prescribed three-dimensional structures. It is accepted at present that the three-dimensional structure of protein molecules is predetermined by their primary structure. That is, all the information required for the formation of the secondary and tertiary structure is contained in the specific sequence of amino acids.

Following the development of techniques which led to the determi-

FIGURE 18-1

The primary, or covalent, structure of ribonuclease. Standard three-letter abbreviations are used to indicate the unique sequence of individual amino acid residues of the molecule. The dark bands connecting across the chains stand for intrachain S-S (disulfide) bonds of the amino acid cystine. Disulfide bonds are the only type of covalent cross-links known to occur in enzymes. Ribonuclease was synthesized from its component amino acids in 1969.

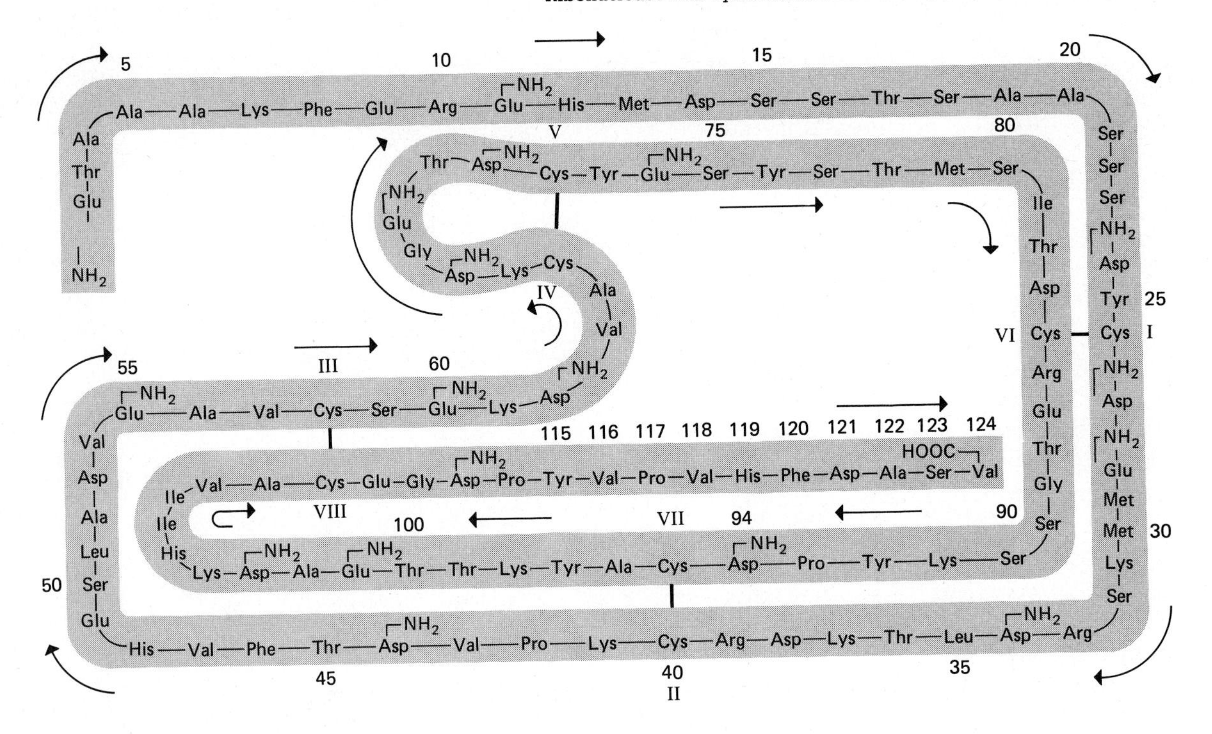

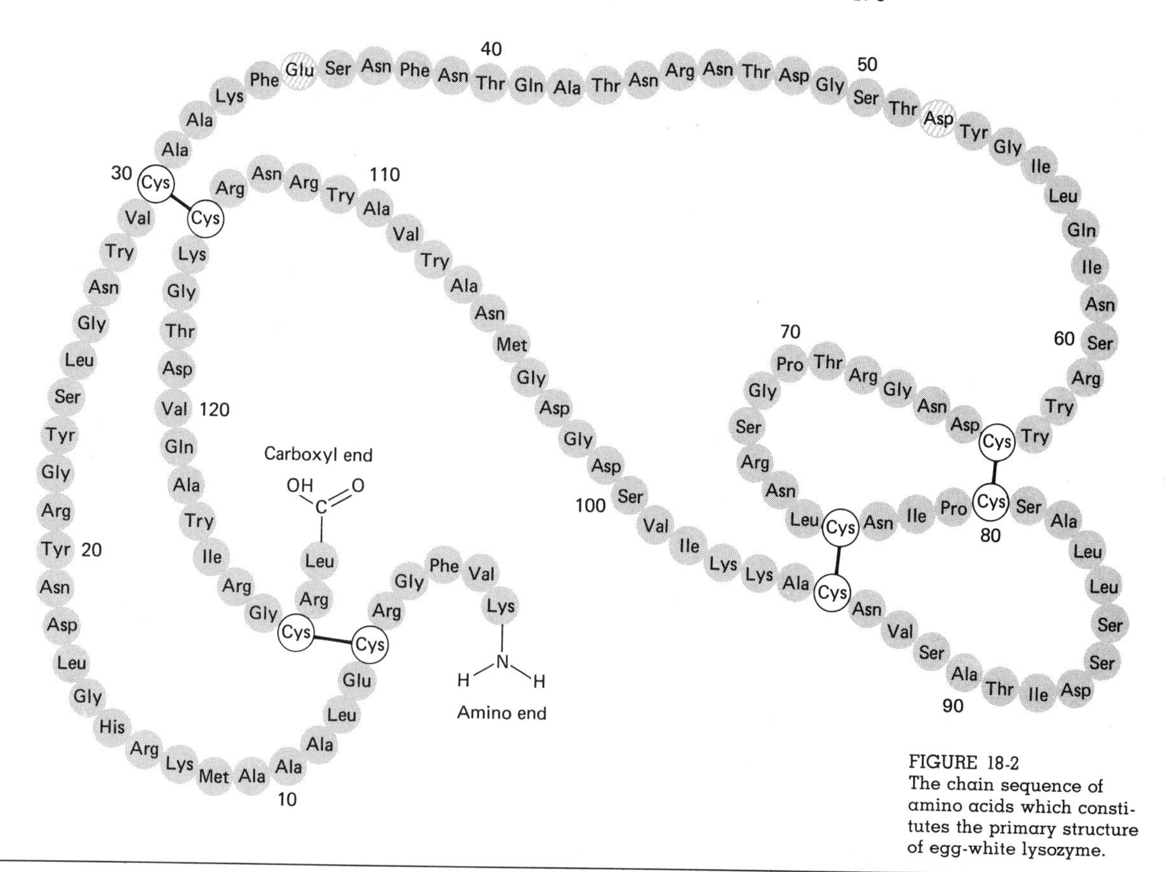

FIGURE 18-2
The chain sequence of amino acids which constitutes the primary structure of egg-white lysozyme.

nation of the three-dimensional structure of myoglobin and hemoglobin, considerable attention has been focused on determining the structures of crystalline enzymes. Electron-density maps of a few enzymes, including egg-white lysozyme, ribonuclease, α-chymotrypsin, carbonic anhydrase, and carboxypeptidase, have been constructed. As a generalization, these molecules have in common a single polypeptide chain, which is highly compact and nearly spherical in shape, and a cavity into which the substrate might be fitted during catalysis (Figs. 18-3 and 18-4).

Other enzymes are known which, instead of being a single polypeptide chain, are composed of a small number of chains (subunits) of less than 30,000 mol. wt. Some of these enzymes have only one type of subunit, but others contain two or more kinds of subunits. It is common to discover that an enzyme is a dimer, trimer, or tetramer of such subunits. Several large proteins, such as glutamic dehydrogenase, urease, and β-galactosidase, with molecular weights of 5×10^5 or greater, have been shown to consist of several subunits, and only the entire aggregation is enzymatically active.

There often may be several proteins from a single tissue source that

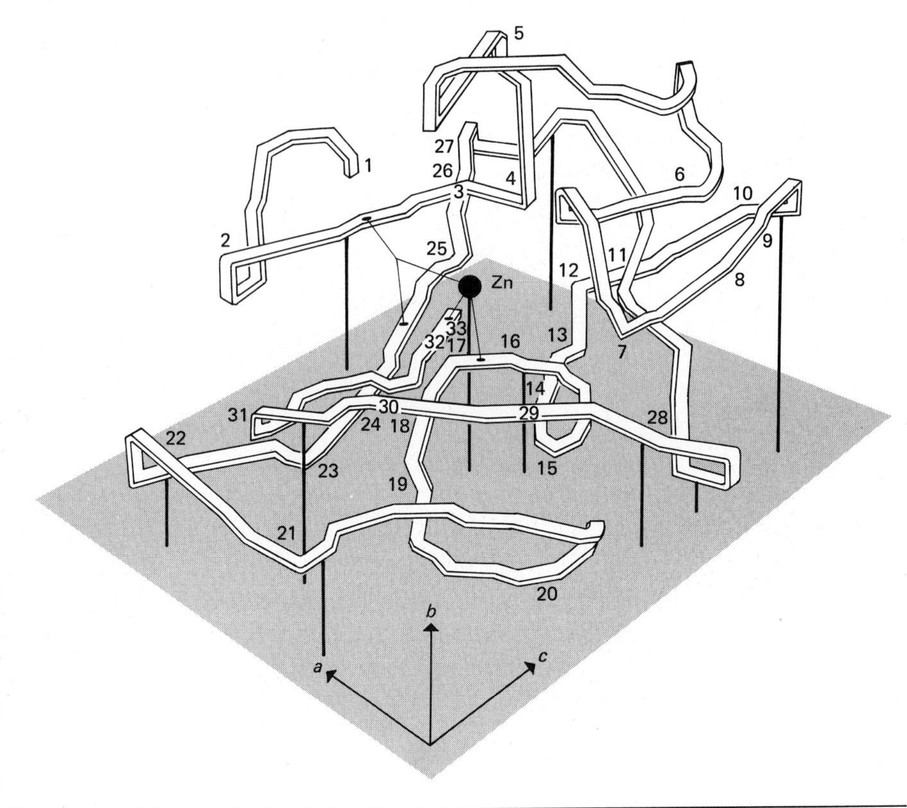

FIGURE 18-3
The chain conformation of carbonic anhydrase. Note the zinc atom near the center of this metalloenzyme. (*After S. A. Bernhard, 1968, The Structure and Function of Enzymes, W. A. Benjamin, Inc., New York, p. 298.*)

have the same enzymatic activity but are not identical structurally. The term "**isoenzyme**" is recommended by the International Union of Biochemistry (I.U.B) to describe these multiple enzyme forms. Such enzymes result from the association of subunits in different combinations. Isoenzymes can be separated by electrophoretic methods, and they often differ in solubility, pH optima, and other features. Among the several enzymes which show this behavior, lactic dehydrogenase (LDH) is the best known. This enzyme (mol. wt. 140,000) has two types of subunits, A and B, which associate to yield five electrophoretically distinguishable tetramer isoenzymes. They are, designated by their subunit composition, B_4, AB_3, A_2B_2, A_3B, and A_4. Both subunits of beef-heart LDH have molecular weights of about 35,000. Dissociation and recombination of LDH subunits to yield active enzyme have been accomplished.

Many of the enzymes involved in oxidation-reduction (e.g., in mitochondria) are bound to membranes and are not readily extracted. Some can be solubilized by detergents, which indicates that lipids are involved in binding them to intracellular structures. Molecular weight determinations are meaningless for such enzymes. The ATPase activity of myosin, mentioned above, demonstrates that enzymatic activity also may be displayed by fibrous proteins.

Efficiency

Under optimal conditions most enzymatic reactions proceed at 10^8 to 10^{11} times more rapidly than the corresponding nonenzymatic reaction. Only small amounts of enzymes are necessary for rapid reaction because they have high *turnover numbers;* i.e., the number of substrate molecules metabolized per enzyme molecule per minute is very high. For most enzymes it is 1000 and, in a few instances, may be more than 1,000,000. For peroxidase it is 2,680,000! Furthermore, enzymes act under mild conditions: 37 to 38°C for most mammals and birds or at lower temperatures in poikilotherms, and in neutral, aqueous solutions (although perhaps there

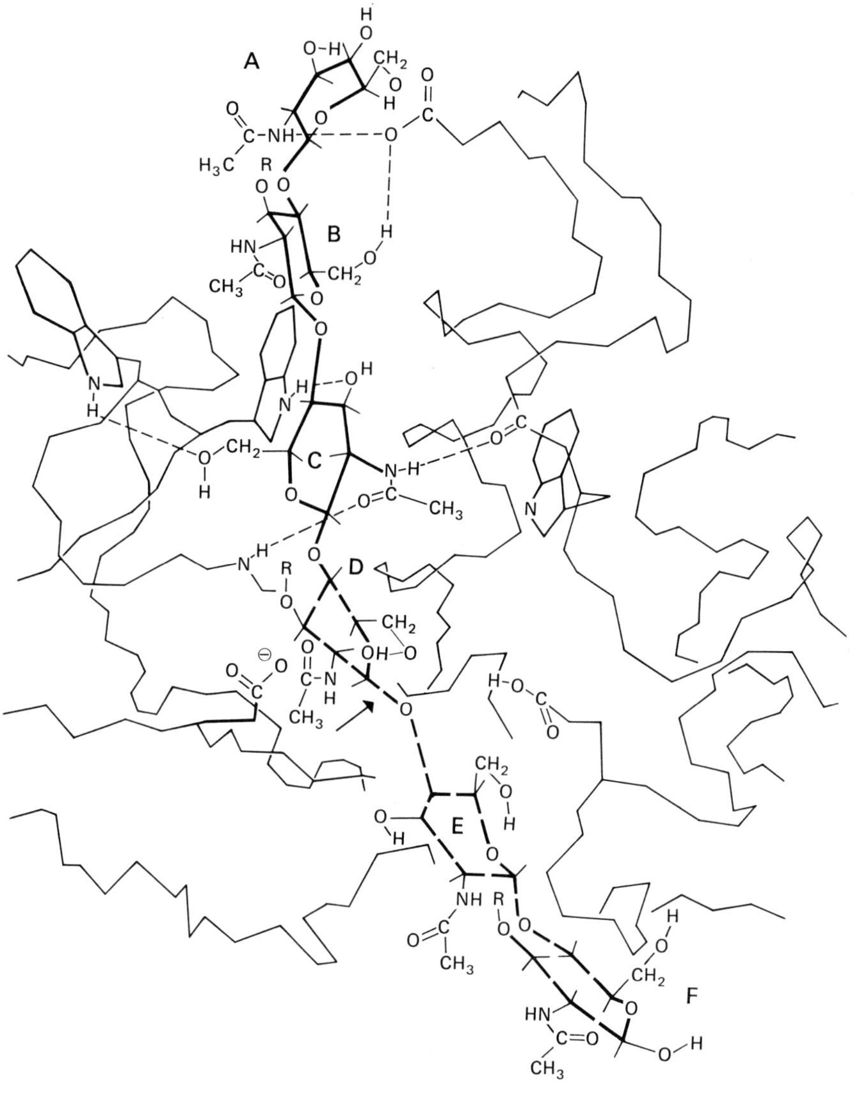

FIGURE 18-4
Three-dimensional model of the region around the active site of the egg-white lysosome-substrate complex. A polysaccharide substrate, whose pyranose rings are shown in heavy lines (sites *A* through *F*), lies in the cleft of the enzyme. The six light dashed lines are hydrogen bonds between the enzyme and substrate. The peptide backbone of the enzyme is represented by the light, continuous lines. (*After M. Chipman and N. Sharon,* 1969, *Science, 165,* p. 456, *based on D. C. Phillips,* 1966, *Sci. Am., 215,* p. 78.)

are nonaqueous compartments where enzymes may act, such as in mitochondrial membranes).

The degree of acceleration effected by some enzymes is far greater than kinetic information would predict it to be. A plausible explanation for the enormous velocities of such enzyme-catalyzed reactions has yet to be formulated, and this is one of the major subjects of investigation in enzymology today.

Yield

Enzymes produce no by-products; i.e., the yield in products is always 100 percent, in contrast to many nonbiological catalysts. For every substrate acted upon by an enzyme a specific product appears.

Incidentally, it should be kept in mind that an enzyme cannot be detected unless at least one of its substrates is known.

Specificity

Most enzymatic reactions are specific in terms of the nature of the reaction catalyzed and the structure of the substrate utilized. *Intracellular enzymes*, especially, are found which act only on their specific substrates and not on closely related substances, even being specific to an optical isomer in some instances. This means that they often can be specifically inhibited also.

It is necessary in describing an enzyme to state the species and organ or tissue of origin, because enzymes catalyzing the same chemical reaction and having the same substrate specificity, but derived from different sources, may differ in other properties. An amylase from human pancreas and one from hog pancreas, although catalytically identical, may differ in solubility, pH optima, and other features. Also, because some enzymes are isoenzymatic there may be differences in rates of reaction between the same kind of enzyme extracted from liver and kidney of any one species.

Sir Frederic Hopkins, an eminent biochemist speaking in 1933, considered the specificity of enzymes as perhaps the "most fundamental and significant phenomenon in nature," and suggested that "control of events by intracellular enzymes . . . by itself secures the status of the cell as a system which can maintain itself in dynamic equilibrium with its environment. . . ."

It is apparent that many thermodynamically possible reactions are available to the cell at all times. Enzymes are used like switches or valves to determine which reactions will proceed at any one time. That is, they are vital parts of the *control systems* of cells.

Reversibility

Enzymes catalyze the reverse reaction to the same extent as the forward reaction; i.e., *both starting materials (reactants) and products are substrates*. Thus, enzymes have at least two substrates. It follows, then, that enzymes do not determine the direction of reactions. That is determined by concentrations, thermodynamic requirements, electron potentials, and other factors. Some reactions are not reversible, but this is not the fault of the enzyme.

Stability

Some enzymes are very stable, retaining their potential for catalysis for years if stored in a dry, cool, and dark place. Others are very labile, quickly losing their activity despite standard precautions to protect them.

Most enzymes undergo changes in activity and become insoluble after heating. This is called **denaturation** and consists of an alteration of chain conformation; i.e., the folding of the polypeptide chain is somehow altered. Enzymes differ in their sensitivity to heating; some denature even when kept near freezing, but others, like the ribonucleases, are remarkably insensitive to heating. Denaturation can be effected in most soluble proteins and, in addition to heat, may be caused by vigorous agitation, ultraviolet radiation, and the actions of acids, bases, organic solvents, salts of heavy metals, urea, detergents, as well as other compounds.

18-2 KINDS OF ENZYMES AND THEIR ACTIONS

Enzymes were formerly classified on the basis of their substrate specificities. The hydrolyzing enzymes, for example, often are still classified as proteases, carbohydrases, and esterases (Sec. 5-4). However, the general use of such trivial names is untenable because certain enzymes do not have such limited substrate specificity. Trypsin and chymotrypsin, for example, catalyze the hydrolysis not only of peptide bonds but also of ester bonds. A systematic classification and nomenclature of enzyme-catalyzed reactions has been established by the Commission on Enzymes of the I.U.B. In this classification, enzymes are divided into six general groups:

1. Oxidoreductases: catalyzing oxidation-reduction reactions; concerned with electron and hydrogen transfers; e.g., dehydrogenases, oxidases, reductases, catalases

2. Transferases: catalyzing group transfer reactions; e.g., methyl transferases, phosphorylases, amino transferases, kinases

3. Hydrolases: catalyzing hydrolytic reactions; e.g., esterases, acid and alkaline phosphatases, peptidases, deaminases, carbohydrases, ureases, ATPases

4. Lyases: catalyzing the addition of groups to double bonds or vice versa; e.g., decarboxylases

5. Isomerases: catalyzing isomerizations; e.g., racemases and epimerases

6. Ligases (synthetases): catalyzing the condensation of two molecules coupled with the cleavage of pyrophosphate bond of ATP or similar triphosphate; e.g., acetyl-CoA synthetase

Each of these broad classes is divided into subclasses on the basis of the nature of the particular reaction involved. Every enzyme has been assigned a systematic name, as well as a number, but is commonly referred to by the trivial name. For example, enzyme number 2.7.1.2 is ATP: glucose-6-phosphotransferase, but is generally called *glucokinase*. In many cases, enzymes are named by simply adding the suffix *-ase* to the name of the substrate, as will be noted among the examples above.

The *oxidoreductases* either transfer electrons alone or electrons accompanied by protons. If the acceptor is oxygen, the transfer is called *aerobic*, and if the acceptor is some other molecule it is called *anaerobic*. As will be discussed in the next section, these enzymes are always associated with a functional molecule known as a *coenzyme*. The next chapter deals largely with the actions of these enzymes.

The transferring enzymes, or *transferases*, mediate the passage of a *functional group* from one compound to another, and in some instances, they also transfer the energy of the "bond" with which the group is attached. The latter are importantly involved in bioenergetics.

A few words about the terms *bond* and *functional group* will aid in understanding much of what follows in this and the next several chapters. The chemical bond between atoms in a covalent compound consists of a shared pair of electrons. By this sharing an atom achieves a stable electronic grouping. The work necessary to dissociate two covalent atoms is a measure of the strength of that bond. We need to be reminded from time to time that the dash (—) in a formula is used to denote this pair of electrons of the outer orbitals shared by two atoms and that other electrons are *understood* to be present.

Molecules have groups or centers that are characteristically more "reactive" than others. A chemist can generally predict where in a given molecule an action will take place in a particular set of circumstances, because of the relative propensity of the bond attaching certain groups to be "broken." Furthermore, bonds generally determine the specific properties of a molecule, the rest of the molecule being rather inert. There are several functional groups of biological importance, such as OH, Cl, NO_2, SH, NH_2, COOH, CO, and CHO. They are the major sites at which small molecules are polymerized to form larger ones, they form hydrogen bonds or disulfide bonds, they are involved in chemical group or energy transfers, or they participate in oxidation-reduction reactions (Fig. 18-5).

The *hydrolases* may be considered as a special group of transferring enzymes in which water serves as the specific acceptor. All the digestive enzymes are hydrolases (described in Chap. 5, Sec. 5-4), and there are numerous intracellular ones. Some examples will illustrate their mode of hydrolytic action:

Carbohydrases: $\mathrm{R{-}O{-}R' + H_2O \rightleftharpoons ROH + R'OH}$

Proteases: $\mathrm{R{-}\overset{\overset{\displaystyle O}{\|}}{C}{-}NHR' + H_2O \rightleftharpoons R{-}\overset{\overset{\displaystyle O}{\|}}{C}{-}OH + NH_2R'}$

FUNCTIONAL GROUP	EXAMPLE
$-COOH$ Carboxyl	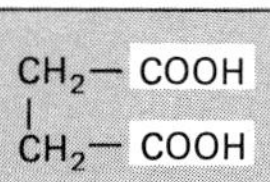 CH_2-COOH / CH_2-COOH Succinic acid (important in energy metabolism)
$-NH_2$ Amino	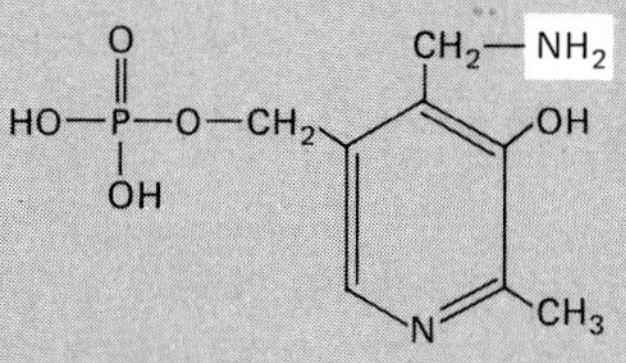 Pyridoxamine phosphate (a form of vitamin B_6)
$-OH$ Hydroxyl	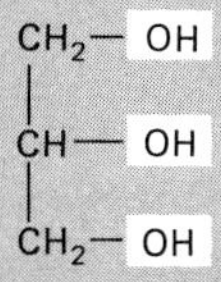 CH_2-OH / $CH-OH$ / CH_2-OH Glycerol (a structural component of fat)
$-SH$ Sulfhydryl	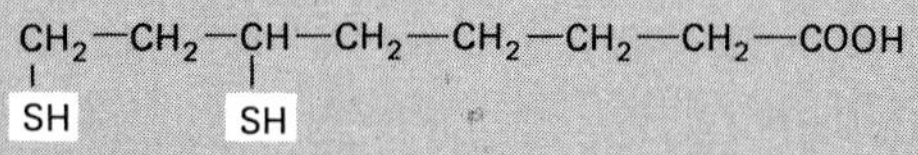 $CH_2(SH)-CH_2-CH(SH)-CH_2-CH_2-CH_2-CH_2-COOH$ Lipoic acid (important in energy metabolism)
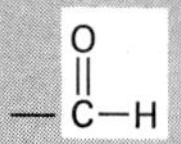 $-C(=O)-H$ Aldehyde	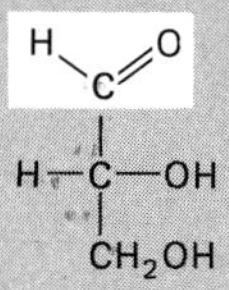 $H-C(=O)$ / $H-C-OH$ / CH_2OH D-Glyceraldehyde (a sugar whose derivatives play a key metabolic role)
 $-C(=O)-$ Keto	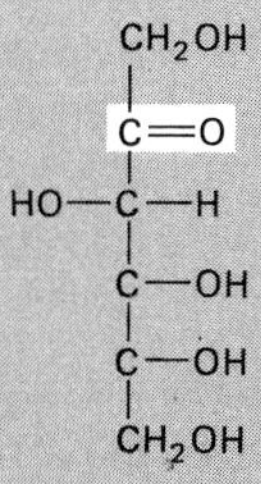 CH_2OH / $C=O$ / $HO-C-H$ / $C-OH$ / $C-OH$ / CH_2OH D-Fructose (a keto sugar important in energy metabolism)

FIGURE 18-5
Some important functional groups in biological systems.

Esterases: $R{-}\overset{\overset{\displaystyle O}{\|}}{C}{-}OR' + H_2O \rightleftharpoons R{-}\overset{\overset{\displaystyle O}{\|}}{C}{-}OH + R'OH$

Phosphatases: $R{-}O{-}PO_3H_2 + H_2O \rightleftharpoons ROH + H_3PO_4$

Whereas the enzymes in the preceding categories catalyze reactions of the type

$$A + B \rightleftharpoons C + D$$

the *lyases* and *ligases* engage in reactions of the type

$$A \rightleftharpoons B + C$$

For example, the action of a decarboxylase:

$$RCOCOOH \rightleftharpoons RCOH + CO_2$$

Lyases remove groups from their substrates, but not by hydrolysis, and leave double bonds (or add groups to double bonds). Ligases catalyze the joining of two molecules and couple the breakdown of an ATP or similar triphosphate.

Isomerases catalyze molecular rearrangements, either by modifying the structure of part of a molecule or by displacing part of it. *Racemases,* isomerases which catalyze the conversion of an optically active substance into its racemate, have been identified in bacteria. For example, there are racemases that convert certain D-amino acids into the L-form.

18-3 COENZYMES, COFACTORS, AND ACTIVATORS

Certain enzymes require association with a nonprotein moiety, or **cocatalyst,** before they are active. This is especially true of certain of the oxidative enzymes. Hydrolytic enzymes, by contrast, generally do not require cocatalysts. As we will be concerned in the next chapter with many enzymes that require these factors, it will be useful to examine their nature and function at this time.

Some of the enzymes possess cocatalyst groups, which are essential for enzymatic activity but which can be removed by dialysis or by passing the enzyme solution through a Sephadex[1] column. Such dissociable moieties are called **prosthetic groups** in biochemistry. The nondialyzable protein portion is called an **apoenzyme,** and the conjugated apoenzyme and cocatalyst is known as the **holoenzyme.** The prosthetic group alone frequently has a slight catalytic activity, but only the holoenzyme has the full activity. Evidently the prosthetic group has a protective action on the apoenzymes, because holoenzymes are much more stable than apoenzymes.

[1] Sephadex is a commercial name for one of the polymeric carbohydrates. It is available as tiny beads containing pores of known size and is used as a molecular sieve, separating molecules according to size and charge.

Terms applying to cocatalysts have become very confused and, therefore, it will be useful to define them here in terms of the catalytic functions they mediate. **Cofactors** are cocatalysts which form part of an active site (see later), and are regenerated upon each turnover of substrate. This is illustrated in the following diagram,

$$E + C \rightleftharpoons EC$$
$$EC + S \rightleftharpoons ECS \rightleftharpoons EC + P$$
$$EC \rightleftharpoons E + C$$

where E = the enzyme
C = the cofactor
S = the substrate
P = the product

Good examples of cofactors are *pyridoxal phosphate* and *thiamine pyrophosphate* (Fig. 18-6).

In some reactions the cocatalyst and substrate form a new chemical compound. With each turnover of substrate the cocatalyst is converted to a new product. Thus, this type of cocatalyst, called a **coenzyme,** does not serve as a stable constituent of the enzyme site, and is in fact a reactant. However, it is regenerated in the overall metabolic process by other enzyme-catalyzed reactions which utilize the coenzyme-product as a reactant. This can be illustrated as:

$$E_1 = S_1^* + C \rightleftharpoons E_1 + C^* + P_1$$
$$C^* + E_2 + S_2 \rightleftharpoons E_2 + P_2^* + C$$

The action of the second enzyme reverses the action of the first enzyme, and the coenzyme thus serves to transfer a chemically reactive intermediate (*) from the first to the second catalytic process.

The *pyridine nucleotides* are coenzymes for a great variety of hydrogenation-dehydrogenation reactions. They serve alternately as an oxidant and reductant in two different catalyzed reactions. There are two

FIGURE 18-6
Enzyme cofactors. Pyridoxal phosphate is associated with the decarboxylation of amino acids, in transaminations, condensations by synthetases, and other reactions. The ring structure is modified pyridoxine, one of the B vitamins. Thiamine pyrophosphate, containing the thiazolium ring (the vitamin thiamine), is a cofactor for carboxylase and the decarboxylation of a series of α-keto acids.

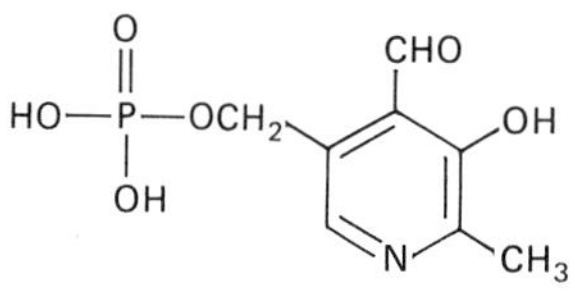

Pyridoxal-5-phosphate

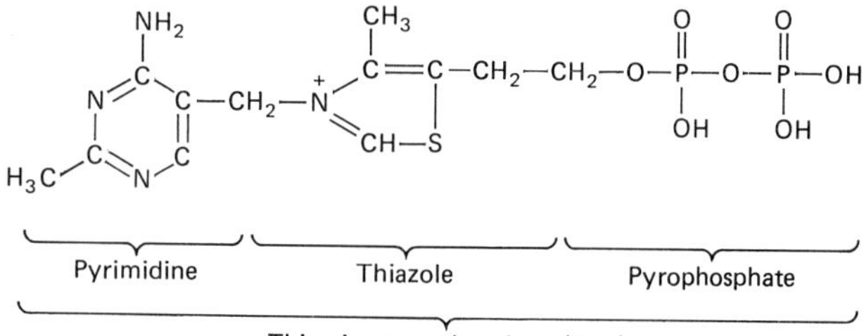

Pyrimidine
Thiazole
Pyrophosphate
Thiamine pyrophosphate (TPP)

pyridine nucleotide coenzymes: *nicotinamide adenine dinucleotide* (NAD) and *nicotinamide adenine dinucleotide phosphate* (NADP) (Fig. 18-7). [In older literature NAD is designated DPN (diphosphopyridine nucleotide) and NADP is TPN (triphosphopyridine nucleotide). The newer names correctly describe the chemical nature of the coenzymes, whereas the older terms do not.]

Lactic dehydrogenase incorporates NAD as a coenzyme. In order for a reaction to proceed there must be a combination of the apoenzyme, the coenzyme, and lactic acid. Two hydrogen atoms are transferred from lactic acid to NAD, and the result is that pyruvic acid and NADH are formed. In a reaction of this type the distinction between *coenzyme* and *substrate* is very subtle and, in fact, could be considered semantic, because both are acted upon (Fig. 18-8).

Coenzyme A is a complex nucleotide which plays an important role in a great variety of acetylation reactions. For brevity the coenzyme is generally written CoA, or sometimes, to indicate the physiologically active part of the molecule, the sulfhydryl group, as CoA-SH (Fig. 18-9).

A group of enzymes which contain prosthetic groups incorporating *riboflavin* are known as the *flavoproteins*. The reoxidation of NADPH is catalyzed by one of these having *riboflavin phosphate*, better known as *flavin mononucleotide* (FMN), as a coenzyme. A number of enzymes catalyzing the oxidations of a variety of substrates have as a coenzyme *flavin adenine dinucleotide* (FAD). Typical substrates for the flavoproteins are pyridine nucleotides, α-amino acids, α-hydroxy acids, and aldehydes. The reduced flavoproteins, in turn, become substrates for

FIGURE 18-7 **Structure of the pyridine nucleotide, nicotinamide adenine dinucleotide (NAD^+). When an additional OPO_3H_2 is present instead of the OH in the 2′ position of the pentose (*) the molecule is nicotinamide adenine dinucleotide phosphate ($NADP^+$).**

Nicotinamide
Adenine
D-Ribose
D-Ribose
Nicotinamide mononucleotide (NMN)
Adenine mononucleotide (AMP)
Pyrophosphate link (PP)
Nicotinamide–adenine dinucleotide (NAD^+)

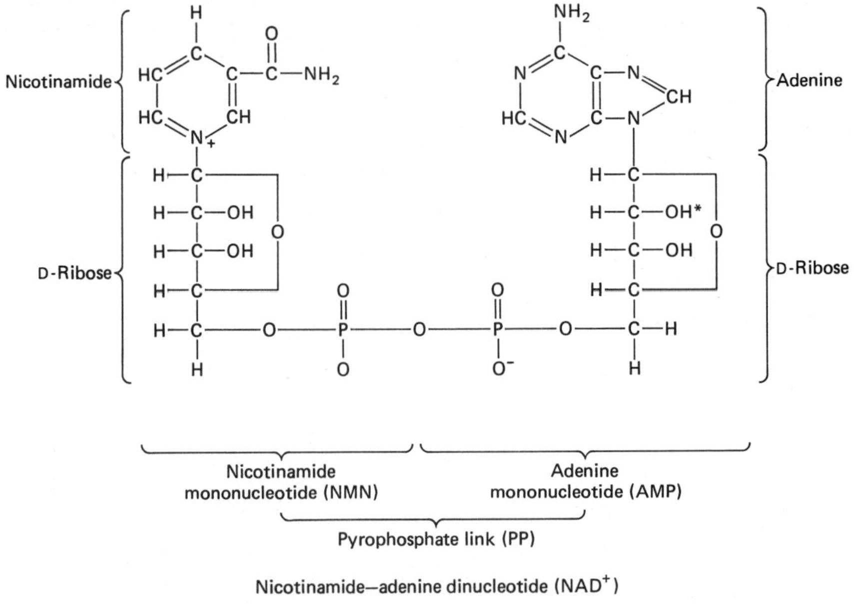

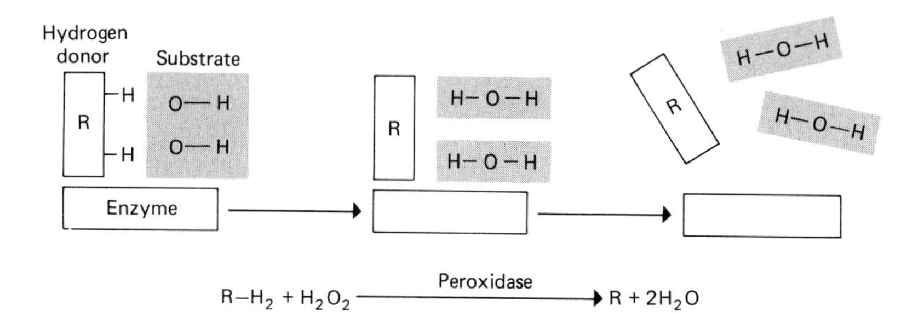

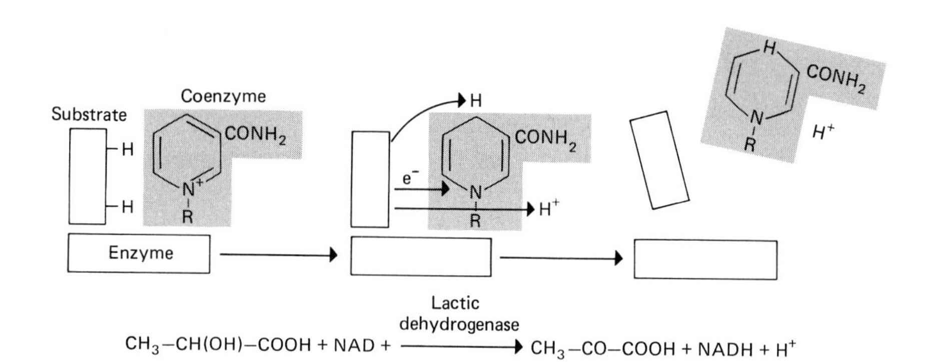

FIGURE 18-8
Diagrams illustrating the similarity of "substrate" and "coenzyme." Peroxidase does not require a coenzyme and can catalyze the hydrogen transfer from several kinds of hydrogen donors to hydrogen peroxide. Lactic dehydrogenase can act only on lactic acid and NAD, and both substrate and coenzyme are modified in the reaction. (*After E. Florey, 1966, An Introduction to General and Comparative Animal Physiology, W. B. Saunders Company, Philadelphia, p. 137.*)

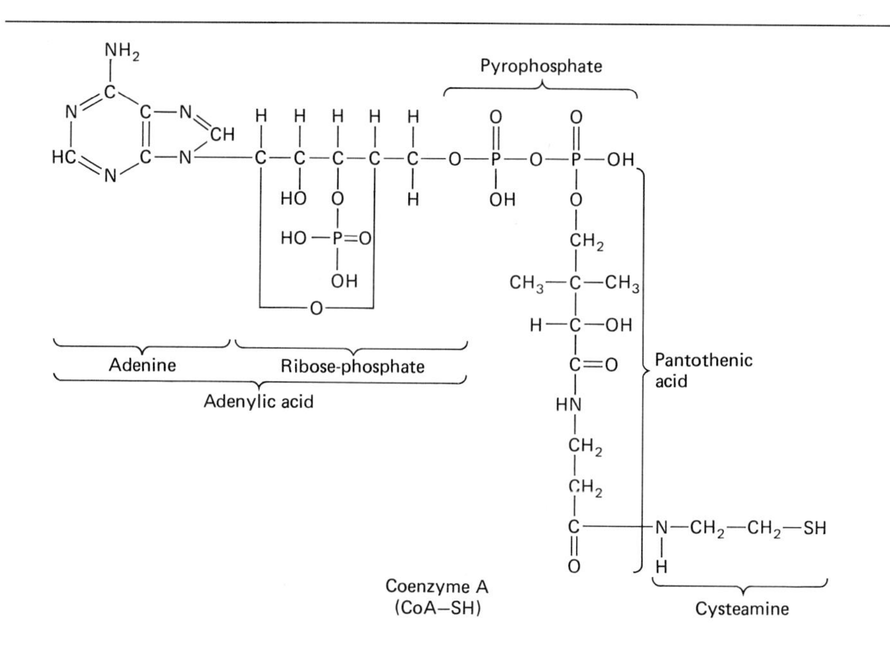

FIGURE 18-9
Coenzyme A. The molecule contains the vitamin pantothenic acid.

reactions involving other electron acceptors and are regenerated to the oxidized form (Fig. 18-10).

Enzymes containing coenzymes derived from *hemes*, which contain a metal ion and the porphyrin nucleus, occupy a central position in primary oxidation processes and in oxidative phosphorylation (the coupling of oxidation to the formation of ATP). For example, *cytochromes*, a class of oxidation-reduction enzymes that are principally concerned with the transfer of electrons from flavoproteins to oxygen or to other electron acceptors, employ heme as a coenzyme. Heme serves as a cofactor for the hydroperoxidases (catalases and peroxidases). It is also involved with the photosynthetic process (ferredoxin) and in the transport of oxygen by hemoglobin and myoglobin.

Biotin, a growth factor of both yeast and humans (vitamin H), serves as a coenzyme for ATP-dependent carboxylations and certain transcarboxylations. Another vitamin, *folic acid*, is incorporated into *tetrahydrofolic acid*, the coenzyme for the transfer for one-carbon fragments. *Vitamin* B_{12} *(cobamide) coenzymes* are known to be involved in five enzymatic reactions with mutases and dehydrases.

Glutathione, a tripeptide made up of cysteine, glutamic acid, and glycine, is a coenzyme for some of the addition reactions of thiols and aldehydes. *Lipoic acid*, a disulfide-containing molecule, is involved with the oxidative decarboxylation of α-keto acids (Fig. 18-11).

The classification of coenzymes is obviously somewhat arbitrary, and just as the pyridine nucleotides may be considered as coenzymes, so

FIGURE 18-10
Structure of the riboflavin-containing coenzymes, FMN and FAD. Riboflavin is a yellow pigment synthesized by plants and microorganisms and serving as a vitamin for several other organisms.

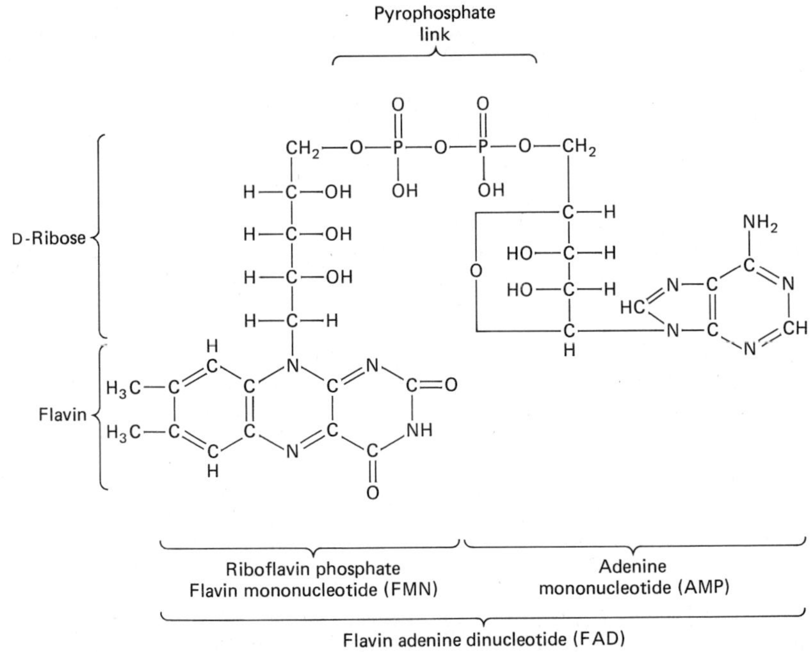

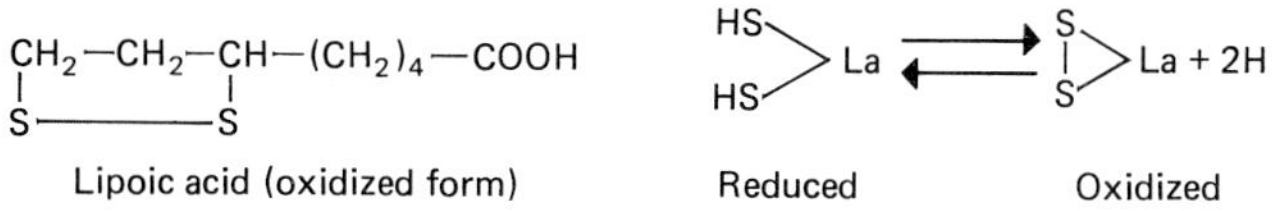

$$\underset{|\qquad\qquad\quad|}{CH_2{-}CH_2{-}CH{-}(CH_2)_4{-}COOH}$$
$$S{-}{-}{-}{-}{-}{-}S$$

Lipoic acid (oxidized form)

$$\text{Reduced (HS)}_2\text{La} \rightleftarrows \text{Oxidized (S–S)La} + 2H$$

FIGURE 18-11
Structure of the coenzyme lipoic acid, showing the functional sulfhydryl group in oxidized and reduced forms.

also may the *adenine nucleotides.* ADP can serve as an acceptor molecule in phosphate transfer; ATP in turn serves as a phosphate donor, and ADP is regenerated. In a somewhat analogous way *uridine nucleotides* serve as coenzymes in carbohydrate metabolism. For example, uridine diphosphate (UDP) complexes with glucose (UDPG) prior to transfer of the glucose to fructose to yield sucrose. Similarly, UDP serves to form intermediates in the synthesis of complex polysaccharides (see Chap. 16; Sec. 16-4).

Cytidine nucleotides are involved in the group transfer reactions leading to the synthesis of lecithins, an important class of lipids. *Guanosine* and *inosine nucleotides* are involved in a few transfer reactions, including carbohydrate syntheses.

Many enzymes require a *divalent* metal ion for activity. In several cases the requirement is specific for a particular metal (e.g., Zn^{2+} in carbonic anhydrase), but in others more than one metal can bring about activation. The metal may either function as a bridging group, to bind substrate and enzyme together through formation of a coordination complex, or it may serve as the catalytic group itself. An example of the latter situation is the iron atoms of catalase, which catalyzes the decomposition of hydrogen peroxide. In a few cases it appears that two metals may be required by an enzyme, as is the situation with pyruvate phosphokinase which requires both Mg^{2+} and K^{+} (Table 18-1).

A large number of enzymes catalyzing a wide variety of unrelated reactions are activated by *monovalent* cations. However, in contrast to

Table 18-1
Some Enzymes That Either Contain a Divalent Cation or Are Activated by One

Metal	Enzyme	Metal	Enzyme
Mo	Xanthine oxidase	Zn	Carbonic anhydrase
	Nitrate reductase		Lactic dehydrogenase
Cu	Tyrosinase		Carboxypeptidase
	Phenolase	Mg	Peptidases
	Ascorbic acid oxidase		Phosphatases
	Cytochrome oxidase		ATP-enzymes, such as the hexokinases
Fe	Cytochrome enzymes	Mn	Arginase
	Catalase		Phosphoglucomutase
	Peroxidases		Dipeptidases
	Tryptophan oxidase	Co	Peptidases
	Homogentisicase		
Ca	Lecithinases A and C		
	Lipases		

the rather specific functions that can be assigned to certain divalent cations, the mode of action of the monovalent cations K^+ and Na^+ is less well established. Those enzymes activated by K^+ generally are little activated by Na^+ and vice versa. (The Na^+, K^+-activated ATPases of active transport, which also require Mg^{2+}, present a special case.) It recently has been proposed that the monovalent ion interacts with the substrate and enzyme to form a functional ternary complex. However, it has proved very difficult to isolate such theoretical intermediates. Just how monovalent ions activate enzymes remains to be clearly defined.

As noted in Sec. 5-5, pepsin, a digestive enzyme, occurs in an inactive form known as *pepsinogen*. The general name for a dormant enzyme is **zymogen.** Pepsinogen is activated to form the active enzyme by H^+ (Fig. 9-5). Reducing agents, such as cysteine and glutathione, serve as activators of -SH enzymes. Enzymes themselves may activate other enzymes or proenzymes; e.g., enterokinase activates trypsinogen to form the active digestive enzyme, trypsin. This activation involves the hydrolysis of a peptide bond in the zymogen molecule to liberate a peptide and thereby make available the catalytically active site.

18-4 ACTIVE SITE

That part of an enzyme which binds a substrate is called an **active site** or **active center.** There is commonly just one such site per enzyme molecule, but there are some—dehydrogenases, for example—which have more than one active site. That active sites are small and geometrically discrete regions is suggested by the relatively small size of the cofactors that form a part of them and by the fact that enzymes are inactivated by combination with molecules that are quite small. The substrate of most enzymes is usually a molecule of relatively low molecular weight or a small portion of a molecule of high molecular weight (Fig. 18-4).

Appropriate molecular models predict that substrates could be in contact with only a few—10 percent or less—of the amino acid residues. It has been possible to "label" the active site of many enzymes by means of a specific chemical reagent—e.g., with a compound that mimics the reactivity of a substrate but lacks the configuration specific to any particular enzyme—which forms metastable intermediates or stable products. The particular residue to which such an analog is attached can be established by procedures for degrading and isolating peptides. The active site of phosphoglucomutase is a part of the peptide chain with the sequence -Thr.Ala.Ser.His.Asp-. A large number of proteolytic enzymes contain the common site sequence -Gly.Asp.Ser.Gly-. Some of the amino acids in such a sequence will make direct contact with the substrate, while others have only indirect roles (Fig. 18-12).

In many enzymes the active site contains either a *sulfhydryl group of cysteine* or a *bivalent metal ion* (Table 18-1). The latter frequently is bound to the sulfur atom of a cysteine residue.

FIGURE 18-12
A schematic representation of an active site. *Solid circles:* "contact" amino acid residues whose fit with substrate determines specificity; *triangles:* catalytic residues acting on substrate bond (jagged line); *open circles:* nonessential residues on the surface; *squares:* residues whose interaction with each other maintains the three-dimensional structure of the protein.

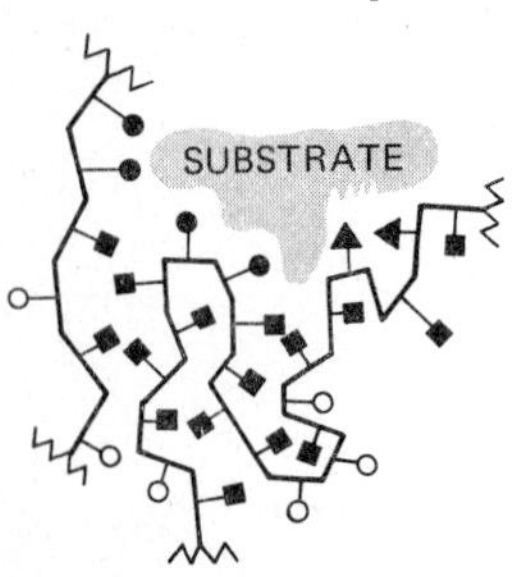

In some cases attempts to localize active sites have failed. This is true for ribonuclease, where different parts of the folded peptide chain are responsible for its catalytic activity. If any of these parts is altered, its enzymatic activity is impaired or destroyed. Thus, in some enzymes the concept of an "active site" does not apply, the unique conformation of most or all of the molecule apparently being essential for proper association with a substrate, as well as for binding of a cocatalyst where used.

There is considerable evidence for the existence of multisited enzyme molecules. The polypeptide subunits of some such enzymes have been shown to be composed of identical amino acid sequences, and each site may be a part of one, or more than one, polypeptide chain. In some cases the number of sites corresponds to the number of chains, as in glyceraldehyde-3-phosphate dehydrogenase, where both are four in number.

There is often multisite interaction. Binding at one site may influence the configuration of the chain bearing it and thereby exert an influence on the conformations of the other chains. There are other enzymes in which it is postulated that catalytic activity at one site is dependent on the state of binding not only of the substrate but of *effector* as well. Effectors are molecules such as amino acids, adenylic acid, ATP, and CTP. They may act as *inhibitors* in some cases and as *activators* in others. It is assumed that multisited enzymes can exist in at least two states, but that the conformations of all subunits within a single aggregate are identical. The role of the effector is to drive the equilibrium toward one or the other of these states—activators toward catalytic activity, inhibitors in the direction of inactivity. Enzymes in which combinations are made at sites other than at the substrate binding site to cause protein subunit interaction are called **allosteric enzymes** (more fully described in Chap. 20, Sec. 20-2). It is believed that allosteric enzymes play an important role in regulation of metabolic processes.

18-5 ENZYME-SUBSTRATE COMPLEXES

Concepts pertaining to chemical reactions and catalysis are based on the supposition that collisions between reacting molecules are a prerequisite for chemical reaction. Similarly, it is postulated that enzymes must form temporary complexes with their substrates. An enzyme-substrate complex, usually denoted by ES, is short-lived and unstable. If a substance combines with an enzyme to give a stable complex there will be no further reaction, as in the case of some enzyme *inhibitors* which combine irreversibly with the enzyme. If an inhibitor can be reversibly bound to the same site of an enzyme as the substrate, or if it prevents the formation of ES, it is called a *competitive inhibitor,* since it competes with the substrate for the active site.

The existence of ES complexes was predicted first from results of kinetic experiments. However, some enzymes that are colored and can be detected photometrically by their characteristic absorption spectra, will show a change in their absorption spectra when they form ES complexes. For example, catalase and peroxidase undergo typical changes in their colors and absorbancies when they combine with the substrate hydrogen peroxide; a green complex is formed first, which is later converted into a red complex. Some enzymes and cocatalysts undergo measurable changes of fluorescence when they combine with a substrate, as is true of the pyridine nucleotides. Photometric methods, then, allow measurement of the concentrations of ES and, in some cases, even the amount of apoenzyme-coenzyme complex present.

It is evident that catalysis requires that the enzyme and substrate molecules fit closely together in a complex. This means that in addition to spatial and geometric fit, there has to be electrostatic complementarity—pairing of oppositely charged groups in enzyme and substrate—as well, to permit close nesting of otherwise complementary surfaces. Furthermore, there must be an exact positioning of catalytic elements in the enzyme relative to the substrate; i.e., the enzyme must impinge upon one highly localized region of the substrate.

In addition to these close-range fits, there must be a long-range fit also. The chemical properties of the substrate molecule, once it is bound to the active site of the enzyme, are significantly influenced by the electrostatic environment in other parts of the enzyme. It is this environment, provided by the enzyme as a whole, that determines which substrate can have access to the active site, and this in a sense helps in determining specificity. Taken all together, the features of "fitness" of the enzyme-substrate combination determine what *specificity* means for each enzyme.

Once the ES complex is formed, what is the specific role of the enzyme? Two kinds of mechanisms have been postulated. In one category, called the *double-displacement mechanism,* two successive attacks occur: the first by the enzyme on one substrate to produce a covalent intermediate (i.e., an amino acid residue of the enzyme forms a covalent bond with the substrate), the second by the second substrate on the intermediate to produce a final product. This mode of action is characteristic of sucrose phosphorylase (where it was first discovered, providing a landmark in the understanding of enzyme action), chymotrypsin, phosphoglucomutase, and other enzymes. In the other category of action, called the *single-displacement mechanism,* no covalent intermediate is formed between the enzyme and substrate. Instead the second substrate makes a direct attack on the first substrate. This demonstrates that enzymes can catalyze chemical transformations by affecting the environment around the bonds in a substrate, probably by polarizing the electron distribution. Electron polarization also occurs in the double-displacement mechanism, but the enzyme serves in addition as a stoichiometric reactant during the catalytic action.

When two molecules approach one another, the inherent "wriggling" (vibration) of all the rotating groups in each molecule imposes a barrier to interaction, the magnitude of which can be evaluated quantitatively in terms of *entropy*. A wriggling molecule is, in the thermodynamic sense, a disordered system. This brings us to the energetic advantages of enzyme-catalyzed reactions.

In the first place, *spatial constraints* imposed within the ES complex prevent oscillatory fluctuations and thus minimize or eliminate the entropy barrier to subsequent chemical reaction. Secondly, as soon as the complex is formed the *effective concentration* of enzyme is increased by a factor of 10,000 or more, because the reaction to follow is *intra*molecular (at least in the case of covalent intermediates), not *inter*molecular. This is a major thermodynamic advantage. The third important advantage is that enzymes break down reactions into a *series of component processes*, each of which requires a relatively small activation energy. Since in any chemical reaction a large activation energy is a great deterrent, a multistep catalysis has considerable kinetic advantage over a one-step catalysis (Fig. 18-13).

Finally, the close fit of substrate and enzyme provides for the *simultaneous withdrawal* of protons and electrons from one group of atoms in the substrate *and delivery* to another group of atoms separated in space from the first group. Such synchronization has great catalytic advantage as compared to two sequential, independent steps, since it leads to a major reduction of the activation barrier.

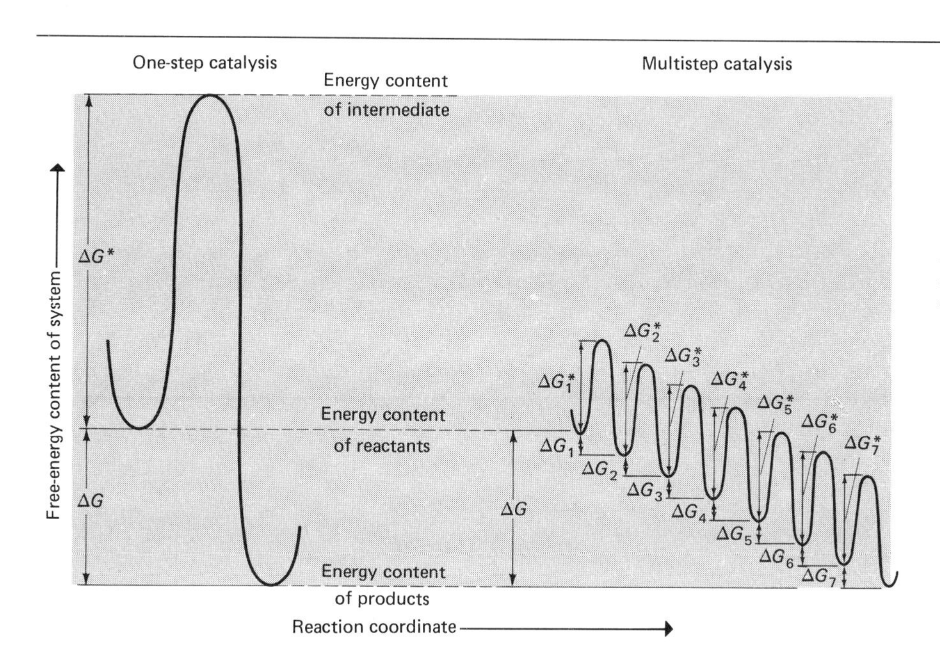

FIGURE 18-13
The activation process of a one-step versus a multistep catalysis. ΔG^* = activation energy; ΔG = free-energy change between initial and final states for each step. (*After D. E. Green and R. F. Goldberger, 1967, Molecular Insights into the Living Process, Academic Press, Inc., New York, p. 84.*)

18-6 FACTORS AFFECTING RATE OF ENZYME ACTIONS

If enzymatic hydrolysis is compared with an analogous reaction brought about by the action of strong acids or bases, it will be found that the former proceeds at a much lower temperature. The reaction paths in the two situations are different, and this is reflected in their activation energies. For acid- or base-effected hydrolysis of an ester the value is about 10 to 13 kcal/mole, and for the hydrolysis of peptide bonds it is about 20 kcal/mole. The values for the analogous enzyme-catalyzed hydrolyses are about 4 and 12 to 14 kcal/mol, respectively.

The activation process may be considered from the view of thermodynamics like any other chemical reaction, and we can write an equation for it analogous to Eq. 17-6:

$$\Delta G^* = \Delta H^* - T\Delta S^*$$

where ΔG^* = the free energy of activation
ΔH^* = the heat of activation
ΔS^* = the entropy of activation

Thus, it can be seen that because ΔH is lower in enzyme-catalyzed reactions, as in the example above, the free energy of activation also decreases. In addition, there are good reasons for believing that enzyme action involves an increase in the entropy of activation.

A useful comparison of rates of a reaction is the **temperature coefficient,** Q_{10}—the ratio of reaction rates at $T°$ + 10°C and at T°C. (See Chap. 14, Sec. 14-3, for a better definition of Q_{10}.) Thermobiological reactions, like thermochemical reactions in general, have Q_{10}'s that are usually 2 or more. The Q_{10} values of enzymatic reactions generally lie between 1.1 to 5.3, and most are 3 or less.

The reason for the twofold or greater increase of thermochemical reactions is probably that, with increasing thermal energy, a greater number of molecules with varying amounts of energy will attain sufficient energy to undergo a thermochemical reaction. It is assumed that for a molecule to undergo decomposition or for two molecules to collide and react, the molecules must possess a certain energy of activation, or *critical increment*. As mentioned earlier, enzymes lower the activation energy of reactions they catalyze.

Temperature also has an effect on the enzyme itself. It can be shown with any enzyme that, like other chemical reactions, the rate of enzymatic activity increases with increasing temperature (but only to a maximum for enzymes). Above a certain temperature there is inactivation (denaturation) of the native protein, and the rate of reaction will decline with increasing temperature. The Q_{10} for denaturation of enzymes between 70 and 80°C is several hundred (Fig. 18-14).

A thermodynamic analysis, however, does not provide much insight into the mechanism of enzymatic catalysis. It is chiefly kinetic data that reveal factors affecting rates of enzyme action.

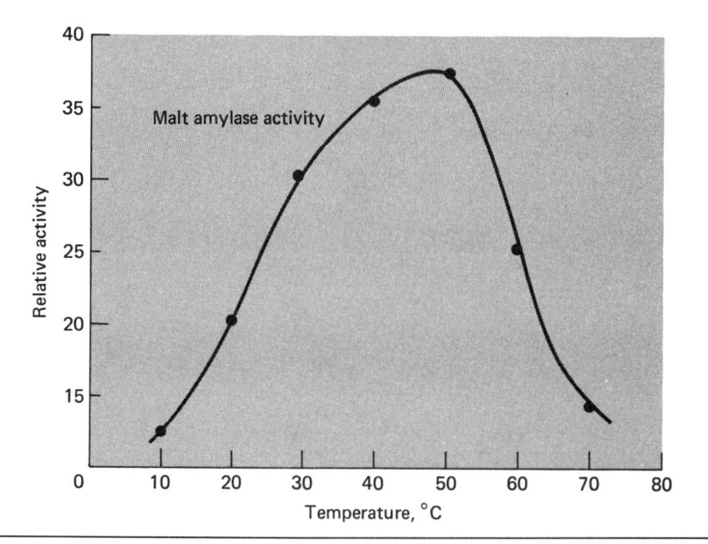

FIGURE 18-14
Effect of temperature on the rate of enzyme activity. (*After A. C. Giese, 1962, Cell Physiology, 2d ed., W. B. Saunders Company, Philadelphia, p. 317.*)

Let us consider the hydrolysis of sucrose, in which a molecule of sucrose is cleaved and a molecule of water reacts (hydrolysis). Because the concentration of water is so much greater than that of the sugar, there is little perceptible change in water concentration accompanying the large change in sucrose concentration. So, for practical purposes only one substance, sucrose, can be considered as reacting, and the rate of its reaction, within limits, will be proportional to its concentration. This, in chemical reaction kinetics, is known as a *first-order reaction*. It is generally found only in reactions that go essentially to completion and in reactions with constant enzyme concentration (Fig. 18-15).

More precisely, the velocity of an enzyme action parallels the substrate concentration only when the amount of substrate is low and other variables are controlled. Furthermore, with sufficiently high substrate concentration the rate of catalysis actually may be slowed. In general,

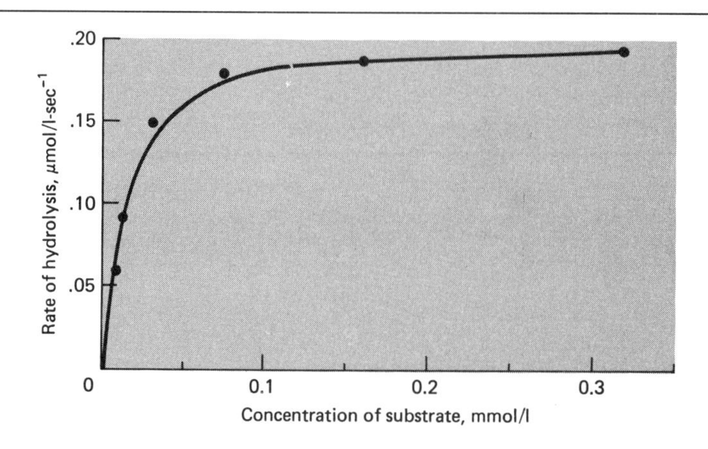

FIGURE 18-15
Effect of substrate concentration on the rate of reaction when a constant amount of enzyme is present. The first part of the curve, where the rate of reaction is proportional to the substrate concentration, is called a *first-order* reaction. As the enzyme molecules become saturated with substrate there is no further increase in rate. The reaction is then known as zero-order. (*After A. C. Giese, 1962, Cell Physiology, 2d ed., W. B. Saunders Company, Philadelphia, p. 316.*)

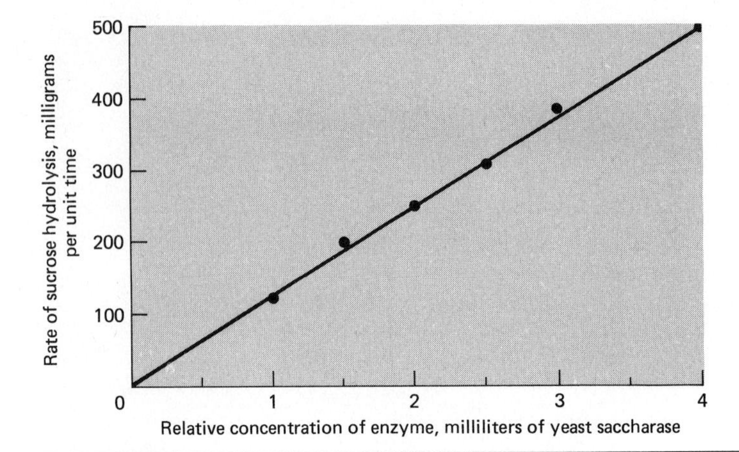

FIGURE 18-16
Effect of enzyme concentration on the rate of a reaction when an excess of substrate is present. (*After E. Baldwin, 1967, Dynamic Aspects of Biochemistry, 5th ed., Cambridge University Press, London, p. 27.*)

however, the velocity of the overall reaction tends to increase with increasing concentration of substrate as proportionately more free enzyme is converted to ES. The maximum velocity is attained when the enzyme is operating at full speed and practically all the enzyme is converted to ES. The rate of reaction is then limited by the concentration of enzyme, and thus the reaction velocity is independent of substrate concentration. This is called a *zero-order reaction* (Figs. 18-15 and 18-16).

In truth, most catalytic reactions are neither strictly first order nor zero order, but proceed according to some intermediate kinetics.

A mathematical treatment of the theory underlying enzyme kinetics cannot be well handled in an account as brief as would be appropriate here; the interested student would be well advised to consult the references at the end of this chapter for information on the subject.

The products of an enzyme action retard the reaction rate, as can be demonstrated by adding products to a system containing purified enzyme. An explanation proposed for this effect is that the product forms a complex with the enzyme that is more stable than the ES complex. Also, it

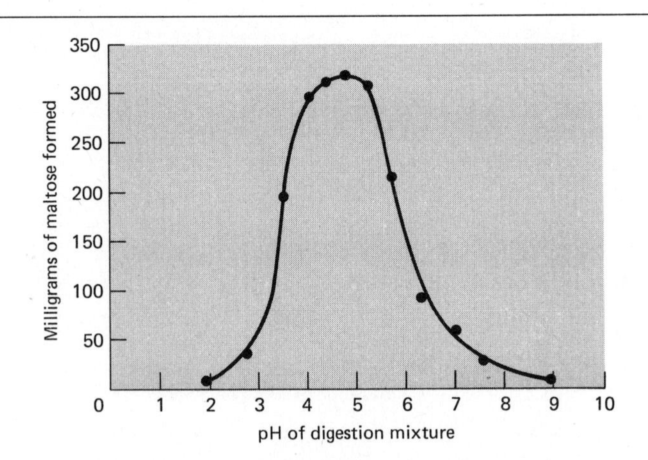

FIGURE 18-17
The influence of hydrogen-ion concentration on the activity of wheat-flour amylase. Ten grams of wheat flour autodigested at 27° for 1 hour. (*After R. A. Gortner, 1938, Outlines of Biochemistry, 2d ed., John Wiley & Sons, Inc., New York, p. 942.*)

can be assumed that in the presence of high concentrations of products, the enzyme may catalyze resynthesis, which would cause an apparent decrease in the rate of decomposition. (Feedback inhibition of enzymes will be discussed in Chap. 20, Sec. 20-2.)

Enzymes are quite sensitive to $[H^+]$ and show maximum activity at a certain pH. Activity declines on either side of this optimum. The pH optimum varies widely among enzymes, and the optimum for any one enzyme will vary somewhat, depending on what substrate the enzyme is acting upon (Figs. 18-17 and 18-18).

The pH sensitivity of enzymes may result from the necessity of a particular charge distribution about the active site for optimal activity, but other factors also may be involved, such as water of dehydration, ionic strength effects, and the relation between apoenzyme and cocatalyst.

18-7 ENZYME COUPLING

Intracellular enzymes often seem to be geographically close together, forming sets of several kinds of enzymes, or multienzyme complexes. The advantage for efficient catalysis by such an arrangement is obvious. These associations are made feasible by forming them into *systems* with some functional connection which will couple them together—for example, by *common-substrate coupling:*

$$A \underset{E_1}{\rightleftharpoons} B \underset{E_2}{\rightleftharpoons} C$$

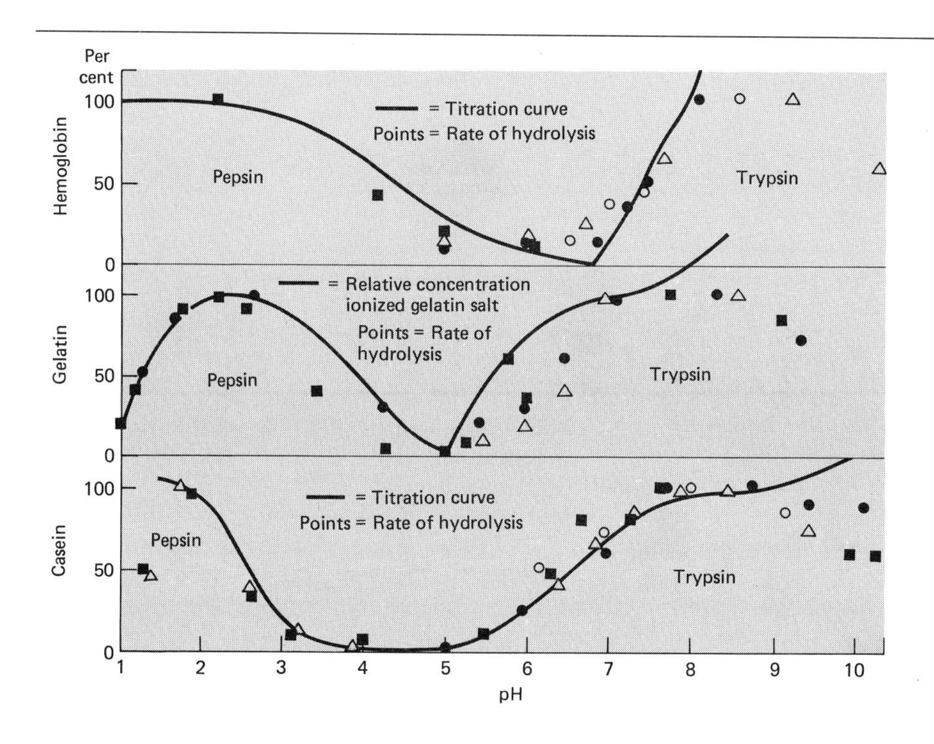

FIGURE 18-18 The effect of pH on the activity of pepsin and trypsin. The relative rate of hydrolysis and percent of protein present as salt are indicated. Note that the pH effect varies with the protein. (*After J. H. Northrop, 1939, Crystalline Enzymes, Columbia University Press, New York, p. 8.*)

Here it is seen that each enzyme has two substrates—B is common for both enzymes (E_1 and E_2). However, A is a substrate only for E_1, and C is a substrate only for E_2. You will see many examples of this type of coupling in metabolic sequences, as detailed in the next chapter.

A second kind of substrate coupling is effected by *reversing reaction*. This can be illustrated by the following oxidation-reduction reaction:

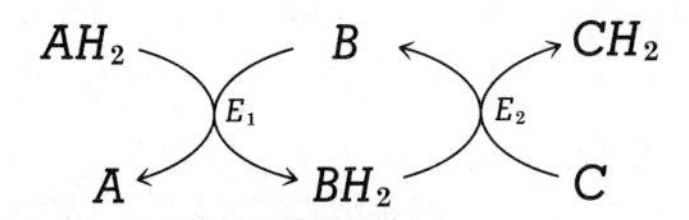

Note that B is reduced by the first enzyme (E_1) and oxidized by the second (E_2). This is a very common mechanism.

In both the examples just cited the enzymes can be considered to be coupled because of a dependence of the second upon the first for *substrate*. Chains of successive enzyme reactions are formed using these simple associations. There is a second kind of coupling of reactions: *energy* coupling. Energy released by one reaction may be transferred to drive a second reaction; i.e., the free energy liberated by one enzyme reaction (e.g., an oxidation) may be utilized by an energy-requiring reaction (e.g., a synthesis), as noted in the preceding chapter (Sec. 17-4). That energy is preserved by some intermediate in the biochemical pathway. In fact, the *preservation of energy in intermediates* is the basis of all biological energetics. ATP is a common intermediate in most reactions, but the preservation of energy also is fundamentally a cellular enzyme function. That is, *enzymes preserve energy* by forming ES complexes which can act as intermediates. In fact, coenzymes are uniquely adapted for directly energizing certain synthetic reactions. For example, saccharide polymerization involves nucleotide coenzyme (Sec. 16-4). Thus, enzymes provide the opportunity to carefully mete out the energy of a food molecule, preserving free energy and eventually making quantal transfers, generally to form ATP. ATP then transfers energy to other molecules through enzyme-complexed intermediates. The involved series of chemical reactions known totally by the name of "metabolism" is, in very large part, a sequential series of such energy transfers. These will be examined in detail in the following chapter.

READINGS

Bernhard, S. A., 1968, *The Structure and Function of Enzymes*, W. A. Benjamin, Inc., New York. A comprehensive, up-to-date treatment of enzyme chemistry for the student who is seeking detailed information. The index, unfortunately, is inadequate.

Dickerson, R.E., and I. Geis, 1969, *The Structure and Action of Proteins*, W. A. Benjamin, Inc., Menlo Park, California. An introduction to the structure and function of proteins, with outstanding illustrations of enzyme construction.

Green, D. E., and R. F. Goldberger, 1967, *Molecular Insights into the Living Process*, Academic Press, Inc., New York. An interestingly composed book which includes a discussion of enzymes (Chaps. 4 and 5) in the context of cellular bioenergetics.

CONTROLLED ENERGY TRANS-FORMATIONS 19

When we speak of "bond formation" we are really referring to the establishment of certain electron configurations, and when we talk in terms of "bond energy" or "breaking of bonds" we actually are dealing with electron energies. In the final analysis, the *energy* of life is derived from electrons, in a way that is rather analogous to the derivation from an electric current of the power that drives man-made machines. Cells obtain energy only by manipulating electrons, moving them in a controlled way from a higher energy state, or potential, to a lower one. This is merely one way of saying that energy transfers in cells are oxidations and reductions, for oxidation is the removal of electrons and reduction is the addition of electrons. By emphasizing that it is from electron transitions that free energy is realized, we place the matter of biological energy in proper perspective.

This simple concept is difficult to see in most biological systems because *energy generation* is so complexly interwoven into the *metabolic processes* of organisms. That is, the few essential reactions which lead to the release of energy are interwoven with and enmeshed among numerous metabolic processes that have nothing to do directly with energy generation.

There are, however, some instructive examples among microorganisms in which the energy-yielding processes are segregated from general metabolism. The bacterium *Hydrogenomonas* carries out the following sample oxidation-reduction reaction:

$$\underset{\text{Electron donor}}{H_2} + \underset{\text{Electron acceptor}}{1/2\ O_2} \xrightarrow{ADP + P_i \ \rightarrow \ ATP} H_2O$$

The drop in electron potential, or energy, occurring with the formation of water is used to form ATP. Other bacteria oxidize sulfur, ammonia, or nitrite with oxygen as the electron acceptor. There are still other species which use sulfate, sulfite, or thiosulfite as an acceptor. All these organisms are called *chemosynthetic autotrophs*. (They will be mentioned again later in Sec. 19-8.)

Segregation of energy generation and metabolism is the exception and not the rule. It appears that metabolic systems have been fashioned around the energy-coupling reactions, which suggests that the metabolic pathways were later evolutionary developments than were the coupling reactions themselves.

Our objective in this chapter will be to examine the complex systems of metabolic and energy-generating machinery of organisms, so that we may identify the sources and the destinies of energetic electrons as well as the mechanisms by which their energy is conserved.

19-1 CENTRAL ROLE OF ATP

Oxidation reactions provide an organism with a means for supplying its energy demand. There is primarily one form to which the utilizable energy released by oxidation is transformed, and that is the terminal phosphate bond of ATP. All the available dimensions of energy appear to be funneled into this *one denomination of common currency*, and all biochemical machines are designed to be energized at some stage by ATP.

Formation of high-energy phosphate bonds results mainly from three biochemical devices: (1) glycolysis, or fermentation, (2) oxidative phosphorylation involving the tricarboxylic acid (TCA) cycle, and (3) photosynthesis.

These three processes have much in common, as becomes apparent when we subdivide the overall events into sequential steps. First, they all generate an electron donor in a set of preparatory or priming reactions. The donor is triose phosphate in glycolysis; NAD and succinate in the TCA cycle; and probably reduced ferredoxin in chloroplast photosynthesis. The second common step is the interaction of the donor with an electron-transfer system, and the resulting oxidoreduction is coupled to the formation of a high-energy intermediate. In step three, the energy of the electrons forming the bond in the intermediate is transferred from one pair of linked molecules (those forming the high-energy intermediate) to another pair of molecules, namely ADP and inorganic phosphate (P_i). Thus, the high-energy bond finally comes to reside in the P-O-P link formed between ADP and P_i. Chemically, this reaction is an esterification of phosphate, but its exact mechanism is not known. The final event would be the manipulation of the P-O-P bond of ATP by replacement reactions to give rise to bonds essential in synthetic processes (Fig. 19-1).

Much attention has been devoted in the last 30 years to an explanation of the factors which contribute to the high-transfer potential of the phosphate of ATP. One factor is that the resonance energy characteristic of the hydrolysis products is higher than that of ATP, meaning that the former are more stable. If hydrolysis converts a substance to products with

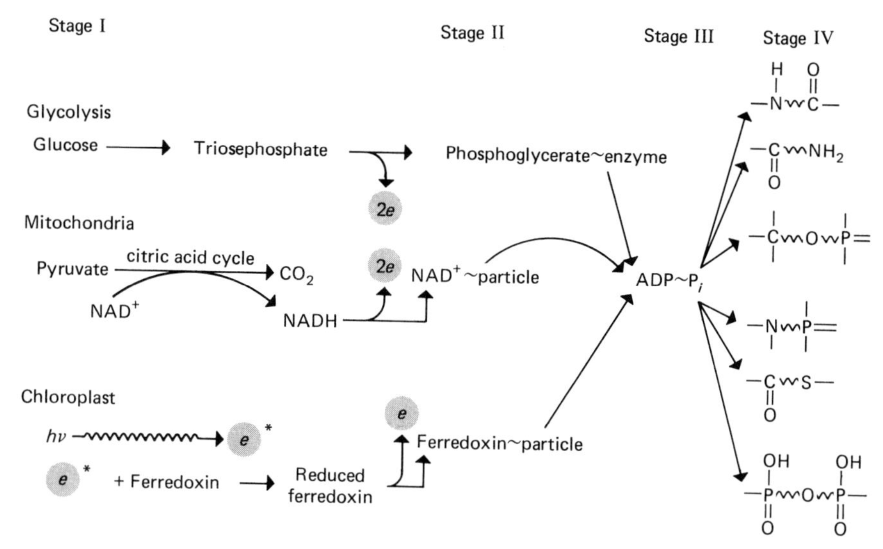

FIGURE 19-1
The four stages in the formation and utilization of phosphate-bond energy. Stage I: preparatory stage during which electron donors are generated; stage II: formation of the high-energy bond by interaction of electron donor with coupling system; stage III: transfer of bond from pair of molecules in first high-energy intermediate to ADP-P_i pair; stage IV: formation of new bonds during ATP-energized synthetic reactions. The exact form of the high-energy intermediate (represented as NAD ~ particle in the mitochondrion and ferredoxin ~ particle in the chloroplast) is unknown, and there are multiple forms. $h\nu$ = radiant energy; e = electron. (*After D. E. Green and R. F. Goldberger,* 1967, *Molecular Insights into the Living Process, Academic Press, Inc., New York,* p. 132.)

increased resonance possibilities, the reaction will be promoted to a more stable state and there will be a decrease in free energy. The loss of resonance energy with the formation of ATP results from a situation termed *competing resonance:* two phosphorus atoms are competing for the electrons of a single oxygen atom. Additionally, ATP and related substances exist as either the trianion or tetraanion at neutral pH, and, consequently, electrostatic repulsions may tend to destabilize the polyphosphate configuration. To overcome charge density, energy is required to bring the four oxygens together in ATP, and their separation during hydrolysis releases a corresponding amount of energy.

The free-energy decreases occurring with the hydrolysis of the magnesium chelate of ATP at pH 7 are as follows:

$$\text{A-R-P}\sim\text{P}\,\vdots\,\text{P} \longrightarrow \text{A-R-P}\sim\text{P} + \text{P}_i \qquad \Delta G^{\circ\prime} = -7.4 \text{ kcal/mol}$$
$$\text{A-R-P}\,\vdots\,\text{P}\sim\text{P} \longrightarrow \text{A-R-P} + \text{PP}_i \qquad \Delta G^{\circ\prime} = -7.6 \text{ kcal/mol}$$

The actual available free energy from the hydrolysis of 1 mol of ATP under intracellular conditions is greater than the standard free-energy change and is probably in the range of − 10 to − 12 kcal/mol.

All this does not mean that ATP is a kinetically unstable molecule; it is in fact surprisingly stable and requires considerable energy to hydrolyze. However, it is thermodynamically a potentially vigorous agent and in the presence of specific enzymes it serves to energize many reactions, both as a phosphorylating and adenylating agent.

19-2 CENTRAL METABOLIC PATHWAYS

Metabolism may be defined as the totality of the chemical processes that an organism is capable of performing. At any one time an organism may not be expressing all the metabolic aspects of which it is capable. Metabolic control or regulation, which is a function of genetic and extrinsic (environmental, physiological) factors, determines what biochemical events take place. We will restrict our inquiry to certain metabolic routes only, as our interest at this point is in those which make energy available for biological use.

Catabolic processes are degradative ones in which large organic molecules, serving as food for the organism, are broken down (generally with the participation of oxidative reactions) to simple cellular constituents, with the concomitant release of free energy. This energy is utilized by the organism for synthesis, mechanical work, transport, and heat. *Anabolic* processes are those which lead to synthesis; these result in simple molecules being incorporated into more complex structures, and frequently involve reductive reactions in addition to the investment of free energy.

Despite the diversity of metabolic routes among organisms and the magnitude of the biochemical events going on within any one of them, there is a *central area*, or *main line of metabolism*, linking catabolic and anabolic processes. It is remarkably similar in the great majority of organisms examined. Large foodstuff molecules yield a very restricted group of small organic molecules, each characteristic of a particular class of substance. Carbohydrates are utilized as monosaccharides and degraded to triose phosphates or pyruvic acid (pyruvate). Lipids are cleaved into glycerol and two-carbon (acetyl) or three-carbon (propionyl) fragments which are then condensed with coenzyme A (CoA). Proteins yield amino acids which are converted to simple acids such as pyruvate, acetate, oxaloacetate, and α-ketoglutarate (Fig. 19-2).

These small molecules are very *dynamic*, serving as the units which are polymerized in anabolic processes, being interconverted as metabolic equivalents, or being completely oxidized to CO_2 and H_2O. Many of the reactions that they engage in are freely reversible, but others are not. Also, anabolic and catabolic routes rarely, if ever, follow the same pathways in detail.

19-3 GLYCOLYSIS

Glycolysis, also called the Embden-Meyerhof sequence, is a series of coupled reactions in which glycogen, starch, or glucose is anaerobically broken down to two molecules of the three-carbon pyruvate, yielding the energy for a gain of two ATP molecules. Carbohydrates are a primary energy source for the majority of organisms. The glycolytic pathway, the first series of energy-yielding reactions to be worked out in

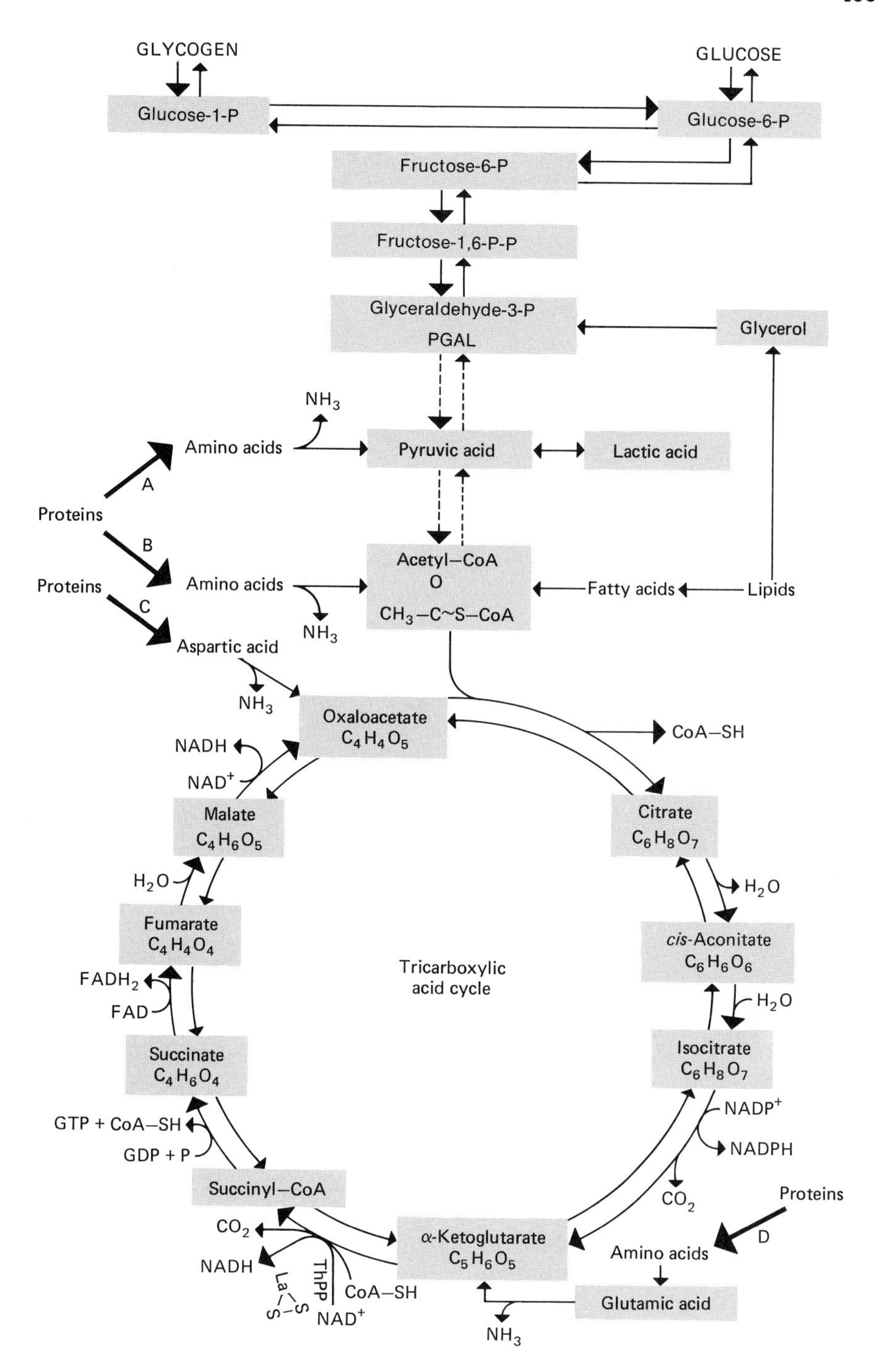

FIGURE 19-2
Simplified version of the common metabolic pathway for different foods. Carbohydrates are utilized as monosaccharides; glucose, the conventional form used, is shown here, but others can be modified and enter this pathway. Lipids largely are consumed after being hydrolyzed as glycerol and two carbon fragments. Amino acids may enter the pathway at the four points shown, depending upon their chemical nature: *A–D* (see Sec. 16-2 for explanation).

detail, is the major route of glucose catabolism in most organisms. It is the main line for the first phases of carbohydrate metabolism in aerobic organisms, and in obligate anaerobes it provides all the intermediates and most or all of the energy for their various functions (Fig. 19-3).

A complete discussion of the glycolytic sequence is beyond the scope of this book. Our intent here will be to direct attention to the manner in which an energy donor is generated and how its energy is transferred and stored for subsequent use.

The glycolytic reactions, as noted earlier, may be divided into three steps. The first, which we shall call *generation of a donor*, begins with the phosphorylation of glucose and extends to the formation of the phosphorylated triose, D-glyceraldehyde-3-phosphate (also called phosphoglyceraldehyde, or PGAL) (Fig. 19-3).

Glucose, when it enters the cell, is in a low energy state and is made more reactive through phosphorylation by ATP. (Glucose derived from glycogen is esterified by P_i.) This is called a *priming reaction*. Glucose, now phosphorylated at C-6, undergoes an internal rearrangement (isomerization) to become fructose-6-P. A second priming reaction with another ATP creates fructose-1,6-diP. This molecule is then cleaved in the middle by a hydrolytic enzyme, fructoaldolase, to form two C_3 units, dihydroxyacetone phosphate and PGAL. The latter occupies a most important position not only in glucose catabolism but also in metabolism of other carbohydrates and lipids and in photosynthesis. (Refer again to Fig. 19-3.)

Glycolysis now proceeds through PGAL into the second step of glycolysis, *formation of a high-energy intermediate*. In a coupled reaction PGAL is oxidized in the presence of P_i to form a diphosphate. This, in turn, in the third step, engages in a transphosphorylation with ADP, becoming 3-phosphoglyceric acid (PGA) in a reaction coupled to the *formation of ATP*. Thus, the two PGAL molecules (C_3) coming from one molecule of glucose (C_6) compensate for the two ATPs invested in the priming steps.

In subsequent reactions, which are variants of the second and third steps of energy conservation, the remaining phosphate on the 3-carbon acid is prepared for transfer to ADP. With the formation of pyruvic acid, ATP is produced. Again there are two phosphorylations per each starting glucose, so that a net gain of two ATP units is realized. When glycogen is the starting material there is a net gain of three ATPs, because inorganic phosphate is used to phosphorylate glucose instead of ATP (see diagrams at bottom of Fig. 19-3).

It will be noted (Fig. 19-3) that two units of reduced NAD (NADH) are produced. These energy-transferring molecules also can be used to form high-energy phosphate bonds in systems where oxygen is used as an acceptor of hydrogens and electrons. They are potential high-energy intermediates and serve in that capacity in *aerobic* systems. In *anaerobic* systems, however, organic compounds generally serve as acceptors,

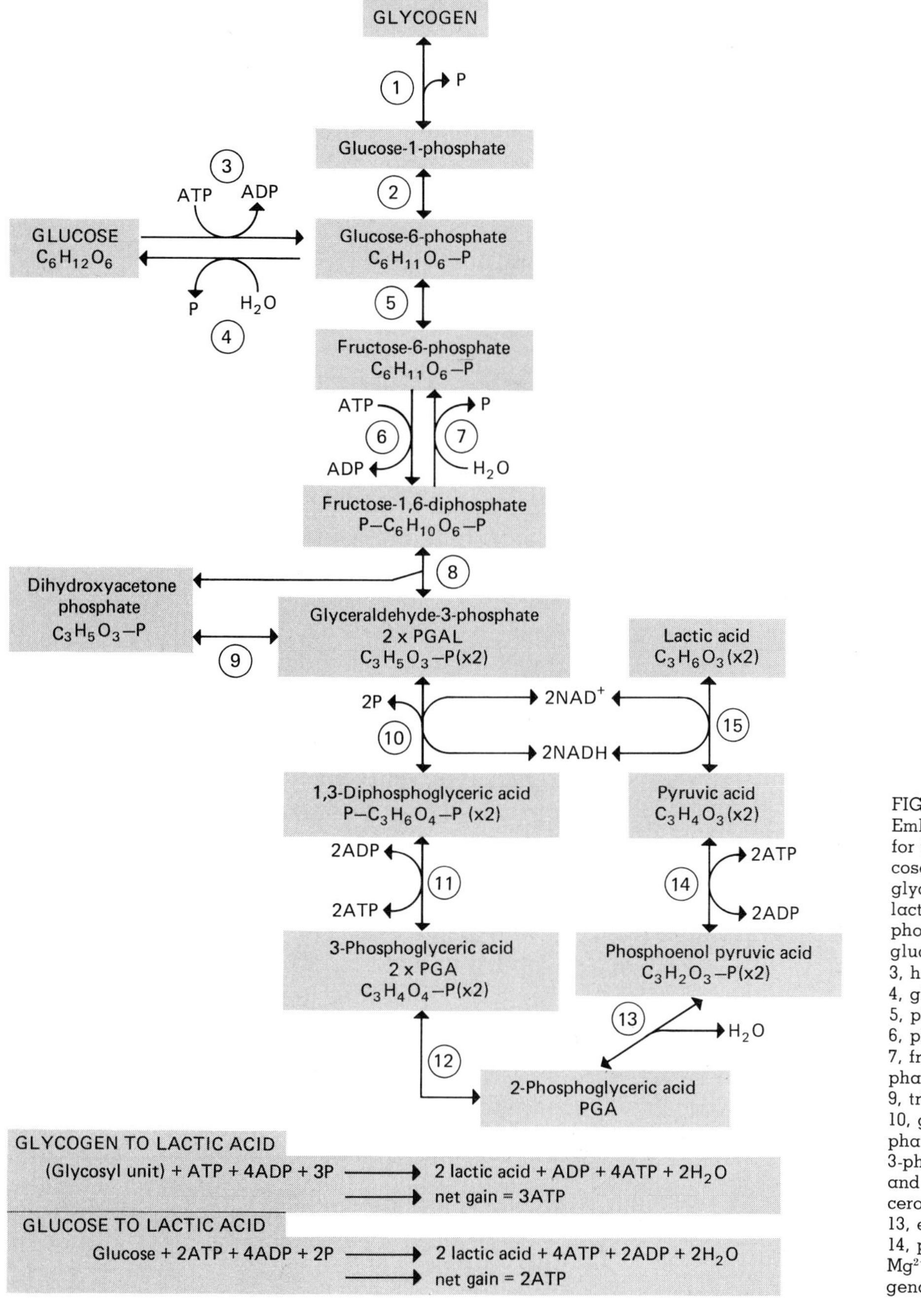

FIGURE 19-3
Embden-Meyerhof sequence for the metabolism of glucose or the glycosyl units of glycogen to pyruvic or lactic acid. Enzymes: 1, phosphorylase; 2, phosphoglucomutase and Mg^{2+}; 3, hexokinase and Mg^{2+}; 4, glucose-6-phosphatase; 5, phosphohexoisomerase; 6, phosphofructo-1-kinase; 7, fructose-1, 6-diphosphatase; 8, fructoaldolase; 9, triose isomerase; 10, glyceraldehyde-3-phosphate dehydrogenase; 11, 3-phosphoglyceric 1-kinase and Mg^{2+}; 12, phosphoglyceromutase and Mg^{2+}; 13, enolase and Mg^{2+}; 14, pyruvic kinase and Mg^{2+}; 15, lactic dehydrogenase and Zn^{2+}.

and the free-energy changes of these reactions are not sufficient to form phosphate bonds. In glycolysis the interaction between the aldotriose phosphate and pyruvate involves two half-reactions, catalyzed by different enzymes but sharing the coenzyme in common. These reactions regenerate the coenzyme and thus serve as *balancing reactions;* i.e., the reactive products of oxidation are combined into stable organic structures.

Pyruvic acid is shown in Fig. 19-3 as the acceptor, forming lactic acid, a process that occurs in muscle cells, for example, but is by no means universal. Insect muscles, for instance, do not form lactate but make glycerol from dihydroxyacetone phosphate. In yeast cells, growing under anaerobic conditions, alcohol is produced:

$$\text{Pyruvate} \xrightarrow[\searrow CO_2]{} \text{acetaldehyde} \xrightarrow{NADH \;\;\; NAD^+} \text{ethanol}$$

Other kinds of microorganisms produce different products, such as acetic and butyric acid. Many of these reactions have been exploited as commercial sources of chemicals and are employed in the production of foods and beverages. Such reactions are examples of a **fermentation**—an oxidoreductive, energy-yielding, coenzyme-linked series of reactions in which oxidized and oxidizing substances are organic compounds.

Let us now examine the details of the energy-yielding reactions of glycolysis more closely. These are the first ones:

$$\underset{\text{3-PGAL}}{\begin{matrix} CH_2OP \\ | \\ CHOH \\ | \\ C{=}O \\ | \\ H \end{matrix}} \;\underset{2H^+}{\overset{P_i}{\rightleftarrows}}\; \underset{\text{1,3-Diphosphoglyceric acid}}{\begin{matrix} CH_2OP \\ | \\ CHOH \\ | \\ C{=}O \\ | \\ OP \end{matrix}} \;\underset{ADP \;\;\; ATP}{\longleftrightarrow}\; \underset{\text{3-PGA}}{\begin{matrix} CH_2OP \\ | \\ CHOH \\ | \\ C{=}O \\ | \\ OH \end{matrix}}$$

Oxidation and phosphorylation occur simultaneously in the first step. The energy of oxidation is conserved in the form of a high-energy bond between the enzyme (glyceraldehyde-3-phosphate dehydrogenase) and the triose. This bond is then manipulated to form a high-energy phosphate and free the enzyme.

1,3-Diphosphoglycerate has a high free energy of hydrolysis (−11,800 cal/mol) because, like ATP, there is a close juxtaposition of two phosphate groups, each with a double negative charge. In addition, there is an unusually high density of electrons in the oxygen-rich anhydride linkage formed between the carboxyl and phosphate groups.

The second reaction goes readily to completion because the repulsive forces between the two negatively charged products prevent recombination, and the electrons assume more stable positions with greater resonance freedom in the carboxyl and phosphate groups. The carboxyl phosphate group of 1,3-diphosphoglycerate, because of its very high phosphate-transfer potential (Table 17-4), is donated completely to ADP with a high thermodynamic driving force.

The second set of energy-yielding reactions is:

$$\underset{\text{2-PGA}}{\begin{array}{c}CH_2OH\\|\\CHOP\\|\\C{=}O\\|\\OH\end{array}} \xrightleftharpoons[H_2O]{} \underset{\text{Phosphoenolpyruvate}}{\begin{array}{c}CH_2\\\|\\COP\\|\\C{=}O\\|\\OH\end{array}} \xrightleftharpoons[ADP \to ATP]{} \underset{\text{Pyruvate}}{\begin{array}{c}CH_2\\\|\\COH\\|\\C{=}O\\|\\OH\end{array}}$$

The first reaction is a dehydration of the phosphate ester of a dihydroxy acid, which can also be regarded as an intramolecular oxidation of C-2 and reduction of C-3. As a result, electrons become more concentrated about C-2 because of the formation of a double bond (two pairs of electrons) and the constriction of the bond distances and angles around this carbon. This causes the negatively charged phosphate and carboxyl groups to be pulled closer together, resulting in a shift in the internal energy of the molecule so that phosphopyruvate becomes a high-energy compound ($\Delta G^{\circ\prime}$ of hydrolysis = $-12{,}800$ cal/mol). Phosphoenolpyruvate readily donates its phosphate group to ADP. The turnover number of the enzyme catalyzing this step, pyruvic kinase, is 6×10^3 for pyruvate formation, as compared with 12 for the reverse direction. Thus, the equilibrium is thermodynamically very favorable for pyruvate formation and would provide a formidable block against use of this reaction in carbohydrate synthesis.

The phosphorylation of ADP in these reactions is called *substrate level* to distinguish it from photophosphorylation and oxidative phosphorylation. In substrate phosphorylation the phosphate group is transferred to ADP from an organic molecule which also provides the energy for the high-energy bond. In the other two processes inorganic phosphate is used to phosphorylate ADP and electrons transferred from several substrates provide the energy to drive the reaction.

19-4 THE PENTOSE PHOSPHATE CYCLE

There is a set of enzymes, widespread in nature, which catabolize hexose by a pathway that involves neither glycolysis nor the citric acid cycle. A number of different names have been assigned to the reaction sequence involved. Sometimes it is called the "Warburg-Dickens pathway" after two of its principal investigators, or the "pentose cycle" because C_5 sugars are involved, or the "hexose-monophosphate shunt" because one of its pathways diverges from glycolysis at the level of glucose-6-P (Fig. 19-4).

None of the reactions of this cycle leads to the synthesis of ATP. Does

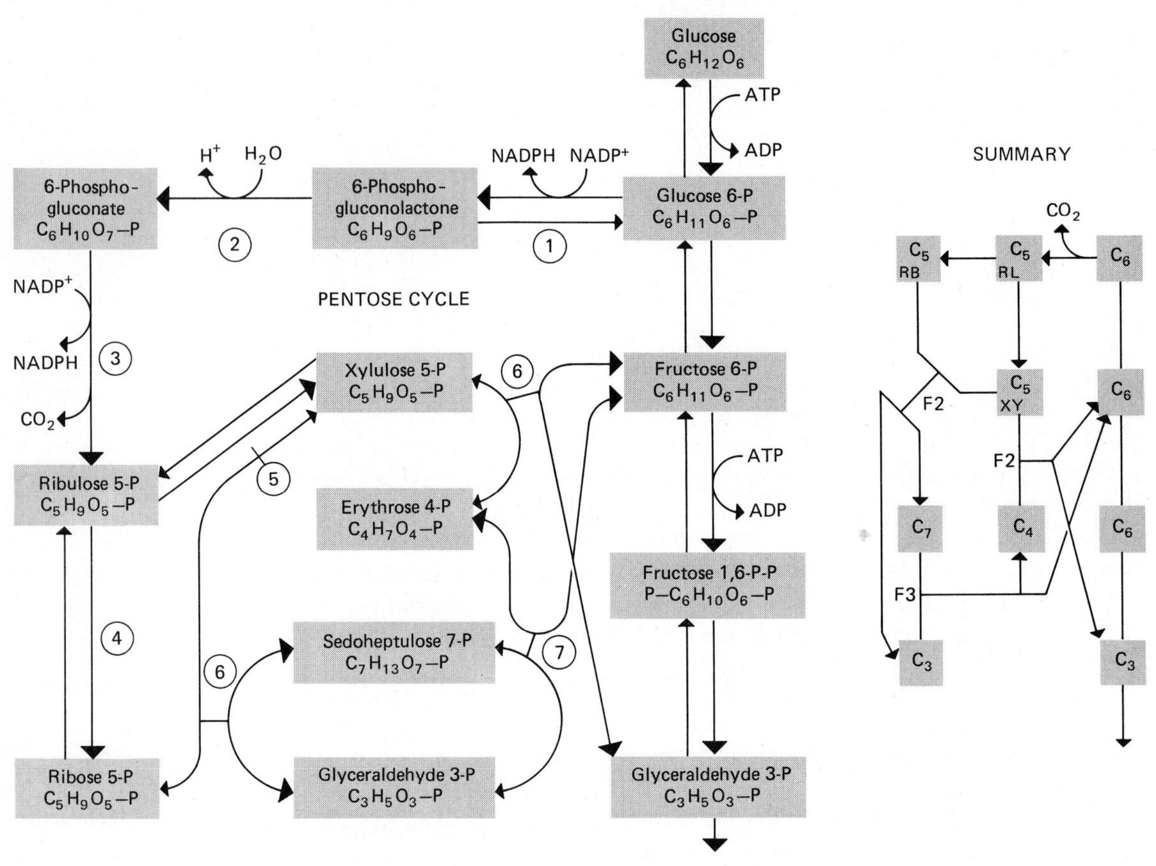

FIGURE 19-4
Major steps in the pentose phosphate cycle. Larger arrowheads show course of reactions in glucose breakdown. Enzymes: 1, glucose-6-phosphate dehydrogenase; 2, gluconolactonase and Mg^{2+}; 3, 6-phosphogluconic dehydrogenase and Mg^{2+}; 4, phosphoriboisomerase; 5, phosphoketopentoepimerase; 6, transketolase and thiamine pyrophosphate; 7, transaldolase. *F2*, Enzymatic transfer of C_2 fragment; *F3*, transfer of a C_3 fragment. *RL*, Ribulose; *RB*, ribose; *XY*, xylose. In summary, three pentose molecules produce two hexose molecules and one triose molecule.

this mean that these reactions are not energy-conserving? Not at all. While the central role of ATP in conservation of energy has been stressed, it will be recalled that coenzymes also can serve as energy-conserving intermediates. Whereas the transferable energy of ATP is in the form of a highly reactive phosphoryl group, that of coenzymes is in the form of a pair of electrons. Most synthetic processes require one to several reductive steps, and coenzymes can directly transfer reducing electrons in certain of these reactions. Thus, the coenzyme of this cycle, NADP, can be regarded as a source of chemical energy.

Theoretically, since the cell can oxidize the NADPH produced in the pentose pathway using O_2 as an electron acceptor (as explained in the next section), a mechanism is provided for the oxidation of glucose to CO_2 and H_2O. The reactions can be summarized as:

6 Glucose-6-P + 12 $NADP^+$ $\longrightarrow$ 6 pentose-P + $6CO_2$ + $6H_2O$ + 12 NADPH
4 Pentose-P $\longrightarrow$ 2 fructose-6-P + 2 tetrose-P
2 Pentose-P + 2 tetrose-P $\longrightarrow$ 2 fructose-6-P + 2 triose-P
2 Triose-P $\longrightarrow$ fructose-1,6-P

Sum: 6 Glucose-6-P + 12 $NADP^+$ $\longrightarrow$
$6CO_2$ + $6H_2O$ + 12 NADPH + 4 fructose-6-P + fructose-1,6-diP

Thus, for every six molecules of glucose-6-P entering the cycle, one sugar is completely oxidized to CO_2 and H_2O and five are regenerated. Note that glucose is not directly oxidized; there are but two oxidations (reactions 1 and 3 in Fig. 19-4), and these are the energy-yielding steps.

The pentose cycle serves the additional functions of providing a pool of enzymes for interconnecting and reshuffling the carbon atoms of monosaccharides, as well as serving as a source of precursors for carbohydrate biosyntheses. Also, it has been suggested that the reactions of the pentose cycle and certain ones of the glycolytic sequence may interact and lead to the oxidative conversion of glucose either to triose-P or to CO_2.

The reactions of the pentose cycle occur almost universally in organisms as an essential adjunct to aerobic metabolism. They have been found in many bacteria, plants, and various animals.

The relative contributions of the glycolytic and pentose pathways in glucose metabolism can be compared by tracer techniques using the radioisotope ^{14}C. Whereas $^{14}CO_2$ results from the metabolism of either glucose 1-^{14}C or glucose 6-^{14}C in glycolysis, that produced in the pentose cycle has its origin in glucose 1-^{14}C only. These techniques have demonstrated that the pentose cycle is more active in insects and certain other tissues than glycolysis.

19-5 THE TRICARBOXYLIC ACID CYCLE

The major pathway for the catabolism of the end product of glycolysis is a sequence of reactions called the **tricarboxylic acid** (TCA) **cycle** (also known as the citric acid cycle after the first compound formed, or the Krebs cycle after its most illustrious proponent) (Fig. 19-2).

When pyruvate is metabolized by way of the citric acid cycle, the great majority of the potential energy originally residing in glucose can be captured in ATPs. The addition of an aerobic stage to an organism's metabolic machinery, therefore, tremendously enhances its energetic position, as shown by the following:

1 Glucose $\longrightarrow$ 2 lactic acid $\qquad \Delta G^{\circ\prime} = -47$ kcal
1 Glucose $\longrightarrow$ 2 ethanol + $2CO_2$ $\qquad \Delta G^{\circ\prime} = -56$ kcal
1 Glucose + $6O_2$ $\longrightarrow$ $6CO_2$ + $6H_2O$ $\qquad \Delta G^{\circ\prime} = -686$ kcal

Clearly, the energy return from an oxygen-utilizing system is spectacular and surely must have selective value among many organisms. However, possession of such a system is not without cost; a special set of many enzymes, as well as devices for assuring access to oxygen, must be maintained.

The TCA cycle is a major metabolic "hub," serving not only in the combustion of pyruvate but in the oxidation of amino acids and the C_2 and C_3 end products of fatty acid catabolism as well. It also plays a central role in synthesis, providing carbon (C_4, C_5, C_6) skeletons for amino acids, purines, pyrimidines, porphyrins, long-chain fatty acids, and a few other compounds. Our present interest in the cycle, however, is in its role in making energy available for ATP formation, a process that occurs in the tissues of all animals, higher plants, and many microorganisms.

Pyruvate cannot enter the cycle directly, but must first undergo a decarboxylation and form with the cofactor, thiamine pyrophosphate (TPP), an exceedingly active derivative which can be regarded as the metabolically active form of acetaldehyde. The acetyl group is then transferred to CoA, forming a highly reactive combination through a sulfur linkage (Fig. 19-5).

Transfer of an acetyl group (acetylation) is rather common, and that of acetyl-CoA is readily mobile. This complex is an intermediate in a variety of catabolic and synthetic pathways. It also can provide the energy needed to add phosphate (pyrophosphate in this case) and form ATP, as shown in Fig. 19-5. But we are now more interested in the pathway by which the acetyl group is moved into the TCA cycle (Fig. 19-6).

In one complete sequence of the eight reactions contained in this cycle, one acetate equivalent is oxidized to CO_2 and H_2O. Also, the intermediates are regenerated. In the first step (1) the acetyl group is condensed with oxaloacetate (C_4) to form citrate (C_6). Free CoA is regenerated in this reaction. In the next two steps (2), catalyzed by aconitase, the citric acid is reversibly converted into two other tricarboxylic acids, namely *cis*-aconitic and isocitric. Removal and addition of water molecules occur in these conversions. Next, (3) isocitrate dehydrogenase catalyzes the removal of two hydrogen atoms and electrons which reduce NAD^+. A molecule of CO_2 is evolved, the first of two which result from each cycle. (There is evidence that, in systems employing NADP as the coenzyme, oxalosuccinic acid is formed, which is then decarboxylated.) The next reaction (4) is another decarboxylation, but quite different from the preceding one. Note that the resulting succinyl (C_4) is combined with CoA. This step is catalyzed by an enzyme complex, α-ketoglutarate dehydrogenase complex, containing FAD, NAD, lipoic acid, TPP, a decarboxylase, reductase-transacetylase, and dehydrogenase. Succinyl CoA, in contrast to acetyl-CoA, has limited reactivity. Mostly it continues in the cycle, the energy of the thioester bond forming a nucleoside phosphate in the next reaction (5). CoA is regenerated, and succinic acid is formed. The next reaction (6) is a dehydrogenation, and the electron acceptor for suc-

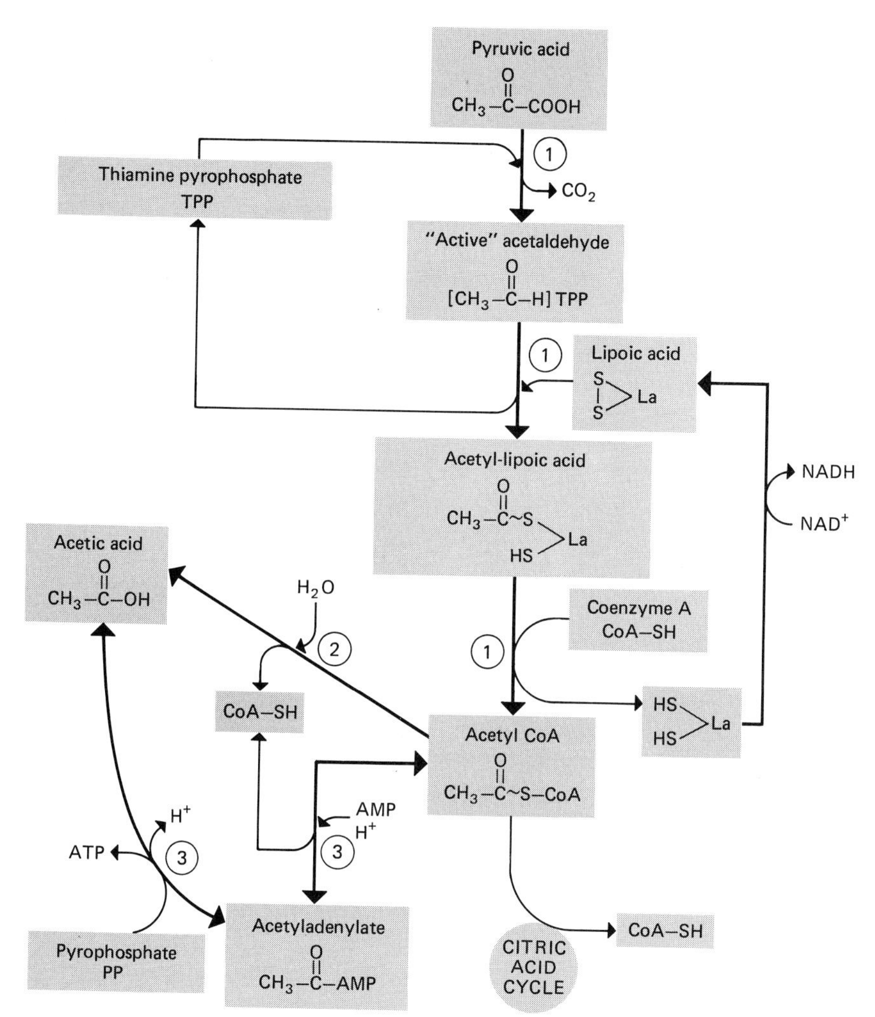

FIGURE 19-5
Metabolism of pyruvic acid to acetyl-CoA and acetic acid. Enzymes: 1, pyruvic dehydrogenase; 2, acetyl-CoA deacylase; 3, acetic thiokinase.

cinic dehydrogenase is FAD. The next two steps (7 and 8), involving first a hydration and then a reduction, complete one turn of the cycle.

Now, one might well ask what has been accomplished. One C_2 (as acetyl-CoA) and one C_4 (oxaloacetate) have been fed into the cycle, and two molecules of CO_2 and a molecule of the C_4 acid again have appeared. Most important of all, in four of the reactions, highly energetic electrons are transferred to coenzymes. These are a potential source of much energy.

19-6 OXIDATION AND ELECTRON TRANSPORT

The several electrons accepted by the coenzymes for the oxidations we have considered so far are transferred to oxygen through a terminal sequence of "carriers" that are collectively referred to as the *respiratory*

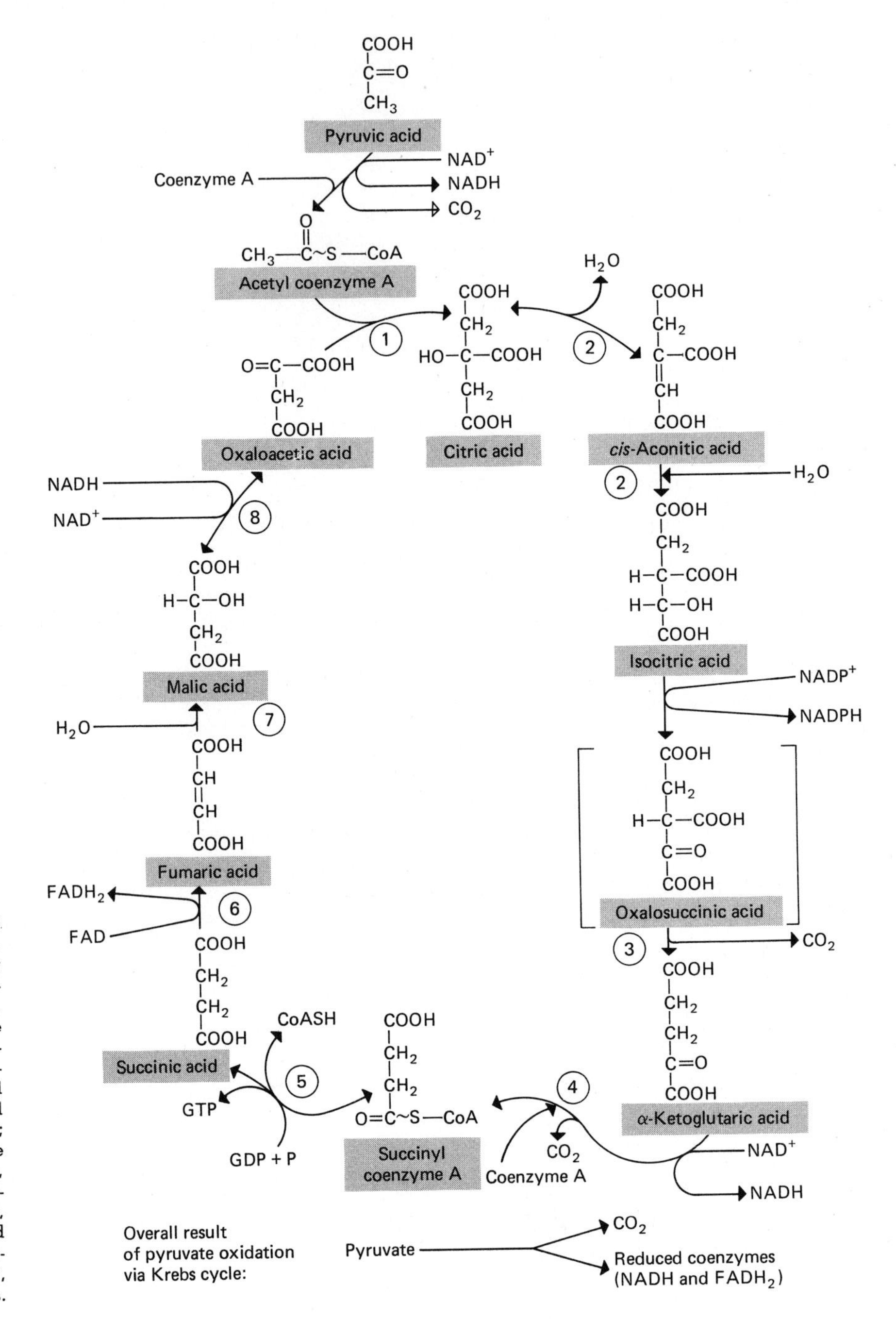

FIGURE 19-6
The tricarboxylic acid cycle. Enzymes: 1, condensing enzyme; 2, aconitase (inhibited by fluorocitrate); 3, isocitrate dehydrogenase and Mg^{2+} or Mn^{2+}; 4, α-ketoglutarate dehydrogenase (complex) (inhibited by arsenite); 5, succinyl CoA synthetase and Mg^{2+}; 6, succinic dehydrogenase (inhibited by malonate); 7, fumarase; 8, malate dehydrogenase. Oxalosuccinate, if formed, is enzyme-bound and therefore not an intermediate. GDP and GTP, guanosine nucleotides.

chain or *electron-transport sequence* or *system* (ETS). The carriers are proteins of four kinds, each associated with a nonprotein functional group:

Protein	Functional group(s)
Flavoproteins	Flavins
Cytochromes	Heme
Nonheme iron proteins	Iron linked to sulfur
Copper proteins	Copper

These complexes are successively reduced and oxidized as electrons move along them (Fig. 19-7).

Reversibility of the oxidation-reduction states of cytochromes can be dramatically demonstrated with light-absorption methods. This fact was discovered in 1925 by Keilin, who, in some classical studies of respiration of insect wing muscle, vertebrate heart muscle, and yeast cells, observed the ubiquity of the respiratory pigment which he named *cytochrome*. Using a spectrometric device fitted to a microscope, he found that the reduced pigment had four characteristic absorption bands, whereas when the pigment was oxidized there were no distinct absorption bands. By 1930, Keilin had concluded that there were three distinct kinds of cytochromes in his preparations. He called them *a*, *b*, and *c*, in the order of their absorption maxima, *a* absorbing the longest and *c* the shortest wavelengths (Fig. 19-8).

Keilin also postulated the existence of another carrier which linked the cytochromes to oxygen, and which he therefore called *cytochrome*

FIGURE 19-7
Flow pattern of hydrogens and electrons in biological oxidation. There are many components in the electron-transport chain, but only NAD, flavoprotein (FP), coenzyme Q (Q), and cytochromes (*b*, *c*, *a*, a_3) are shown. (*After A. L. Lehninger, Bioenergetics, 2d, ed., W. A. Benjamin, Inc., Menlo Park, Calif., p. 74.*)

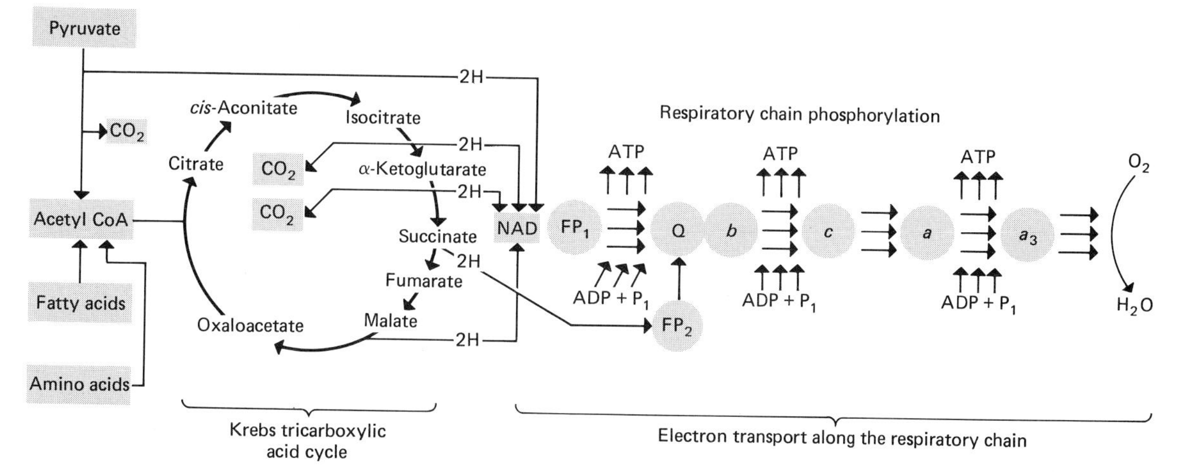

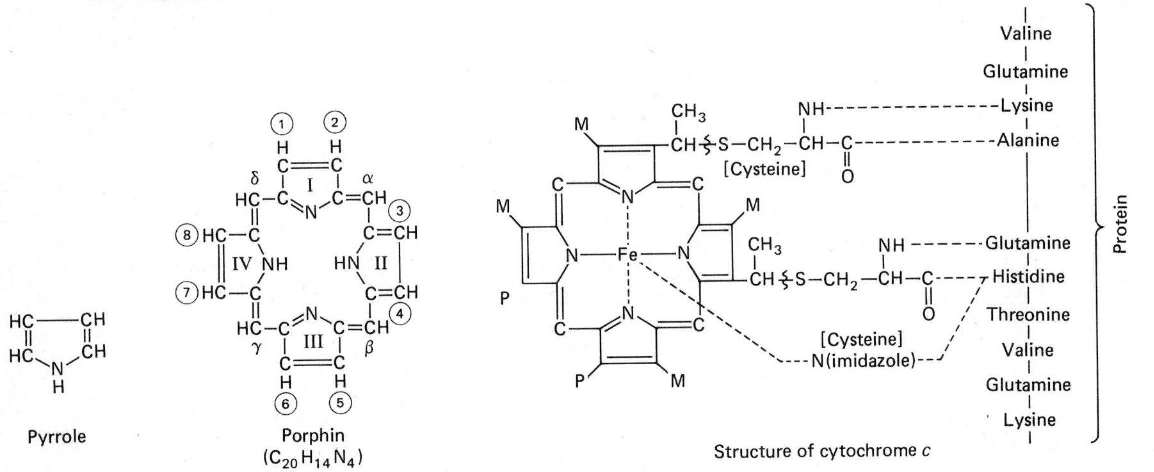

FIGURE 19-8
(a) The pyrrole ring. (b) Porphin, a tetrapyrrole. The position of the double bonds can be only arbitrarily fixed in this kind of structure. The conventional numbering of rings and positions is indicated. (c) Cytochrome c. It contains one iron atom per molecule. The linkage of the porphyrin to protein is shown. M = methyl group (—CH_3); P = propionic group (—CH_2—CH_2—COOH).

oxidase. This substance was shown to be CO-, KCN-, and H_2S-sensitive. It was subsequently found that this enzyme, containing two heme groups and two copper atoms, is an aggregate of cytochrome *a* and ferrocytochrome a_3. The latter is thought to bind oxygen.

When a coenzyme accepts electrons, the substrate acts as an electron donor and is, therefore, a *reducing* agent. The coenzyme acts as an acceptor and is an *oxidizing* agent. In the case of the pyridine nucleotides, a hydride ion ($H:^-$), removed by a dehydrogenase reaction, is acquired. The oxidations in which NAD and NADP participate involve removal of two electrons and two protons from the substrate; both electrons, but only one proton, are transferred to the coenzyme to effect reduction. The proton and electron of the unincorporated hydrogen are separated during transfer. Thus, the accepted hydrogen is equivalent to a hydride ion: a hydrogen with two electrons (H^-). The oxidized forms of the coenzymes will be represented as NAD^+ and $NADP^+$, the reduced forms as NADH and NADPH. Some texts use NADH + H^+ and NADPH + H^+ for the latter, and others write $NADH_2$ and $NADPH_2$. Showing the free hydrogen ion is appropriate, but it is not bound to the coenzyme. The use of $NADH_2$ and $NADPH_2$ is obviously misleading.

The addition of the hydride to NAD^+ causes a shift of electrons within the pyrimidine ring, resulting in an additional electron becoming associated with—but not bonded to—the nitrogen. The shifts of electrons around the ring are associated with profound changes in the shape of the molecule, which appear to favor (in a mechanical way) the passage of electrons through the coenzyme. The nonbound, or unshared, electrons freely move back into the ring when the coenzyme is oxidized (Fig. 19-9).

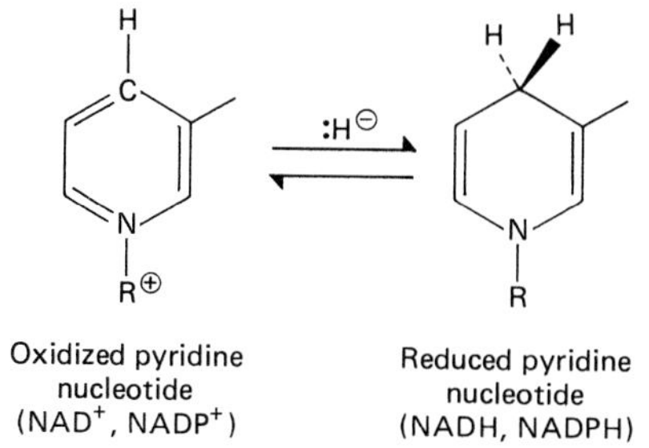

FIGURE 19-9
The ring of pyridine nucleotide coenzymes in the oxidized and reduced states. Reduction (hydrogenation) results in a tetrahedral C-4 atom and a redistribution of electrons, so that a pair of unshared electrons associates with the nitrogen. Upon being oxidized the electrons are redistributed in the ring as indicated by the bond positions.

The reactions catalyzed by the pyridine nucleotides can be readily followed by spectrophotometric methods. The reduced forms absorb at 340 mμ and have a strong fluorescence when excited by light at 440 mμ wavelength.

The flavin coenzymes, FAD and FMN, are tightly bound to protein and thus act as prosthetic groups (hence the term "flavoprotein"), rather than as freely dissociating coenzymes like the pyridine nucleotides. The flavins have characteristic absorption bands in the oxidized state, which diminish with reduction. They also have a strong yellow-green fluorescence. Reduction of flavin enzymes occurs as two sequential one-electron transfers. Thus reduced FAD can be represented as FADH and $FADH_2$, depending upon whether one or two electrons are accepted (Fig. 19-10).

All electron donors and acceptors have a characteristic affinity for electrons. Recall that standard reduction potentials for many biological entities are known (Table 17-3) and that one can predict whether a compound will have a tendency to donate or accept electrons from another compound. An *electron pressure* tends to drive electrons from the more negative to the less negative compound. When the coenzymes of oxidation and the ETS components are arranged on a scale of reduction potential, one can predict the free-energy change of each of the transfers and the total energy change for the entire system (Fig. 19-11).

The change in free energy at each electron transfer is directly re-

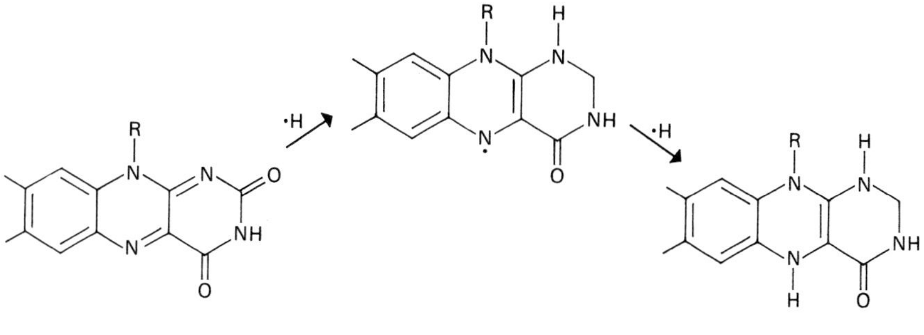

FIGURE 19-10
The two-step acquisition of electrons in the reduction of flavins. The intermediate free radical is referred to as a *semiquinone*. Note that an accepted electron associates with a nitrogen, as in NAD, but also hydrogen bonds to nitrogen, unlike the situation in NAD.

lated to the magnitude of the change in potential, the value of which can be calculated according to an equation similar to Eq. 17-13, viz.:

$$\Delta G^{\circ\prime} = -nF\ \Delta E$$

where ΔE = the difference in the standard reduction potential of the reacting carriers.

In Fig. 19-11 it can be seen that there are three large free-energy decrements alternating with two smaller ones. But for now let us consider only the overall change which occurs when a pair of electrons is transported all the way from NAD to oxygen, the most electropositive acceptor. As indicated in the figure, the total change is −52,000 cal/mol. Now, in the complete oxidation of one glucose to CO_2 and H_2O, 12 pairs of electrons pass through the ETS. Thus, 12 × −52,000, or −624,000 cal/mol of glucose, are made available in this process. Since the $\Delta G^{\circ\prime}$ of combustion of glucose is −686,000, it is obvious that almost all the free energy of glucose oxidation is made available during the electron-transport phase.

Careful stoichiometric measurements indicate that when a pair of electrons passes down the energy gradient of the ETS, an average of three ATPs are formed from ADP and P_i. Since the free energy of hydrolysis of ATP is about 7000 cal/mol, then 3 × 7000, or about 21,000 cal, are captured as ATPs. Thus, 3 moles of ADP conserves approximately 40 percent (3 × 7000/52,000) of the energy released when a pair of electrons is transported by this system. Transporting of 12 pairs should permit conservation of at least 252,000 cal of the 624,000 cal made available by the electrons emanating from glucose.

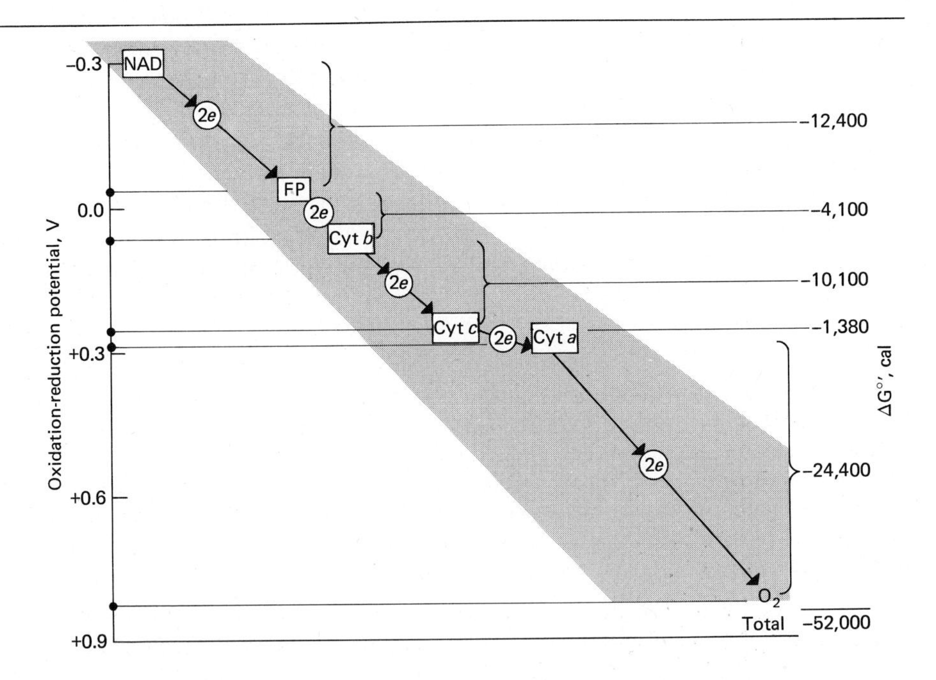

FIGURE 19-11
Some of the electron carriers arranged on a thermodynamic scale of "electron pressure." The free-energy changes occurring as electrons pass from one carrier to another are indicated.

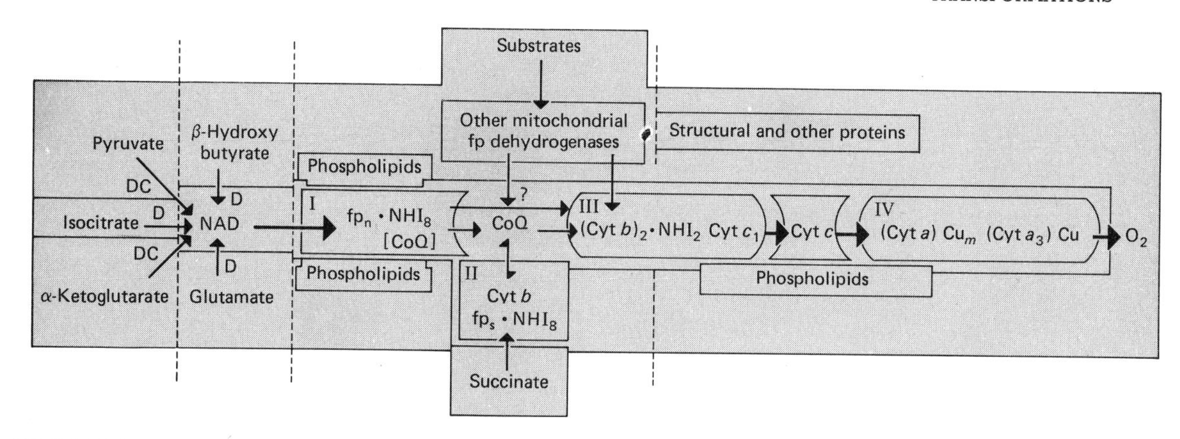

FIGURE 19-12
An electron-transport system of mitochondria from beef heart. The system, linked to the outer mitochondrial membranes, can be stripped away and the intramitochondrial content liberated; electron-transport particles are obtained which are capable of catalyzing the key reactions NADH→O_2 and succinate→O_2, and sometimes also of retaining the capacity for oxidative phosphorylation. By further comminutive procedures there can then be obtained the unit of electron transport, consisting of four complexes, I through IV. Complex I: NADH-CoQ (oxido) reductase; complex II: succinate-CoQ (oxido) reductase; complex III: $CoQH_2$-cytochrome c (oxido) reductase; complex IV: cytochrome oxidase. *D,* dehydrogenase (NAD-linked); *DC,* dehydrogenase complex (NAD-linked); *fp,* flavoprotein; *fp$_n$,* NADH dehydrogenase; *fp$_s$,* succinate dehydrogenase; *NHI,* nonheme iron. (*After H. R. Mahler and E. H. Cordes,* 1968, *Basic Biological Chemistry, Harper & Row, Publishers, Incorporated, New York,* p. 363.)

As noted earlier in this chapter, the free-energy change that accompanies the hydrolysis of ATP intracellularly is probably considerably greater than 7000 cal/mol. If this is so, the above estimate of energy conserved is too low. The chief difficulty in determining the most accurate value for $\Delta G'$s is that the hydrolysis of ATP is quite sensitive to changes in temperature, pH, and Mg^{2+} concentration. In the living cell, where the environment does change and the concentrations are unlikely to be those that apply to standard conditions, it is likely that a value of −11,000 or more may be more realistic. Whatever the value, the important point is that ATP is a very energetic substance.

Several details concerning the transport of electrons will be omitted here. Many things are unknown about the ETS, and only a few of the better-known facts have been presented. Before leaving this system, however, attention needs to be focused on a few more points, and this can be done best by reference to a diagram (Fig. 19-12).

1. All the reactions we have examined are by necessity shown as two-dimensional constructs. The enzymes, cocatalysts, and supporting elements of a system are commonly organized into complexes or sets, having specific three-dimensional configurations. This is well established for the components of the TCA cycle and ETS. As shown in Fig. 19-12, the highly organized sets of enzymes are assembled as complexes, which in turn have a precise order in the membranes of mitochondria (as will be discussed in Chap. 20).

2. Some 80 molecules are believed to constitute one complete electron-transfer chain, but not all of them participate in the transfers. Perhaps about 15

kinds of protein are directly engaged in the process. Each is associated with some oxidation-reduction prosthetic group such as flavin, heme, nonheme iron, or copper.

3. Whereas the electrons from the TCA pool are mainly transferred through pyridine nucleotides to flavoprotein (complex I), note that those coming from succinate enter the carrier system by way of another flavoprotein (complex II).

4. NAD, coenzyme Q (also called ubiquinone), and cytochrome c appear to serve as mobile carriers, shuttling electrons into or between transport complexes.

5. As might be expected, the various carrier proteins appear to be arranged in an order of decreasing negative potential (from NAD to cytochrome a_3, in beef-heart mitochondria).

6. Cytochromes transport electrons singly, but flavoprotein and CoQ transport electrons as pairs. Oxygen accepts two pairs of electrons in forming H_2O.

7. Mitochondria contain a relatively large amount of lipid, more than 90 percent of which is phospholipid. Its function appears to be to stabilize the active conformations of the various proteins and permit their interaction with other essential components, such as the mobile CoQ.

8. Most mitochondria contain vitamin K and other related naphthoquinones, as well as vitamin E or related compounds. The functions of these molecules remain obscure in most cases.

9. The cytochrome system is the predominant terminal pathway in aerobic organisms, probably the only one in free-living animals, but other systems are known which also use oxygen as an electron acceptor. Two possible alternative pathways, one involving phenol oxidase and the other ascorbic acid oxidase (both copper-containing enzymes), are particularly common in plants.

19-7 OXIDATIVE PHOSPHORYLATION

It will be recalled that there are three places in the ETS where there is a relatively large free-energy change (Fig. 19-11; see also Fig. 19-7). It is at these positions in the transport chain that high-energy intermediates are generated during electron transfer. These, in turn, can donate high-energy phosphoryl groups to ADP.

The true nature of the coupling reactions of the respiratory sequence is not known, but it appears that energy is delivered in packets of appropriate size for forming phosphate bonds as the transferred electrons are lowered in potential through a series of small steps.

Perhaps significantly, phosphorylation appears to occur only at sites in the ETS where there is a mobile carrier: NAD associated with complex I, CoQ between complexes I, II, and III, and cytochrome c between complexes III and IV. The mobile components may provide the phosphoryl group, joined by a high-energy bond, which can be trans-

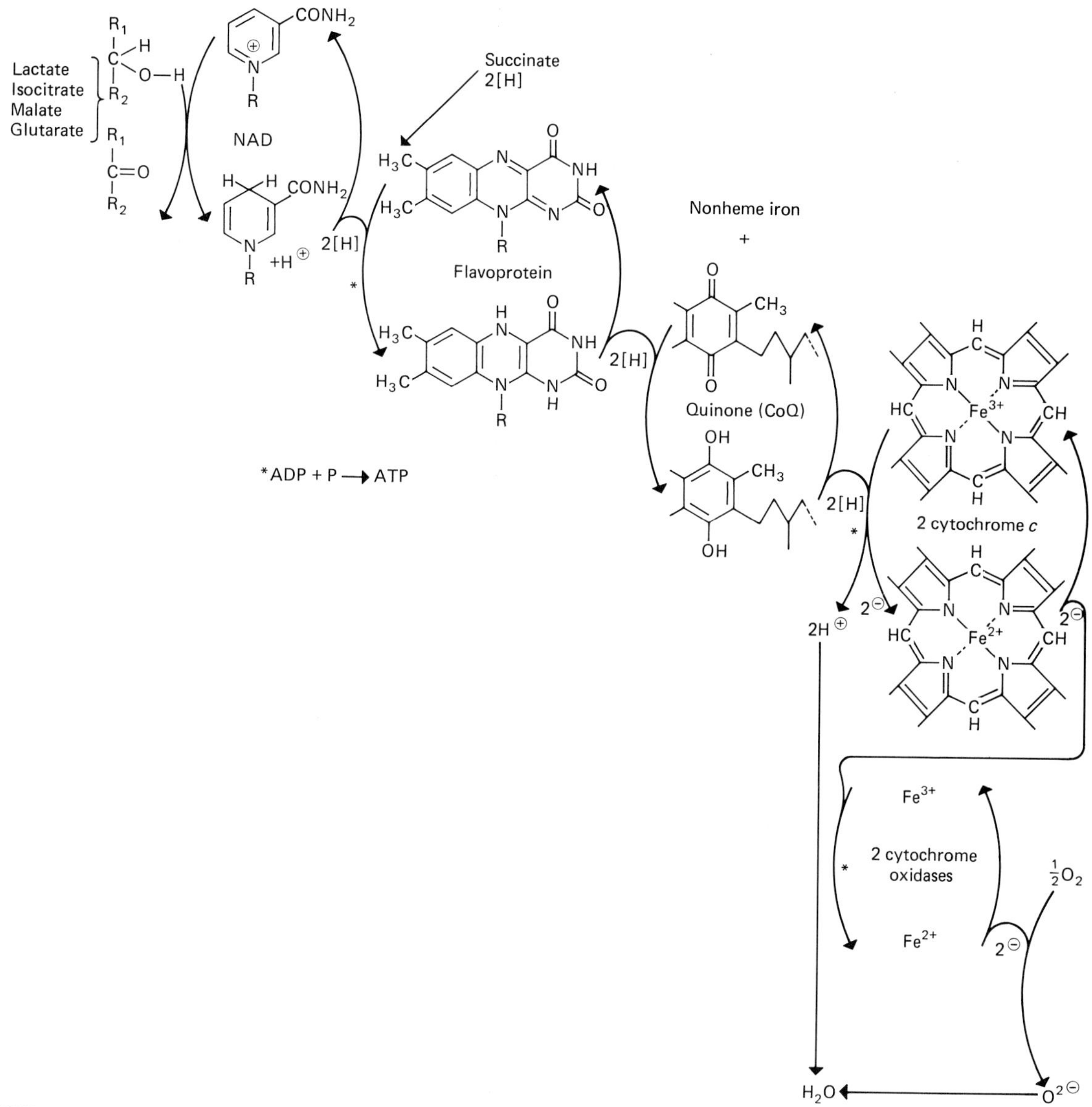

FIGURE 19-13
Simplified diagram of oxidative phosphorylation. Suggested sites of phosphorylation are shown by asterisks. Note that two cytochrome *c* are required to pass two electrons, since each transports only one at a time, and that H^+ is not accepted.

ferred to ADP with the formation of ATP (Fig. 19-13).

The efficiency of the coupling process is experimentally determined by measuring the P/O ratio; i.e., the number of phosphate bonds that are generated when one oxygen accepts a pair of electrons. A generally accepted ratio is three. This is an average value because the transported electrons are not energetically uniform. Those transported from succinate support two phosphorylations, whereas those from α-ketoglutarate provide four, and those of isocitrate and malonate three each. Thus, the *average* is three phosphorylations per pair of electrons transported through the entire sequence. However, some investigators have found conditions under which ratios approaching six can be measured. The efficiency of ATP formation, therefore, may be higher than is generally indicated.

Respiratory inhibitors, especially cyanide, carbon monoxide, and certain narcotics, have been especially useful in advancing our understanding of the respiratory chain. Some of them can block the chain at specific points, thereby revealing the sequence of components and their mode of interaction. For example, HCN and CO specifically inhibit electron transfer by cytochrome c, and barbiturates block transfers between flavin and CoQ.

A standard method for preventing phosphorylation, while permitting electron transport to continue, is to use an "uncoupling" agent, 2,4-dinitrophenol being the classical one. By use of such agents it has been shown that phosphorylation is almost exclusively due to the electron-transfer process in mitochondria, rather than to substrate-level oxidations.

The overall process of cellular respiration is commonly indicated to yield a net gain of 38 mol ATP per mole of glucose oxidized, the balance sheet being as follows:

$$1\ \text{Glucose unit} \xrightarrow{\text{glycolysis}} 2\ \text{pyruvate units} + 2\ \text{ATP} + 4\text{H}$$

$$2\ \text{Pyruvate units} \xrightarrow{\text{oxidation}} 2\ \text{acetyl-CoA units} + 2CO_2 + 4\text{H}$$

$$2\ \text{Acetyl-CoA units} \xrightarrow{\text{TCA cycle}} 4CO_2 + 16\text{H}$$

$$24\text{H} \xrightarrow{\text{oxidative phosphorylation}} 36\ \text{ATP}$$

$$\text{Sum: Glucose} + 6O_2 \longrightarrow 6CO_2 + 6H_2O + 38\ \text{ATP}$$

The efficiency of the process in energy conservation can be calculated thus:

$$\frac{38 \times 7{,}000}{686{,}000}\,100 = 39\%$$

When compared to man-made machines, the efficiency of the respiratory system is quite superior. However, this is only an estimate based on "standard" conditions; in the cell, as already noted, the free-energy changes may be influenced by local pH conditions, salt concentrations, and compartmentation of enzymes, so that equilibria of reactions may be quite different than they are in laboratory experiments. Therefore, the efficiency actually may be considerably higher—perhaps as high as 60 percent in the living system.

19-8 AUTOTROPHY

Life cannot exist solely by "dissecting" molecules; i.e., living forms cannot be only a collection of energy "consumers." In order that the supply not be exhausted, some organisms must assemble organic molecules. Those organisms which can produce a *net* synthesis of organic molecules from inorganic precursors provide a source of energy for all the remainder of the biomass. How these "producers" gather to themselves exogenous energy and use it to power the endergonic reactions that form organic molecules will be considered briefly. Although this book is about the functions of animals, it behooves us to gain some appreciation of the energetic strategies of those nonanimals that make high-energy compounds available to animals. That life is very largely "a cyclic procession of matter driven by sunlight" becomes much better comprehended when one has the larger view of bioenergetics.

Organisms which are capable of using carbon dioxide as their sole source of carbon are called **autotrophs,** meaning literally "self-feeders." Before it can be used in the synthesis of organic molecules carbon dioxide must be reduced, and this is a strongly endergonic reaction. An organism which uses light as the required energy is called a **photoautotroph.** If the energy is obtained from the oxidation of an oxidizable compound or element the organism is called a **chemoautotroph.** The autotrophs serve as the primary energy sources for all other organisms, known collectively as **heterotrophs** (Fig. 5-3).

All known chemoautotrophs are bacteria. In terms of their contribution to bioenergetics, considered on a world scale, the chemoautotrophic bacteria are not a highly significant group. However, they challenge our conventional ideas about the means by which living forms secure energy. This small, aberrant group of microorganisms has coupled their metabolic machinery to the energy releases of "inorganic" reactions. In its relentless search for potential sources of energy, life seems to have left no stones unturned.

Because the unique feature of chemoautotrophic bacteria is their ability to synthesize their substances utilizing carbon dioxide as their only source of carbon and the energy derived from simple oxidations, it has become customary to refer to them as **chemosynthetic bacteria.** Different species of bacteria have been found which can utilize the energy derived from the oxidation of sulfur, ammonia, nitrite, hydrogen, carbon monoxide, methane, and ferrous carbonate.

There are many things not known about the energetics of chemoautotrophs, but the fundamental unknown is the manner by which the energy liberated in the primary oxidations is made available for biosynthesis. Specifically, the problem is how the energy of oxidation is used in the reduction of carbon dioxide. It is assumed that hydrogen is transported through coenzymes in a manner similar to the process in other tissues. If this is true, chemosynthetic bacteria must have mechanisms for using the free energy released by oxidations to form reduced carrier and ATP, but these processes have not been demonstrated.

By using labeled carbon dioxide ($^{14}CO_2$) in investigations of chemosynthesis it has been found that the first stable product of chemosynthesis is 3-phosphoglyceric acid. The same substance is also the first stable product of photosynthesis and, as in photosynthesis, it is thought that phosphoglycerate is formed in a reaction between CO_2 and ribulose diphosphate.

It is most interesting that a number of heterotrophic organisms are known to oxidize inorganic substances, but only the chemoautotrophs seem to have devised mechanisms for coupling the released energy to their energy-requiring metabolic processes. Their efficiency in energy utilization compares favorably with that of green plants in the field using light. Rabinowitch[1] has commented that "if chemoautotrophic organisms did not succeed in spreading over the whole surface of the earth, as did the green plant, it was not for lack of efficiency, but merely because chemical energy is available only in a few non-equilibrated spots while sunlight flows abundantly everywhere."

19-9 PHOTOSYNTHESIS

Except for the chemoautotrophic bacteria, photosynthesis is the fundamental process upon which all life on earth depends. Energy "harvested" from the sun supports the biomass, for it is electromagnetic energy, within a segment of the spectrum called *light*, that is absorbed by pigments and transduced into the chemical energy which drives bioenergetics. Only a very small portion of the continuous stream of energy that flows from the sun is captured by *green* plants and funneled into the biomass.

Photosynthesis is the photochemical reduction of carbon dioxide with the concomitant oxidation (or dehydrogenation) of water and evolution of gaseous oxygen. Water serves as the reducing agent, and carbohydrates are the primary products. The process is strongly endothermic, and the net gain in energy is provided by light absorbed by pigments. The overall equation is usually written as:

$$CO_2 + H_2O \xrightarrow{h\nu} (CH_2O) + O_2$$

This equation is a great oversimplification of what actually occurs, for not only are carbohydrate and oxygen formed, but amino acids, proteins, fats, and a variety of other compounds result as well. Also it fails even to suggest that photochemical reactions occur which provide energy-rich compounds to plants without the necessity of reducing carbon dioxide. Photosynthesis, therefore, is not one simple chain of sequential reactions but is, in fact, a multitude of reactions.

[1] E. I. Rabinowitch, 1945, *Photosynthesis and Related Processes*, vol. 1, Interscience Publishers, Inc., New York.

The simplified equation also does not reveal that there are some photosynthetic organisms which do not produce oxygen. For example, a group of pigmented photosynthetic bacteria, called *purple sulfur bacteria,* use H_2S, instead of H_2O, as a reductant, and they produce elementary sulfur as a by-product. Other kinds of photosynthetic bacteria use hydrogen gas as the hydrogen donor for photoreduction. Thus, it is immediately apparent that photosynthesis involves several variations of the basic process.

Photosynthetic Pigments

It has been known for about 70 years that photosynthesis takes place within aggregations of certain pigmented molecules, principally the tetrapyrrolic **chlorophylls.** In higher plants and some of the algae these and other pigments are incorporated into membrane configurations known as **chloroplasts.** Although many pigments have been found associated with the photosynthetic apparatus of the various green plants, only two of them seem to be indispensable: chlorophyll *a* in higher green plants, and *bacteriochlorophyll a* in photosynthetic bacteria. In addition to chlorophyll *a*, there also is found chlorophyll *b* in all higher green plants and the green algae. Chlorophylls *a* and *c* are found in brown algae and diatoms, and chlorophylls *a* and *d* in red algae. In addition to bacteriochlorophyll, the green bacteria also have a second related pigment called *bacterioviridin.*

Another group of photosynthetic pigments are called **carotenoids.** They are $C_{40}H_{56}$ aliphatic polyene chains. Each taxonomic group of photosynthetic plants has its own characteristic carotenoids. They absorb light in the blue and green spectral regions. The best-known ones are the orange-yellow *carotenes.* (The precursor of vitamin A is β-carotene.) The yellow carotenoid alcohols, *xanthophylls,* belong also to this group.

In certain algae the **phycobilins,** a group of water-soluble proteins with a tetrapyrrolic prosthetic group, are found. *Phycoerythrin,* of most red and some blue-green algae, and *phycocyanin,* of most blue-green and some red algae, are the best characterized of these. The prosthetic groups are structurally similar to animal bile pigments.

The Photosynthetic Unit

Much evidence indicates that many chlorophyll and other pigment molecules cooperate to collect the energy of light quanta and deliver it to a smaller number of chemical reaction sites. When light excites the pigment molecules, there is resonance transfer of excitation to neighbor molecules and eventually to a final "sink" or "reaction center." At this point an electron of an acceptor molecule is excited to move into the transport system (Fig. 19-14).

The size of the aggregate associated with one reaction center can be computed from results obtained by an intermittent-illumination technique. If the photochemical event at the reaction center is driven by the energy of one quantum, the unit of the chloroplasts of higher plants is calculated to contain about 300 chlorophyll molecules. In the purple photosynthetic bacteria the unit is thought to contain approximately 50 molecules of bacteriochlorophyll. Electron micrographs have shown structural entities in chloroplast lamellae which may be the morphological counterpart of these functional units (Fig. 20-16).

What determines that the absorbed energy will move to the reaction center, and how is the energy trapped there? The answer appears to be that in the chloroplast lamellae the pigments are densely packed and arranged in arrays around a reaction center. Pigment molecules in high density will allow *energy migration*. That is, electronic excitation energy, generated upon absorption of light by one pigment molecule, can hop to a neighboring pigment molecule, and then another, etc., until it is dissipated as heat, is reemitted as light (fluorescence), or does chemical work.

Excitation may be transferred with good efficiency to pigment molecules at distances up to 5 nm. By a random-walk process excitation moves from molecule to molecule until it arrives at the reaction center, where it is trapped. The migration occurs during a state of electronic excitation which persists for no longer than 10^{-9} second, but several hundred pigment molecules may be traversed. The overall direction of travel is from pigments with a shorter-wavelength absorption band to ones with a longer band, but whose energy levels overlap. Excitation energy can be transferred to molecules with electrons capable of being elevated to the same or lower (but not higher) energy states. The wavelengths absorbed indicate the permissible energy states of the absorbing molecule. Shorter wavelengths have higher energy photons than longer wavelengths. For example, energy absorbed by chlorophyll *b* can be transferred to chlorophyll *a*, as would be predicted from the relative positions of their major absorption peaks (about 430 and 660 nm for *a* versus 480 and 760 nm for *b*).

Absorption spectrophotometry of whole green algae reveals dif-

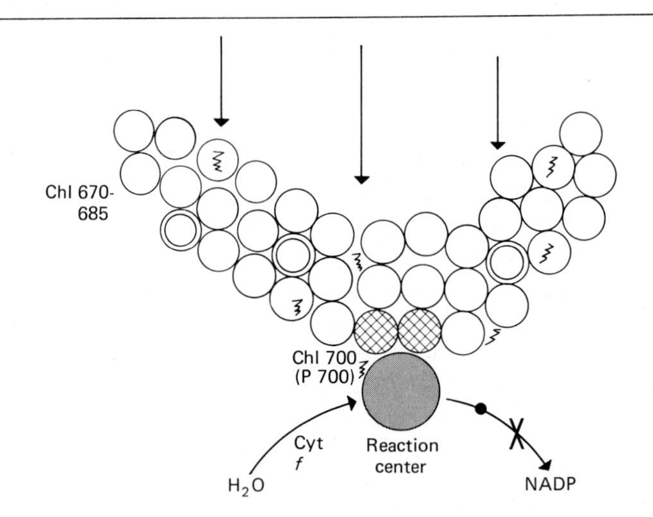

FIGURE 19-14
A scheme for the funneling of energy by several chlorophyll (Chl 670–685) and other pigments (double circles) of plastids into a few sinks or traps [cross-hatched circles (Chl 700 or P 700)] which are involved in energy conversion necessary for electron transfer from H_2O to NADP. The dark sphere below the pigment molecules represents a "reaction center" where electron transfers from H_2O through NADP to organic compounds occur. X is an unknown substance that is alternately reduced and oxidized. cyt *f*, cytochrome *f*.

ferent absorption bands that are considered to be forms of chlorophyll *a*. Though in only one form when isolated from chloroplasts, chlorophyll *a* seems to occur in different forms in situ. It is thought that the pigment molecules may be bound to the lipoprotein lamellae of chloroplasts in different ways which establish slightly different conformations, thus accounting for the variants revealed by absorption bands.

One form of chlorophyll *a* is detected as a minor band positioned at about 700 nm and appears to be a small amount of a special form. Indeed, there may be only one of these molecules, which is called P700, per photosynthetic unit. Its very long wavelength absorption indicates a relatively low excitation state, and it is considered to be the final trap for migrating excitation (Fig. 19-14). (In other organisms the trapping pigment may be different. For example, in purple bacteria the analogous pigment is P890.)

Electron Transport and the Two-pigment System

"Far-red" light (λ greater than 685 nm) is curiously inefficient in promoting photosynthesis even though it is absorbed by chlorophyll *a*. The low efficiency can be improved by superimposing light of shorter wavelength. That is, there is an "enhancement effect," so that the rate of photosynthesis with the combined long and short wavelengths is greater than the sum of the rates produced by the two wavelengths separately. The discovery of this **Emerson effect**, and related observations (including the identification and characterization of cytochromes in green tissues), provided the basis for construction of a model by Hill and Bendall in 1960 to explain photosynthetic transport of electrons (Fig. 19-15).

According to the Hill-Bendall model, system I, considered to be mainly one form of chlorophyll *a*, absorbs far-red light. System II absorbs shorter wavelengths and appears to be a second form of chlorophyll *a* plus accessory pigments. The energy of the absorbed photons is used to transport electrons against a thermochemical gradient. Supposedly, the two photoacts function in tandem, partially and collaboratively, to favor electron flow. The systems can function independently but are more efficient when acting cooperatively. That is, there is "enhancement" when both pigment systems are absorbing photons. The exact pigment composition of the two systems varies with species.

System II is assumed to transport electrons from a potential of about +0.8 to 0 V, and system I spans from about +0.4 to −0.4 V. The net change in potential is therefore about 1.2 V. It is apparent then that the two systems have overlapping potentials. Electron flow in the region of overlap will result in a release of energy, as it is *with* the potential gradient. Some of this energy is thought to be trapped as ATP.

Several oxidation-reduction carriers (cytochromes and quinones) have been identified in the photosynthetic apparatus, and some of them are thought to transport electrons in the pathway which links the two photosystems. These are placed in the scheme of Fig. 19-15 according to their

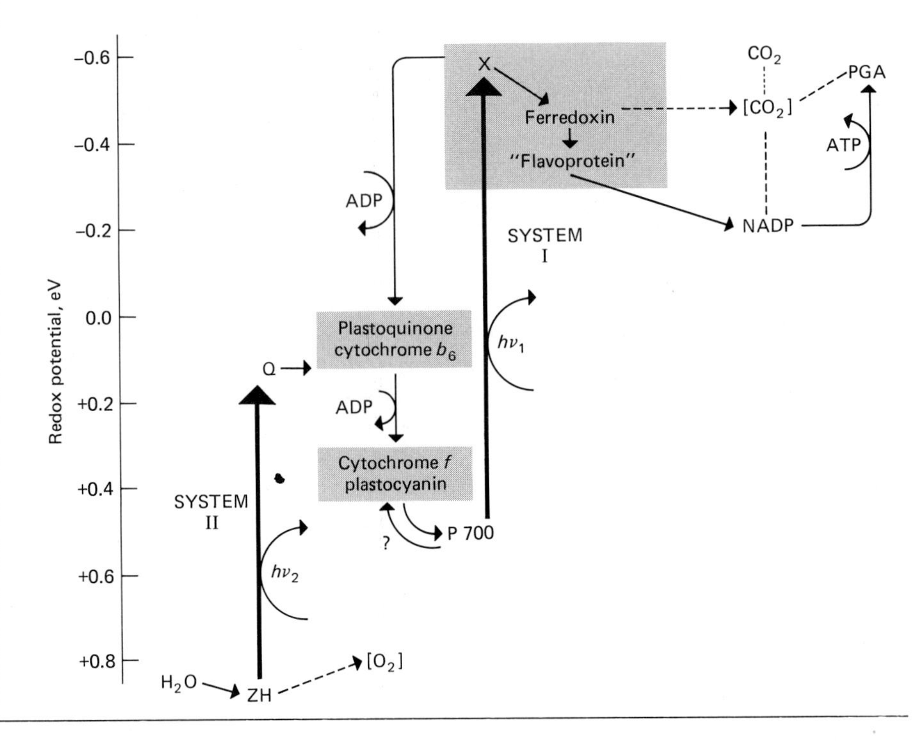

FIGURE 19-15
Scheme for photosynthesis in terms of two light reactions, as proposed largely by Hill and Bendall. The direction of hydrogen or electron transport of oxygen-evolving photosynthetic organisms is indicated by the direction of the arrows. The redox substances surrounded by a rectangle may be partly bypassed. The two light reactions are represented by the large arrows. "Flavoprotein" denotes ferredoxin-NADP reductase. *PGA*, 3-phosphoglyceric acid. (See text for further explanation.)

characteristic potentials. For example, plastoquinone and cytochrome b_6 have potentials near 0 V and are thought to be associated with systems II, whereas plastocyanin (a copper protein) and cytochrome f are close to 0.4 V and probably are a part of system I. These carriers are oxidized when system I absorbs light and reduced when light is absorbed by system II, as is predicted by the Hill-Bendall model.

Now we can reasonably well integrate the scheme for funneling energy, shown in Fig. 19-14, with the scheme for electron transfer shown in Fig. 19-15. The latter is a more detailed conception of what transpires in the "reaction center," wherein reside systems I and II. Not only are P700 molecules at an appropriate low excitation energy, but it appears that they are oxidized when light is absorbed by system I pigments and reduced when light is absorbed by system II pigments. Because of special binding, P700 may be not just a normal chlorophyll but could be considered a photocatalyst. Upon trapping excitation it is oxidized and an unknown substance, called *X*, is reduced (Figs. 19-14 and 19-15). Thus, the postulated role of P700 is to collect energy from the other pigment molecules of the photosynthetic unit of system I, transfer an electron to an acceptor, and recover the electron from a cytochrome (Fig. 19-15).

Following the formation of the unknown primary photoreductant, XH (in algae and higher plants), reduced coenzyme, NADPH, is produced. The formation of XH marks the end of the *light reactions*, and the reduction of NADP is the beginning of the *dark reactions* of photosynthesis —the latter so called because they can occur in the dark after reducing power is built up by the light reactions. Reduction of NADP and the formation of NADPH involves two enzymes. The first is an iron-containing, nonheme reductase called *ferredoxin*. The other is a flavoprotein, *ferredoxin-NADP reductase*. (The latter enzyme is called by other specific names and is designated in Fig. 19-15 simply as "flavoprotein.").

There is some evidence that CO_2 also may be reduced by a more direct pathway involving reductive carboxylation of acetyl-CoA and succinyl-CoA, and which requires reduced ferredoxin but not reduced coenzyme (Fig. 19-15, dashed lines).

The formation of reduced coenzyme is the prelude to the fixation and reduction of CO_2, as discussed below. But before considering the dark reactions let us look at the primary reactions in system II of the two-system, oxygen-evolving plants. Pigments in system II presumably transfer their energy to a reaction center in a manner analogous to those of system I. A pigment which absorbs at 680 nm may serve as the energy trap for this system. A postulated substance, ZH, is oxidized to Z. Several molecules of Z cooperate to dehydrogenate water and become reduced to ZH. The reduction of Z and the concomitant production of oxygen are complicated and poorly defined dark reactions involving at least two water molecules and the transport of four electrons (Fig. 19-15).

Reduction and Fixation of Carbon Dioxide

Carbon dioxide is incorporated by an elaborate cyclic process, the **Calvin cycle,** or the **carbon reduction cycle,** involving as one of its steps the acceptance of CO_2 by the five-carbon sugar *ribulose-1,5-diphosphate*. The product of this reaction is two molecules of the three-carbon 3-phosphoglyceric acid. The latter step is independent of light and the products of the light reaction (ATP and NADPH), but other parts of the Calvin cycle require products of the light reaction (Fig. 19-16).

FIGURE 19-16
Diagram of the reductive carbohydrate cycle in chloroplasts. The cycle consists of three phases. In the carboxylative phase (I) ribulose-5-phosphate (Ru-5-P) is phosphorylated to ribulose diphosphate (RuDP), which then accepts a molecule of CO_2 and is cleaved to two molecules of phosphoglyceric acid (PGA). In the reductive phase (II) PGA is reduced and converted to hexose phosphate. In the regenerative phase (III) hexose phosphate is converted into storage carbohydrates (starch) and into the pentose monophosphate needed for the carboxylative phase. All the reactions of the cycle occur in the dark. The reactions of the carboxylative and reductive phases are driven by ATP and $NADPH_2$ formed in the light. Products of the light phase are circled.

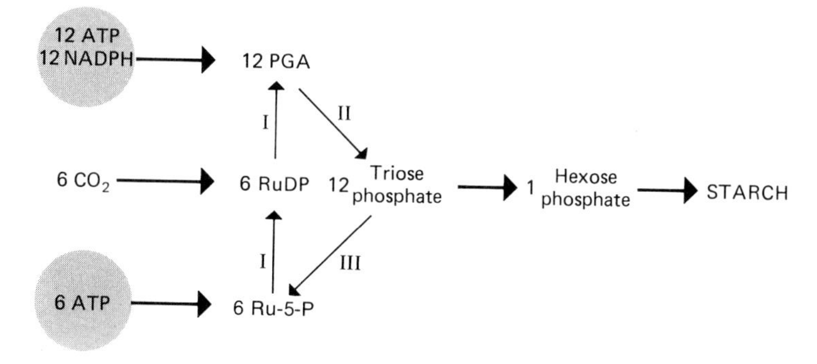

There are some workers in the field of photosynthesis who are critical of the postulated carbon cycle, and it would be misleading to give the impression that all is known about carbon reduction in green plants. However, the cycle is widely accepted as essentially correct.

Photophosphorylation

Two kinds of phosphorylations associated with photosynthetic systems have been distinguished. One is **noncyclic phosphorylation.** It is associated with reactions producing, in addition to ATP, oxidized and reduced products, such as O_2 and NADPH. It is called noncyclic because the electrons expelled from the excited system do not return to it but instead are removed by way of NADP to be used in reduction of CO_2. They are replenished by electrons from water in this "open" type of system. The amount of ATP formed in this way is probably not sufficient for the reduction of CO_2 by the Calvin cycle, and definitely is not enough for supporting the simultaneous synthesis of lipids and proteins. And so it appears that light is used to drive a second kind of phosphorylation called **cyclic phosphorylation.** It is not accompanied by net production of oxidized and reduced products. It is cyclic in that excited electrons move only in a closed transport pathway through several carriers within a pigment system. The carriers are considered to be cofactors, including vitamin K and FMN, and cytochromes.

The site or sites of phosphorylation in the photosynthetic chain are not known with certainty. A probable site is the energy drop between plastoquinone and cytochrome *f* (or plastocyanin) (Fig. 19-15). Phosphorylation here could be of the noncyclic type, as the reactions which lead to O_2 evolution and coenzyme reduction would use this electron pathway. There may be other phosphorylation sites; one of these, presumed to operate during cyclic phosphorylation, is indicated in Fig. 19-15 as between X and plastoquinone. (Electron carriers are not shown in the figure.)

In both cases of photophosphorylation there is absorption of a photon which brings an electron to a high energy level. (In truth, it is the *molecule* which is brought to an excited state.) Capture of the energy upon return of the electron to a lower energy state is a thermochemical reaction that can proceed in the dark. The unique features of cyclic photophosphorylation, in contrast to phosphorylation by mitochondria, are that it is a closed system not requiring an exogenous substrate as the donor of electrons, it does not reduce oxygen, and the pigment system generates both the donor and the acceptor of the electron by using light.

Photosynthesis is but one of many photobiological processes. The fact that it channels light energy into chemical events is not unique, for this occurs also in phototropic and photoperiodic processes, in vision, and other biological phenomena. Photosynthesis is unique, however, in that it alone is the process through which large amounts of energy are captured and fixed as chemical potential. At the center of the energy-trapping system is the remarkable molecule, chlorophyll. As Timiryazev has remarked, "Chlorophyll . . . is the real Prometheus, stealing fire from the heavens."

The play seems out for an almost infinite run.
Don't mind a little thing like the actors fighting.
The only thing I worry about is the sun.
We'll be all right if nothing goes wrong with the lighting.[1]

READINGS

Green, D. E., and H. Baum, 1970, *Energy and the Mitochondrion*, Academic Press, Inc., New York. An interpretation of the mitochondrion as a functional unit and as a prototype of membrane systems.

Gregory, R. P. F., 1971, *Biochemistry of Photosynthesis*, Interscience Publishers, a division of John Wiley & Sons, Inc., New York. A clear and comprehensive summary of photosynthesis.

Lehninger, A. L., 1971, *Bioenergetics*, 2d ed., W. A. Benjamin, Inc., Menlo Park, Calif. The best introduction to molecular mechanisms of energy transformation to be had.

[1] Robert Frost, 1969, "It Bids Pretty Fair," in *The Poetry of Robert Frost*, edited by Edward Connery Lathem, Holt, Rinehart, & Winston, New York, p. 392.

VI

EPILOGUE

Starting with the functional processes common to all organisms, and then continuing with the specialized activities of cells, the basic outlines of physiology have been sketched. It is appropriate to conclude this exploration of the intricacies of the workings of the parts of animals with some further attention to the unique organization of living matter and to speculate upon its evolutionary origins.

To perceive the functional order that provides the basis for life is to experience a special kind of beauty. Living materials are the ultimate expression of the interactive properties of matter. Within the basic physical properties of elements and molecules is embedded the potential for inevitably evolving into the living state of matter.

FUNCTIONAL ORDER 20

It has been, and remains, one of the goals of biological research to resolve the complex structures and processes of living things into more and more elementary parts and analyze them. This attitude, of course, has been the immensely successful one applied by chemists and physicists. Obviously, it also has been very successful in biology; with progressive resolution has come new insights and deeper understanding. Most of the information displayed in the preceding chapters has come from such a "reductionist" approach to problems in biology. There is faith that when its elementary units are understood, an explanation of the real nature of life will be at hand.

However, analysis of this kind is not sufficient to deal with the phenomena of life completely, for as von Bertalanffy[1] says

> . . . each individual part and each individual event depends not only on conditions within itself, but also to a greater or lesser extent on the conditions within the whole, or within super-ordinate units of which it is a part. Hence the behaviour of an isolated part is, in general, different from its behaviour within the context of the whole. . . . Secondly, the actual whole shows properties that are absent from its isolated parts. The problem of life is *organization.*

Life resides in organized systems, and when they are destroyed, so is the living state. Here is the uniqueness of life, and to understand the essential nature of its order is the central object of the science of biology.

It is now quite apparent that the fabric of life, or **structure,** and the processes of life, or **function,** are manifestations of the same thing. Von Bertalanffy has termed this concept **dynamic morphology.** The molecules which constitute the "morphology" of an organism are those which simultaneously give it "physiology." Indeed, it is impossible to state whether it is the particular arrangements of atoms and molecules in an organism that determine their functions or functions that determine their arrangements.

From the elementary units of matter to the complete organism, there are ascending **levels of organization,** each of which has unique characteristics not found at lower levels. The biologist must attempt to fit his in-

[1] L. von Bertalanffy, 1952, *Problems of Life,* C. A. Watts & Co., Ltd., London, p. 12.

formation from one level into an interpretation at other levels, and he must conduct his investigation within various of these levels.

We know proportionately less about progressively higher levels of organization; with increasing complexity, simple laws of matter have increasingly less predictive value and the number of variables becomes enormously large. Nevertheless, encouraging progress is being made in analyzing functional organization a both the subcellular and cellular levels.

The energetic processes which we have been considering have a structural and functional order. They occur as orderly sequences in space and time. The nature of that order and the control it endows is the subject of this chapter.

20-1
TEMPORAL AND SPATIAL ORDER

A living system makes a multitude of different molecules which support a great variety of functions. It cannot be stated that one kind of molecule is more important than another, as they are all essential to the organism that makes them, but as a class of molecules, the *enzymes* must be singled out as the one that endows life with its unique status. What enzymes do in supporting life functions is dictated by a number of circumstances. It is not sufficient that their modes of action be determined only by principles of chemical reaction, nor is it appropriate that they should be present in fixed amounts. To respond appropriately to perturbations in its surroundings, an organism makes adjustments in its functional mechanisms. Central to these adjustments are changes in enzymes, both qualitative and quantitative. Metabolic pathways can be modulated, altered, or suppressed, within certain genetic limits, as demanded by prevailing circumstances. Thus, a temporal orderliness characterizes the living state.

There is another kind of orderliness also—spatial. Within any living entity there occur chemical events that have a high degree of improbability. Their improbability may become a certainty because a special microcosm is created in which the reaction becomes inevitable. Furthermore, by anchoring together sets of enzymes in membrane configurations and thus confining a series of reactions to a specific locus, a variety of quite different reactions, appropriately spaced, can go on simultaneously without disturbing one another. If it were not for such "compartmentations" the chemistry of cells would simply follow the tendency toward chaos and maximum entropy. Also, by fixing enzymes in specific arrays, only certain ones of several thermodynamically possible reactions are permitted to occur. In addition, since the chemistry of biological processes, like all reactions, must utilize ephemeral excited states, there must be intimate associations between the reactants to effect the efficient energy use and transfer that is demanded by the living state. There is, then, orderliness of time and space underlying the functions of organisms, an imposed orderliness that begins at the molecular level and endures as a continuum through the totality of the organism.

The major outlines of metabolism are now rather clearly visualized, and most of the participants, at least for the major metabolic events, have been identified. Much effort is currently being directed toward elucidating the control mechanisms that regulate the interactions of the components of the metabolic machinery.

20-2 CONTROL OF ENZYME ACTIVITY

Commonly a metabolic sequence involves a series of enzymes acting sequentially until an end product is formed. Schematically this could be represented as:

$$A \xrightarrow{E_{ab}} B \xrightarrow{E_{bc}} C \xrightarrow{E_{cd}} D \xrightarrow{E_{de}} E \xrightarrow{E_{ef}} F$$

One kind of control mechanism for such a sequence was first demonstrated in the 1950s in enzymes of biosynthetic pathways of the intestinal bacterium, *Escherichia coli*. Several cases are known in which a mutant of the bacterium lacks one of the enzymes of a pathway and, thus, cannot synthesize certain compounds. If, for example, the enzyme for the conversion of C to D were lacking, the organism would accumulate C, but not D, E, or F. Now if the mutant is grown in a medium containing F, an interesting thing happens: it severely reduces the rate of synthesis of compound C. The inhibition by the normal end product of the synthesis of an intermediate is an example of **negative feedback.**

Let us consider what happens in other bacterial strains where the entire sequence can be synthesized. When the end product F is added to the mediùm, the rate of production of C is again reduced. Obviously, if C is produced at a slower rate, all subsequent compounds, D, E, and F, also will be produced at a proportionally slower rate. Thus, the enzyme, E_{bc}, is **rate-limiting,** or one can say the conversion of B to C is a **pacemaker reaction.** This is a kind of enzyme inhibition known to occur in several amino acid and nucleic acid synthetic pathways of *E. coli* (Fig. 20-1).

Similar feedback inhibitions of enzymes have been demonstrated in nucleotide synthetic pathways in tissues of birds and mammals. Generally, the enzymes inhibited act in early stages of a pathway, often the first or second enzyme. The effect of such a mechanism is to prevent the buildup of an excess of the product.

Much less common is feedback of the opposite kind, **positive feedback.** It has little use in biological systems because a positive-feedback situation has the potential for uncontrolled reaction, like an explosion. No good examples of use of positive feedback in metabolic control are known. (Recall, nevertheless, that positive feedback is important in blood clotting; Chap. 8, Sec. 8-4.)

It has been demonstrated that some reactions are inhibited by the

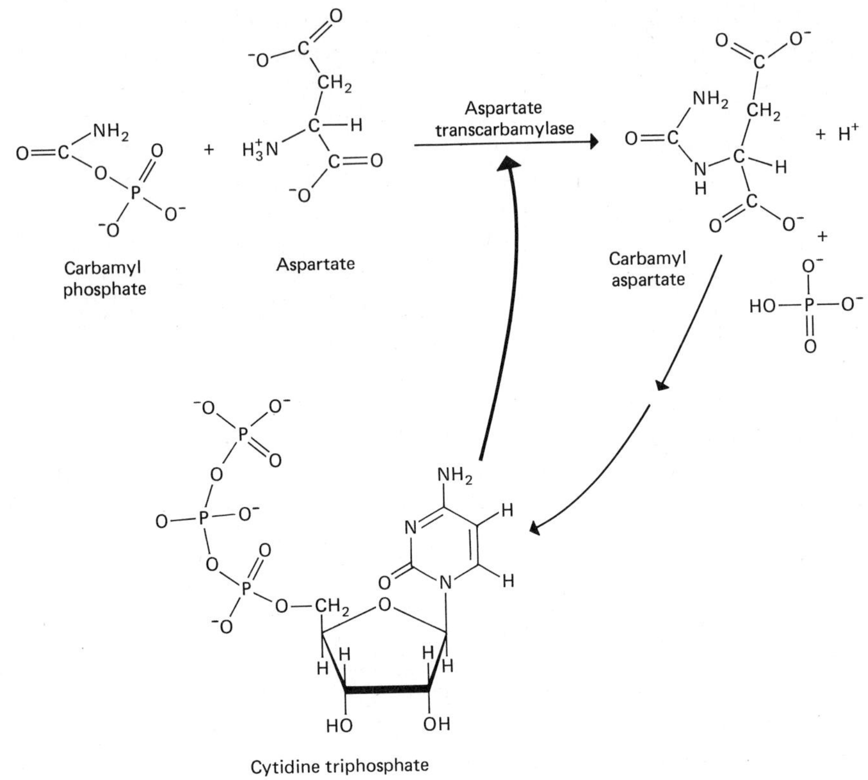

FIGURE 20-1
Feedback inhibition of aspartate transcarbamylase in the pyrimidine pathway of *E. coli*. The enzyme inhibited is the first of a series of six in a pathway leading to cytidine triphosphate formation.

immediate product of that reaction to an extent greater than would be predicted by the laws of mass action. This is called **product inhibition,** and it may act as a control mechanism, although good in vivo demonstrations of the fact are lacking. There is some reason to believe that in one instance, the inhibition by glucose-6-phosphate of glucose phosphorylation by hexokinase, such a mechanism serves as a metabolic control.

There appear to be mechanisms that regulate the relative rates of alternative pathways in metabolic schemes. One of the best-studied examples of this is the phenomenon called the **Pasteur effect.** In 1876 the French scientist Louis Pasteur published the observation that fermentation is inhibited by respiration. Thus, the regulation of the relative rates of glycolysis and aerobic respiration are adjusted to the availability of oxygen. Its adaptive value, however, is that aerobic respiration provides energy much more efficiently than glycolysis. The mechanism for this effect remains to be revealed.

There is a class of mechanisms for controlling the amounts of active enzyme that can be called **activations.** For example, hydrogenases in certain bacteria and algae are activated by hydrogen, and mammalian liver produces a tryptophan pyrrolase that is activated by tryptophan.

The exact natures of activations are not as yet well characterized and seem to be quite diverse. In some instances the inactive enzyme may be in the form of a typical zymogen which is activated by a second enzyme, as noted in Chap. 18, Sec. 18-3. Others are activated by the removal of a tightly bound, specific inhibitor.

Inquiry into possible molecular mechanisms of inhibition has revealed some remarkable facts and led to some ingenious explanations. If an inhibitor product is similar to the initial substrate, there is little difficulty in understanding the mechanism of operation. It may act as a competitive inhibitor and bind the catalytic sites in an **isosteric** manner; i.e., it does not cause conformational changes in the enzyme. However, in a few well-studied cases the inhibiting product and substrate are quite dissimilar, and the assumption has been made that there are two different sites on the enzyme—one catalytic and the other regulatory. This idea is based upon several kinds of observations, one of which is that susceptibility to inhibition can be abolished without affecting the catalytic activity of the enzyme. It is believed that the action of the inhibitor is to change the conformation of the enzyme in the way that the substrate would if it were bound, and thus it cannot complex with any more substrate. Such enzymes are called **allosteric** (Fig. 20-2).

Isocitrate dehydrogenase, which catalyzes the NAD-mediated oxidation of isocitrate to oxalosuccinate (Fig. 19-6), is one of the enzymes which show allosteric properties. The reaction rate, at low substrate concentrations, can be shown to be dependent upon Mg^{2+} and adenylic acid (AMP), two species which play no apparent role in the stoichiometric reaction. One explanation is that these substances serve as allosteric activators (or positive effectors).

20-3 CONTROL OF ENZYME SYNTHESIS

A straightforward way of controlling enzyme activities is to regulate the amount of enzyme present at any one time. However, there are many

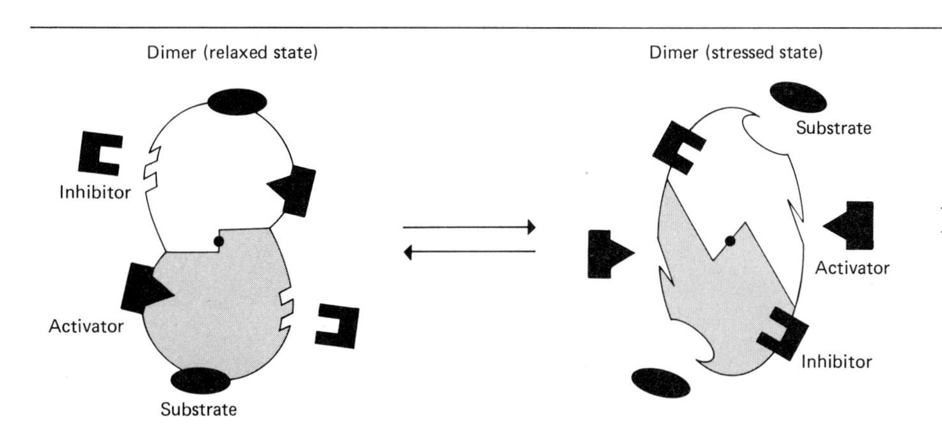

FIGURE 20-2
Model of allosteric transition of an enzyme. In one conformation it can attach substrate and activator, but in the other only inhibitor bonds can be formed. (*After J. Monod, 1966, Science, 154, p. 481.*)

cells in which certain enzymes appear to be produced at a very constant rate, regardless of variations in their substrates. These unregulated enzymes are known as **constitutive enzymes.** On the other hand, there are enzymes that are produced in varying amounts, depending upon prevailing circumstances.

Many enzymes are produced only when certain specific compounds, usually substrates, are present. This is known as **enzyme induction.** Sometimes more than just one enzyme is induced; synthesis of a set of enzymes for a particular sequence may be evoked by just one specific compound. Furthermore, induction in bacteria is not limited to catabolic enzymes. Enzymes involved in the active transport of substances into the cell are sometimes inducible. For example, certain sugars enter bacteria only by diffusion when they are first introduced, but after a brief lag period the cells acquire a system specifically for active transport of that sugar.

A second method of control of enzyme synthesis, called **repression,** was discovered when it was observed that addition of arginine to the medium of *E. coli* prevented the formation of an enzyme, acetylornithinase, which is in the synthetic path for arginine. Repression refers, then, to those situations in which the product of an enzymatic reaction, or the final product of a sequence of reactions, interferes with the continued *production* of an enzyme. (This is not the same as feedback inhibition, where the *action* of the enzyme is affected.)

The major steps in the making of enzymes are transcription of deoxyribonucleotide sequences of structural genes and translation of their specific information into particular amino acid assemblies. The control of enzyme production, therefore, is a matter of regulating protein biosynthesis. An overall control of protein synthesis, however, is not an adequate way to adjust the availability of specific enzymes with time and conditions. Specific genes, or blocks of genes, which control individual reactions, must be regulated.

It is largely from the work of Jacob, Monod, and their colleagues that the mechanisms of induction and repression of specific enzymes have come to be better understood. Their mapping of the genetic region for control of β-galactosidase, an inducible enzyme, in certain strains of *E. coli* provided the basis for postulating a control mechanism. In the *lac* locus, adjacent to the β-galactosidase structural gene (*z*), are the genes (*y*), for an associated "permease" needed for the uptake of the galactoside lactose (the transport aspects of this system are discussed in Sec. 14-6), and (*x*), for galactoside transacetylase. Also in the *lac* region are three other genes: (1) a gene *i* (for inducer), which in the wild type (i^+) suppresses the formation of both β-galactosidase and the permease, but which in its mutant form (i^-) allows the constitutive formation of the enzymes, even when the inducer (lactose or certain other galactosides) is absent; (2) a gene o (for operator), which is thought to respond to a hypo-

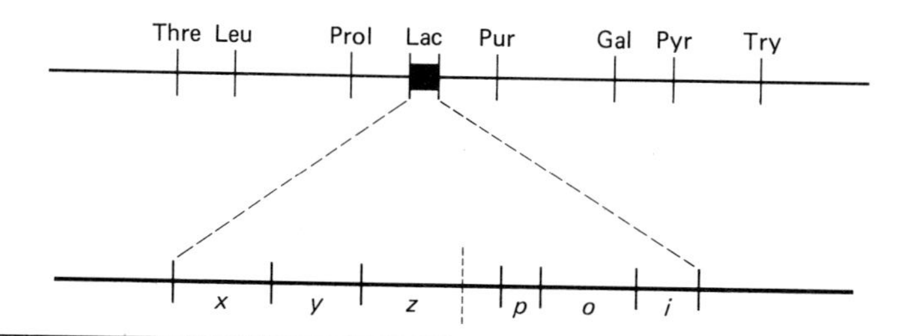

FIGURE 20-3
Genes for the utilization of lactose by *E. coli*. The position of the *lac* region on the chromosome is shown at the top; the lower line is an enlargement of this region. See text for explanation.

thetical repressor substance produced by *i*; and (3) a gene *p* (for promoter) which initiates transcription (Fig. 20-3).

The gene for inducibility (*i*) is now called a *regulator* gene, and the complex of structural genes (z, *y*, *x*) plus the related operator gene (o) for controlling the transcription is called an *operon*. The operon is hypothesized to be a genetic unit encompassing a DNA region which can be transcribed into a single mRNA molecule. The unidirectional synthesis of mRNA on the *lac* locus does not occur if the repressor produced by *i* is "clogging" the operator, and thus the transcription of structural information cannot occur. Just how the repressor does this is unknown. Recent evidence suggests that transcription of mRNA in the *lac* operon begins at the z gene and is initiated by the *promoter* (p) gene, which also appears to determine the maximum rate at which mRNA can be formed. An interesting feature of this mechanism is that the genes for β-galactosidase, "permease," and transacetylase, all connected to one another but functionally unique, are controlled simultaneously by one operon.

Most cases of enzyme induction are now regarded as being due to enzyme **derepression.** Chromosomal DNA is considered to be normally linked with a histone-type protein in a state where it cannot undergo transcription. Chromosomal RNA is postulated to give the histone its specificity for base sequences for DNA. The repressor, according to these ideas, is an RNA-histone complex. When an inducer (substrate) is added, it is thought to combine with and inactivate the repressor, allowing the formation of an enzyme. Thus, we have a hypothetical explanation of why formation of an enzyme is evoked only when its substrate is available.

To explain repression by the same basic mechanism requires an additional assumption: that some regulator genes form incomplete repressors—i.e., an **aporepressor**—which alone cannot act on the structural gene. The repressor is completed and becomes functional when combined with the end product of the biosynthetic chain, the smaller molecule serving as a **corepressor.** Thus, we now have a hypothetical explanation of why an enzyme is not formed when enough of its product is on hand.

The reality of these ingenious mechanisms has become more probable with the isolation of two of the postulated repressor molecules, *lac* repressor and λ phage repressor. They are reported to be protein and specific for the appropriate operator region of DNA. It will be reassuring if such mechanisms are found to have a general phyletic distribution; in addition to bacteria, they have been demonstrated, so far, only in fungi and seed plants.

20-4 MULTIENZYME COMPLEXES

Proteins can vary in the number and kind of amino acids composing them, in the number and kind of associated polypeptide chains, in conformational details, and in subunit assemblages. The number of combinations is infinite. It seems beyond question that it is the protein macromolecule which endows genetic specificity to a species and the individual organism. Let us look at the structural organization of proteins and then give special consideration to some of the assemblies that enzyme proteins form.

Amino acids are united through an acid-amide type of linkage called a *peptide bond* to form peptides and proteins (cf. pp. 115–116). The formation of each bond results in the loss of two charged groups, ammonium and carboxylate. A "backbone" is formed by amide linkages with amino acids projecting to the side (Fig. 20-4).

There are a number of naturally occurring peptides. The hormones of the posterior lobe of the pituitary are octapeptides. Adrenocorticotropic hormone (ACTH) consists of 39 amino acid residues. Most peptides, however, are high-molecular-weight compounds and are grouped under the name *protein*. Most proteins are large enough to be called *macromole-*

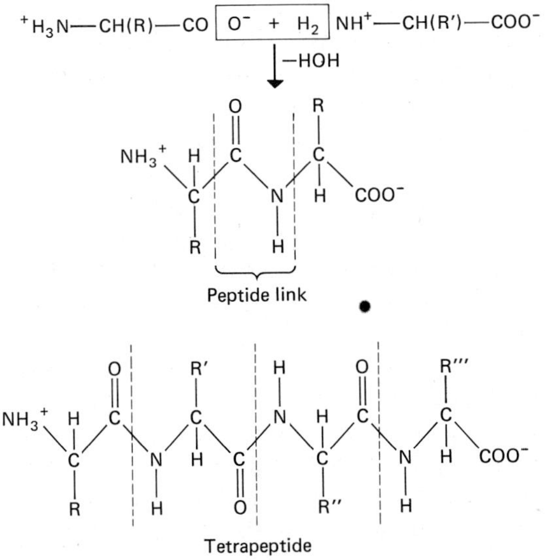

FIGURE 20-4
The "peptide linkage," an acid-amide type of bond, through which the amino acids of proteins are united. (See Fig. 5-6 also.)

$$^{+}H_3N{-}CH(R){-}CO\,\boxed{O^- + H_2}\,NH^+{-}CH(R'){-}COO^-$$

$\downarrow -HOH$

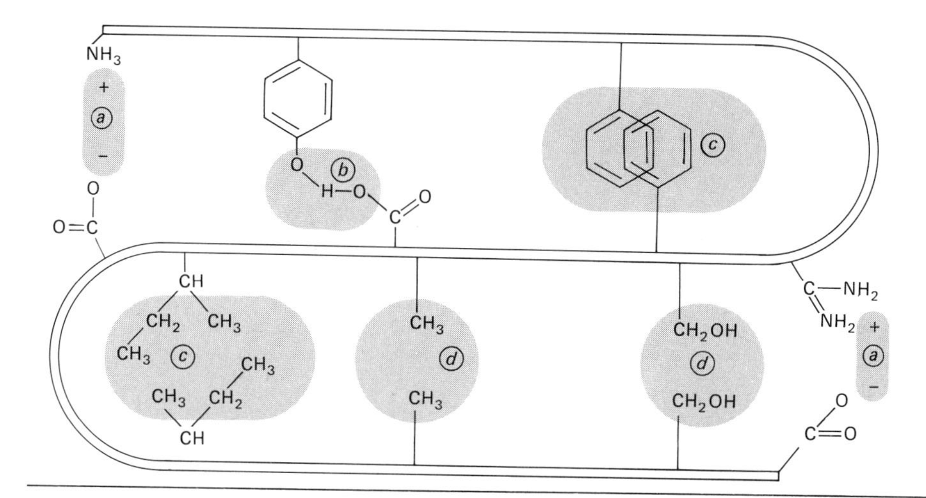

FIGURE 20-5
Some types of noncovalent bonds that stabilize protein structure: (*a*) electrostatic interaction; (*b*) hydrogen bonding between tyrosine residues and carboxylate groups on side chains; (c) interaction of nonpolar side chains caused by the mutual repulsion of solvent; (*d*) van der Waals interactions. (*After C. B. Anfinsen,* 1959, *The Molecular Basis of Evolution, John Wiley & Sons, Inc., New York, p.* 102.)

cules. [Any chemical compound whose molecular weight is above 10,000 (some definitions use 5000 as the lower limit) and in which covalent forces are effective in all available space is a macromolecule.] For an insoluble protein, however, molecular-weight measurements usually reflect preparation procedure, and the true size of the molecule cannot be determined.

One of the great breakthroughs in biochemistry was the development of techniques and methods for determining the sequence of amino acids in proteins. Sanger's elucidation of the complete sequence for the protein hormone, insulin, earned him the Nobel Prize in 1960. Insulin is composed of two chains of amino acids, linked together by two disulfide bridges (Fig. 9-6). One chain contains 21 and the other 30 amino acids, but the exact sequence of residues varies with mammal species.

As in the case of insulin, long-chain polymers can be linked with other polymers through covalent bonds such as S-S linkages. Polypeptides also can be coiled, looped, and interwoven. The shape of long-chain polymers, to a very large degree, determines their biological function. The three-dimensional structure of such macromolecules is stabilized by several kinds of attractive and repulsive forces between amino acid residues and between the residues and surrounding solvent (Fig. 20-5).

Largely through x-ray diffraction patterns it has been shown that polypeptide chains twist and coil upon themselves. Although such chains could have an infinite number of configurations, it is assumed that the most stable are those in which —NH groups of peptide bonds are maximally bonded to —C═O groups through *hydrogen bonds* (cf. Fig. 7-1). The hydrogen bond, $>C{=}O \ldots H{-}N<$, gives order to the flexible polypeptide backbone (Fig. 20-6).

The number and sequence of amino acid residues linked by peptide bonds define the *primary* structure of proteins. The structure due to the formation of hydrogen bonds is called *secondary* structure. It may be imagined that long, helical segments of polypeptides might assume a

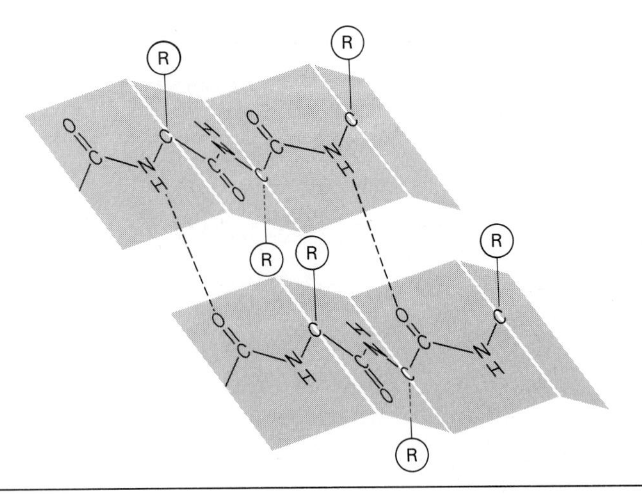

FIGURE 20-6
Two polypeptide chains, each in the proposed structure of Pauling and Corey, and hydrogen-bonded together.

variety of conformations because of various forces and linkages interacting between the various amino acid side chains. Structure resulting from such interactions is called *tertiary* structure.

Certain proteins are aggregates of several polypeptide chains which are not covalently linked to one another. Hemoglobin is an example of such a protein, being composed of four chains (Fig. 6-25). Each chain may be considered a subunit of structure. It appears that most, if not all, large proteins are composed of such subunits.

Enzymes sometimes are composed of polypeptide subunits, each having its own primary, secondary, and tertiary structure. Some of the multichain enzymes can be dissociated into monomeric units, and often these individual units are found to be enzymatically active. For example, a glutamic dehydrogenase with a molecular weight of about 1.2×10^6 has been dissociated into enzymatically active subunits of about 300,000 mol wt. Thus, the fully associated dehydrogenase would appear to be a tetramer.

Not only do similar or identical monomers aggregate to make larger enzymes, but different kinds of fully associated enzymes can be found in still larger aggregates. A good example is the pyruvate dehydrogenase complex of *E. coli*. It is a particle which can be separated into three enzymes, pyruvate decarboxylase, dihydrolipoyl transacetylase, and a flavoprotein, dihydrolipoyl dehydrogenase. The complete particle can be resolved into 88 separate polypeptides. If the component parts are mixed together, they show a high degree of self-organization, reassociating themselves to yield a large number of complexes just like the original dehydrogenase.

The enzymes involved in fatty acid synthesis also are organized into isolable particles that are of a multienzyme nature. Fatty acid synthetase from bird liver has been resolved into two components, while that of *E. coli* has several, including two kinds of transferases, a condensing enzyme, two reductases, and a hydrase (cf. p. 430). A similar multienzyme complex has been isolated from yeast (Fig. 20-7).

The respiratory-chain system of mitochondria is the best-studied of the multienzyme complexes. Particles isolated from mitochondria have complete sets of enzymes for transporting electrons (see below). These electron-transport particles, in turn, are composed of four kinds of enzyme complexes, as discussed in Chap. 19 (see Fig. 19-12).

The study of aggregates of enzymes has led to radical changes in concepts of enzyme function. For one thing, laws based on mass action and ideal collision do not account for the behavior of enzymes in "solid-state" arrays. Reaction kinetics and mechanisms do not presently consider interaction of individual catalysts where degrees of freedom would seem to be considerably diminished. Enzymes bound to membranes almost invariably have activities that are not shown by the same enzymes in solution. The conformational changes undergone by reacting enzymes probably alter the relation of one enzyme to another, as well as to other functional components. Thus, "mechanical" features probably limit some enzymes to specific and irreversible actions.

The analysis of enzyme complexes is an outgrowth of improvements in methods of cell disruption and of isolation as well as purification of particulate systems. The fact that multienzyme complexes are isolable from cell homogenates probably does not mean that within cells they exist as unattached particles. It seems highly likely that such aggregates are bound within themselves more tightly than they are bound to something else in the cell. Fractionation procedures take advantage of the fact that cell components are organized into diverse assemblages held together by forces of different energies. Relatively mild chemical or mechanical treatments are used to partially dissociate cell components when the objective is to obtain intact nuclei, mitochondria, chloroplasts, or other so-called *organelles*. More rigorous or more specific methods may be employed to break these particles into subunits.

These are several methods for disrupting or homogenizing cells, including ultrasonic vibrations, mechanical grinding, and repeated

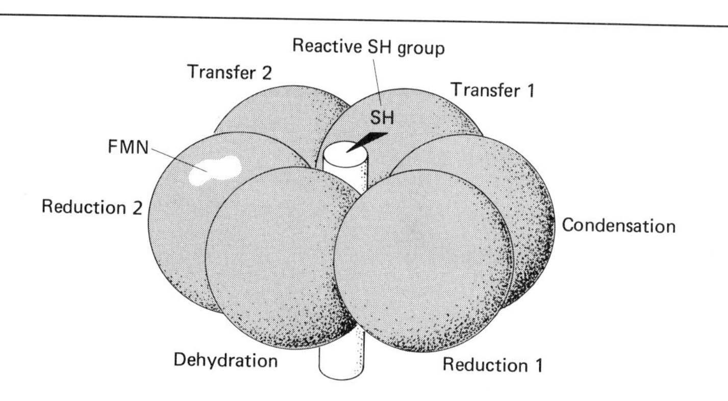

FIGURE 20-7
A multienzyme complex isolated from yeast by Lynen which carries out the entire synthesis of long-chain fatty acids.

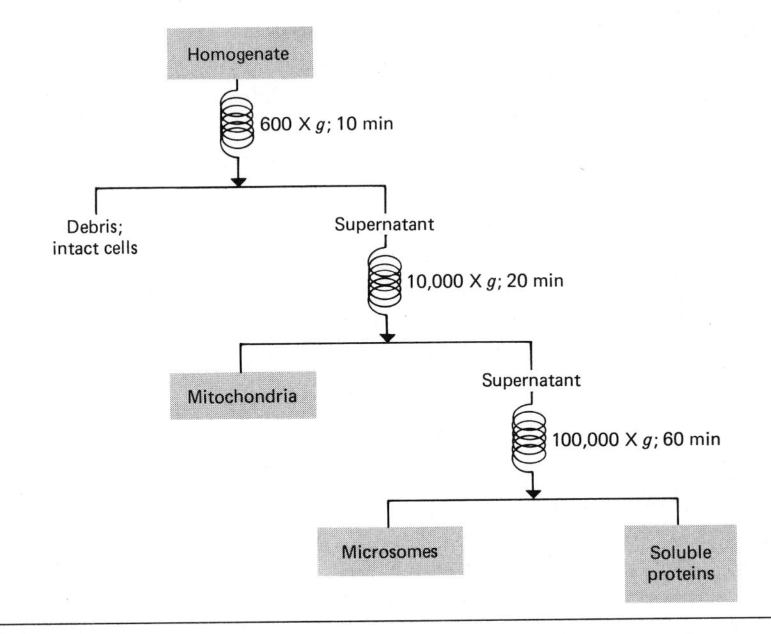

FIGURE 20-8
Sequential procedure commonly employed in differential centrifugation of cell homogenates. Force and duration of centrifugations are indicated.

freezing and thawing. The suspensions produced by these methods are generally subjected to conventional centrifugation in order to separate particles differentially by their densities (Fig. 20-8).

The materials remaining suspended in solution after high-speed centrifugation can be resolved by additional methods, such as various kinds of chromatography, electrophoresis, and density-gradient sedimentation in an ultracentrifuge. Also, the larger particles, differentiated by centrifugation, sometimes can be disrupted and be resolved by these same methods into their component parts.

The information obtained from enzymatic and physicochemical analysis of the cell entities separated by such methods has been joined partially with discoveries emanating from electron microscopy to give a far better understanding of dynamic morphology than was possible just a few years ago. It now seems justifiable to speculate that what is visualized as "structure" in cells is, in fact, biochemical machinery entirely, there being no structural elements per se. It may no longer be correct, then, to think of a cell as having a "skeleton" on which are hung its functional parts. The functional parts probably are, at the same time, the supporting structure. The lamellae of a chloroplast or mitochondrion and the variously fashioned endoplasmic reticula probably should be regarded entirely as "solid-state" arrays of enzymes plus other molecules which create appropriate environments for functioning of the enzymes.

20-5 MEMBRANE CHEMISTRY AND STRUCTURE

Within a dramatically short time, and largely because of the advent of electron microscopy, our conception of how a cell is structured has changed radically. Nebulous ideas about the colloid nature of "living material," and use of the terms "protoplasm" and "ground substance," have been largely discarded. The textbook picture of the "typical" cell changed drastically during the 1950s and the 1960s. The new emphasis is on *membranes*—many, many membranes, and in a variety of configurations (Fig. 20-9).

There have been concomitant changes in the emphases of investigations in cell biochemistry. What might be called an "antireductionist" trend is evident. Systems of a complexity that would have been baffling a few years ago are being analyzed today. A new field of biochemical investigation, membrane biochemistry, has emerged.

An historically important hypothesis for the structure of membranes

FIGURE 20-9
Diagramatic generalized cell to show its various membrane systems as seen in electron micrographs.

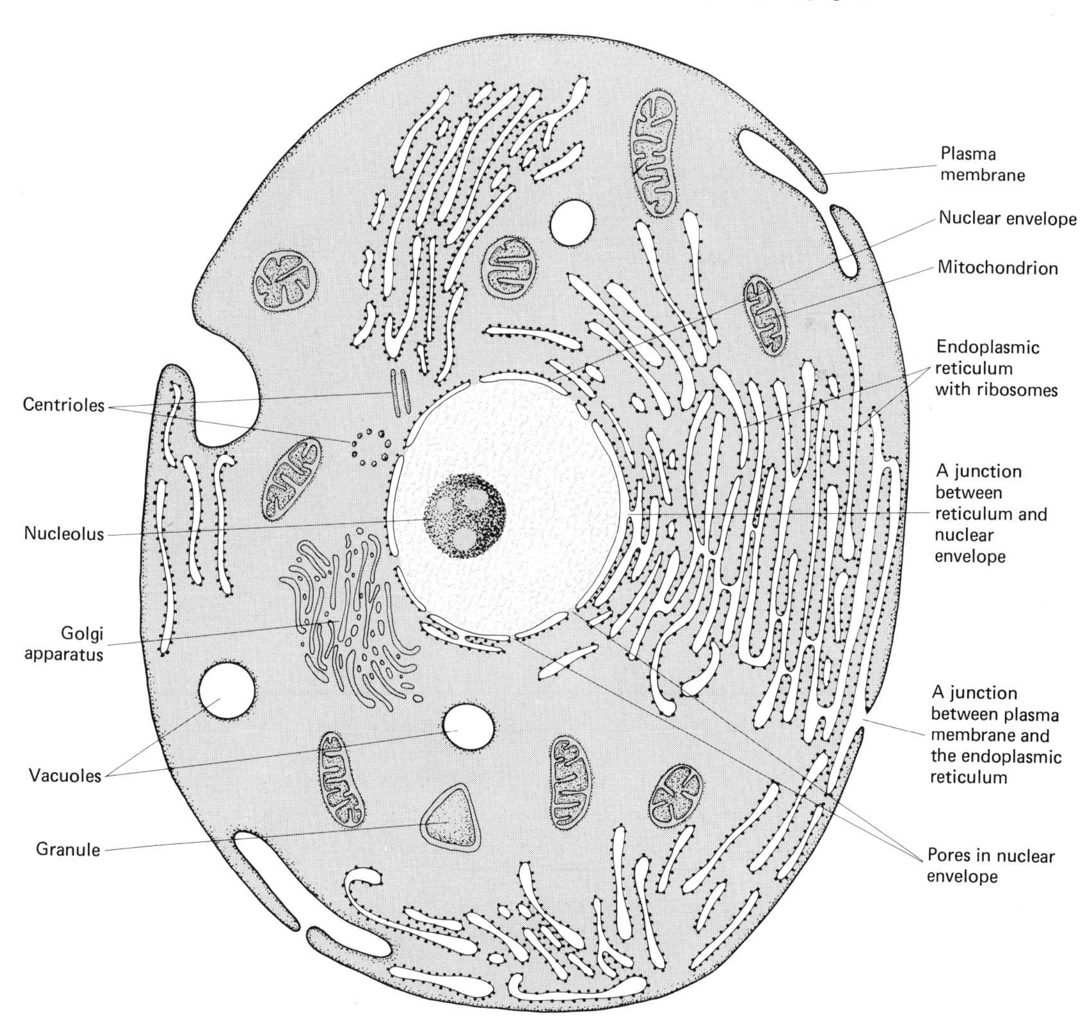

is the concept of the **unit membrane,** as proposed by Robertson in 1959. It was formulated to account for electron-microscopic and x-ray diffraction data, while at the same time accommodating much of the information accumulated over many years about the physicochemical nature of membranes, particularly plasma membranes.

The unit-membrane theory proposes that there is one basic structure common to all membranes or to most portions of all membranes: a bimolecular leaflet of phospholipids, whose nonpolar moieties are oriented inward and perpendicular to the plane of the membrane, and whose polar groups project outward toward the surfaces and associate with proteins (and, perhaps, some carbohydrate). Thus, a sequence of three kinds of layers is formed: protein, bimolecular phospholipid, protein. In electron micrographs three zones are seen: dense, light, dark—supposedly corresponding to the three kinds of chemical zones—with a total thickness of about 7.5 nm. This dimension is remarkably similar to that predicted from a membrane model proposed by Danielli and Davson in 1935, and the specific chemical sequence they proposed influenced the later interpretations of electron micrographs (Fig. 20-10).

The electron microscope shows the triple-layered configuration to be present (after fixation in $KMnO_4$, and often after fixation in OsO_4) in Schwann cells (which surround axons of neurons), myelin sheaths (which may be internal extensions of the plasma membranes of Schwann cells) (cf. Fig. 11-14), endoplasmic reticula, the two membranes of the nuclear envelope, mitochondrial membranes, chloroplast membranes, and others. All tend to be about 7.5 nm thick, but can vary between 5 and 10 nm.

The unit-membrane theory has been criticized on the grounds that (1) chemical data have been obtained from only very few kinds of membranes, and mostly from exploded (hemolysed) erythrocytes which leave behind thick cell coverings, called "ghosts" (which may be atypical membranes); (2) electron micrographs do not reveal chemical information, so that the chemical nature of dense and light zones is not known; also; methods employed in their preparation can produce artifacts; and (3) the

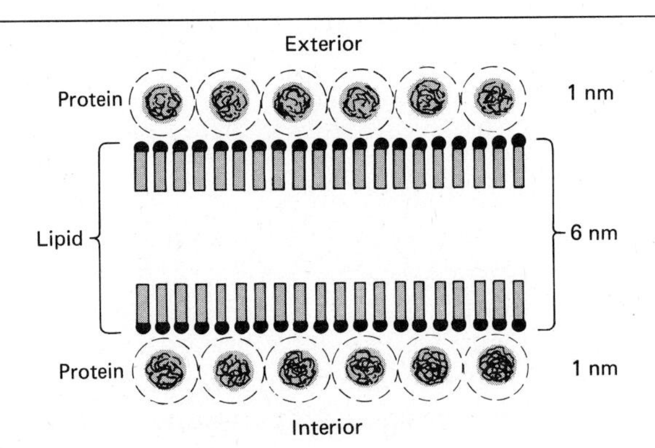

FIGURE 20-10
Model of plasma membrane proposed by Danielli and Davson in 1935.

available x-ray diffraction data are from model systems and only one kind of biological membrane, the myelin sheath of neurons.

Recent information about the structure of proteins and other macromolecules also casts doubt upon the validity of the Danielli-Davson-Robertson model. Conformations of proteins are now considered to be largely determined by hydrophobic interactions, rather than the electrostatic ones called for by their model. Recent enzyme-digestion studies suggest a more exposed position for phospholipid than would be allowed by a bimolecular leaflet. Other models have been proposed lately which place phospholipid and protein in intermixed positions and suggest that they are organized into subunits.

Mitochondria are regarded by some investigators as model systems for the study of membranes and, from what has been learned of their structure-function, a **subunit** or **elementary-particle** theory has been developed about the nature of biological membranes in general. Some of the ubiquitous features of membranes are, according to this view, the following. (1) Their outer structured layer is a composite of nesting, repeating subunits or elementary particles; in any one kind of membrane these units are distinctive and identical. (2) Repeating units are composed of two or more parts: a base piece, serving as the membrane-forming subunit, and a free, projecting piece attached to the base piece. (3) Although the repeating units appear to be identical, they are not necessarily identical chemically, but each has a characteristic complement of proteins and a set of enzymatic activities. (4) Repeating units that have been dissociated (or separated base pieces alone), under proper conditions, spontaneously assemble themselves into vesicles; thus, external information is not required for membrane formation (Fig. 20-11).

According to Green, the chief architect of this subunit theory of membrane construction, most of the phospholipid of membranes is associated with the base pieces, which join to form a continuum of all phospholipids within a membrane. The membrane protein is of two types: "structural" and "catalytic." (In mitochondria the two types occur in about a 1:1 ratio by weight.) Both kinds are distinctive for any given membrane, but all known structural proteins have in common the properties of combining with phospholipids and binding inorganic phosphate, ATP, and NAD^+.

The subunits assume a geometric form which achieves physiological stability by satisfying the principle of minimal free energy of conformation. The function of a membrane, according to this concept, is the integration of the functions of the individual, oriented, repeating units which make up membranes. All aspects of membrane function are, then, the sum of the properties of the individual units.

One of the essential points of departure between the repeating-unit theory and the unit-membrane theory is the mode of membrane formation. Green visualizes the process as a great number of preformed particles assembling themselves in two-dimensional arrays, whereas Rob-

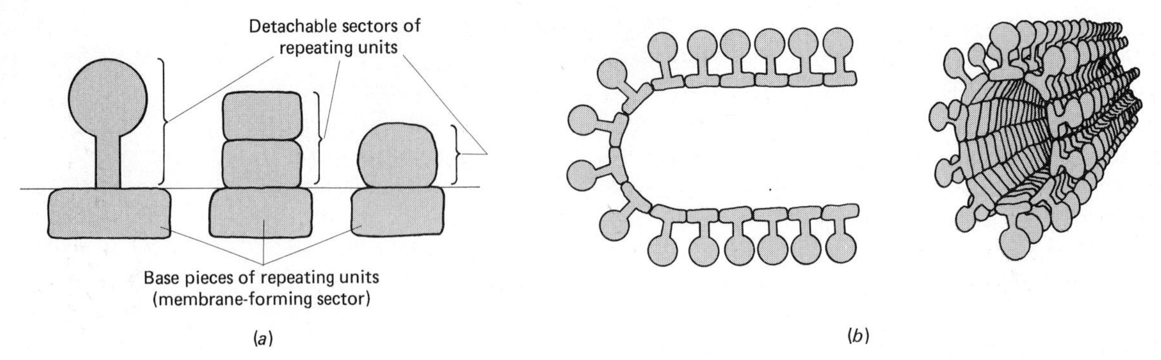

FIGURE 20-11
(a) Diagramatic representation of various kinds of repeating units and of the two sectors of each repeating unit. (b) Diagramatic representation of the membrane as a fusion of repeating units. These particular repeating units are the elementary particles of the mitochondrial inner membrane. (*After D. E. Green and R. E. Goldberger, 1967, Molecular Insights into the Living Process, Academic Press, Inc., New York, p. 224.*)

ertson believes that phospholipids form bimolecular leaflets to which metabolically functional proteins subsequently attach.

Cinematographic studies of living cells reveal that their membrane systems change constantly. For example, chloroplasts and mitochondria increase and decrease in size; they break up and they form from coalescing subunits. Thus, there is a *dynamic turnover* of subcellular bodies. They must not be regarded as rigid, immutable, static structures, but rather as plastic, ephemeral forms of matter which are constantly being disassembled and reassembled.

Perhaps the most nonstatic model for membrane structure to date is the **fluid-mosaic** concept. Biological membranes are conceived to be mosaics of alternating globular proteins and phospholipid bilayer that is fluid or dynamic and probably is a two-dimensional, oriented, viscous solution. The theory applies to the so-called *functional* membranes, such as the bounding plasmalemmal and intracellular membranes, including the membranes of mitochondria, chloroplasts, and other organelles. However, it does not necessarily describe other membranelike systems which may be rigid, such as myelin and the lipoproteins membrane of certain viruses.

Central to the fluid-mosaic model are thermodynamic considerations. *Hydrophobic* interactions are responsible for the sequestering of nonpolar groups away from water. The sum of the free-energy contributions of the many nonpolar amino acid residues of soluble proteins is regarded as of sufficient magnitude to result in the nonpolar residues being sequestered in the interior of the molecules away from contact with water. *Hydrophilic* interactions, on the other hand, result in essentially all the ionic residues of the protein molecules being in contact with water, usually on the outer surface of the molecule. Although other forces are influential, it is the hydrophobic and hydrophilic interactions which are regarded in this theory as the primary determinants of gross structure. These interactions are thought to be maximized, and thus the lowest free-energy state is attained. If this theory is correct, then the trilaminar model of Danielli-Davson-Robertson is rendered unlikely; it would be thermodynamically unstable because not only are the nonpolar amino acid residues of the proteins largely exposed to water, but the ionic and polar groups of the lipid are sequestered by a layer of protein from contact with water. Thus, neither hydrophobic nor hydrophilic interactions are maximized in the classical model.

Of the three major classes of membrane components—proteins, lipids, and oligosaccharides—the proteins are predominant. Two kinds are distinguished in the fluid-mosaic model: *peripheral* and *integral*. Peripheral proteins can be dissociated comparatively easily from the membrane, they dissociate free of lipids, and they are relatively soluble in neutral aqueous buffers, all of which suggests that they are held to the membrane by rather weak noncovalent interactions and are not strongly associated with membrane lipid. Cytochrome *c*, of mitochondrial membranes, and the protein, spectrin, of erythrocyte membranes, are examples of proteins which fit this category. The peripheral proteins may be unique to certain kinds of membranes. (One might suggest that they correspond to the detachable sectors of Green's model.) The integral proteins, the major portion, in contrast, require much more severe treatments to dissociate them from the membrane. They commonly remain associated with lipids when isolated, and they usually are highly insoluble, or aggregated in neutral, aqueous buffers. Thus, it appears that the integral proteins are the essential proteins for structural integrity, and they may be largely common to all or most membranes (Fig. 20-12).

The membrane phospholipids are considered to be in a bilayer. The globular molecules of the integral proteins (perhaps in some instances attached to oligosaccharides to form glycoproteins or to lipids to form lipoproteins) are conceived to alternate with sections of the phospholipid bilayer in a mosaic pattern. The protein molecule, or an aggregate of protein molecules, may be embedded in one side of the membrane, or it may traverse the entire membrane and have regions in contact with the aqueous solvent on both sides of the membrane.

The phospholipids, with their ionic and polar heads in contact with the aqueous phase, are thought to constitute the *matrix*. The lipids are in a fluid rather than a crystalline state, and the integral proteins are dispersed in this fluid lipid matrix (Fig. 20-13).

20-6 MEMBRANE SYSTEMS

It would be premature to conclude that any totally satisfactory understanding of the construction of biological membranes exists. This is reflected in the fact that there are numerous models, in addition to the three just described, to explain membrane structure (see Hendler, listed in the Readings, for an account of several of these). Membranes are known to vary widely in their function and chemical composition, so it seems

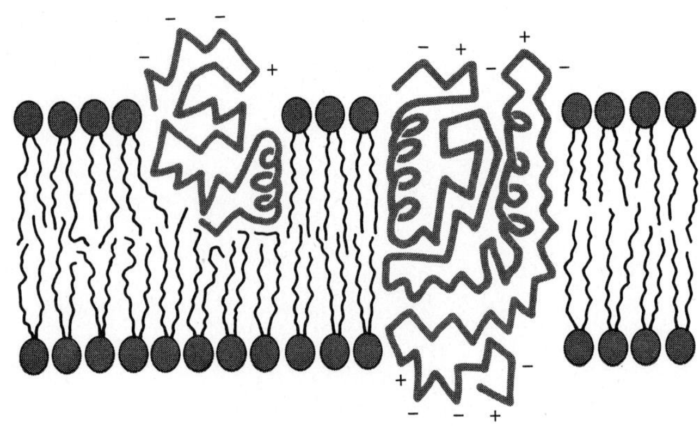

FIGURE 20-12
The lipid-globular protein mosaic model of membrane structure in schematic cross-sectional view. The phospholipids are in a discontinuous bilayer with their ionic and bipolar heads in contact with the aqueous medium along both surfaces. The integral proteins, with the heavy lines representing the folded polypeptide chains, are shown as globular molecules partially embedded in and partially protruding from the membrane. The protruding parts have on their surfaces the ionic residues (– and +) of the protein, while the nonpolar residues are largely in the embedded parts. The degree to which the proteins are embedded and, in particular, whether they span the entire membrane thickness depend on the size and structure of the molecules. (*After S. J. Singer and G. L. Nicolson,* 1972, *Science, 175,* p. 722.)

improbable that all of them have exactly the same appearance and fundamental organization. Also, only a few kinds of membranes have been closely studied—too few to develop substantial generalizations about them. A brief account of some of the information available about certain kinds of membrane systems will serve to emphasize their similarities and

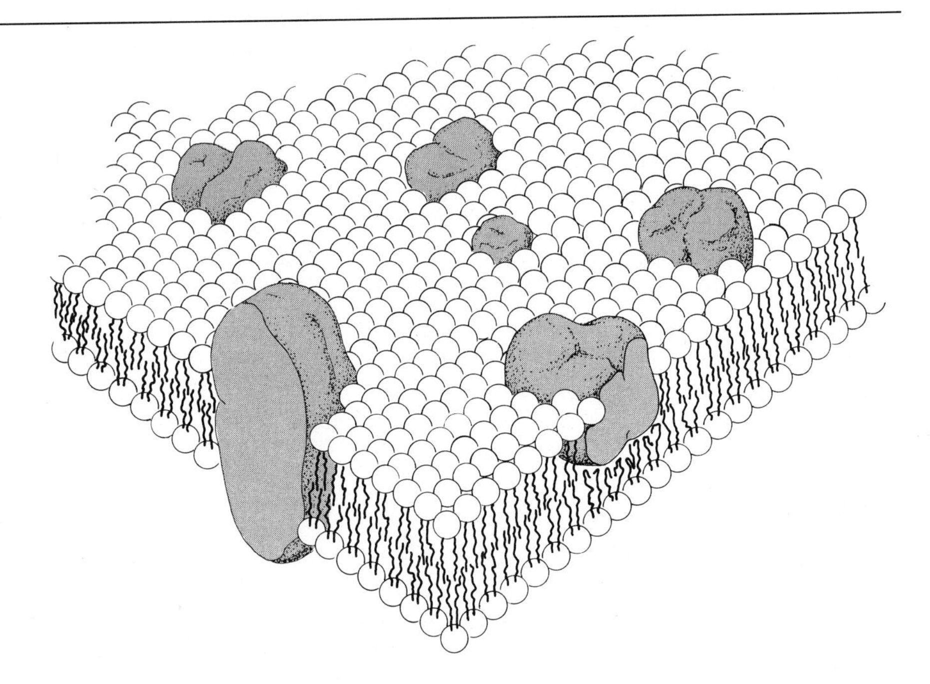

FIGURE 20-13
The lipid-globular protein mosaic model with a lipid matrix (the fluid-mosaic model). The solid bodies with stippled surfaces represent the globular integral proteins, which are distributed in the lipid bilayer. (*After S. J. Singer and G. L. Nicolson,* 1972, *Science, 175,* p. 723.)

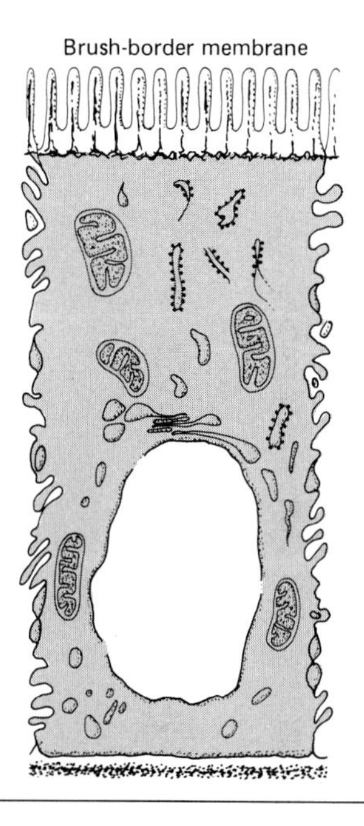

FIGURE 20-14
Diagramatic representation of the various membranous components seen in electron micrographs of sections through intestinal epithelial cells. Note especially the brush border.

differences, and to underscore the inadequacy of the information we now have about membranes.

Plasma Membranes

The outermost membrane enclosing a cell is called the **cell membrane** or **plasma membrane.** A plasma membrane should not be regarded as a simple physical barrier between the outside and the inside of a cell, but should be thought of as one of the essential pieces of living machinery, with prescribed functions and a dynamic morphology. In many cases cells are surrounded by further constructions of their own making—cellulose walls, collagen, bone, or siliceous material—but these are nonliving.

In addition to being specialized for exchange processes, plasma membranes have structural and enzymatic specializations reflecting their specific adaptations. Epithelial cells of the intestinal mucosa, for example, have, on their mucosal surfaces, the plasma membrane projecting as minute processes. There may be as many as 3000 of these **microvilli** per cell. A similar membrane is seen on cells of kidney tubules. It is believed that these processes increase the efficiency of transmembrane exchange by increasing the cell surface area. Such a configuration is called a **brush border** (Fig. 20-14).

Some electron micrographs show what appears to be a continuity of the endoplasmic reticulum (see below) with the cell membrane. If this is so, it may mean that the exterior of the cell can extend deep within the cytoplasm and even may be connected by the reticulum to the nuclear envelope (note suggestion of this in Fig. 20-9). Thus, the idea of a continuous, peripheral barrier around the cell may be untenable. Concepts of modes of permeation of molecules to and from cells must take into account these observations.

At places where the plasma membranes of adjacent cells are apposed, the membranes of the two cells generally appear to remain separated by what in electron micrographs looks like a low-density region about 20 nm wide. It has been speculated that this space is maintained rather constant by the interplay of electrostatic repulsion forces resulting from the ionic nature of the membrane and attractions due to van der Waals interactions.

Endoplasmic Reticula and Associated Organelles

In the cells of higher organisms the electron microscope reveals an extensive, lamellar set of membranes, called collectively the **endoplasmic reticulum.** Though highly folded, forming numerous flattened sacs and tubules, and morphologically diverse, it is thought to be a continuum. One form of it is coated with RNA-containing granules called **ribosomes.** Biochemical evidence suggests that this kind of membrane is the locus for the synthesis of proteins, fatty acids, phospholipids, and steroids, and, in addition, is a site for "detoxification" processes.

A second form of endoplasmic reticulum lacks ribosomes and has a smooth appearance in electron micrographs. The smooth variety has not been seen in plant cells. In animal cells it is often found along with the rough form.

The equivalent of the endoplasmic reticulum takes a special form in striated muscle cells and is called a **sarcoplasmic reticulum** (see Chap. 15, Sec. 15-3). Another set of membranes arranged in a transverse reticulum, and also probably invaginating from the plasma membrane, is present (see Fig. 15-14).

Ribosomal particles are generally collected by differential centrifugation. If a suspension of broken metazoan cells is centrifuged to remove debris, nuclei, and mitochondria, there is left a supernatant which can be centrifuged at 50,000 to 100,000 times gravity to yield a sediment which is referred to as **microsomes** (Fig. 20-8). This is an operational term, because the sediment is a mixture of membrane fragments, ribosomes, proteins, and other particles and subunits from several membrane systems. The microsomal fraction will incorporate amino acids into protein in cell-free systems, as noted in Chap. 16, Sec. 16-6.

Among the membranes of endoplasmic reticula there is frequently seen a special set of membranes forming flattened sacs and vacuoles called the **Golgi apparatus** (Fig. 20-9). Its size and position vary from one cell type to another, and its function is not known with certainty. There is strong circumstantial evidence that it has a secretory function, the apparatus being particularly well developed in secretory cells and neurons (which are secretory cells, it will be recalled).

Striking morphological modifications of the endoplasmic reticulum in relation to functional states of cells have been reported, suggesting that it participates in certain phases of metabolism. For example, in starved mammals the membranes of exocrine cells of the pancreas have a very different configuration than in animals which have been fed recently, as if food intake triggers a rapid modification in the reticulum. The endoplasmic reticulum is believed to participate in intracellular transport of enzymes, which are discharged as zymogen granules. The Golgi membranes may encapsulate such secretions in membranes.

The endoplasmic reticulum, where present, assumes many forms and participates in a variety of functions. No other membrane system seems so variable. Its morphologic characteristics and specific activities depend upon the special nature of the cell in which it occurs. All cells engage in the same basic metabolic transactions, but there are variants in bioenergetic pathways. Exactly which transactions, and to which membranes they are assigned, may determine the nature of a cell's structure.

A class of particles with centrifugal properties between mitochondria and ribosomes is called **lysosomes.** Lysosomes have diameters of 5.5 to 8 nm, are covered by a membrane, and are distinguished from other particles by their hydrolytic enzymes. Lysosomes provide lytic enzymes for dissolution of injured or dying cells, digest materials taken in by phagocytic cells, and may play a role in developing organisms by eroding defunct structures, such as the tail of a metamorphosing tadpole.

Nuclear Envelopes

Electron micrographs often reveal a double set of membranes surrounding nuclei with a space of about 50 nm between them. Because it is a complex of membranes, many workers prefer to call the lamellae around the nucleus a **nuclear envelope.** Some evidence supports the view that the outer nuclear membrane is continuous with endoplasmic reticulum, and like the rough form of reticulum, it is granular, whereas the inner membrane is smooth. Usually the envelope appears to be penetrated by numerous openings or pores that are 50 to 70 nm in diameter (Fig. 20-9).

From studies with isolated nuclei and with nuclei in situ much evidence has been collected to suggest that the envelope is quite porous. Macromolecules, such as albumin, hemoglobin, DNase, RNase, and fluorescent antibodies (protein), have been demonstrated to enter the nuclei of several different kinds of cells. These large particles may enter by some active process, such as pinocytosis, but the presence of pores would seem to make it unnecessary to invoke such a mechanism.

In plant and animal cells an important inclusion of the nuclear envelope is chromosomal DNA. In bacteria, however, chromosomes are not surrounded by such a membrane. This may not be a very important morphological distinction, since their chromosomes may be surrounded by some kind of individual membrane capsule, whereas metazoans have a common capsule for all the chromosomes. Furthermore, at the time of karyokinesis the envelope is absent temporarily, although some kind of membrane may still invest the chromosomes individually.

Chloroplast Membranes

A mature functional **chloroplast** is often about 4 to 6 μm in diameter and 2 to 3 μm thick. It is composed of layers of numerous flattened, hollow, tubelike unit membranes. At fairly regular intervals the lamellae form dense regions called **grana,** which are seen in electron micrographs to be composed of closed sacs, or **thylakoids** (Gr. *thylakos,* sack). In these membranes are concentrated the pigments of photosynthesis (Fig. 20-15).

Dark-grown seedlings of higher plants develop a precursor of chlorophyll called **protochlorophyll.** A one-quantum light reaction is required to reduce it to the active, dark-green chlorophyll. Protochlorophyll is the light-absorbing agent for its own transformation. This remarkable molecule exhibits another kind of structure and function relationship that is reflected in the plastid with which it associates; the characteristic lamellae of the chloroplast do not develop until chlorophyll is synthesized. This suggests that chlorophyll is not simply attached to a membrane; rather, chlorophyll is an essential "structural" part of the membrane.

By a dissolution process it has been shown that the grana are composed of functional subunits, containing an estimated 200 to 400 densely packed molecules of chlorophyll. These isolated particles carry on photosynthetic reactions, evolving oxygen when illuminated. Electron micrographs have revealed an ordered array of polygonal elements in grana which may be the photosynthetic particles. Each polygon, which Park and Biggins called a **quantasome,** is calculated to contain 230 chlorophyll molecules each (Fig. 20-16).

In several nonmetazoan organisms, neither mitochondria nor chloroplasts are found. Nevertheless, the unicellular forms are functionally

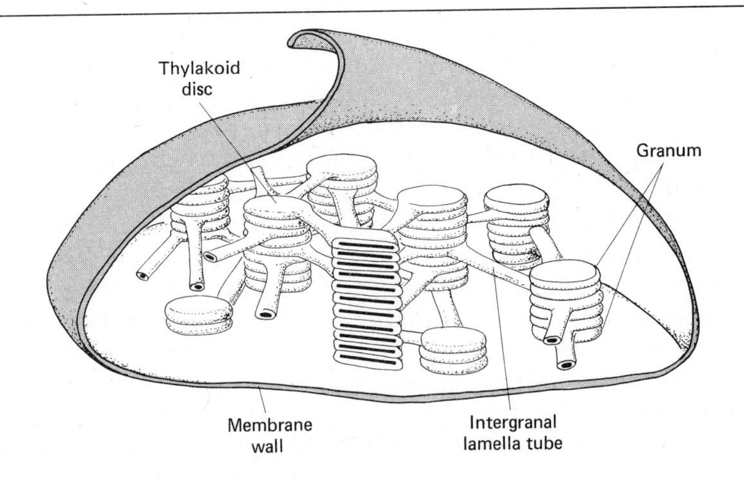

FIGURE 20-15
Three-dimensional reconstruction of the chloroplast of corn.

FIGURE 20-16
Electron micrograph of a single lamella and part of another one in a chloroplast. Where the membrane is torn away an ordered array of units called *quantasomes*, can be seen. Chromium shadowed. (*Courtesy of R. B. Park.*)

equivalent. It seems that their energy transductions occur on extensions of plasma membranes. Blue-green algae and purple bacteria show concentrations of photosynthetic pigments, but without the high degree of lamellar arrangements found in higher plants. Obviously, then, structure-function relationships are not fixed; the same function can be conducted within different matrices.

Mitochondrial Membranes

Almost all aerobic cells have mitochondria, but they are absent in cells that are obligatory anaerobes. Mitochondria vary considerably in shape from one cell type to another. Their size varies also, but common dimensions are 2 to 3 μm long and 1 μm thick. Chemical analysis shows them to be protein and phospholipid mainly, like other membrane systems. Their detailed structure has been studied extensively by electron microscopy. They are seen to have two distinct sets of membranes. The outer one is smooth, whereas the inner one is folded inward extensively to form a series of flattened tubes or plates known as **cristae** (Fig. 20-9).

Mitochondria are disrupted with detergents such as cholate or digitonin, or by physical methods such as sonic radiation. Analysis of the fractions which result reveals that the cytochromes and flavoproteins of the respiratory chain are found exclusively in the insoluble, particulate fraction. The particles, presumably derived from membranes, if properly prepared, will transport electrons, coupled with phosphorylation of ADP. Most of the enzymes of the TCA cycle are contained in the soluble fraction, and for this reason are generally considered to be organized as complexes in the intermembrane space or attached loosely to the cristae.

Electron micrographs reveal configurations of the inner membrane and cristae that differ, depending upon the energy state of the mitochondrion. When substrate (succinate or pyruvate plus malate) is added to a suspension of mitochondria, all are found to have their membranes in a swollen condition that is called the *energized state*. When inhibitors of electron transfer (antimycin and rotenone) are added to the suspension, the membranes assume a thinner, sheetlike configuration which is referred to as the *nonenergized state*. Here again is an example of the oneness of structure and function.

As mentioned previously, elementary particles of the inner membrane of mitochondria have been characterized. They were first seen by Fernández-Morán in beef-heart mitochondria. The particles forming the inner and outer membranes may be distinctly different morphologically. The inner ones have been rather well characterized and are visualized as having stalks between the base piece and head, whereas the outer units are thought to be stalkless (Figs. 20-17 and 20-18).

What enzymes compose the elementary particles is not settled at the time of this writing. One group of investigators proposes that the particle is tripartite (Fig. 20-18) and contains a respiratory assembly (Fig. 19-12). An opposing view, held by other groups, is that elementary particles are coupling factors for ATP formation. There also is the possibility that elementary particles of electron micrographs are artifacts of staining methods.

Although the mitochondrion is a special structure with certain unique respiratory enzymes, its general morphologic characteristics are not very different from those of other energy-transforming membrane systems. All subcellular systems seem to have a double membrane separated by a space, a largely protein and phospholipid composition, ele-

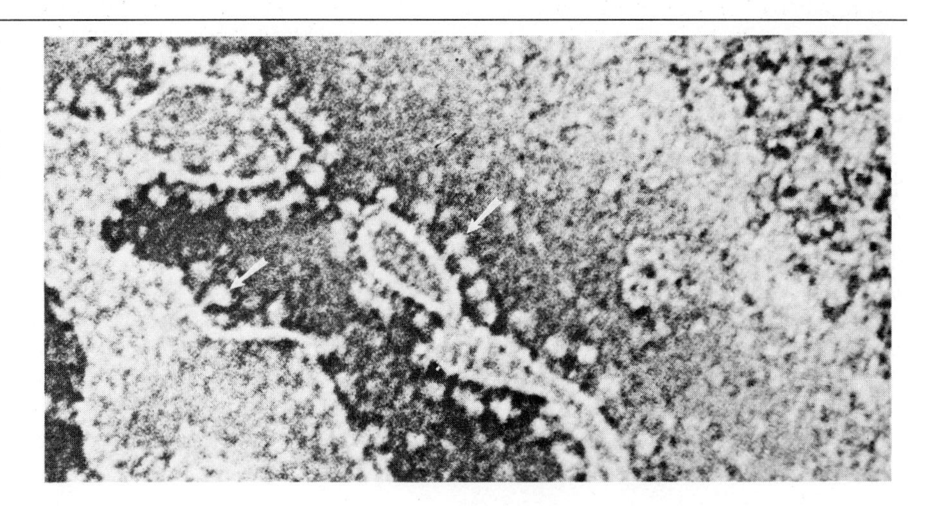

FIGURE 20-17 Electron micrograph of a whole mitochondrion isolated from beef-heart muscle. Arrows point to repeating units. (*Courtesy of H. Fernández-Morán*)

mentary particles that are functional units, and mechanisms for ion translocations. Membrane systems which superficially appear to be different may have more similarities than is even now apparent.

In terms of function, membrane systems may be regarded as machinelike. Their compartmentalized, closely regulated enzyme sets give a specific output for a specific input. We have seen that they employ "feedback" principles in their operations. Certain of these machines even regulate other machines; e.g., the chromosomes regulate the ribosomes. Thus, they are fully "automated" machines. Humans have stumbled onto these same principles and employed them in their machines only very recently.

There is much to be learned about organized enzymes, but we can assume that there are inherent advantages in having enzymes in complexes or as membrane-bound arrays. Probably it is increased efficiency and control of biochemical events that are the *raison d'être* for such functional order.

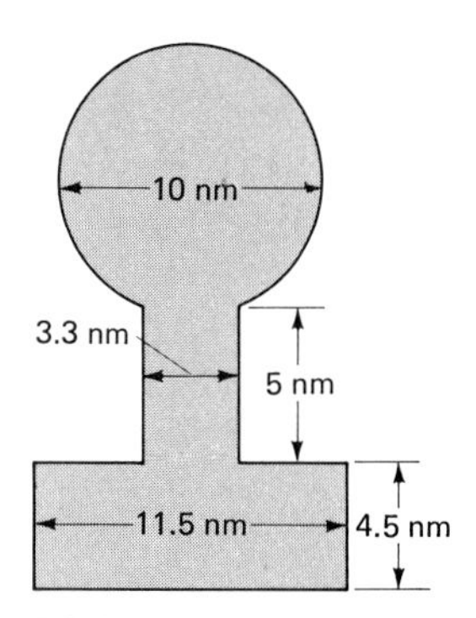

FIGURE 20-18
Dimensions of the elementary particle of the inner membrane of beef-heart mitochondria. See Fig. 20-11*b* for a suggestion about how the particles may be assembled into membranes of mitochondria. (*After D. E. Green and R. F. Goldberger,* 1967, *Molecular Insights into the Living Process, Academic Press, Inc., New York,* p. 247.)

READINGS

Hendler, R. W., 1971, Biological membrane ultrastructure, in *Physiol. Rev.*, **51**, 66–97. A comprehensive examination of membrane models from the 1935 model of Danielli-Davson through the 1970 model of Hybl-Dorset.

Monod, J., 1966, From enzymatic adaptation to allosteric transitions, *Science*, **154**, 475–483. A personal account by Monod of his discoveries of mechanisms controlling enzyme synthesis, presented as a lecture on the occasion of his being awarded a Nobel prize.

Nystrom, R. A., 1973, *Membrane Physiology*, Prentice-Hall, Inc., Englewood Cliffs, N.J. Chapter 2 presents an account of the arguments about the nature of the plasma membrane. Chapters 3 and 4 deal with permeation and excitability, respectively, and Chap. 5 describes artificial membrane systems.

Singer, S. J., and G. L. Nicolson, 1972, The fluid mosaic model of the structure of cell membranes, *Science*, **175**, 720–731. An argument for a current membrane model.

von Bertalanffy, L., 1952, *Problems of Life*, C. A. Watts & Co., Ltd., London. A provocative inquiry into the uniqueness of the living state, with special emphasis on structure-function interdependencies.

BEGINNINGS OF LIFE 21

Almost everyone sooner or later comes to be interested in speculation about how life began on the spaceship Earth. However, for various reasons, many would not consider it a subject worth the serious concern of scientists. Some have been persuaded by religious dogmas that all life was "created" by a special act of a superbeing or, at least, the establishment of human beings on earth was some kind of supernatural event. These ideas are accepted as articles of faith and thus are little influenced by reason. More objective people, influenced by the rigorous demonstrations of Pasteur and others that spontaneous generation of organisms does not occur, find it difficult to accept the possibility of abiogenesis—life originating naturally from lifeless matter. Even among biologists, who are aware of the staggeringly complex nature of present-day organisms, there is a justifiable skepticism about the capacity of matter to organize itself spontaneously into the complicated life state.

Close examination of speculations about the origin of life and related laboratory experimentation is, nevertheless, a most worthwhile endeavor. The central problems in this field, which are the domain of a small but growing body of theoretical and experimental scientists, are at the very core of understanding the true nature of living things. The focus is on the transition between nonliving and living states of matter; at this "interface" life should be in simplest form, yet have unique properties that make it living matter.

21-1 SPECULATIONS

Although there is disagreement on how life began originally, there is no problem in accepting the premise that it did arise at least once in the past. Various terms are used to refer to this event. **Abiogenesis** will be used here, and it specifically means the generation of life from nonliving matter. The terms **autogenesis** and **neobiogenesis** are sometimes used as synonyms for abiogenesis. All these terms admit the possibility that life was "generated" more than once—in fact, that such generation may still occur. A more restrictive term is **eobiogenesis,** which refers to the *first* living form.

Far from regarding an abiogenic event as something having the probability of a "miracle," there are many of sound mind who regard it as almost inevitable. Theoretical speculations and experimental evidence can be drawn together effectively to support the contention that life is the consequence of the natural, spontaneous development of a set of particular substances under a specific set of circumstances. That is to say, once certain conditions were attained in the evolution of the earth, formation of a life state became highly probable. The progression of life as an autonomous process, trending toward increased complexity, may be regarded as an extrapolation of the same conditions that drove matter through the initial transformation to life state.

A variant speculation about the origin of life on earth is that it was transported here from another planet, probably one in another system. According to one such hypothesis, there came to the earth some space voyagers at a time when the earth was without life and perhaps rather inhospitable. During their visit the visitors "contaminated" the sterile earth with living forms; perhaps they dumped some of their accumulated wastes in which were viable organisms. These forms—probably fungi- or algaelike organisms—survived and evolved into the multitude of living things we have today. Thus, earth life had its origin in a garbage dump.

Serious consideration once was given to the hypothesis that spores, carried through space as the result of radiation pressure, could have "seeded" the earth with life.[1] This is now regarded as extremely unlikely because intense ultraviolet radiations (which are destructive to living systems—even in the absence of oxygen, contrary to what was then believed) and heating by absorbed radiations would have destroyed any such units in outer space.

These speculations, though amusing, avoid the question of how nonlife begets life. They simply move the great problem from this planet to another. Furthermore, it probably is safe to say they provide no explanation about the way life began on earth.

The first comprehensive statement of a hypothesis of the origin of life on earth was published in Russia by A. I. Oparin in 1936. Drawing together diverse information from astronomy, geology, and chemistry, Oparin synthesized a brilliant argument that has continued to shape ideas about abiogenesis and the nature of life. There have been revisions of some of Oparin's ideas, partly because of changes in astronomical theory and partly because of the evolution of new concepts, but much of what follows can be directly attributed to his original ideas.

It is now generally held that life evolved in three major steps: (1) the accumulation of molecular complexes and energy sources for maintaining the first systems definable as living; this step is called *chemical evolution;* (2) the actual emergence of these first primitive systems, or **eo-**

[1] S. Arrhenius, 1908, *Worlds in the Making*, Harper & Brothers, New York.

bionts; (3) the evolution of one or more eobionts into what we now call cells. Let us look now at the reasoning and the evidence which may make such a scheme tenable.

21-2 FITNESS OF THE ENVIRONMENT

The entirety of those parts of the earth occupied by life is called the **biosphere.** It is a discontinuous entity, being composed of the space inhabited by all individual organisms. It extends into the atmosphere no more than 20 km, below the surface of continents to about 10 m, and in the oceans down to perhaps 10,000 m. Life, therefore, inhabits a rather thin veneer on and near the surface of this planet. But we can marvel at how suitable a place this veneer is for life. It is difficult to imagine a more appropriate set of conditions in which life could flourish. However, it is probable that life has been "shaped" to "fit" the preexisting physical conditions—hence, this explains in part the remarkable compatibility of life with earth conditions. On the other hand, there seems to have evolved a physical environment on this planet that might be called "preadapted" for life, a concept that has been called "fitness of the environment" by Henderson.[1] A brief account of some of the physical features of the earth, as it may have been at the time primordial life began and as it is now, will illustrate the basis for such a concept.

Atmosphere

Astronomers generally agree that the primeval atmosphere was largely hydrogen and methane with a little carbon dioxide, but little if any free oxygen. Consequently it was of reducing character. Oxygen would have been present on the early earth in chemical compounds, of which water was the most important. An oxygen-free atmosphere, if such occurred, would have permitted the possibility of formation and survival of compounds which today are forbidden by our oxidizing environment. Relatively simple compounds of C, H, O, N, S, and P, which are now regarded as natural organic compounds, may have been formed and remained stable for prolonged periods of time.

A second consequence of an atmosphere having no gaseous oxygen is that there would have been no ozone layer to filter short-wave ultraviolet sunlight. Such light is of very high energy and may have served to drive photochemical reactions at the surface of the earth, leading to the formation of molecular compounds—a kind of inorganic photosynthesis, perhaps. However, present-day life is destroyed by such high-energy radiations, and it seems equally possible that ultraviolet light also could have had decomposing effects.

There is disagreement concerning the evolution of the atmosphere, but commonly it is held that much of the hydrogen, being too light, was

[1] L. J. Henderson, 1927, *Fitness of the Environment,* The Macmillan Company, New York.

not held by the nascent earth's gravity. This was followed by a stage—a steady state perhaps as long as 100 million years—in which there was a good deal more carbon dioxide than at present. Methane, water vapor, ammonia, hydrogen sulfide, carbon monoxide, and nitrogen possibly were constituents at that time also.

The origin of the molecular oxygen which makes up one-fifth of the present atmosphere is variously ascribed. It has been proposed that O_2 is formed by physical processes; e.g., by photochemical and ionization processes in the stratosphere from oxides of carbon and nitrogen. Others contend that the great majority of the O_2 was released through the process of photosynthesis. (Photosynthesis also would have lowered atmospheric carbon dioxide by reducing it to carbohydrates and other compounds.) Possibly atmospheric oxygen originated from more than one source.

Regardless of its evolution, the atmosphere and its present steady-state composition are intimately related to living things. Many organisms are adapted for exchanges of CO_2 and O_2 at the levels at which they occur in the atmosphere (0.03 percent and 21 percent by volume, respectively) and can tolerate only minor changes in their concentrations. However, there is a significant number of organisms, the anaerobes, that do not use oxygen metabolically. Curiously, molecular oxygen, so vital to so many organisms, is toxic to all living matter. It was, perhaps, the need for development of antioxidant mechanisms, as the atmosphere took on more and more oxygen, that led to the utilization of oxygen as an electron acceptor in respiration. Anaerobic organisms that have not developed antioxidant mechanisms must take refuge in niches which are kept relatively free of oxygen.

Thus, organisms have contributed to the atmosphere's evolution, and simultaneously the composition of the atmosphere has profoundly shaped their respiratory functions. Even the density of the atmosphere, which permits flight, determines the aerodynamic characteristics of those forms that traverse it. And so the atmosphere seems eminently suited for life.

Lithosphere

As related in Chap. 5 (Sec. 5-1), the predominant elements in the earth's crust, plus nitrogen, the most abundant constituent of the atmosphere, account for virtually the total mass of organisms. Living things are composed of the lightest, most water-soluble, most ionic, most mobile, and most abundant elements of the earth. Life is not chemically exotic. It was compounded from the most readily available materials at the earth's surface.

Hydrosphere

The nature of life and the manner of its origin are intimately related to the colligative properties of water. The unique thermal properties of water make it the main factor in temperature stabilization or buffering of the physical environment. Temperatures in the biosphere remain within remarkably narrow limits (0 to 40°C). Because it comprises 75 percent or more of their volume, water likewise serves to thermally stabilize the internal environment of organisms. A brief review of water's thermal properties is given in Chap. 4 (Sec. 4-4). (See Table 21-1.)

Water is the best solvent known, as described in Chap. 7 (Sec. 7-1). This has paramount importance in determining the composition of living forms. Its properties, especially that of hydrogen bonding, also make important contributions to biological configurations.

The *pH* of water bodies reflects their composition of solutes, which includes atmospheric gases. The pH of seawater in equilibrium with air is 8.16. There is reason to believe that the earliest sea was at first alkaline, as the air contained much ammonia and little carbon dioxide. As hydrogen was lost from the atmosphere, the relative concentration of carbon dioxide increased and the sea may have become more acid. Then as the atmosphere evolved to its present composition the pH rose again.

Finally, it should be noted that water is quite *transparent*. Light, particularly blue, penetrates water to several meters, permitting vision and photosynthesis, and thereby extending the biosphere into the hydrosphere.

There are many more aspects of "fitness" of the environment that have been recorded elsewhere. The reader is referred especially to Henderson's book, cited earlier, from which the following is taken:

The properties of matter and the course of cosmic evolution are now seen to be intimately related to the structure of the living being and to its activities; they become, therefore, far more important in biology than has been previously suspected. For the whole evolutionary process, both cosmic and organic, is one and the biologist may now rightly regard the universe in its very essence as biocentric.[1]

[1] Ibid., p. 312.

Table 21-1
Some Physical Properties of a Series of Isoelectric Substances

Substance	No. atoms	Nucleus (+)	Hydrogen (+)	M.P., °C	B.P., °C	Molal heat of vaporization,* cal/mol
CH_4	5	6	4	−184	−161	2200
NH_3	4	7	3	−78	−33	5550
H_2O	3	8	2	0	+100	9750
HF	2	9	1	−92	+19	7220
Ne	1	10	0	−249	−246	415

* Heat required to change 1 mol from liquid to vapor.

Source: From J. T. Edsall and J. Wyman, 1958, *Biophysical Chemistry*, Academic Press, Inc., New York, p. 28.

21-3 CHEMICAL EVOLUTION

All those basic properties which we recognize as common to living organisms probably were established or selected during the early phases of life's origin. In fact, the most important and complex mechanisms of life may have been established before there was "life." These would include:

1. Mechanisms of energy storage and transfer (e.g., the "high-energy" phosphate bonds)
2. Oxidation-reduction mechanisms involving electron transfers
3. Specific composition and use of certain ions (e.g., metals in porphyrins, as in respiratory pigments and chlorophylls)
4. Catalytic mechanisms to provide direction and to increase the magnitude of prebiological reactions

These mechanisms probably were established in the latter part of the *chemical evolution* phase—that period in the evolutionary history of the earth during which the chemical components on its surface were changed from their primeval form into chemicals upon which living organisms, or from which living organisms, could develop. The abiogenic production of "organic" molecules is now regarded as not a rare and mysterious phenomenon but rather as a highly probable process. This is in contrast to views on the subject a few decades ago. As we shall see in the next section, laboratory experiments have dispelled the aura of sanctity long associated with organic synthesis.

Let us turn briefly to the problem of how, and under what conditions, the abiogenic synthesis of organic compounds might have taken place. Oparin based his speculations on the then accepted theory of the fiery origin of the planets, as proposed by Sir James Jeans. The first carbon compounds, Oparin contended, were hydrocarbons, as would be expected in a reducing atmosphere. As the atmosphere cooled to 1000°C and lower, reactive free radicals like CH and CH_2 would combine to form simple hydrocarbons:

$$\vdots CH + \vdots CH \longrightarrow \underset{\text{Acetylene}}{HC{\equiv}CH}$$

$$:CH_2 + :CH_2 \longrightarrow \underset{\text{Ethylene}}{H_2C{=}CH_2}$$

$$:CH_2 + :CH_2 \longrightarrow \underset{\text{Methane}}{CH_4} + C$$

Condensations, polymerizations, oxidations, and then splitting, reduction, and other reactions could follow to form a variety of compounds. However, the Jeans theory is no longer generally accepted, and other modes of synthesis are believed to have occurred. This is not to say that thermal energy was not employed in organic synthesis, but perhaps the initial high-temperature chemistry did not occur.

Other methods are regarded as more likely to have produced metastable molecules when the earth had a reducing atmosphere; e.g., electric discharge through mixtures of hydrogen, methane, ammonia, and water vapor (i.e., the constituents of the primitive atmosphere). Potential reactants may have been activated by other energies, including solar radiations, cosmic radiations, radioactive emissions, and sonic vibrations (as caused by meteorites traversing the atmosphere).

Whatever the mechanisms, it is commonly held that carbon compounds would have been formed on the nascent earth. The first simple molecular aggregates would be expected to have evolved into more complex ones with the continued input of energy. Increasing complexity implies increasing instability, but many organic molecules have half-lives of the order of a thousand years or more. A variety of such metastable compounds may have formed an accumulation from which a living unit, or several living units, evolved. How such eobionts were assembled, modified, replicated, and coordinated to form functional units are questions about which there can only be speculation.

When did chemical evolution occur? Presumably during a period of about 2 billion (2×10^9) years and sometime prior to 2 billion years ago. Geochemists estimate the age of the earth to be some 4.5×10^9 years. A few reference points will aid in grasping the magnitude of such a span of time. The oldest unequivocal fossils are probably the laminated algal stromatolites whose geological relations suggest that they are at least 2.7 billion years old. Recently fossil evidence of 3.1-billion-year-old bacteria has been found. Since bacteria and, especially, algae are quite complex forms, it can be speculated that the origin of life may be dated as early as 4 billion years ago. The oldest animal fossils (Cambrian) are at least 500 million years old. (Human beings have been extant for perhaps 20 million years; if biological time is represented as a 24-hour span of time, man appears in the last 10 minutes.) Thus, chemical evolution may have occupied a period of time at least as long as that from the beginning of simple life forms to the present (Fig. 21-1).

The great amount of time available between the origin of the earth and the first evidence of life has been used to argue that highly improbable events could have occurred which led to the appearance of life. In other words, given sufficient time, improbable events become probable. However, Blum[1] has pointed out that:

" . . . the greater the time elapsed the greater should be the approach to equilibrium, the most probable state, and it seems that this ought to take precedence in our thinking over the idea that time provides the possibility for the occurrence of the highly improbable. The more we study living systems the more we marvel at their beautifully ordered complexity; and we may estimate that the forming of such systems (or even much simpler ones) by a single chance act would have an improbability of the order of a miracle, that could have happened only once in our universe. On the other hand, the more we

[1] H. F. Blum, 1962, *Time's Arrow and Evolution*, 2d ed., Harper & Bros., New York (Torchbook Edition), p.178A.

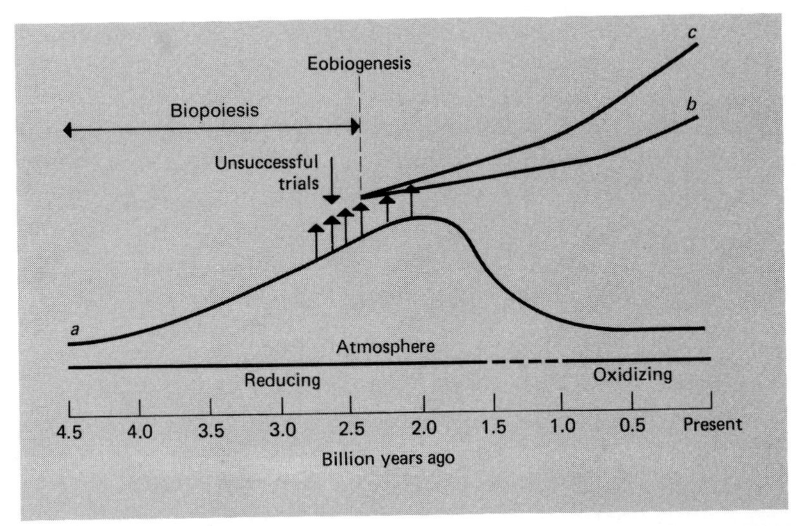

FIGURE 21-1
Significant events in the origin and evolution of life. Line *a* suggests the progress of abiogenic production of organic chemicals. *Biopoiesis* is that span of chemical history that preceded the origin of the first living thing. The vertical *arrows* indicate the appearance and disappearance of less-developed subvital entities. With the advent of the first successful eobionts (*eobiogenesis*) there began the period of biogenic synthesis of organic compounds, indicated by line *b*, which has supplanted abiogenic synthesis. The increasing complexity of organisms through biological evolution is represented by line *c*. Note the transition of the atmosphere from *reducing* to *oxidizing* sometime after the origin of life, presumably because of the advent of autotrophic organisms. (*After J. Keosian*, 1964, *The Origin of Life, 2d ed., Reinhold Book Corporation, New York*, p. 90.)

study mechanisms that operate in the living organism, the more satisfactorily are we able to explain them in terms of physical principles; and this may lead us to the opposite position; that is, that life originated here and at many other places in the universe as a matter of course owing to the straightforward functioning of physical principles, among which the second law of thermodynamics would play an outstanding role.

21-4 LABORATORY CHEMISTRY

Melvin Calvin, in the 1950s, was among the first to consider it feasible and purposeful to do experiments that were designed to simulate conditions of the prebiological earth. Subsequently a number of similar and related experiments were done. Given the premise that the composition of the primeval atmosphere at the surface of the earth was some mixture of

H—O(—H)	C≡O	O=C=O	H—C(H)(H)—H	H—H	H—N(H)—H
Water	Carbon monoxide	Carbon dioxide	Methane	Hydrogen	Ammonia

the question was asked: By introducing energy into an aggregation of these molecules, would any kind of significant evolutionary development of molecular structures occur? Several kinds of mixtures have been exposed to a variety of energy sources—ionizing radiation (particulate and gamma radiation), ultraviolet radiation, and electrical discharge —with the result that indeed radical or ionic fragments form, some of which recombine in new metastable forms. The formation of the following compounds, for example, has been demonstrated:

$H{-}C{\equiv}N$	$HN(C{\equiv}N)_2$	$H{-}C({=}O){-}OH$	$H{-}C(H){=}O$
Hydrocyanic acid	Dicyanamide	Formic acid	Formaldehyde

$HOCH_2{-}C(H){=}O$	$CH_3{-}C({=}O){-}OH$	$HO{-}C({=}O){-}CH_2{-}CH_2{-}C({=}O){-}OH$
Glycolaldehyde	Acetic acid	Succinic acid

$H_2N{-}CH_2{-}C({=}O){-}OH$	$CH_3{-}CH(NH_2){-}C({=}O){-}OH$	$HO{-}C({=}O){-}CH_2{-}CH(NH_2){-}C({=}O){-}OH$
Glycine	Alanine	Aspartic acid

The design of the apparatus and the conditions required for such syntheses are comparatively simple. One of the first such experiments is explained in Figs. 21-2 and 21-3.

Even more complex molecules have been made abiogenically. Sugars and heterocyclic compounds have been formed. Adenine has

FIGURE 21-2
Spark-discharge apparatus used by Miller in 1953 to synthesize organic compounds from a reducing mixture by simple gases. Methane, ammonia, and hydrogen were circulated past electric discharges from tungsten electrodes energized by a high-frequency Tesla coil. Water in the small flask was boiled to add steam to mix with the other gases. From the condenser the water flows back into the boiling flask, bringing down some of the less-volatile products. The U tube at the base of the apparatus traps liquids and thus promotes circulation in one direction. After 1 week of continuous running, analysis of the fluid in the small flask indicated the presence of glycine, alanine, β-alanine, aspartic acid, and α-amino-n-butyric acid. (*After S. L. Miller*, 1955, *J. Am. Chem. Soc.*, *77*, p. 2351.)

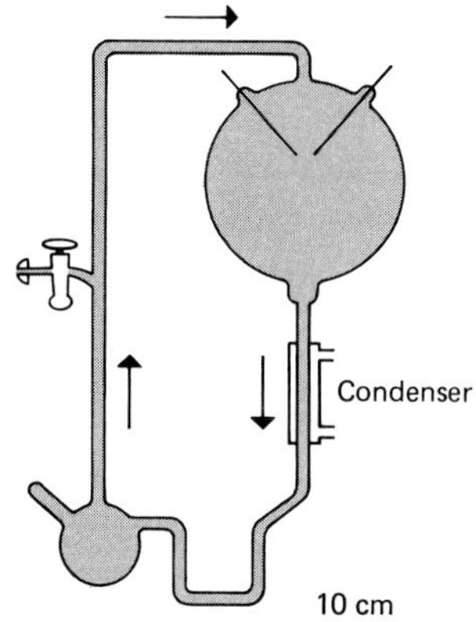

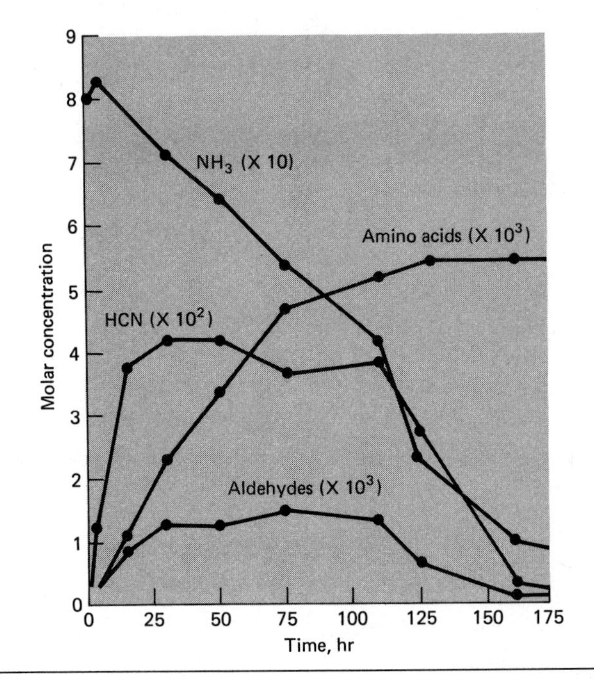

FIGURE 21-3
Changes in concentrations of ammonia, hydrogen cyanide, and aldehydes in the U tube and of amino acids in the flask of Miller's apparatus while sparking a mixture of methane, ammonia, hydrogen, and water. (*After S. L. Miller, 1957, Biochem. Biophys. Acta, 23, p. 484.*)

been produced from ammonia and HCN. This, of course, is of paramount significance, since adenine is a part of the ubiquitous energy-transfer molecule, adenosine triphosphate, as well as being a component of nucleic acids. Other purines and pyrimidines, as well, have been synthesized under simulated primitive earth conditions.

Primitive life was probably protein-centered—as life is today—hence, it is important to be able to suggest the spontaneous origin of precellular protein. It has been demonstrated that amino acids can give rise in chemical systems to protein-type molecules. One method has used an aqueous solution of amino acids exposed to ultraviolet light. Dipeptides (i.e., two amino acids joined by a peptide linkage) are readily formed in this system. Another method, that of S. W. Fox, is to heat dry or slightly moist amino acid mixtures. When two or more amino acids, one being aspartic acid, glutamic acid, or lysine, are heated together, mixtures of genuine peptides result. Thermal polymers or proteinoids containing as many as the 18 amino acids common to all organisms have been obtained. The conditions simulated by such thermal polymerizations are perhaps those of some primitive mud flat at the edge of the sea.

Fox has reported that when 15 mg of his proteinoid is treated with 2.5 ml hot seawater and allowed to cool, microspheres having diameters about like those of small bacteria are formed. This has been interpreted as indicating that self-organization of primitive cells from molecular matrices by a process analogous to crystallization may have occurred on innumerable occasions. These microspheres act as if their envelopes had osmotic properties; i.e., they shrink in hypertonic solutions. Perhaps this suggests the beginnings of biological membranes. However, it is several giant steps from proteinoid spheres to a real cell.

21-5 RESIDUAL PROBLEMS

The events attending the beginnings of life, which formerly seemed to be so completely shrouded in mystery, at least now have been brought into the realm where reasoned speculations about them can be formulated. Obviously, there are a host of unsettled and unsettling problems.

One of these is the nature of the energetics of the first living system—or eobiont. According to one group of arguments, the first form of terrestrial life must have been an *autotroph;* i.e., an organism that could manufacture its own organic substance out of inorganic substances, as can the contemporary green plants. The logic of such a hypothesis is obvious and persuasive because it accounts for the energy requirement of the living state as well as the suitability of an environment in which free oxygen may have been absent. Later, according to this thesis, organisms evolved which could consume the organic molecules made by autotrophs and extract from them energy; these were what we now call *heterotrophs.*

Others argue that the earliest form of metabolism was heterotrophic. They assume that since chemical evolution would have produced a complex set of organic compounds which formed living entities, the same set of compounds also could have been used as the original energy source for the eobionts. Later, as these organic molecules were being depleted, autotrophic forms evolved and continued to supply the heterotrophs with energetic molecules (Fig. 21-4).

Regardless of the specific sequence of events, as Blum points out, the greatest problem is *what* was the first organism. It is difficult to imagine a sharp transition. When did a collection of molecules cease to be just molecules and become life? In particular, since modern organisms are strictly dependent upon protein catalysts (enzymes), then how, when no life existed, did substances come into being which today are absolutely essential to living systems, yet which can be formed only by those systems?

FIGURE 21-4
A hypothetical scheme for the monophyletic proliferation of life.

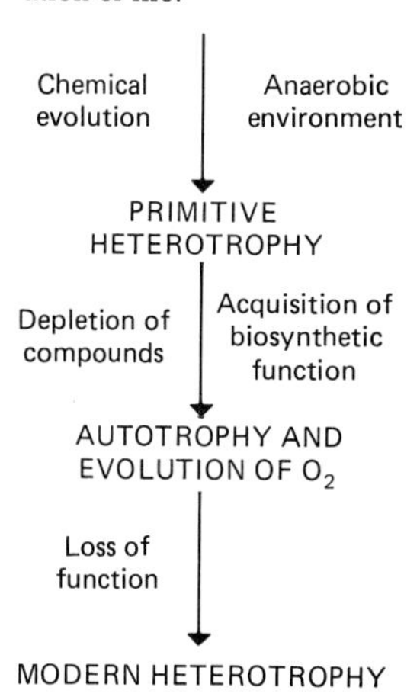

A hypothesis by Horowitz,[1] though not solving the dilemma posed by Blum, does aid in thinking about this kind of problem. He suggests that the first living entity was a completely heterotrophic unit, reproducing itself at the expense of prefabricated organic molecules in its environment. Depletion of the environment continued until a point was reached where the supply of essential substances limited further multiplication. By a process of mutation a means was eventually discovered for utilizing other, simpler substances. With this event the evolution of biosynthesis

[1] M. H. Horowitz, 1945, On the evolution of biochemical syntheses, *Proc. Natl. Acad. Sci.*, **31**, 157.

began. Further mutations provided additional steps until there were organisms which could synthesize all the essential complex compounds. This evolution was probably based on the chance combination of genes, resulting to a large extent in the development of short reaction chains utilizing substances whose synthesis had been previously acquired in different mutant forms.

It probably is apparent now that the character of the eobiont has been approached mainly from two different directions—the "metabolically alive" system and the "genetically alive" system. Some of those whose criteria for "aliveness" fall in the former category regard linked endergonic and exergonic reactions as satisfying the definition of an organism, as noted in Chap. 1 (Sec. 1-3). To them life can be an abstraction, resident in a single molecule that is potentially able to engage in the next reaction in a series, or, at the other extreme, the whole of the primordial seas, "the metabolizing ocean,"[1] may be considered to have been the only one living thing at an early stage of life.

In contrast, others do not hold with such an abstraction and demand that life be embodied in a morphological component. The importance of energy utilization is recognized, but the chemical reactions constituting the "metabolism" of the system produce morphological components as the "fabric" of the system which has permanence and continuity. Different efficiencies of formation of end products would provide a basis for selection, and thus would such a system evolve.

The living forms which confront us now, the products of eons of evolution, with all their perfections for replicating and energy transformation, lead us to think that life processes must be orderly and efficient. It is likely that this was not the case in the beginning; the first forms may have been quite inefficient. The kinds of selection pressures existing today undoubtedly were imposed gradually. For instance, the emergence of nucleic acid–based genetic mechanisms may have been a late arrival in living things. It has been found that when peptides are formed abiogenically there is a statistical preference for peptide sequences of known proteins over those sequences not common in nature. Furthermore, the addition of a peptide to an amino acid mixture increases the extent to which such sequences are formed. Still more relevant is the finding of a virus which produces a disease called *scrapie* in sheep that, from circumstantial evidence, appears to be free of either RNA or DNA. Thus, molecular replication may occur even today without participation of nucleic acid.

If we accept the thesis that spontaneous generation of life occurred on this earth, we are then confronted with two interesting problems: How many times has the event happened, and is it still occurring? Biologists generally believe that abiogenesis has not occurred during at least the last billion years. One reason is that the accumulation of appropriate and sufficient molecules by chemical evolution is improbable in more recent times because oxidations and biological attrition would destroy them. Furthermore, all attempts at demonstration of spontaneous generation

[1] M. Ycas, 1955, A note on the origin of life, *Proc. Natl. Acad. Sci.*, **41**, 715.

have failed. It is generally believed that even if it were to occur, the unit would be wiped out almost immediately by existing living organisms.

Several eobionts may have appeared near the end of the chemical evolution phase (Fig. 21-1). They may have replicated over many years and then most of them may have failed to continue. How many eobionts survived to give rise to present forms is unknown, of course, but the striking biochemical similarities of all living things suggest a common ancestor. However, they may only reflect the homogeneity that existed when the potential for life still resided in the "metabolizing ocean."

These arguments notwithstanding, some biologists consider it possible that life can begin now. Fox argues that at the beginning of its own evolution a de novo form of life would be inconspicuous because of the many characteristics it shares already with existing life. He considers the possibility that if life were to begin in hot springs (considered to be favorable for thermal synthesis) one would not be able to distinguish de novo forms from established forms, such as the sulfur bacteria or the thermophilic blue-green algae, the lineal, unevolved descendants of ancestors that first appeared millions of years ago.

There was never naught,
There was always thought . . .
Matter was begun-
And in fact complete,
One and yet discrete
To conflict and pair.
Everything was there,
Every single thing
Waiting was to bring,
Clear from hydrogen
All the way to men. . . .[1]

READINGS

Blum, H. F., 1968, *Time's Arrow and Evolution*, 3d ed., Princeton University Press, Princeton, N.J. A classic work, the theme of which is the relationship between the second law of thermodynamics and organic evolution. The nature and the origin of life are considered in a profound and stimulating set of arguments.

Fox, S. W., and K. Dose, 1972, *Molecular Evolution and the Origin of Life*, W. H. Freeman and Company, San Francisco. A synthesis of information from astronomy, geology, physics, chemistry, and biology about the origin of life.

Keosian, J., 1968, *The Origin of Life*, 2d ed., Reinhold Book Corporation, New York. A concise little book which sets forth in a very readable way a summary of thinking on this subject.

Oparin, A. I., 1938, *The Origin of Life* (translated by Sergius Morgulis), The Macmillan Company, New York. This historic work was published in Russia in 1936. It contains a comprehensive statement of his ideas of the gradual evolution of primary organic substances.

[1] From "A Never Naught Song," by Robert Frost, 1969, in *The Poetry of Robert Frost*, edited by Edward Connery Lathem, Holt, Rinehart and Winston, New York, p. 426.

Index